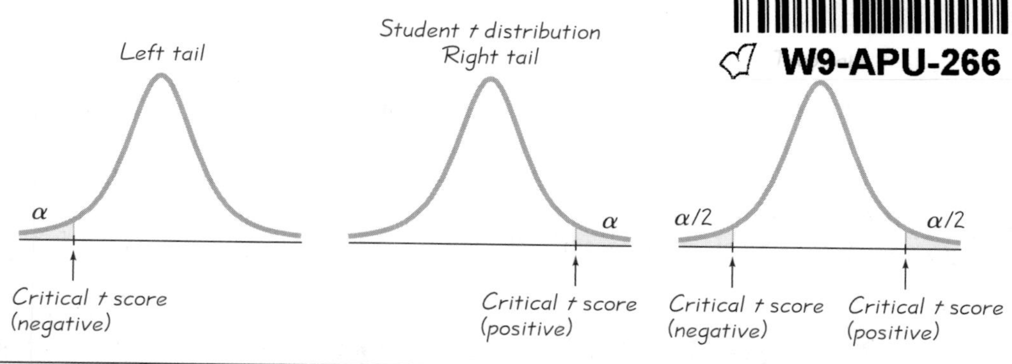

TABLE A-3 t Distribution

	α					
Degrees of Freedom	.005 (one tail) .01 99% (two tails)	.01 (one tail) .02 98% (two tails)	.025 (one tail) .05 95% (two tails)	.05 (one tail) .10 90% (two tails)	.10 (one tail) .20 80% (two tails)	.25 (one tail) .50 50% (two tails)
1	63.657	31.821	12.706	6.314	3.078	1.000
2	9.925	6.965	4.303	2.920	1.886	.816
3	5.841	4.541	3.182	2.353	1.638	.765
4	4.604	3.747	2.776	2.132	1.533	.741
5	4.032	3.365	2.571	2.015	1.476	.727
6	3.707	3.143	2.447	1.943	1.440	.718
7	3.500	2.998	2.365	1.895	1.415	.711
8	3.355	2.896	2.306	1.860	1.397	.706
9	3.250	2.821	2.262	1.833	1.383	.703
10	3.169	2.764	2.228	1.812	1.372	.700
11	3.106	2.718	2.201	1.796	1.363	.697
12	3.054	2.681	2.179	1.782	1.356	.696
13	3.012	2.650	2.160	1.771	1.350	.694
14	2.977	2.625	2.145	1.761	1.345	.692
15	2.947	2.602	2.132	1.753	1.341	.691
16	2.921	2.584	2.120	1.746	1.337	.690
17	2.898	2.567	2.110	1.740	1.333	.689
18	2.878	2.552	2.101	1.734	1.330	.688
19	2.861	2.540	2.093	1.729	1.328	.688
20	2.845	2.528	2.086	1.725	1.325	.687
21	2.831	2.518	2.080	1.721	1.323	.686
22	2.819	2.508	2.074	1.717	1.321	.686
23	2.807	2.500	2.069	1.714	1.320	.685
24	2.797	2.492	2.064	1.711	1.318	.685
25	2.787	2.485	2.060	1.708	1.316	.684
26	2.779	2.479	2.056	1.706	1.315	.684
27	2.771	2.473	2.052	1.703	1.314	.684
28	2.763	2.467	2.048	1.701	1.313	.683
29	2.756	2.462	2.045	1.699	1.311	.683
Large (z)	2.575	2.326	1.960	1.645	1.282	.675

ELEMENTARY STATISTICS

EIGHTH EDITION

ELEMENTARY STATISTICS

EIGHTH EDITION

MARIO F. TRIOLA

Boston San Francisco New York
London Toronto Sydney Tokyo Singapore Madrid
Mexico City Munich Paris Cape Town Hong Kong Montreal

Sponsoring Editor: Deirdre Lynch
Editorial Assistant: Rebecca Martin
Development Editor: Elka Block
Executive Marketing Manager: Brenda Bravener
Senior Marketing Coordinator: Laura Potter Walton
Managing Editor: Karen Guardino
Senior Production Supervisor: Peggy McMahon
Text and Cover Designer: Barbara T. Atkinson
Cover photo: ©Kelly-Mooney Photography/CORBIS
Art Editor: Meredith Nightingale
Photo Research: Beth Anderson
Prepress Service Manager: Caroline Fell
Manufacturing Buyer: Evelyn Beaton
Technical Illustrator: Academy Artworks
Composition: WestWords, Inc.
Printer: VonHoffman Press

Library of Congress Cataloging-in-Publication Data

Triola, Mario F.
Elementary statistics / Mario F. Triola.—8th ed.
p. cm.
Includes bibliographical references.
ISBN 0-201-61477-4 (hc)
1. Statistics. I. Title

QA276.12 .T76 2000
519.5—dc21 99-087305

2 3 4 5 6 7 8 9 10—VHP—03020100

To Marc and Scott

About the Author

Mario F. Triola is a Professor of Mathematics at Dutchess Community College, where he has taught statistics for over 30 years. Marty is the author of *Mathematics and the Modern World* and *A Survey of Mathematics* and is coauthor of *Introduction to Technical Mathematics*, and *Business Statistics*. Outside of the classroom, Marty's consulting work includes the mathematical design of casino slot machines and fishing rods, and he has worked with attorneys in determining probabilities in paternity lawsuits, identifying salary inequities based on gender, and analyzing disputed election results. Marty was recently a writing team member of the Project Coalition writing team with NASA and the American Mathematics Association of Two-Year Colleges.

When he's not working, Marty enjoys playing golf and tennis, running, hiking, and anything that flies. He has a commercial pilot's license with an instrument rating, and has flown airplanes, helicopters, sail planes, hang gliders, and hot air balloons. His passion for flying has included parachute jumps, flying in a Goodyear blimp, and parasailing.

The Text and Academic Authors Association has awarded Mario F. Triola a "Texty" for Excellence for his work on *Elementary Statistics*.

Preface

ABOUT THIS BOOK

The author and publisher of this award-winning book share a common goal: Provide the best possible introductory statistics book for both students and professors. ***Elementary Statistics***, Eighth Edition, incorporates the latest technological and teaching advances, an abundance of pedagogical features, and real, meaningful, and interesting applications, such as those related to these questions:

- Does it rain more on weekends (as reported by the media)?
- Is the mean body temperature really 98.6° F, as is commonly believed?
- When the Titanic sank, did passengers follow the rule of "women and children first"?
- Would you be surprised to learn that regular Coke weighs more than diet Coke?
- Can analysis of discarded garbage be used to predict the size of a population?
- Do fighter jet ejection seats need to be designed to better accommodate female pilots?

AUDIENCE/PREREQUISITES

Elementary Statistics is written for students majoring in any field. Although the use of algebra is minimal, students should have completed at least an elementary algebra course. In many cases, underlying theory is included, but this book does not stress the mathematical rigor more suitable for mathematics majors. Because the many examples and exercises cover a wide variety of different and interesting statistical applications, ***Elementary Statistics*** is appropriate for students pursuing careers drawing from a wide variety of disciplines ranging from the social sciences of psychology and sociology to areas such as education, the allied health fields, business, economics, engineering, the humanities, the physical sciences, journalism, communications, and liberal arts.

CONTENT AND ORGANIZATION CHANGES IN THE EIGHTH EDITION

- Section 1-5 (Statistics with Calculators and Computers) from the 7th edition has been deleted, because references to calculators and computers are now included throughout this 8th edition. The Cumulative Review Exercises for Chapter 1 continue to include calculator "warm-up" exercises.

- The 7th edition section on the Multiplication Rule is now divided into two separate sections: Section 3-4 (Multiplication Rule: Basics) and Section 3-5 (Multiplication Rule: Complements and Conditional Probability). This provides greater flexibility for instructors who prefer to include less probability, and it also helps students by allowing them to focus on the basics of the multiplication rule before studying the more difficult applications of that rule.
- Section 5-7 (Determining Normality) is a new section that describes methods for determining whether a set of sample data appears to come from a population having a normal distribution.
- The 7th edition section on "Estimating a Population Mean: Large Samples" has been reorganized so that the subsection of "Determining Sample Size" is now a separate section (Section 6-4). It is our experience that the former Section 6-2 could not be covered in a single class hour. The new organization makes it easier to cover confidence interval estimates for means before considering the separate issue of determining sample size.
- The sections of Chapter 8 (Inferences from Two Samples) have been rearranged to better fit the preferences of instructors who would like to include some, but not all, sections of this chapter.
- Section 9-6 (Modeling) is a new section that uses some of the correlation/regression tools in developing a mathematical function that describes real-world data. It includes nonlinear functions, such as quadratic, exponential, and power functions.
- Throughout the book there is even greater emphasis on *describing, exploring,* and *comparing* data. For examples, see Section 2-3 or 8-2, or the Chapter Problem for Chapter 11.
- Throughout the book there is even greater emphasis on *interpreting* results. Instead of simply obtaining an answer, the consequences of the answer are considered. For example, when discussing probability in Chapter 3, instead of simply finding probability values, we interpret them by differentiating between events that are usual and those that are unusual. In Section 4-4, after we find the values of means and standard deviations in binomial probability distributions, we use those results to distinguish between results that are usual and those that are unusual. With hypothesis testing, we don't simply end with a conclusion of rejecting or failing to reject a null hypothesis; we proceed to state a practical conclusion that addresses the real issue. Students are encouraged to *think* about the implications of results instead of cranking out cookbook results that make no real sense.

FLEXIBLE SYLLABUS

The organization of this book reflects the preferences of most statistics instructors, but there are two common variations that can be used easily with this 8th edition:

- **Early coverage of correlation/regression:** Some instructors prefer to cover the basics of correlation and regression early in the course, such as immediately following the topics of Chapter 2. *Sections 9-2 (Correlation) and 9-3 (Regression)* *can be covered early in the course.* Simply omit the subsection in Section 9-2 clearly identified as "Formal Hypothesis Test (Requires Coverage of Chapter 7)."

- **Minimum probability:** Some instructors feel strongly that coverage of probability should be extensive, while others feel just as strongly that coverage should be kept to a bare minimum. Instructors preferring minimum coverage can include Section 3-2 while skipping the remaining sections of Chapter 3. Sections 3-3 through 3-7 are not essential for the chapters that follow. Many instructors prefer to cover only the fundamentals of probability along with the basics of the addition rule and multiplication rule. The reorganized coverage of the multiplication rule (Sections 3-4 and 3-5) now offers that flexibility.

NEW FEATURES

- **Statistics in the News**, a feature included at the end of each chapter, provides real-life illustrations and activities that relate to the use of statistics in current and newsworthy applications.
- **Internet Projects**, also included at the end of each chapter, involve the student with applications using data found on the Internet. The projects can be found at http://www.awlonline.com/triola.

- **TI-83 Plus** instructions and screen displays are included throughout the book.
- **Excel** instructions and screen displays are now included throughout the book.
- **Titles for Examples and Exercises** are included so that instructors and students can quickly identify the topic and/or methodology involved.
- **Full-Color Design** increases reader interest through a more functional and visually appealing presentation.
- **ActivStats® icons** throughout the text encourage students to go to the ActivStats CD when encountering difficult concepts. AS
- **CD-ROM,** packaged free with the text, contains the STATDISK statistical software, the Excel add-in (Data Desk/XL), and the Appendix B data sets in formats for different technologies, as well as a TI-83 Plus data application.

EXERCISES

There are **1500** exercises—*more than 60 percent of them new!* Many more of the exercises require interpretation of results. Because exercises are of such critical importance to any statistics book, great care has been taken to ensure their usefulness, relevance, and accuracy. Three statisticians have read carefully through the final stages of the book to verify accuracy of the text material and exercise answers. Exercises are arranged in order of increasing difficulty by dividing them into two groups: (1) Basic Skills and Concepts and (2) Beyond the Basics. The Beyond the Basics exercises address more difficult concepts or require a somewhat stronger mathematical background. In some cases, these exercises also introduce a new concept.

Real data: *More than half of the exercises use real data.* Because the use of real data is such an important consideration for students, hundreds of hours have been devoted to finding real, meaningful, and interesting data. In addition to the real data included throughout the book, many exercises refer to the 20 data sets listed in Appendix B.

HALLMARK FEATURES

Beyond an interesting and accessible (and sometimes humorous) writing style, great care has been taken to ensure that each chapter of ***Elementary Statistics*** will help students understand the concepts presented. The following features are designed to help meet that objective:

- **Chapter-opening features:** A list of chapter sections previews the chapter for the student; a chapter-opening problem, using real data, then motivates the chapter material; and the first section is a chapter overview that provides a statement of the chapter's objectives.
- **End-of-chapter features:**
  Statistics in the News encourages analysis of timely applications of statistics.
  A **Vocabulary List** of important terms (a full glossary is found in Appendix D);
  A **Chapter Review** summarizes the key concepts and topics of the chapter;
  Review Exercises provide practice on chapter concepts and procedures;
  Cumulative Review Exercises reinforce earlier material;
  From Data to Decision: Critical Thinking, a capstone problem, requires critical thinking and a writing component;
  Cooperative Group Activities encourage active learning in groups;
  Technology Projects are specifically designed to incorporate an application using technology;
  Internet Project that applies important chapter concepts.

- **Margin Essays:** The text includes 106 margin essays, which illustrate uses and abuses of statistics in real, practical, and interesting applications. Topics include "Six Degrees of Separation," "Statistics and Land Mines," "Using Statistics to Identify Thieves," and "Choosing Personal Security Codes."
- **STATDISK, Minitab, Excel, and TI-83 Plus** instructions and output appear throughout the book.
- **Flowcharts:** These appear throughout the text to simplify and clarify more complex concepts and procedures.
- **Real Data Sets:** Real data are used extensively throughout the entire book. Appendix B lists 20 data sets, 10 of which are new. These data sets are provided in printed form in Appendix B at the back of the book, and in electronic form on the CD packaged with the book. The data sets include such varied topics as eruptions of the Old Faithful geyser, nicotine contents of cigarettes, diamond prices and characteristics, weights of diet and regular Coke and Pepsi, movie financial and rating data, and rainfall amounts.
- **Interviews:** Every chapter of the text now includes interviews with professional men and women in a variety of fields who use statistics in their day-to-day work.
- **Appendices:** Appendix A contains tables; Appendix B lists 20 data sets; Appendix C is a summary of important TI-83 Plus procedures; Appendix D is a glossary of important terms; Appendix E is a bibliography of recommended text and reference books; Appendix F contains answers to all the odd-numbered section exercises, as well as *all* answers to Review Exercises and Cumulative Review Exercises.
- **Quick-Reference Endpapers:** Tables A-2 and A-3 (the normal and t distributions) are reproduced on the front inside cover pages. A symbol table is included on the last pages for quick and easy reference to key symbols.

- **Detachable Formula/Table Card:** This insert, organized by chapter, gives students a quick reference for studying, or for use when taking tests (if allowed by the instructor). The formula/table card is also available on the book Web site.

TECHNOLOGY

Elementary Statistics, Eighth Edition, can be easily used without reference to any specific technology. Many instructors have successfully used past editions of this book with students using only scientific calculators. However, for those who choose to supplement the course with specific technology, the following tools are available:

TECHNOLOGY SUPPLEMENTS

- **STATDISK:** An easy-to-use statistical software package developed specifically for use with ***Elementary Statistics***, included on the CD-ROM provided with this book. The new Version 8.1 of STATDISK is available for Windows and Macintosh systems. In addition to this free software, there are references to STATDISK throughout this book, and there is a separate STATDISK manual/workbook designed specifically as a supplement to this book.

STATDISK

- **Minitab:** There are references to Minitab (Release 12) throughout this book, and there is a separate Minitab manual/workbook designed specifically as a supplement. The Appendix B data sets are included on the CD-ROM in a Minitab format.

MINITAB

- **Excel:** References to Excel are included throughout this book, and there is a separate Excel manual/workbook designed specifically as a supplement to this text. The Appendix B data sets are included on the CD-ROM in an Excel Format.

EXCEL

- **Excel Add-In:** Data Desk/XL is a software supplement on the CD-ROM that enhances the statistics capabilities of the Excel program.
- **TI-83 Plus:** References to the TI-83 Plus calculator appear throughout, and there is a separate manual/workbook designed specifically as a supplement to this book. The CD-ROM includes a data application for the TI-83 Plus. This application includes the Appendix B data sets.

TI-83 Plus

- **SPSS:** There is a separate SPSS manual/workbook designed specifically as a supplement to this book. The Appendix B data sets are included on the CD-ROM in an SPSS format.
- **Data CD:** The Appendix B data sets included in formats for Minitab, Excel, SPSS, the TI-83 Plus calculator, and text files are included on the CD and packaged with the book.
- **Web Site:** www.awlonline.com/triola includes further information and links to other sites with useful data sets.

SUPPLEMENTS

The student and instructor supplements packages are intended to be the most complete and helpful learning system available for an introductory statistics course. The following supplements are available from Addison Wesley Longman Publishing Company. Instructors should contact the local AWL sales consultant, or e-mail the company directly at exam@awl.com for examination copies.

FOR THE INSTRUCTOR

- ***Annotated Instructor's Edition***, by Mario F. Triola, contains answers to all exercises in the margin, recommended assignments, and teaching suggestions. ISBN: 0-201-61480-4. (Student for-sale edition ISBN: 0-201-61477-4).
- ***Instructor's Guide and Solutions Manual***, by Mario F. Triola and Milton Loyer (Penn State University), contains solutions to all the exercises, quizzes (with answers), transparency masters, and sample course syllabi. ISBN: 0-201-70460-9.
- **Testing System**: Great care has been taken to ensure the strongest possible testing system for the new edition of ***Elementary Statistics***. Not only is there a printed test bank, but there is also a computerized test generator, **TestGen-EQ 3.0 and Quizmaster 2.0**, that lets you view and edit testbank questions, transfer them to tests, and print in a variety of formats. The program offers many options for organizing and displaying test banks and tests. A built-in random number and test generator makes **TestGen-EQ** ideal for creating multiple versions of tests and provides more possible test items than print testbank questions. Powerful search and sort functions let the instructor easily locate questions and arrange them in the preferred order. Users can export tests as HTML text files so they can be viewed with a web browser. Printed Testbank ISBN: 0-201-70463-3; TestGen-EQ for Mac and Windows ISBN: 0-201-70468-4.
- **PowerPoint® Lecture Presentation CD:** Free classroom lecture presentation software geared specifically to the sequence and philosophy of ***Elementary Statistics***, has been prepared by Cheryl Slayden of Pellissippi State Technical Community College. Key graphics from the book are also included. Mac and Windows ISBN: 0-201-70462-5.
- *New* **Triola Web Site:** The new Triola Web site can be accessed at **http://www.awlonline.com/triola.** It provides dynamic resources for instructors and students. Some of the resources include internet projects keyed to every chapter of the text, data downloads, sample syllabi, and more. Many student resources are also available on the web site, including practice quizzes, study tips, instructional video clips and more.

FOR THE STUDENT

- *New* **Videos**, designed to supplement the sections in the book, with many topics presented by Mario F. Triola. The videos feature professor and student interaction, and all technologies in the book are represented on the videos. This is an excellent resource for students who have missed class or wish to review a topic. It is also a valuable resource for instructors involved with distance learning, individual study, or self-paced learning programs. Videotapes ISBN: 0-201-70470-6. Available on CD-ROM. CD-ROM ISBN: 0-201-70966-X
- **Data Software**, prepared by Mario F. Triola, includes the data sets (except for Data Set 6) from Appendix B in the textbook. These data sets are stored as text files, Minitab worksheets, TI Data application for the TI-83 Plus, STATDISK files, SPSS files, and Excel workbooks and are included on the CD-ROM located in the back of the book. The CD-ROM also includes programs for the TI-83 Plus® graphing calculator.
- **STATDISK Statistical Software**, specifically for the Triola text, is a statistical software package (for both Windows and Macintosh) licensed free to

adopters of ***Elementary Statistics***, Eighth edition. This software is included on the CD-ROM in the back of the book.

- *New* **Excel Add-In**: Free software, called Data Desk/XL, designed to enhance the capabilities of Excel's statistics programs. Available on the CD-ROM bound in the back of the book.
- *New* **CD-ROM**: A free CD-ROM, packaged with every new copy of the text, contains the text data sets in formats for various technologies, a TI-83 Plus data application, TI-83 Plus programs, STATDISK software, and the Excel Add-In.
- *New* **Triola *Elementary Statistics* Web Site**: This Web site may be accessed at http://www.awlonline.com/triola, and provides dynamic resources for students. Some of the resources include internet projects keyed to every chapter of the text, data downloads, study tips, instructional video clips, and more.
- ***Student Solutions Manual***, Eighth edition, by Milton Loyer (Penn State University), provides detailed, worked-out solutions to odd-numbered exercises. ISBN: 0-201-70465-X.
- ***STATDISK Student Laboratory Manual and Workbook,*** Eighth edition, written by Mario F. Triola, includes instructions and experiments to be conducted by students using STATDISK software, either in the computer lab, or for out-of-class assignments. ISBN: 0-201-70466-8.
- ***Excel® Student Laboratory Manual and Workbook***, written by Johanna Halsey and Ellena Reda (Dutchess Community College), includes instructions on and examples of Excel in use. The manual includes many examples and problems from the book as well as other appropriate exercises, and it encourages further exploration of statistical concepts. ISBN: 0-201-70459-5.
- ***Minitab® Student Laboratory Manual and Workbook***, Eighth edition, written by Mario F. Triola, includes instructions on and examples of Minitab use. It also supplies many computer experiments and allows further exploration of statistical concepts. ISBN: 0-201-70461-7.
- *New* ***SPSS® Student Laboratory Manual and Workbook***, written by Roger Peck (California State University-Bakersfield), includes instructions on and examples of SPSS use. It also includes appropriate exercises from the text. ISBN: 0-201-70464-1.
- ***TI-83 Plus® Companion to Elementary Statistics***, Eighth edition, by Larry Morgan (Montgomery County Community College) is organized to follow the sequence of topics in the text, and it is an easy-to-follow, step-by-step guide on how to use the TI-83 Plus® graphing calculator. It provides worked-out examples to help students fully understand and use the graphing calculator. This supplement is also suitable for use with a TI-83 calculator. ISBN: 0-201-70469-2.
- **Triola Version of *ActivStats*®**, developed by Paul Velleman and Data Description, Inc., provides complete coverage of introductory statistics topics on CD-ROM. ActivStats integrates video, simulation, animation, narration, text, interactive experiments, World Wide Web access, and Data Desk®, a statistical software package. Homework problems and data sets from the text are included on the CD-ROM. Also available are *ActivStats for Excel* and *ActivStats for SPSS*. *ActivStats for Excel:* Windows and Macintosh ISBN: 0-201-70861-2. *ActivStats for SPSS:* Windows ISBN: 0-201-70860-4. Windows and Macintosh ISBN: 0-201-70859-0. **AS**

- ***AWL Tutor Center***: Free tutoring is available to students who purchase a new copy of the eighth edition of ***Elementary Statistics***. The Addison Wesley Longman Tutor Center (AWLTC) is staffed by qualified statistics and mathematics instructors who provide students with tutoring on text examples, problems, and odd-numbered exercises. Tutoring assistance is provided by telephone, fax, and e-mail and is available five days a week, seven hours a day. Each new book can be bundled with a registration number that provides each student with a free six-month subscription to the service. Request ISBN 0-201-61357-3 (text bundled with AWLTC registration). Students who already have their text may purchase a subscription to the AWLTC by having their bookstore order ISBN 0-201-44461-5. For more information, please contact your Addison Wesley Longman sales representative.

ALSO AVAILABLE FROM THE PUBLISHER . . .

For more than a decade, Addison Wesley Longman Publishing Company has enjoyed an important partnership with Minitab®, Inc., through our publication of the popular Student Editions of Minitab. The Student Editions of Minitab provide a reduced version of Minitab professional software (which allows the user access to 3500 data points and to a full range of Minitab functionalities and graphics capabilities), and an accompanying manual (which includes case studies and hands-on tutorials). Currently available Student Editions are:

Student Edition of Minitab, Release 12 for Windows 95/98 NT
ISBN: 0-201-39715-3
Student Edition of Minitab, Release 8 for DOS ISBN: 0-201-83590-8
Student Edition of Minitab, Release 8 for the Macintosh
ISBN: 0-201-83591-6

Addison Wesley Longman is pleased to offer another statistical software product, ***Data Desk Version 5.0 for Macintosh*** and ***Data Desk 6.0 for Windows***. Data Desk® is an interactive and highly graphical statistics software program originally developed by Paul Velleman (Cornell University).

Data Desk Student Version 6.0 for Windows ISBN: 0-201-25831-5
Data Desk Student Version 5.0 for the Macintosh ISBN: 0-201-57124-2

Any of these products may be purchased separately or bundled with Addison Wesley Longman texts. Instructors can contact local AWL sales consultants for details or contact the company at exam@awl.com for examination copies of any of these items.

M.F.T.
LaGrange, New York

ACKNOWLEDGMENTS

It is an honor and a privilege to be part of a team so committed to the goal of publishing the best possible introductory statistics textbook. It is a pleasure working with the truly exceptional Addison Wesley Longman publishing team, and I thank Deirdre Lynch, Greg Tobin, Elka Block, Rebecca Martin, Peggy McMahon, Barbara Atkinson, Meredith Nightingale, Brenda Bravener, Laura Potter Walton, Joe Vetere, Trish Mescall, Holly Rioux, and the entire Addison Wesley Longman staff.

I thank the professionals who agreed to be interviewed for this book, and I thank Heather Carielli and Ashley McConnaughhay for their help in constructing new data sets. I thank Karen Estes for her work with icon placement, and I thank Paul Lorczak for his work on the Internet Projects. I am grateful for the many suggestions and comments provided by instructors and students, and I am grateful to the hundreds of researchers who did studies, conducted surveys, and compiled interesting and meaningful data used in examples and exercises throughout this book.

I also take great pride in thanking my wife, Ginny, and my sons, Marc and Scott, for their help and encouragement.

I would also like to thank the following for their help with the Eighth Edition:

Text Accuracy Reviewers:
Milton Loyer
David Lund, University of Wisconsin at Eau Claire
Tommy Leavelle, Mississippi College
John Ritschdorff, Marist College

Reviewers of the Eighth Edition:
William A. Ahroon, Plattsburgh State
Scott Albert, College of Du Page
John Bray, Broward Community College–Central
Patricia Buchanan, Pennsylvania State University
Paul Cox, Ricks College
Nirmal Devi, Embry Riddle Aeronautical University
Joseph DeMaio, Kennesaw State University
Dennis Doverspike, University of Akron
Lauren Johnson, Inver Hills Community College
John Klages, County College of Morris
Marlene Kovaly, Florida Community College at Jacksonville
Christopher Jay Lacke, Rowan University

Tommy Leavelle, Mississippi College
Sergio Loch, Grand View College
Rhonda Magel, North Dakota State University–Fargo
Kathleen Mittag, University of Texas–San Antonio
Lyn Noble, Florida Community College at Jacksonville–South
Keith Oberlander, Pasadena City College
Lindsay Packer, College of Charleston
Fabio Santos, LaGuardia Community College
Laura Snook, Blackhawk Community College
Tom Sutton, Mohawk College

For Submitting Student Reviews:
Laura Kincaid, Baldwin–Wallace College

For their work on the videos:
Randall Allbritton, Daytona Beach Community College
Elka Block, Twin Prime Editorial
Nick Gellar, Collin County Community College
Len Groeneveld, Springfield Technical Community College
Denise Heban
Sergio Loch, Grand View College
Frank Purcell, Twin Prime Editorial
John Ritschdorff, Marist College
Jacci White, St. Leo University

I extend my sincere thanks for the suggestions made by the following reviewers and users of previous editions of the book:

Mary Abkemeier, Fontbonne College
Jules Albertini, Ulster County Community College
Tim Allen, Delta College
Stu Anderson, College of Du Page
Jeff Andrews, TSG Associates, Inc.
Mary Anne Anthony, Rancho Santiago Community College
William Applebaugh, University of Wisconsin–Eau Claire
James Baker, Jefferson Community College
Anna Bampton, Christopher Newport University
James Beatty, Burlington County College

Philip M. Beckman, Black Hawk College
Marian Bedee, BGSU, Firelands College
Don Benbow, Marshalltown Community College
Michelle Benedict, Augusta College
Kathryn Benjamin, Suffolk County Community College
Ronald Bensema, Joliet Junior College
David Bernklau, Long Island University
Maria Betkowski, Middlesex Community College
Shirley Blatchley, Brookdale Community College
David Blaueuer, University of Findlay
Randy Boan, Aims Community College
Denise Brown, Collin County Community College
John Buchl, John Wood Community College
Michael Butler, Mt. San Antonio College
Jerome J. Cardell, Brevard Community College
Don Chambless, Auburn University
Rodney Chase, Oakland Community College
Bob Chow, Grossmont College
Philip S. Clarke, Los Angeles Valley College
Darrell Clevidence, Carl Sandburg College
Susan Cribelli, Aims Community College
Imad Dakka, Oakland Community College
Arthur Daniel, Macomb Community College
Gregory Davis, University of Wisconsin, Green Bay
Tom E. Davis, III, Daytona Beach Community College
Charles Deeter, Texas Christian University
Joe Dennin, Fairfield University
Richard Dilling, Grace College
Rose Dios, New Jersey Institute of Technology
Paul Duchow, Pasadena City College
Bill Dunn, Las Positas College
Marie Dupuis, Milwaukee Area Technical College
Evelyn Dwyer, Walters State Community College
Jane Early, Manatee Community College
Sharon Emerson-Stonnell, Longwood College
P. Teresa Farnum, Franklin Pierce College
Ruth Feigenbaum, Bergen Community College

Vince Ferlini, Keene State College
Maggie Flint, Northeast State Technical Community College
Bob France, Edmonds Community College
Christine Franklin, University of Georgia
Richard Fritz, Moraine Valley Community College
Maureen Gallagher, Hartwick College
Mahmood Ghamsary, Long Beach City College
Tena Golding, Southeastern Louisiana University
Elizabeth Gray, Southeastern Louisiana University
David Gurney, Southeastern Louisiana University
Francis Hannick, Mankato State University
Sr. Joan Harnett, Molloy College
Leonard Heath, Pikes Peak Community College
Peter Herron, Suffolk County Community College
Mary Hill, College of Du Page
Larry Howe, Rowan College of New Jersey
Lloyd Jaisingh, Morehead State University
Martin Johnson, Gavilan College
Roger Johnson, Carleton College
Herb Jolliff, Oregon Institute of Technology
Francis Jones, Huntington College
Toni Kasper, Borough of Manhattan Community College
Alvin Kaumeyer, Pueblo Community College
William Keane, Boston College
Robert Keever, SUNY, Plattsburgh
Alice J. Kelly, Santa Clara University
Dave Kender, Wright State University
Michael Kern, Bismarck State College
Marlene Kovaly, Florida Community College at Jacksonville
Tomas Kozubowski, University of Tennessee
Shantra Krishnamachari, Borough of Manhattan Community College
Richard Kulp, David Lipscomb University
Linda Kurz, SUNY College of Technology
Tommy Leavelle, Mississippi College
R. E. Lentz, Mankato State University
Timothy Lesnick, Grand Valley State University
Dawn Lindquist, College of St. Francis
George Litman, National-Louis University

Benny Lo, Ohlone College
Sergio Loch, Grand View College
Vincent Long, Gaston College
Barbara Loughead, National-Louis University
David Lund, University of Wisconsin–Eau Claire
Rhonda Magel, North Dakota State University
Gene Majors, Fullerton College
Hossein Mansouri, Texas State Technical College
Virgil Marco, Eastern New Mexico University
Joseph Mazonec, Delta College
Caren McClure, Rancho Santiago Community College
Phillip McGill, Illinois Central College
Marjorie McLean, University of Tennessee
Austen Meek, Cañada College
Robert Mignone, College of Charleston
Glen Miller, Borough of Manhattan Community College
Kermit Miller, Florida Community College at Jacksonville
Mitra Moassessi, Santa Monica College
Charlene Moeckel, Polk Community College
Theodore Moore, Mohawk Valley Community College
Gerald Mueller, Columbus State Community College
Sandra Murrell, Shelby State Community College
Faye Muse, Asheville–Buncombe Technical Community College
Gale Nash, Western State College
Felix D. Nieves, Antillean Adventist University
DeWayne Nymann, University of Tennessee
Patricia Oakley, Seattle Pacific University
Patricia Odell, Bryant College
James O'Donnell, Bergen Community College
Alan Olinsky, Bryant College
Ron Pacheco, Harding University
Kwadwo Paku, Los Medanos College
Deborah Paschal, Sacramento City College
S. A. Patil, Tennessee Technological University
Robin Pepper, Tri-County Technical College
David C. Perkins, Texas A&M University–Corpus Christi
Anthony Piccolino, Montclair State University
Richard J. Pulskamp, Xavier University

Vance Revennaugh, Northwestern College
C. Richard, Southeastern Michigan College
Sylvester Roebuck, Jr., Olive Harvey College
Kenneth Ross, Broward Community College
Charles M. Roy, Camden County College
Kara Ryan, College of Notre Dame
Richard Schoenecker, University of Wisconsin, Stevens Point
Nancy Schoeps, University of North Carolina, Charlotte
Jean Schrader, Jamestown Community College
A. L. Schroeder, Long Beach City College
Phyllis Schumacher, Bryant College
Sankar Sethuraman, Augusta College
Rosa Seyfried, Harrisburg Area Community College
Calvin Shad, Barstow College
Carole Shapero, Oakton Community College
Lewis Shoemaker, Millersville University
Joan Sholars, Mt. San Antonio College
Galen Shorack, University of Washington
Teresa Siak, Davidson County Community College
Cheryl Slayden, Pellissippi State Technical Community College
Arthur Smith, Rhode Island College
Marty Smith, East Texas Baptist University
Aileen Solomon, Trident Technical College
Sandra Spain, Thomas Nelson Community College
Maria Spinacia, Pasco–Hernandez Community College
Paulette St. Ours, University of New England
W. A. Stanback, Norfolk State University
Carol Stanton, Contra Costa College
Richard Stephens, Western Carolina College
W. E. Stephens, McNeese State University
Terry Stephenson, Spartanburg Methodist College
Consuelo Stewart, Howard Community College
Ellen Stutes, Louisiana State University at Eunice
Sr. Loretta Sullivan, University of Detroit Mercy
Andrew Thomas, Triton College
Evan Thweatt, American River College
Judith A. Tully, Bunker Hill Community College

Paul Velleman, Cornell University
Gary Van Velsir, Anne Arundel Community College
Randy Villa, Napa Valley College
Hugh Walker, Chattanooga State Technical Community College
Charles Wall, Trident Technical College
Glen Weber, Christopher Newport College
David Weiner, Beaver College
Sue Welsch, Sierra Nevada College
Roger Willig, Montgomery County Community College
Gail Wiltse, St. Johns River Community College
Odell Witherspoon, Western Piedmont Community College
Jean Woody, Tulsa Junior College
Thomas Zachariah, Loyola Marymount University
Elyse Zois, Kean College of New Jersey

Contents

CHAPTER 10 Multinomial Experiments and Contingency Tables 572

CHAPTER 11 Analysis of Variance 612

CHAPTER 12 Statistical Process Control 652

CHAPTER 13 Nonparametric Statistics 682

CHAPTER 14 Projects, Procedures, Perspectives 753

Appendices 760

ELEMENTARY STATISTICS

EIGHTH EDITION

CHAPTER 1

Introduction to Statistics

As a student, are you in the most dangerous profession?

Two employees were installing aluminum siding on a farmhouse when it became necessary to remove a 36-foot CB antenna fastened at the top of the house. One employee disconnected the antenna and handed it to the other employee who was standing on the ground. When the second employee attempted to lay the antenna on the ground, it made contact with a 7200-volt power transimssion line. This second employee received a fatal shock (based on information from the Occupational Safety and Health Administration).

We all know that some professions are inherently more dangerous than others. Fatal injuries are sometimes suffered by police officers involved in car crashes or gunfights. Firefighters sometimes die as a result of smoke inhalation, collapsing buildings, or explosions. Taxicab drivers are sometimes killed by passengers determined to steal the money collected as fares. Coal miners have suffered shortened lives as a result of their exposure to dangerous levels of coal dust typically found in mines. Some seemingly safe jobs, such as those in the U.S. Postal Service, have gained notoriety from deaths that resulted from employee rage.

The Swiss physician H. C. Lombard once compiled longevity data for different professions. He used death certificates that included name, age at death, and profession. He then proceeded to compute the average (mean) length of life for the different professions, and he found that students were lowest with a mean of only 20.7 years! (See "A Selection of Selection Anomolies" by Wainer, Palmer, and Bradlow in *Chance,* Volume 11, No. 2.) Similar results would be obtained if the same data were collected today in the United States. Is being a student really more dangerous than working near power lines, or being a police officer, a taxicab driver, or a postal employee? We will address this issue later in the chapter.

1-1 Overview

We begin our study of *statistics* by noting that the word has two basic meanings. In one sense, the word refers to specific numbers, such as this published result: "Twenty-three percent of people polled believed that there are too many polls." A second meaning refers to statistics as a method of analysis. It is this second meaning that will concern us here.

DEFINITION

Statistics is a collection of methods for planning experiments, obtaining data, and then organizing, summarizing, presenting, analyzing, interpreting, and drawing conclusions based on the data.

In this book we introduce the methods of statistics and show their application in many fields of study. You will see that statistics is much more than the mere calculation of averages accompanied by colorful graphs. It is a powerful analytical tool that you can use to develop more generalized and meaningful conclusions about populations that go beyond the original sample data.

In statistics we commonly use the terms *population* and *sample.* Because these terms are central to our study, we define them now.

DEFINITIONS

A **population** is the complete collection of all elements (scores, people, measurements, and so on) to be studied. The collection is complete in the sense that it includes all subjects to be studied.

A **census** is the collection of data from *every* element in a population.

A **sample** is a subcollection of elements drawn from a population.

For example, a typical Nielsen television survey uses a *sample* of 4000 households, and the results are used to form conclusions about the *population* of all 103,215,027 households in the United States. Or the *New York Times* might commission a survey of 1000 people to estimate the proportion of the population favoring legalization of marijuana for medical purposes. The 1000 people constitute the *sample,* and all of us constitute the *population.* Every 10 years our government tries to obtain a *census*, but fails because it is impossible to reach everyone.

In this chapter we present some basic concepts about the nature of data, and some examples of uses and abuses of statistics. We conclude with a discussion of the design of experiments. The following important points will be emphasized in this chapter:

- Sample data must be collected in an appropriate way, such as through a process of *random* selection.
- If sample data are not collected in an appropriate way, no amount of statistical torturing can salvage the data.

What we ask is that you begin your study of statistics with an open mind. We are convinced that by the time you complete this introductory course, you will be firm in your belief that statistics is an interesting and rich subject with applications that are extensive, real, and meaningful. We are also convinced that with regular class attendance and appropriate diligence, you will succeed in mastering the basic concepts presented in this course.

The State of Statistics

The word *statistics* is derived from the Latin word *status* (meaning "state"). Early uses of statistics involved compilations of data and graphs describing various aspects of a state or country. In 1662, John Graunt published statistical information about births and deaths. Graunt's work was followed by studies of mortality and disease rates, population sizes, incomes, and unemployment rates. Households, governments, and businesses rely heavily on statistical data for guidance. For example, unemployment rates, inflation rates, consumer indexes, and birth and death rates are carefully compiled on a regular basis, and the resulting data are used by business leaders to make decisions affecting future hiring, production levels, and expansion into new markets.

1-2 The Nature of Data

Data are observations (such as measurements, genders, and survey responses) that have been collected. Data are sometimes used to find statistics. An isolated list of lifeless numbers might appear to be a data set awaiting some statistical manipulation, but the effective use of statistics requires that we know the context of the data, how the data were obtained, and the population from which the data were obtained. We will demonstrate later the importance of sampling methods and the principle of randomness. Now, however, we will introduce some key terms.

DEFINITIONS

A **parameter** is a numerical measurement describing some characteristic of a *population.*

A **statistic** is a numerical measurement describing some characteristic of a *sample*.

EXAMPLES

1. **Parameter:** When Lincoln was first elected to the presidency, he received 39.82% of the 1,865,908 votes cast. If we consider the collection of all of those votes to be the population being considered, then the 39.82% is a *parameter,* not a statistic.
2. **Statistic:** Based on a sample of 877 surveyed executives, it was found that 45% of them would not hire anyone whose job application contained a typographical error. The figure of 45% is a *statistic* because it is based on a sample, not the entire population of all executives.

Some data sets consist of numbers (such as heights), and others are nonnumerical (such as eye colors). The terms *quantitative data* and *qualitative data* are often used to distinguish between these types.

Interesting Statistics

The following claims were circulated by e-mail. They can be tested using methods presented in this book.

- Women blink nearly twice as much as men.
- Right-handed people live, on average, nine years longer than left-handed people.
- You share your birthday with at least 9 million other people.
- China has more English-speaking people than the United States.
- American Airlines saves $40,000 in a year by eliminating one olive from each salad served in first-class.
- The three most valuable brand names on earth are Marlboro, Coca-Cola, and Budweiser, in that order.

Do you think that all of these claims are true?

DEFINITIONS

Quantitative data consist of numbers representing counts or measurements.

Qualitative (or **categorical** or **attribute**) **data** can be separated into different categories that are distinguished by some nonnumeric characteristic.

EXAMPLES

1. **Quantitative data:** The incomes of college graduates
2. **Qualitative data:** The genders (male/female) of college graduates

When working with quantitative data, it is important to use the appropriate units of measurement, such as dollars, feet, hours, and so on. In some cases, quantitative data are scaled to be larger or smaller so that we can work with more convenient numbers. We should be especially careful to observe such notations as "all amounts are in thousands of dollars," or "all times are in hundredths of a second," or "units are in kilograms." Ignoring such units of measurement can lead to very wrong conclusions.

We can further describe quantitative data by distinguishing between *discrete* and *continuous* types.

DEFINITIONS

Discrete data result when the number of possible values is either a finite number or a "countable" number. (That is, the number of possible values is 0 or 1 or 2 and so on.)

Continuous (numerical) data result from infinitely many possible values that correspond to some continuous scale that covers a range of values without gaps, interruptions, or jumps.

EXAMPLES

1. **Discrete data:** The numbers of eggs that hens lay are *discrete* data because they represent *counts.*
2. **Continuous data:** The amounts of milk that cows produce are *continuous* data because they are *measurements* that can assume any value over a continuous span. During a given time interval, a cow might yield an amount of milk that can be any value between 0 gallons and

5 gallons. It would be possible to get 2.343115 gallons, because the cow is not restricted to the discrete amounts of 0, 1, 2, 3, 4, or 5 gallons.

Another common way of classifying data is to use four levels of measurement: nominal, ordinal, interval, and ratio. In applying statistics to real problems, the level of measurement of the data is an important factor in determining which procedure to use. (See Figure 14-1 on page 758.) There will be some references to these levels of measurement in this book, but the important point here is based on common sense: Never do computations and never use statistical methods with data that are not appropriate. For example, it would not make sense to compute an average of social security numbers, because those numbers are data that are used for identification; they don't represent measurements or counts of anything.

DEFINITION

The **nominal level of measurement** is characterized by data that consist of names, labels, or categories only. The data cannot be arranged in an ordering scheme (such as low to high).

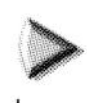

EXAMPLES The following are examples of sample data at the nominal level of measurement.

1. Survey responses of yes, no, and undecided
2. The names of the television shows watched at 10:00 P.M. tonight

Because nominal data lack any ordering or numerical significance, they cannot be used for calculations. Numbers are sometimes assigned to the different categories (especially when data are computerized), but these numbers have no real computational significance and any average calculated with them is meaningless.

DEFINITION

Data at the **ordinal level of measurement** can be arranged in some order, but differences between data values either cannot be determined or are meaningless.

Measuring Disobedience

How are data collected about something that doesn't seem to be measurable, such as people's level of disobedience? Psychologist Stanley Milgram devised the following experiment: A researcher instructed a volunteer subject to operate a control board that gave increasingly painful "electrical shocks" to a third person. Actually, no real shocks were given, and the third person was an actor. The volunteer began with 15 volts and was instructed to increase the shocks by increments of 15 volts. The disobedience level was the point at which the subject refused to increase the voltage. Surprisingly, two-thirds of the subjects obeyed orders even though the actor screamed and faked a heart attack.

EXAMPLES The following are examples of sample data at the ordinal level of measurement.

1. **Course grades:** A college professor assigns grades of A, B, C, D, or F. These grades can be arranged in order, but we can't determine differences between the grades. Thus we know, for example, that A is higher than B (so there is an ordering), but we cannot subtract B from A (so the difference cannot be found).
2. **Rankings:** Based on several criteria, a magazine ranks cities according to their "livability." Those rankings (first, second, third, and so on) determine an ordering. However, the differences between rankings are meaningless. For example, a difference of "second minus first" might suggest $2 - 1 = 1$, but this difference of 1 is meaningless because it is not an exact quantity that can be compared to other such differences. The difference between the first city and the second city is not the same as the difference between the second city and the third city. Using the magazine rankings, the *difference* between New York City and Boston cannot be quantitatively compared to the *difference* between St. Louis and Philadelphia.

Ordinal data provide information about relative comparisons, but not the magnitudes of the differences. They should not be used for calculations.

DEFINITION

The **interval level of measurement** is like the ordinal level, with the additional property that the difference between any two data values is meaningful. However, there is no natural zero starting point (where *none* of the quantity is present).

EXAMPLES The following are examples of data at the interval level of measurement.

1. **Temperatures:** Body temperatures of 98.2°F and 98.6°F. Those values are ordered, and we can determine their difference (often called the *distance* between the two values). However, there is no natural starting point. The value of 0°F might seem like a starting point, but it is arbitrary and does not represent the total absence of heat. It is wrong to say that 50°F is twice as hot as 25°F. (Temperature readings on the Kelvin scale are at the ratio level of measurement; that scale has an absolute zero.)
2. **Years:** The years 1000, 2000, 1776, and 1492. (Time did not begin in the year 0, so the year 0 is arbitrary instead of being a natural zero starting point.)

DEFINITION

The **ratio level of measurement** is the interval level modified to include the natural zero starting point (where zero indicates that *none* of the quantity is present). For values at this level, differences and ratios are both meaningful.

EXAMPLES The following are examples of data at the ratio level of measurement. Note the presence of the natural zero value, and note the use of meaningful ratios of "twice" and "three times."

1. **Weights:** Weights (in carats) of diamond engagement rings (0 does represent no weight, and 4 carats is twice as heavy as 2 carats)
2. **Prices:** Prices of college textbooks ($0 does represent no cost, and a $90 book is three times as costly as a $30 book)

This level of measurement is called the ratio level because the starting point makes ratios meaningful. Because a 200-lb weight is twice as heavy as a 100-lb weight, but 50°F is *not* twice as hot as 25°F, weights are at the ratio level while Fahrenheit temperatures are at the interval level. For a concise comparison and review, study Table 1-1 to see the differences among the four levels of measurement.

TABLE 1-1 Levels of Measurement of Data

Level	Summary	Example	
Nominal	Categories only. Data cannot be arranged in an ordering scheme.	Student states: 5 Californians 20 Texans 40 New Yorkers	Categories or names only
Ordinal	Categories are ordered, but differences cannot be determined or they are meaningless.	Student cars: 5 compact 20 mid-size 40 full-size	An order is determined by "compact, mid-size, full-size."
Interval	Differences between values are meaningful, but there is no natural starting point. Ratios are meaningless.	Campus temperatures: 5°F 20°F 40°F	"0°F doesn't mean no heat." 40°F is not twice as hot as 20°F.
Ratio	Like interval level, but there is a natural zero starting point. Ratios are meaningful.	Student commuting distances: 5 mi 20 mi 40 mi	40 mi is *twice* as far as 20 mi.

1-2 Basic Skills and Concepts

Understanding a Statistic Versus a Parameter. *In Exercises 1–4, determine whether the given value is a statistic or a parameter.*

1. A sample of students is selected, and the average (mean) age is 20.7 years.

2. After checking computer records for every commercial movie made last year, the longest running time is found to be 187 minutes.

3. All of the state governors are surveyed, and 30 of them are found to be Democrats.

4. Several Domino sugar packs are randomly selected, and the average (mean) weight of the contents is 3.647 g.

Understanding Discrete Versus Continuous. *In Exercises 5–8, determine whether the given values are from a discrete or continuous data set.*

5. A statistics professor counts 3 absent students.

6. A statistics professor finds that on the first test, the first paper is turned in 39.627 minutes after the test began.

7. In a survey of 1068 Americans, 673 state that they own answering machines.

8. A manufacturer of rechargeable calculator batteries finds that one batch consists of 850 good batteries and 7 that are defective.

Determining Appropriate Measurements. *In Exercises 9–18, determine which of the four levels of measurement (nominal, ordinal, interval, ratio) is most appropriate.*

9. Heights of women basketball players in the WNBA

10. Ratings of superior, above average, average, below average, or poor for blind dates

11. Noon temperatures (in degrees Fahrenheit) in Death Valley this week

12. Social security numbers

13. The years in which new editions of this book were published

14. Zip codes

15. *Consumer Reports* magazine ratings of "best buy, recommended, not recommended"

16. Distances traveled by students who commute to college

17. The actual contents (in ounces) of cola in Coke cans labeled 12 oz

18. The number of errors made when a Circuit City store scans 8000 different purchases

1-2 Beyond the Basics

19. Temperature Increase In the "Born Loser" cartoon strip by Art Sansom, Brutus expresses joy over an increase in temperature from 1° to 2°. When asked what is so good about 2°, he answers that "It's twice as warm as this morning." Explain why Brutus is wrong yet again.

20. Interpreting Political Polling A pollster surveys 200 people and asks them their preference of political party. He codes the responses as 0 (for Democrat), 1 (for Republican), 2 (for Independent), or 3 (for any other responses). He then calculates the average (mean of the numbers) and gets 0.95. How can that value be interpreted?

1-3 Uses and Abuses of Statistics

Uses of Statistics

Many students study statistics because they know that it impresses employers to see a statistics course on a job applicant's transcript, and because virtually any field of study benefits from the application of statistical methods. In this book we include an abundance of practical uses of statistics. Here are a few brief examples:

- Poll results are used to determine the television shows we watch and the products we buy.
- Manufacturers provide better products at lower costs by using statistical quality control tools, such as control charts.
- Diseases are controlled through analyses designed to anticipate epidemics.
- Endangered species of fish and other wildlife are protected through regulations and laws that react to statistical estimates of changing population sizes.
- Through statistical analysis of fatality rates, legislators can better justify laws, such as those governing air pollution, auto inspections, seat belt and air bag use, and drunk driving.

We cite only these few examples, because a complete compilation of the uses of statistics would easily fill the remainder of this book (a prospect not totally unpleasant to some readers). As you proceed with this course, you will encounter many different applications of statistics.

Abuses of Statistics

Abuses of statistics are abundant. About a century ago, statesman Benjamin Disraeli famously said, "There are three kinds of lies: lies, damned lies, and statistics." It has also been said that "figures don't lie; liars figure." Historian Andrew Lang said that some people use statistics "as a drunken man uses lampposts—for support rather than illumination." Political cartoonist Don Wright encourages us to "bring back the mystery of life: lie to a pollster." These statements refer to uses of statistics in which data are presented in ways that are designed to be misleading. Some abusers of statistics are simply ignorant or careless, but others have personal objectives and are willing to suppress unfavorable data while emphasizing supportive data. We will now present several examples of ways in which data can be distorted.

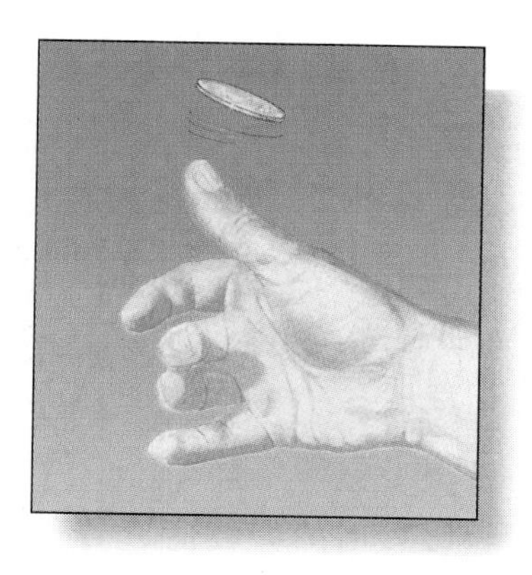

Detecting Phony Data

A class is given the homework assignment of recording the results when a coin is tossed 500 times. One dishonest student decides to save time by just making up the results instead of actually flipping a coin. Because people generally cannot make up results that are really random, we can often identify such phony data. With 500 tosses of an actual coin, it is extremely likely that you will get a run of six heads or six tails, but people almost never include such a run when they make up results.

AS Another way to detect fabricated data is to establish that the results violate Benford's Law: For many collections of data, the leading digits are not uniformly distributed. Instead, the leading digits of 1, 2, . . . , 9 occur with rates of 30%, 18%, 12%, 10%, 8%, 7%, 6%, 5%, and 5% respectively. (See "The Difficulty of Faking Data" by Theodore Hill, Chance, Vol. 12, No. 3.)

Statistics and Land Mines

The International Campaign to Ban Land Mines and the executive director of the Vietnam Veterans of America Foundation (VVAF) were recently awarded the Nobel Peace Prize. When VVAF asked for help in collecting data about land mines, a team of notable statisticians was assembled. Instead of working with intangible data, such as the value of a human life, they worked with tangible raw data, such as the area that a minefield makes unusable, and the cost of crops that cannot be grown. The data were included in *After the Guns Fall Silent: The Enduring Legacy of Landmines*, which became a key resource book in discussions of the land mine issue. The *AMSTAT News* quoted one of the book's editors: "... this data-gathering and analysis effort is what made it possible to put the issue before policy-makers. This work really made a difference."

Bad Samples A major source of deceptive statistics is the use of inappropriate methods to collect data. One very common sampling method somehow allows the sample subjects to decide for themselves whether to be included.

DEFINITION

A **self-selected survey** (or **voluntary response sample**) is one in which the respondents themselves decide whether to be included.

In such surveys, people with strong opinions are more likely to participate, so the responses obtained are not necessarily representative of the whole population. *Self-selected samples* are obtained when a television news program asks viewers to call in with their preference for a political candidate or when a "researcher" mails a survey to people who can decide either to ignore it or to respond.

A self-selected sample is only one way in which the method of collecting data can be seriously flawed. As another example, consider our Chapter Problem, which described how Swiss physician H. C. Lombard once compiled longevity data for different professions by referring to death certificates that included name, age at death, and profession. He then computed the average (mean) length of life for the different professions and found that students ranked lowest with a mean of only 20.7 years. The problem is that the sample is not appropriate, because most people are students when they are relatively young; they are no longer students when they become a little older and are employed in some other profession. Also, all of the students included in the Lombard study *died*, so the 20.7-year longevity figure merely suggests that when students die, they are likely to be young. A similar error would be made if you computed the average age of teenagers who died. That figure could not possibly be more than 19 years, but this does not really mean that being a teenager is more dangerous than being a test pilot, a profession in which the members would surely die at ages with a mean considerably higher than 19 years.

Small Samples It can be very misleading to make broad conclusions or inferences based on samples that are far too small. An example is the Children's Defense Fund's publication of *Children Out of School in America,* which reported that among secondary school students suspended in one region, 67% were suspended at least three times. That figure was based on a sample of only *three* students, and media reports failed to mention that the sample size was so small. Sometimes a sample might seem relatively large (as in a survey of "2000 randomly selected adult Americans"), but if conclusions are based on subgroups, such as the Catholic male Republicans from Maine, such conclusions might be based on samples that are too small. Although it is important to have a sample that is sufficiently large, size is not the only issue; it is just as important to have sample data that have been collected in an appropriate way, such as through random selection. Even large samples can be bad samples.

Loaded Questions Survey questions can be worded to elicit a desired response. In a recent survey, different wordings of a question resulted in different responses:

- 97% yes: "Should the president have the line item veto to eliminate waste?"
- 57% yes: "Should the president have the line item veto, or not?"

Sometimes questions are unintentionally loaded by such factors as the order of the items being considered. Consider these questions from a poll conducted in Germany:

- Would you say that traffic contributes more or less to air pollution than industry?
- Would you say that industry contributes more or less to air pollution than traffic?

When traffic was presented first, 45% blamed traffic and 27% blamed industry; when industry was presented first, 24% blamed traffic and 57% blamed industry.

Misleading Graphs Many visual devices—such as bar graphs and pie charts—can be used to exaggerate or diminish the true import of the data. (Such devices will be discussed in Chapter 2.) The two graphs in Figure 1-1 depict the *same data* from the Bureau of Labor Statistics, but part (a) is designed to exaggerate the difference between the salaries of people who have bachelor's degrees and the salaries of those who have only high school diplomas. By not starting the horizontal axis at zero, the graph in part (a) produces a misleading subjective impression. Figure 1-1 carries an important lesson: To interpret a graph correctly, we must analyze the *numerical* information provided with the graph, so that we won't be misled by its general shape.

AS

Pictographs Drawings of objects, called *pictographs,* may also be misleading. Some objects commonly used to depict data include three-dimensional

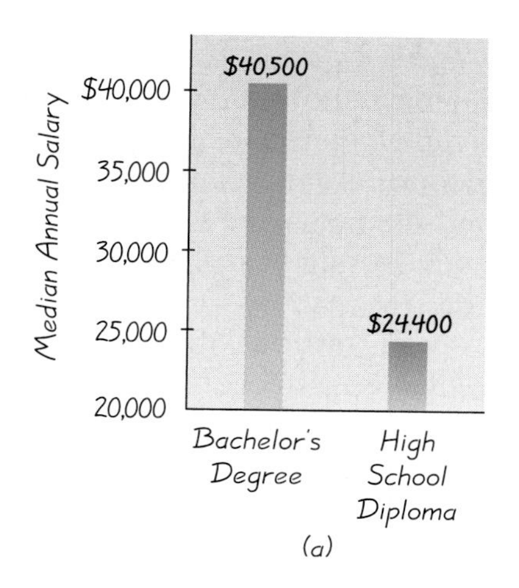

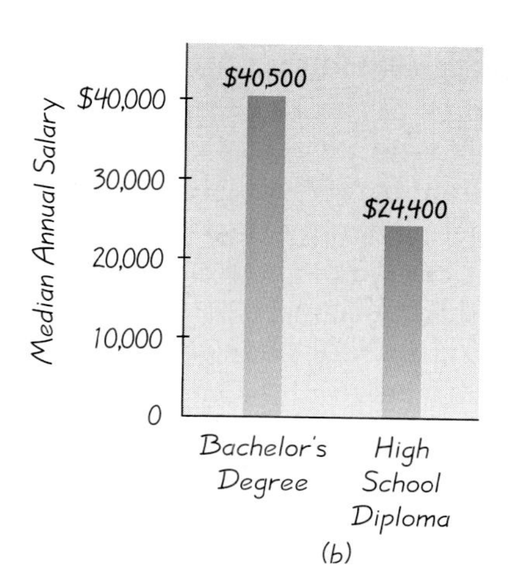

FIGURE 1-1
Salaries of People with Bachelor's Degrees and with High School Diplomas

FIGURE 1-2
Double the length, width, and height of a cube, and the volume increases by a factor of eight.

objects, such as moneybags, stacks of coins, army tanks (for military expenditures), barrels (for oil production), and houses (for home construction). When drawing such objects, artists can create false impressions that distort differences. If you double each side of a square, the area doesn't merely double; it increases by a factor of four. If you double each side of a cube, the volume doesn't merely double; it increases by a factor of eight. If taxes double over a decade, an artist may depict tax amounts with one moneybag for the first year and a second moneybag that is twice as deep, twice as tall, and twice as wide. Instead of appearing to double, taxes will appear to increase by a factor of eight, so the truth will be distorted by the drawing. See Figure 1-2.

Precise Numbers When a statement includes a very precise figure, such as "There are now 103,215,027 households in the United States," many people assume incorrectly that because it is precise it is also *accurate*. In this case, the number is an estimate, and it would be better to state that the number of households is about 103 million.

Distorted Percentages Misleading or unclear percentages are sometimes used. Continental Airlines ran ads boasting better service. In referring to lost baggage, the ads claimed that this was "an area where we've already improved 100% in the last six months." In an editorial criticizing this statistic, the *New York Times* interpreted the 100% improvement figure to mean that no baggage is now being lost—an accomplishment not yet achieved by Continental Airlines.

Partial Pictures "Ninety percent of all our cars sold in this country in the last 10 years are still on the road." Millions of consumers heard that impressive commercial message. What they were not told was that 90% of the cars the advertiser sold in this country were sold within the last three years. The claim was technically correct, but it was very misleading in that it did not present the complete story.

Deliberate Distortions In the book *Tainted Truth,* Cynthia Crossen cites an example of the magazine *Corporate Travel,* which published results showing that among car rental companies, Avis was the winner in a survey of people who rent cars. When Hertz requested detailed information about the survey, the actual survey responses disappeared and the magazine's survey coordinator resigned. Hertz sued Avis (for false advertising based on the survey) and the magazine; a settlement was reached.

Entire books have been devoted to abuses of statistics, including Darrell Huff's classic *How to Lie with Statistics,* Robert Reichard's *The Figure Finaglers,* and Cynthia Crossen's *Tainted Truth.* Understanding these practices will be extremely helpful in evaluating the statistical data found in everyday situations.

1-3 Basic Skills and Concepts

1. "900" Numbers In an ABC "Nightline" poll, 186,000 viewers each paid 50 cents to call a "900" telephone number with their opinion about keeping the United Nations headquarters in the United States. The results showed that 67% of those who called were in favor of moving the United Nations headquarters out of the United States. Interpret the results by identifying what we can conclude about the way the general population feels about keeping the United Nations headquarters in the United States.

2. Telephone Surveys The Hartford Insurance Company has hired you to poll a sample of adults about their car purchases. What is wrong with using telephone directories as the population from which the sample is drawn?

3. CPI The Consumer Price Index (CPI) is based on the cost of goods and services purchased by typical consumers. Assume that the cost is $500 this year.
   a. If there is inflation so that all costs rise 5% next year, what is next year's cost?
   b. Assume that in the year after next year, all costs drop by 5%. What is that year's cost? When the 5% increase is followed by the 5% decrease, do costs return to the original $500 level?

4. Campus Crime In a study of college campus crimes committed by students high on alcohol or drugs, a mail survey of 1875 students was conducted. A *USA Today* article noted, "Eight percent of the students responding anonymously say they've committed a campus crime. And 62% of that group say they did so under the influence of alcohol or drugs." Assuming that the number of students responding anonymously is 1875, how many actually committed a campus crime while under the influence of alcohol or drugs?

5. Healthy Mothers The *Newport Chronicle* claims that pregnant mothers can increase their chances of having healthy babies by eating lobsters. That claim is based on a study showing that babies born to lobster-eating mothers have fewer health problems than babies born to mothers who don't eat lobster. What is wrong with the claim?

6. Job Applicants "According to a nationwide survey of 250 hiring professionals, scuffed shoes was the most common reason for a male job seeker's failure to make a good first impression." Newspapers carried this statement based on a poll commissioned by Kiwi Brands, producers of shoe polish. Comment on why the results of this survey might be questionable.

7. Motorcycle Helmets The Hawaii State Senate held hearings when it was considering a law requiring that motorcyclists wear helmets. Some motorcyclists testified that they had been in crashes in which helmets would not have been helpful. Which important group was not able to testify? (See "A Selection of Selection Anomalies" by Wainer, Palmer, and Bradlow in *Chance*, Volume 11, No. 2.)

8. Longevity You need to conduct a study of longevity for people who were born after the end of World War II in 1945. If you were to visit graveyards and use the birth and death dates listed on tombstones, would you get good results? Why or why not?

9. Bad Question A survey includes this item: "Enter your height in inches." It is expected that actual heights of respondents can be obtained and analyzed, but there are two different major problems with this item. Identify them.

10. Mail Survey When author Shere Hite wrote *Women and Love: A Cultural Revolution in Progress*, she based her conclusions on 4500 replies that she received after mailing 100,000 questionnaires to various women's groups. Are her conclusions likely to be valid in the sense that they can be applied to the general population of all women?

11. Understanding Percentages Is a 10% price cut the same as two consecutive 5% price cuts? Why or why not?

12. Car Weights Refer to Data Set 18 in Appendix B and consider the weights of the randomly selected car models. If you were to calculate the average (mean) weight of those cars, is the result likely to be a good estimate of the weight of the cars being driven in the United States? Why or why not?

1-3 Beyond the Basics

13. Phony Data A researcher at the Sloan-Kettering Cancer Research Center was once criticized for falsifying data. Among his data were figures obtained from six groups of mice, with 20 individual mice in each group. These values were given for the percentage of successes in each group: 53%, 58%, 63%, 46%, 48%, 67%. What is the major flaw?

14. What's Wrong with This Picture? Try to identify each of the four major flaws in the following. A daily newspaper ran a survey by asking readers to call in their response to this question: "Do you support the development of atomic weapons that could kill millions of innocent people?" It was reported that 20 readers responded, and 87% said "no," while 13% said "yes."

15. Understanding Percentages A *New York Times* editorial criticized a chart caption that described a dental rinse as one that "reduces plaque on teeth by over 300%."

a. If you remove 100% of some quantity, how much is left?
b. What does it mean to reduce plaque by over 300%?

16. Interpreting Surveys In an angry letter to the Associated Press, a company president complained about a poll being unfair. He claimed that because the sample size of 1223 people was used to represent 120 million people, his letter represented 98,000 people who have the same views. Is his arithmetic correct? Is his claim valid?

1-4 Design of Experiments

Sometimes we have an interesting and important data set, but we don't have a particular objective in mind, so we want to *explore* the data to see what insights we might acquire. (In Chapter 2 we will introduce basic tools that can be used to explore or describe data sets.) More often, though, we do have a specific objective, and we want to collect the data and do the analysis that will help us to meet that objective. We typically get our data from two common sources: *observational studies* (such as polls) and *experiments* (such as using a treatment to improve hair growth).

DEFINITIONS

In an **observational study,** we observe and measure specific characteristics, but we don't attempt to *modify* the subjects being studied.

In an **experiment,** we apply some *treatment* and then proceed to observe its effects on the subjects.

An example of an observational study is a Nielsen poll. Each night, the Nielsen organization polls people to determine the percentage of the population that is watching NBC at 9:00 P.M. These people are merely counted; they do not receive any type of treatment. An example of an experiment was the testing of the Salk vaccine. In the famous 1954 experiment, some children were given the actual Salk vaccine and others were given a placebo that contained no medicine or drug. This was an experiment, as opposed to an observational study, because the subjects were given a treatment. In a well-designed experiment, like this one, we compare effects in the treatment group to effects in a group not given the treatment. The use of two groups (treatment group and placebo group) made it possible to evaluate the effectiveness of the Salk vaccine in preventing polio.

There are a few basic steps that should be followed in designing an experiment that is capable of yielding valid results.

1. *Identify your objective:* Identify the exact question to be answered, and clearly identify the relevant population. (For example, "Is the Salk vaccine effective in reducing the incidence of polio in the population of children?")
2. *Collect sample data:* The way in which sample data are collected is absolutely critical to the success of the experiment. The sample data must be representative of the population in question, the sample must be large enough so that the effects of the treatment can be known, and the question should be addressed without interference from extraneous factors.
3. *Use a random procedure that avoids bias.* (In the Salk vaccine experiment, for example, children were assigned to the two groups through a process of random selection.)
4. *Analyze the data and form conclusions.*

Controlling Effects of Variables

When conducting an experiment, it is all too easy to receive interference from variable factors that are not relevant to the issue being studied. These effects can be controlled through good experimental design. In the 1954 polio experiment, for example, a treatment group of children was given the actual vaccine and a control group was given a placebo that contained no

Sampling Rejected for the Census

It has been estimated that in the 1990 Census, roughly 10 million people were not counted, while 4 million others were counted twice. These errors can be largely corrected by applying the same sampling techniques used by pollsters. But there is a political issue. Population counts affect the number of seats in the House of Representatives, so Republicans oppose sampling because undercounted regions tend to be largely Democratic, while overcounted regions tend to be largely Republican. Democrats favor the use of sampling methods. Some people argue that the Constitution specifies that the census be an "actual enumeration," or a head count which does not allow for sampling methods, and the Supreme Court upheld that position. Will modern sampling methods be eventually used, or will the traditional census head count prevail?

Hawthorne and Experimenter Effects

The well-known placebo effect occurs when an untreated subject incorrectly believes that he or she is receiving a real treatment and reports an improvement in symptoms. The Hawthorne effect occurs when treated subjects somehow respond differently, simply because they are part of an experiment. (This phenomenon was called the "Hawthorne effect" because it was first observed in a study of factory workers at Western Electric's Hawthorne plant.) An experimenter effect (sometimes called a Rosenthall effect) occurs when the researcher or experimenter unintentionally influences subjects through such factors as facial expression, tone of voice, or attitude.

drug at all. In experiments of this type, a **placebo effect** occurs when an untreated subject incorrectly believes that he or she is receiving a treatment and reports an improvement in symptoms. The placebo effect can be countered by using **blinding,** a technique in which the subject doesn't know whether he or she is receiving a treatment or a placebo. Blinding is used so that we can determine whether the treatment effect is significantly greater than the placebo effect. The polio experiment was *double-blind*, meaning that blinding occurred at two levels: (1) The children being injected didn't know whether they were getting the Salk vaccine or a placebo, and (2) the doctors who gave the injections and evaluated the results didn't know either.

When designing an experiment to test the effectiveness of one or more treatments, it is important to put the subjects (often called *experimental units*) in different groups (or *blocks*) in such a way that those groups are very similar. A **block** is a group of subjects (or experimental units) that are similar. (They need to be similar only in the ways that might affect the outcome of the experiment.)

When testing one or more different treatments, form blocks so that each one consists of subjects that are similar.

When deciding how to assign the subjects to different blocks, you can use random selection or you can try to carefully control the assignment so that the subjects within each block are similar. One approach is to use a **completely randomized experimental design,** in which subjects are put into different blocks through a process of *random selection*. An example of a completely randomized experimental design is, again, the polio experiment. Children were assigned to the treatment group or the placebo group through a process of random selection (equivalent to flipping a coin). Another approach is to use a **rigorously controlled design,** in which experimental units are very *carefully chosen* so that the subjects in each block are similar in the ways that are important. With a rigorously controlled design, you might start by giving the Salk vaccine to a healthy seven-year-old girl from Texas, while another healthy seven-year-old girl from Texas would be given the placebo. To make this work, you would have to go beyond health, age, and state of residence and identify other relevant factors to be considered.

When conducting experiments, the results are sometimes ruined because of *confounding*.

DEFINITION

Confounding occurs in an experiment when the effects from two or more variables cannot be distinguished from each other.

For example, suppose a professor in Vermont experiments with a new attendance policy ("your course average drops one point for each class cut"), but an exceptionally mild winter moderates the discomforts (snow and cold temperatures) that have reduced attendance in the past. If attendance does improve, we can't tell whether the improvement is attributable to the new attendance policy or to the mild winter—the effects of the attendance policy and the weather have been confounded.

Sample Size

Another important consideration in conducting experiments is the size of your sample. It must be large enough so that erratic behavior of very small samples will not produce misleading results. Repetition of an experiment is called *replication*.

A large sample is not necessarily a good sample. Although it is important to have a sample that is sufficiently large, it is more important to have a sample in which the elements have been chosen in an appropriate way, such as random selection.

Use a sample size large enough so that we can see the true nature of any effects, and obtain the sample using an appropriate method, such as one based on *randomness*.

In the testing of the Salk vaccine, a sample size of only four people in the treatment group and four other people in the control group would not reveal whether the vaccine is effective. Fortunately, the actual experiment involved large samples: 200,000 children were given the actual Salk vaccine and 200,000 other children were given a placebo. Because the actual experiment used sufficiently large sample sizes, the effectiveness of the vaccine could be seen. Nevertheless, even though the treatment and placebo groups were very large, the experiment would have failed if subjects had not been assigned to the two groups in an appropriate way.

Randomization

One of the worst mistakes is to collect data in a way that is inappropriate. We cannot overstress this very important point:

Data carelessly collected may be so completely useless that no amount of statistical torturing can salvage them.

We noted in Section 1-3 that a self-selected survey (or voluntary response sample) is one in which people decide themselves whether to respond. Self-selected samples are very common, but their results are generally useless for making valid inferences about larger populations.

We now define and describe some of the more common methods of sampling.

DEFINITIONS

In a **random sample** members of the population are selected in such a way that each has an *equal chance* of being selected.

A **simple random sample** of size n subjects is selected in such a way that every possible sample of size n has the same chance of being chosen.

Meta-Analysis

The term *meta-analysis* refers to a technique of doing a study that essentially combines results of other studies. It has the advantage that separate smaller samples can be combined into one big sample, making the collective results more meaningful. It also has the advantage of using work that has already been done. Meta-analysis has the disadvantage of being only as good as the studies that are used. If the previous studies are flawed, the "garbage in, garbage out" phenomenon can occur. The use of meta-analysis is currently popular in medical research and psychological research. This study used data from 46 other studies: "Meta-Analysis of Migraine Headache Treatments: Combining Information from Heterogeneous Designs" by Dominici et al, *Journal of the American Statistical Association*, Vol. 94, No. 445.

The Literary Digest Poll

In the 1936 presidential race, *Literary Digest* magazine ran a poll and predicted an Alf Landon victory, but Franklin D. Roosevelt won by a landslide. Maurice Bryson notes, "Ten million sample ballots were mailed to prospective voters, but only 2.3 million were returned. As everyone ought to know, such samples are practically always biased." He also states, "Voluntary response to mailed questionnaires is perhaps the most common method of social science data collection encountered by statisticians, and perhaps also the worst." (See Bryson's "The *Literary Digest* Poll: Making of a Statistical Myth," *The American Statistician*, Vol. 30, No. 4.)

With random sampling we expect all groups of the population to be (approximately) proportionately represented. Random samples are selected by many different methods, including using computers to generate random numbers. (Before computers, tables of random numbers were often used instead. For truly exciting reading, see this book consisting of 1 million digits that were randomly generated: *A Million Random Digits*, published by Free Press. The *Cliff Notes* summary of the plot is not yet available.) Careless or haphazard sampling can easily result in a biased sample with characteristics unlike the population from which the sample was drawn. Random sampling, in contrast, is very carefully planned and executed. There are other sampling techniques in use, and we describe the common ones here, but *only random sampling and simple random sampling will be used throughout the remainder of the book.*

DEFINITION

In **systematic sampling,** we select some starting point and then select every kth (such as every 50th) element in the population.

For example, if Coca Cola managers wanted to poll the 29,500 employees, they could begin with a complete employee roster, then select every 50th person to obtain a sample of size 590. This method is simple and is often used. (If we randomly pick the starting point, then randomly pick every 50th person on an alphabetized list, we have a *random sample* because everyone has the same chance of being chosen, but we do not have a *simple random sample* because not every possible sample of 590 employees has the same chance. For example, there is no chance of getting the first 590 people on the alphabetized list.)

DEFINITION

With **convenience sampling,** we simply use results that are readily available.

In some cases, results from convenience sampling may be quite good, but in many other cases they may be seriously biased. In investigating the proportion of left-handed people, it would be convenient for a student to survey his or her classmates, because they are readily available. Even though such a sample is not random, the results should be quite good. In contrast, it might be convenient (and perhaps also profitable) for ABC News to conduct a poll by asking audience members to call a "900" telephone number to register their opinions, but this would be a self-selected sample and the results would likely be biased.

DEFINITION

With **stratified sampling,** we subdivide the population into at least two different subgroups (or strata) that share the same characteristics (such as gender or age bracket), then we draw a sample from each stratum.

For example, using the 50 states as strata, we might select a random sample of voters in each state. If the different strata have sample sizes that are in the same proportion as in the population, we say that we have *proportionate* sampling. If it should happen that some strata are not represented in the proper proportion, then the results can be adjusted or weighted accordingly. For a fixed sample size, if you randomly select subjects from different strata, you are likely to get more consistent (and less variable) results than by simply selecting a random sample from the general population. For that reason, stratified sampling is often used to reduce the variation in the results.

DEFINITION

In **cluster sampling,** we first divide the population area into sections (or clusters), then randomly select some of those clusters, and then choose *all* the members from those selected clusters.

Note that stratified sampling and cluster sampling both involve the formation of subgroups, but cluster sampling uses *all* members from a sample of clusters, whereas stratified sampling uses a *sample* of members from all strata. An example of cluster sampling can be found in a pre-election poll, in which we randomly select 30 election precincts and then survey all the people from each of those precincts. This would be much faster and much less expensive than selecting one person from each of the many precincts in the population area. The results can be adjusted or weighted to correct for any disproportionate representations of groups. Cluster sampling is used extensively by government and private research organizations.

Figure 1-3 on page 22 illustrates the five common methods of sampling just described. In practice, professionals often collect data by using some combination of the five methods. Here is one typical example of what is called a *multistage sample design*: First randomly select a sample of counties from all 50 states, then randomly select cities and towns in those counties, then randomly select residential blocks in each city or town, then randomly select households in each block, then randomly select someone from each household. We will not use such a sample design in this book. We should again stress that the methods of this book typically require that we have a *simple random sample*.

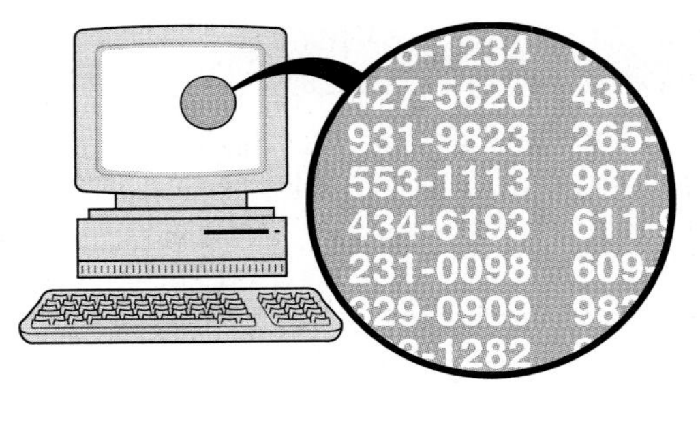

Random Sampling:
Each member of the population has an equal chance of being selected. Computers are often used to generate random telephone numbers.

Systematic Sampling:
Select every kth member.

Convenience Sampling:
Use results that are readily available.

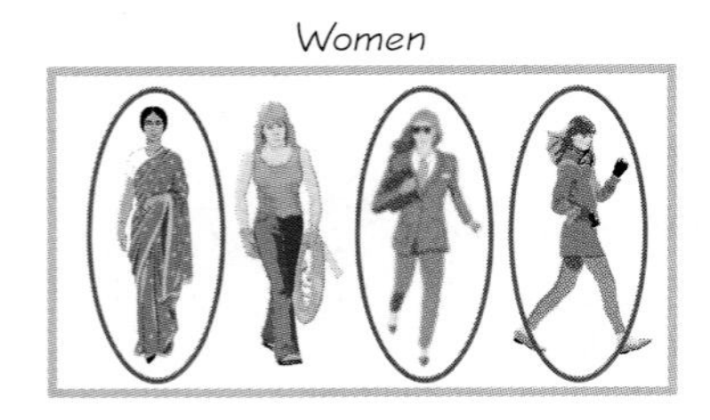

Stratified Sampling:
Classify the population into at least two strata, then draw a sample from each.

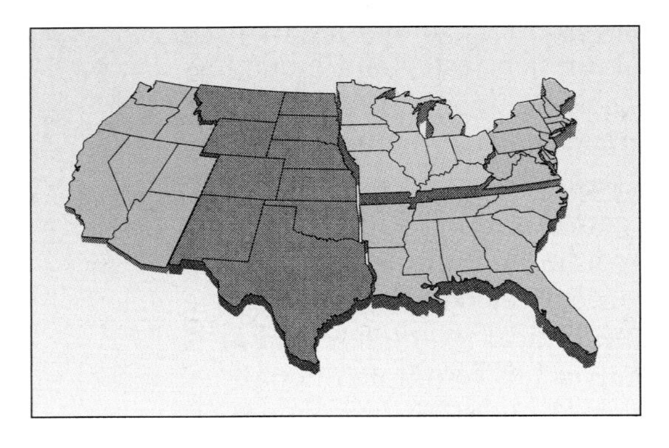

Cluster Sampling:
Divide the population area into sections, randomly select a few of those sections, and then choose all members in them.

FIGURE 1-3
Common Sampling Methods

No matter how well you plan and execute the sample collection process, there is likely to be some error in the results. For example, you might randomly select 1000 adults, ask them if they graduated from high school, and record the sample percentage of "yes" responses. If you randomly select another sample of 1000 adults, it is likely that you will obtain a *different* sample percentage.

DEFINITIONS

A **sampling error** is the difference between a sample result and the true population result; such an error results from chance sample fluctuations.

A **nonsampling error** occurs when the sample data are incorrectly collected, recorded, or analyzed (such as by selecting a biased sample, using a defective measurement instrument, or copying the data incorrectly).

If we carefully collect a sample so that it is representative of the population, we can use the methods in this book to analyze the sampling error, but we must exercise extreme care so that nonsampling error is minimized.

1-4 Basic Skills and Concepts

Distinguishing Between Observational Study and Experiment. *In Exercises 1–4, determine whether the given description corresponds to an observational study or an experiment.*

1. Cans of Coke are opened, and the volumes (in ounces) of the contents are measured.

2. The new drug Statisticzene is tested by recording its effects on the students who are given the drug.

3. The effectiveness of multimedia teaching is tested with a sample of students who complete a course of study using the multimedia approach.

4. Much controversy arose over a study of patients with syphilis who were *not* given a treatment that could have cured them.

Identifying Types of Sampling. *In Exercises 5–16, identify which of these types of sampling is used: random, systematic, convenience, stratified, or cluster.*

5. Telephone Surveys The Gallup Organization plans to conduct a poll of New York City residents with the "212" area code. Computers are used to randomly generate telephone numbers that are automatically called.

6. MTV Survey A marketing expert for MTV is planning a survey in which 500 people will be randomly selected from each age group of 10–19, 20–29, and so on.

7. News Reporting An ABC news reporter polls people as they pass him on the street.

8. Medical Research A Johns Hopkins University researcher surveys all cardiac patients in each of 30 randomly selected hospitals.

9. Car Ownership A General Motors researcher has partitioned all registered cars into categories of subcompact, compact, mid-size, intermediate, and full-size. She is surveying 200 randomly selected car owners from each category.

10. Jury Selection The Dutchess County Commissioner of Jurors obtains a list of 42,763 car owners and constructs a pool of jurors by selecting every 100th name on that list.

11. Student Drinking The College of Newport conducts a study of student drinking by randomly selecting 10 different classes and interviewing all of the students in each of those classes.

12. Lobbying A lobbyist for the tobacco industry obtains a sample of members of Congress by writing the 535 names on individual index cards, putting them in a box, mixing them, then selecting 50 different names.

13. Education and Salary An economist is studying the effect of education on salary and conducts a survey of 150 randomly selected workers from each of these categories: less than a high school diploma, high school diploma, more than a high school diploma.

14. Prisoners A prison guard surveys all prisoners in his ward.

15. Sobriety Checkpoint The author was an observer at a police sobriety checkpoint at which every fifth driver was stopped and interviewed. (He witnessed the arrest of a former student.)

16. Exit Poll CNN is planning an exit poll in which 100 polling stations will be randomly selected and all voters will be interviewed as they leave the premises.

Obtaining a Simple Random Sample. *In Exercises 17–20, describe a procedure for obtaining a simple random sample.*

17. Select a simple random sample of 200 students from the population of full-time students at your college.

18. Select a simple random sample of 50 Bufferin aspirin tablets to be tested for the precise amount of aspirin that they contain.

19. Select a simple random sample of 250 college textbooks from the population of all college textbooks currently being used in your state.

20. Select a simple random sample of 100 households in your county.

1-4 Beyond the Basics

21. Reducing Crime Two categories of survey questions are *open* and *closed.* An open question allows a free response, while a closed question allows only a fixed response. The following examples are based on Gallup surveys.

    Open question: What do you think can be done to reduce crime?

    Closed question: Which of the following approaches would be most effective in reducing crime?

    - Hire more police officers.
    - Get parents to discipline children more.
    - Correct social and economic conditions in poor neighborhoods.
    - Improve rehabilitation efforts in jails.
    - Give convicted criminals tougher sentences.
    - Reform courts.

    a. What are the advantages and disadvantages of open questions?
    b. What are the advantages and disadvantages of closed questions?
    c. Which type is easier to analyze with formal statistical procedures, and why is that type easier?

22. Sampling Design The Addison Wesley Longman Publishing Company has commissioned you to survey 100 students who use this book. Describe procedures for obtaining a sample of each type: random, systematic, convenience, stratified, cluster.

23. Confounding Give an example (different than the one in the text) illustrating how confounding occurs.

24. Random Selection Among the 50 states, one state is randomly selected. Then, a statewide voter registration list is obtained and one name is randomly selected. Does this procedure result in a randomly selected voter?

25. Simple Random Samples We noted that systematic sampling could result in a random sample, but not a simple random sample.

a. Does stratified sampling result in a random sample? A simple random sample? Explain.

b. Does cluster sampling result in a random sample? A simple random sample? Explain.

Misleading Graphs

1. Refer to the accompanying graphs reproduced from a front-page story appearing in the *Poughkeepsie Journal*. Identify two different flaws in those graphs.

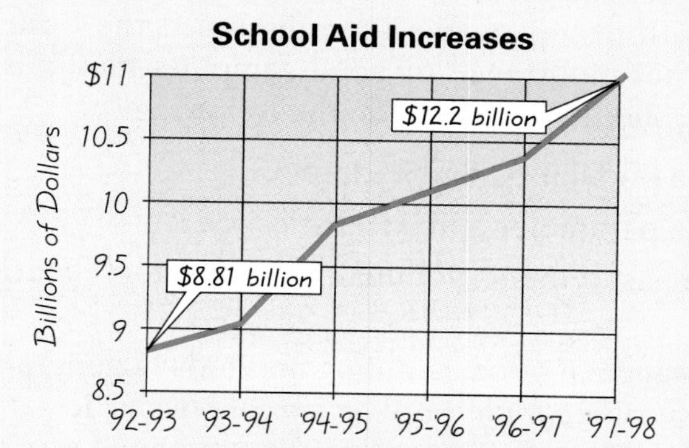

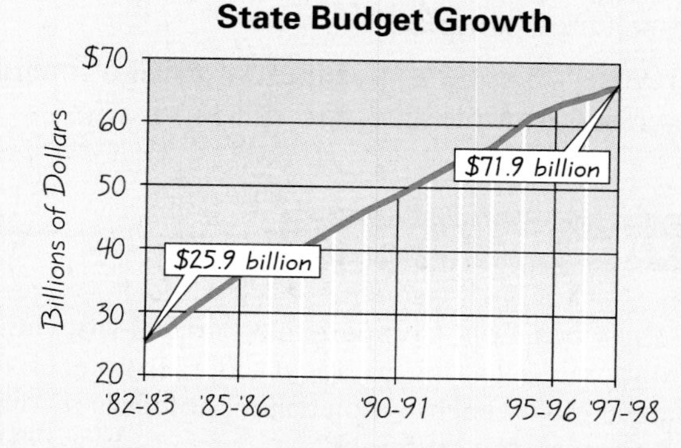

2. Refer to the *USA Today* illustration in which percentages are represented by volumes of a portion of someone's head. Are the data presented in a format that makes them easy to understand and compare? Could the same information be presented in a better way? If so, construct a graph of your own that better depicts the given information.

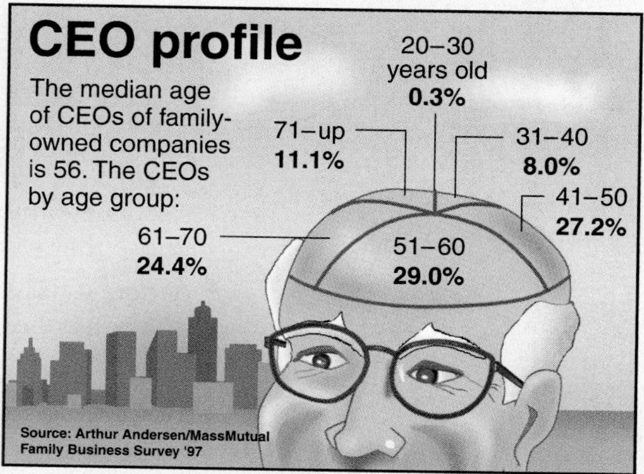

USA Today Snapshot. "A look at statistics that shape your finances."

VOCABULARY LIST

statistics
population
census
sample
data
parameter
statistic
quantitative data
qualitative data
discrete data
continuous data
nominal level of measurement
ordinal level of measurement
interval level of measurement
ratio level of measurement
self-selected survey (or voluntary reponse sample)
observational study
experiment
placebo effect
blinding
block
completely randomized experimental design
rigorously controlled design
confounding
replication
random sample
simple random sample
systematic sampling
convenience sampling
stratified sampling
cluster sampling
sampling error
nonsampling error

REVIEW

We began this chapter with a general description of the nature of statistics, then discussed different aspects of the nature of data. Uses and abuses of statistics were illustrated with examples. We then discussed the design of experiments, emphasizing the importance of good sampling methods. On completing this chapter, you should be able to do the following:

- Distinguish between a population and a sample.
- Distinguish between a parameter and a statistic.
- Identify the level of measurement (nominal, ordinal, interval, ratio) of a set of data.
- Recognize the importance of good sampling methods in general; recognize the importance of a *simple random sample* in particular; and recognize the serious deficiency of poor sampling methods, especially those that result in self-selected samples.
- Understand the importance of good experimental design, including the control of variable effects, sample size, and randomization

REVIEW EXERCISES

1. Understanding Coke Data The Coca Cola Company has 366,000 stockholders, and a poll is conducted by randomly selecting 30 stockholders from each of the 50 states. The number of shares held by each sampled stockholder is recorded.

a. Are the values obtained discrete or continuous?

b. Identify the level of measurement (nominal, ordinal, interval, ratio) for the sample data.
c. Which type of sampling (random, systematic, convenience, stratified, cluster) is being used?
d. If the average (mean) number of shares is calculated from the sample group, is the result a statistic or a parameter?
e. If you are the chief executive officer of the Coca Cola Company, what characteristic of the data set would you consider to be extremely important?
f. What is wrong with gauging stockholder views by mailing a questionnaire that stockholders could complete and mail back?

2. More Coke Identify the type of sampling (random, systematic, convenience, stratified, cluster) used when a sample of the 366,000 Coca Cola shareholders is obtained as described.
   a. A complete list of all stockholders is compiled, and every 500th name is selected.
   b. At the annual stockholders' meeting, a survey is conducted of all who attend.
   c. Fifty different stockbrokers are randomly selected, and a survey is made of all their clients who own shares of Coca Cola.
   d. A computer file of all stockholders is compiled so that they are all numbered consecutively, then random numbers generated by computer are used to select the sample of stockholders.
   e. All of the stockholders' zip codes are collected, and five stockholders are randomly selected from each zip code.

3. Sampling Design You plan to conduct a poll of the full-time students who attend your college. Describe a procedure for obtaining a sample of each type: random, systematic, convenience, stratified, cluster.

4. Design of Experiment You plan to conduct an experiment to test the effectiveness of Sleepeze, a new drug that allegedly cures insomnia. You will use a sample of subjects who are treated with the drug and another sample of subjects who are given a placebo.
   a. What is "blinding," and how might it be used in this experiment?
   b. Why is it important to use blinding in this experiment?
   c. What is a completely randomized block design?
   d. What is a rigorously controlled block design?
   e. What is replication, and why is it important?

5. Dropout Rate Your college dean has commissioned you to do a study of student dropout rates, and you plan to poll every 50th student who passes a central campus location. What is fundamentally wrong with this sampling plan?

6. Identify the level of measurement (nominal, ordinal, interval, ratio) used in each of the following.
   a. The amounts of tar (in milligrams) in a sample of cigarettes
   b. A movie critic's ratings of "must see, recommended, not recommended, don't even think about going"
   c. A movie critic's classification of "drama, comedy, adventure"
   d. IQ scores, where the score is considered to be a measure of intelligence (not the number of points scored on the IQ test)
   e. IQ scores, where the score is considered to be the number of points scored on the IQ test (not a measure of intelligence)

Cumulative Review Exercises

This book's Cumulative Review Exercises are designed to include topics from preceding chapters. For this chapter, we present calculator warm-up exercises with expressions similar to those found throughout the book. Use your calculator to obtain the indicated values.

1. $\dfrac{1.23 + 4.56 + 7.89}{3}$

2. $\sqrt{\dfrac{(5-7)^2 + (12-7)^2 + (4-7)^2}{3-1}}$

3. $\dfrac{1.96^2 \times (0.4)(0.6)}{0.025^2}$

4. $\dfrac{98.20 - 98.60}{\dfrac{0.62}{\sqrt{106}}}$

5. $\dfrac{25!}{16!\,9!}$

6. $\sqrt{\dfrac{10(513.27) - 71.5^2}{10(10-1)}}$

7. $\dfrac{8(151{,}879) - (516.5)(2176)}{\sqrt{8(34{,}525.75) - 516.5^2}\sqrt{8(728{,}520) - 2176^2}}$

8. $\dfrac{(183 - 137.09)^2}{137.09} + \dfrac{(30 - 41.68)^2}{41.68}$

In Exercises 9–12, some of the given expressions are designed to yield results expressed in a form of scientific notation. For example, the result of 1.23E5 can be expressed as 123,000, and the result of 4.56E − 4 can be expressed as 0.000456. Perform the indicated operation and express the result as an ordinary number that is not in scientific notation.

9. 0.95^{150} **10.** 25^8 **11.** 52^6 **12.** 0.25^5

Cooperative Group Activities

1. *In-class activity:* From the cafeteria, obtain 18 straws. Cut 6 of them in half, cut 6 of them into quarters, and leave the other 6 as they are. There should now be 42 straws of different lengths. Put them in a bag, mix them up, then select one straw, find its length, then replace it. Repeat this until 20 straws have been selected. Important: Select the straws without looking into the bag, and select the first straw that is touched. Find the average (mean) length of the sample of 20 straws. Now remove all of the straws and find the mean length of the population. Did the sample provide an average that was close to the true population average? Why or why not?

2. *Out-of-class activity:* The following survey question raised concerns when responses suggested that about 22% of Americans thought that the Holocaust might not have occurred.

"Does it seem possible or does it seem impossible to you that the Nazi extermination of the Jews never happened?"

A subsequent poll showed that respondents were probably confused by the double negative in the wording of the question. Here is the wording used in a subsequent Roper poll:

"Does it seem possible to you that the Nazi extermination of the Jews never happened, or do you feel certain that it happened?"

Is this second version substantially less confusing? Write the question in a way that is clearer than both of these versions.

TECHNOLOGY PROJECT

The objective of this project is to introduce the technology resources that you will be using in your statistics course. Refer to Data Set 1 in Appendix B and use the first 10 volumes (in ounces) of regular Coke. Using your statistics software package or a TI-83 Plus calculator, enter those 10 values, then obtain a printout of them.

STATDISK: Click on **Data** at the top of the screen, then select **Sample Editor** and proceed to enter the data. To obtain a printout, click on **File,** then select **Print.**

Minitab: Enter the data in the column C1, then click on **File,** and select **Print Worksheet.**

Excel: Enter the data in column A, then click on **File,** and select **Print.**

TI-83 Plus: Printing a TI-83 Plus screen display is possible only if you are using a Graphlink connection to a computer.

Critical Thinking

Exercise 1 in Section 1-3 included a reference to an ABC "Nightline" poll in which 186,000 viewers each paid 50 cents to call a "900" telephone number with their opinion about keeping the United Nations in the United States. Of those who called, 67% were in favor of moving the United Nations out of the United States. The show's host, Ted Koppel, reported that a "scientific" poll of 500 people showed that 72% of us want the United Nations to *stay* in the United States.

Analyzing the results

Write a response to someone who claims that the first result is likely to be better because it is based on 186,000 people, whereas the second result is based on only 500 people.

Internet Project

In this section of each chapter, you will be directed to the Web page at

http://www.awlonline.com/triola

from which you can reach the page for all the Internet Projects given for *Elementary Statistics, Eighth Edition*. Go to this Web site now and familiarize yourself with all of the available features for the book.

Each Internet Project includes activities, such as exploring data sets, performing simulations, and researching true-to-life examples found at various Web sites. These activities will help you explore and understand the rich nature of statistics and its importance in our world. So visit this Web site often and enjoy the activities!

Statistics *at work*

"We use statistics to determine the degree of isolation between putative stocks."

Sarah Mesnick

Behavioral and Molecular Ecologist

Sarah Mesnick is a National Research Council postdoctoral fellow. In her work as a marine mammal biologist, she conducts research at sea as well as in the Laboratory of Molecular Ecology. Her research focuses on the social organization and population structure of sperm whales. She received her doctorate in evolutionary biology at the University of Arizona.

What do you do?

My research focuses on the relationship between sociality and population structure in sperm whales. We use this information to build better management models for the conservation of this, and other, endangered marine mammal species.

What concepts of statistics do you use?

Currently, I use chi-square and *F*-statistics to examine population structure and regression measures to estimate the degree of relatedness among individuals within whale pods. We use the chi-square and *F*-statistics to determine how discrete populations of whales are in the Pacific. Discrete populations are managed as independent stocks. The regression analysis of relatedness is used to determine kinship within groups.

Could you cite a specific example illustrating the use of statistics?

I'm currently working with tissue samples obtained from three mass strandings of sperm whales. We use genetic markers to determine the degree of relatedness among individuals within the strandings. This is a striking behavior—entire pods swam up onto the beach following a young female calf, stranded, and subsequently all died. We thought that to do something as dramatic as this, the individuals involved must be very closely related. We're finding, however, that they are not. The statistics enable us to determine the probability that two individuals are related given the number alleles that they share. Also, sperm whale—and many other marine mammal, bird, and turtle species—are injured or killed incidentally in fishing operations. We need to know the size of the population from which these animals are taken. If the population is small, and the incidental kill large, the marine mammal population may be threatened. We use statistics to determine the degree of isolation between putative stocks. If stocks are found to be isolated, we would use this information to prepare management plans specifically designed to conserve the marine mammals of the region. Human activities may need to protect the health of the marine environment and its inhabitants.

How do you approach your research?

We try not to have preconceived notions about how the animals are dispersed in their environment. In marine mammals in particular, because they are so difficult to study, there are generally accepted notions about what the animals are doing, yet these have not been critically investigated. In the case of relatedness among individuals within sperm whale groups, they were once thought to be matrilineal and accompanied by a "harem master." With the advent of genetic techniques, and dedicated field work, more open minds and more critical analyses—the statistics come in here—we're able to reassess these notions.

CHAPTER 2

Describing, Exploring, and Comparing Data

CHAPTER PROBLEM

Is there a keyboard configuration that is more efficient than the one that most of us now use?

The traditional typewriter keyboard configuration is called a *Qwerty* keyboard because of the position of the letters QWERTY in the top row of letters. Developed in 1872, the Qwerty configuration was supposed to force typists to slow down so that their machines would be less likely to jam. The Dvorak keyboard, developed in 1936, positioned the keys most frequently used in the middle (or "home") row, a move intended to improve efficiency. Both keyboard configurations are shown in the accompanying illustration.

QWERTY Keyboard

Dvorak Keyboard

An article in the magazine *Discover* suggests that you can measure the ease of typing by using this point rating system: Count each letter on the home row as 0, each letter on the top row as 1, and each letter on the bottom row as 2. (See "Typecasting" by Scott Kim, *Discover*.) For example, the word *statistics* would have a rating of 7 on the Qwerty keyboard and a rating of 1 on the Dvorak keyboard:

	s	t	a	t	i	s	t	i	c	s	
Qwerty keyboard	0	1	0	1	1	0	1	1	2	0	(sum = 7)
Dvorak keyboard	0	0	0	0	0	0	0	0	1	0	(sum = 1)

Using this rating system with each of the 52 words in the Preamble to the Constitution, we get the rating values shown in Tables 2-1 and 2-2.

Interpreting Results: A visual comparison of Tables 2-1 and 2-2 might not reveal much at first, but in this chapter we will present methods for gaining meaningful insights, which will enable us to make intelligent and productive comparisons.

TABLE 2-1 Qwerty Keyboard Word Ratings

2	2	5	1	2	6	3	3	4	2
4	0	5	7	7	5	6	6	8	10
7	2	2	10	5	8	2	5	4	2
6	2	6	1	7	2	7	2	3	8
1	5	2	5	2	14	2	2	6	3
1	7								

TABLE 2-2 Dvorak Keyboard Word Ratings

2	0	3	1	0	0	0	0	2	0
4	0	3	4	0	3	3	1	3	5
4	2	0	5	1	4	0	3	5	0
2	0	4	1	5	0	4	0	1	3
0	1	0	3	0	1	2	0	0	0
1	4								

2-1 Overview

The title of this chapter—Describing, Exploring, and Comparing Data—is intended to suggest that we sometimes describe a data set, sometimes explore a data set, and sometimes compare data sets. The following are examples of each function.

- *Describing a data set:* A statistics professor has graded the tests in her class, and she wants to better understand how the class performed. She describes the test scores by computing values that measure the center and spread of the data and constructing a graph showing the distribution of the scores.
- *Exploring a data set:* A researcher has obtained the annual incomes of people who have purchased new cars. He wants to explore that data set in an attempt to identify any notable characteristics, such as the average income and the range of incomes.
- *Comparing data sets:* A sociologist has obtained annual incomes of female and male statistics professors. She wants to identify similarities and differences in an attempt to determine whether both genders are compensated equally.

To work with data in any of the ways just listed, we need a variety of tools to help us *understand* the data set. This chapter describes those tools.

When analyzing a data set, we must first determine whether we have a *sample* or a complete *population.* That determination will affect both the methods we use and the conclusions we form. We use methods of **descriptive statistics** to summarize or describe the important characteristics of a set of data, and we use methods of **inferential statistics** when we use sample data to make inferences (or generalizations) about a population. When your professor calculates the final exam average for your statistics class, that result is an example of a descriptive statistic. However, if we state that the result is an estimate of the final exam average for all statistics classes, we are making an inference that goes beyond the known data.

Descriptive statistics and inferential statistics are the two general divisions of the subject of statistics. This chapter deals with the basic concepts of descriptive statistics.

Important Characteristics of Data

When describing, exploring, and comparing data sets, the following characteristics of data are usually most important:

1. *Center:* A representative or average value that indicates where the middle of the data set is located
2. *Variation:* A measure of the amount that the values vary among themselves
3. *Distribution:* The nature or shape of the distribution of the data (such as bell-shaped, uniform, or skewed)

4. *Outliers:* Sample values that lie very far away from the vast majority of the other sample values
5. *Time:* Changing characteristics of the data over time

In this book we will show how the tools of statistics can be applied to these characteristics of data to provide instructive insights.

Manual Calculations: Although this chapter includes detailed steps for important procedures, it is not necessary to master all those steps in all cases. Technology now makes it easy for us to obtain results by calculator or computer so that we can focus on meaning and interpretation. Nevertheless, we recommend that in each case you perform a few manual calculations before using your computer or calculator. Your understanding will be enhanced, and you will acquire a better appreciation for the technology.

2-2 Summarizing Data with Frequency Tables

When working with large data sets, it is generally helpful to organize and summarize the data by constructing a frequency table.

DEFINITION

A **frequency table** lists classes (or categories) of values, along with frequencies (or counts) of the number of values that fall into each class.

Table 2-3 is a frequency table summarizing the Qwerty word ratings from Table 2-1. The **frequency** for a particular class is the number of original scores that fall into that class. For example, the first class in Table 2-3 has a frequency of 20, indicating that there are 20 values between 0 and 2 inclusive.

We will first present some standard terms used in discussing frequency tables, and then we will describe how to construct and interpret them.

TABLE 2-3

Frequency Table of Qwerty Word Ratings

Rating	Frequency
0–2	20
3–5	14
6–8	15
9–11	2
12–14	1

DEFINITIONS

Lower class limits are the smallest numbers that can belong to the different classes. (Table 2-3 has lower class limits of 0, 3, 6, 9, and 12.)

Upper class limits are the largest numbers that can belong to the different classes. (Table 2-3 has upper class limits of 2, 5, 8, 11, and 14.)

Class boundaries are the numbers used to separate classes, but without the gaps created by class limits. They are obtained as follows: Find the size of the gap between the upper class limit of one class and the lower class limit of the next class. Add half of that amount to each upper class limit to find

Authors Identified

In 1787–88 Alexander Hamilton, John Jay, and James Madison anonymously published the famous *Federalist* papers in an attempt to convince New Yorkers that they should ratify the Constitution. The identity of most of the papers' authors became known, but the authorship of 12 of the papers was contested. Through statistical analysis of the frequencies of various words, we can now conclude that James Madison is the *likely* author of these 12 papers. For many of the disputed papers, the evidence in favor of Madison's authorship is overwhelming to the degree that we can be almost certain of being correct.

the upper class boundaries; subtract half of that amount from each lower class limit to find the lower class boundaries. (Table 2-3 has gaps between classes of exactly 1 unit, so 0.5 is added to the upper class limits and subtracted from the lower class limits. The class boundaries are −0.5, 2.5, 5.5, 8.5, 11.5, and 14.5.)

Class midpoints are the midpoints of the classes. (Table 2-3 has class midpoints of 1, 4, 7, 10, and 13.) Each class midpoint can be found by adding the lower class limit to the upper class limit and dividing the sum by 2.

Class width is the difference between two consecutive lower class limits or two consecutive lower class boundaries. (Table 2-3 uses a class width of 3.)

The definitions of class width and class boundaries are tricky. Be careful to avoid the easy mistake of making the class width the difference between the lower class limit and the upper class limit. See Table 2-3 and note that the class width is 3, not 2. You can simplify the process of finding class boundaries by understanding that they basically fill the gaps between classes by splitting the difference between the end of one class and the beginning of the next class. Carefully examine the definition of class boundaries, and spend some time until you understand it well.

Constructing Frequency Tables

The main reason for constructing a frequency table is to use it for constructing a graph that effectively shows the distribution of the data (for example, see *histograms*, introduced in the following section). Figure 2-1 describes in detail the procedure for constructing a frequency table. Some instructors include those detailed steps, while others skip them. It is important to observe the following guidelines when constructing a frequency table.

1. *Be sure that the classes are mutually exclusive.* In other words, each of the original values must belong to only one class.
2. *Include all classes, even if the frequency is zero.*
3. *Try to use the same width for all classes.* Sometimes open-ended intervals, such as "65 years or older," are impossible to avoid.
4. *Select convenient numbers for class limits.* Round up to use fewer decimal places or use numbers relevant to the situation.
5. *Use between 5 and 20 classes.*
6. *The sum of the class frequencies must equal the number of original data values.*

EXAMPLE **Qwerty Frequency Table** Use the 52 values in Table 2-1 for the Qwerty keyboard and follow the procedure shown in Figure 2-1 to construct the frequency table shown in Table 2-4, which appears on page 38. Assume that you want 5 classes.

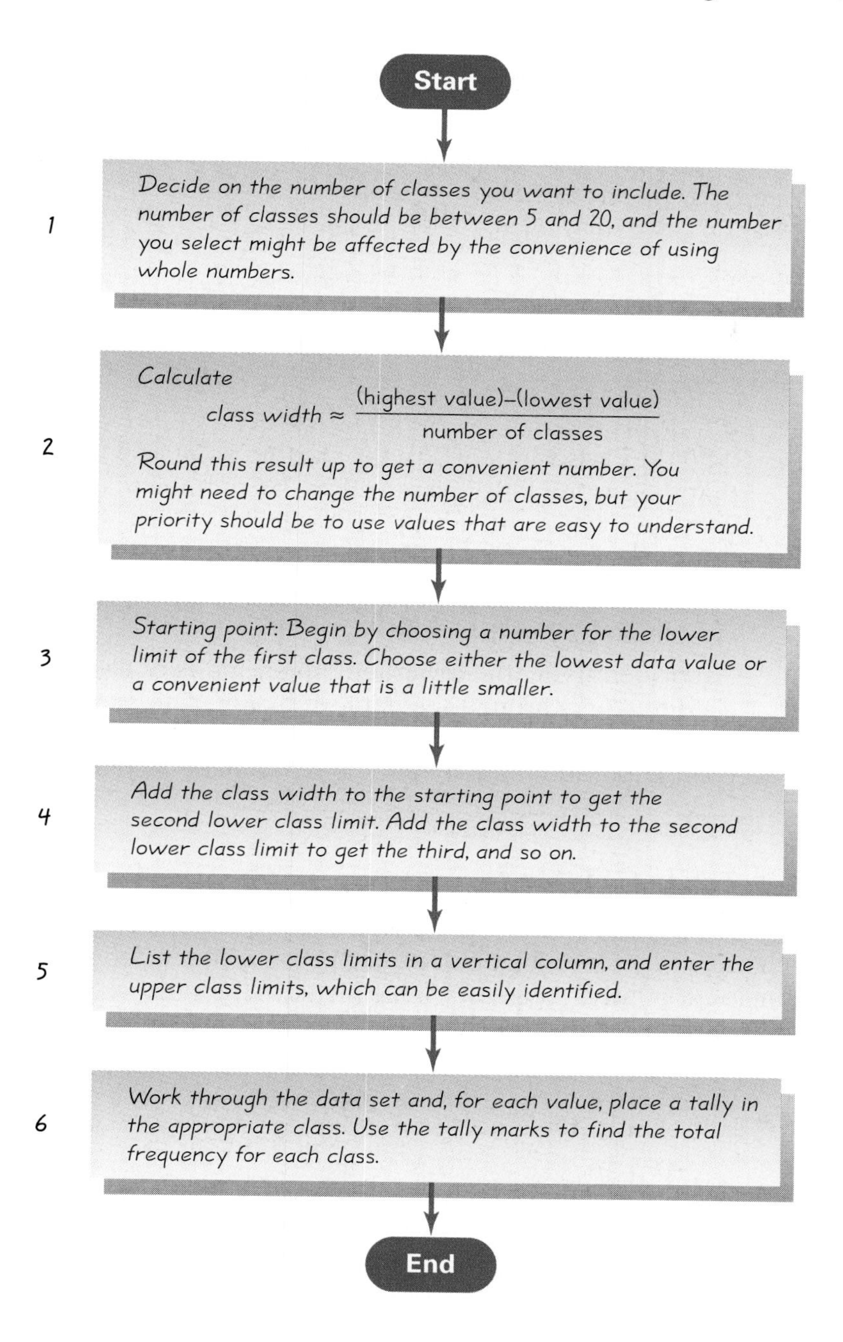

FIGURE 2-1
Procedure for Constructing a Frequency Table

SOLUTION

Step 1: Begin by selecting 5 as the number of desired classes.

Step 2: Calculate the class width.

$$\text{class width} \approx \frac{(\text{highest value}) - (\text{lowest value})}{\text{number of classes}} = \frac{14 - 0}{5} = 2.8 \approx 3$$

Step 3: Choose a starting point of 0, which is the lowest value in the list.

TABLE 2-4 Frequency Table of Qwerty Word Ratings

x	Tally	f
0–2	||||||||||||||||||||	20
3–5	||||||||||||||	14
6–8	|||||||||||||||	15
9–11	||	2
12–14	|	1

Step 4: Add the class width of 3 to the starting point of 0 to determine that the second lower class limit is 3. Continue to add the class width of 3 to get the remaining lower class limits of 6, 9, and 12.

Step 5: List the lower class limits vertically as shown in the margin. From this list, we can easily identify the corresponding upper class limits as 2, 5, 8, 11, and 14.

0–
3–
6–
9–
12–

Step 6: After identifying the lower and upper limits of each class, proceed to work through the data set by entering a tally mark for each value, as shown in Table 2-4. When the tally marks are completed, add them to find the frequencies as shown.

TABLE 2-5 Relative Frequency Table of Qwerty Word Ratings

Rating	Relative Frequency
0–2	38.5%
3–5	26.9%
6–8	28.8%
9–11	3.8%
12–14	1.9%

Relative Frequency Table

An important variation of the basic frequency table uses **relative frequencies,** which are easily found by dividing each class frequency by the total of all frequencies. A **relative frequency table** includes the same class limits as a frequency table, but relative frequencies are used instead of actual frequencies. The relative frequencies are often expressed as percents.

$$\text{relative frequency} = \frac{\text{class frequency}}{\text{sum of all frequencies}}$$

In Table 2-5 the actual frequencies shown in Table 2-4 are replaced by the corresponding relative frequencies expressed as percents. The first class has a relative frequency of $20/52 = 0.385$, or 38.5%. The second class has a relative frequency of $14/52 = 0.269$, or 26.9%, and so on. If constructed correctly, the sum of the relative frequencies should total 1 (or 100%), with some small discrepancies allowed for rounding errors.

Relative frequency tables make it easier for us to understand the distribution of the data and to compare different sets of data.

Cumulative Frequency Table

Another variation of the standard frequency table is used when cumulative totals are desired. The **cumulative frequency** for a class is the sum of the frequencies for that class and all previous classes. Table 2-6 is an example of a **cumulative frequency table,** with cumulative frequencies used instead of the

TABLE 2-6 Cumulative Frequency Table of Qwerty Word Ratings

Rating	Cumulative Frequency
Less than 3	20
Less than 6	34
Less than 9	49
Less than 12	51
Less than 15	52

individual class frequencies. Using the original frequencies of 20, 14, 15, 2, and 1 from Table 2-3 or 2-4, we add 20 + 14 = 34 to get the second cumulative sum, then we add 20 + 14 + 15 = 49 to get the third, and so on.

Interpreting Frequency Tables

The following examples show how frequency tables can be used as tools for describing, exploring, and comparing data.

EXAMPLE **Describing Data** Mark McGwire hit 70 home runs in the 1998 season, and *USA Today* published the distances of those home runs. Table 2-7 summarizes the *last digits* of those distances. When such distances are actually measured, we usually find that the last digits occur with frequencies that are roughly the same. Because of its very uneven distribution, Table 2-7 strongly suggests that the home run distances were not actually measured, but were estimated instead.

TABLE 2-7 Last Digits of Home-Run Distances

Last Digit	Frequency
0	55
1	2
2	1
3	1
4	0
5	3
6	0
7	2
8	4
9	2

EXAMPLE **Exploring Data** In studying the behavior of the Old Faithful geyser in Yellowstone National Park, geologists collect data for the times (in minutes) between eruptions. Table 2-8 summarizes actual data that were obtained. Examination of the frequency table reveals unexpected behavior: The distribution of times has two different peaks. This distribution led geologists to consider various possible explanations.

TABLE 2-8 Times Between Old Faithful Eruptions

Time	Frequency
40–49	8
50–59	44
60–69	23
70–79	6
80–89	107
90–99	11
100–109	1

EXAMPLE **Comparing Data Sets** The Chapter Problem at the beginning of this chapter includes two different data sets collected from the Qwerty and Dvorak keyboard arrangements. The Preamble to the Constitution was used as a sample, and each word was rated as described in the Chapter Problem. Table 2-9 combines the relative frequency tables for those two data sets. From Table 2-9 we see that there is a substantial

TABLE 2-9 Qwerty and Dvorak Keyboard Ratings

Rating	Qwerty Relative Frequency	Dvorak Relative Frequency
0–2	38.5%	63.5%
3–5	26.9%	36.5%
6–8	28.8%	0%
9–11	3.8%	0%
12–14	1.9%	0%

difference between the two data sets. The Dvorak ratings are considerably lower, suggesting that the Dvorak keyboard is considerably more efficient than the Qwerty keyboard. This ability to create meaningful order from apparently chaotic data is truly wonderful.

Frequency tables can be used to identify the general nature of the distribution of data, and they can also be used to construct graphs that visually display the distribution of data. Such graphs are discussed in the next section.

2-2 Basic Skills and Concepts

Describing Data. *In Exercises 1–4, identify the class width, class midpoints, and class boundaries for the given frequency table.*

1.

Height of Men (in.)	Frequency
55–59	1
60–64	3
65–69	49
70–74	46
75–79	1

2.

Height of Women (in.)	Frequency
55–59	11
60–64	121
65–69	63
70–74	4
75–79	1

3.

GPA	Frequency
0.00–0.49	72
0.50–0.99	23
1.00–1.49	47
1.50–1.99	135
2.00–2.49	288
2.50–2.99	276
3.00–3.49	202
3.50–3.99	97

4.

Absences	Frequency
0–9	21
10–19	10
20–29	3
30–39	2
40–49	14

Describing Data. *In Exercises 5–8, construct the relative frequency table that corresponds to the frequency table in the exercise indicated.*

5. Exercise 1 **6.** Exercise 2 **7.** Exercise 3 **8.** Exercise 4

Describing Data. *In Exercises 9–12, construct the cumulative frequency table that corresponds to the frequency table in the exercise indicated.*

9. Exercise 1 **10.** Exercise 2 **11.** Exercise 3 **12.** Exercise 4

Describing and Comparing Data. *In Exercises 13–16, use the data from Data Set 1 in Appendix B.*

13. Weights of Coke Construct a frequency table for the weights of regular Coke. Start the first class at 0.7900 lb and use a class width of 0.0050 lb. Does the result violate any of the guidelines for constructing frequency tables?

14. Weights of Diet Coke Construct a frequency table for the weights of diet Coke. Start the first class at 0.7750 lb and use a class width of 0.0050 lb. Does the result violate any of the guidelines for constructing frequency tables?

15. Comparing Weights of Pepsi and Coke Construct a frequency table for the weights of regular Pepsi. Start the first class at 0.8100 lb and use a class width of 0.0050 lb. Compare the frequency table to the result from Exercise 13. Is there a notable difference?

16. Comparing Weights of Diet Pepsi and Pepsi Construct a frequency table for the weights of diet Pepsi. Start the first class at 0.7700 lb and use a class width of 0.0050 lb. Compare the frequency table to the result from Exercise 15. Is there a notable difference?

17. Bears Refer to Data Set 7 in Appendix B and construct a frequency table of the weights of bears. Use 11 classes, beginning with a lower class limit of 0.

18. Exploring Body Temperature Data Refer to Data Set 6 in Appendix B and construct a frequency table of the body temperatures for midnight on the second day. Use eight classes, beginning with a lower class limit of 96.5. Describe two different notable features of the result.

19. Strengths of Aluminum Cans Refer to Data Set 12 in Appendix B. The axial loads are the weights that were applied before the cans collapsed. Construct a frequency table for the cans that are 0.0109 in. thick. Start the first class at 200 lb and use a class width of 20 lb. When the top lids of the cans are pressed into place, the applied pressure varies between 158 lb and 165 lb. Will these cans withstand that pressure?

20. Exploring Strengths of Aluminum Cans Refer to Data Set 12 in Appendix B. The axial loads are the weights that were applied before the cans collapsed. Construct a frequency table for the cans that are 0.0111 in. thick, but do not include the value of 504 lb. (See Exercise 21.) Start the first class at 200 lb and use a class width of 20 lb. Compare the result to the frequency table for Exercise 19. Determine whether the thicker cans appear to be stronger.

2-2 Beyond the Basics

21. Interpreting Outlier Effect Refer to Data Set 12 for the axial loads of the cans that are 0.0111 in. thick. The load of 504 pounds is called an *outlier* because it is far away from all of the other values. Repeat Exercise 20, but include the value of 504 pounds. Interpret the result by stating a generalization about how much of an effect an outlier might have on a frequency table.

22. Analyzing Effect of Classes Is it possible that the number of classes in a frequency table can have a dramatic effect on the apparent distribution of the data? If so, construct a data set for which a change from five classes to six results in a dramatic change in the distribution of the frequencies.

23. Phony Data Listed below are two sets of values that are supposed to be heights (in inches) of randomly selected adult males. One of the sets consists of heights actually obtained from randomly selected adult males, but the other set consists of numbers that were fabricated. Construct a frequency table for each set of heights. Compare the two frequency tables and interpret your results by identifying the set of data that you believe to be false and stating your reason.

a. 70 73 70 72 71 73 71 67 68 72 67 72 71 73
72 70 72 68 71 71 71 73 69 73 71 66 77 67

b. 70 73 70 72 71 66 74 76 68 75 67 68 71 77
66 69 72 67 77 75 66 76 76 77 73 74 69 67

24. Analyzing the Number of Classes In constructing a frequency table, Sturges' guideline suggests that the ideal number of classes can be approximated by $1 + (\log n)/(\log 2)$, where n is the number of data values. Use this guideline to complete the table in the margin.

Number of Values	Ideal Number of Classes
16–22	5
23–45	6
	7
	8
	9
	10
	11
	12

2-3 Pictures of Data

In Section 2-2 we showed how frequency tables can be used to describe, explore, or compare distributions of data sets. In this section we continue the study of distributions by introducing graphs that show the distributions of data in pictorial form. As you read through this section, keep in mind that the objective is not simply to construct a graph, but rather to learn something about a data set—that is, to understand the nature of the distribution.

Histograms

A common and important graphic device for presenting data is the histogram.

DEFINITION

A **histogram** is a bar graph in which the horizontal scale represents classes and the vertical scale represents frequencies. The heights of the bars correspond to the frequency values, and the bars are drawn adjacent to each other (without gaps).

We can construct a histogram after we have first completed a frequency table representing a data set. The histogram in Figure 2-2 corresponds directly to the frequency values in Table 2-3 (in the preceding section). Each bar of this histogram is marked with its lower class boundary at the left and its upper class boundary at the right. It is often more practical, however, to use class midpoint values instead of class boundaries. The use of class midpoint values is common in software packages that automatically generate histograms.

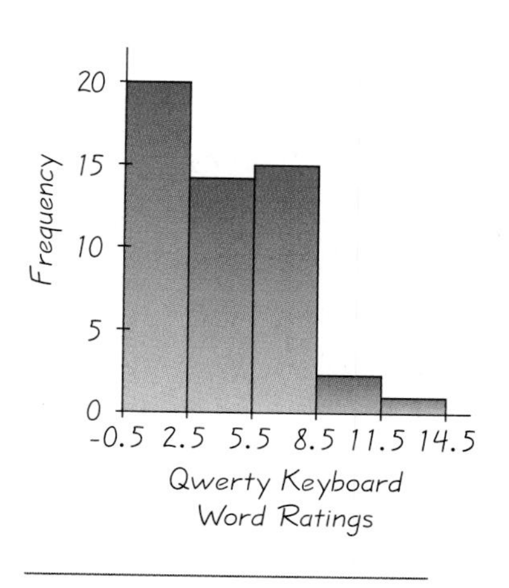

FIGURE 2-2
Histogram of Qwerty Word Ratings

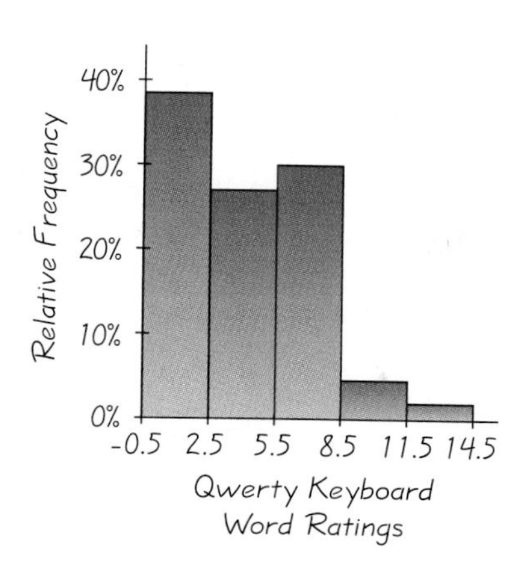

FIGURE 2-3
Relative Frequency Histogram of Qwerty Word Ratings

Before constructing a histogram from a completed frequency table, we must consider the scales used on the vertical and horizontal axes. The maximum frequency (or the next highest convenient number) should suggest a value for the top of the vertical scale; 0 should be at the bottom. In Figure 2-2 we designed the vertical scale to run from 0 to 20. The horizontal scale should be subdivided in a way that allows all the classes to fit well. Ideally, we should try to follow the rule of thumb that the vertical height of the histogram should be about three-fourths of the total width. Both axes should be clearly labeled. **AS**

Relative Frequency Histogram

A **relative frequency histogram** has the same shape and horizontal scale as a histogram, but the vertical scale is marked with *relative frequencies* instead of actual frequencies, as in Figure 2-3.

Frequency Polygon

A **frequency polygon** uses line segments connected to points located directly above class midpoint values. Figure 2-4 shows the frequency polygon corresponding to Table 2-3. The heights of the points correspond to the class frequencies, and the line segments are extended to the right and left so that the graph begins and ends on the x-axis.

Ogive

An **ogive** (pronounced "oh-jive") is a line graph that depicts *cumulative* frequencies, just as the cumulative frequency table (see Table 2-6 in the preceding section) lists cumulative frequencies. Figure 2-5 is an ogive corresponding to Table 2-6. Note that the ogive uses class boundaries along the horizontal scale and that the graph begins with the lower boundary of the first class and

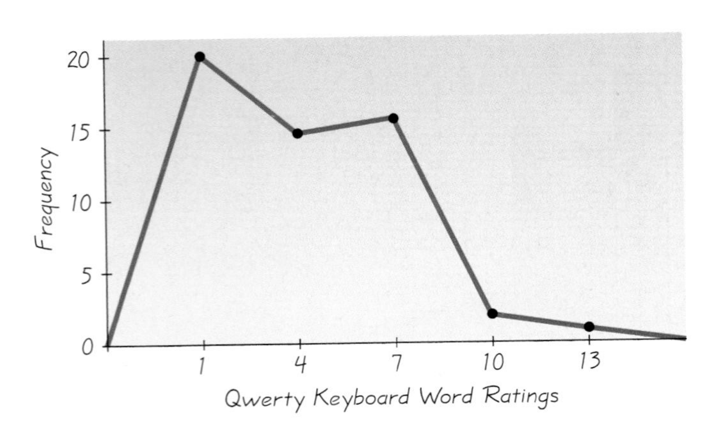

FIGURE 2-4
Frequency Polygon of Qwerty Word Ratings

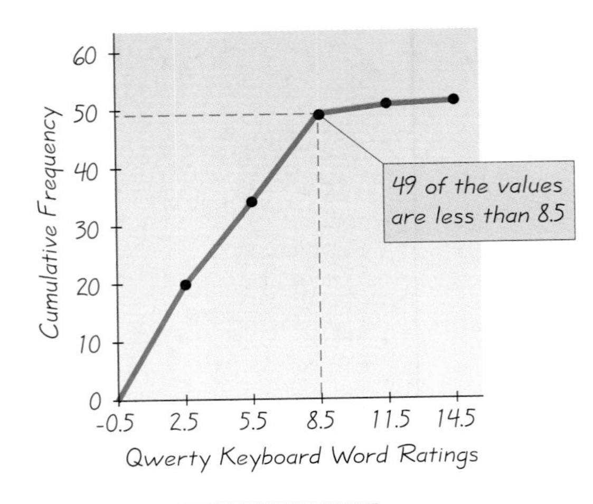

FIGURE 2-5
Ogive of Qwerty Word Ratings

ends with the upper boundary of the last class. Ogives are useful for determining the number of values less than some particular class boundary. For example, Figure 2-5 shows that 49 of the values are less than 8.5.

Dotplots

A **dotplot** consists of a graph in which each data value is plotted as a point (or dot) along a scale of values. Dots representing the same values are stacked. See Figure 2-6, which represents the 52 Qwerty keyboard word ratings listed in Table 2-1. The leftmost dot, for example, represents the value of 0. The four dots stacked above 1 show that there are four words with ratings of 1, and so on. The dotplot is similar to the histogram in that it shows the *distribution* of the data.

AS

Stem-and-Leaf Plots

A **stem-and-leaf plot** represents data by separating each value into two parts: the stem (such as the leftmost digit) and the leaf (such as the rightmost digit). Although we could illustrate a stem-and-leaf plot with the same

FIGURE 2-6
Dotplot of Qwerty Word Ratings

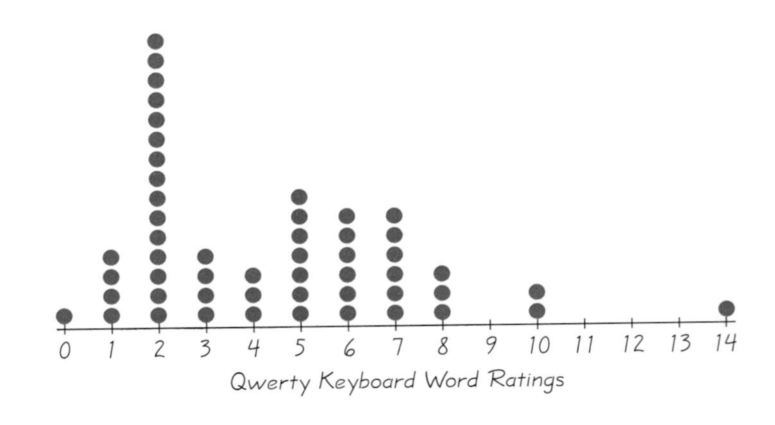

Qwerty keyboard data used in the other examples, the illustration below does a better job of showing just how a stem-and-leaf plot is constructed.

Raw Data (Test Grades)

67	72	85	75	89
89	88	90	99	100

Stem	Leaves	
6	7	← Grade of 67
7	25	← 72, 75
8	5899	
9	09	
10	0	

From this illustration, you can see how 67 is separated into its stem (6) and its leaf (7). Each of the other values is broken up in a similar way. Note that the leaves are arranged in increasing order, *not* the order in which they occur.

By turning the page on its side, we can see a distribution of these data. The great advantage of the stem-and-leaf plot is that we can see the distribution of data and yet keep all the information in the original list. If necessary, we could reconstruct the original list of values.

The rows of digits in a stem-and-leaf plot are similar in nature to the bars in a histogram. One of the guidelines for constructing histograms is that the number of classes should be between 5 and 20, and that same guideline applies to stem-and leaf plots for the same reasons. Stem-and-leaf plots can be *expanded* to include more rows or *condensed* to include fewer rows. The stem-and-leaf plot in our example can be expanded by subdividing rows into those with the digits 0 through 4 and those with digits 5 through 9. This expanded stem-and-leaf plot is shown here.

Stem	Leaves	
6	7	
7	2	← Include leaves of 0, 1, 2, 3, 4
7	5	← Include leaves of 5, 6, 7, 8, 9,
8		← No values between 80 and 84
8	5899	
9	0	
9	9	
10	0	

When it becomes necessary to *reduce* the number of rows, we can condense a stem-and-leaf plot by combining adjacent rows, as in the following illustration. Note that we use an asterisk to separate digits associated with the numbers in each stem. Every row in the condensed plot must include exactly one asterisk so that the shape of the plot is not distorted.

Stem	Leaves	
5-6	*7	← 67
7-8	25*5899	← 72, 75, 85, 88, 89, 89
9-10	09*0	← 90, 99, 100

Another advantage of stem-and-leaf plots is that their construction provides a fast and easy procedure for *sorting* data (arranging data in order). AS

Data must be sorted for a variety of statistical procedures, such as finding the median (discussed in Section 2-4) and finding percentiles or quartiles (discussed in Section 2-6).

Pareto Charts

Consider the following statement: Among 75,200 accidental deaths in the United States in a recent year, 43,500 were attributable to motor vehicles, 12,200 to falls, 6400 to poison, 4600 to drowning, 4200 to fire, 2900 to ingestion of food or an object, and 1400 to firearms (based on data from the National Safety Council). The information contained in this statement concerning the relationships among the data would be much more effectively conveyed in a Pareto chart.

A **Pareto chart** is a bar graph for qualitative data, with the bars arranged in order according to frequencies. As in histograms, vertical scales in Pareto charts can represent frequencies or relative frequencies. The tallest bar is at the left, and the smaller bars are farther to the right. By arranging the bars in order of frequency, the Pareto chart focuses attention on the more important categories. Figure 2-7 shows clearly that the number of motor vehicle accidental deaths is far greater than the numbers in the other categories. Although firearm accidental deaths attract considerable media attention, they are a relatively minor occurrence when compared to the other categories.

Pie Charts

Pie charts are yet another method used to represent data in pictorial form. Figure 2-8 is a **pie chart** that graphically represents as slices of pie the same accidental death data used for the Pareto chart. Construction of a pie chart involves slicing up the pie into the proper proportions. If the category of motor vehicles represents 57.8% of the total, then the wedge representing motor vehicles should be 57.8% of the total (with a central angle of $0.578 \times 360° = 208°$).

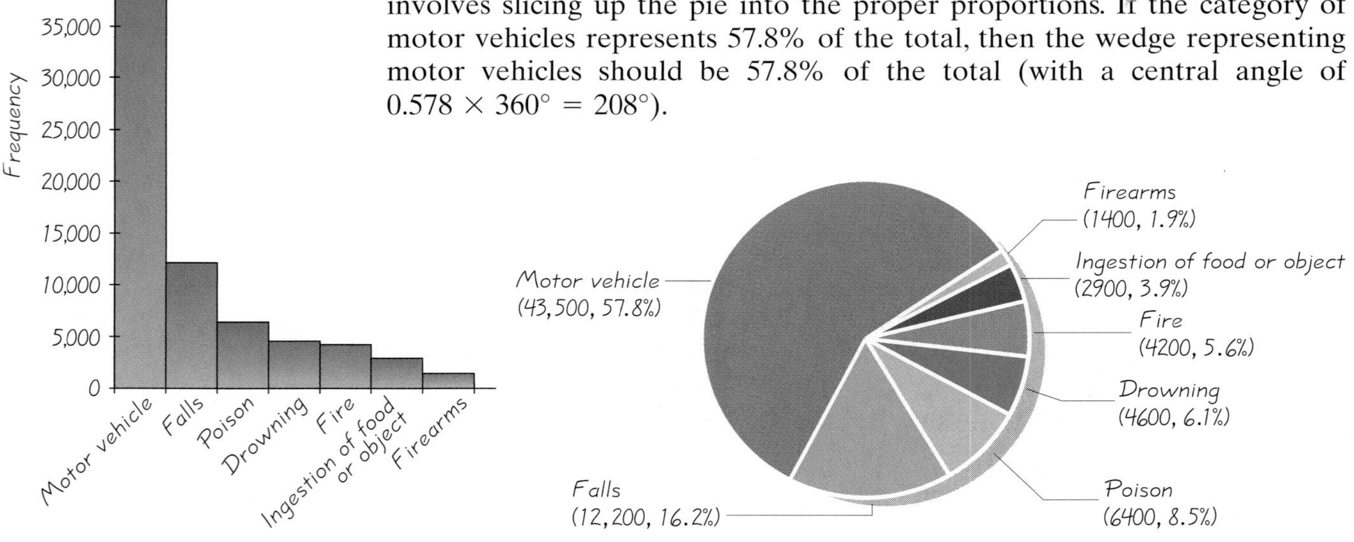

FIGURE 2-7
Pareto Chart of Accidental Deaths by Type

FIGURE 2-8
Pie Chart of Accidental Deaths by Type

The Pareto chart (Figure 2-7) and the pie chart (Figure 2-8) depict the same data in different ways, but a comparison will probably show that the Pareto chart does a better job of showing the relative sizes of the different components.

Scatter Diagrams

A **scatter diagram** is a plot of the paired (x, y) data with a horizontal x-axis and a vertical y-axis. The data are paired in a way that matches each value from one data set with a corresponding value from a second data set. To manually construct a scatter diagram, construct a horizontal axis for the values of the first variable, construct a vertical axis for the values of the second variable, then plot the points. The pattern of the plotted points is often helpful in determining whether there is some relationship between the two variables. (This issue is discussed at length when the topic of correlation is considered in Section 9-2.) Using the cigarette nicotine and tar data from Data Set 8 in Appendix B, we used Minitab to generate the scatter diagram shown here. On the basis of that graph, there does appear to be a relationship between the nicotine and tar contents in cigarettes, as shown by the pattern of the points.

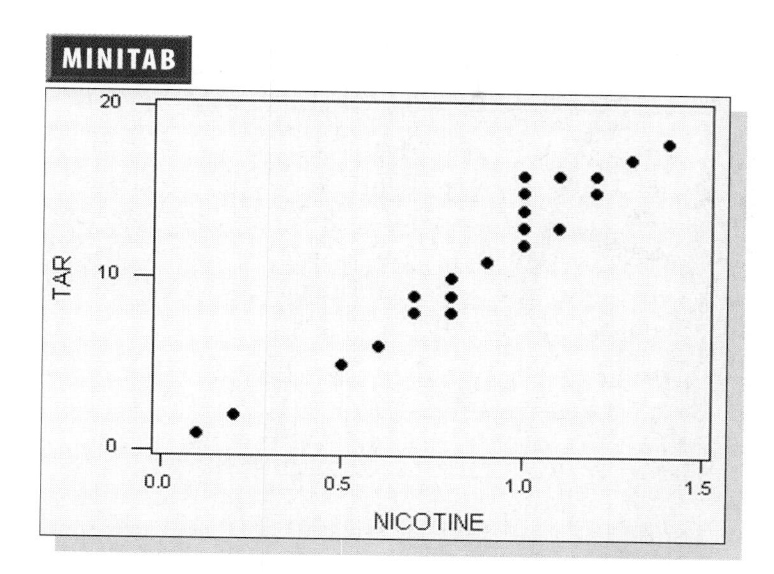

Other Graphs

Numerous pictorial displays other than the ones just described can be used to represent data dramatically and effectively. In Section 2-7 we present boxplots, which are very useful for revealing the spread of data. Pictographs depict data by using pictures of objects, such as soldiers, tanks, airplanes, stacks of coins, or moneybags. Various graphs in Chapter 12 depict patterns of data over time.

The figure on page 48 has been described as possibly "the best statistical graphic ever drawn." It includes six different variables relevant to the march

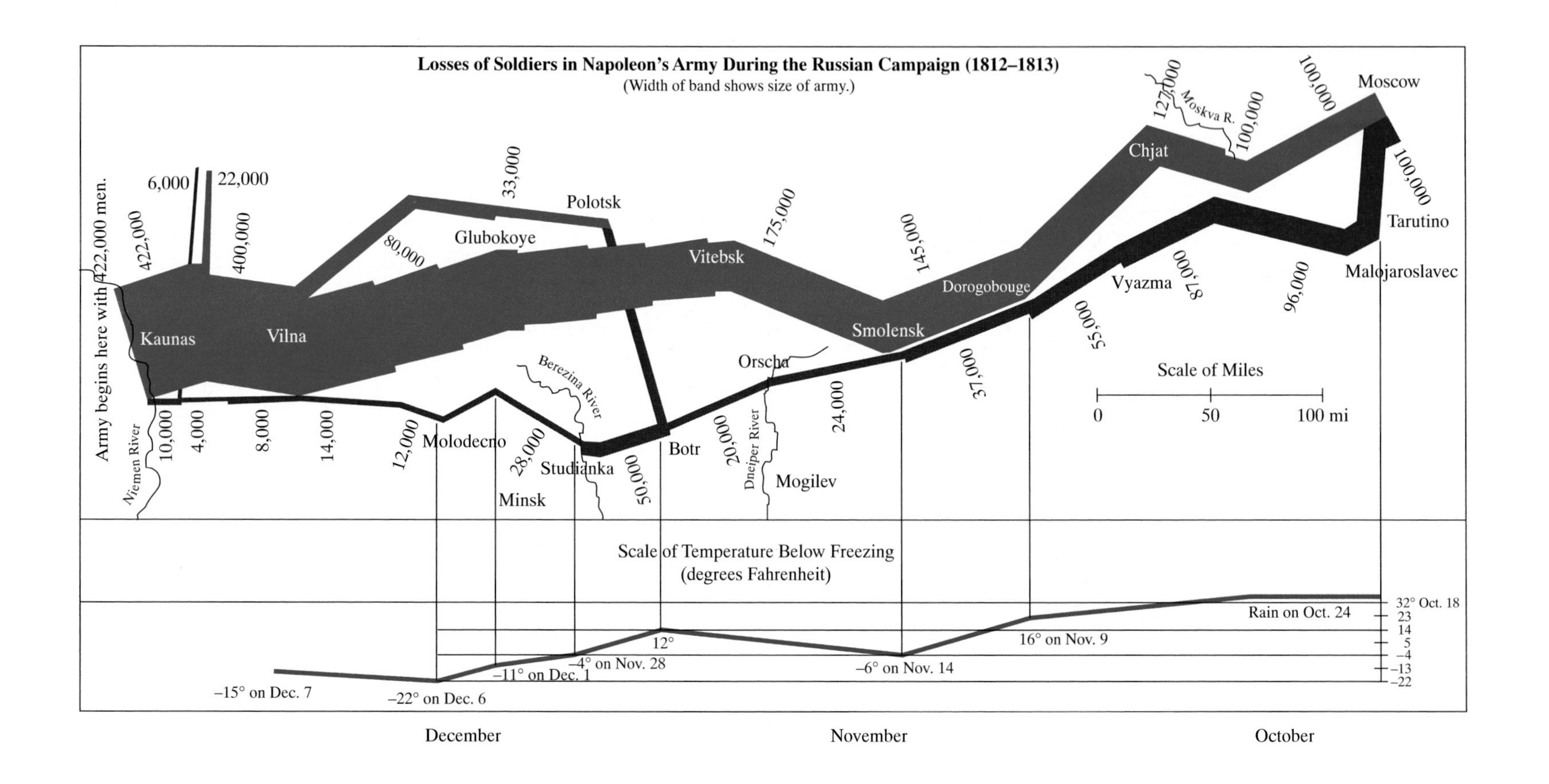

Losses of Soldiers in Napoleon's Army During the Russian Campaign (1812–1813)
(Width of band shows size of army.)
Army begins here with 422,000 men.
422,000
Kaunas
6,000
22,000
400,000
Vilna
80,000
Glubokoye
33,000
Polotsk
Vitebsk
175,000
Smolensk
145,000
Dorogobouge
Chjat
127,000
Moskva R.
100,000
100,000
Moscow
100,000
Tarutino
Malojaroslavec
96,000
87,000
Vyazma
55,000
37,000
24,000
Orscha
Dneiper River
Mogilev
20,000
Botr
50,000
Studianka
Berezina River
28,000
Minsk
Molodecno
12,000
14,000
8,000
4,000
10,000
Niemen River
Scale of Miles
0
50
100 mi
Scale of Temperature Below Freezing
(degrees Fahrenheit)
32° Oct. 18
23
14
5
−4
−13
−22
Rain on Oct. 24
16° on Nov. 9
−6° on Nov. 14
12°
−4° on Nov. 28
−11° on Dec. 1
−22° on Dec. 6
−15° on Dec. 7
October
November
December

of Napoleon's army to Moscow and back during 1812–1813. The thick band at the left depicts the size of the army when it began its invasion of Russia from Poland. The lower band shows its size during the retreat, along with corresponding temperatures and dates. Although first developed in 1861 by Charles Joseph Minard, this graph is ingenious even by today's standards.

Another historically important graph is one developed by the world's most famous nurse, Florence Nightingale. This graph, shown in Figure 2-9, is particularly interesting because it actually saved lives when Nightingale used it to convince British officials that military hospitals needed to improve sanitary conditions, treatment, and supplies. It is drawn somewhat like a pie chart, except that the central angles are all the same and different radii are used to show changes in the numbers of deaths each month. The outermost regions represent deaths due to preventable diseases, the innermost regions represent deaths from wounds, and the middle regions represent deaths from other causes.

Florence Nightingale

Florence Nightingale (1820–1910) is known to many as the founder of the nursing profession, but she also saved thousands of lives by using statistics. When she encountered an unsanitary and undersupplied hospital, she improved those conditions and then used statistics to convince others of the need for more widespread medical reform. She developed original graphs to illustrate that, during the Crimean War, more soldiers died as a result of unsanitary conditions than were killed in combat. Florence Nightingale pioneered the use of social statistics as well as graphics techniques.

Conclusion

The effectiveness of Florence Nightingale's graph illustrates well this important point: A graph is not in itself an end result; it is a tool for describing, exploring, and comparing data. When we create meaningful pictures of data, we are really doing the following:

Describing data: In a histogram, for example, we consider the shape of the distribution, whether there are extreme values, and any other notable char-

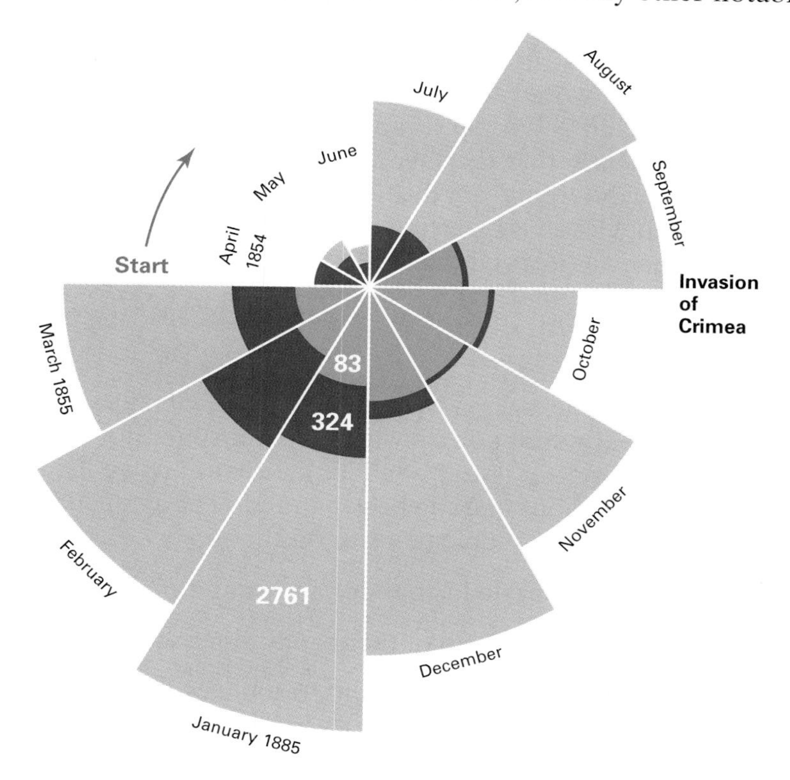

FIGURE 2-9
Deaths in British Military Hospitals During the Crimean War

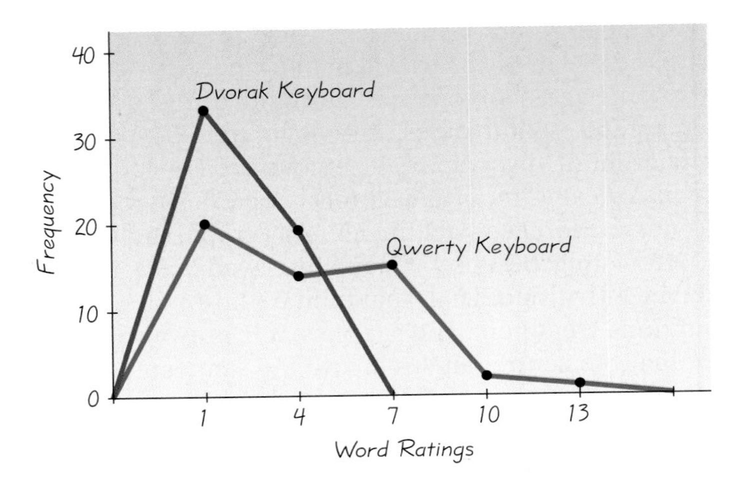

FIGURE 2-10
Frequency Polygons for Qwerty and Dvorak Keyboards

acteristics. We note the overall shape of the distribution. Are the values evenly distributed? Is the distribution skewed (lopsided) to the right or left? Does the distribution peak in the middle? We observe any outliers (values located far away from most of the other values) that might hide the true nature of the distribution. (See Exercise 29.)

Exploring data: We look for any features of the graph that reveal useful and/or interesting characteristics of the data set. In Figure 2-9, for example, we see that more soldiers were dying from inadequate hospital care than were dying from battle wounds.

Comparing data: Figure 2-10, which combines the frequency polygons for the Qwerty keyboard word ratings and the Dvorak keyboard word ratings listed in Tables 2-1 and 2-2 in the Chapter Problem, makes the comparison easy. We can see clearly that the Dvorak word ratings have a distribution that is considerably different from the Qwerty word ratings. The Dvorak values are considerably lower, suggesting that the Dvorak keyboard is considerably more efficient, as it was designed to be.

Using Technology

Powerful software packages now exist which are capable of generating impressive graphs. In this book we will make frequent reference to STATDISK, Minitab, Excel, and the TI-83 Plus calculator, so we list the graphs (discussed in this section) that can be generated. (For detailed procedures, see the manuals that are supplements to this book.)

STATDISK: Can generate histograms and scatter diagrams

Minitab: Can generate all of the graphs discussed in this section

Excel: Can generate histograms, frequency polygons, pie charts, and scatter diagrams

TI-83 Plus: Can generate histograms and scatter diagrams

2-3 Basic Skills and Concepts

Analyzing Graphs. *In Exercises 1–4, refer to the STATDISK-generated histogram, which represents Boston rainfall amounts (in inches) on the Mondays in a recent year.*

1. How many Mondays had no rain at all?

2. How much rain fell on the Monday with the most rain?

3. How many Mondays had at least 3/4 in. of rain?

4. What is the class width?

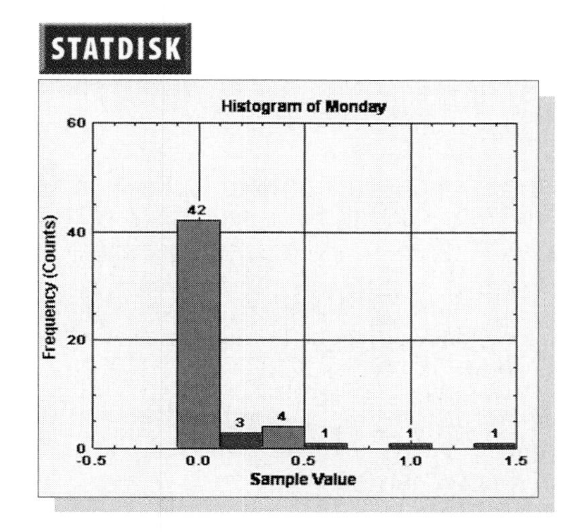

5. Planning to See Old Faithful Visitors to Yellowstone National Park consider an eruption of the Old Faithful geyser to be a major attraction that should not be missed. The accompanying frequency table summarizes a sample of times (in minutes) between eruptions. Construct a histogram corresponding to the frequency table. If you're scheduling a bus tour of Yellowstone, what is the minimum time you should allocate to Old Faithful if you want to be reasonably sure that your tourists will see an eruption?

Table for Exercise 5

Time	Frequency
40–49	8
50–59	44
60–69	23
70–79	6
80–89	107
90–99	11
100–109	1

6. Comparing Student and Faculty Cars Samples of student cars and faculty/staff cars were obtained at the author's college, and their ages (in years) are summarized in the accompanying frequency table. Construct a relative frequency histogram for student cars and another relative frequency histogram for faculty cars. Compare the two frequency histograms. What are the noticeable differences between the two samples?

Table for Exercise 6

Age	Students	Faculty/ Staff
0–2	23	30
3–5	33	47
6–8	63	36
9–11	68	30
12–14	19	8
15–17	10	0
18–20	1	0
21–23	0	1

7. Speeding Tickets The accompanying frequency table describes the speeds of drivers ticketed by the Town of Poughkeepsie police. These drivers were traveling through a 30 mi/h speed zone on Creek Road, which passes the author's college. Construct a histogram corresponding to the frequency table. What does the distribution suggest about the enforced speed limit compared to the posted speed limit?

Table for Exercise 7

Speed	Frequency
42–45	25
46–49	14
50–53	7
54–57	3
58–61	1

8. Exploring Body Temperature Data The accompanying frequency table summarizes a sample of human body temperatures. (See the temperatures for midnight on the second day, as listed in Data Set 6 in Appendix B.) Explore this data set by constructing a histogram corresponding to the frequency table. What does the distribution suggest about the common belief that the average body temperature is 98.6°F? If the subjects are randomly selected, the temperatures should have a distribution that is approximately bell-shaped. Is it? Why is the distribution of human temperatures so important?

Temperature	Frequency
96.5–96.8	1
96.9–97.2	8
97.3–97.6	14
97.7–98.0	22
98.1–98.4	19
98.5–98.8	32
98.9–99.2	6
99.3–99.6	4

Comparing Data. *In Exercises 9–12, make the comparisons by using the frequency tables developed for Data Set 1 in Appendix B. (See Exercises 13–16 in Section 2-2.)*

9. Regular/Diet Coke Compare the weights of regular Coke and diet Coke by constructing two frequency polygons on the same set of axes, as in Figure 2-10. (See Exercises 13 and 14 in Section 2-2.)

10. Coke/Pepsi Compare the weights of regular Coke and regular Pepsi by constructing two frequency polygons on the same set of axes, as in Figure 2-10. (See Exercises 13 and 15 in Section 2-2.)

11. Diet Coke/Diet Pepsi Compare the weights of diet Coke and diet Pepsi by constructing two frequency polygons on the same set of axes, as in Figure 2-10. (See Exercises 14 and 16 in Section 2-2.)

12. Regular/Diet Pepsi Compare the weights of regular Pepsi and diet Pepsi by constructing two frequency polygons on the same set of axes, as in Figure 2-10. (See Exercises 15 and 16 in Section 2-2.)

Describing Data. *In Exercises 13 and 14, list the original data represented by the given stem-and-leaf plots.*

13.

Stem	Leaves
60	0117
61	02889
62	13577
63	
64	0099

14.

Stem	Leaves
25	36 55 89
26	01 17 27 36
27	37 42 67
28	92

Describing Data. *In Exercises 15 and 16, construct the dotplot for the data represented by the stem-and-leaf plot in the given exercise.*

15. Exercise 13

16. Exercise 14

Describing and Interpreting Data. *In Exercises 17 and 18, construct the stem-and-leaf plots for the given data sets found in Appendix B.*

17. Bears The lengths (in inches) of the bears in Data Set 7. (*Hint:* First round the lengths to the nearest inch.)

18. Weights of Discarded Plastic Weights (in pounds) of plastic discarded by 62 households: Refer to Data Set 5, and start by rounding the listed weights to the nearest tenth of a pound (or one decimal place). (Use an expanded stem-and-leaf plot with about 11 rows.)

19. Job Hunting A study was conducted to determine how people get jobs. The table lists data from 400 randomly selected subjects. The data are based on results from the National Center for Career Strategies. Construct a Pareto chart that corresponds to the given data. If someone would like to get a job, what seems to be the most effective approach?

Job Sources of Survey Respondents	Frequency
Help-wanted ads	56
Executive search firms	44
Networking	280
Mass mailing	20

20. Job Sources Refer to the data given in Exercise 19, and construct a pie chart. Compare the pie chart to the Pareto chart. Can you determine which graph is more effective in showing the relative importance of job sources?

21. Causes of Train Derailments An analysis of train derailment incidents showed that 23 derailments were caused by bad track, 9 were due to faulty equipment, 12 were attributable to human error, and 6 had other causes (based on data from the Federal Railroad Administration). Construct a pie chart representing the given data.

22. Analyzing Causes of Train Derailments Refer to the data given in Exercise 21, and construct a Pareto chart. Compare the Pareto chart to the pie chart. Can you determine which graph is more effective in showing the relative importance of the causes of train derailments?

Describing and Interpreting Data. *In Exercises 23 and 24, use the given paired data from Appendix B to construct a scatter diagram.*

23. Cigarette Tar/CO In Data Set 8, use tar for the horizontal scale and use carbon monoxide for the vertical scale. Interpret the results by determining whether there appears to be a relationship between cigarette tar and carbon monoxide. If so, describe the relationship.

24. Bear Neck/Weight In Data Set 7, use the distances around bear necks for the horizontal scale and use the bear weights for the vertical scale. Based on the result, what is the relationship between a bear's neck size and its weight?

In Exercises 25–28, refer to the figure on page 48, which describes Napoleon's 1812–1813 campaign to Moscow and back. The thick band at the left depicts the size of the army when it began its invasion of Russia from Poland, and the lower band describes its retreat.

25. Find the percentage of men who survived the entire campaign.

26. Find the number of men and the percentage of men who died crossing the Berezina River.

27. How many men died on the return from Moscow during the time when the temperature dropped from 16°F to −6°F?

28. Of the men who made it to Moscow, how many died on the return trip between Moscow and Botr? (Note that 33,000 men did not go to Moscow, but they joined the returning men who did.)

2-3 Beyond the Basics

29. a. Constructing Histograms and Comparing Data Refer to Data Set 12 in Appendix B and construct a histogram for the axial loads of cans that are 0.0111 in. thick. That data set includes an outlier of 504 lb. (An outlier is a value that is far away from the other values.)
b. Repeat part (a) after excluding the outlier of 504 lb.
c. How much of an effect does an outlier have on the shape of the histogram?

30. Constructing and Analyzing Histograms Using a collection of sample data, we construct a frequency table with 10 classes and then construct the corresponding histogram. How is the histogram affected if the number of classes is doubled but the same vertical scale is used?

31. Comparing Ages of Oscar Winners In "Ages of Oscar-winning Best Actors and Actresses" (*Mathematics Teacher* magazine) by Richard Brown and Gretchen Davis, stem-and-leaf plots are used to compare the ages of actors and actresses at the time they won Oscars. Here are the results for recent winners from each category.

Actors:	32	37	36	32	51	53	33	61	35	45	55	39
	76	37	42	40	32	60	38	56	48	48	40	43
	62	43	42	44	41	56	39	46	31	47	45	60
Actresses:	50	44	35	80	26	28	41	21	61	38	49	33
	74	30	33	41	31	35	41	42	37	26	34	34
	35	26	61	60	34	24	30	37	31	27	39	34

Actors' Ages	Stem	Actresses' Ages
	2	
72	3	
	4	4
	5	0
	6	
	7	
	8	

a. Construct a back-to-back stem-and-leaf plot for the above data. The first two ages from each group have been entered in the margin.
b. Compare the stem-and-leaf plots for the two different sets of data and explain any differences.

32. Analyzing Graphs Figure 2-9, a *polar-area diagram*, has been called misleading because it uses varying areas to represent information that is one-dimensional. How might the data be graphed so that the information can be better understood?

2-4 Measures of Center

The main objectives of this section are to present the important measures of center and to show how to compute them.

DEFINITION

A **measure of center** is a value at the center or middle of a data set.

In Sections 2-2 and 2-3 we considered frequency tables and graphs that reveal the nature or shape of the *distribution* of a data set. In this section we focus on finding values that are at the *center* of a data set. There are several different ways to determine the center, so we have different definitions of measures of center, including the mean, median, mode, and midrange. We begin with the mean.

Mean

The (arithmetic) mean is generally the most important of all numerical descriptive measurements, and it is what most people call an average.

DEFINITION

The **arithmetic mean** of a set of values is the number obtained by adding the values and dividing the total by the number of values. This measure of center will be used often throughout the remainder of this text, and it will be referred to simply as the **mean.**

This definition can be expressed as Formula 2-1, in which the Greek letter Σ (uppercase Greek sigma) indicates that the data values should be added, Σx represents the sum of all data values, and n denotes the **sample size** (the number of values in the data set).

Formula 2-1

$$\text{mean} = \frac{\Sigma x}{n}$$

The mean is denoted by $\overline{x}$ (pronounced "x-bar") if the data set is a *sample* from a larger population; if all values of the population are used, then we denote the mean by μ (lowercase Greek mu). (Sample statistics are usually represented by English letters, such as $\overline{x}$, and population parameters are usually represented by Greek letters, such as μ.)

Notation

Σ denotes the *addition* of a set of values.

x is the *variable* usually used to represent the individual data values.

n represents the *number of values in a sample.*

(continued)

Marion: Average City USA

Because it closely parallels the makeup of the nation, Marion, Indiana is a favorite choice of researchers who must conduct market testing. When researchers wanted to gauge reaction to potato chips made with the fat substitute olestra, they went to Marion in search of opinions that are likely to reflect the sense of the nation.

Marion has a median household income of $37,396, compared to the national median of $38,738. Marion is also ethnically diverse with more than 40 nationalities represented.

N represents the *number of values in a population.*

$\bar{x} = \dfrac{\Sigma x}{n}$ is the *mean of a set of sample values.*

$\mu = \dfrac{\Sigma x}{N}$ denotes the mean of all values in a *population.*

EXAMPLE **Volumes of Coke** Listed below are the volumes (in ounces) of the Coke in five different cans. Find the mean for this sample.

12.3 12.1 12.2 12.3 12.2

SOLUTION

The mean is computed by using Formula 2-1. First add the values, then divide by the number of values:

$$\bar{x} = \frac{\sum x}{n} = \frac{12.3 + 12.1 + 12.2 + 12.3 + 12.2}{5} = \frac{61.1}{5} = 12.22$$

The mean volume is 12.22 oz.

One disadvantage of the mean is that it is sensitive to every value, so one exceptional value can affect the mean dramatically. The median largely overcomes that disadvantage.

Median

DEFINITION

The **median** of a data set is the middle value when the original data values are arranged in order of increasing (or decreasing) magnitude. The median is often denoted by $\tilde{x}$ (pronounced "x-tilde").

To find the median, first sort the values (arrange them in order), then follow one of these two procedures:

1. If the number of values is odd, the median is the number located in the exact middle of the list.
2. If the number of values is even, the median is found by computing the mean of the two middle numbers.

Figure 2-11 summarizes this procedure for finding the median.

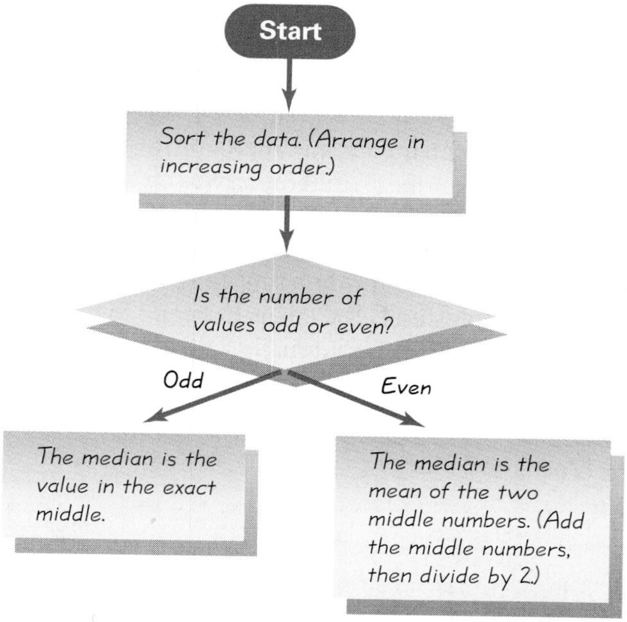

Start

Sort the data. (Arrange in increasing order.)

Is the number of values odd or even?

Odd: The median is the value in the exact middle.

Even: The median is the mean of the two middle numbers. (Add the middle numbers, then divide by 2.)

FIGURE 2-11
Procedure for Finding the Median

EXAMPLE Executive Women Find the median of the following salaries (in millions of dollars) paid to female executives (based on data from *Working Woman* magazine):

6.72 3.46 3.60 6.44

SOLUTION First arrange the values in order:

3.46 3.60 6.44 6.72

Because the number of values is an even number (4), the median is found by computing the mean of the two middle values, 3.60 and 6.44.

$$\text{median} = \frac{3.60 + 6.44}{2} = \frac{10.04}{2} = 5.02$$

The median is \$5.02 million.

EXAMPLE Executive Women Repeat the preceding example, this time including another salary of \$26.70 million. That is, find the median of the following salaries (in millions of dollars):

6.72 3.46 3.60 6.44 26.70

SOLUTION First arrange the values in order:

3.46 3.60 6.44 6.72 26.70

Class Size Paradox

There are at least two ways to obtain the mean class size, and they can have very different results. At one college, if we take the numbers of students in 737 classes, we get a mean of 40 students. But if we were to compile a list of the class sizes for each student and use this list, we would get a mean class size of 147. This large discrepancy is due to the fact that there are many students in large classes, while there are few students in small classes. Without changing the number of classes or faculty, we could reduce the mean class size experienced by students by making all classes about the same size. This would also improve attendance, which is better in smaller classes.

Because the number of values is an odd number (5), the median is in the exact middle of the list. The median is therefore $6.44 million.

In Table 2-10, we show the means and medians for the data sets used in the two preceding examples. Examining that table, we see that when we include the extreme value of $26.70 million, the mean changes dramatically, but the median changes by only a small amount. Because the median is relatively insensitive to extreme values, it is often used for data sets with a relatively small number of extreme values. For example, the U.S. Census Bureau recently reported that the *median* household income is $36,078 annually. The median is used here because there is a small number of households with really high incomes.

TABLE 2-10 Means and Medians for Two Data Sets

Data Set	Mean	Median
3.46 3.60 6.44 6.72	5.055	5.02
3.46 3.60 6.44 6.72 26.70	9.384	6.44

Mode

DEFINITION

The **mode** of a data set is the value that occurs most frequently. When two values occur with the same greatest frequency, each one is a mode and the data set is **bimodal.** When more than two values occur with the same greatest frequency, each is a mode and the data set is said to be **multimodal.** When no value is repeated, we say that there is no mode. The mode is often denoted by M.

EXAMPLE Find the modes of the following data sets.

a. 5 5 5 3 1 5 1 4 3 5
b. 1 2 2 2 3 4 5 6 6 6 7 9
c. 1 2 3 6 7 8 9 10

SOLUTION

a. The number 5 is the mode because it is the value that occurs most often.

b. The numbers 2 and 6 are both modes because they occur with the same greatest frequency. This data set is bimodal.

c. There is no mode because no value is repeated.

In fact, the mode is not often used with numerical data. But among the different measures of center we are considering, the mode is the only one that can be used with data at the nominal level of measurement. (Recall that the nominal level of measurement applies to data that consist of names, labels, or categories only.) The following example illustrates this point.

EXAMPLE **College Majors** A study of college students included 10 psychology majors, 20 English majors, and 5 math majors. We can't compute a mean or median of these majors, but we can report that the mode is English major, because that is the one occurring with the greatest frequency.

Midrange

DEFINITION

The **midrange** is the value midway between the highest and lowest values in the original data set. It is found by adding the highest data value to the lowest data value and then dividing the sum by 2, as in the following formula.

$$\text{midrange} = \frac{\text{highest value} + \text{lowest value}}{2}$$

The midrange is seldom used. Because it is derived from only the highest and the lowest values, it is too sensitive to those extremes. Nevertheless, it does have two redeeming features: (1) It is easy to compute; and (2) it helps to reinforce the important point that there are several different ways to define the center of a data set. Figure 2-12 illustrates the differences among the mean, median, mode, and midrange.

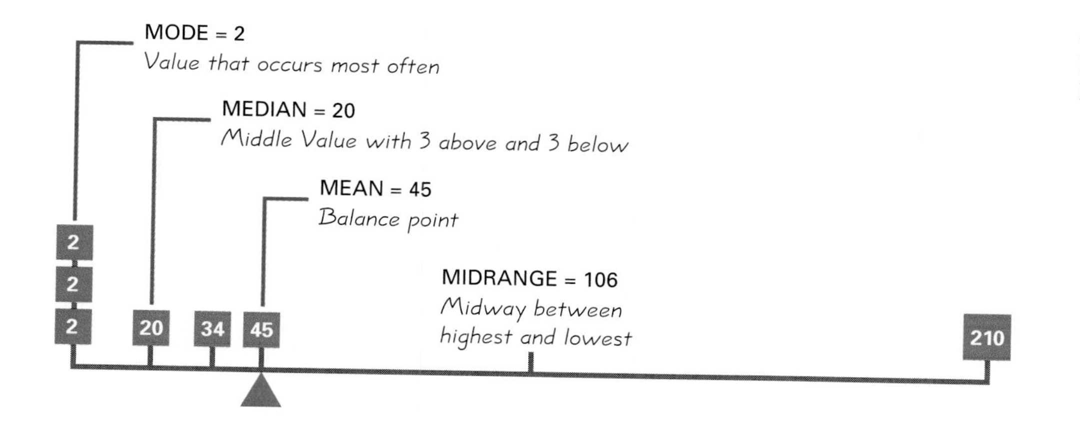

FIGURE 2-12
Measures of Center

EXAMPLE **Ages of Thieves** Find the midrange of the ages of people arrested on theft charges at the Dutchess County jail (based on data from a student of the author).

18 16 23 25 19 18 20 38

SOLUTION The midrange is found as follows:

$$\frac{\text{highest value} + \text{lowest value}}{2} = \frac{38 + 16}{2} = 27 \text{ years}$$

Unfortunately, the term *average* is sometimes used for any measure of center, and it is sometimes used for the mean. Because of this ambiguity, we should avoid using the term *average* when referring to a particular measure of center and use instead the specific term, such as mean, median, mode, or midrange. When encountering a value reported as being an *average*, we should be aware that the value could have been computed by any one of several different approaches.

In the spirit of describing, exploring, and comparing data, we provide Table 2-11, which summarizes the different measures of center for the Qwerty and Dvorak keyboard word ratings listed in Tables 2-1 and 2-2 in the Chapter Problem. Recall that these two data sets consist of the ratings for each of the 52 words in the Preamble to the Constitution. Lower ratings represent words that are easier to type; higher ratings result from words that are more difficult to type. A comparison of the measures of center suggests that the Qwerty keyboard is more difficult to use. Its mean of 4.4 appears to be substantially greater than the mean of 1.7 obtained with the Dvorak keyboard. In later chapters of this text we will introduce methods for objectively determining whether such apparent differences are actually significant.

TABLE 2-11 Comparison of Qwerty and Dvorak Keyboards

	Qwerty	Dvorak
Mean	4.4	1.7
Median	4.0	1.0
Mode	2	0
Midrange	7.0	2.5

Round-Off Rule

A simple rule for rounding answers is this:

Carry one more decimal place than is present in the original set of values.

When implementing this rule, we round only the final answer, not the intermediate values. Thus the mean of 2, 3, 5 is 3.333333. . . , which is rounded to 3.3. Because the original values were whole numbers, we rounded to the nearest tenth. As another example, the mean of 2.1, 3.4, 5.7 is 3.7333333. . . , which is rounded to 3.73 (one more decimal place than was used for the original values).

Mean from a Frequency Table

When data are summarized in a frequency table, we don't know the exact values falling in a particular class. To make calculations possible, we pretend that within each class, all sample values are equal to the class midpoint. Because each class midpoint is repeated a number of times equal to the class frequency, the sum of all sample values becomes $\Sigma(f \cdot x)$, where f denotes frequency and x represents the class midpoint. The total number of sample values is the sum of frequencies Σf. Formula 2-2 is used to compute the mean when the sample data are summarized in a frequency table. Formula 2-2 is not really a new concept; it is simply a variation of Formula 2-1.

First multiply each frequency and class midpoint, then add the products.

↓

Formula 2-2

$$\bar{x} = \frac{\Sigma(f \cdot x)}{\Sigma f} \quad \text{mean from frequency table}$$

↑

sum of frequencies

In Table 2-12, the first two columns from Frequency Table 2-3 (on page 35) have been entered. Table 2-12 illustrates the procedure for using Formula 2-2 when calculating a mean from data summarized in a frequency table. (In reality, software or a calculator are generally used in place of manual calculations.

TABLE 2-12 Finding the Mean from a Frequency Table

Rating	Frequency f	Class Midpoint x	$f \cdot x$
0–2	20	1	20
3–5	14	4	56
6–8	15	7	105
9–11	2	10	20
12–14	1	13	13
Totals:	$\Sigma f = 52$		$\Sigma(f \cdot x) = 214$

$$\bar{x} = \frac{\Sigma(f \cdot x)}{\Sigma f} = \frac{214}{52} = 4.1$$

Six Degrees of Separation

Social psychologists, historians, political scientists, and communications specialists are among those interested in "The Small World Problem": Given any two people in the world, how many intermediate links are required in order to connect the two original people? Social psychologist Stanley Milgram conducted an experiment using the U.S. mail system. Subjects were instructed to try and contact other target people by mailing an information folder to an acquaintance who they thought would be closer to the target. Among 160 such chains that were initiated, only 44 were completed. The number of intermediate acquaintances varied from 2 to 10, with a median of 5. A mathematical model was used to show that if those missing chains were completed, the median would be slightly greater than 5. (See "The Small World Problem," by Stanley Milgram, Psychology Today, May 1967.)

Not At Home

Pollsters cannot simply ignore those who were not at home when called the first time. One solution is to make repeated callback attempts until the person can be reached. Alfred Politz and Willard Simmons describe a way to compensate for those missing results without making repeated callbacks. They suggest weighting results based on how often people are not at home. For example, a person at home only two days out of six will have a 2/6 or 1/3 probability of being at home when called the first time. When such a person is reached the first time, his or her results are weighted to count three times as much as someone who is always home. This weighting is a compensation for the other similar people who are home two days out of six and were not at home when called the first time. This clever solution was first presented in 1949.

To compute the mean using a TI-83 Plus calculator, enter the class midpoints in list L1, enter the frequencies in list L2, then press **STAT, CALC,** and select **1-Var Stats** and enter L1, L2 with the comma.)

Using Table 2-12 we get $\overline{x} = 4.1$, but we get $\overline{x} = 4.4$ if we use the original list of 52 values. Remember, the frequency table yields an *approximation* of $\overline{x}$, because it is not based on the exact original list of sample values.

Weighted Mean

In some situations, the values vary in their degree of importance, so we may want to compute a **weighted mean,** which is a mean computed with the different scores assigned different weights. In such cases, we can calculate the weighted mean by assigning different weights to different values, as shown in Formula 2-3.

Formula 2-3 weighted mean: $\overline{x} = \dfrac{\Sigma(w \cdot x)}{\Sigma w}$

For example, suppose we need a mean of three test scores (85, 90, 75), but the first test counts for 20%, the second test counts for 30%, and the third test counts for 50% of the final grade. We can assign weights of 20, 30, and 50 to the test scores, then proceed to calculate the mean by using Formula 2-3 as follows:

$$\begin{aligned}\overline{x} &= \frac{\Sigma(w \cdot x)}{\Sigma w} \\ &= \frac{(20 \times 85) + (30 \times 90) + (50 \times 75)}{20 + 30 + 50} \\ &= \frac{8150}{100} = 81.5\end{aligned}$$

As another example, college grade-point averages can be computed by assigning each letter grade the appropriate number of points (A = 4, B = 3, etc.), then assigning to each number a weight equal to the number of credit hours. Again, Formula 2-3 can be used to compute the grade-point average.

AS The Best Measure of Center

So far, we have considered the mean, median, mode, and midrange as measures of center. Which one of these is best? Unfortunately, there is no single best answer to that question, because there are no objective criteria for determining the most representative measure for all data sets. The different measures of center have different advantages and disadvantages, some of which are summarized in Table 2-13. An important advantage of the mean is that it takes every value into account, but an important disadvantage is that it is sometimes dramatically affected by a few extreme values. This disadvantage can be overcome by using a trimmed mean, as described in Exercise 23.

TABLE 2-13 Comparison of Mean, Median, Mode, and Midrange

Average	Definition	How Common?	Existence	Takes Every Value into Account?	Affected by Extreme Values?	Advantages and Disadvantages
Mean	$\bar{x} = \frac{\Sigma x}{n}$	most familiar "average"	always exists	yes	yes	used throughout this book; works well with many statistical methods
Median	middle score	commonly used	always exists	no	no	often a good choice if there are some extreme values
Mode	most frequent score	sometimes used	might not exist; may be more than one mode	no	no	appropriate for data at the nominal level
Midrange	$\frac{\text{high} + \text{low}}{2}$	rarely used	always exists	no	yes	very sensitive to extreme values

General comments:

- For a data collection that is approximately symmetric with one mode, the mean, median, mode, and midrange tend to be about the same.
- For a data collection that is obviously asymmetric, it would be good to report both the mean and median.
- The mean is relatively *reliable*. That is, when samples are drawn from the same population, the sample means tend to be more consistent than the other averages (consistent in the sense that the means of samples drawn from the same population don't vary as much as the other averages).

Skewness

A comparison of the mean, median, and mode can reveal information about the characteristic of skewness, defined below and illustrated in Figure 2-13.

DEFINITION

A distribution of data is **skewed** if it is not symmetric and if it extends more to one side than the other. (A distribution of data is **symmetric** if the left half of its histogram is roughly a mirror image of its right half.)

Data skewed to the *left* are said to be **negatively skewed;** the mean and median are to the left of the mode. Although not always predictable, negatively skewed data generally have the mean to the left of the median. [See Figure 2-13(a).] Data skewed to the *right* are said to be **positively skewed;** the mean and median are to the right of the mode. Again, although not

FIGURE 2-13
Skewness

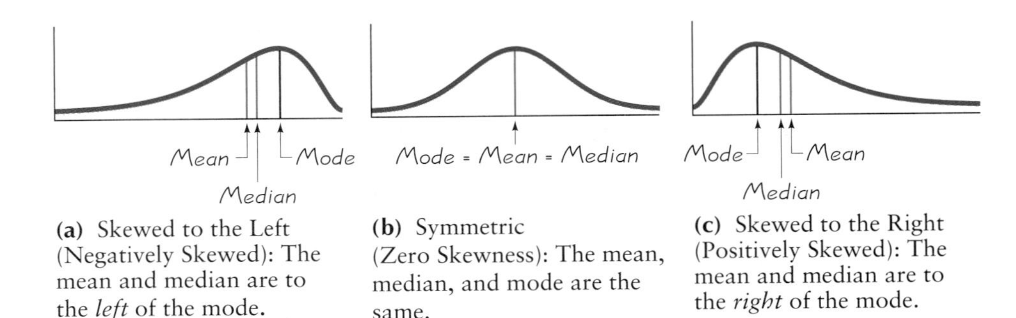

(a) Skewed to the Left (Negatively Skewed): The mean and median are to the *left* of the mode.

(b) Symmetric (Zero Skewness): The mean, median, and mode are the same.

(c) Skewed to the Right (Positively Skewed): The mean and median are to the *right* of the mode.

always predictable, positively skewed data generally have the mean to the right of the median. [See Figure 2-13(c).]

Lopsided to the right = Skewed to left = Negatively skewed

Lopsided to the left = Skewed to right = Positively skewed

Data not lopsided = Symmetric = Zero skewness

If we examine the histogram in Figure 2-2 for the word ratings using the Qwerty keyboard, we see a graph that appears to be skewed to the right. In practice, many distributions of data are symmetric and without skewness. Distributions skewed to the right are more common than those skewed to the left because it's often easier to get exceptionally large values than values that are exceptionally small. With annual incomes, for example, it's impossible to get values below the lower limit of zero, but there are a few people who earn millions of dollars in a year. Annual incomes therefore tend to be skewed to the right, as in Figure 2-13(c).

Using Technology

The calculations in this section are fairly simple, but some of the calculations in the following sections are more rigorous. Many computer software programs allow you to enter a data set and use one operation to get several different sample statistics, referred to as *descriptive statistics.* (See Section 2-6 for sample displays resulting from STATDISK, Minitab, Excel, and the TI-83 Plus calculator.) The following are some of the procedures for obtaining such displays.

STATDISK: Choose the main menu item of Data, and use the Sample Editor to enter the data. Click on **Copy,** then click on **Data** once again, but now select **Descriptive Statistics.** Click on **Paste** to get the data set that was entered. Now click on **Evaluate** to get the various descriptive statistics, including the mean, median, midrange, and other statistics to be discussed in the following sections.

Minitab: Enter the data in the column with the heading C1. Click on **Stat,** select **Basic Statistics,** then select **Descriptive**

Statistics. The results will include the mean and median as well as other statistics.

Excel: Enter the sample data in column A. Select **Tools,** then **Data Analysis,** then select **Descriptive Statistics** and click **OK**. In the dialog box, enter the input range (such as A1:A20 for 20 values in column A), click on **Summary Statistics,** then click **OK.** If Data Analysis does not appear in the Tools menu, it must be installed using Add-Ins.

TI-83 Plus: First enter the data in list L1 by pressing **STAT,** then selecting **Edit** and pressing the **ENTER** key. After the data values have been entered, press **STAT** and select **CALC,** then select **1-Var Stats** and press the **ENTER** key twice. The display will include the mean $\overline{x}$, the median, the minimum value, and the maximum value. Use the down-arrow key to see the results that don't fit on the initial display.

2-4 Basic Skills and Concepts

Finding Measures of Centers of Data. *In Exercises 1–4, find the (a) mean, (b) median, (c) mode, and (d) midrange for the given sample data. These data sets were provided by students of the author.*

1. Car Crashes The given values are the numbers of Dutchess County car crashes for each month of a recent year.

   27 8 17 11 15 25 16 14 14 14 13 18

2. Eyeglass Fittings The given values are the measurements of distances (in millimeters) between the pupils of adult patients being fitted for eyeglasses.

   67 66 59 62 63 66 66 55

3. McDonald's Service Times The given values are the service times (in seconds) of McDonald's drive-through customers. McDonald's wants to maintain a mean of 90 sec or less. Are these results acceptable?

   88 107 35 93 65 55 119 83 99 74 46 108

4. Ordering Steaks by Weight The given values are the weights (in ounces) of steaks listed on a restaurant menu as "20-ounce Porterhouse" steaks. The weights are supposed to be 21 oz because the steaks supposedly lose an ounce when cooked. Do these steaks appear to weigh enough?

   17 20 21 18 20 20 20 18 19 19

   20 19 21 20 18 20 20 19 18 19

Finding and Comparing Measures of Centers of Data. *In Exercises 5–8, find the mean, median, mode, and midrange for each of the two samples, then compare the two sets of results.*

5. Customer Waiting Times Waiting times of customers (in minutes) at the Jefferson Valley Bank (where all customers enter a single waiting line) and the Bank of Providence (where customers wait in individual lines at three different teller windows):

Jefferson Valley:	6.5	6.6	6.7	6.8	7.1	7.3	7.4	7.7	7.7	7.7
Providence:	4.2	5.4	5.8	6.2	6.7	7.7	7.7	8.5	9.3	10.0

Interpret the results by determining whether there is a difference between the two data sets that is not apparent from a comparison of the measures of center. If so, how are the data sets different?

6. Regular/Diet Coke Weights (in pounds) of samples of the contents in cans of regular Coke and diet Coke:

Regular:	0.8192	0.8150	0.8163	0.8211	0.8181	0.8247
Diet:	0.7773	0.7758	0.7896	0.7868	0.7844	0.7861

Does there appear to be a significant difference between the two data sets? How might such a difference be explained?

7. Comparing Coke and Pepsi Weights (in pounds) of samples of the contents in cans of regular Coke and regular Pepsi:

Coke:	0.8192	0.8150	0.8163	0.8211	0.8181	0.8247
Pepsi:	0.8258	0.8156	0.8211	0.8170	0.8216	0.8302

Does there appear to be a significant difference in the weights of the two different brands?

8. Male Skull Breadths Maximum breadth of samples of male Egyptian skulls from 4000 B.C. and 150 A.D.:

4000 B.C.:	131	119	138	125	129	126	131	132	126	128	128	131
150 A.D.:	136	130	126	126	139	141	137	138	133	131	134	129

(Based on data from *Ancient Races of the Thebaid* by Thomson and Randall-Maciver.)

Changes in head sizes over time suggest interbreeding with people from other regions. Do the head sizes appear to have changed from 4000 B.C. to 150 A.D.?

Finding the Mean of Data. *In Exercises 9–12, find the mean of the data summarized in the accompanying frequency table.*

9. Old Faithful Eruptions Visitors to Yellowstone National Park consider an eruption of the Old Faithful geyser to be a major attraction that should not be missed. The accompanying frequency table summarizes a sample of times (in minutes) between eruptions.

Table for Exercise 9

Time	Frequency
40–49	8
50–59	44
60–69	23
70–79	6
80–89	107
90–99	11
100–109	1

10. Comparing Student and Faculty Cars Samples of student cars and faculty/staff cars were obtained at the author's college, and their ages (in years) are summarized in the accompanying frequency table. Compare the two means.

Table for Exercise 10

Age	Students	Faculty/Staff
0–2	23	30
3–5	33	47
6–8	63	36
9–11	68	30
12–14	19	8
15–17	10	0
18–20	1	0
21–23	0	1

11. How Fast Were They Driving? The accompanying frequency table describes the speeds of drivers ticketed by the Town of Poughkeepsie police. These drivers were traveling through a 30 mi/h speed zone on Creek Road, which passes the author's college. How does the mean compare to the posted speed limit of 30 mi/h?

Table for Exercise 11

Speed	Frequency
42–45	25
46–49	14
50–53	7
54–57	3
58–61	1

12. Interpreting Body Temperature Differences The accompanying frequency table summarizes a sample of human body temperatures. (See the temperatures for midnight on the second day, as listed in Data Set 6 in Appendix B.) How does the mean compare to the value of 98.6°F, which is the value that most people assume to be the mean?

Table for Exercise 12

Temperature	Frequency
96.5–96.8	1
96.9–97.2	8
97.3–97.6	14
97.7–98.0	22
98.1–98.4	19
98.5–98.8	32
98.9–99.2	6
99.3–99.6	4

Finding the Mean and Median of Data. *In Exercises 13–16, refer to the data set in Appendix B. Use computer software or a calculator to find the mean and median.*

13. Comparing Textbook Prices Data Set 2 in Appendix B: Find the mean and median of the new textbook prices at the author's college and find the mean and median of new textbook prices at the University of Massachusetts. Compare and interpret the results.

14. Cigarette Nicotine Content Data Set 8 in Appendix B: Find the mean and median of the nicotine contents of all cigarettes listed.

15. Interpreting Boston Rainfall Amounts Data Set 17 in Appendix B: Find the mean and median of the rainfall amounts in Boston on Thursday and find the mean and median of the rainfall amounts in Boston on Sunday. One report claimed that it rains more on weekends than during the week. Do these results appear to support that claim?

16. Comparing Home-Run Distances Data Set 19 in Appendix B: Find the mean and median for the home-run distances of Mark McGwire and find the mean and median for the home-run distances of Sammy Sosa. Compare and interpret the results.

2-4 Beyond the Basics

17. The **harmonic mean** is often used as a measure of center for data sets consisting of rates of change, such as speeds. It is found by dividing the number of values (n) by the sum of the *reciprocals* of all values, expressed as

$$\frac{n}{\Sigma \frac{1}{x}}$$

(No value can be zero.) Four students drive from New York to Florida (1200 miles) at a speed of 40 mi/h (yeah, right!) and return at a speed of 60 mi/h. What is their average speed for the round trip? (The harmonic mean is used in averaging speeds.)

18. The **geometric mean** is often used in business and economics for finding average rates of change, average rates of growth, or average ratios. Given n values (all of which are positive), the geometric mean is the nth root of their product. The *average growth factor* for money compounded at annual interest rates of 10%, 8%, 9%, 12%, and 7% can be found by computing the geometric mean of 1.10, 1.08, 1.09, 1.12, and 1.07. Find that average growth factor.

19. The **quadratic mean** (or **root mean square,** or **R.M.S.**) is usually used in physical applications. In power distribution systems, for example, voltages and currents are usually referred to in terms of their R.M.S. values. The quadratic mean of a set of values is obtained by squaring each value, adding the results, dividing by the number of values (n), and then taking the square root of that result, expressed as

$$\text{quadratic mean} = \sqrt{\frac{\Sigma x^2}{n}}$$

Find the R.M.S. of these power supplies (in volts): 110, 0, −60, 12.

20. Finding the Weighted Mean Kelly Bell gets quiz grades of 65, 83, 80, and 90. She gets a 92 on her final exam. Find the weighted mean if the quizzes each count for 15% and the final counts for 40% of the final grade.

21. Analyzing Transformed Data In each of the following, describe how the mean, median, mode, and midrange of a data set are affected.

a. The same constant k is added to each value of the data set.

b. Each value of the data set is multiplied by the same constant k.

22. Finding the Median When data are summarized in a frequency table, the median can be found by first identifying the *median class* (the class that contains the median). We then assume that the values in that class are evenly distributed and we can interpolate. This process can be described by

$$\text{(lower limit of median class)} + \text{(class width)}\left(\frac{\left(\frac{n+1}{2}\right) - (m+1)}{\text{frequency of median class}}\right)$$

where n is the sum of all class frequencies and m is the sum of the class frequencies that *precede* the median class. Use this procedure to find the median of the data set summarized in Frequency Table 2-3.

23. Finding and Comparing the Trimmed Mean Because the mean is very sensitive to extreme values, we say that it is not a *resistant* measure of center. The **trimmed mean** is more resistant. To find the 10% trimmed mean for a data set, first arrange the data in order, then delete the bottom 10% of the values and the top 10% of the values, and calculate the mean of the remaining values. For the weights of the bears in Data Set 7 from Appendix B, find (a) the mean, (b) the 10% trimmed mean, (c) the 20% trimmed mean. How do the results compare?

24. Average Teacher's Salary Using an almanac, a researcher finds the average teacher's salary for each state. He adds those 50 values, then divides by 50 to obtain their mean. Is the result equal to the national average teacher's salary? Why or why not?

2-5 Measures of Variation

In this section we introduce the characteristic of variation. Because variation is so important in statistics, this is one of the most important sections in the entire book. We will discuss in detail the following key concepts: (1) Variation refers to the amount that values vary among themselves, and it can be measured with specific numbers; (2) values that are relatively close together have lower measures of variation, and values that are spread farther apart have measures of variation that are larger; (3) the standard deviation, which is a particularly important measure of variation, can be computed; and (4) the values of standard deviation must be *interpreted* correctly.

Many banks once required that customers wait in separate lines at each teller's window, but most have now changed to a single main waiting line. Why did they make that change? The mean waiting time didn't change, because the waiting-line configuration doesn't affect the efficiency of the tellers. They changed to the single line because customers prefer waiting times that are more *consistent* with less variation. Thus thousands of banks made a change that resulted in lower variation (and happier customers), even though the mean was not affected. Let's now consider the same bank sample data used in Exercise 5 in the preceding section. The listed values are waiting times (in minutes) of customers.

Jefferson Valley Bank (Single waiting line)	6.5	6.6	6.7	6.8	7.1	7.3	7.4	7.7	7.7	7.7
Bank of Providence (Multiple waiting lines)	4.2	5.4	5.8	6.2	6.7	7.7	7.7	8.5	9.3	10.0

In Figure 2-14 we show dotplots of both data sets, and we include the measures of center. A comparison of the measures of center does not reveal any differences between the two data sets, but a comparison of the dotplots provides clear and strong visual evidence that the Jefferson Valley Bank has waiting times with substantially less variation than the times for the Bank of Providence. Looking at Figure 2-14, we see that the dots for Jefferson Valley are closer together (indicating less variation) than the dots for the Bank of Providence.

Let's now proceed to develop some specific ways of actually *measuring* variation. We begin with the range.

Range

The **range** of a set of data is the difference between the highest value and the lowest value.

$$\textbf{range} = \textbf{(highest value)} - \textbf{(lowest value)}$$

To compute it, simply subtract the lowest value from the highest value. For the Jefferson Valley Bank customer waiting times, the range is $7.7 - 6.5 = 1.2$ min. The Bank of Providence has waiting times with a range of 5.8 min, and this larger value suggests greater variation.

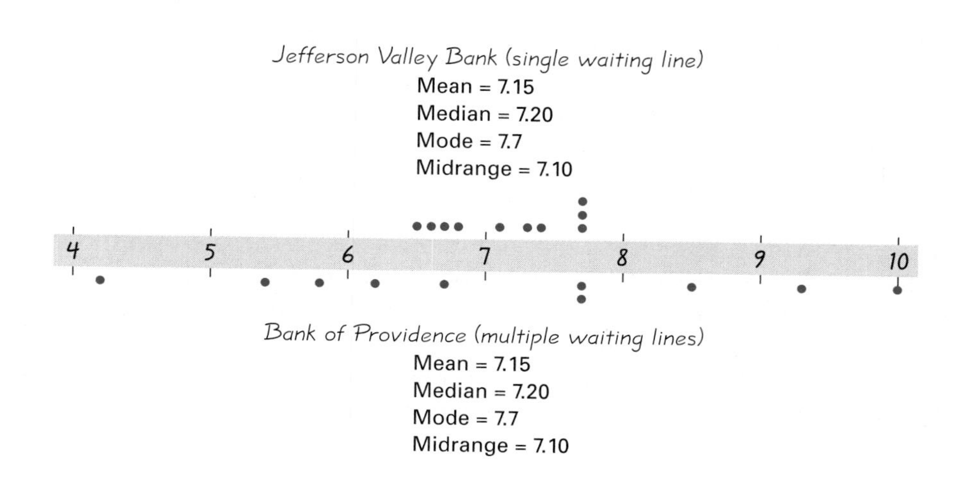

FIGURE 2-14
Dotplots of Waiting Times

More Stocks, Less Risk

In their book *Investments*, authors Zvi Bodie, Alex Kane, and Alan Marcus state that "the average standard deviation for returns of portfolios composed of only one stock was 0.554. The average portfolio risk fell rapidly as the number of stocks included in the portfolio increased." They note that with 32 stocks, the standard deviation is 0.325, indicating much less variation and risk. They make the point that with only a few stocks, a portfolio has a high degree of "firm-specific" risk, meaning that the risk is attributable to the few stocks involved. With more than 30 stocks, there is very little firm-specific risk; instead, almost all of the risk is "market risk," attributable to the stock market as a whole. They note that these principles are "just an application of the well-known law of averages."

The range is very easy to compute, but because it depends on only the highest and the lowest values, it isn't as useful as the other measures of variation that use every value. (See Exercise 29 for an example in which the range is misleading.)

Standard Deviation of a Sample

The standard deviation is the measure of variation that is generally the most important and useful. We define the standard deviation now, but in order to understand it fully you will need to read the remainder of this section very carefully.

DEFINITION

The **standard deviation** of a set of sample values is a measure of variation of values about the mean. It is calculated by using Formula 2-4 or 2-5.

Formula 2-4 $$s = \sqrt{\frac{\Sigma(x - \bar{x})^2}{n - 1}}$$ sample standard deviation

Formula 2-5 $$s = \sqrt{\frac{n\Sigma(x^2) - (\Sigma x)^2}{n(n - 1)}}$$ shortcut formula for standard deviation

Formulas 2-4 and 2-5 are equivalent in the sense that they will always yield the same result. Formula 2-4 has the advantage of reinforcing the concept that the standard deviation is a type of average deviation. Formula 2-5 has the advantage of being easier to use when you must calculate standard deviations on your own. Formula 2-5 also eliminates the intermediate rounding errors introduced in Formula 2-4 when the exact value of the mean is not used. Formula 2-5 is used in calculators and software programs because it requires only three memory locations (for n, Σx, and Σx^2) instead of a memory location for every value in the data set. So which formula should you use? We advise the following: Use Formula 2-4 for a few examples, then learn how to find standard deviations on your calculator and by using a software program.

Most scientific calculators are programmed so that you can enter a list of values and get the standard deviation automatically. If you are using a TI-83 Plus calculator, see the procedure described in Section 2-4. The TI-83 Plus results will include the standard deviation, identified as Sx.

AS

We define a measure of variation in the way described by Formula 2-4 for the following reason: Let's suppose that we have a sample with the three numbers, 2, 4, and 15. The mean of those three values is 7, and we will measure variation away from that mean. We begin with the individual amounts by which values deviate from the mean. For a particular value x, the amount of **deviation** is $x - \bar{x}$, which is the difference between the value and the mean. For the sample of 2, 4, and 15, the deviations away from the mean of 7 are

-5, -3, and 8. However, the sum of all such deviations is always zero, which doesn't really help us. To avoid always getting zero, we could take absolute values, as in $\Sigma|x - \bar{x}|$. For our sample of 2, 4, and 15, we get $\Sigma|x - \bar{x}| = 5 + 3 + 8 = 16$. If we divide that sum by the number of sample values, we get the **mean absolute deviation,** described as follows:

$$\text{mean absolute deviation} = \frac{\Sigma|x - \bar{x}|}{n}$$

The mean absolute deviation of 2, 4, and 15 is 16/3. Instead of using absolute values, we can get a better measure of variation by making all deviations $(x - \bar{x})$ nonnegative by squaring them. The deviations of -5, -3, and 8 become 25, 9, and 64 when squared. We add all of the squared deviations, then divide by $n - 1$ to get an average. The average of the squared deviations 25, 9, and 64 is 98/2, or 49. (We divide by $n - 1$ because there are only $n - 1$ independent values. That is, with a given mean, only $n - 1$ values can be assigned any number before the nth value is determined.) Finally, we take the square root to compensate for that squaring, as in $\sqrt{49} = 7$. As a result, the standard deviation has the same units of measurement as the original values. For example, if customer waiting times are in minutes, the standard deviation of those times will also be in minutes.

When applying Formula 2-4, it is best to take an approach that is often successful in working with mathematical expressions: Start on the inside and proceed outward. The procedure for calculating the standard deviation can be summarized as follows.

Procedure for Finding the Standard Deviation with Formula 2-4

Step 1: Find the mean of the values $(\bar{x})$.

Step 2: Subtract the mean from each individual value to get a list of deviations of the form $(x - \bar{x})$.

Step 3: Square each of the differences obtained from Step 2. [This produces numbers of the form $(x - \bar{x})^2$.]

Step 4: Add all of the squares obtained from Step 3 to get $\Sigma(x - \bar{x})^2$.

Step 5: Divide the total from Step 4 by the number $(n - 1)$; that is, 1 less than the total number of values present.

Step 6: Find the square root of the result of Step 5.

EXAMPLE **Using Formula 2-4** Use Formula 2-4 to find the standard deviation of the Jefferson Valley Bank customer waiting times. Those times (in minutes) are reproduced below:

6.5 6.6 6.7 6.8 7.1 7.3 7.4 7.7 7.7 7.7

TABLE 2-14 Calculating Standard Deviation for Jefferson Valley Bank Customers

x	$x - \overline{x}$	$(x - \overline{x})^2$
6.5	−0.65	0.4225
6.6	−0.55	0.3025
6.7	−0.45	0.2025
6.8	−0.35	0.1225
7.1	−0.05	0.0025
7.3	0.15	0.0225
7.4	0.25	0.0625
7.7	0.55	0.3025
7.7	0.55	0.3025
7.7	0.55	0.3025
Totals: 71.5		2.0450
↓		↓
$\overline{x} = \frac{71.5}{10} = 7.15$ min		$s = \sqrt{\frac{2.0450}{10 - 1}} = \sqrt{0.2272} = 0.48$ min

SOLUTION We will follow the six steps in the procedure just given. Refer to those steps and to Table 2-14, which shows the detailed calculations.

Step 1: Obtain the mean of 7.15 by adding the values and then dividing by the number of values:

$$\overline{x} = \frac{\Sigma x}{n} = \frac{71.5}{10} = 7.15 \text{ min}$$

Step 2: Subtract the mean of 7.15 from each value to get these values of $(x - \overline{x})$: −0.65, −0.55, . . . , 0.55.

Step 3: Square each value obtained in Step 2 to get these values of $(x - \overline{x})^2$: 0.4225, 0.3025, . . . , 0.3025.

Step 4: Sum all of the preceding values to get the value of

$$\Sigma(x - \overline{x})^2 = 2.0450$$

Step 5: With $n = 10$ values, divide by 1 less than 10:

$$\frac{2.0450}{9} = 0.2272$$

Step 6: Find the square root of 0.2272. The standard deviation is

$$\sqrt{0.2272} = 0.48 \text{ min}$$

Ideally, we would now interpret the meaning of the result, but such interpretations will be discussed later in this section.

Where Are the 0.400 Hitters?

The last baseball player to hit above 0.400 was Ted Williams, who hit 0.406 in 1941. There were averages above 0.400 in 1876, 1879, 1887, 1894, 1895, 1896, 1897, 1899, 1901, 1911, 1920, 1922, 1924, 1925, and 1930, but none since 1941. Are there no longer great hitters? Harvard's Stephen Jay Gould notes that the mean batting average has been steady at 0.260 for about 100 years, but the standard deviation has been decreasing from 0.049 in the 1870s to 0.031, where it is now. He argues that today's stars are as good as those from the past, but consistently better pitchers now keep averages below 0.400. Dr. Gould discusses this in Program 4 of the series *Against All Odds: Inside Statistics.*

EXAMPLE **Using Formula 2-5** The preceding example used Formula 2-4 for finding the standard deviation of the Jefferson Valley Bank customer waiting times. Using the same data set, find the standard deviation by applying Formula 2-5.

SOLUTION To use Formula 2-5 we first must find values for n, Σx, and Σx^2.

$n = 10$	(because there are 10 values in the sample)
$\Sigma x = 71.5$	(the sum of the 10 sample values)
$\Sigma x^2 = 513.27$	(the sum of the squares of the sample values: $6.5^2 + 6.6^2 + 6.7^2 + \cdots + 7.7^2$)

Formula 2-5 can now be used to find the standard deviation.

$$s = \sqrt{\frac{n(\Sigma x^2) - (\Sigma x)^2}{n(n-1)}} = \sqrt{\frac{10(513.27) - (71.5)^2}{10(10-1)}} = \sqrt{\frac{20.45}{90}} = 0.48 \text{ min}$$

A great self-test is to stop here and calculate the standard deviation of the waiting times for the Bank of Providence. Follow the same procedures used in the preceding two examples and verify that, for the Bank of Providence, $s = 1.82$ min. Although the interpretations of these standard deviations will be discussed later, we can now compare them to see that the standard deviation of the times for the Jefferson Valley Bank (0.48 min) is much lower than the standard deviation for the Bank of Providence (1.82 min). This supports our subjective conclusion that the waiting times at the Jefferson Valley Bank have much less variation than those at the Bank of Providence.

Standard Deviation of a Population

In our definition of standard deviation, we referred to the standard deviation of *sample* data. A slightly different formula is used to calculate the standard deviation σ (lowercase Greek sigma) of a *population:* Instead of dividing by $n - 1$, divide by the population size N, as in the following expression.

$$\sigma = \sqrt{\frac{\Sigma(x-\mu)^2}{N}} \quad \text{population standard deviation}$$

Because we generally deal with sample data, we will usually use Formula 2-4, in which we divide by $n - 1$. Many calculators do both the sample standard deviation and the population standard deviation, but they use a variety of different notations. Be sure to identify the notation used by your calculator, so that you get the correct result. (The TI-83 Plus uses Sx for the sample standard deviation and σx for the population standard deviation.)

Variance of a Sample and Population

We are using the term *variation* as a general description of the amount that values vary among themselves. The term *variance* refers to a specific definition.

DEFINITION

The **variance** of a set of values is a measure of variation equal to the square of the standard deviation.

Sample variance: Square of the standard deviation s.

Population variance: Square of the population standard deviation σ.

The sample variance s^2 is an **unbiased estimator** of the population variance σ^2, which means that values of s^2 tend to target the value of σ^2 instead of systematically tending to overestimate or underestimate σ^2. (See Exercise 35.)

EXAMPLE In the preceding example, we used the Jefferson Valley Bank customer waiting times to find that the standard deviation is given by $s = 0.48$ min. Find the variance of that same sample.

SOLUTION Instead of using the rounded value of $s = 0.48$ min, we will use the more precise value of $s = 0.4767$ min. Because the variance is the square of the standard deviation, we get this result:

$$\text{sample variance} = s^2 = (0.4767 \text{ min})^2 = 0.23 \text{ min}^2$$

The variance is an important statistic used in some important statistical calculations, such as analysis of variance (discussed in Chapter 11). For our present purposes, the variance has a serious disadvantage: The units of variance are different than the units of the original data set. For example, if the original customer waiting times are in minutes, the units of the variance are square minutes (min^2). What is a square minute? (Have some fun constructing a creative answer to that question.) Because the variance uses different units, it is extremely difficult to understand variance by relating it to the original data set. Because of this property, we will focus on the standard deviation as we try to develop an understanding of variation.

We now present the notation and the round-off rule we are using.

Notation

s = *sample* standard deviation
s^2 = *sample* variance
σ = *population* standard deviation
σ^2 = *population* variance

Note: Articles in professional journals and reports often use SD for standard deviation and VAR for variance.

Round-Off Rule

As discussed in Section 2-4, we use this rule for rounding final results:

> **Carry one more decimal place than is present in the original set of values.**

Round only the final answer, never in the middle of a calculation. (If it becomes absolutely necessary to round in the middle, carry at least twice as many decimal places as will be used in the final answer.)

Finding Standard Deviation from a Frequency Table

We sometimes need to compute the standard deviation of a data set that is summarized in the form of a frequency table, such as Table 2-3 in Section 2-2. If the original list of sample values is available, use those values with Formula 2-4 or 2-5 so that the result will be more exact. If the original data are not available, use one of these two methods:

1. If the total number of values is not too large, use your calculator or software program by entering each class midpoint a number of times equal to the class frequency.
2. Calculate the standard deviation using Formula 2-6.

Formula 2-6 $$s = \sqrt{\frac{n[\Sigma(f \cdot x^2)] - [\Sigma(f \cdot x)]^2}{n(n-1)}}$$ standard deviation for frequency table

EXAMPLE **Qwerty Keyboard Ratings** Find the standard deviation of the 52 values summarized in Table 2-3, assuming that the original data set is not available.

SOLUTION

Method 1: Table 2-3 has class midpoints of 1, 4, 7, 10, and 13. Using a calculator or software program, enter the value of 1 twenty

Data Mining

The term *data mining* is commonly used to describe the now popular practice of analyzing an existing large set of data for the purpose of finding relationships, patterns, or any interesting results that were not found in the original studies of the data set. Some statisticians express concern about ad hoc inference—a practice in which a researcher goes on a fishing expedition through old data, finds something significant, and then identifies an important question that has already been answered. Robert Gentleman, a column editor for *Chance* magazine, writes that "there are some interesting and fundamental statistical issues that data mining can address. We simply hope that its current success and hype don't do our discipline (statistics) too much damage before its limitations are discussed."

TABLE 2-15 Calculating Standard Deviation from Frequency Table

Word Rating	Frequency, f	Class Midpoint, x	$f \cdot x$	$f \cdot x^2$
0–2	20	1	20	20
3–5	14	4	56	224
6–8	15	7	105	735
9–11	2	10	20	200
12–14	1	13	13	169
Totals:	$\Sigma f = 52$		$\Sigma(f \cdot x) = 214$	$\Sigma(f \cdot x^2) = 1348$

times (because the frequency of the first class is 20), enter 4 fourteen times, and so on. Find the standard deviation of this set of 52 class midpoints. The result should be 3.0.

Method 2: Use Formula 2-6. To do this, we first need to find the values of $n, \Sigma(f \cdot x)$, and $\Sigma(f \cdot x^2)$. After obtaining those values from Table 2-15, we apply Formula 2-6 as follows:

$$s = \sqrt{\frac{n[\Sigma(f \cdot x^2)] - [\Sigma(f \cdot x)]^2}{n(n-1)}} = \sqrt{\frac{52[1348] - [214]^2}{52(52-1)}}$$

$$= \sqrt{\frac{24{,}300}{2652}} = \sqrt{9.1628959} = 3.0$$

Unlike most calculators, the TI-83 Plus is designed to compute the standard deviation of values summarized in a frequency table. First enter the class midpoints in list L1, then enter the frequencies in list L2. Now press **STAT,** select **CALC,** select **1-VarStats,** and enter L1, L2 (including the comma) to obtain results that include the mean and standard deviation. Again, the sample standard deviation is identified as Sx, and the population standard deviation is identified as σx.

Interpreting and Understanding Standard Deviation

We will now attempt to make some intuitive sense of standard deviation. First, we should clearly understand that the standard deviation measures the variation among values. Values close together will yield a small standard deviation, whereas values spread farther apart will yield a larger standard deviation. Refer again to Figure 2-14, which is a dotplot showing different amounts of variation.

Because variation is such an important concept and because the standard deviation is such an important tool in measuring variation, we will consider three different ways of developing a sense for values of standard deviations. The first one, the **range rule of thumb,** is based on the principle that for many data sets, the vast majority (such as 95%) of sample values lie within 2 standard deviations of the mean. (We could improve the accuracy of this rule by taking into account such factors as the size of the sample and the nature of the distribution, but we prefer to sacrifice accuracy for the sake of simplicity. We want a simple rule that will help us interpret values of standard deviations; later methods will produce more accurate results.)

Range Rule of Thumb

For Estimation: To obtain a rough estimate of the standard deviation s, use the equation

$$s \approx \frac{\text{range}}{4}$$

where range = (highest value) − (lowest value).

For Interpretation: If the standard deviation s is known, we can use it to find rough estimates of the minimum and maximum "usual" sample values as follows.

minimum "usual" value ≈ (mean) − 2 × (standard deviation)
maximum "usual" value ≈ (mean) + 2 × (standard deviation)

When calculating a standard deviation using Formula 2-4 or 2-5, you can use the range rule of thumb as a check on your result, but you must realize that although the approximation will get you in the general vicinity of the answer, it can be off by a fairly large amount.

EXAMPLE **Customer Waiting Times** For the Jefferson Valley Bank customer waiting times (6.5, 6.6, 6.7, 6.8, 7.1, 7.3, 7.4, 7.7, 7.7, 7.7) we used Formulas 2-4 and 2-5 to compute the standard deviation as $s = 0.48$ min. Find the estimated standard deviation by using the range rule of thumb.

SOLUTION Scanning the list of values, we find a high of 7.7 and a low of 6.5, so the range is $7.7 - 6.5 = 1.2$. We use the range rule of thumb to get a rough estimate of s as follows:

$$s \approx \frac{\text{range}}{4} = \frac{1.2}{4} = 0.3 \text{ min}$$

INTERPRETATION Previously, we found that the standard deviation is 0.48, so the range-rule-of-thumb estimate of 0.3 is a bit too low. Nevertheless, this estimate confirms that we are in the ballpark, and we would know that a value for s such as 7 is probably not correct.

Mail Consistency

A recent survey of 29,000 people who use the U.S. Postal Service revealed that they wanted better consistency in the time it takes to make a delivery. Now, a local letter could take one day or several days. *USA Today* reported a common complaint: "Just tell me how many days ahead I have to mail my mother's birthday card."

The level of consistency can be measured by the standard deviation of the delivery times. A lower standard deviation reflects more consistency. The standard deviation is often a critically important tool used to monitor and control the quality of goods and services.

EXAMPLE **Qwerty Keyboard Ratings** Use the range rule of thumb to find a rough estimate of the standard deviation of the sample of 52 word ratings from the Qwerty keyboard, as listed in Table 2-1.

SOLUTION In using the range rule of thumb to estimate the standard deviation of sample data, we find the range and divide by 4. By scanning the list of ratings, we can see that the lowest is 0 and the highest is 14, so the range is 14. The standard deviation s is estimated as follows:

$$s \approx \frac{\text{range}}{4} = \frac{14}{4} = 3.5$$

INTERPRETATION This result is in the ballpark of the correct value of 2.8 that is obtained by calculating the exact value of the standard deviation with Formula 2-4 or 2-5.

The preceding examples illustrated how we can use known information about the range to estimate the standard deviation. The following example is particularly important as an illustration of one way to *interpret* the value of a standard deviation.

EXAMPLE **Heights of Men** Previous results from the National Health Survey show that the heights of men have a mean of 69.0 in. and a standard deviation of 2.8 in. Use the range rule of thumb to find the minimum and maximum "usual" heights.

SOLUTION With a mean of 69.0 in. and a standard deviation of 2.8 in., we use the range rule of thumb to find the minimum and maximum usual heights as follows:

$$\begin{aligned}\text{minimum} &\approx (\text{mean}) - 2 \times (\text{standard deviation})\\ &= 69.0 - 2(2.8) = 69.0 - 5.6 = 63.4 \text{ in.}\\ \text{maximum} &\approx (\text{mean}) + 2 \times (\text{standard deviation})\\ &= 69.0 + 2(2.8) = 69.0 + 5.6 = 74.6 \text{ in.}\end{aligned}$$

INTERPRETATION Based on these results, we expect that typical men will range in height between 63.4 in. and 74.6 in. The heights of some men do not fall between those limits, but those men are unusually short or unusually tall. We now have a much better understanding of how men's heights vary.

Empirical (or 68–95–99.7) Rule for Data with a Bell-Shaped Distribution

Another rule that is helpful in interpreting values for a standard deviation is the **empirical rule.** This rule states that for data sets having a distribution that is approximately bell-shaped, the following properties apply. (See Figure 2-15.)

- About 68% of all values fall within 1 standard deviation of the mean.
- About 95% of all values fall within 2 standard deviations of the mean.
- About 99.7% of all values fall within 3 standard deviations of the mean.

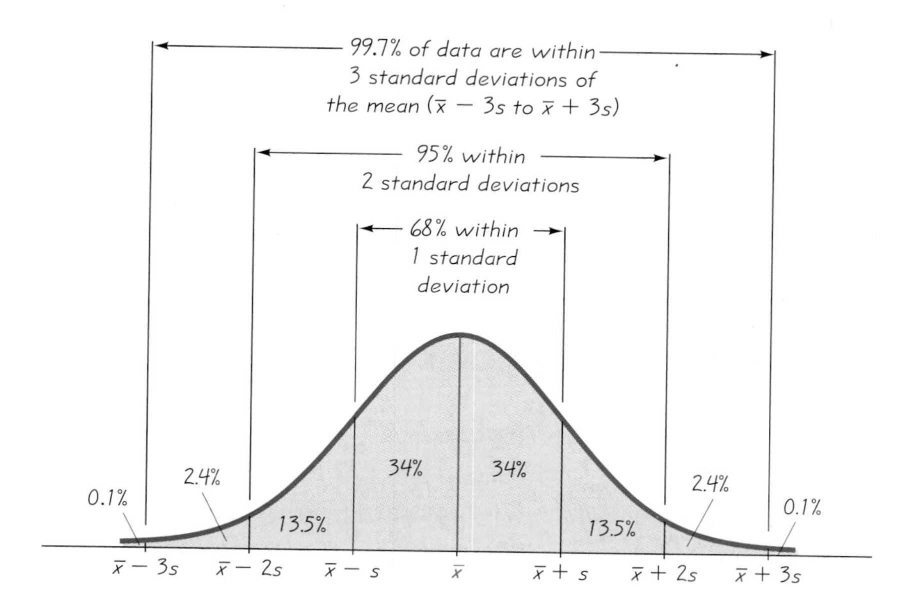

FIGURE 2-15
The Empirical Rule

AS

EXAMPLE **Heights of Men** The heights of men have a bell-shaped distribution with a mean of 69.0 in. and a standard deviation of 2.8 in. (based on data from the National Health Survey). What percentage of men have heights between 60.6 in. and 77.4 in.?

SOLUTION The key to solving this problem is to recognize that 60.6 in. and 77.4 in. are each exactly 3 standard deviations away from the mean of 69.0 in., as shown below.

$$3 \text{ standard deviations} = 3s = 3(2.8) = 8.4$$

Therefore, 3 standard deviations from the mean is

$$69.0 - 8.4 = 60.6$$

or

$$69.0 + 8.4 = 77.4$$

The empirical rule tells us that about 99.7% of all values are within 3 standard deviations of the mean, so about 99.7% of all men's heights are between 60.6 in. and 77.4 in.

Buying Cars

For buying a new or used car, an excellent reference is the reliability data compiled and reported by *Consumer Reports* magazine. Frequency-of-repair data are based on 10 million pieces of data collected from thousands of readers. Statisticians analyze the data for patterns that lead to lists of both reliable cars and cars that should be avoided. Consumers Union President Rhoda Karpatkin writes, "Because numbers describe so much of our work, it should be no surprise that statisticians are key to that process."

Hint: Difficulty in applying the empirical rule usually stems from confusion over the meaning of such phrases as "within 3 standard deviations of the mean." If you are at all uncertain, you should stop here and review the preceding example. The following table should help as well.

Phrase	**Meaning**
Within 1 standard deviation of the mean	Between $(\overline{x} - s)$ and $(\overline{x} + s)$
Within 2 standard deviations of the mean	Between $(\overline{x} - 2s)$ and $(\overline{x} + 2s)$
Within 3 standard deviations of the mean	Between $(\overline{x} - 3s)$ and $(\overline{x} + 3s)$

A third concept that is helpful in understanding or interpreting a value of a standard deviation is **Chebyshev's theorem.** The empirical rule applies only to data sets with a bell-shaped distribution. Chebyshev's theorem applies to *any* data set, but its results are very approximate.

Chebyshev's Theorem

The proportion (or fraction) of any set of data lying within K standard deviations of the mean is always *at least* $1 - 1/K^2$, where K is any positive number greater than 1. For $K = 2$ and $K = 3$, we get the following results:

- At least 3/4 (or 75%) of all values lie within 2 standard deviations of the mean.
- At least 8/9 (or 89%) of all values lie within 3 standard deviations of the mean.

EXAMPLE **Heights of Men** Heights of men have a mean of 69.0 in. and a standard deviation of 2.8 in. What can we conclude from Chebyshev's theorem?

SOLUTION Applying Chebyshev's theorem with a mean of 69.0 and a standard deviation of 2.8, we can reach the following conclusions:

- At least 3/4 (or 75%) of all men have heights within 2 standard deviations of the mean (between 63.4 in. and 74.6 in.).
- At least 8/9 (or 89%) of all men have heights within 3 standard deviations of the mean (between 60.6 in. and 77.4 in.).

After studying this section, you should understand that the standard deviation is a measure of variation among scores. Given sample data, you should be able to compute the value of the standard deviation. You should be able to interpret the values of standard deviations that you compute. You should recognize that for typical data sets, it is unusual for a score to differ from the mean by more than 2 or 3 standard deviations.

2-5 Basic Skills and Concepts

Finding the Measures of Variation. *In Exercises 1–4, find the range, variance, and standard deviation for the given data. (The same data were used in Section 2-4 where we found measures of center. Here we find measures of variation.)*

1. Car Crashes The given values are the numbers of Dutchess County car crashes for each month of a recent year.

27 8 17 11 15 25 16 14 14 14 13 18

2. Eyeglass Fittings The given values are the measurements of distances (in millimeters) between the pupils of adult patients being fitted for eyeglasses.

67 66 59 62 63 66 66 55

3. McDonald's Service Times The given values are the service times (in seconds) of McDonald's drive-through customers.

88 107 35 93 65 55 119 83 99 74 46 108

4. Actual Steak Weights The given values are the weights (in ounces) of steaks listed on a restaurant menu as "20-ounce Porterhouse" steaks.

17 20 21 18 20 20 20 18 19 19
20 19 21 20 18 20 20 19 18 19

Comparing Measures of Variation. *In Exercises 5–8, find the range, variance, and standard deviation for each of the two samples, then compare the two sets of results. (The same data were used in Section 2-4.)*

5. Customer Waiting Times Examples in this section included the following waiting times of customers at the Jefferson Valley Bank (where all customers enter a single waiting line) and the Bank of Providence (where customers wait in individual lines at three different teller windows):

Jefferson Valley:	6.5	6.6	6.7	6.8	7.1	7.3	7.4	7.7	7.7	7.7
Providence:	4.2	5.4	5.8	6.2	6.7	7.7	7.7	8.5	9.3	10.0

6. Comparing Regular and Diet Coke Data Weights (in pounds) of samples of the contents in cans of regular Coke and diet Coke:

Regular:	0.8192	0.8150	0.8163	0.8211	0.8181	0.8247
Diet:	0.7773	0.7758	0.7896	0.7868	0.7844	0.7861

7. Comparing Coke and Pepsi Data Weights (in pounds) of samples of the contents in cans of regular Coke and regular Pepsi:

Coke:	0.8192	0.8150	0.8163	0.8211	0.8181	0.8247
Pepsi:	0.8258	0.8156	0.8211	0.8170	0.8216	0.8302

8. Ancient Skull Breadths Maximum breadth of samples of male Egyptian skulls from 4000 B.C. and 150 A.D.:

4000 B.C.:	131	119	138	125	129	126	131	132	126	128	128	131
150 A.D.:	136	130	126	126	139	141	137	138	133	131	134	129

(Based on data from *Ancient Races of the Thebaid* by Thomson and Randall-Maciver.)

Table for Exercise 9

Time	Frequency
40–49	8
50–59	44
60–69	23
70–79	6
80–89	107
90–99	11
100–109	1

Table for Exercise 10

Age	Students	Faculty/ Staff
0–2	23	30
3–5	33	47
6–8	63	36
9–11	68	30
12–14	19	8
15–17	10	0
18–20	1	0
21–23	0	1

Table for Exercise 11

Speed	Frequency
42–45	25
46–49	14
50–53	7
54–57	3
58–61	1

Table for Exercise 12

Temperature	Frequency
96.5–96.8	1
96.9–97.2	8
97.3–97.6	14
97.7–98.0	22
98.1–98.4	19
98.5–98.8	32
98.9–99.2	6
99.3–99.6	4

Finding the Standard Deviation. *In Exercises 9–12, find the standard deviation of the data summarized in the given frequency table. (The same frequency tables were used in Section 2-4.)*

9. Old Faithful Eruption The accompanying frequency table summarizes a sample of times (in minutes) between eruptions of the Old Faithful geyser in Yellowstone National Park.

10. Comparing Student and Faculty Cars Samples of students cars and faculty/staff cars were obtained at the author's college, and their ages (in years) are summarized in the accompanying frequency table.

11. Speeds of Ticketed Drivers The accompanying frequency table describes the speeds of drivers ticketed by the Town of Poughkeepsie police.

12. Body Temperature Differences The accompanying frequency table summarizes a sample of human body temperatures (from the temperatures for midnight on the second day, as listed in Data Set 6 in Appendix B).

Finding Standard Deviations Using Technology. *In Exercises 13–16, refer to the data set in Appendix B. Use computer software or a calculator to find the standard deviation. (The same data sets were used in Section 2-4.)*

13. Comparing Textbook Prices Data Set 2 in Appendix B: Find the standard deviation of new textbook prices at the author's college and find the standard deviation of new textbook prices at the University of Massachusetts. Compare and interpret the standard deviations.

14. Nicotine Content Data Set 8 in Appendix B: Find the standard deviation of the nicotine contents of all cigarettes listed.

15. Interpreting Boston Rainfall Amounts Data Set 17 in Appendix B: Find the standard deviation of the rainfall amounts in Boston on Thursday and find the standard deviation of the rainfall amounts in Boston on Sunday. Compare and interpret the standard deviations.

16. Using the Range Rule of Thumb Data Set 19 in Appendix B: Find the standard deviation for the home-run distances of Mark McGwire and find the standard deviation for the home-run distances of Sammy Sosa. Compare and interpret the standard deviations.

17. Using the Range Rule of Thumb Use the range rule of thumb to estimate the standard deviation of ages of your classmates at the time of your high school graduation.

18. Interpreting Test Scores Use the range rule of thumb to estimate the standard deviation of the scores on the first statistics test in your class.

19. Interpreting Test Scores The test scores in a statistics class have a mean of 75 and a standard deviation of 12. Use the range rule of thumb to estimate the minimum and maximum "usual" scores. Is a score of 50 considered unusual in this context?

20. Interpreting Heights of Women Heights of women have a mean of 63.6 in. and a standard deviation of 2.5 in. (based on data from the National Health Survey). Use the range rule of thumb to estimate the minimum and maximum "usual" heights of women. In this context, is it unusual for a woman to be 6 ft tall?

21. Estimating Heights of Women Heights of women have a bell-shaped distribution with a mean of 63.6 in. and a standard deviation of 2.5 in. Using the empirical rule, what is the approximate percentage of women with heights between

a. 61.1 in. and 66.1 in.?
b. 56.1 in. and 71.1 in.?

22. Estimating IQ Scores IQ scores have a bell-shaped distribution with a mean of 100 and a standard deviation of 15. Using the empirical rule, what is the approximate percentage of people with IQ scores between

a. 70 and 130?
b. 85 and 115?

23. Using Chebyshev's Theorem If heights of women have a mean of 63.6 in. and a standard deviation of 2.5 in., what can you conclude from Chebyshev's theorem about the percentage of women between 58.6 in. and 68.6 in. tall?

24. Using Chebyshev's Theorem If IQ scores have a mean of 100 and a standard deviation of 15, what can you conclude from Chebyshev's theorem about the percentage of IQ scores between 55 and 145?

25. Understanding Standard Deviation What do you know about the values in a data set having a standard deviation of $s = 0$?

26. Understanding Units of Measurement If a data set consists of the prices of textbooks expressed in dollars, what are the units used for standard deviation? What are the units used for variance?

27. Comparing Car Battery Lives The Everlast and Endurance brands of car battery are both labeled as lasting 48 months. In reality, they both have a mean life of 50 months, but the Everlast batteries have a standard deviation of 2 months, while the Endurance batteries have a standard deviation of 6 months. Which brand is the better choice? Why?

28. Interpreting Outliers A data set consists of 20 values that are fairly close together. Another value is included, but this new value is an outlier (very far away from the other values). How is the standard deviation affected by the outlier? No effect? A small effect? A large effect?

2-5 Beyond the Basics

29. Comparing Data Sets Two different sections of a statistics class take the same quiz and the scores are recorded below. Find the range and standard deviation for each section. What do the range values lead you to conclude about the variation in the two sections? Why is the range misleading in this case? What do the standard deviation values lead you to conclude about the variation in the two sections?

Section 1:	1	20	20	20	20	20	20	20	20	20	20
Section 2:	2	3	4	5	6	14	15	16	17	18	19

30. Transforming Data For each of the following, describe how the range and standard deviation of a data set are affected.

a. The same constant k is added to each value of the data set.
b. Each value of the data set is multiplied by the same constant k.

(Continued on next page.)

c. For the body temperature data listed in Data Set 6 of Appendix B (12 A.M. on day 2), $\bar{x} = 98.20°$ F and $s = 0.62°$ F. Find the values of $\bar{x}$ and s after each temperature has been converted to the Celsius scale. [*Hint:* $C = 5(F - 32)/9$.]

31. a. The **coefficient of variation,** expressed as a percent, is used to describe the standard deviation relative to the mean. It allows us to compare variability of data sets with different measurement units (such as feet versus minutes), and it is calculated as follows:

$$\frac{s}{\bar{x}} \cdot 100 \quad \text{or} \quad \frac{\sigma}{\mu} \cdot 100$$

Find the coefficient of variation for the following sample of car ages (in years):

0 1 3 3 5 6 6 6 6 8 12

b. Genichi Taguchi developed a method of improving quality and reducing manufacturing costs through a combination of engineering and statistics. A key tool in the Taguchi method is the **signal-to-noise ratio.** The simplest way to calculate this ratio is to divide the mean by the standard deviation. Find the signal-to-noise ratio for the sample data given in part (a).

32. In Section 2-4 we introduced the general concept of skewness. Skewness can be measured by **Pearson's index of skewness:**

$$I = \frac{3(\bar{x} - \text{median})}{s}$$

If $I \geq 1.00$ or $I \leq -1.00$, the data can be considered to be *significantly skewed.* Find Pearson's index of skewness for the Qwerty keyboard word ratings listed in Table 2-1, and then determine whether there is significant skewness.

33. Understanding Standard Deviation A sample consists of six values that fall between 1 and 9 inclusive. What is the largest possible standard deviation?

34. Phony Data? For any data set of n values with standard deviation s, every value must be within $s\sqrt{n-1}$ of the mean. A statistics teacher reports that the test scores in her class of 17 students had a mean of 75.0 and a standard deviation of 5.0. Kelly, the class's self-proclaimed best student, claims that she received a grade of 97. Could Kelly be telling the truth?

35. Understanding Variance Let a population consist of the values 1, 2, and 3. Assume that samples of two different values are randomly selected *with replacement.*

a. Find the variance σ^2 of the population $\{1, 2, 3\}$.

b. List the nine different possible samples and find the sample variance s^2 for each of them. If you repeatedly select two different items, what is the mean value of the sample variances s^2?

c. For each of the nine samples, find the variance by treating each sample as if it is a population. (Be sure to use the formula for population variance.) If you repeatedly select two different items, what is the mean value of the population variances?

d. Which approach results in values that are better estimates of σ^2: part (b) or part (c)? Why? When computing variances of samples, should you use division by n or $n - 1$?

e. The preceding parts show that s^2 is an unbiased estimator of σ^2. Is s an unbiased estimator of σ?

2-6 Measures of Position

In this section we introduce measures that can be used to compare values from different data sets or to compare values within the same data set. The basic tools are z scores, quartiles, deciles, and percentiles. We begin with z scores.

z Scores

A z score (or standard score) is found by converting a value to a standardized scale, as given in the following definition.

DEFINITION

A **standard score,** or **z score,** is the number of standard deviations that a given value x is above or below the mean. It is found using the following expressions.

Sample		Population
$z = \dfrac{x - \overline{x}}{s}$	or	$z = \dfrac{x - \mu}{\sigma}$

(Round z to two decimal places.)

The following example illustrates how z scores can be used to compare values, even though they might come from different populations.

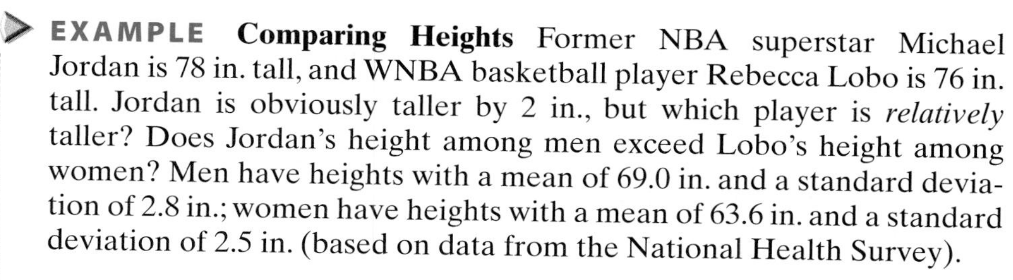

EXAMPLE **Comparing Heights** Former NBA superstar Michael Jordan is 78 in. tall, and WNBA basketball player Rebecca Lobo is 76 in. tall. Jordan is obviously taller by 2 in., but which player is *relatively* taller? Does Jordan's height among men exceed Lobo's height among women? Men have heights with a mean of 69.0 in. and a standard deviation of 2.8 in.; women have heights with a mean of 63.6 in. and a standard deviation of 2.5 in. (based on data from the National Health Survey).

SOLUTION To compare the heights of Michael Jordan and Rebecca Lobo relative to the populations of men and women, we need to standardize those heights by converting them to z scores.

$$\text{Jordan:} \quad z = \frac{x - \mu}{\sigma} = \frac{78 - 69.0}{2.8} = 3.21$$

$$\text{Lobo:} \quad z = \frac{x - \mu}{\sigma} = \frac{76 - 63.6}{2.5} = 4.96$$

INTERPRETATION Michael Jordan's height is 3.21 standard deviations above the mean, but Rebecca Lobo's height is a whopping 4.96 standard deviations above the mean. Rebecca Lobo's height among women is greater than Michael Jordan's height among men.

Market Testing Olestra

In a Chicago study involving the use of olestra as a substitute for fat in potato chips, Procter & Gamble researchers reported that on a 9-point taste scale, olestra chips got a mean of 5.6 and regular chips got a mean of 6.4. When the author requested the two sets of sample data that resulted in those means, Greg Allgood, Procter & Gamble's Associate Director of Regulatory and Clinical Development, wrote that "we respectfully decline your request and do not feel that providing this business proprietary information will be beneficial to us." Procter & Gamble published the summary statistics, but would not reveal the raw data that were used. Is this ethical?

FIGURE 2-16
Interpreting z Scores

Unusual values are those less than $z = -2.00$ or greater than $z = 2.00$.

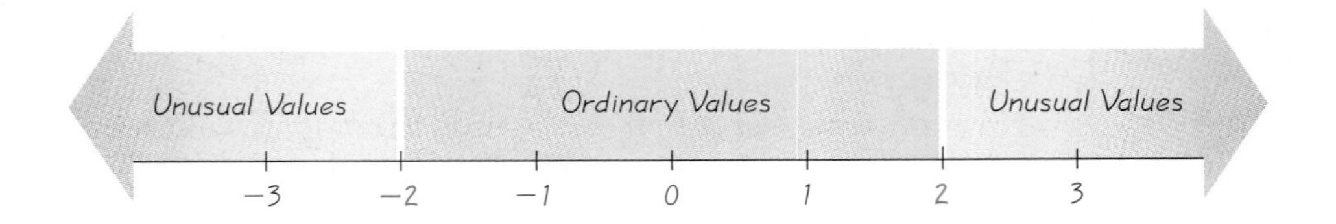

In Section 2-5 we used the range rule of thumb to conclude that a value is "unusual" if it is more than 2 standard deviations away from the mean. It follows that unusual values have z scores less than -2 or greater than $+2$. (See Figure 2-16.) Using this criterion, both Michael Jordan and Rebecca Lobo are unusually tall because they both have heights with z scores greater than 2.

While considering basketball players with exceptional heights, another player is Mugsy Bogues, a successful player who is only 5 ft 3 in. tall. After converting 5 ft 3 in. to 63 in., we convert his height to a z score as follows.

$$\text{Bogues:} \quad z = \frac{x - \mu}{\sigma} = \frac{63 - 69.0}{2.8} = -2.14$$

Let's thank Mugsy Bogues for his years of inspired play and for illustrating this principle:

Whenever a value is below the mean, the corresponding z score is negative.

Ordinary values: $-2 \le z \text{ score} \le 2$

Unusual values: $z \text{ score} < -2$ *or* $z \text{ score} > 2$

As the preceding example showed, z scores provide useful measurements for making comparisons between different sets of data. Likewise, quartiles, deciles, and percentiles are measures of position useful for comparing values within one set of data or between different sets of data.

Quartiles, Deciles, and Percentiles

Recall from Section 2-4 that the median of a data set is the middle value, so 50% of the values are equal to or less than the median and 50% of the values are greater than or equal to the median. Just as the median divides the data into two equal parts, the three **quartiles,** denoted by Q_1, Q_2, and Q_3, divide the sorted values into four equal parts. (Values are sorted when they are arranged in order.) Roughly speaking, Q_1 separates the bottom 25% of the sorted values from the top 75%; Q_2 is the median, which separates the bottom 50% from the top 50%; and Q_3 separates the bottom 75% from the top 25%. More precisely, at least 25% of the sorted values will be less than or equal to Q_1, and at least 75% will be greater than or equal to Q_1. At least 75% of the data will be less than or equal to Q_3, while at least 25% will be equal to or greater than Q_3.

Quartiles are useful for investigating the nature of the distribution of data. We will describe a procedure for finding quartiles after we discuss percentiles. There is no universally approved procedure for calculating quartiles, and different computer programs often yield different results. For example, if you use the data set of 1, 3, 6, 10, 15, 21, 28, and 36, you will get the following results.

	Q_1	Q_2	Q_3
STATDISK	4.5	12.5	24.5
Minitab	3.75	12.5	26.25
Excel	5.25	12.5	22.75
TI-83 Plus	4.5	12.5	24.5

For this particular data set, STATDISK and the TI-83 Plus calculator agree, but they do not always agree. If you use a calculator or computer software for exercises involving quartiles, you may get results that differ slightly from the answers given in the back of the book.

Just as there are three quartiles separating a data set into four parts, there are nine **deciles,** denoted by $D_1, D_2, D_3, \ldots, D_9$, which separate the data into 10 groups with about 10% of the values in each group. There are also 99 **percentiles,** denoted by $P_1, P_2, \ldots, P_{99}$, which partition the data into 100 groups with about 1% of the values in each group. (Quartiles, deciles, and percentiles are examples of *quantiles*—or *fractiles*—which partition data into parts that are approximately equal.)

The process of finding the percentile that corresponds to a particular value x is fairly simple, as indicated in the following expression.

$$\text{percentile of value } x = \frac{\text{number of values less than } x}{\text{total number of values}} \cdot 100$$

EXAMPLE **Weights of Coke** Table 2-16 lists 36 weights (in pounds) of the contents of 36 cans of regular Coke. Find the percentile corresponding to the weight of 0.8143 lb.

SOLUTION From Table 2-16 we see that there are 8 values less than 0.8143, so

$$\text{percentile of } 0.8143 = \frac{8}{36} \cdot 100 = 22 \text{ (rounded)}$$

INTERPRETATION The weight of 0.8143 lb is the 22nd percentile.

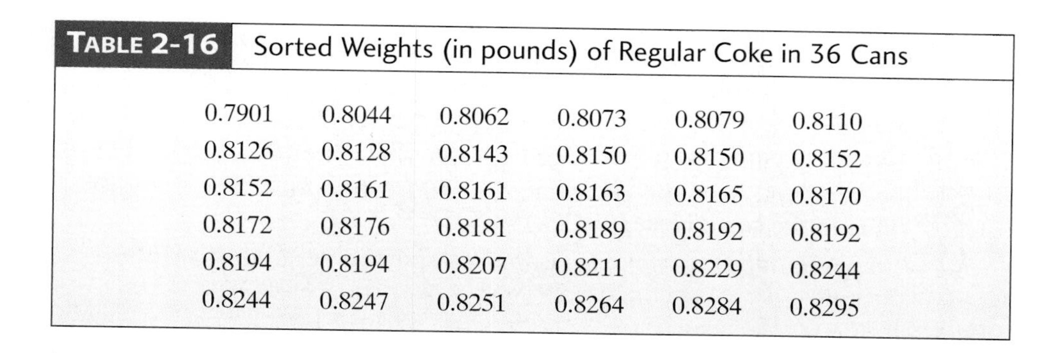

TABLE 2-16 Sorted Weights (in pounds) of Regular Coke in 36 Cans

0.7901	0.8044	0.8062	0.8073	0.8079	0.8110
0.8126	0.8128	0.8143	0.8150	0.8150	0.8152
0.8152	0.8161	0.8161	0.8163	0.8165	0.8170
0.8172	0.8176	0.8181	0.8189	0.8192	0.8192
0.8194	0.8194	0.8207	0.8211	0.8229	0.8244
0.8244	0.8247	0.8251	0.8264	0.8284	0.8295

Cost of Laughing Index

There really is a Cost of Laughing Index (CLI), which tracks costs of such items as rubber chickens, Groucho Marx glasses, admission to comedy clubs, and 13 other leading humor indicators. This is the same basic approach used in developing the Consumer Price Index (CPI), which is based on a weighted average of goods and services purchased by typical consumers. While standard scores and percentiles allow us to compare different values, they ignore any element of time. Index numbers, such as the CLI and CPI, allow us to compare the value of some variable to its value at some base time period. The value of an index number is the current value, divided by the base value, multiplied by 100.

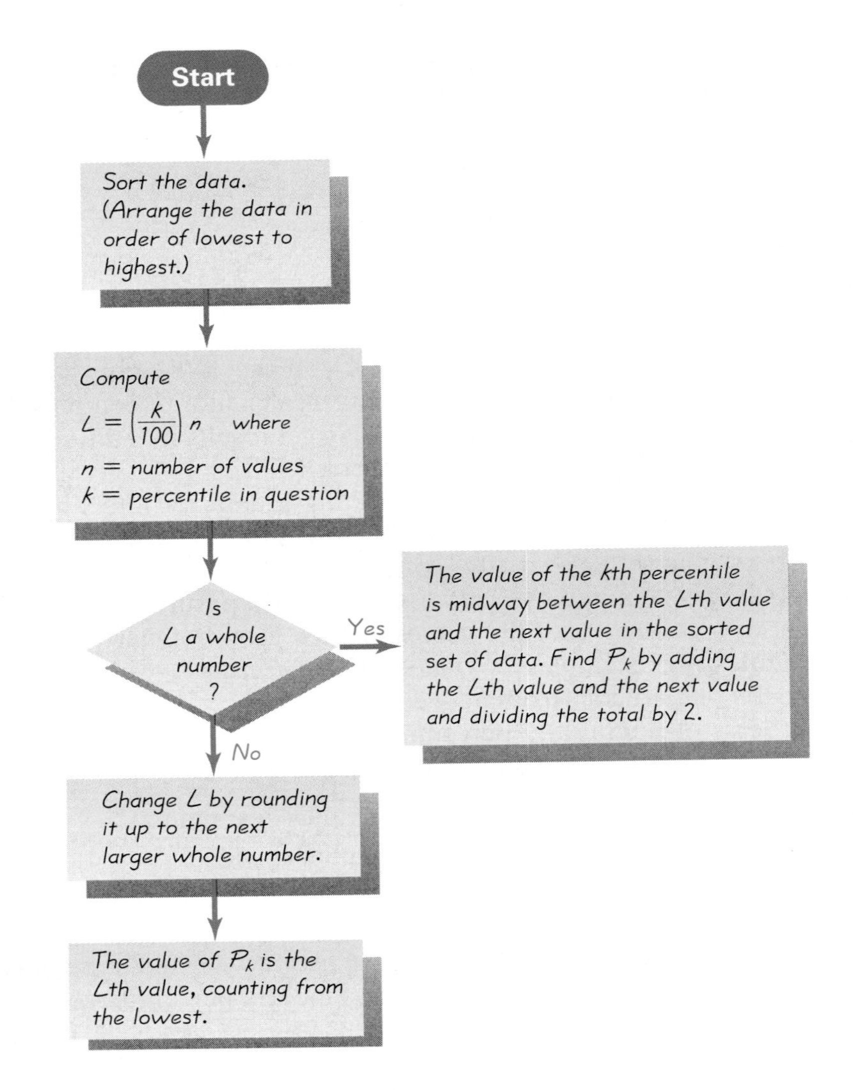

FIGURE 2-17
Finding the Value of the *k*th Percentile

The preceding example shows how to use a given sample value to find the corresponding percentile. There are several different methods for performing the reverse procedure, taking a given percentile and finding the corresponding sample value. The one we will use is summarized in Figure 2-17, which uses the following notation.

Notation

n = total number of values in the data set

k = percentile being used (Example: For the 25th percentile, $k = 25$.)

L = locator that gives the *position* of a value (Example: For the 25th value in the sorted list, $L = 25$.)

P_k = kth percentile (Example: P_{25} is the 25th percentile.)

EXAMPLE **Weights of Coke** Refer to the sample of Coke weights given in Table 2-16. Use Figure 2-17 to find the value of the 25th percentile, P_{25}.

SOLUTION Referring to Figure 2-17, we see that the sample data are already sorted, so we can proceed to compute the value of the locator L. In this computation, we use $k = 25$ because we are attempting to find the value of the 25th percentile, and we use $n = 36$ because there are 36 sample values:

$$L = \frac{k}{100} \cdot n = \frac{25}{100} \cdot 36 = 9$$

Next, we are asked if L is a whole number. The answer is yes, so we proceed to the box located at the right. We now see that the value of the kth (25th) percentile is midway between the Lth (9th) value and the next value in the original set of data. That is, the value of the 25th percentile is midway between the 9th value and the 10th value. The 9th value is 0.8143 and the 10th value is 0.8150, so the value midway between them is 0.81465. We conclude that the 25th percentile is $P_{25} = 0.81465$ lb.

EXAMPLE **Decile** For the sample of Coke weights in Table 2-16, find the value of the 4th decile, D_4.

SOLUTION First, we should note that the 4th decile, D_4, is the same as the 40th percentile, P_{40}, so we can proceed with the objective of finding the value of the 40th percentile.

Referring to Figure 2-17, the data are already sorted, so we compute the locator $L = (40/100)36 = 14.4$. Next, L is not a whole number, so we proceed straight down to where we change L by rounding it up to the next larger whole number: 15. The value of the 40th percentile (or the 4th decile) is the 15th value in the list, which is 0.8161 lb.

The preceding example showed that $D_4 = P_{40} = 0.8161$ lb, but you can verify that 0.8161 lb converts to the 36th percentile, not the 40th percentile. There is a discrepancy here, but such discrepancies become smaller as the sample size increases. We could eliminate the discrepancy by using a more exact and more complicated procedure that includes interpolation instead of rounding.

The preceding example showed that when finding a decile value (such as D_4), we can use the equivalent percentile value (such as P_{40}) instead. The table in the margin shows relationships relating quartiles and deciles to equivalent percentiles.

Quartiles	Deciles
$Q_1 = P_{25}$	$D_1 = P_{10}$
$Q_2 = P_{50}$	$D_2 = P_{20}$
$Q_3 = P_{75}$	$\vdots$
	$D_9 = P_{90}$

In earlier sections of this chapter we described several statistics, including the mean, median, mode, range, and standard deviation. Other statistics are sometimes defined using quartiles and percentiles, as in the following:

a. NBA basketball player Shaquille O'Neal, who is 7 ft 1 in. tall
b. Bob Jenkins, who is 5 ft 4 in. tall
c. The author, who is a 69.72-in.-tall golf and tennis "player"

4. Body Temperatures Human body temperatures have a mean of 98.20° and a standard deviation of 0.62°. Convert the given temperatures to z scores.
a. 100° **b.** 96.96° **c.** 98.20°

Interpreting z Scores. *In Exercises 5–8, express all z scores with two decimal places. Consider a value to be unusual if its z score is less than −2.00 or greater than 2.00.*

5. Heights of Women The Beanstalk Club is limited to women and men who are very tall. The minimum height requirement for women is 70 in. Women's heights have a mean of 63.6 in. and a standard deviation of 2.5 in. Find the z score corresponding to a woman with a height of 70 in. and determine whether that height is unusual.

6. Length of Pregnancy A woman wrote to *Dear Abby* and claimed that she gave birth 308 days after a visit from her husband, who was in the Navy. Lengths of pregnancies have a mean of 268 days and a standard deviation of 15 days. Find the z score for 308 days. Is such a length unusual? What do you conclude?

7. Body Temperatures Human body temperatures have a mean of 98.20° and a standard deviation of 0.62°. An emergency room patient is found to have a temperature of 101°. Convert 101° to a z score. Is that temperature unusually high? What does it suggest?

8. Cholesterol Levels in Men For men aged between 18 and 24 years, serum cholesterol levels (in mg/100 mL) have a mean of 178.1 and a standard deviation of 40.7 (based on data from the National Health Survey). Find the z score corresponding to a male, aged 18–24 years, who has a serum cholesterol level of 259.0 mg/100 mL. Is this level unusually high?

9. Comparing Test Scores Which is relatively better: a score of 75 on a history test or a score of 27 on a psychology test? Scores on the history test have a mean of 80 and a standard deviation of 12. Scores on the psychology test have a mean of 30 and a standard deviation of 8.

10. Comparing Scores Three students take equivalent tests of a sense of humor and, after the laughter dies down, their scores are calculated. Which is the highest relative score?
a. A score of 144 on a test with a mean of 128 and a standard deviation of 34
b. A score of 90 on a test with a mean of 86 and a standard deviation of 18
c. A score of 18 on a test with a mean of 15 and a standard deviation of 5

11. Weights of Coke Refer to Data Set 1 in Appendix B for the sample of 36 weights of regular Coke. Convert the weight of 0.7901 to a z score. Is 0.7901 an unusual weight for regular Coke?

12. Weights of Diet Coke Refer to Data Set 1 in Appendix B for the sample of 36 weights of diet Coke. Convert the weight of 0.7907 to a z score. Is 0.7907 an unusual weight for diet Coke?

Finding Percentiles. *In Exercises 13–16, use the 36 sorted weights of regular Coke listed in Table 2-16. Find the percentile corresponding to the given value.*

13. 0.8264 **14.** 0.8172 **15.** 0.8192 **16.** 0.8079

Finding Percentiles, Quartiles, and Deciles. In Exercises 17–24, use the 36 sorted weights of regular Coke listed in Table 2-16. Find the indicated percentile, quartile, or decile.

17. P_{80} **18.** D_8 **19.** D_6 **20.** D_3

21. Q_3 **22.** P_{10} **23.** D_1 **24.** P_{33}

Finding Percentiles. In Exercises 25–28, use the weights (in pounds) of bears listed in Data Set 7 of Appendix B. Find the percentile corresponding to the given weight.

25. 144 **26.** 212 **27.** 316 **28.** 90

Finding Percentiles, Quartiles, and Deciles. In Exercises 29–36, use the weights (in pounds) of bears listed in Data Set 7 of Appendix B. Find the indicated percentile, quartile, or decile.

29. P_{85} **30.** P_{35} **31.** Q_1 **32.** Q_3

33. D_9 **34.** D_3 **35.** P_{50} **36.** P_{95}

2-6 Beyond the Basics

37. Understanding Units of Measurement When finding a z score for the height of a basketball player in the NBA, how is the result affected if, instead of using inches, all values are expressed in centimeters? In general, how are z scores affected by the particular unit of measurement that is used?

38. Interpreting a z Score Heights of women have a mean of 63.6 in. and a standard deviation of 2.5 in. Erin Zito has a height described by $z = -1.25$. What is her height in inches?

39. Distribution of z Scores

a. A data set has a distribution that is uniform. If all of the values are converted to z scores, what is the shape of the distribution of the z scores?

b. A data set has a distribution that is bell-shaped. If all of the values are converted to z scores, what is the shape of the distribution of the z scores?

c. In general, how is the shape of a distribution affected if all values are converted to z scores?

40. Fibonacci Sequence The first several terms of the famous Fibonacci sequence are 1, 1, 2, 3, 5, 8, 13.

a. Find the mean $\bar{x}$ and standard deviation s, then convert each value to a z score. Don't round the z scores; carry as many places as your calculator can handle.

b. Find the mean and standard deviation of the z scores found in part (a).

c. If you use any other data set, will you get the same results obtained in part (b)?

41. Weights of Coke Use the sorted weights of regular Coke listed in Table 2-16.

a. Find the interquartile range.

b. Find the midquartile.

c. Find the 10–90 percentile range.

d. Does $P_{50} = Q_2$? If so, does P_{50} *always* equal Q_2?

e. Does $Q_2 = (Q_1 + Q_3)/2$? If so, does Q_2 *always* equal $(Q_1 + Q_3)/2$?

42. Interpolation When finding percentiles using Figure 2-17, if the locator L is not a whole number, we round it up to the next larger whole number. An alternative to this procedure is to interpolate so that a locator of 23.75 leads to a value that is 0.75 (or 3/4) of the way between the 23rd and 24th scores. Use this method of interpolation to find P_{35}, Q_1, and D_3 for the weights of bears listed in Data Set 7 of Appendix B.

2-7 Exploratory Data Analysis (EDA)

The general theme of this chapter is describing, exploring, and comparing data, and the focus of this section is exploration. We will first define exploratory data analysis, then introduce some new tools—outliers, 5-number summaries, and boxplots. We can then add these to the techniques presented earlier in this chapter.

DEFINITION

Exploratory data analysis is the process of using statistical tools (such as graphs, measures of center, and measures of variation) to investigate data sets in order to understand their important characteristics.

Recall that in Section 2-1 we listed five important characteristics of data beginning with these three: (1) *center,* a representative or average value; (2) *variation,* a measure of the amount of spread among the sample values; and (3) *distribution*, the nature or shape of the distribution of the data. When exploring a data set, we usually want to calculate the mean and the standard deviation and to generate a histogram, but we should not stop there. It is important to further examine the data set to identify any notable features, especially those that could have a strong effect on results and conclusions. One such feature is the presence of outliers.

Outliers

An **outlier** is a value that is located very far away from almost all of the other values. Relative to the other data, an outlier is an *extreme* value. When exploring a data set, outliers should be considered because they may reveal important information. The following example illustrates the effects of an outlier.

EXAMPLE **Qwerty Keyboard Ratings** When using computer software, it is easy to make typing errors. Refer to the Qwerty word ratings listed in Table 2-1 with the Chapter Problem and assume that the first entry of 2 was incorrectly entered as 222 because the key was pressed too long when you were distracted by a meteorite landing on your porch. How does that error affect your exploration of the data set?

SOLUTION After the first entry of 2 is replaced with the incorrect value of 222, the data in Table 2-1 will result in a mean of 8.6 (instead of the correct value of 4.4) and a standard deviation of 30.3 (instead of the correct value of 2.8). The accompanying STATDISK display shows the histogram for this modified data set.

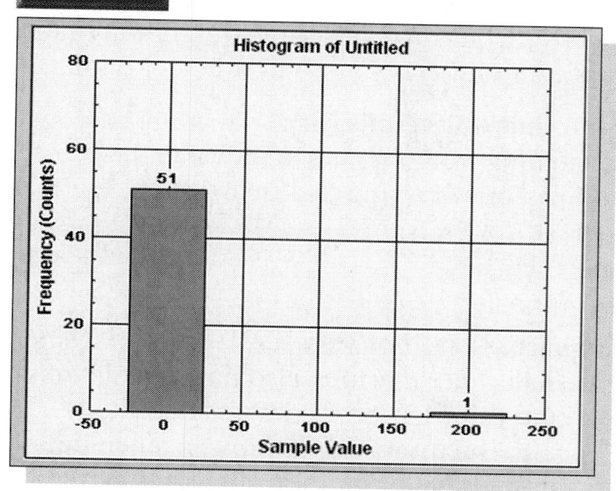

Compare this histogram to the correct histogram shown in Figure 2-2 of Section 2-3 and you will see that this histogram is substantially different.

Our example demonstrates these important principles:

1. An outlier can have a dramatic effect on the mean.
2. An outlier can have a dramatic effect on the standard deviation.
3. An outlier can have a dramatic effect on the scale of the histogram so that the true nature of the distribution is totally obscured.

An easy way to find outliers is to examine a *sorted* list of the data. In particular, look at the minimum and maximum sample values and determine whether they are very far away from the other typical values. In the preceding example, the outlier of 222 is an error. When such an error is identified, we should either correct it or delete it. Remember, however, that some data sets include outliers that are correct values, not errors. When exploring data, we might study the effects of outliers by constructing graphs and calculating statistics with and without the outliers included. (See Exercise 11 for a way to depict outliers on boxplots.)

Boxplots

Another frequently used graph is the boxplot. Boxplots are useful for revealing the center of the data, the spread of the data, the distribution of the data, and the presence of outliers. To construct a boxplot we first obtain the minimum value, the maximum value, and quartiles, as defined in the *5-number summary.*

DEFINITIONS

For a set of data, the **5-number summary** consists of the minimum value; the first quartile, Q_1; the median, or second quartile, Q_2; the third quartile, Q_3; and the maximum value.

A **boxplot** (or **box-and-whisker diagram**) is a graph of a data set that consists of a line extending from the minimum value to the maximum value, and a box with lines drawn at the first quartile, Q_1; the median; and the third quartile, Q_3.

Medians and quartiles are not very sensitive to extreme values. So boxplots, which use medians and quartiles, also have the advantage of not being as sensitive to extreme values as other devices based on the mean and standard deviation. Because boxplots don't show as much detailed information as histograms or stem-and-leaf plots, they might not be the best choice when dealing with a single data set. They are often terrific, though, for comparing two or more data sets. When using two or more boxplots for comparing different data sets, it is important to use the same scale so that correct comparisons can be made.

AS

EXAMPLE **Qwerty Keyboard Ratings** Refer to the 52 Qwerty word ratings in Table 2-1 (without the error of 222 used in the preceding example).

a. Find the values constituting the 5-number summary.

b. Construct a boxplot.

SOLUTION

a. The 5-number summary consists of the minimum, Q_1, the median, Q_3, and the maximum. To find those values, first sort the data (by arranging them in order from lowest to highest). The minimum of 0 and the maximum of 14 are easy to identify from the sorted list. Now find the quartiles. Using the flowchart of Figure 2-17, we get $Q_1 = P_{25} = 2$, which is located by calculating the locator $L = (25/100)52 = 13$ and finding the value midway between the 13th and 14th values in the sorted list. The median is 4, which is the value midway between the 26th and 27th values. We also find that $Q_3 = 6$ by using Figure 2-17 for the 75th percentile. The 5-number summary is therefore 0, 2, 4, 6, and 14.

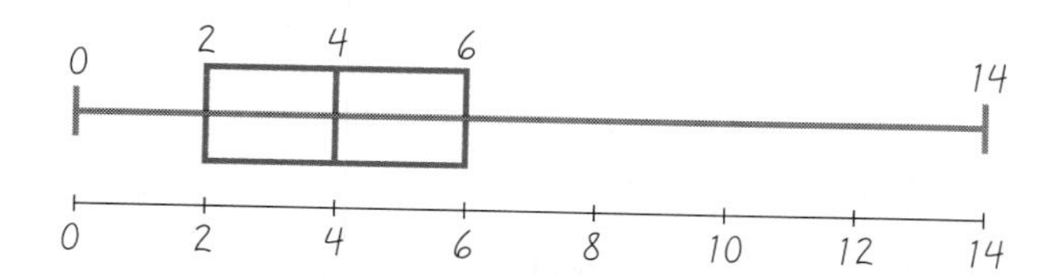

FIGURE 2-18
Boxplot of Qwerty Word Ratings

b. In Figure 2-18 we graph the boxplot for the data. We use the minimum (0) and the maximum (14) to determine a scale of values, then we plot the values from the 5-number summary as shown.

In Figure 2-19 we show some generic boxplots along with common distribution shapes. It appears that the Qwerty word ratings have a skewed distribution.

In the accompanying STATDISK display, we show the Qwerty word ratings in the top boxplot and the Dvorak word ratings in the lower boxplot. A comparison of the two boxplots reveals that the Dvorak ratings are grouped considerably farther to the left, suggesting that the Dvorak keyboard is considerably easier to use. Also, the Qwerty ratings appear to have more spread because the Qwerty boxplot stretches farther to the right.

STATDISK

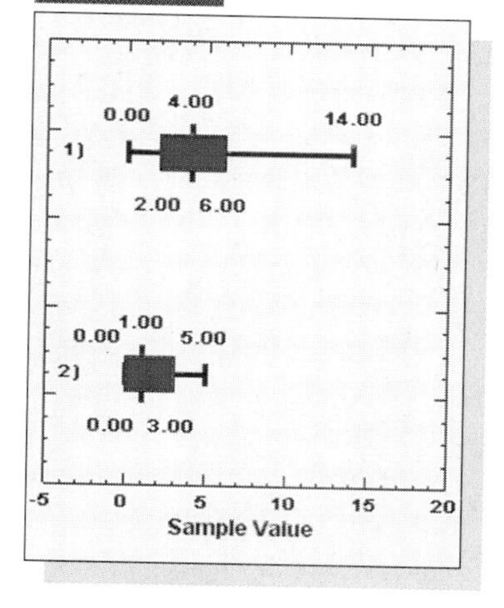

Exploring

In this section we have added outliers, 5-number summaries, and boxplots to our list of tools useful for exploring data sets. We now have the following arsenal of exploration tools:

- *Measures of center*: mean, median, and mode
- *Measures of variation*: standard deviation and range
- *Measures of spread and relative location*: minimum value, maximum value, and quartiles
- *Unusual values*: outliers
- *Distribution*: histograms, stem-and-leaf plots, and boxplots

Now that you have all these tools in hand, you need to begin to *think creatively*! Rather than simply cranking out statistics and graphs, try to identify those that are particularly interesting or important. As a first step, investigate outliers and consider their effects by finding measures and graphs with and without the outliers included.

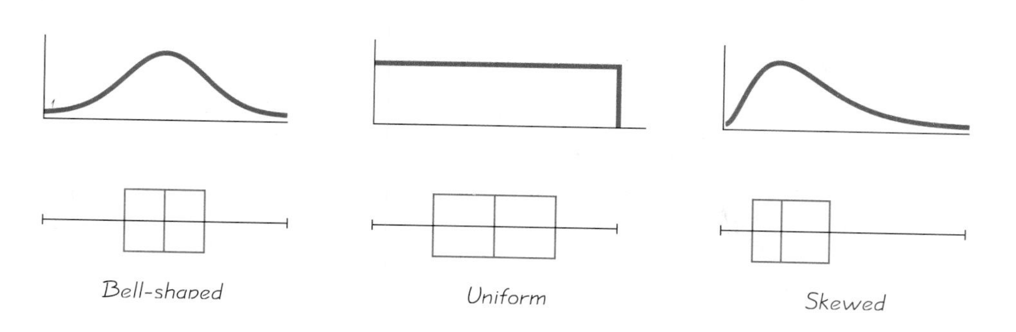

FIGURE 2-19
Boxplots Corresponding to Bell-Shaped, Uniform, and Skewed Distributions

The following examples ilustrate how these exploration tools work.

EXAMPLE Qwerty and Dvorak Keyboards: Did We Miss Anything? The Chapter Problem included 52 word ratings from the Qwerty keyboard and 52 word ratings from the Dvorak keyboard. In preceding sections we have compared those data sets using relative frequency tables (Table 2-9), frequency polygons (Figure 2-10), measures of center (Table 2-11), and boxplots (STATDISK display in this section). That seems to be a reasonably thorough comparison, but apart from formal methods to be considered later in this book, have we missed anything that is really important? Yes! Our comparisons treated the two data sets as if they were separate and independent, but the word ratings came from the *same* 52 words in the Preamble to the Constitution. We should therefore explore the *differences* between the pairs of ratings corresponding to each of the 52 words. For example, the first word of the Preamble is "we," which has a Qwerty rating of 2 and a Dvorak rating of 2, so the difference is $2 - 2 = 0$. Explore the set of data consisting of the 52 differences listed here:

0 2 2 0 2 6 3 3 2 2 0 0 2 3 7 2 3 5 5 5 3 0 2 5 4 4

2 2 −1 2 4 2 2 0 2 2 3 2 2 5 1 4 2 2 2 13 0 2 6 3 0 3

SOLUTION Keeping in mind our list of exploration tools, we find the mean, standard deviation, 5-number summary, and outliers, and we generate a histogram and boxplot. The results are shown below.

Mean: 2.7

Standard deviation: 2.3

5-number summary: minimum $= -1$, $Q_1 = 2$, median $= 2$, $Q_3 = 3.5$, maximum $= 13$

Outlier: 13

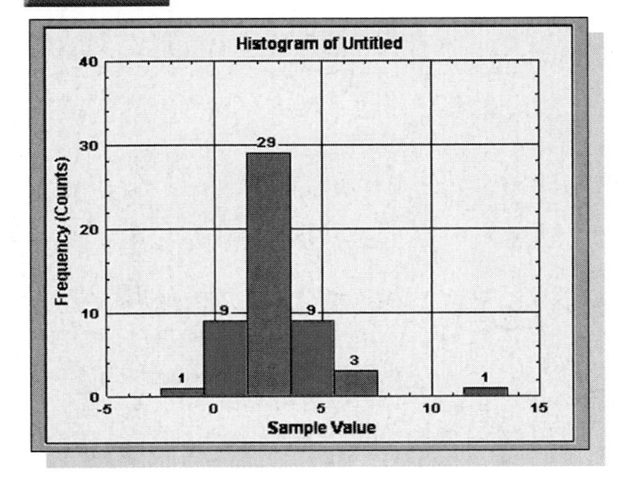

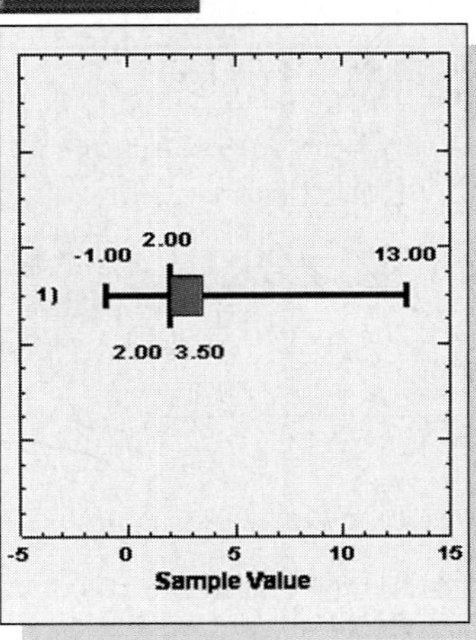

Formulas and Tables

for *Elementary Statistics, Eighth Edition*, by Mario F. Triola

TABLE A-4 Chi-Square (χ^2) Distribution

Degrees of Freedom	Area to the Right of the Critical Value									
	0.995	0.99	0.975	0.95	0.90	0.10	0.05	0.025	0.01	0.005
1	—	—	0.001	0.004	0.016	2.706	3.841	5.024	6.635	7.879
2	0.010	0.020	0.051	0.103	0.211	4.605	5.991	7.378	9.210	10.597
3	0.072	0.115	0.216	0.352	0.584	6.251	7.815	9.348	11.345	12.838
4	0.207	0.297	0.484	0.711	1.064	7.779	9.488	11.143	13.277	14.860
5	0.412	0.554	0.831	1.145	1.610	9.236	11.071	12.833	15.086	16.750
6	0.676	0.872	1.237	1.635	2.204	10.645	12.592	14.449	16.812	18.548
7	0.989	1.239	1.690	2.167	2.833	12.017	14.067	16.013	18.475	20.278
8	1.344	1.646	2.180	2.733	3.490	13.362	15.507	17.535	20.090	21.955
9	1.735	2.088	2.700	3.325	4.168	14.684	16.919	19.023	21.666	23.589
10	2.156	2.558	3.247	3.940	4.865	15.987	18.307	20.483	23.209	25.188
11	2.603	3.053	3.816	4.575	5.578	17.275	19.675	21.920	24.725	26.757
12	3.074	3.571	4.404	5.226	6.304	18.549	21.026	23.337	26.217	28.299
13	3.565	4.107	5.009	5.892	7.042	19.812	22.362	24.736	27.688	29.819
14	4.075	4.660	5.629	6.571	7.790	21.064	23.685	26.119	29.141	31.319
15	4.601	5.229	6.262	7.261	8.547	22.307	24.996	27.488	30.578	32.801
16	5.142	5.812	6.908	7.962	9.312	23.542	26.296	28.845	32.000	34.267
17	5.697	6.408	7.564	8.672	10.085	24.769	27.587	30.191	33.409	35.718
18	6.265	7.015	8.231	9.390	10.865	25.989	28.869	31.526	34.805	37.156
19	6.844	7.633	8.907	10.117	11.651	27.204	30.144	32.852	36.191	38.582
20	7.434	8.260	9.591	10.851	12.443	28.412	31.410	34.170	37.566	39.997
21	8.034	8.897	10.283	11.591	13.240	29.615	32.671	35.479	38.932	41.401
22	8.643	9.542	10.982	12.338	14.042	30.813	33.924	36.781	40.289	42.796
23	9.260	10.196	11.689	13.091	14.848	32.007	35.172	38.076	41.638	44.181
24	9.886	10.856	12.401	13.848	15.659	33.196	36.415	39.364	42.980	45.559
25	10.520	11.524	13.120	14.611	16.473	34.382	37.652	40.646	44.314	46.928
26	11.160	12.198	13.844	15.379	17.292	35.563	38.885	41.923	45.642	48.290
27	11.808	12.879	14.573	16.151	18.114	36.741	40.113	43.194	46.963	49.645
28	12.461	13.565	15.308	16.928	18.939	37.916	41.337	44.461	48.278	50.993
29	13.121	14.257	16.047	17.708	19.768	39.087	42.557	45.722	49.588	52.336
30	13.787	14.954	16.791	18.493	20.599	40.256	43.773	46.979	50.892	53.672
40	20.707	22.164	24.433	26.509	29.051	51.805	55.758	59.342	63.691	66.766
50	27.991	29.707	32.357	34.764	37.689	63.167	67.505	71.420	76.154	79.490
60	35.534	37.485	40.482	43.188	46.459	74.397	79.082	83.298	88.379	91.952
70	43.275	45.442	48.758	51.739	55.329	85.527	90.531	95.023	100.425	104.215
80	51.172	53.540	57.153	60.391	64.278	96.578	101.879	106.629	112.329	116.321
90	59.196	61.754	65.647	69.126	73.291	107.565	113.145	118.136	124.116	128.299
100	67.328	70.065	74.222	77.929	82.358	118.498	124.342	129.561	135.807	140.169

From Donald B. Owen, *Handbook of Statistical Tables*, ©1962 Addison-Wesley Publishing Co., Reading, MA. Reprinted with permission of the publisher.

HYPOTHESIS TEST: WORDING OF FINAL CONCLUSION

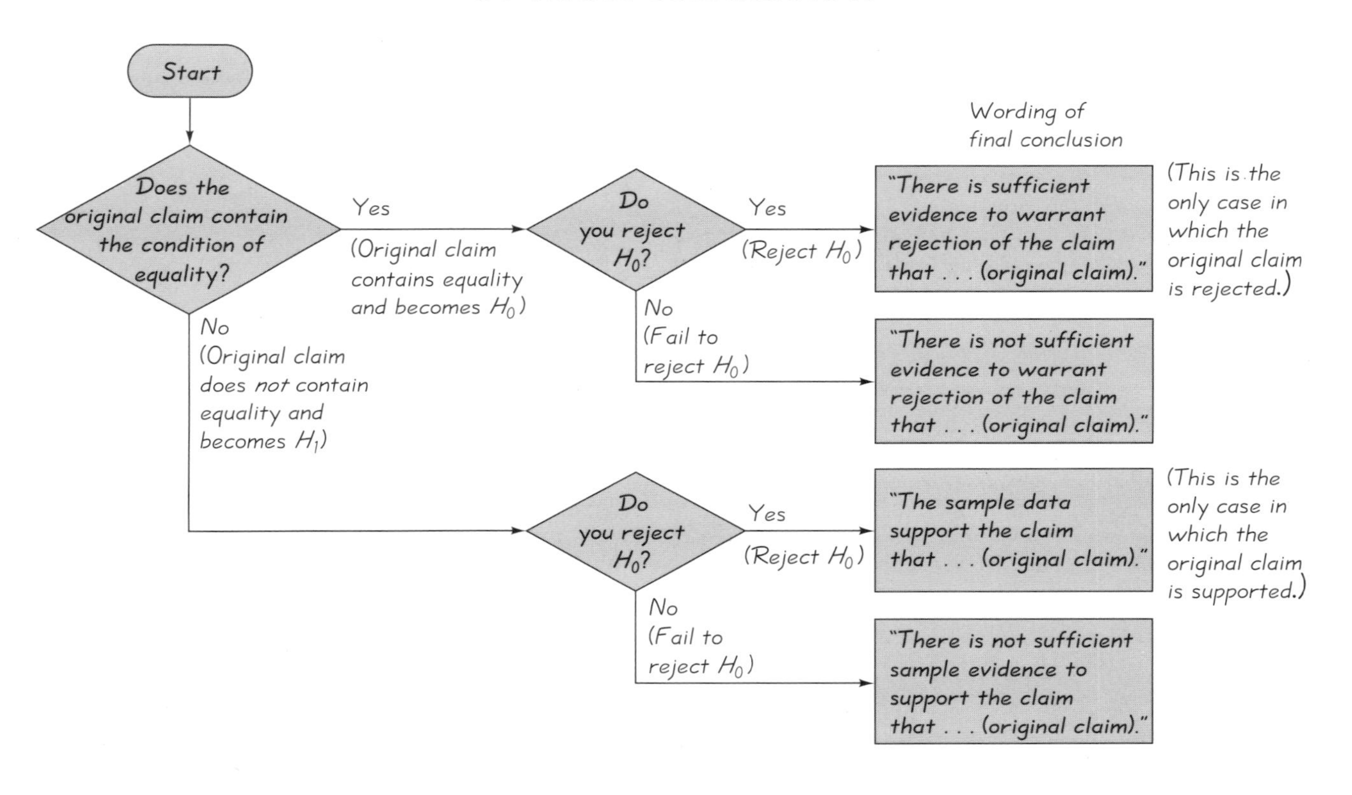

INTERPRETATION: If the Qwerty and Dvorak keyboards had the same level of typing difficulty, we would expect the differences between word ratings to average around 0, but the mean difference is 2.7, and the histogram shows that the distribution is not centered about 0. Because the differences between Qwerty and Dvorak appear to center around a value well above 0, it appears that the Qwerty configuration results in greater typing difficulty. If we were to exclude the single outlier of 13, we would find that the mean does not change by much (it becomes 2.5). Excluding the outlier does not affect the histogram very much. And excluding the outlier does not affect the boxplot very much, either; the right tail is shortened so that it ends at 7 instead of 13. The apparent difference is present with or without the outlier. This is convincing evidence that the Dvorak keyboard does in fact have a configuration that makes typing easier. (See also the end-of-chapter project "From Data to Decision: Critical Thinking.")

Good Advice for Journalists

Columnist Max Frankel wrote in *The New York Times* that "Most schools of journalism give statistics short shrift and some let students graduate without any numbers training at all. How can such reporters write sensibly about trade and welfare and crime, or air fares, health care and nutrition? The media's sloppy use of numbers about the incidence of accidents or disease frightens people and leaves them vulnerable to journalistic hype, political demagoguery, and commercial fraud." He cites several cases, including an example of a full-page article about New York City's deficit with a promise by the mayor of New York City to close a budget gap of $2.7 billion; the entire article never once mentioned the *total* size of the budget, so the $2.7 billion figure had no context.

EXAMPLE **Does It Rain More on Weekends?** Refer to Data Set 17 in Appendix B, which lists rainfall amounts (in inches) in Boston for every day of a recent year. The collection of this data set was inspired by media reports claiming that it rains more on weekends (Saturdays and Sundays) than on weekdays. Later in this book we will describe statistical methods that can be used to formally address that claim, but for now, let's explore the data set to see what can be learned. (Even if we already knew how to apply those formal statistical methods, we should first explore the data before proceeding with the formal analysis.)

SOLUTION Referring again to our list of exploration tools, we find the mean and standard deviation for each day of the week, and we also find the 5-number summary for each day of the week. The results are summarized in the following table. The STATDISK display shows boxplots for each of the seven days of the week, starting with Monday at the top. Because the histograms for all seven days are pretty much the same, we show only the histogram for the Monday rainfall amounts.

	Mean	Standard Deviation	Minimum	Q_1	Median	Q_3	Maximum
Monday	0.100	0.263	0.00	0.000	0.000	0.010	1.410
Tuesday	0.058	0.157	0.00	0.000	0.000	0.015	0.740
Wednesday	0.051	0.135	0.00	0.000	0.000	0.010	0.640
Thursday	0.069	0.167	0.00	0.000	0.000	0.040	0.850
Friday	0.095	0.228	0.00	0.000	0.000	0.040	0.960
Saturday	0.143	0.290	0.00	0.000	0.000	0.100	1.480
Sunday	0.068	0.200	0.00	0.000	0.000	0.010	1.280

INTERPRETATION Examining and comparing the statistics and graphs, we make the following important observations.

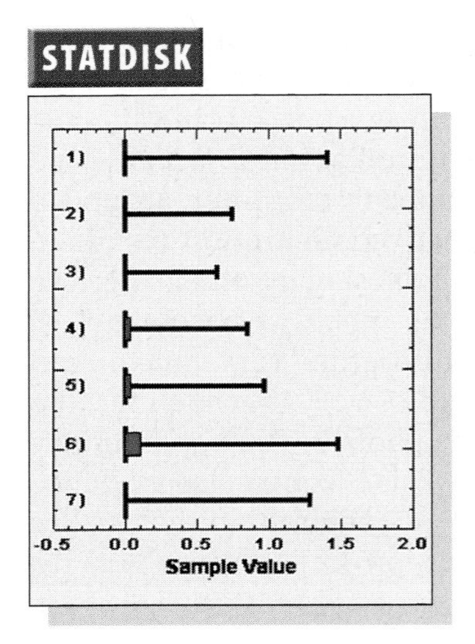

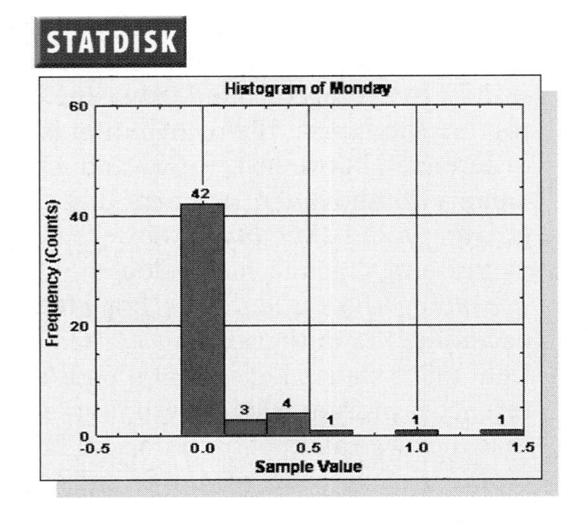

- *Means:* The means vary from a low of 0.051 in. to a high of 0.143 in. The seven means vary by considerable amounts, and in later chapters of this book we will pursue whether these differences are *significant*. (Later methods will show that the means do not differ by significant amounts.) If we list the means in order from low to high, we get this sequence of days: Wednesday, Tuesday, Sunday, Thursday, Friday, Monday, Saturday. There does not appear to be a pattern of higher rainfall on weekends. Also, see the Excel graph of the seven means. In that graph, Monday is the first day, Tuesday is the second day, and so on. The Excel graph does not support the claim of more rainfall on Saturday and Sunday.

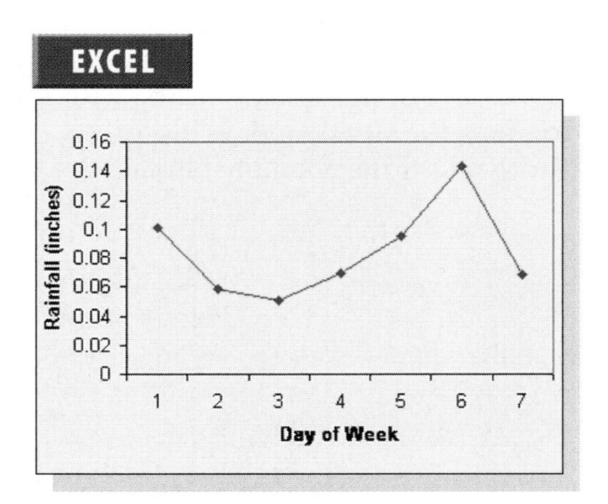

- *Variation:* The seven standard deviations vary from 0.135 in. to 0.290 in., but those values are not dramatically different. There does not appear to be anything highly unusual about the amounts of variation.

- The *minimums, first quartiles,* and *medians* are all 0.00 for each of the seven days. This is explained by the fact that for each day of the week, there are many days with no rain. The abundance of zeros is also seen in the boxplots and histograms, which show that the data have distributions that are heavy toward the low end (skewed right).
- *Outliers:* There are no outliers or unusual values. At the low end, there are many rainfall amounts of zero. At the high end, the sorted list of all 365 rainfall amounts ends with the high values of 0.92, 0.96, 1.28, 1.41, and 1.48.
- *Distributions:* The distributions of the rainfall amounts are skewed to the right. They are not bell-shaped as we might have expected. If the application of a particular method of statistics requires normally distributed (bell-shaped) populations, that requirement is not satisfied.

We have now acquired considerable insight into the nature of Boston rainfall amounts for different days of the week. Based on our exploration, we can conclude that Boston does not experience more rain on weekends than on the other days of the week.

Using Technology

This section introduced outliers, 5-number summaries, and boxplots. To find outliers, sort the data in order from the lowest value to the highest, then examine the lowest and highest values to determine whether they are far away from the other sample values. STATDISK, Minitab, Excel, and the TI-83 Plus calculator can provide values of quartiles, so the 5-number summary is easy to find. STATDISK, Minitab, Excel, and the TI-83 Plus calculator can be used to create boxplots, and we will now describe the different procedures.

Caution: Remember that quartile values calculated by Minitab and the TI-83 Plus calculator may differ slightly from those calculated by applying Figure 2-17, so the boxplots may differ slightly as well.

STATDISK: Choose the main menu item of **Data** and use the **Sample Editor** to enter the data, then click on **COPY.** Now select **Data,** then **Boxplot** and click on **PASTE,** then **Evaluate.**

Minitab: Enter the data in column C1, then select **Graph,** then **Boxplot.** Enter C1 in the first cell under the Y column, then click OK.

Excel: Although Excel is not designed to generate boxplots, they can be generated using the Data Desk XL add-in that is a supplement to this book. First enter the data in column A. Click on **DDXL** and select **Charts and Plots.** Under Function Type, select the option of **Boxplot.** In the dialog box, click on the pencil icon and enter the range of data, such as A1:A52 if you have 52 values listed in column A. Click on **OK.** The result is a modified boxplot as described in Exercise 11. The values of the 5-number summary are also displayed.

TI-83 Plus: Enter the sample data in list L1. Now select **STAT PLOT** by pressing the 2nd key followed by the key labeled Y =. Press the **ENTER** key, then select the option of **ON,** and select the boxplot type that is positioned in the middle of the second row. The Xlist should indicate L1, and the Freq value should be 1. Now press the **ZOOM** key and select the option 9 for **ZoomStat.** Press the **ENTER** key, and the boxplot should be displayed. You can use the arrow keys to move right or left so that values can be read from the horizontal scale.

2-7 Basic Skills and Concepts

Exploring Data Using Boxplots. *In Exercises 1–6, find the 5-number summary and construct a boxplot. Explore the data set and identify at least one characteristic that is particularly noteworthy.*

1. Employee Ages The following list shows the ages of most of the employees at the Vita Needle Company.

76 45 72 77 63 65 87 73 84 86 79 86 75
87 74 39 75 41 82 34 88 85 79 73 53

(Based on data from "Where Retirement Became a Dirty Word" by Julie Flaherty, *New York Times.*)

2. Ages of Faculty Cars The following list shows the ages of cars in the faculty/staff parking lot at the author's college.

4 9 9 8 7 1 5 4 4 4 7 6 7 7 23

3. Pulse Rates Refer to Data Set 14 in Appendix B and explore the pulse rates.

4. Old Faithful Eruptions Refer to Data Set 11 in Appendix B and explore the time intervals between eruptions of the Old Faithful geyser.

5. Strengths of Aluminum Cans Refer to Data Set 12 in Appendix B and explore the axial loads of the cans that are 0.0111 in. thick.

6. Body Temperatures Refer to Data Set 6 in Appendix B and explore the body temperatures for 12 A.M. on day 2.

Comparing Data Sets. *In Exercises 7–10, find 5-number summaries, construct boxplots, and compare the data sets.*

7. Ages of Oscar Winners In "Ages of Oscar-Winning Best Actors and Actresses" (*Mathematics Teacher* magazine) by Richard Brown and Gretchen Davis, the authors compare the ages of actors and actresses at the time they won Oscars. The results for recent winners from each category are listed in the following table. Use boxplots to compare the two data sets.

Actors: 32 37 36 32 51 53 33 61 35 45 55 39
76 37 42 40 32 60 38 56 48 48 40 43
62 43 42 44 41 56 39 46 31 47 45 60

Actresses: 50 44 35 80 26 28 41 21 61 38 49 33
74 30 33 41 31 35 41 42 37 26 34 34
35 26 61 60 34 24 30 37 31 27 39 34

8. Garbage Refer to Data Set 5 in Appendix B and use the weights of discarded paper and the weights of discarded plastic.

9. Textbook Prices Refer to Data Set 2 in Appendix B and use the new textbook prices at the author's college and the new textbook prices at the University of Massachusetts.

10. Regular/Diet Coke Refer to Data Set 1 in Appendix B and use the weights of regular Coke and the weights of diet Coke.

2-7 Beyond the Basics

11. The boxplots discussed in this section are often called *skeletal* (or *regular*) boxplots. **Modified boxplots** are constructed as follows:
   a. Find the *IQR,* which denotes the interquartile range defined by $IQR = Q_3 - Q_1$.
   b. Draw the box with the median and quartiles as usual, but when drawing the lines to the right and left of the box, draw the lines only as far as the points corresponding to the largest and smallest values that are within 1.5 *IQR* of the box.
   c. **Mild outliers** are sample values above Q_3 or below Q_1 by an amount that is greater than 1.5 *IQR* but not greater than 3 *IQR*. Plot mild outliers as *solid* dots.
   d. **Extreme outliers** are values that are either above Q_3 by more than 3 *IQR* or below Q_1 by more than 3 *IQR*. Plot extreme outliers as small *hollow* circles.

   The accompanying figure is an example of a modified boxplot. Refer to Data Set 12 in Appendix B for the axial loads of the cans that are 0.0111 in. thick. For that data set, the quartiles are 275, 285, and 295. Identify any mild outliers and extreme outliers.

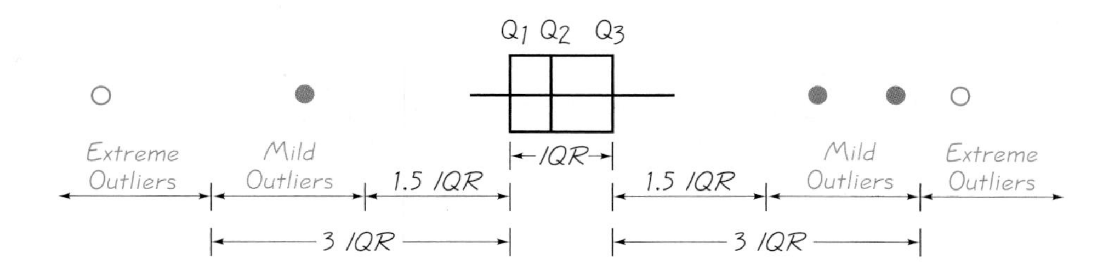

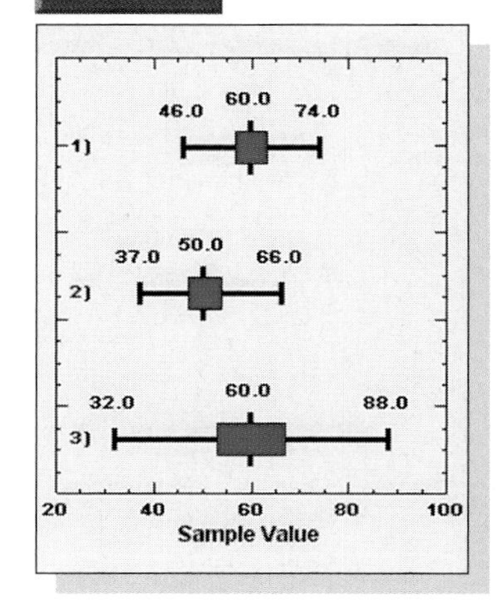

12. Comparing Boxplots Refer to the accompanying STATDISK display of three boxplots that represent the longevity (in months) of samples of three different car batteries. If you are the manager of a fleet of cars and you must select one of the three brands, which boxplot represents the brand you should choose? Why?

The Plague of Unwanted Calls

BOSTON All of us have had our dinners interrupted by telemarketers who seem to have honed the ability to call at the most inconvenient times. Legislators have already imposed some controls, so that calls in the middle of the night are now prohibited. Revised controls are constantly being considered. Instead of basing our analysis of the problem on anecdotal evidence, it is important to collect objective data that truly represent existing conditions. See the accompanying illustration from *USA Today.*

USA Today Snapshot. "A look at statistics that shape the nation" from 3/30/99

1. Refer to the *USA Today* illustration and assume that it is based on a survey of 1000 people. Construct the frequency table corresponding to the given data.

2. A major problem is caused by the two vague classes "don't know" and "6–up." Because those classes are not specific values, we cannot perform exact calculations. Delete the "don't know" class, and assume that the class of "6–up" is actually "6–10." Calculate the mean and standard deviation.

3. Repeat Exercise 2, but assume that the class of "6-up" is actually "6–12." Are the results dramatically affected?

4. Can the *USA Today* illustration be used to obtain reasonably good estimates of the mean and standard deviation?

5. Do the results suggest that unwanted calls really are a problem?

Vocabulary List

descriptive statistics
inferential statistics
frequency table
frequency
lower class limits
upper class limits
class boundaries
class midpoints
class width
relative frequencies
relative frequency table
cumulative frequency
cumulative frequency table
histogram
relative frequency histogram
frequency polygon
ogive
dotplot
stem-and-leaf plot
Pareto chart
pie chart
scatter diagram

measure of center
arithmetic mean
mean
sample size
median
mode
bimodal
multimodal
midrange
weighted mean
skewed
symmetric
negatively skewed
positively skewed
range
standard deviation
deviation
mean absolute deviation
variance
unbiased estimator
range rule of thumb
empirical (or 68–95–99.7) rule
Chebyshev's theorem
standard score
z score
quartiles
deciles
percentiles
exploratory data analysis (EDA)
outlier
5-number summary
boxplot (or box-and-whisker diagram)

Review

In this chapter we considered methods for describing, exploring, and comparing data sets. When investigating a data set, the following characteristics are generally very important:

1. *Center:* A representative or average value
2. *Variation:* A measure of the amount that the values vary
3. *Distribution:* The nature or shape of the distribution of the data (such as bell-shaped, uniform, or skewed)
4. *Outliers:* Sample values that lie very far away from the vast majority of the other sample values
5. *Time:* Changing characteristics of the data over time

After completing this chapter, you should be able to do the following:

- Summarize the data by constructing a frequency table or relative frequency table (Section 2-2)
- Visually display the nature of the distribution by constructing a histogram, dotplot, stem-and-leaf plot, pie chart, or Pareto chart (Section 2-3)
- Calculate measures of center by finding the mean, median, mode, and midrange (Section 2-4)
- Calculate measures of variation by finding the standard deviation, variance, and range (Section 2-5)
- Compare individual values by using z scores, quartiles, deciles, or percentiles (Section 2-6)
- Investigate and explore the spread of data, the center of the data, and the range of values by constructing a boxplot (Section 2-7)

In addition to creating these tables, graphs, and measures, you should be able to understand and interpret the results. For example, you should clearly understand that the standard deviation is a measure of how much the data vary, and you should be able to use the standard deviation to distinguish between values that are usual and values that are unusual.

REVIEW EXERCISES

1. Ages of Presidents Given below are the ages of U.S. presidents when they were inaugurated. Find the (a) mean, (b) median, (c) mode, (d) midrange, (e) range, (f) standard deviation, (g) variance, (h) Q_1, (i) P_{30}, (j) D_7.

57	61	57	57	58	57	61	54	68	51	49	64	50	48
65	52	56	46	54	49	51	47	55	55	54	42	51	56
55	51	54	51	60	62	43	55	56	61	52	69	64	46

2. **a.** Teddy Roosevelt was 42 years of age when he was inaugurated. Using the results from Exercise 1, convert his age to a z score.
   b. Is Teddy Roosevelt's age of 42 years "unusual"? Why or why not?
   c. Using the range rule of thumb, identify any other listed ages that are unusual.

3. Constructing a Frequency Table Using the same ages listed in Exercise 1, construct a frequency table. Use six classes with 40 as the lower limit of the first class, and use a class width of 5.

4. Constructing a Histogram Using the frequency table from Exercise 3, construct a histogram and identify the general nature of the distribution (such as uniform, bell-shaped, or skewed).

5. Constructing a Boxplot Using the same ages listed in Exercise 1, construct a boxplot and identify the values constituting the 5-number summary.

6. Using the Empirical Rule Assume the ages of past, present, and future presidents have a bell-shaped distribution with a mean of 54.9 years and a standard deviation of 6.3 years.
   a. What does the empirical rule say about the percentage of ages between 48.6 years and 61.2 years (or within 1 standard deviation of the mean)?
   b. What does the empirical rule say about the percentage of ages between 42.3 years and 67.5 years?

7. Comparing Scores An industrial psychologist for the Citation Corporation develops two different tests to measure job satisfaction. Which score is better: a score of 72 on the management test, which has a mean of 80 and a standard deviation of 12, or a score of 19 on the test for production employees, which has a mean of 20 and a standard deviation of 5? Explain.

8. Using the Range Rule of Thumb
   a. Make a rough estimate of the mean age of textbooks currently required at your college.
   b. Use the range rule of thumb to make a rough estimate of the standard deviation of the ages of textbooks currently required at your college.

9. Transforming Data A data set has a mean of 12.3 minutes and a standard deviation of 4.0 minutes.
   a. What is the mean after 5.0 minutes has been added to every value?
   b. What is the standard deviation after 5.0 minutes has been added to every value?
   c. What is the variance after 5.0 minutes has been added to every value?

10. Constructing a Pareto Chart The United States Coast Guard collected data on serious boating accidents and listed the categories as shown below, with their frequencies given in parentheses. Construct a Pareto chart summarizing the given data.

Colliding with another boat (2203)
Colliding with a fixed object (839)
Running aground (341)
Person falling overboard (431)
Capsizing (458)

Cumulative Review Exercises

1. Ages of Commercial Jets Safety concerns grow as jets in the U.S. commercial fleet become older. The ages (in years) of 16 randomly selected Boeing 737 jets are listed below.

12	25	7	4	13	7	11	12
23	0	19	7	12	30	18	1

Based on data from the Federal Aviation Administration.

a. Find the mean, median, mode, and midrange.
b. Find the standard deviation, variance, and range.
c. Are the given ages drawn from a population that is discrete or continuous?
d. What is the level of measurement of these values (nominal, ordinal, interval, ratio)?

2. **a.** A set of data is at the nominal level of measurement and you want to obtain a representative data value. Which of the following is most appropriate: mean, median, mode, or midrange? Why?
b. A sample is obtained by telephoning the first 250 people listed in the local telephone directory. What type of sampling is being used (random, stratified, systematic, cluster, convenience)?
c. An exit poll is conducted by surveying everyone who leaves the polling booth at 50 randomly selected election precincts. What type of sampling is being used (random, stratified, systematic, cluster, convenience)?

3. Energy Consumption Each year, the United States Department of Energy publishes an *Annual Energy Review* that includes the per capita energy consumption (in millions of Btu) for each of the 50 states. If you calculate the mean of these 50 values, is the result the mean per capita energy consumption for the population in all 50 states combined? If it is not, explain how you would calculate the mean per capita energy consumption for all 50 states combined.

Cooperative Group Activities

1. *Out-of-class activity:* Are estimates influenced by "anchoring" numbers? In the article "Weighing Anchors" in *Omni* magazine, author John Rubin observed that when people estimate a value, their estimate is often "anchored" to (or influenced by) a preceding number, even if that preceding number is totally unrelated to the quantity being estimated. To demonstrate this, he asked people to give a quick estimate of the value of $8 \times 7 \times 6 \times 5 \times 4 \times 3 \times 2 \times 1$. The average answer given was 2250, but when the order of the numbers was reversed, the average became 512. Rubin explained that when we begin calculations with larger numbers (as in $8 \times 7 \times 6$), our estimates tend to be larger. He noted that both 2250 and 512 are far below the correct product, 40,320. The article suggests that irrelevant numbers can play a role in influencing real estate appraisals, estimates of car values, and estimates of the likelihood of nuclear war.

Conduct an experiment to test this theory. Select some subjects and ask them to quickly estimate the value of

$$8 \times 7 \times 6 \times 5 \times 4 \times 3 \times 2 \times 1$$

Then select other subjects and ask them to quickly estimate the value of

$$1 \times 2 \times 3 \times 4 \times 5 \times 6 \times 7 \times 8$$

Record the estimates along with the particular order used. Carefully design the experiment so that conditions are uniform and the two sample groups are selected in a way that minimizes any bias. Don't describe the theory to subjects until after they have provided their estimates. Compare the two sets of sample results by using the methods of this chapter. Provide a printed report that includes the data collected, the detailed methods used, the method of analysis, any relevant graphs and/or statistics, and a statement of conclusions. Include a critique of reasons why the results might not be correct and describe ways in which the experiment could be improved.

A variation of the preceding experiment is to survey people about their knowledge of the population of Kenya. First ask half of the subjects whether they think the population is above 5 million or below 5 million, then ask them to estimate the population with an actual number. Ask the other half of the subjects whether they think the population is above 80 million or below 80 million, then ask them to estimate the population. (Kenya's population is 28 million.) Compare the two sets of results and identify the "anchoring" effect of the initial number that the survey subjects are given.

2. *Out-of-class activity:* In each group of three or four students, collect an original data set of values at the interval or ratio level of measurement. Provide the following: (1) a list of sample values; (2) printed computer results of descriptive statistics and graphs; and (3) a written description of the nature of the data, the method of collection, and important characteristics.

3. *In-class activity:* Given below are the ages of motorcyclists at the time they were fatally injured in traffic accidents. If your objective is to dramatize the dangers of motorcycles for young people, which would be most effective: histogram, Pareto chart, pie chart, dotplot, mean, median . . . ? Construct the graph and find the statistic that best meets that objective. Is it okay to deliberately distort data if the objective is a positive one, such as saving the lives of motorcyclists?

17	38	27	14	18	34	16	42	28	24	40	20	23	31
37	21	30	25	17	28	33	25	23	19	51	18	29	

Based on data from the U.S. Department of Transportation.

4. *Out-of-class activity:* In each group of three or four students, select one of the following items and construct a graph that is effective in addressing the question.
   a. Is there a difference between the set of home-run distances for Mark McGwire and Sammy Sosa? (See Data Set 19 in Appendix B.)
   b. Is there a relationship between the selling prices of homes and the asking prices? (See Data Set 16 in Appendix B.)
   c. When comparing weights of M&M candies to weights of quarters, what characteristics are similar? (See Data Sets 10 and 13 in Appendix B.)

TECHNOLOGY PROJECT

It is commonly believed that the mean body temperature of healthy adults is 98.6° F. Refer to Data Set 6 in Appendix B and consider the body temperatures taken at midnight on the second day. (Data Set 6 is not stored as a STATDISK or Minitab or

Excel file.) Use STATDISK, Minitab, Excel, or a TI-83 Plus calculator to enter the 106 temperatures and save them as a file named BODYTEMP. Proceed to obtain a histogram, boxplot, measures of center, measures of variation, Q_1, Q_3, the minimum, and the maximum values. Use the results to describe important characteristics of the data set. Based on this sample, what do you conclude about the common belief that the mean body temperature is 98.6° F? Is this the result you would have expected?

from DATA to DECISION

Critical Thinking

The Chapter Problem included data in Tables 2-1 and 2-2. Both tables listed ratings for the 52 words in the Preamble to the Constitution, but Table 2-1 is based on the Qwerty keyboard, whereas Table 2-2 is based on the Dvorak keyboard. In this chapter we constructed frequency tables and graphs, and calculated descriptive measures that included the mean and standard deviation. Based on comparisons, it appeared that the Dvorak keyboard is more efficient than the Qwerty keyboard used by most of us. Instead of blindly accepting that conclusion, however, we might question the basic premises.

Analyzing the results

Address at least one of the following questions by considering the suggested alternative. Collect sample data and use the methods of this chapter to compare the efficiency of the Qwerty and Dvorak keyboards. Does it appear that the Dvorak keyboard really is more efficient?

a. Are the words in the Preamble representative? Would the same conclusion be reached with a sample of words from this textbook, which was written much more recently?

b. Each sample value is a total rating for the letters in a *word,* but should we be using words, or should we simply use the individual letters?

c. Is the rating system appropriate? (Letters in the top row are assigned 1, letters in the middle row are assigned 0, and letters in the bottom row are assigned 2.) Is there a different rating system that would better reflect the difficulty of typing?

Internet Project

Data on the Internet

The Internet is host to a wealth of information and much of that information comes in the form of raw data, which can be studied and summarized using the statistics introduced in this chapter. For example, we found the following information with just a few clicks:

- The value of stock in Walt Disney Corporation tends to rise during the winter months but varies throughout the year. In 1998 the maximum and minimum stock prices differed by nearly 20 points.
- Average attendance during the 1998 regular football season at the new 80,166 seat capacity Fed Ex Field in Washington, D.C. was 67,762.
- The populations of California, New York and Texas account for over 25% of the total U.S. population.

The Internet Project for this chapter, located at

http://www.awlonline.com/triola

will point you to data sets in the areas of sports, finance and the weather. Once you have assembled a data set you will apply the methods of this chapter to summarize and classify the data.

Statistics *at work*

"The journalist must be able to cast a critical eye on research, divine the true context and significance of the work."

Mark Fenton

Editor at Large, Walking Magazine

Mark Fenton is also a walking advocate and is a champion walker. He was on the U.S. national racewalking team five times and has represented the U.S. in numerous international competitions. He has studied biomechanics and exercise physiology at the Olympic Training Center's Sports Science Laboratory in Colorado Springs, Colorado.

What is your job?

I'm editor-at-large for a health and fitness magazine, *Walking Magazine*. I spend a good deal of time as a public spokesman and public health advocate.

What concepts of statistics do you use?

I have to be familiar with all the common tools used in exercise science and public health research statistical analyses; from means and standard deviations, to statistical significance of differences, confidence intervals, and analyses of variance, as examples.

How do you use statistics in your job?

I regularly read medical research journals (*Medicine and Science in Sports and Exercise* and the *Journal of the American Medical Association* are two most notable) for exercise physiology, public health, and epidemiological research. Thus, I must be conversant in how studies are controlled and analyzed, and understand the relative power or value of a study's outcome. This information is used in my composition of articles and speeches on the findings of such work.

Please describe one specific example illustrating how the use of statistics was successful in improving a product or service.

I regularly read research papers, and have to be cautious of holes in the evidence—dangerously small sample sizes; statistical significance but small absolute differences between groups; extremely large standard deviations, which can obscure the relevance of a finding for the average person.

A simple example is the problem of saying that the AVERAGE maximum heart rate for a 36-year-old woman is 190 beats per minute (226 minus age). That's helpful in calculating target heart rate, until you find that the deviation is quite high, and so the resulting figure can be off by as much as 10 beats per minute for roughly 1/3 of the population—an error large enough to be problematic when exercising. The study that pointed this out has altered how I offer exercise intensity recommendations.

Is your use of probability and statistics increasing, decreasing, or remaining stable?

With my growing work in public health, and the increasing complexity of the statistical tools being used in large population analyses, my conversance in the field MUST continue growing if I am to keep up.

Do you recommend statistics for today's college students? Why?

I absolutely recommend at least one course in statistics, as it is a useful tool in assessing the information with which we are bombarded—much of it without context—every day. At very least a comprehensive introductory course in statistics is a must for anyone interested in science journalism. The journalist must be able to cast a critical eye on research, divine the true context and significance of the work.

Do you feel job applicants are viewed more favorably if they have studied some statistics?

I suspect it is rarely asked of science journalists, but frankly should be one of the specific questions in an interview: Do you have the skills to assess the accuracy and power of the work you'll be reporting on? Keep in mind, one doesn't have to be a statistics expert. Rather, one should be able to ask and understand the answers to basic questions on analytical techniques, the true power, confidence, magnitude, and significance of findings.

Which other skills are important for today's college students?

Very high on my list of necessary attributes are excellent oral and written communications skills. These are a must in the workplace, and I am particularly concerned at the number of people unable to quickly and cogently construct and present an oral argument or presentation. It is the basis of a great deal of decision making in the business environment.

Please describe how you try to ensure objectivity.

I am usually assessing other's works, so I try to look for the simplest and clearest explanations of statistical work. It's my suspicion that the more convoluted an analysis, the further the researcher was from obtaining the desired result, and the more they felt obliged to "contrive" some positive outcome in the research. So, I think it's critical that I be able to knowledgeably question them about their methods. For example, in public health research it's common to analyze the effect of one condition on a population (obesity, for example) while controlling for others (smoking, diet, exercise habits, alcohol consumption), and, eventually, to try to assess the relative risks of disease and death for these various states. So, these are fairly involved processes, and it's important to at least be able to understand the premise and the balance of the researcher's approach.

CHAPTER 3

Probability

CHAPTER PROBLEM

Women and children first?

It is a well-known rule of the sea that if the sinking of a ship is imminent, then the lifeboats are filled first with women and children. Was this rule followed when the *Titanic* sank on Monday, April 15, 1912? Let's consider the data in Table 3-1.

After doing a little basic arithmetic, we see that 19.6% of the men (332 out of 1692) survived, and 70.4% of the women and children (or 374 out of 531) survived. Such simple calculations are the basis for finding important values of probabilities of events, such as these:

- The probability of getting a survivor if a *Titanic* passenger is randomly selected
- The probability of getting a woman or child if a *Titanic* survivor is randomly selected

We will find these probabilities later in this chapter.

TABLE 3-1 *Titanic* Mortality

	Men	Women	Boys	Girls	**Total**
Survived	332	318	29	27	**706**
Died	1360	104	35	18	**1517**
Total	**1692**	**422**	**64**	**45**	**2223**

3-1 Overview

In later chapters of this book, we will use sample data to make inferences (or conclusions) about populations. Many of those inferences will be based on *probabilities* of events. As a simple example, suppose that you were to win the top prize in a state lottery five consecutive times. There would be an uproar, and everyone would suspect that you were somehow cheating. They would know that even though there is a chance of someone winning five consecutive times, that chance is so incredibly low that they would reject it as a reasonable explanation. This is exactly how statisticians think: They reject explanations based on very low probabilities. Statisticians use the *rare event rule.*

Rare Event Rule for Inferential Statistics

If, under a given assumption (such as a lottery being fair), the probability of a particular observed event (such as five consecutive lottery wins) is extremely small, we conclude that the assumption is probably not correct.

Our primary objective in this chapter is to develop a sound understanding of probability values, which we will build upon in subsequent chapters. A secondary objective is to develop the basic skills necessary to solve probability problems.

3-2 Fundamentals

In considering probability, we deal with procedures that produce outcomes.

DEFINITIONS

An **event** is any collection of results or outcomes of a procedure.

A **simple event** is an outcome or an event that cannot be further broken down into simpler components.

The **sample space** for a procedure consists of all possible *simple* events. That is, the sample space consists of all outcomes that cannot be broken down any further.

EXAMPLES

Procedure	Example of Event	Sample Space
Roll one die	5 (simple event)	{1, 2, 3, 4, 5, 6}
Roll two dice	7 (not a simple event)	{1-1, 1-2, . . . , 6-6}

When rolling one die, 5 is a *simple event* because it cannot be broken down any further. When rolling two dice, 7 is *not a simple event* because it can be broken down into simpler events, such as 3-4 and 6-1. When rolling two dice, the *sample space* consists of 36 simple events: 1-1, 1-2, . . . , 6-6. When rolling two dice, the outcome of 3-4 is considered a simple event, because it is an outcome that cannot be broken down any further. We might incorrectly think that 3-4 can be further broken down into the individual results of 3 and 4, but 3 and 4 are not individual outcomes when two dice are rolled. When two dice are rolled, there are exactly 36 outcomes that are simple events: 1-1, 1-2, . . . , 6-6.

Subjective Probabilities at the Racetrack

Researchers studied the ability of racetrack bettors to develop realistic subjective probabilities. (See "Racetrack Betting: Do Bettors Understand the Odds?," by Brown, D'Amato, and Gertner, *Chance* magazine, Vol. 7, No. 3.) After analyzing results for 4400 races, they concluded that although bettors slightly overestimate the winning probabilities of "longshots" and slightly underestimate the winning probabilities of "favorites," their general performance is quite good. The subjective probabilities were calculated from the payoffs, which are based on the amounts bet, and the actual probabilities were calculated from the actual race results.

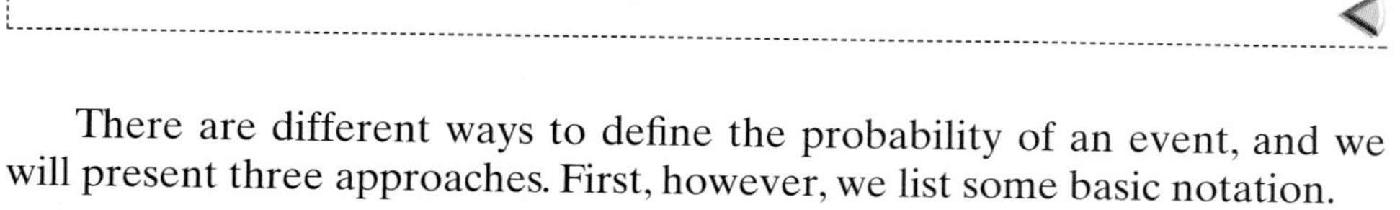

There are different ways to define the probability of an event, and we will present three approaches. First, however, we list some basic notation.

Notation for Probabilities

P denotes a probability.
A, B, and C denote specific events.
$P(A)$ denotes the probability of event A occurring.

Rule 1: Relative Frequency Approximation of Probability

Conduct (or observe) a procedure a large number of times, and count the number of times that event A actually occurs. Based on these actual results, $P(A)$ is *estimated* as follows:

$$P(A) = \frac{\text{number of times } A \text{ occurred}}{\text{number of times trial was repeated}}$$

Rule 2: Classical Approach to Probability (Requires Equally Likely Outcomes)

Assume that a given procedure has n different simple events and that each of those simple events has an *equal chance* of occurring. If event A can occur in s of these n ways, then

$$P(A) = \frac{\text{number of ways } A \text{ can occur}}{\text{number of different simple events}} = \frac{s}{n}$$

Rule 3: Subjective Probabilities

$P(A)$, the probability of event A, is found by simply guessing or estimating its value based on knowledge of the relevant circumstances.

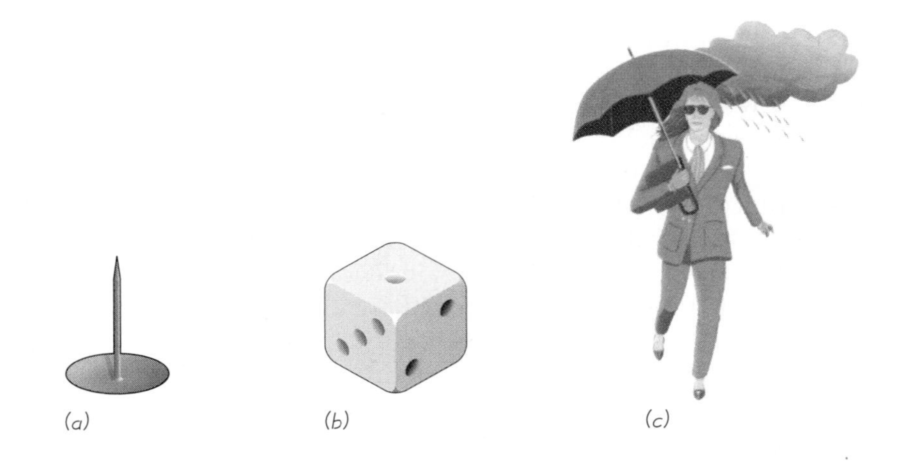

FIGURE 3-1
Three Approaches to Finding Probability

(a) Relative Frequency Approach (Rule 1): When trying to determine: *P*(tack lands point up), we must repeat the procedure of tossing the tack many times and then find the ratio of the number of times the tack lands with the point up to the number of tosses.
(b) Classical Approach (Rule 2): When trying to determine *P*(2) with a balanced and fair die, each of the six faces has an equal chance of occurring.

$$P(2) = \frac{\text{number of ways 2 can occur}}{\text{total number of simple events}} = \frac{1}{6}$$

(c) Subjective Probability (Rule 3): When trying to estimate the probability of rain tomorrow, meteorologists use their expert knowledge of weather conditions to develop an estimate of the probability.

It is very important to note that *the classical approach* (*Rule 2*) *requires equally likely outcomes.* If the outcomes are not equally likely, either we must use the relative frequency estimate or we must rely on our knowledge of the circumstances to make an *educated guess*. Figure 3-1 illustrates the three approaches.

When finding probabilities with the relative frequency approach (Rule 1), we obtain an *approximation* instead of an exact value. As the total number of observations increases, the corresponding approximations tend to get closer to the actual probability. This property is stated as a theorem commonly referred to as the *law of large numbers.*

Law of Large Numbers

As a procedure is repeated again and again, the relative frequency probability (from Rule 1) of an event tends to approach the actual probability.

The law of large numbers tells us that the relative frequency approximations from Rule 1 tend to get better with more observations. This law reflects a simple notion supported by common sense: A probability estimate based on only a few trials can be off by substantial amounts, but with a very large number of trials, the estimate tends to be much more accurate. For example, an opinion poll of only a dozen people could easily be in error by large amounts, but a poll of thousands of *randomly selected* people will be much closer to the true population values.

The TI-83 Plus display in Figure 3-2 illustrates the law of large numbers by showing results simulated on the calculator. (See Larry Morgan's *TI-83 Plus Companion to Elementary Statistics*, a supplement to this book.) The horizontal scale represents the number of births, which increases from left to right. The vertical scale represents the proportion of girls, with 0.5 corresponding to the horizontal line in the middle. Note that as the number of births increases, the proportion of girls approaches the 0.5 value.

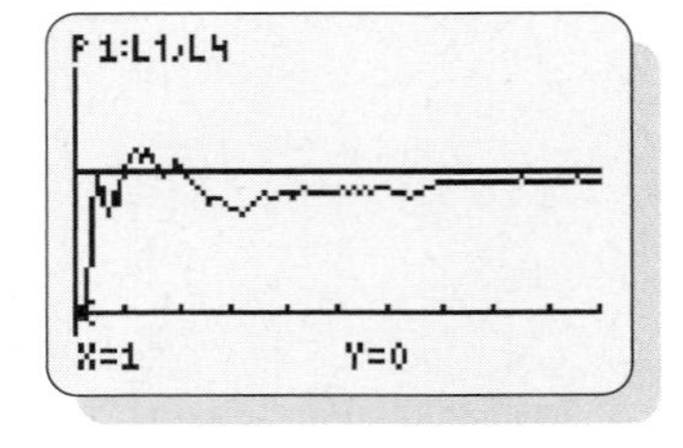

FIGURE 3-2
Illustration of the Law of Large Numbers

After examining the preceding three rules, it might seem that we should always use Rule 2 when a procedure has equally likely outcomes. Unfortunately, many procedures are so complicated that the classical approach (Rule 2) is impractical to use. In the game of solitaire, for example, the outcomes (hands dealt) are all equally likely, but it is extremely frustrating to try to use Rule 2 to find the probability of winning. In such cases we can more easily get good estimates by using the relative frequency approach (Rule 1). Simulations are often helpful when using this approach. (A **simulation** of a procedure is a process that behaves in the same ways as the procedure itself, so that similar results are produced.) For example, it's much easier to use Rule 1 for estimating the probability of winning at solitaire—that is, to play the game many times (or to run a computer simulation)—than to perform the extremely complex calculations required with Rule 2.

The following examples illustrate how the three rules work. In some of the examples we use the term *random.* Recall these definitions from Section 1-4: In a **random sample of one element** from a population, all elements available for selection have the same chance of being chosen; a sample of n items is a **random sample** (or a simple random sample) if it is selected in such a way that every possible sample of n items from the population has the same chance of being chosen. The general concept of randomness is of critical importance in statistics. When making inferences based on samples, we must have a sampling process that is representative, impartial, and unbiased. If a sample is not carefully selected, it may be totally worthless.

EXAMPLE **Lightning Strikes** Find the probability that a randomly selected person will be struck by lightning this year.

SOLUTION The sample space consists of two simple events: The selected person is struck by lightning this year or is not. Because the sample space consists of events that are not equally likely, we can't use the classical approach (Rule 2). We must either use a relative frequency approximation (Rule 1) or subjectively estimate the probability (Rule 3). Before continuing, stop and try to guess the probability of a person being struck by lightning this year. Many people would make a reasonable guess that the probability is quite low, such as "one chance in a million." If we elect to use Rule 1, it isn't practical to conduct actual trials, but we can research past events. In a recent year, 377 people were struck by lightning in the United States. With a population of 274,037,295, the probability of being struck by lightning in a year is estimated to be

$$\frac{377}{274{,}037{,}295} \approx \frac{1}{727{,}000}$$

How Probable?

How do we interpret such terms as *probable*, *improbable*, or *extremely probable*? The FAA interprets these terms as follows. *Probable:* A probability on the order of 0.00001 or greater for each hour of flight. Such events are expected to occur several times during the operational life of each airplane. *Improbable:* A probability on the order of 0.00001 or less. Such events are not expected to occur during the total operational life of a single airplane of a particular type, but may occur during the total operational life of all airplanes of a particular type. *Extremely improbable:* A probabilty on the order of 0.000000001 or less. Such events are so unlikely that they need not be considered to ever occur.

EXAMPLE **Guessing Answers** On an SAT or ACT test, a typical multiple-choice question has 5 possible answers. If you make a random guess on one such question, what is the probability that your response is wrong?

SOLUTION There are 5 possible outcomes or answers, and there are 4 ways to answer incorrectly. Random guessing implies that the outcomes in the sample space are equally likely, so we apply the classical approach (Rule 2) to get

$$P(\text{wrong answer}) = \frac{4}{5} = 0.8$$

In basic probability problems of the type we are now considering, it is very important to examine the available information carefully and to identify the total number of possible outcomes correctly. In some cases, the total number of possible outcomes is directly available, but in other cases we must process the available information to determine that total. The preceding example included the information that the total number of outcomes is 5, but the following example requires us to calculate the total number of possible outcomes.

EXAMPLE **Computer Ownership** In a statistics class, 18 students own their own computers and 7 do not. If one of the students is randomly selected, find the probability of getting one who does *not* own a computer.

SOLUTION *Hint:* Instead of trying to formulate an answer directly from the written statement, summarize the given information in a format that allows you to better understand it. For example:

18 students own computers
7 students do not own computers
25 total number of students

With random selection, the 25 students are equally likely to be selected, and Rule 2 applies as follows:

$$P(\text{student does not own a computer}) = \frac{\text{number who do not own a computer}}{\text{total number of students}} = \frac{7}{25} = 0.28$$

There is a 0.28 probability that when one of the students is randomly selected, he or she does not own a computer. The way this solution was reached illustrates two important principles: First, you should list the given information in a clear and understandable format. Second, instead of finding probability values by merely dividing the smaller value by the larger value, you need to evaluate the available information carefully to identify the total number of possible outcomes and the number of ways in which the event can occur.

EXAMPLE **Gender of Children** Find the probability that a couple with 3 children will have exactly 2 boys. Assume that boys and girls are equally likely and that the gender of any child is not influenced by the gender of any other child.

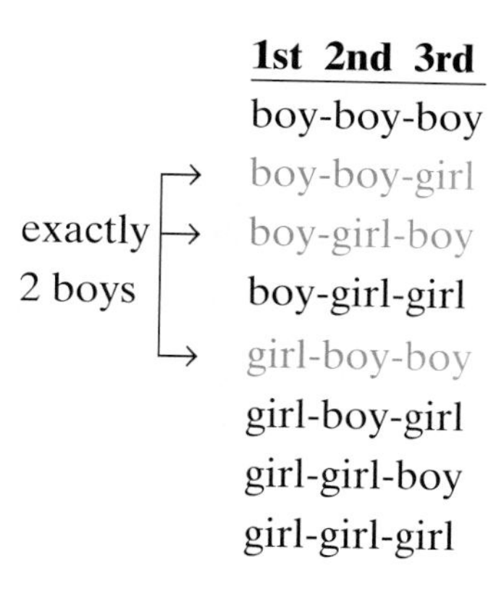

SOLUTION We first list the sample space that identifies the 8 outcomes. Those 8 outcomes are equally likely, so we use Rule 2. Of those 8 different possible outcomes, 3 correspond to exactly 2 boys, so

$$P(\text{2 boys in 3 births}) = \frac{3}{8} = 0.375$$

INTERPRETATION There is a 0.375 probability that if a couple has 3 children, exactly 2 will be boys.

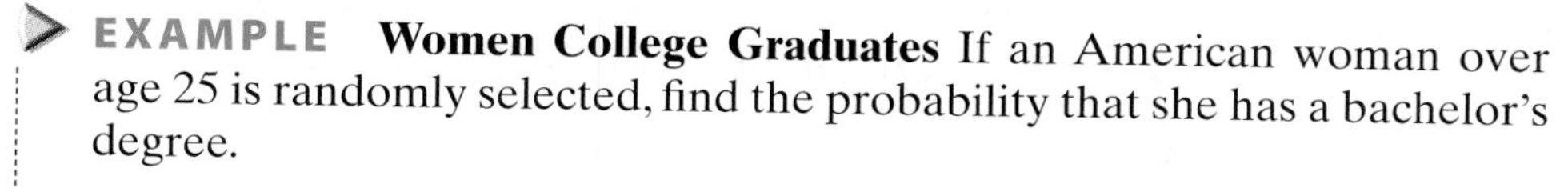

EXAMPLE **Women College Graduates** If an American woman over age 25 is randomly selected, find the probability that she has a bachelor's degree.

SOLUTION There are only two possible outcomes: Either the selected woman has a bachelor's degree or she does not. Because those two outcomes are not equally likely, we must either use the relative frequency approximation or settle for a subjective estimate of the probability. A study of 2500 randomly selected women over age 25 showed that 545 of them have bachelor's degrees (based on data from the Census Bureau). Based on that result, we *estimate* that the probability is 545/2500, or 0.218.

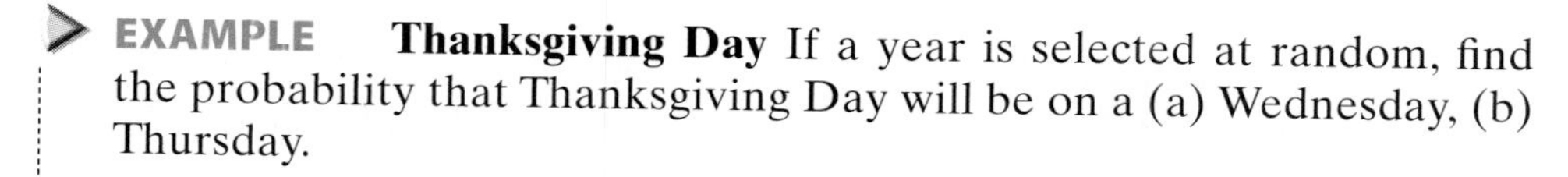

EXAMPLE **Thanksgiving Day** If a year is selected at random, find the probability that Thanksgiving Day will be on a (a) Wednesday, (b) Thursday.

SOLUTION

a. Thanksgiving Day always falls on the fourth Thursday in November. It is therefore impossible for Thanksgiving to be on a Wednesday. When an event is impossible, we say that its probability is 0.

b. It is certain that Thanksgiving will be on a Thursday. When an event is certain to occur, we say that its probability is 1.

Because any event imaginable is impossible, certain, or somewhere in between, it is reasonable to conclude that the mathematical probability of any event is 0, 1, or a number between 0 and 1 (see Figure 3-3).

- **The probability of an impossible event is 0.**
- **The probability of an event that is certain to occur is 1.**
- $0 \leq P(A) \leq 1$ **for any event** A.

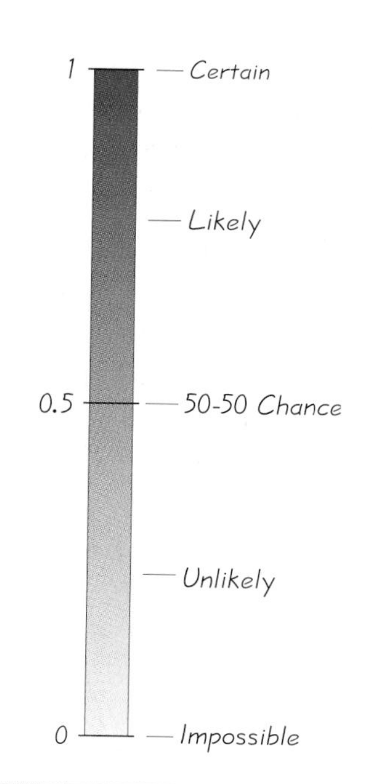

FIGURE 3-3
Possible Values for Probabilities

In Figure 3-3, the scale of 0 through 1 is shown on the left, and the more familiar and common expressions of likelihood are shown on the right.

Complementary Events

Sometimes we need to find the probability that an event A does *not* occur.

DEFINITION

The **complement** of event A, denoted by $\overline{A}$, consists of all outcomes in which event A does *not* occur.

EXAMPLE **Testing Corvettes** The General Motors Corporation wants to conduct tests of a new model of Corvette. A pool of 50 drivers has been recruited, 20 of whom are men. When the first person is selected from this pool, what is the probability of *not* getting a male driver?

SOLUTION Because 20 of the 50 subjects are men, it follows that 30 of the 50 subjects are women, so

$$
\begin{aligned}
P(\text{not selecting a man}) &= P(\overline{\text{man}}) \\
&= P(\text{woman}) \\
&= \frac{30}{50} = 0.6
\end{aligned}
$$

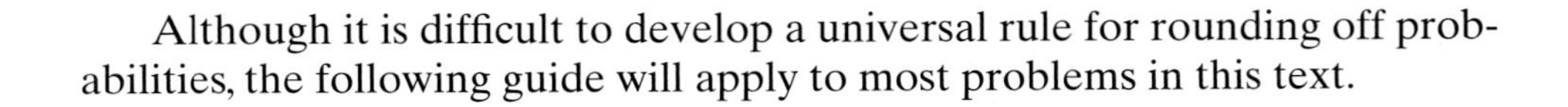

Although it is difficult to develop a universal rule for rounding off probabilities, the following guide will apply to most problems in this text.

Rounding Off Probabilities

When expressing the value of a probability, either give the *exact* fraction or decimal or round off final decimal results to three significant digits. (*Suggestion:* When a probability is not a simple fraction such as 2/3 or 5/9, express it as a decimal so that the number can be better understood.)

All digits in a number are significant except for the zeros that are included for proper placement of the decimal point.

EXAMPLES

- The probability of 0.000072098 has five significant digits (72098), and it can be rounded to three significant digits as 0.0000721.
- The probability of 1/3 can be left as a fraction, or rounded to 0.333. Do *not* round to 0.3.

Probabilities that Challenge Intuition

In certain cases, our subjective estimates of probability values are dramatically different from the actual probabilities. Here is a classic example: If you take a deep breath, there is better than a 99% chance that you will inhale a molecule that was exhaled in dying Caesar's last breath. In that same morbid and unintuitive spirit, if Socrates' fatal cup of hemlock was mostly water, then the next glass of water you drink will likely contain one of those same molecules. Here's another less morbid example that can be verified: In classes of 25 students, there is better than a 50% chance that at least 2 students will share the same birthday.

- The probability of heads in a coin toss can be expressed as $1/2$ or 0.5; because 0.5 is exact, there's no need to express it as 0.500.
- The fraction $7723/93725$ is exact, but its value isn't obvious, so express it as the decimal 0.0824.

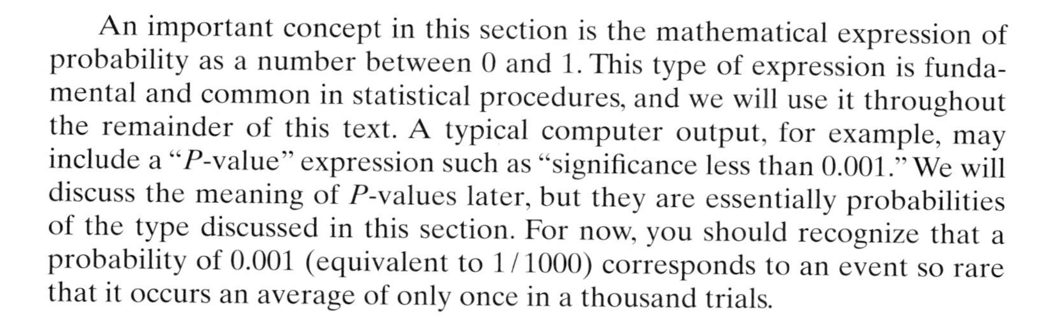

An important concept in this section is the mathematical expression of probability as a number between 0 and 1. This type of expression is fundamental and common in statistical procedures, and we will use it throughout the remainder of this text. A typical computer output, for example, may include a "*P*-value" expression such as "significance less than 0.001." We will discuss the meaning of *P*-values later, but they are essentially probabilities of the type discussed in this section. For now, you should recognize that a probability of 0.001 (equivalent to $1/1000$) corresponds to an event so rare that it occurs an average of only once in a thousand trials.

Odds

Expressions of likelihood are often given as *odds,* such as 50:1 (or "50 to 1"). A serious disadvantage of odds is that they make many calculations extremely difficult. As a result, statisticians, mathematicians, and scientists prefer to use probabilities. The advantage of odds is that they make it easier to deal with money transfers associated with gambling, so they tend to be used in casinos, lotteries, and racetracks. There are three definitions that apply: the actual odds *against* event A occurring, the actual odds *in favor* of event A occurring, and the *payoff* odds. Note that in the three definitions that follow, the actual odds against and the actual odds in favor describe the actual likelihood of some event, but the payoff odds describe the relationship between the bet and the amount of the payoff. Racetracks and casinos are in business to make a profit, so the payoff odds will not be the same as the actual odds.

DEFINITION

The **actual odds against** event A occurring are the ratio $P(\overline{A})/P(A)$, usually expressed in the form of $a:b$ (or "a to b"), where a and b are integers having no common factors.

The **actual odds in favor** of event A are the reciprocal of the actual odds against that event. If the odds against A are $a:b$, then the odds in favor of A are $b:a$.

The **payoff odds** against event A represent the ratio of net profit (if you win) to the amount bet.

$$\text{payoff odds against event } A = (\text{net profit}):(\text{amount bet})$$

Using Probability to Identify Unusual Events. *In Exercises 9–16, consider an event to be "unusual" if its probability is less than or equal to 0.05.*

9. a. Find the probability of getting a total of 2 when a pair of dice is rolled. (*Hint:* List the sample space of 36 outcomes, consisting of 1-1, 1-2, . . . , 6-6.)

b. Is it "unusual" to get a total of 2 when a pair of dice is rolled?

10. a. Find the probability of getting a 7 when a pair of dice is rolled. (See Exercise 9.)

b. Is it "unusual" to get a 7 when a pair of dice is rolled?

11. a. An instructor is obsessed with the metric system and insists that all multiple-choice questions must have 10 different possible answers, one of which is correct. What is the probability of answering a question correctly if a random guess is made?

b. Is it "unusual" to answer a question correctly by guessing?

12. a. A roulette wheel has 38 slots: One slot is 0, another slot is 00, and the other slots are numbered 1 through 36. If you bet all of your textbook money on the number 13, what is the probability that you will win?

b. Is it "unusual" to win when you bet on a single number in roulette?

13. a. A study of 400 randomly selected American Airlines flights showed that 344 arrived on time (based on data from the Department of Transportation). What is the estimated probability of an American Airlines flight arriving late?

b. Is it "unusual" for an American Airlines flight to arrive late?

14. Guessing Birthdays On their first date, Kelly asks Mike to guess the date of her birth, not including the year.

a. What is the probability that Mike will guess correctly? (Ignore leap years.)

b. Would it be "unusual" for him to guess correctly on his first try?

c. If you were Kelly, and Mike did guess correctly on his first try, would you believe his claim that he made a lucky guess, or would you be convinced that he already knew when you were born?

d. If Kelly asks Mike to guess her age, and Mike's guess is too high by 15 years, what is the probability that Mike and Kelly will have a second date?

15. Lottery In the old New York State Lottery, you had to select six numbers between 1 and 54 inclusive. There were 25,827,165 different possible six-number combinations, and you had to select the correct combination of all six digits to win the grand prize. For a $1 bet, you selected two different six-number combinations. (You could not select a single six-number combination; you had to select two.)

a. If you placed a $1 bet and selected two different six-number combinations, what was the probability of winning the grand prize?

b. Was it unusual to win the grand prize?

16. Breaking Codes A code breaker for the National Security Agency is trying to decipher an intercepted message, and she is experimenting with different letter-substitution schemes.

a. If she randomly selects a letter from the alphabet, what is the probability that she will select the correct letter, which is *e*?

b. Would it be considered unusual to select *e* on her first try and find that it is the correct choice?

17. Probability of Alzheimer's Disease In a study of Americans over 65 years of age, it is found that 255 have Alzheimer's disease and 2302 do not (based on data from the Alzheimer's Association). If an American over 65 years of age is randomly selected, what is the estimated probability that he or she has Alzheimer's disease? Based on that probability, is Alzheimer's disease a major concern for those over 65 years of age?

18. Probability of Computer Failures A *PC World* survey of 4000 personal computer owners showed that 992 of them broke down during the first two years. (The computers broke down, not the owners.) In choosing among several computer suppliers, a purchasing agent wants to know the probability of a personal computer breaking down during the first two years. Use the survey results to estimate that probability.

19. Blood Groups In a study of blood donors, 225 were classified as group O and 275 had a classification other than group O (based on data from the Greater New York Blood Program). What is the approximate probability that a person will have group O blood?

20. TV Surveys A Nielsen survey of 3857 households shows that 463 have their televisions tuned to CBS on Monday night between 10:00 P.M. and 10:30 P.M. If a household is randomly selected, estimate the probability of getting one tuned to CBS in that time slot.

21. Probability of a Birthday

a. If a person is randomly selected, find the probability that his or her birthday is October 18, which is National Statistics Day in Japan. Ignore leap years.

b. If a person is randomly selected, find the probability that his or her birthday is in October. Ignore leap years.

22. Probability of Brand Recognition

a. In a study of brand recognition, 831 consumers knew of Campbell's Soup, and 18 did not (based on data from Total Research Corporation). Use these results to estimate the probability that a randomly selected consumer will recognize Campbell's Soup.

b. *Estimate* the probability that a randomly selected adult American consumer will recognize the brand name of Bose. (Bose is a manufacturer of high-quality speakers and components.)

23. Fruitcake Survey In a Bruskin-Goldring Research poll, respondents were asked how a fruitcake should be used. One hundred thirty-two respondents indicated that it should be used for a doorstop, and 880 other respondents cited other uses, including birdfeed, landfill, and a gift. If one of these respondents is randomly selected, what is the probability of getting someone who would use the fruitcake as a doorstop?

24. Probability of IRS Correctness The U.S. General Accounting Office recently tested the IRS for correctness of answers to taxpayers' questions. For 1733 trials, the IRS was correct 1107 times. Use these results to estimate the probability that a random taxpayer's question will be answered correctly. Based on the result, would you say that the IRS does a good job answering taxpayers' questions correctly?

25. Probability of a Car Crash Among 400 randomly selected drivers in the 20–24 age bracket, 136 were in a car accident during the last year (based on data from the National Safety Council). If a driver in that age bracket is randomly selected, what is the approximate probability that he or she will be in a car accident during the next year? Is the resulting value high enough to be of concern to those in the 20–24 age bracket?

26. Probability of a Jail Sentence Data provided by the Bureau of Justice Statistics revealed that for a representative sample of convicted burglars, 76,000 were jailed, 25,000 were put on probation, and 2000 received other sentences. Use these results to estimate the probability that a convicted burglar will serve jail time. Does it seem that the result is high enough to deter burglars?

27. Probability of an Adverse Drug Reaction When the drug Viagra was clinically tested, 117 patients reported headaches and 617 did not (based on data from Pfizer, Inc.). Use this sample to estimate the probability that a Viagra user will experience a headache.

Method of Fraud	Number
Stolen card	243
Counterfeit card	85
Mail/phone order	52
Other	46

28. Probability of Credit-Card Fraud A study of credit-card fraud was conducted by Master Card International, and the accompanying table is based on the results. If one case of credit-card fraud is randomly selected from the cases summarized in the table, find the probability that the fraud resulted from a counterfeit card.

29. Gender of Children: Constructing Sample Space This section included a table summarizing the gender outcomes for a couple planning to have three children.

a. Construct a similar table for a couple planning to have two children.
b. Assuming that the outcomes listed in part (a) are equally likely, find the probability of getting two girls.
c. Find the probability of getting exactly one child of each gender.

30. Genetics: Constructing Sample Space Both parents have the brown/blue pair of eye-color genes, and each parent contributes one gene to a child. Assume that if the child has at least one brown gene, that color will dominate and the eyes will be brown. (The actual determination of eye color is somewhat more complicated.)

a. List the different possible outcomes. Assume that these outcomes are equally likely.
b. What is the probability that a child of these parents will have the blue/blue pair of genes?
c. What is the probability that the child will have brown eyes?

31. Finding Odds and Payoffs The American Statistical Association decides to invest some of its member revenue by buying a racehorse named Mean. Mean is entered in a race in which the actual probability of winning is 3/17.

a. Find the actual odds against Mean winning.
b. If the payoff odds are listed as 4:1, how much profit do you make if you bet $4 and Mean wins?

32. Finding Odds in Roulette A roulette wheel has 38 slots. One slot is 0, another is 00, and the others are numbered 1 through 36, respectively. You are placing a bet that the outcome is an odd number.

a. What is your probability of winning?
b. What are the actual odds against winning?
c. When you bet that the outcome is an odd number, the payoff odds are 1:1. How much profit do you make if you bet $18 and win?
d. How much profit would you make on the $18 bet if you could somehow convince the casino to change its payoff odds so that they are the same as the actual odds against winning? (*Recommendation:* Don't actually try to convince any casino of this; their sense of humor is remarkably absent when it comes to things of this sort.)

3-2 Beyond the Basics

33. Interpreting Effectiveness A double-blind experiment is designed to test the effectiveness of the drug Statisticzene as a treatment for number blindness. When treated with Statisticzene, subjects seem to show improvement. Researchers calculate that there is a 0.04 probability that the treatment group would show improvement if the drug has no effect. What should you conclude about the effectiveness of Statisticzene?

34. Determining Whether a Jury Is Random An attorney is defending a client accused of not meeting his alimony obligations. The pool of 20 potential jurors consists of all women, and the attorney calculates that there is a probability of 1/1,048,576 that 20 randomly selected people will be all women. Is there justification for arguing that the jury pool is unfair to his client?

35. Finding Probability from Odds If the actual odds against event A are $a:b$, then $P(A) = b/(a + b)$. Find the probability of Millenium winning his next race, given that the actual odds against his winning are 3 : 5.

36. Finding Probability from Stem-and-Leaf Plot The stem-and-leaf plot summarizes the time (in hours) that managers spend on paperwork in one day (based on data from Adia Personnel Services). Use this sample to estimate the probability that a randomly selected manager spends more than 2.0 hours per day on paperwork.

Stem	Leaves
0.	00
1.	0578
2.	00113449
3.	347
4.	445

37. Using Boxplot of IQ Scores After collecting IQ scores from hundreds of subjects, a boxplot is constructed with this five-number summary: 82, 91, 100, 109, 118. If one of the subjects is randomly selected, find the probability that his or her IQ score is greater than 109.

38. Leap Years and Guessing Birthdays In part (a) of Exercise 21, leap years were ignored in finding the probability that a randomly selected person will have a birthday on October 18.

a. Recalculate this probability, assuming that a leap year occurs every four years. (Express your answer as an exact fraction.)

b. Leap years occur in years evenly divisible by 4, except they are skipped in three of every four centesimal years (years ending in 00). The years 1700, 1800, and 1900 were not leap years, but 2000 is a leap year. Find the exact probability for this case, and express it as an exact fraction.

39. Flies on an Orange If two flies land on an orange, find the probability that they are on points that are within the same hemisphere.

40. Points on a Stick Two points along a straight stick are randomly selected. The stick is then broken at those two points. Find the probability that the three resulting pieces can be arranged to form a triangle. (This is a very difficult problem.)

3-3 Addition Rule

The main objective of this section is to introduce the *addition rule* as a device for finding probabilities that can be expressed as $P(A \text{ or } B)$, the probability that either event A occurs or event B occurs (or they both occur) as the single outcome of a procedure. The key word to remember is *or.* Throughout this text we use the *inclusive or,* which means either one or the other or both. (Except for Exercise 27, we will not consider the *exclusive or,* which means either one or the other but not both.)

In the previous section we presented the fundamentals of probability and considered events categorized as *simple.* In this and the following section we consider *compound events.*

Guess on SATs?

Students preparing for multiple-choice test questions are often told not to guess, but that's not necessarily good advice. Standardized tests with multiple-choice questions typically *compensate* for guessing, but they don't *penalize* it. For questions with five answer choices, one-fourth of a point is usually subtracted for each incorrect response. Principles of probability show that in the long run, pure random guessing will neither raise nor lower the exam score. Definitely guess if you can eliminate at least one choice or if you have a sense for the right answer, but avoid tricky questions with attractive wrong answers. Also, don't waste too much time on such questions.

DEFINITION

A **compound event** is any event combining two or more simple events.

Notation for Addition Rule

$P(A \text{ or } B) = P(\text{event } A \text{ occurs or event } B \text{ occurs or they both occur})$

See Figure 3-4, which depicts a sample of 5 men and 10 women. Refer to that figure and answer this question: How many of the 15 subjects are men or red? (Remember, "men or red" really means "men, or red, or both.") Examination of Figure 3-4 should show that a total of 9 subjects are men or red. (*Important note*: It is *wrong* to add the 5 men to the 7 red people, because this total of 12 would have counted the 3 red men twice, but they are individual subjects that should be counted once each.) Because 9 of the 15 subjects are "men or red," the probability of randomly selecting someone who is a man or is red can be expressed as $P(\text{man or red}) = 9/15$.

This example suggests a general rule whereby we add the number of outcomes corresponding to each of the events in question:

When finding the probability that event *A* occurs or event *B* occurs, find the total of the number of ways *A* can occur and the number of ways *B* can occur, but *find that total in such a way that no outcome is counted more than once.*

FIGURE 3-4

One approach is to combine the number of ways event A can occur with the number of ways event B can occur and, if there is any overlap, compensate by subtracting the number of outcomes that are counted twice, as in the following rule.

Formal Addition Rule

$$P(A \text{ or } B) = P(A) + P(B) - P(A \text{ and } B)$$

where $P(A \text{ and } B)$ denotes the probability that A and B both occur at the same time as an outcome in a trial of a procedure.

Although the formal addition rule is presented as a formula, it is generally better to understand the spirit of the rule and apply it intuitively, as follows.

Intuitive Addition Rule

To find $P(A \text{ or } B)$, find the sum of the number of ways event A can occur and the number of ways event B can occur, *adding in such a way that every outcome is counted only once.* $P(A \text{ or } B)$ is equal to that sum, divided by the total number of outcomes.

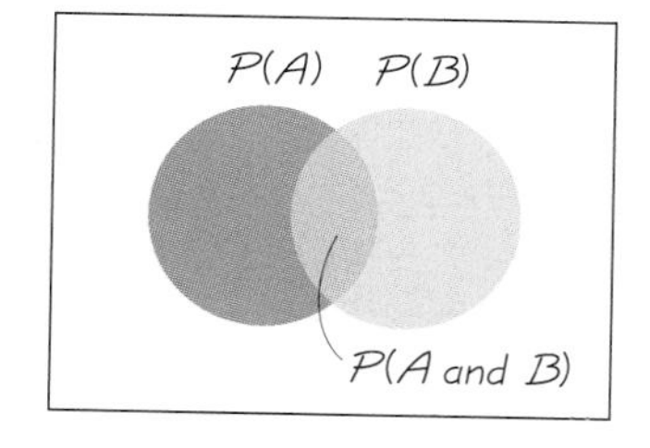

FIGURE 3-5
Venn Diagram Showing Overlapping Events

Figure 3-5 shows a Venn diagram that provides a visual illustration of the formal addition rule. In this figure we can see that the probability of A or B equals the probability of A (left circle) plus the probability of B (right circle) minus the probability of A and B (football-shaped middle region). This figure shows that the addition of the areas of the two circles will cause double-counting of the football-shaped middle region. This is the basic concept that underlies the addition rule. Because of the relationship between the addition rule and the Venn diagram shown in Figure 3-5, the notation $P(A \cup B)$ is sometimes used in place of $P(A \text{ or } B)$. Similarly, the notation $P(A \cap B)$ is sometimes used in place of $P(A \text{ and } B)$, so the formal addition rule can be expressed as

$$P(A \cup B) = P(A) + P(B) - P(A \cap B)$$

The addition rule is simplified whenever A and B cannot occur simultaneously, so $P(A \text{ and } B)$ becomes zero. Figure 3-6 illustrates that with no overlapping of A and B, we have $P(A \text{ or } B) = P(A) + P(B)$. The following definition formalizes the lack of overlapping shown in Figure 3-6.

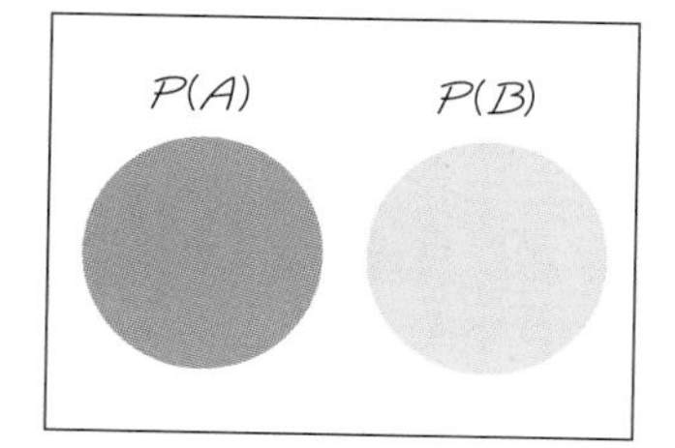

FIGURE 3-6
Venn Diagram Showing Nonoverlapping Events

DEFINITION

Events A and B are **mutually exclusive** if they cannot occur simultaneously.

The flowchart of Figure 3-7 shows how mutually exclusive events affect the addition rule.

FIGURE 3-7
Applying the Addition Rule

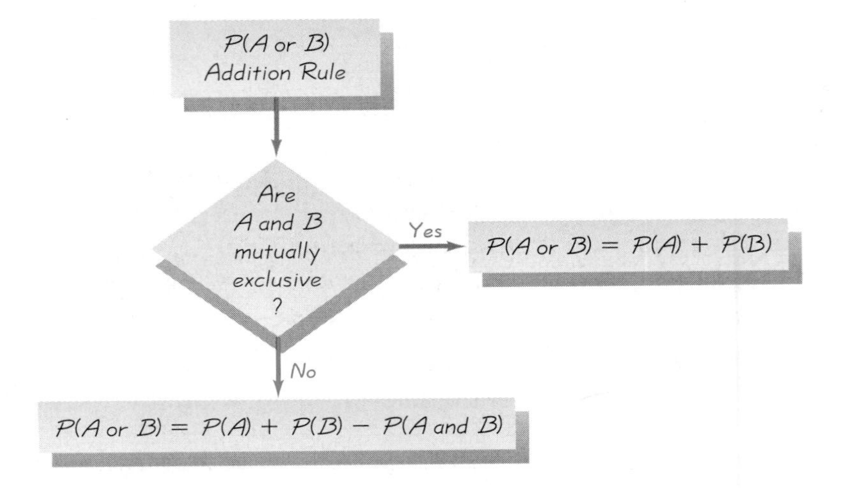

EXAMPLE **Titanic Passengers** Refer to Table 3-1, reproduced here for your convenience. Assuming that 1 person is randomly selected from the 2223 people aboard the *Titanic,* apply the addition rule to do the following.

a. Find P(selecting a man or a boy).

b. Find P(selecting a man or someone who survived).

SOLUTION

a. From the table we see that among the 2223 people, the total number of men and boys is $1692 + 64 = 1756$. Because 1756 of the people are men or boys and the total number of people is 2223, we get $P(\text{man or boy}) = 1756/2223 = 0.789924 = 0.790$ (rounded). Here, the event of selecting a man and the event of selecting a boy are mutually exclusive events, so we can add without concern for double-counting.

b. From the table we see that among the 2223 people, there are 1692 men and 706 people who survived, but we should not add 1692 and 706

TABLE 3-1 *Titanic* Mortality

	Men	Women	Boys	Girls	**Total**
Survived	332	318	29	27	**706**
Died	1360	104	35	18	**1517**
Total	**1692**	**422**	**64**	**45**	**2223**

because those events (men, survived) overlap and are not mutually exclusive. Using the intuitive addition rule, we must add in such a way that we do not double-count the 332 men who survived. The result should be 2066 individual people who were men or survived. We therefore get this result: $P(\text{man or survived}) = 2066/2223 = 0.929$.

There are several strategies you could use for counting people who are men or survivors. Any of the following would work.

- Color the cells representing people that are men or survivors, then add the numbers in those colored cells, being careful to add each number only once. This approach yields

$$1360 + 332 + 318 + 29 + 27 = 2066$$

- Add the 1692 men to the 706 survivors, but compensate for the double-counting by subtracting the 332 men survivors. This approach yields a result of

$$1692 + 706 - 332 = 2066$$

- Start with the total of 1692 men, then add the survivors who were not yet included in that total, to get a result of

$$1692 + 318 + 29 + 27 = 2066$$

Carefully study the preceding example, because it makes clear this essential feature of the addition rule: "Or" suggests addition, and the addition must be done without double-counting.

We can summarize the key points of this section as follows:

1. To find $P(A \text{ or } B)$, begin by associating "or" with addition.
2. Consider whether events A and B are mutually exclusive; in other words, can they happen at the same time? If they are not mutually exclusive (that is, if they can happen at the same time), be sure to avoid (or at least compensate for) double-counting when adding the relevant probabilities. If you understand the importance of not double-counting when you find $P(A \text{ or } B)$, you don't necessarily have to calculate the value of $P(A) + P(B) - P(A \text{ and } B)$.

Errors made when applying the addition rule often involve double-counting; that is, events that are not mutually exclusive are treated as if they were. One indication of such an error is a total probability that exceeds 1; however, errors involving the addition rule do not always cause the total probability to exceed 1.

Complementary Events

In Section 3-2 we defined the complement of event A and denoted it by $\overline{A}$. We said that $\overline{A}$ consists of all the outcomes in which event A does *not* occur.

Shakespeare's Vocabulary

According to Bradley Efron and Ronald Thisted, Shakespeare's writings included 31,534 different words. They used probability theory to conclude that Shakespeare probably knew at least another 35,000 words that he didn't use in his writings. The problem of estimating the size of a population is an important problem often encountered in ecology studies, but the result given here is another interesting application. (See "Estimating the Number of Unseen Species: How Many Words Did Shakespeare Know?", in *Biometrika*, Vol. 63, No. 3.)

Events A and $\overline{A}$ must be mutually exclusive, because it is impossible for an event and its complement to occur at the same time. Also, we can be absolutely certain that A either does or does not occur, which implies that either A or $\overline{A}$ must occur. These observations enable us to apply the addition rule for mutually exclusive events as follows:

$$P(A \text{ or } \overline{A}) = P(A) + P(\overline{A}) = 1$$

We justify $P(A \text{ or } \overline{A}) = P(A) + P(\overline{A})$ by noting that A and $\overline{A}$ are mutually exclusive; we justify the total of 1 by our certainty that A either does or does not occur. This result of the addition rule leads to the following three equivalent forms.

Rule of Complementary Events

$$P(A) + P(\overline{A}) = 1$$
$$P(\overline{A}) = 1 - P(A)$$
$$P(A) = 1 - P(\overline{A})$$

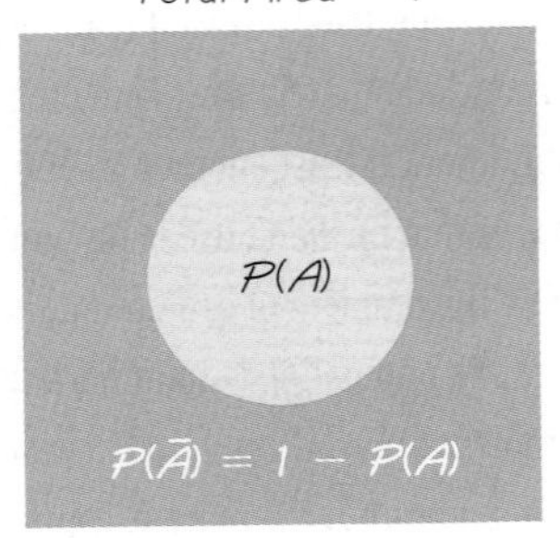

FIGURE 3-8
Venn Diagram for the Complement of Event A

Figure 3-8 visually displays the relationship between $P(A)$ and $P(\overline{A})$.

EXAMPLE If $P(A) = 0.3$, find $P(\overline{A})$.

SOLUTION Using the rule of complementary events, we get

$$P(\overline{A}) = 1 - P(A) = 1 - 0.3 = 0.7$$

A major advantage of the *rule of complementary events* is that its use can greatly simplify certain problems. We will illustrate this advantage in Section 3-5.

3-3 Basic Skills and Concepts

Determining Whether Events Are Mutually Exclusive. *For each part of Exercises 1 and 2, are the two events mutually exclusive for a single trial? (Hint: Consider "mutually exclusive" to be equivalent to "separate" or "not overlapping.")*

1. a. Selecting a voter who is under the age of 30 No
Selecting a voter whose principal news source is MTV

b. Selecting someone treated with an experimental drug
Selecting someone who experiences improved symptoms

c. Getting an odd number when a roulette wheel is spun
Getting an even number when a roulette wheel is spun

2. a. Selecting an ace from a deck of cards
Selecting a card that is a spade

b. Selecting a survey subject who is a registered Democrat
Selecting a survey subject who is not a registered voter

c. Selecting a survey subject who is watching CNN
Selecting a survey subject who is not watching television

3. Finding Complements

a. If $P(A) = 0.05$, find $P(\overline{A})$.

b. Based on recent data from the U.S. National Center for Health Statistics, the probability of a baby being a boy is 0.513. Find the probability of a baby being a girl.

4. Finding Complements

a. Find $P(\overline{A})$, given that $P(A) = 0.228$.

b. Based on data from the National Health Examination, the probability of a randomly selected adult male being taller than 6 ft is 0.14. Find the probability of randomly selecting an adult male and getting someone with a height of 6 ft or less.

5. Using Addition Rule Refer to Figure 3-4. Assume that one person is randomly selected. Find the probability of getting a female or someone coded green.

6. Using Addition Rule Refer to Figure 3-4. Assume that one person is randomly selected. Find the probability of getting a female or someone coded red.

7. *Titanic* Passengers Refer to Table 3-1. Assume that one person aboard the *Titanic* is randomly selected.

a. Find the probability of selecting a woman or a girl.

b. Find the probability of selecting a woman or someone who survived.

8. *Titanic* Passengers Refer to Table 3-1. Assume that one person aboard the *Titanic* is randomly selected.

a. Find the probability of selecting a woman or boy or girl.

b. Find the probability of selecting a woman or someone who died in the sinking of the ship.

9. National Statistics Day If someone is randomly selected, find the probability that his or her birthday is not October 18, which is National Statistics Day in Japan. Ignore leap years.

10. Birthday and Complement If someone is randomly selected, find the probability that his or her birthday is not in October. Ignore leap years.

11. Age and Tickets In a study of 82 young (under the age of 32) drivers, 39 were men who were ticketed, 11 were men who were not ticketed, 8 were women who were ticketed, and 24 were women who were not ticketed (based on data from the Department of Transportation). If one of these subjects is randomly selected, find the probability of getting a man or someone who was ticketed.

12. Age and Tickets Refer to the same data set as in Exercise 11. If one of the subjects is randomly selected, find the probability of getting a woman or someone who was not ticketed.

Using the Addition Rule with Blood Categories. *In Exercises 13–20, refer to the accompanying figure, which describes the blood groups and Rh types of 100 people (based on data from the Greater New York Blood Program). In each case, assume that 1 of the 100 subjects is randomly selected, and find the indicated probability.*

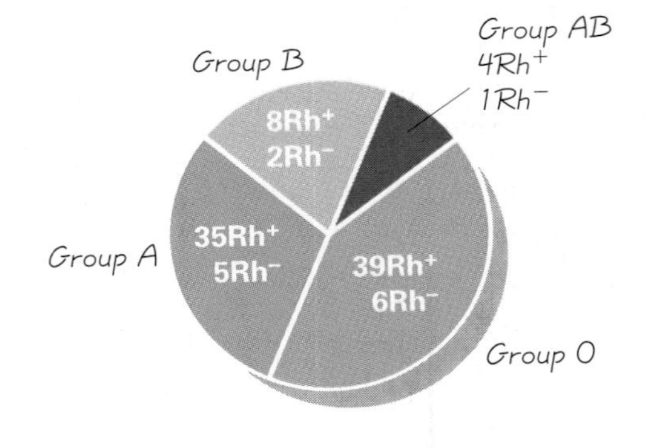

13. $P(\text{not group A})$

14. $P(\text{type Rh}^-)$

15. $P(\text{group A or type Rh}^-)$

16. $P(\text{group A or group B})$

17. $P(\text{not type Rh}^+)$

18. $P(\text{group B or type Rh}^+)$

19. $P(\text{group AB or type Rh}^+)$

20. $P(\text{group A or O or type Rh}^+)$

21. Poll Resistance Pollsters are concerned about declining levels of cooperation among persons contacted in surveys. A pollster contacts 84 people in the 18–21 age bracket and finds that 73 of them respond and 11 refuse to respond. When 275 people in the 22–29 age bracket are contacted, 255 respond and 20 refuse to respond (based on data from "I Hear You Knocking but You Can't Come In," by Fitzgerald and Fuller, *Sociological Methods and Research,* Vol. 11, No. 1). Assume that 1 of the 359 people is randomly selected. Find the probability of getting someone in the 18–21 age bracket or someone who refused to respond.

22. Poll Resistance Refer to the same data set as in Exercise 21. Assume that 1 of the 359 people is randomly selected, and find the probability of getting someone who is in the 18–21 age bracket or someone who responded.

23. Sexual Harassment Problems of sexual harassment have received much attention in recent years. In one survey, 420 workers (240 of whom are men) considered a friendly pat on the shoulder to be a form of harassment, whereas 580 workers (380 of whom are men) did not consider that to be a form of harassment (based on data from Bruskin/Goldring Research). If one of the surveyed workers is randomly selected, find the probability of getting someone who does not consider a pat on the shoulder to be a form of harassment.

24. Sexual Harassment Refer to the same data set as in Exercise 23, and find the probability of randomly selecting a man or someone who does not consider a pat on the shoulder to be a form of harassment.

3-3 Beyond the Basics

25. Determining Whether Events Are Mutually Exclusive

a. If $P(A) = 3/11$, $P(B) = 4/11$, and $P(A \text{ or } B) = 7/11$, what do you know about events A and B?

b. If $P(A) = 5/18$, $P(B) = 11/18$, and $P(A \text{ or } B) = 13/18$, what do you know about events A and B?

26. Mutually Exclusive Events If events A and B are mutually exclusive and events B and C are mutually exclusive, must events A and C be mutually exclusive? Give an example supporting your answer.

27. Exclusive *Or* How is the addition rule changed if the *exclusive or* is used instead of the *inclusive or?* Recall that the *exclusive or* means either one or the other, but not both.

28. Extending the Addition Rule Given that $P(A \text{ or } B) = P(A) + P(B) - P(A \text{ and } B)$, develop a formal rule for $P(A \text{ or } B \text{ or } C)$. (*Hint:* Draw a Venn diagram.)

3-4 Multiplication Rule: Basics

In Section 3-3 we presented the addition rule for finding $P(A \text{ or } B)$, the probability that a trial has an outcome of A or B or both. The objective of this section is to develop a rule for finding $P(A \text{ and } B)$, the probability that event A occurs in a first trial and event B occurs in a second trial.

Notation

$P(A \text{ and } B) = P(\text{event } A \text{ occurs in a first trial and event } B \text{ occurs in a second trial})$

In Section 3-3 we associated *or* with addition; in this section we will associate *and* with multiplication. We will see that $P(A \text{ and } B)$ involves multiplication of probabilities and that we must sometimes adjust the probability of event B to reflect the outcome of event A.

Probability theory is used extensively in the analysis and design of standardized tests, such as the SAT, ACT, LSAT (for law), and MCAT (for medicine). For ease of grading, such tests typically use true/false or multiple-choice questions. Let's assume that the first question on a test is a true/false type, while the second question is a multiple-choice type with five possible answers (a, b, c, d, e). We will use the following two questions. Try them!

1. True or false: The probability of being struck by lightning is greater than the probability of winning a state lottery.

2. The Pearson correlation coefficient is named after
   a. Karl Marx
   b. Carl Friedrich Gauss
   c. Karl Pearson
   d. Carly Simon
   e. Mario Triola

The answers to the two questions are T (for "true") and c. Let's find the probability that if someone makes random guesses for both answers, the first answer will be correct *and* the second answer will be correct. One way to find that probability is to list the sample space as follows.

T,a	T,b	T,c	T,d	T,e
F,a	F,b	F,c	F,d	F,e

Redundancy

Reliability of systems can be greatly improved with redundancy of critical components. Airplanes have two independent electrical systems, and aircraft used for instrument flight typically have two separate radios. The following is from a *Popular Science* article about stealth aircraft: "One plane built largely of carbon fiber was the Lear Fan 2100 which had to carry two radar transponders. That's because if a single transponder failed, the plane was nearly invisible to radar." Such redundancy is an application of the multiplication rule in probability theory. If one component has a 0.001 probability of failure, the probability of two independent components both failing is only 0.000001.

If the answers are random guesses, then the 10 possible outcomes are equally likely, so

$$P(\text{both correct}) = P(T \text{ and } c) = \frac{1}{10} = 0.1$$

Now note that $P(T \text{ and } c) = 1/10$, $P(T) = 1/2$ and $P(c) = 1/5$, from which we see that

$$\frac{1}{10} = \frac{1}{2} \times \frac{1}{5}$$

so that

$$P(T \text{ and } c) = P(T) \times P(c)$$

This suggests that, in general, $P(A \text{ and } B) = P(A) \cdot P(B)$, but let's consider another example before making that generalization.

For now, we note that tree diagrams are sometimes helpful in determining the number of possible outcomes in a sample space. A **tree diagram** is a picture of the possible outcomes of a procedure, shown as line segments emanating from one starting point. These diagrams are helpful in counting the number of possible outcomes if the number of possibilities is not too large. The tree diagram shown in Figure 3-9 summarizes the outcomes of the true/false and multiple-choice questions. From Figure 3-9 we see that if both answers are random guesses, all 10 branches are equally likely and the probability of getting the correct pair (T, c) is 1/10. For each response to the first question, there are 5 responses to the second. The total number of outcomes is 5 taken 2 times, or 10. The tree diagram in Figure 3-9 illustrates the reason for the use of multiplication.

Our first example of the true/false and multiple-choice questions suggested that $P(A \text{ and } B) = P(A) \cdot P(B)$, but the next example will introduce another important element.

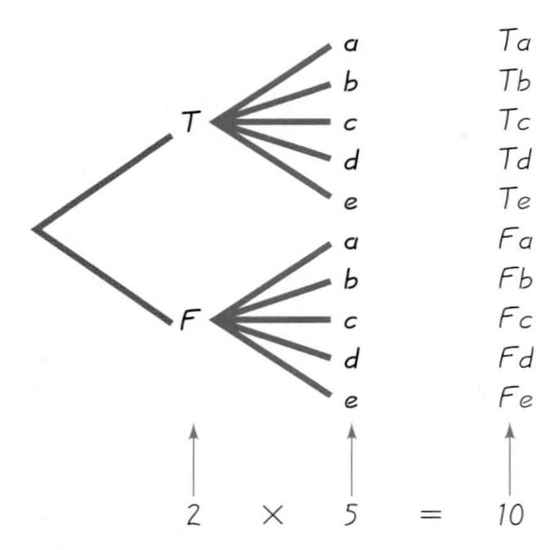

FIGURE 3-9
Tree Diagram of Test Answers

EXAMPLE Traffic Signals A box contains glass lenses used for traffic signals. Five of the lenses are red, 4 are yellow, and 3 are green. If 2 of the lenses are randomly selected from the box, find the probability that the first lens is red *and* the second lens is green. Assume that the first lens is *not replaced* before the second lens is selected.

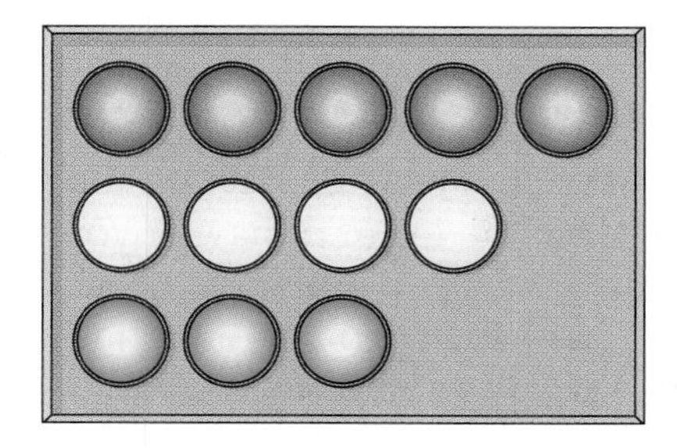

SOLUTION

First selection: $P(\text{red}) = 5/12$ (because there are 12 lenses, 5 of which are red)

Second selection: $P(\text{green}) = 3/11$ (because there are 11 lenses remaining, 3 of which are green)

With $P(\text{red}) = 5/12$ and $P(\text{green}) = 3/11$, we have

$$P(\text{red and green}) = \frac{5}{12} \times \frac{3}{11} = 0.114$$

The key point is that we have to adjust the probability of event B to reflect the outcome of event A. Because the second selection is made without replacement of the first selected lens, the second probability must take into account the result of "red" for the first selection. After the red lens has been selected on the first trial, only 11 lenses remain and 3 of them are green, so $P(\text{green}) = 3/11$.

This example illustrates the important principle that *the probability for event B should take into account the fact that event A has already occurred.* This principle is often expressed using the following notation.

Notation for Conditional Probability

$P(B|A)$ represents the probability of event B occurring after it is assumed that event A has already occurred. (We can read $B|A$ as "B given A.")

DEFINITIONS

Two events A and B are **independent** if the occurrence of one does not affect the probability of the occurrence of the other. (Several events are similarly independent if the occurrence of any does not affect the proba-

Independent Jet Engines

A three-engine jet departed from Miami International Airport en route to South America, but one engine failed immediately after takeoff. While the plane was turning back to the runway, the other two engines also failed, but the pilot was able to make a safe landing. With independent jet engines, the probability of all three failing is only 0.0001^3, or about one chance in a trillion. The FAA found that the same mechanic who replaced the oil in all three engines incorrectly positioned the oil plug sealing rings. A goal in using three separate engines is to increase safety with independent engines, but the use of a single mechanic caused their operation to become dependent. Maintenance procedures now require that the engines be serviced by different mechanics.

bilities of the occurrence of the others.) If A and B are not independent, they are said to be **dependent.**

For example, flipping a coin and then tossing a die are *independent* events because the outcome of the coin has no effect on the probabilities of the outcomes of the die. In contrast, the event of having your car start and the event of getting to class on time are *dependent* events, because the outcome of trying to start your car does affect the probability of getting to class on time.

Using the preceding notation and definitions, along with the principles illustrated in the preceding examples, we summarize the key concept of this section as the *formal multiplication rule.*

Formal Multiplication Rule

$$P(A \text{ and } B) = P(A) \cdot P(B|A)$$

If A and B are independent events, $P(B|A)$ is really the same as $P(B)$. (For further discussion about determining whether events are independent or dependent, see the subsection "Testing for Independence" in Section 3-5. For now, try to understand the basic concept of independence and how it affects the computed probabilities.) See the following *intuitive multiplication rule.* (Also see Figure 3-10.)

Intuitive Multiplication Rule

When finding the probability that event A occurs in one trial and event B occurs in the next trial, multiply the probability of event A by the probability of event B, but be sure that the probability of event B takes into account the previous occurrence of event A.

FIGURE 3-10
Applying the Multiplication Rule

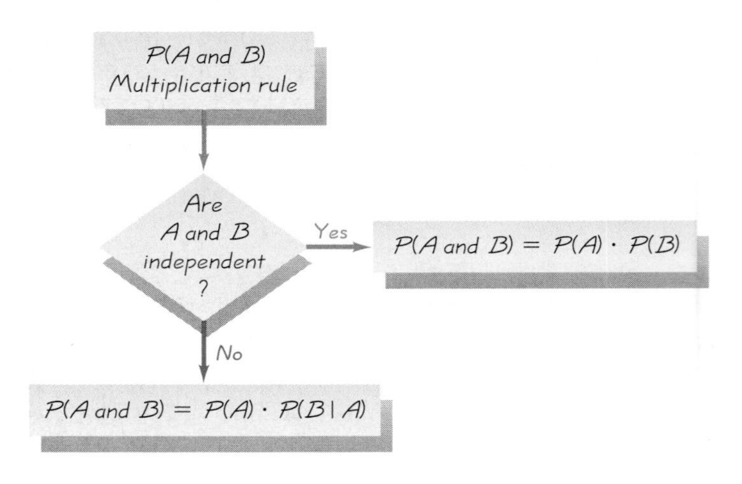

EXAMPLE **Jury Selection** A pool of potential jurors consists of 10 men and 15 women. The Commissioner of Jurors randomly selects two names from this pool. Find the probability that the first is a man and the second is a man if the two people are selected (a) with replacement; (b) without replacement.

SOLUTION *Hint:* Before beginning to attempt a solution, organize the given information in a format that can be visualized much better than a paragraph of words. Here is such a model:

10 men
15 women

} Select 2 people.

25 total people

a. If the two people are selected with replacement, the two selections are independent because the second event is not affected by the first outcome. We therefore get

$$P(\text{first is a man and second is a man}) = \frac{10}{25} \cdot \frac{10}{25} = 0.160$$

b. If the two people are selected without replacement, the two selections are dependent because the second event is affected by the first outcome. We therefore get

$$P(\text{first is a man and second is a man}) = \frac{10}{25} \cdot \frac{9}{24} = 0.150$$

Note that in this case, we adjust the second probability to take into account the selection of a man in the first outcome. After selecting a man the first time, there would be 9 men among the 24 people that remain.

So far we have discussed two events, but the multiplication rule can be easily extended to several events. In general, the probability of any sequence of independent events is simply the product of their corresponding probabilities. For example, the probability of tossing a coin three times and getting all heads is $0.5 \cdot 0.5 \cdot 0.5 = 0.125$. We can also extend the multiplication rule so that it applies to several dependent events; simply adjust the probabilities as you go along. If three different people are randomly selected without replacement from a poll of 10 men and 15 women (as in the preceding example), the probability of getting three men is

$$\frac{10}{25} \cdot \frac{9}{24} \cdot \frac{8}{23} = 0.0522$$

In this last example involving the selection of three men, we assumed that the events were dependent because the selections were made without replacement. However, it is a common practice to treat events as independent when *small samples* are drawn from *large populations*. In such cases, it is rare to select the same item twice. Here is a common guideline:

Convicted by Probability

A witness described a Los Angeles robber as a Caucasian woman with blond hair in a ponytail who escaped in a yellow car driven by an African-American male with a mustache and beard. Janet and Malcolm Collins fit this description, and they were convicted based on testimony that there is only about 1 chance in 12 million that any couple would have these characteristics. It was estimated that the probability of a yellow car is 1/10, and the other probabilities were estimated to be 1/4, 1/10, 1/3, 1/10, and 1/1000. The convictions were later overturned when it was noted that no evidence was presented to support the estimated probabilities or the independence of the events. However, because the couple was not randomly selected, a serious error was made in not considering the probability of *other* couples being in the same region with the same characteristics.

If a sample size is no more than 5% of the size of the population, treat the selections as being *independent* (even if the selections are made without replacement, so they are technically dependent).

When pollsters survey 1200 adults from a population of millions, they typically assume independence, even though they sample without replacement.

EXAMPLE **Quality Control** A quality-control manager claims that a new process for manufacturing digital cameras is better because the rate of defects is lower than 5%, which had been the rate of defects in the past. A thousand cameras are manufactured, and 12 of them are randomly selected and tested, with the result that there are no defects. Assuming that the new method has the same 5% defect rate as in the past, find the probability of getting no defects among the 12 digital cameras. Based on the result, is there strong evidence to conclude that the new process is better?

SOLUTION We want to find P(all 12 cameras are good), assuming a defect rate of 5%. If the defect rate is 5%, we have $P(\text{good camera}) = 0.95$. We can now apply the multiplication rule for independent events as follows:

$$\begin{aligned} &P(\text{all 12 cameras are good}) \\ &\quad = P(\text{1st is good and 2nd is good and 3rd is good} \ldots \text{and 12th is good}) \\ &\quad = P(\text{good camera}) \cdot P(\text{good camera}) \cdot \cdots \cdot P(\text{good camera}) \\ &\quad = 0.95 \cdot 0.95 \cdot 0.95 \cdot 0.95 \cdot 0.95 \cdot 0.95 \cdot 0.95 \cdot 0.95 \cdot 0.95 \cdot 0.95 \cdot 0.95 \cdot 0.95 \\ &\quad = 0.95^{12} = 0.540 \end{aligned}$$

Because there is an excellent chance (0.540) of getting 12 good cameras with a defect rate of 5%, we do not have sufficient evidence to conclude that the new method is better. The new method might be better, but this evidence doesn't justify such a conclusion.

We can summarize the fundamentals of the addition and multiplication rules as follows.

- In the addition rule, the word "or" in $P(A \text{ or } B)$ suggests addition. Add $P(A)$ and $P(B)$, being careful to add in such a way that every outcome is counted only once.
- In the multiplication rule, the word "and" in $P(A \text{ and } B)$ suggests multiplication. Multiply $P(A)$ and $P(B)$, but be sure that the probability of event B takes into account the previous occurrence of event A.

3-4 Basic Skills and Concepts

Identifying Events as Independent or Dependent. *In Exercises 1 and 2, for each given pair of events, classify the two events as independent or dependent. Some of the other exercises are based on concepts from earlier sections of this chapter.*

1. a. Flipping a coin and getting heads
Flipping a coin a second time and getting heads

b. Speeding while driving to class
Getting a traffic ticket while driving to class

c. Finding that your car will not start
Finding that your kitchen light will not work

2. a. Finding that your kitchen light is not working
Finding that your refrigerator is not working

b. Drinking until your driving ability is impaired
Being involved in a car crash

c. Testing positive for a virus infection
Being left-handed

3. Applying the Multiplication Rule A box contains glass lenses used for traffic signals. Five of the lenses are red, four are yellow, and three are green. If two of the lenses are randomly selected from the box, find the probability that they are both yellow.
   a. Assume that the first lens is replaced before the second lens is selected.
   b. Assume that the first lens is not replaced before the second lens is selected.

4. Applying the Multiplication Rule Using the box of lenses from Exercise 3, find the probability of randomly selecting three lenses and getting red on the first selection, green on the second, and yellow on the third.
   a. Assume that each lens is replaced before the next lens is selected.
   b. Assume that none of the selected lenses is replaced before the others are selected.

5. Probability of Two Women Jurors A pool of potential jurors consists of 10 men and 15 women. If two *different* people are randomly selected from this pool, find the probability that they are both women.

6. Coin and Die Find the probability of getting the outcome of heads and 5 when a coin is tossed and a single die is rolled.

7. Probability of All Wrong Guesses A quick quiz consists of three multiple-choice questions, each with five possible answers, only one of which is correct. If you make random guesses for each answer, what is the probability that all three of your answers are wrong?

8. Selecting U.S. Senators In the 105th Congress, the Senate consists of 9 women and 91 men. If a lobbyist for the tobacco industry randomly selects three different Senators, what is the probability that they are all men?

9. Coincidental Birthdays
   a. What is the probability that two randomly selected people are both born on July 4? (Ignore leap years.)
   b. What is the probability that two randomly selected people have the same birthday? (Ignore leap years.)

10. Coincidental Birthdays
    a. One couple attracted media attention when their three children, born in different years, were all born on July 4. Ignoring leap years, find the probability that three randomly selected people were all born on July 4.
    b. Ignoring leap years, find the probability that three randomly selected people all have the same birthday.

11. Acceptance Sampling With one method of a procedure called *acceptance sampling,* a sample of items is randomly selected without replacement and the entire batch is

accepted if every item in the sample is OK. The Niko Electronics Company has just manufactured 5000 CDs, and 3% are defective. If 12 of these CDs are randomly selected for testing, what is the probability that the entire batch will be accepted?

12. Poll Confidence Level It is common for public opinion polls to have a "confidence level" of 95%, meaning that there is a 0.95 probability that the poll results are accurate within the claimed margins of error. If five different organizations conduct independent polls, what is the probability that all five of them are accurate within the claimed margins of error?

13. Testing Effectiveness of Gender-Selection Method Recent developments appear to make it possible for couples to dramatically increase the likelihood that they will conceive a child with the gender of their choice. In a test of a gender-selection method, 10 couples try to have baby girls. If this gender-selection method has no effect, what is the probability that the 10 babies will be all girls? If there are actually 10 girls among 10 children, does this gender-selection method appear to be effective? Why?

14. Selecting U.S. Senators In the 105th Congress, the Senate consists of 55 Republicans and 45 Democrats. If a committee is formed by randomly selecting four different Senators, what is the probability that they are all Republicans?

15. Selecting Left-Handed People Ten percent of us are left-handed.

a. What is the probability of randomly selecting three people who are all left-handed?

b. If a survey of the 20 students in a statistics class shows that all of them are left-handed, what do you conclude? Why?

16. Nuclear Reactor Reliability Remote sensors are used to control each of two separate and independent valves, denoted by p and q, that open to provide water for emergency cooling of a nuclear reactor. Each valve has a 0.95 probability of opening when triggered. For the given configuration, find the probability that when both sensors are triggered, water will get through the system so that cooling can occur.

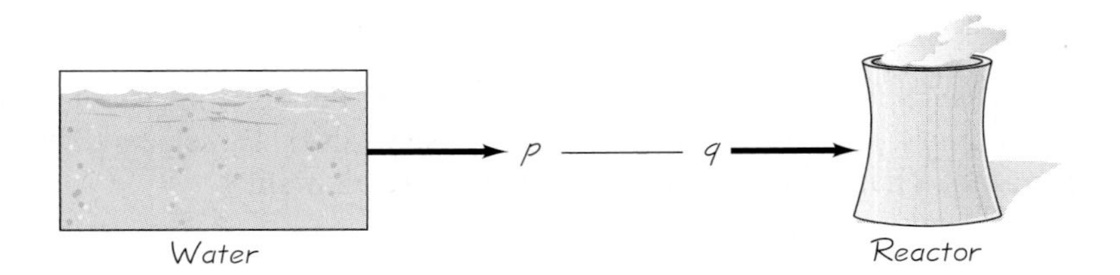

17. Flat Tire Excuse A classic excuse for a missed test is offered by four students who claim that their car had a flat tire. On the makeup test, the instructor asks the students to identify the particular tire that went flat. If they really didn't have a flat tire and randomly select one that supposedly went flat, what is the probability that they will all select the same tire?

18. Voice Identification of Criminal In a Riverhead, New York, case, nine different crime victims listened to voice recordings of five different men. All nine victims identified the same voice as that of the criminal. If the voice identifications were made by random guesses, find the probability that all nine victims would select the same person. Does this constitute reasonable doubt?

19. Quality Control: Is the New Process Better? A quality-control manager claims that a new process for manufacturing cell phones is better because the rate of defects is lower than 4%, which had been the rate of defects in the past. When 25 cell phones are manufactured with the new process, there are no defects. Assuming that the new method has the same 4% defect rate as in the past, find the probability of getting no defects among the 25 cell phones. Based on the result, is there strong evidence to conclude that the new process is better?

20. Coincidences: Give Me a Brake The Rollins Trucking Company has a fleet of 16 trucks used to transport produce from Kansas. When 4 of the trucks are inspected at a police safety checkpoint, it is found that they all have defective brakes. The owner of the company, Ray Rollins, claims that the other 12 trucks have good brakes, and it is just a coincidence that police happened to select the faulty trucks. Assuming that Ray Rollins' claim is true, find the probability of randomly selecting 4 of his trucks and getting the 4 with defective brakes. Based on the result, does his claim seem plausible?

Accuracy of HIV Test. *Based on data from the New York State Health Department, there is a 0.3% HIV rate for the general population. Under certain conditions, a preliminary screening test for HIV is correct 95% of the time. (Subjects are not told that they are HIV infected until later tests verify the preliminary results.) If we randomly select 20,000 people from the general population, we expect results as summarized in the accompanying table. Use this table in Exercises 21–24.*

Sample from the General Population: Preliminary Screening Test

	Positive	Negative
HIV virus infected	57	3
Not HIV infected	997	18,943

21. HIV Positive If two different people are randomly selected, find the probability that they both test positive for HIV.

22. HIV If someone is randomly selected, find the probability of getting someone who tests positive or someone who is HIV infected.

23. HIV If two different people are randomly selected, find the probability that they are both HIV infected.

24. HIV Negative If three different people are randomly selected, find the probability that they all test negative.

3-4 Beyond the Basics

25. Same Birthdays Find the probability that no two people have the same birthday when the number of randomly selected people is

a. 3 **b.** 5 **c.** 25

26. Gender of Children If a couple plans to have eight children, find the probability that they are all of the same gender.

27. Drawing Cards Two cards are to be randomly selected without replacement from a shuffled deck. Find the probability of getting an ace on the first card and a spade on the second card.

28. Complements and the Addition Rule

a. Develop a formula for the probability of not getting either A or B on a single trial. That is, find an expression for $P(\overline{A \text{ or } B})$.

b. Develop a formula for the probability of not getting A or not getting B on a single trial. That is, find an expression for $P(\overline{A} \text{ or } \overline{B})$.

c. Compare the results from parts (a) and (b). Does $P(\overline{A \text{ or } B}) = P(\overline{A} \text{ or } \overline{B})$?

3-5 Multiplication Rule: Complements and Conditional Probability

In this section we continue our discussion of the multiplication rule and show how the basic rule can be used in two other applications. We begin with the use of complements of events as a way to simplify certain calculations.

Complements: The Probability of "At Least One"

The multiplication rule and the rule of complements can be used together to greatly simplify certain types of calculations, such as those in which we want to find the probability that among several trials, *at least one* will result in some specified outcome. In such cases, the meaning of the language must be clearly understood:

- "At least one" is equivalent to "one or more."
- The complement of getting at least one item of a particular type is that you get no items of that type.

Suppose a couple plans to have three children and they want to know the probability of getting at least one girl. See the following interpretations.

At least 1 girl among 3 = 1 or more girls.
The complement of "at least 1 girl" = no girls = all 3 children are boys.

We could easily find the probability from a list of the entire sample space of eight outcomes, but we want to illustrate the use of complements, which can be used in many other problems that cannot be solved so easily.

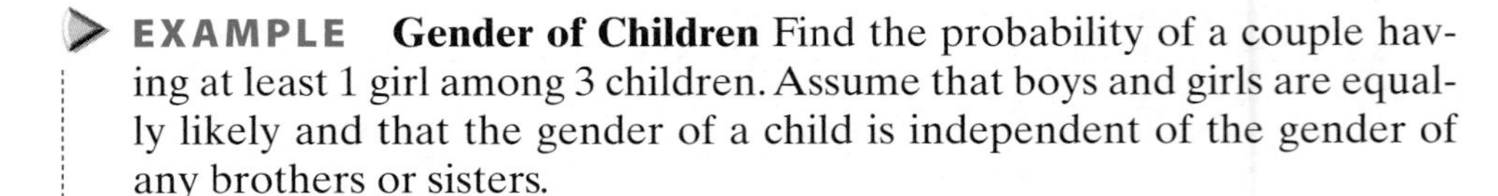

EXAMPLE **Gender of Children** Find the probability of a couple having at least 1 girl among 3 children. Assume that boys and girls are equally likely and that the gender of a child is independent of the gender of any brothers or sisters.

SOLUTION

Step 1: Use a symbol to represent the event desired. In this case, let A = at least 1 of the 3 children is a girl.

Step 2: Identify the event that is the complement of A.

$$\overline{A} = \textit{not} \text{ getting at least 1 girl among 3 children}$$
$$= \text{all 3 children are boys}$$
$$= \text{boy and boy and boy}$$

Step 3: Find the probability of the complement.

$$P(\overline{A}) = P(\text{boy and boy and boy})$$
$$= \frac{1}{2} \cdot \frac{1}{2} \cdot \frac{1}{2} = \frac{1}{8}$$

Step 4: Find $P(A)$ by evaluating $1 - P(\overline{A})$.

$$P(A) = 1 - P(\overline{A}) = 1 - \frac{1}{8} = \frac{7}{8}$$

INTERPRETATION There is a 7/8 probability that if a couple has 3 children, at least 1 of them is a girl.

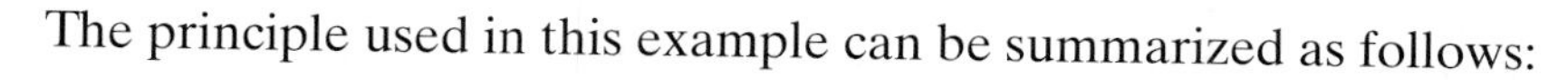

The principle used in this example can be summarized as follows:

To find the probability of *at least one* of something, calculate the probability of *none*, then subtract that result from 1.

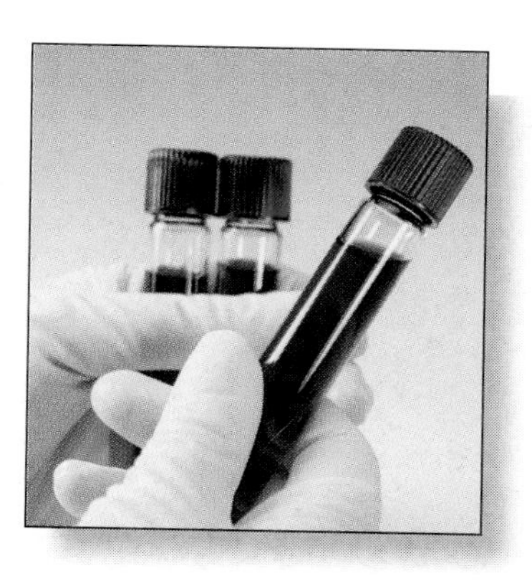

Composite Sampling

The U.S. Army once tested for syphilis by giving each inductee an individual blood test that was analyzed separately. One researcher suggested mixing pairs of blood samples. After the mixed pairs were tested, syphilitic inductees could be identified by retesting the few blood samples that were in the pairs that tested positive. The total number of analyses was reduced by pairing blood specimens, so why not put them in groups of three or four or more? Probability theory was used to find the most efficient group size, and a general theory was developed for detecting the defects in any population. This technique is known as *composite sampling*.

Conditional Probability

The probability of an event is often affected by knowledge of circumstances. For example, if you randomly select someone from the general population, the probability of getting a male is 0.5, but if you know that the selected person frequently changes TV channels with a remote control, the probability is 0.999 (OK, that might be a slight exaggeration). A *conditional probability* of an event occurs when the probability is affected by the knowledge of other circumstances. The conditional probability of event B occurring, given that event A has already occurred, can be found by using the multiplication rule $[P(A \text{ and } B) = P(A) \cdot P(B|A)]$ and solving for $P(B|A)$ by dividing both sides of the equation by $P(A)$.

DEFINITION

The **conditional probability** of event B occurring, given that event A has already occurred, can be found by dividing the probability of events A and B both occurring by the probability of event A:

$$P(B|A) = \frac{P(A \text{ and } B)}{P(A)}$$

Bayes' Theorem

Thomas Bayes (1702–1761) said that probabilities should be revised when we learn more about an event. Here's one form of Bayes' theorem:

$$P(A|B) = \frac{P(A) \cdot P(B|A)}{P(A) \cdot P(B|A) + P(\bar{A}) \cdot P(B|\bar{A})}$$

Suppose 60% of a company's computer chips are made in one factory (denoted by A) and 40% are made in its other factory (denoted by $\bar{A}$). For a randomly selected chip, the probability it came from factory A is 0.60. Suppose we learn that the chip is defective and the defect rates for the two factories are 35% (for A) and 25% (for $\bar{A}$). We can use the above formula to find that there is a 0.677 probability the defective chip came from factory A.

This formula is a formal expression of conditional probability, but we can often use this intuitive approach:

Intuitive Approach to Conditional Probability

The conditional probability of B given A can be found by assuming that event A has occurred and, working under that assumption, calculating the probability that event B will occur.

EXAMPLE ***Titanic* Passengers** Refer to Table 3-1, reproduced here. Find the following:

a. If 1 of the 2223 people is randomly selected, what is the probability that this person survived, given that the selected person is a man?

b. If 1 of the 2223 people is randomly selected, what is the probability of getting a man, given that the selected person survived?

SOLUTION **a.** We want $P(\text{survived}|\text{man})$, the probability of getting someone who survived, *given that the selected person is a man.* If we assume that the selected person is a man, we are dealing with the 1692 people in the first column of Table 3-1. Among those 1692 people, 332 survived, so

$$P(\text{survived}|\text{man}) = \frac{332}{1692} = 0.196$$

The same result can be found by using the formula given with the definition of conditional probability:

$$P(\text{survived}|\text{man}) = \frac{P(\text{man and survived})}{P(\text{man})} = \frac{332/2223}{1692/2223} = 0.196$$

b. Here we want $P(\text{man}|\text{survived})$. If we assume that the person selected is a survivor, we are dealing with the 706 people in the first row of Table 3-1. Among those 706 people, 332 are men, so

$$P(\text{man}|\text{survived}) = \frac{332}{706} = 0.470$$

TABLE 3-1 *Titanic* Mortality

	Men	Women	Boys	Girls	**Total**
Survived	332	318	29	27	**706**
Died	1360	104	35	18	**1517**
Total	**1692**	**422**	**64**	**45**	**2223**

Again, the same result can be found by applying the formula for conditional probability:

$$P(\text{man}|\text{survived}) = \frac{P(\text{survived and man})}{P(\text{survived})}$$
$$= \frac{332/2223}{706/2223} = 0.470$$

By comparing the results from parts (a) and (b), we see that $P(\text{survived}|\text{man})$ is very different from $P(\text{man}|\text{survived})$.

Does the probability of surviving the sinking of the *Titanic depend* on whether the person is a man, woman, or child? Let's list the relevant probabilities:

$$P(\text{survived}|\text{man}) = 0.196$$
$$P(\text{survived}|\text{woman or boy or girl}) = 0.704$$

INTERPRETATION The rule of "women and children first" was not strictly followed on the *Titanic,* because some women and children died while some men survived. But the probabilities we just calculated show that if you were on the *Titanic,* you had a substantially better chance of surviving if you were a woman or child.

The calculation of conditional probabilities is not easy, but using the intuitive approach helps considerably. The next example would be quite difficult if a solution were attempted using the formal rules of probability, but if we carefully read the question and use the intuitive approach, the solution becomes much easier.

EXAMPLE ***Titanic* Passengers** Refer to the *Titanic* mortality data in Table 3-1 and find the probability of getting a woman or child if a *Titanic* survivor is randomly selected.

SOLUTION The question asks us to find $P(\text{woman or child}|\text{survivor})$. If we assume that a survivor is selected, we can restrict our sample space to the first row of 706 survivors. Among the 706 survivors, the number that are women or children is $318 + 29 + 27 = 374$. $P(\text{woman or child}|\text{survivor})$ is therefore $374/706 = 0.530$. Just try combining the formal addition rule and the formula for conditional probability to arrive at the same solution, and you will see the benefit of using this intuitive approach.

Sometimes the information at hand is confusing in its present format, but it can be simplified by constructing a table similar to Table 3-1. The next example shows the usefulness of such a table.

Coincidences?

John Adams and Thomas Jefferson (the second and third presidents) both died on July 4, 1826. President Lincoln was assassinated in Ford's Theater; President Kennedy was assassinated in a Lincoln car made by the Ford Motor Company. Lincoln and Kennedy were both succeeded by vice presidents named Johnson. Fourteen years *before* the sinking of the Titanic, a novel described the sinking of the Titan, a ship that hit an iceberg; see Martin Gardner's *The Wreck of the Titanic Foretold?* Gardner states, "In most cases of startling coincidences, it is impossible to make even a rough estimate of their probability."

EXAMPLE **HIV** The New York State Health Department reports a 10% rate of the HIV virus for the "at-risk" population. Under certain conditions, a preliminary screening test for the HIV virus is correct 95% of the time. (Subjects are not told that they are HIV infected until additional tests verify the initial screening results.) If someone is randomly selected from the at-risk population, what is the probability that they have the HIV virus if it is known that they have tested positive in the initial screening?

SOLUTION There are enough numbers floating around here to confuse the most able of us. Let's begin by imposing some order. Let's assume that the at-risk population consists of 5000 people, so we can now work with concrete numbers, and we can deduce the following.

1. Because there is a 10% rate of HIV, we expect to have 500 infected people (10% of 5000) and 4500 people (90% of 5000) not infected.
2. For the 500 infected people, the preliminary screening test will be correct 95% of the time, so 475 of them (95% of 500) will test positive and 25 of them (5% of 500) will test negative.
3. For the 4500 people not infected, the preliminary screening test will be correct 95% of the time, so 4275 of them (95% of 4500) will test negative and 225 will test positive.

These results can now be summarized in Table 3-2.

We can now address the key question: If someone from the at-risk population is randomly selected, what is the probability that this person has the HIV virus if it is known that he or she has tested positive in the preliminary screening? That is, find $P(\text{HIV}|\text{positive preliminary test})$. Table 3-2 shows that there are $475 + 225 = 700$ people who tested positive. Among those 700 people, 475 have the virus, so

$$P(\text{HIV}|\text{positive preliminary test}) = 475/700 = 0.679.$$

TABLE 3-2 Results from the At-Risk Population: Preliminary Screening

	Positive	Negative
HIV infected	475	25
Not HIV infected	225	4275

Testing for Independence

In the multiplication rule for dependent events, if $P(B|A) = P(B)$, then the occurrence of event A has no effect on the probability of event B. This is often used as a test for independence. If $P(B|A) = P(B)$, then A and B are independent events; however, if $P(B|A) \neq P(B)$, then A and B are dependent events. Another test for independence involves checking for the equality of $P(A \text{ and } B)$ and $P(A) \cdot P(B)$. If they are equal, events A and B

are independent. If $P(A \text{ and } B) \neq P(A) \cdot P(B)$, then A and B are dependent events. These results are summarized as follows.

Two events A and B are *independent* if	Two events A and B are *dependent* if
$P(B\|A) = P(B)$	$P(B\|A) \neq P(B)$
or	or
$P(A \text{ and } B) = P(A) \cdot P(B)$	$P(A \text{ and } B) \neq P(A) \cdot P(B)$

3-5 Basic Skills and Concepts

Describing Complements. *In Exercises 1–4, provide a written description of the complement of the given event.*

1. Gender of Children When a couple has three children, at least one of them is a girl.

2. Freshness When 12 apples are tested for freshness, all of them pass the test.

3. Statistics Grades In a statistics class, all students receive a grade of A.

4. Missile Hits When five missiles are fired, at least one of them strikes its target.

5. Probability of At Least One Girl If a couple plans to have three children, what is the probability that they will have at least one girl?

6. Probability of At Least One Girl If a couple plans to have 10 children (it could happen), what is the probability that there will be at least one girl?

7. At Least One 6 If you roll a die five times, what is the probability of getting at least one 6?

8. At Least One Correct Answer If you make random guesses for three multiple-choice test questions (each with five possible answers), what is the probability of getting at least one correct?

9. Tossing Coins Find the probability of getting heads on the third toss of a coin, given that the first two tosses resulted in heads.

10. Coin and Die A procedure consists of flipping a coin, then rolling a die. What is the probability of getting a 5, given that the coin flip resulted in heads?

11. Subjective Probability Use subjective probability to estimate the probability of randomly selecting an adult and getting a woman, given that the selected person is known to have given birth to a baby.

12. Subjective Probability Use subjective probability to estimate the probability of randomly selecting an adult and getting a male, given that the selected person owns a motorcycle.

13. Redundancy in Alarm Clocks A student experiences difficulties with malfunctioning alarm clocks. Instead of using one alarm clock, he decides to use three. What is the probability that at least one of his alarm clocks works correctly if each individual alarm clock has a 99% chance of working correctly?

14. Acceptance Sampling With one method of the procedure called *acceptance sampling,* a sample of items is randomly selected without replacement, and the entire batch is rejected if there is at least one defect. The Niko Electronics

Company has just manufactured 5000 CDs, and 3% are defective. If 10 of the CDs are selected and tested, what is the probability that the entire batch will be rejected?

15. Using Composite Blood Samples When doing blood testing for HIV infections, the procedure can be made more efficient and less expensive by combining samples of blood specimens. If samples from three people are combined and the mixture tests negative, we know that all three individual samples are negative. Find the probability of a positive result for three samples combined into one mixture, assuming the probability of an individual blood sample testing positive is 0.1 (the probability for the "at-risk" population, based on data from the New York State Health Department).

16. Using Composite Water Samples The Home Testing Company tests water quality of homes that are in the process of being sold. To reduce laboratory costs, water samples from five homes are combined for one test, and further testing is done only if the combined sample fails. Based on past records, 4% of the homes fail. Find the probability that a combined sample from five homes will fail.

Conditional Probabilities. *In Exercises 17–20, refer to the Titanic mortality data in Table 3-1.*

17. If we randomly select someone who was aboard the *Titanic,* what is the probability of getting a man, given that the selected person died?

18. If we randomly select someone who died, what is the probability of getting a man?

19. What is the probability of getting a boy or girl, given that the randomly selected person is someone who survived?

20. What is the probability of getting a man or woman, given that the randomly selected person is someone who died?

3-5 Beyond the Basics

21. Determining Effect of Assumed Population Size Table 3-2 was constructed by assuming that the population size is 5000. (Recall that the table is based on a 10% rate of the HIV virus for the "at-risk" population, and that the preliminary screening test for the HIV virus is currently correct 95% of the time.) Change the assumed population size from 5000 to 15,000.

a. How are the entries in Table 3-2 affected?

b. Using this new table based on a population size of 15,000, find $P(\text{HIV}|\text{positive})$, and compare the result to the one found in the example of this section.

c. Does the assumed size of the population affect the probability values?

22. Using a Two-Way Table The New York State Health Department reports a 0.3% HIV rate for the general population, and under certain conditions, preliminary screening tests for the HIV virus are correct 95% of the time. Assume that the general population consists of 100,000 people.

a. Construct a table for the general population that is similar to Table 3-2 for the "at-risk" population. (*Hint:* See this section for the procedure used to construct Table 3-2.)

b. Using the table from part (a), find $P(\text{HIV}|\text{positive})$ for someone randomly selected from the general population. That is, find the probability of randomly selecting someone with HIV, given that this person tested positive.

23. Shared Birthdays Find the probability that of 25 randomly selected people,
   a. no 2 share the same birthday.
   b. at least 2 share the same birthday.

24. Unseen Coins A statistics professor tosses two coins that cannot be seen by any students. One student asks if one of the coins turned up heads. Given that the professor's response is "yes," find the probability that both coins turned up heads.

3-6 Probabilities Through Simulations

Finding probabilities of events sometimes seems very difficult. Sometimes a probability problem is important for us to solve, but we are not sure that our theoretical solution is correct. The following alternatives are often useful.

1. We can conduct actual trials or collect real data to find probability values or verify that our theoretical results are correct.
2. We can *simulate* the circumstances to learn how the results would occur in reality.

The first alternative is simply an application of the relative frequency approach to finding probabilities, as discussed in Section 3-2. For example, if we want the probability of rolling a pair of dice and getting a total of 7, we could actually roll a real pair of dice 100 times and record the number of 7s that occur. Very often, however, it is not practical to conduct actual trials, so we try instead to develop a *simulation*.

DEFINITION

A **simulation** of a procedure is a process that behaves the same way as the procedure, so that similar results are produced.

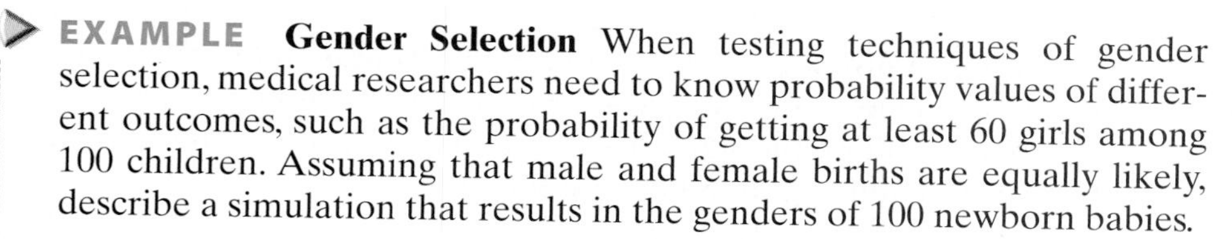

EXAMPLE Gender Selection When testing techniques of gender selection, medical researchers need to know probability values of different outcomes, such as the probability of getting at least 60 girls among 100 children. Assuming that male and female births are equally likely, describe a simulation that results in the genders of 100 newborn babies.

SOLUTION One approach is simply to flip a coin 100 times, with heads representing females and tails representing males. Another approach is to use a calculator or computer to randomly generate 0s and 1s, with 0 representing a male and 1 representing a female. The numbers must be generated in such a way that they are equally likely.

Monkey Typists

A classical claim is that a monkey randomly hitting a keyboard would eventually produce the complete works of Shakespeare, assuming that it continues to type century after century. The multiplication rule for probability has been used to find such estimates. One result of 1,000,000,000,000,000,000,000,000,000,000,000,000 years is considered by some to be too short. In the same spirit, Sir Arthur Eddington wrote this poem: "There once was a brainy baboon, who always breathed down a bassoon. For he said, 'It appears that in billions of years, I shall certainly hit on a tune.'"

To Win, Bet Boldly

The *New York Times* published an article by Andrew Pollack in which he reported lower than expected earnings for the Mirage casino in Las Vegas. He wrote that "winnings for Mirage can be particularly volatile, because it caters to high rollers, gamblers who might bet $100,000 or more on a hand of cards. The law of averages does not work as consistently for a few large bets as it does for thousands of smaller ones..." This reflects the most fundamental principle of gambling: To win, place one big bet instead of many small bets! With the right game, such as craps, you have just under a 50% chance of doubling your money if you place one big bet. With many small bets, your chance of doubling your money drops substantially.

EXAMPLE **Same Birthdays** Exercise 23 in Section 3-5 refers to the birthday problem, in which we find the probability that in a randomly selected group of 25 people, at least 2 share the same birthday. The theoretical solution is difficult. It isn't practical to survey many different groups of 25 people, so we develop a simulation instead.

SOLUTION Begin by representing birthdays by integers from 1 through 365, where 1 = January 1, 2 = January 2, . . . , 365 = December 31. Then use a calculator or computer program to generate 25 random numbers, each between 1 and 365. Those numbers can then be sorted, so it becomes easy to survey the list to determine whether any 2 of the simulated birth dates are the same. We can repeat the process as many times as we like, until we are satisfied that we have a good basis for determining the probability. Our estimate of the probability is the number of times we did get at least 2 birth dates that are the same, divided by the total number of groups of 25 that were generated.

There are several ways of obtaining randomly generated numbers from 1 through 365, including the following.

- **A table of random digits:** Refer, for example, to the *CRC Standard Probability and Statistics Tables and Formulae,* which contains a table of 14,000 digits. (In such a table there are many ways to extract numbers from 1 through 365. One way is by referring to the digits in the first three columns and ignoring 000 as well as anything above 365.)
- **STATDISK:** Select **Data** from the main menu bar, then select **Uniform Generator** and proceed to enter a sample size of 25, a minimum of 1, and a maximum of 365; enter 0 for the number of decimal places. The resulting STATDISK display is shown below. Using **copy/paste,** copy the data set to the **Sample Editor,** where the values can be arranged in increasing order. From the STATDISK display, we see that the first two people have the same birth date, which is the 78th day of the year.
- **Minitab:** Select **Calc** from the main menu bar, then select **Random Data,** and next select **Integer.** In the dialog box, enter 25 for the number of rows, store the results in column C1, and enter a minimum of 1 and a maximum of 365. You can then use **Manip** and **Sort** to arrange the data in increasing order. The result will be as shown below, but the numbers won't be the same. This Minitab result of 25 numbers shows that the 9th and 10th numbers are the same.
- **Excel:** Click on the cell in the upper left corner, then click on the function icon **fx**. Select **Math & Trig,** then select **RANDBETWEEN.** In the dialog box, enter 1 for bottom, and enter 365 for top. After getting the random number in the first cell, click and hold down the mouse button to drag the lower right corner of this first cell, and pull it down the column until 25 cells are highlighted. When you release the mouse button, all 25 random numbers should be present. This display shows that the 1st and 3rd numbers are the same.

STATDISK

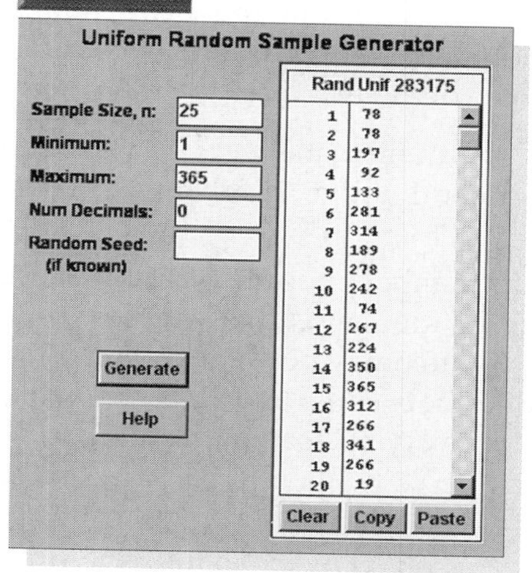

MINITAB

↓	C1	C2
1	38	
2	48	
3	59	
4	71	
5	101	
6	107	
7	122	
8	129	
9	153	
10	153	
11	163	

EXCEL

	A
1	15
2	3
3	15
4	362
5	164
6	184
7	158
8	59
9	143
10	85
11	134

TI-83 Plus

```
randInt(1,365,25
→L1
{79 206 340 133...
SortA(L1)
                Done
L1
{17 34 46 70 79...
```

The Random Secretary

One classic problem of probability goes like this: A secretary addresses 50 different letters and envelopes to 50 different people, but the letters are randomly mixed before being put into envelopes. What is the probability that at least one letter gets into the correct envelope? Although the probability might seem like it should be small, it's actually 0.632. Even with a million letters and a million envelopes, the probability is 0.632. The solution is beyond the scope of this text—way beyond.

- **TI-83 Plus calculator:** Press the **MATH** key, select **PRB,** then choose **randInt(** and proceed to enter the minimum of 1, the maximum of 365, and 25 for the number of scores. See the TI-83 Plus screen display, which shows that we used **randInt** to generate the numbers, which were then stored in list L1, where they were sorted and displayed. This display shows that there are no matching numbers among the first few that can be seen. You can press **STAT** and select **Edit** to see the whole list of generated numbers.

It is extremely important to construct a simulation so that it behaves just like the real procedure. In the next example we demonstrate the right way and the wrong way to construct a simulation.

➤ **EXAMPLE** **Simulating Dice** Describe a procedure for simulating the rolling of a pair of dice.

SOLUTION In the procedure of rolling a pair of dice, each of the two dice yields a number between 1 and 6 (inclusive), and those two numbers are then added. Any simulation should do the same thing. There is a right way and a wrong way to simulate rolling two dice.

The right way: Randomly generate one number between 1 and 6, randomly generate another number between 1 and 6, and then add the two results.

The wrong way: Randomly generate numbers between 2 and 12. This procedure is similar to rolling dice in the sense that the results are always between 2 and 12, but these outcomes between 2 and 12 are equally likely. With real dice, the values between 2 and 12 are *not* equally likely. This simulation would produce very misleading results.

Some probability problems can only be solved by estimating the probability from actual observations or constructing a simulation. The widespread availability of powerful calculators and computers has greatly facilitated the use of simulation methods.

3-6 Basic Skills and Concepts

46196
99438
72113
44044
86763
00151
64703
78907
19155
67640
98746
29910
82855
25259
14752
85446
75260
92532
87333
55848

1. Simulating Families of Four Children The margin contains a brief list of randomly selected digits. Use these random numbers to develop a simulation for finding the probability of getting exactly three girls in a family of four children. Describe the simulation, then estimate the probability based on its results. (*Hint:* Let the odd digits represent girls.)

2. Simulating Families of Five Children Use the random digits in the margin for developing a simulation for finding the probability of getting at least two girls in a family of five children. Describe the simulation, then estimate the probability based on its results.

3. Simulating Three Dice Use the random digits in the margin to develop a simulation for rolling three dice. Describe the simulation, then use it to estimate the probability of getting a total of 10 when three dice are rolled.

4. Simulating Left-Handedness Ten percent of us are left-handed. In a study of dexterity, people are randomly selected in groups of five. Use the random digits in the margin to develop a simulation for finding the probability of getting at least one left-handed person in a group of five. (*Hint:* Because 10% of us are left-handed, let the digit 0 represent someone who is left-handed, and let the other digits represent someone who is not left-handed.)

In Exercises 5–8, use a TI-83 Plus calculator, STATDISK, Minitab, Excel, or any other suitable calculator or program.

5. Simulating Families of Four Children In Exercise 1 we used the digits in the margin to estimate the probability of getting exactly three girls in a family of four children. Develop a simulation of 100 families and estimate the probability. Describe the procedure you used and identify the estimated probability value.

6. Simulating Families of Five Children Develop a simulation for finding the probability of getting at least two girls in a family of five children. Simulate 100 families. Describe the simulation, then estimate the probability based on its results.

7. Simulating Three Dice Develop a simulation for rolling three dice. Simulate the rolling of the three dice 100 times. Describe the simulation, then use it to estimate the probability of getting a total of 10 when three dice are rolled.

8. Simulating Left-Handedness Ten percent of us are left-handed. In a study of dexterity, people are randomly selected in groups of five. Develop a simulation for finding the probability of getting at least one left-handed person in a group of five. Simulate 100 groups of five. (See Exercise 4.)

3-6 Beyond the Basics

9. Simulating the Monty Hall Problem A problem that has attracted much attention in recent years is the *Monty Hall problem,* based on the old television game show "Let's Make a Deal," hosted by Monty Hall. Suppose you are a contestant who has selected one of three doors after being told that two of them conceal nothing, but that a new red Corvette is behind one of the three. Next, the host opens one of the doors you didn't select and shows that there is nothing behind it. He then offers you the choice of sticking with your first selection or switching to the other unopened door. Should you stick with your first choice or should you switch? Develop a simulation of this game and determine whether you should stick or switch. (According to *Chance* magazine, business schools at such institutions as Harvard and Stanford use this problem to help students deal with decision making.)

10. Simulating Birthdays
    - **a.** Develop a simulation for finding the probability that when 50 people are randomly selected, at least two of them have the same birth date. Describe the simulation and estimate the probability.
    - **b.** Develop a simulation for finding the probability that when 50 people are randomly selected, at least three of them have the same birth date. Describe the simulation and estimate the probability.

11. Genetics: Simulating Population Control A classic probability problem involves a king who wanted to increase the proportion of women by decreeing that after a mother gives birth to a son, she is prohibited from having any more children. The king reasons that some families will have just one boy, whereas other families will have a few girls and one boy, so the proportion of girls will be increased. Is his reasoning correct? Will the proportion of girls increase?

3-7 Counting

Let's consider a probability problem that is of interest to many people: What is the probability that you will win the lottery? In New York State's lottery, which is typical, you must choose six numbers between 1 and 51 inclusive. If you get the same six-number combination that is randomly drawn by the lottery officials, you win millions of dollars. There are some lesser prizes, but they

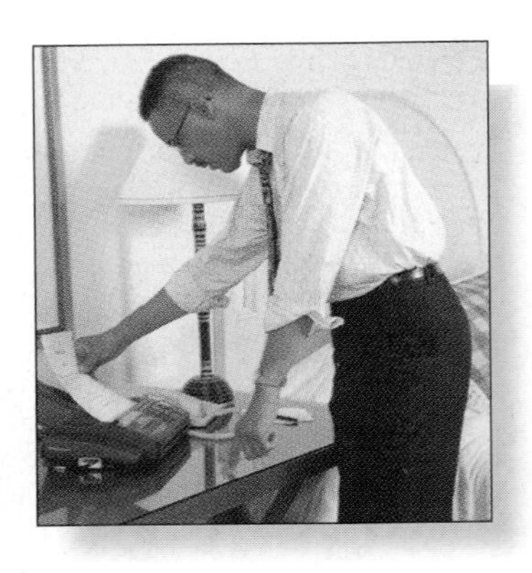

The Number Crunch

Every so often telephone companies split regions with one area code into regions with two or more area codes because the increased number of area fax and internet lines has nearly exhausted the possible numbers that can be listed under a single code. A seven-digit telephone number cannot begin with a 0 or 1, but if we allow all other possibilities, we get $8 \cdot 10 \cdot 10 \cdot 10 \cdot 10 \cdot 10 \cdot 10 = 8{,}000{,}000$ different possible numbers! Even so, after surviving for 80 years with the single area code of 212, New York City was recently partitioned into the two area codes of 212 and 718. Many other regions have also been assigned split area codes. While writing this book, the author's area code also changed.

are relatively insignificant. Using the classical approach to probability (because the outcomes are equally likely), the probability of winning the lottery is found by using $P(\text{win}) = s/n$, where s is the number of ways you can win and n is the total number of possible outcomes. With New York State's lottery $s = 1$ because there is only one way to win the grand prize: Choose the same six-number combination that is drawn in the lottery. Knowing that there is only 1 way to win, we now need to find n, the total number of outcomes; that is, how many six-number combinations are possible when you select numbers from 1 to 51? Writing a list of the possibilities would take about three years of nonstop work, and that's no fun. We could construct a tree diagram, but it would be about 100 miles high and would violate airspace regulations. We need a more practical way of finding the total number of possibilities. This section introduces efficient methods for finding such numbers without directly counting the possibilities. We will return to this lottery problem after we present some basic principles. We begin with the *fundamental counting rule.*

Fundamental Counting Rule

For a sequence of two events in which the first event can occur m ways and the second event can occur n ways, the events together can occur a total of $m \cdot n$ ways.

EXAMPLE **Breaking Codes** You have discovered that all of the gold in Fort Knox is accessible with a two-character code consisting of a letter followed by a digit. Of course, you would never try to actually get any of that gold, but just how many possibilities are there?

SOLUTION There are 26 letters and 10 digits, so the number of different possibilities is $26 \cdot 10 = 260$.

The fundamental counting rule easily extends to situations involving more than two events, as illustrated in the following example.

EXAMPLE **Computer Design** In designing a computer, if a *byte* is defined to be a sequence of 8 bits and each bit must be a 0 or 1, how many different bytes are possible? (A byte is often used to represent an individual character, such as a letter, digit, or punctuation symbol. For example, one coding system represents the letter A as 01000001.)

SOLUTION Because each bit can occur in two ways (0 or 1) and we have a sequence of 8 bits, the total number of different possibilities is given by

$$2 \cdot 2 \cdot 2 \cdot 2 \cdot 2 \cdot 2 \cdot 2 \cdot 2 = 256$$

There are 256 different possible bytes.

EXAMPLE **Chance and Skill** Bob claims that he has the ability to roll a die in such a way that 6 will almost always occur. You test him by giving him a fair die, which he proceeds to roll five times, getting a 6 each time. If Bob has no control over the die, how many outcomes are possible with five rolls of a die? If Bob does get five 6s in five rolls, does it appear that he has control of the die?

SOLUTION There are six possible outcomes with each roll, so the total number of possible outcomes from five rolls is

$$6 \cdot 6 \cdot 6 \cdot 6 \cdot 6 = 7776$$

Only one of those outcomes consists of five 6s, so if Bob does roll a die five times and if he does actually get five 6s, then it would appear that he is able to control the outcome of the die.

How Many Shuffles?

After conducting extensive research, Harvard mathematician Persi Diaconis found that it takes seven shuffles of a deck of cards to get a complete mixture. The mixture is complete in the sense that all possible arrangements are equally likely. More than seven shuffles will not have a significant effect, and fewer than seven are not enough. Casino dealers rarely shuffle as often as seven times, so the decks are not completely mixed. Some expert card players have been able to take advantage of the incomplete mixtures that result from fewer than seven shuffles.

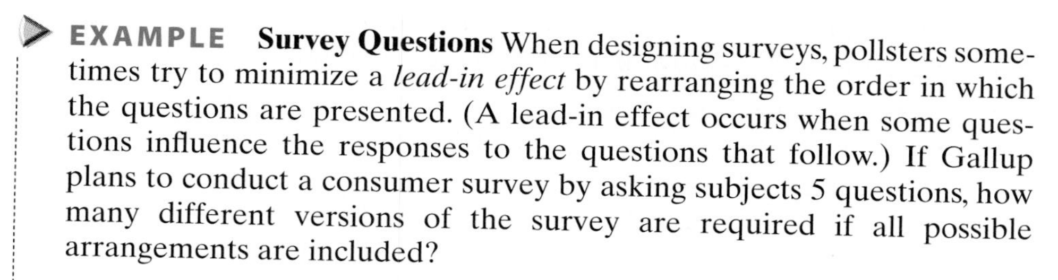

EXAMPLE **Survey Questions** When designing surveys, pollsters sometimes try to minimize a *lead-in effect* by rearranging the order in which the questions are presented. (A lead-in effect occurs when some questions influence the responses to the questions that follow.) If Gallup plans to conduct a consumer survey by asking subjects 5 questions, how many different versions of the survey are required if all possible arrangements are included?

SOLUTION In arranging any individual survey, there are 5 possible choices for the first question, 4 remaining choices for the second question, 3 choices for the third question, 2 choices for the fourth question, and only 1 choice for the fifth question. The total number of possible arrangements is therefore

$$5 \cdot 4 \cdot 3 \cdot 2 \cdot 1 = 120$$

That is, Gallup would need 120 versions of the survey in order to include every possible arrangement.

In the preceding example, we found that 5 survey questions can be arranged $5 \cdot 4 \cdot 3 \cdot 2 \cdot 1 = 120$ different ways. This particular solution can be generalized by using the following notation for the symbol ! and the following *factorial rule*.

Notation

The **factorial symbol !** denotes the product of decreasing positive whole numbers. For example, $4! = 4 \cdot 3 \cdot 2 \cdot 1 = 24$. By special definition, $0! = 1$. (Many calculators have a factorial key. On the TI-83 Plus calculator, first enter the number, then press **MATH** and select **PRB** and menu item 4.)

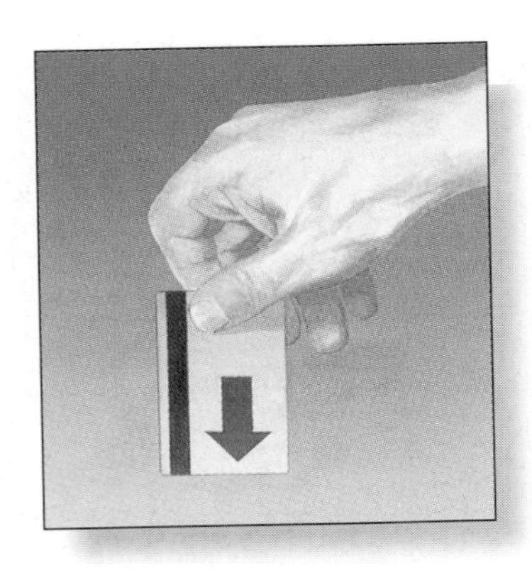

Safety in Numbers

Some hotels have abandoned the traditional room key in favor of an electronic key with a number code. A central computer changes the access code to a room as soon as a guest checks out. A typical electronic key has 32 different positions that are either punched or left untouched. This configuration allows for 2^{32}, or 4,294,967,296, different possible codes, so it is impractical to develop a complete set of keys or try to make an illegal entry by trial and error.

Factorial Rule

A collection of n different items can be arranged in order $n!$ different ways. (This **factorial rule** reflects the fact that the first item may be selected n different ways, the second item may be selected $n - 1$ ways, and so on.)

Routing problems often involve application of the factorial rule. AT&T wants to route telephone calls through the shortest networks. Federal Express wants to find the shortest routes for its deliveries. American Airlines wants to find the shortest route for returning crew members to their homes. See the following example.

EXAMPLE **Air Routes** You have just started your own airline company called Air America (motto: "Where your probability of a safe flight is greater than zero"). You have one plane for a route connecting Austin, Boise, and Chicago. How many routes are possible?

SOLUTION Using the factorial rule, we see that the 3 different cities (Austin, Boise, Chicago) can be arranged in $3! = 6$ different ways. In Figure 3-11 we can see that there are 3 choices for the first city and 2 choices for the second city. This leaves only 1 choice for the third city. The number of possible arrangements for the 3 cities is $3 \cdot 2 \cdot 1 = 6$.

EXAMPLE **Routes to All 50 Capitals** Because of your success in a statistics course, you have been hired by the Gallup Organization, and your first assignment is to conduct a survey in each of the 50 state capitals. As you plan your route of travel, you want to determine the number of different possible routes. How many different routes are possible?

SOLUTION By applying the factorial rule, we know that 50 items can be arranged in order 50! different ways. That is, the 50 state capitals can be arranged 50! ways, so the number of different routes is 50!, or

30,414,093,201,713,378,043,612,608,166,064,768,844,377,641,568,960,512,000,000,000,000

Now there's a large number.

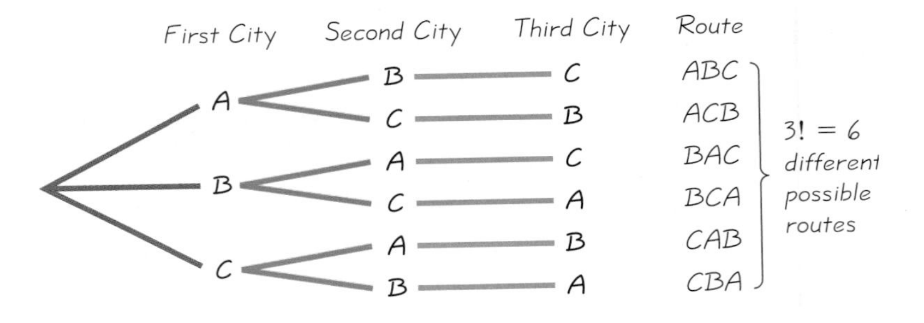

FIGURE 3-11
Tree Diagram of Routes

The preceding example is a variation of a classical problem called the *traveling salesman problem.* It is especially interesting because the large number of possibilities means that we can't use a computer to calculate the distance of each route. The time it would take even the fastest computer to calculate the shortest possible route is about

1,000,000,000,000,000,000,000,000,000,000,000,000,000,000 centuries

Considerable effort is currently being devoted to finding efficient ways of solving such problems.

According to the factorial rule, n different items can be arranged $n!$ different ways. Sometimes we have n different items, but we need to select *some* of them instead of all of them. If we must conduct surveys in state capitals, as in the preceding example, but we have time to visit only four capitals, the number of different possible routes is $50 \cdot 49 \cdot 48 \cdot 47 = 5{,}527{,}200$. Another way to obtain this same result is to evaluate

$$\frac{50!}{46!} = 50 \cdot 49 \cdot 48 \cdot 47 = 5{,}527{,}200$$

In this calculation, note that the factors in the numerator divide out with the factors in the denominator, except for the factors of 50, 49, 48, and 47 that remain. We can generalize this result by noting that if we have n different items available and we want to select r of them, the number of different arrangements possible is $n!/(n-r)!$, as in $50!/46!$. This generalization is commonly called the *permutations rule.*

Permutations Rule (When Items Are All Different)

The number of **permutations** (or sequences) of r items selected from n available items (without replacement) is

$$_nP_r = \frac{n!}{(n-r)!}$$

Many calculators can evaluate expressions of $_nP_r$. With the TI-83 Plus calculator, enter the value for n, then press **MATH, PRB, $_n$P$_r$,** then enter the value for r, and press the **ENTER** key.

It is very important to recognize that the permutations rule requires the following conditions:

- We must have a total of n *different* items available. (This rule does not apply if some of the items are identical to others.)
- We must select r of the n items (without replacement).
- We must consider rearrangements of the same items to be different sequences.

When we use the term *permutations, arrangements,* or *sequences,* we imply that *order is taken into account* in the sense that different orderings of the same items are counted separately. The letters ABC can be arranged six

Choosing Personal Security Codes

All of us use personal security codes for ATM machines, computer internet accounts, and home security systems. The safety of such codes depends on the large number of different possibilities, but hackers now have sophisticated tools that can largely overcome that obstacle. Researchers found that by using variations of the user's first and last names along with 1800 other first names, they could identify 10%-20% of the passwords on typical computer systems. When choosing a password, *do not* use a variation of any name, a word found in a dictionary, a password shorter than seven characters, telephone numbers, or social security numbers. Do include nonalphabetic characters, such as digits or punctuation marks.

different ways: *ABC, ACB, BAC, BCA, CAB, CBA*. (Later, we will refer to *combinations,* which do not count such arrangements separately.) In the following example, we are asked to find the total number of different sequences that are possible. That suggests use of the permutations rule.

EXAMPLE Singing legend Frank Sinatra recorded 381 songs. From a list of his top-10 songs, you must select 3 that will be sung in a medley as a tribute at the next MTV Music Awards ceremony. The order of the songs is important so that they fit together well. If you select 3 of Sinatra's top-10 songs, how many different sequences are possible?

SOLUTION We need to select $r = 3$ songs from $n = 10$ songs on Sinatra's top-10 list. The number of different arrangements is found as shown:

$$_nP_r = \frac{n!}{(n-r)!} = \frac{10!}{(10-3)!} = 720$$

There are 720 different possible arrangements of 3 songs selected from 10.

We sometimes need to find the number of permutations, but some of the items are identical to others. The following variation of the permutations rule applies to such cases.

Permutations Rule (When Some Items Are Identical to Others)

If there are n items with n_1 alike, n_2 alike, . . . , n_k alike, the number of permutations of all n items is

$$\frac{n!}{n_1!\,n_2!\cdots n_k!}$$

EXAMPLE **Letter Arrangements** The classic examples of the permutations rule are those showing that the letters of the word *Mississippi* can be arranged 34,650 different ways and that the letters of the word *statistics* can be arranged 50,400 ways. We will instead consider the letters *DDDDRRRRR*, which are included in a discussion of the runs test for randomness (Section 13-7). Those letters represent a sequence of diet (D) and regular (R) colas. How many ways can we arrange the letters *DDDDRRRRR*?

SOLUTION In the sequence *DDDDRRRRR* we have $n = 9$ items, with $n_1 = 4$ alike and $n_2 = 5$ others that are alike. The number of permutations is computed as follows.

$$\frac{n!}{n_1!\,n_2!} = \frac{9!}{4!\,5!} = \frac{362{,}880}{2880} = 126$$

In Section 13-7 we need the fact that there are 126 different possible sequences of *DDDDRRRRR*, and we can now see how that result is obtained.

The preceding example involved n items, each belonging to one of two categories. When there are only two categories, we can stipulate that x of the items are alike and the other $n - x$ items are alike, so the permutations formula simplifies to

$$\frac{n!}{(n-x)!\,x!}$$

This particular result will be used for the discussion of binomial probabilities, which are introduced in Section 4-3.

When we intend to select r items from n different items but *do not take order into account,* we are really concerned with possible combinations rather than permutations. That is, **when different orderings of the same items are counted separately, we have a permutation problem, but when different orderings of the same items are not counted separately, we have a combination problem** and may apply the following rule.

Combinations Rule

The number of **combinations** of r items selected from n different items is

$${}_nC_r = \frac{n!}{(n-r)!\,r!}$$

Many calculators are designed to evaluate ${}_nC_r$. With the TI-83 Plus calculator, enter the value for n, then press **MATH,** select **PRB** and **${}_nC_r$,** then enter the value for r, and press the **ENTER** key.

It is very important to recognize that in applying the combinations rule, the following conditions apply:

- We must have a total of n different items available.
- We must select r of the n items (without replacement).
- We must consider rearrangements of the same items to be the same. (The combination *ABC* is the same as *CBA*.)

Because choosing between the permutations rule and the combinations rule can be confusing, we provide the following example, which is intended to emphasize the difference between them.

EXAMPLE **Elected Offices** The Board of Trustees at the author's college has 9 members. Each year, they elect a 3-person committee to

Boys and Girls Are Not Equally Likely

In many probability calculations, good results are obtained by assuming that boys and girls are equally likely to be born. In reality, a boy is more likely to be born (with probability 0.5121) than a girl (with probability 0.4879). However, the birth rate for boys is lower than is was 30 years ago, when the probability of a boy was 0.5134. Researchers are not sure why that drop has occurred, but possible explanations include environmental changes and exposure to chemicals. The magnitude of the drop from 0.5134 to 0.5121 might not seem very substantial, but it is significant. Because the drop might be an indicator of some health or environmental hazard, further research is required.

oversee buildings and grounds. Each year, they also elect a chairperson, vice chairperson, and secretary.

a. When the board elects the buildings and grounds committee, how many different 3-person committees are possible?

b. When the board elects the 3 officers (chairperson, vice chairperson, and secretary), how many different slates of candidates are possible?

SOLUTION Note that order is irrelevant when electing the buildings and grounds committee. When electing officers, however, different orders are counted separately.

a. Because order does not count for the committees, we want the number of combinations of $r = 3$ people selected from the $n = 9$ available people. We get

$$_9C_3 = \frac{n!}{(n-r)!\,r!} = \frac{9!}{(9-3)!\,3!} = \frac{362{,}880}{4320} = 84$$

b. Because order does count with the slates of candidates, we want the number of sequences (or permutations) of $r = 3$ people selected from the $n = 9$ available people. We get

$$_9P_3 = \frac{n!}{(n-r)!} = \frac{9!}{(9-3)!} = \frac{362{,}880}{720} = 504$$

There are 84 different possible committees of 3 board members, but there are 504 different possible slates of candidates.

The counting techniques presented in this section are sometimes used in probability problems. The following examples illustrate such applications.

EXAMPLE In the New York State lottery, a player wins first prize by selecting the correct 6-number combination when 6 different numbers from 1 through 51 are drawn. If a player selects one particular 6-number combination, find the probability of winning. (The player need not select the 6 numbers in the same order as they are drawn, so order is irrelevant.)

SOLUTION Because 6 different numbers are selected from 51 different possibilities, the total number of combinations is

$$_{51}C_6 = \frac{51!}{(51-6)!\,6!} = \frac{51!}{45!6!} = 18{,}009{,}460$$

With only one combination selected, the player's probability of winning is only 1/18,009,460.

EXAMPLE Defective Modems The United Components Company ships you a carton of 12 computer modems, 3 of which are defective. If you randomly select 5 of the modems and test them, what is the probability that exactly 2 of them are defective?

SOLUTION First, let's express the given information in a format that allows for better understanding:

3 defective modems
9 good modems
12 total modems

We want P(2 out of 5 are defective), so we might use the multiplication rule for dependent events to get this result (where D represents a defective modem and G represents a good modem):

$$P(\text{D and D and G and G and G}) = \frac{3}{12} \cdot \frac{2}{11} \cdot \frac{9}{10} \cdot \frac{8}{9} \cdot \frac{7}{8} = 0.0318$$

As brilliant as this solution might appear, it is *wrong* because it assumes that the *first* two modems are defective and the last three are good, but there are other ways to get two defective and three good modems. Finding the total number of ways is equivalent to finding the number of different possible arrangements of DDGGG, which is found by using the expression for the number of permutations of n items when some of them are identical to others:

$$\frac{n!}{n_1!\, n_2!} = \frac{5!}{2!\, 3!} = 10$$

Each of those 10 arrangements has a probability of 0.0318, so the probability we seek is $10(0.0318) = 0.318$.

We can summarize the procedure used in the preceding example as follows:

1. Assume that the first two items are defective and the remaining three items are good, then use the multiplication rule to find P(D and D and G and G and G).
2. Use the permutations rule (for some items alike) to find the number of permutations of the items DDGGG.
3. Multiply the probability from Step 1 by the number of permutations from Step 2.

This same procedure can be used for different circumstances by simply changing the numbers as appropriate. For example, we can use this procedure to find the probability of getting exactly 4 boys among 10 births. Replace "defect" with "boy," find $P(\text{BBBBGGGGGG}) = 1/1024$, find the number of arrangements of BBBBGGGGGG [which is $10!/(4!\,6!) = 210$], and then multiply to get 210/1024. This procedure is the basis for finding binomial probabilities, discussed later in Section 4-3.

This section presented five different counting devices. When deciding which particular approach applies, we can reduce confusion by using a systematic approach that addresses key issues. The following summary may be helpful.

- For a sequence of events in which the first can occur m ways, the second can occur n ways, and so on, use the fundamental counting rule: Multiply $m \cdot n$ and so on.
- To find the number of arrangements of n *different* items, with *all* of them to be used in different arrangements, use the factorial rule: Evaluate $n!$.
- For n *different* items with *some* of them to be used in different arrangements, use the permutations rule: Evaluate

$$_nP_r = \frac{n!}{(n-r)!}$$

- To find the total number of different arrangements when there are n items with n_1 of them *identical* to each other, n_2 of them identical to each other, and so on: Evaluate

$$\frac{n!}{n_1!\,n_2!\cdots n_k!}$$

- To find the number of different combinations when there are n different items and r of them are to be selected, use the combinations rule: Evaluate

$$_nC_r = \frac{n!}{(n-r)!\,r!}$$

3-7 Basic Skills and Concepts

Calculating Factorials, Combinations, Permutations. *In Exercises 1–4, evaluate the given expressions and express all results using the usual format for writing numbers (instead of scientific notation).*

1. 14! **2.** $_{20}C_5$ **3.** $_{50}P_3$ **4.** $_{50}P_4$

Probability of Winning the Lottery. *This section included an example showing that the probability of winning the New York State lottery is 1/18,009,460. In Exercises 5–9, find the probability of winning the indicated lottery.*

5. Florida: Select the winning six numbers from 1, 2, . . . , 49.

6. New York regional lottery: Select the winning six numbers from 1, 2, . . . , 31.

7. Connecticut: Select the winning six numbers from 1, 2, . . . , 44.

8. Powerball (12 states): Select the winning five numbers from 1, 2, . . . , 49.

9. New York Lottery The probability of winning the New York State lottery is 1/18,009,460. What is the probability of winning if the rules are changed so that in addition to selecting the correct six numbers from 1 to 51, you must now select them in the same order as they are drawn?

10. Number of Security System Codes The author uses an ADT home security system that has a code consisting of four digits (0, 1, . . . , 9) that must be entered in the correct sequence. The digits can be repeated in the code.

- **a.** How many different possibilities are there?
- **b.** If it takes a burglar five seconds to try a code, how long would it take to try every possibility? Is this time long enough to be a deterrent?

11. Elected Board of Directors There are 12 members on the board of directors for the Newport General Hospital.

- **a.** If they must elect a chairperson, first vice chairperson, second vice chairperson, and secretary, how many different slates of candidates are possible?
- **b.** If they must form an ethics subcommittee of four members, how many different subcommittees are possible?

12. Social Security Numbers Each social security number is a sequence of nine digits. What is the probability of randomly generating nine digits and getting *your* social security number?

13. Scheduling Assignments The starting five players for the New York Knicks basketball team have agreed to make appearances for charity events tomorrow night. If you must send three of them to a United Way event and the other two to a Heart Fund event, how many different ways can you make the assignments?

14. Selection of Treatment Group Walton Pharmaceuticals wants to test the effectiveness of a new drug designed to relieve allergy symptoms. The initial test will be conducted by treating six people chosen from a pool of 15 volunteers. If the treatment group is randomly selected, what is the probability that it consists of the six youngest people in the pool?

15. Delivery Routes A Federal Express delivery route must include stops in four cities.

- **a.** How many different routes are possible?
- **b.** If the route is randomly selected, what is the probability that the cities will be arranged in alphabetical order?

16. Jumble Puzzle Many newspapers carry "Jumble," a puzzle in which the reader must unscramble letters to form words. For example, the letters GRACIT were included in newspapers on the day this exercise was written. How many ways can the letters of GRACIT be arranged?

17. Delivery Routes A UPS dispatcher sends a delivery truck to eight different locations. If the order in which the deliveries are made is randomly determined, find the probability that the resulting route is one of the two shortest possible routes. (The two shortest routes are actually the same route in two opposite directions.)

18. Probabilities of Gender Sequences

- **a.** If a couple plans to have eight children, how many different gender sequences are possible?
- **b.** If a couple has four boys and four girls, how many different gender sequences are possible?

(continued)

c. Based on the results from parts (a) and (b), what is the probability that when a couple has eight children, the result will consist of four boys and four girls?

19. Finding the Number of Possible Melodies In Denys Parsons' *Directory of Tunes and Musical Themes,* melodies for more than 14,000 songs are listed according to the following scheme: The first note of every song is represented by an asterisk *, and successive notes are represented by R (for repeat the previous note), U (for a note that goes up), or D (for a note that goes down). Beethoven's Fifth Symphony begins as $*RRD$. Classical melodies are represented through the first 16 notes. With this scheme, how many different classical melodies are possible?

20. Finding the Number of Routes for Water Samples The Bureau of Fisheries once asked Bell Laboratories for help in finding the shortest route for getting samples from locations in the Gulf of Mexico. How many different routes are possible if samples must be taken from eight locations?

21. Is the Researcher Cheating? You become suspicious when a genetics researcher randomly selects groups of 20 newborn babies and seems to consistently get 10 girls and 10 boys. The researcher explains that it is common to get 10 boys and 10 girls in such cases.

a. If 20 newborn babies are randomly selected, how many different gender sequences are possible?

b. How many different ways can 10 boys and 10 girls be arranged in sequence?

c. What is the probability of getting 10 boys and 10 girls when 20 babies are born?

d. Based on the preceding results, do you agree with the researcher's explanation that it is common to get 10 boys and 10 girls when 20 babies are randomly selected?

22. Determining Age Discrimination In an age-discrimination case against Darmin, Inc., evidence showed that among the last 12 applicants for employment, only the 5 youngest were hired. Find the probability of randomly selecting 5 of 12 people and getting the 5 youngest. Based on the result, does it appear that age discrimination is occurring?

23. Radon Testing After testing 12 homes for the presence of radon, an inspector is concerned that her test equipment is defective because the measured radon level at each home was higher than the reading at the preceding home. That is, the 12 readings were arranged in order from low to high. If the homes were randomly selected, what is the probability of getting this particular arrangement? Based on the result, is her concern about the test equipment justified?

24. Finding the Number of Area Codes *USA Today* reporter Paul Wiseman described the old rules for telephone area codes by writing about "possible area codes with 1 or 0 in the second digit. (Excluded: codes ending in 00 or 11, for toll-free calls, emergency services, and other special uses.)" Codes beginning with 0 or 1 should also be excluded. How many different area codes were possible under these old rules?

25. Cracked Eggs A carton contains 12 eggs, 3 of which are cracked. If we randomly select 5 of the eggs for hard boiling, what is the probability of the following events?

a. All of the cracked eggs are selected.

b. None of the cracked eggs are selected.
c. Two of the cracked eggs are selected.

26. Combination Locks A typical "combination" lock is opened with the correct sequence of three numbers between 0 and 49 inclusive. (A number can be used more than once.) What is the probability of guessing those three numbers and opening the lock with the first try?

27. Five Card Flush A standard deck of cards contains 13 clubs, 13 diamonds, 13 hearts, and 13 spades. If five cards are randomly selected, find the probability of getting a flush. (A flush is obtained when all five cards are of the same suit. That is, they are all clubs, or all diamonds, or all hearts, or all spades.)

28. Guessing Chronological Order A creative history teacher gives a quiz on the first day of class. The quiz consists of 10 historical events, and the students are instructed to make a chronological list of the three events that occurred first. Ben randomly selects three of the events and puts them in a random order (because he wants to finish the quiz quickly so he can get to his statistics class on time). What is the probability that Ben's answer is correct?

3-7 Beyond the Basics

29. Finding the Number of Computer Variable Names A common computer programming rule is that names of variables must be between 1 and 8 characters long. The first character can be any of the 26 letters, while successive characters can be any of the 26 letters or any of the 10 digits. For example, allowable variable names are A, BBB, and M3477K. How many different variable names are possible?

30. Handshakes and Round Tables
 a. Five managers gather for a meeting. If each manager shakes hands with each other manager exactly once, what is the total number of handshakes?
 b. If n managers shake hands with each other exactly once, what is the total number of handshakes?
 c. How many different ways can five managers be seated at a round table? (Assume that if everyone moves to the right, the seating arrangement is the same.)
 d. How many different ways can n managers be seated at a round table?

31. Evaluating Large Factorials Many calculators or computers cannot directly calculate 70! or higher. When n is large, $n!$ can be approximated by $n! = 10^K$, where $K = (n + 0.5) \log n + 0.39908993 - 0.43429448n$.
 a. Evaluate 50! using the factorial key on a calculator and also by using the approximation given here.
 b. The Bureau of Fisheries once asked Bell Laboratories for help finding the shortest route for getting samples from 300 locations in the Gulf of Mexico. If you compute the number of different possible routes, how many digits are used to write that number?

32. Computer Intelligence Can computers "think"? According to the *Turing test,* a computer can be considered to think if, when a person communicates with it, the person believes he or she is communicating with another person instead of a computer. In an experiment at Boston's Computer Museum, each of 10 judges

communicated with four computers and four other people and was asked to distinguish between them.

a. Assume that the first judge cannot distinguish between the four computers and the four people. If this judge makes random guesses, what is the probability of correctly identifying the four computers and the four people?

b. Assume that all 10 judges cannot distinguish between computers and people, so they make random guesses. Based on the result from part (a), what is the probability that all 10 judges make all correct guesses? (That event would lead us to conclude that computers cannot "think" when, according to the Turing test, they can.)

Men, Women, and Other Planets

BOSTON The possibility of intelligent life on other planets intrigues us all. Interest is fueled by sporadic reports of unidentified flying objects—UFOs. Some people strongly believe that the U.S. Government is hiding direct knowledge of the existence of intelligent life from other planets, and that remains from crashed UFOs are being kept at a secret storage facility in New Mexico. If you're going to store UFO remains, New Mexico is probably as good a place as any. In one survey, registered voters were asked if they believe in the existence of intelligent life on other planets. See the *USA Today* illustration summarizing the results.

Refer to the *USA Today* illustration to answer the following questions.

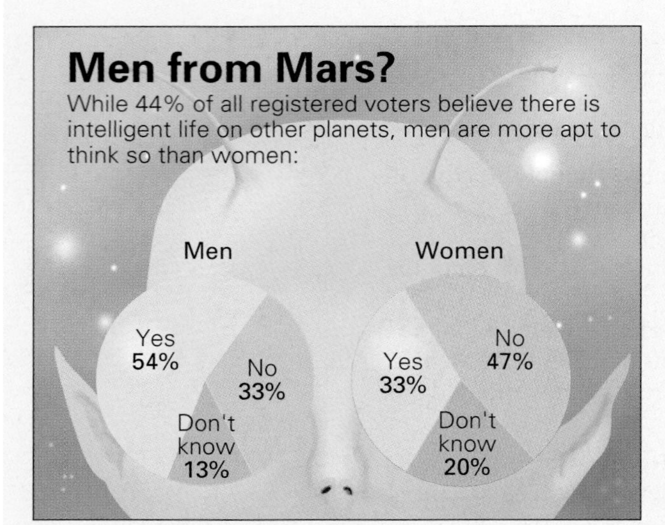

USA Today Snapshot. "A look at statistics that shape the nation"

1. What is the probability that a male voter believes in intelligent life on other planets?
2. What is the probability that a female voter believes in intelligent life on other planets?
3. If two different male voters are randomly selected, what is the probability that they both believe in intelligent life on other planets?
4. Assuming that the population of voters includes an equal number of men and women, what is the probability that a randomly selected voter believes in intelligent life on other planets?
5. If a voter is randomly selected and believes in intelligent life on other planets, what is the probability that the selected person is a male?

Vocabulary List

rare event rule
event
simple event
sample space
relative frequency approximation of probability
classical approach to probability
subjective probabilities
law of large numbers
simulation
random sample of one element
random sample
complement
actual odds against
actual odds in favor
payoff odds
addition rule
compound event
mutually exclusive
rule of complementary events
tree diagram
independent events
dependent events
multiplication rule
conditional probability
simulation
fundamental counting rule
factorial rule
permutations rule
combinations rule

Review

In this chapter we introduced the basic concepts of probability theory. In Section 3-2 we presented the basic definitions and notation, including the representation of events by letters such as A. We defined probabilities of simple events as

$$P(A) = \frac{\text{number of times } A \text{ occurred}}{\text{number of times trial was repeated}} \qquad \text{(relative frequency)}$$

$$P(A) = \frac{\text{number of ways } A \text{ can occur}}{\text{number of different simple events}} = \frac{s}{n} \qquad \text{(for equally likely outcomes)}$$

We noted that the probability of any impossible event is 0, the probability of any certain event is 1, and for any event A, $0 \le P(A) \le 1$. Also, $\overline{A}$ denotes the complement of event A. That is, $\overline{A}$ indicates that event A does *not* occur.

After Section 3-2 we considered compound events, which are events combining two or more simple events. In general, associate *or* with addition and associate *and* with multiplication. Always keep in mind the following key considerations.

- When conducting one trial, do we want the probability of event *A or B?* If so, use the addition rule, but be careful to avoid counting any outcomes more than once.
- When finding the probability that event A occurs on one trial *and* event B occurs on a second trial, use the multiplication rule. Multiply the probability of event A by the probability of event B. *Caution:* When calculating the probability of event B, be sure to take into account the fact that event A has already occurred.

In some probability problems, the biggest obstacle is finding the total number of possible outcomes. The last section of this chapter was devoted to the following counting techniques, which are briefly summarized at the end of Section 3-7.

- Fundamental counting rule
- Factorial rule

- Permutations rule (when items are all different)
- Permutations rule (when some items are identical to others)
- Combinations rule

Most of the material in the following chapters deals with statistical inferences based on probabilities. As an example of the basic approach used, consider a test of someone's claim that a quarter used in a coin toss is fair. If we flip the quarter 10 times and get 10 consecutive heads, we can make one of two inferences from these sample results:

1. The coin is actually fair, and the string of 10 consecutive heads is a fluke.
2. The coin is not fair.

Statisticians use the rare event rule when deciding which inference is correct: In this case, the probability of getting 10 consecutive heads is so small (1/1024) that the inference of unfairness is the better choice. Here we can see the important role played by probability in the standard methods of statistical inference.

REVIEW EXERCISES

Nicorette and Mouth or Throat Soreness. *In Exercises 1–8, use the data obtained from a test of Nicorette, a chewing gum designed to help people stop smoking. The following table is based on data from Merrell Dow Pharmaceuticals, Inc.*

	Nicorette	Placebo
Mouth or throat soreness	43	35
No mouth or throat soreness	109	118

1. If 1 of the 305 subjects is randomly selected, find the probability of getting someone who used Nicorette.

2. If 1 of the 305 subjects is randomly selected, find the probability of getting someone who had no mouth or throat soreness.

3. If 1 of the 305 study subjects is randomly selected, find the probability of getting someone who used Nicorette or had mouth or throat soreness.

4. If 1 of the 305 study subjects is randomly selected, find the probability of getting someone who used the placebo or had no mouth or throat soreness.

5. If two different subjects are randomly selected, find the probability that they both experienced mouth or throat soreness.

6. If two different subjects are randomly selected, find the probability that they are both from the placebo group.

7. If one subject is randomly selected, find the probability that he or she experienced mouth or throat soreness, given that the selected person used a placebo.

8. If one subject is randomly selected, find the probability that he or she experienced mouth or throat soreness or used a placebo.

9. Defective Chips in DVD Players The Binary Computer Company manufactures computer chips used in DVD players. Those chips are made with a 27% yield, meaning that 27% of them are good and the others are defective.

a. If one chip is randomly selected, find the probability that it is not good.

b. If two chips are randomly selected, find the probability that they are both good.
c. If three chips are randomly selected, find the probability that at least one is good.

10. Acceptance Sampling With one method of acceptance sampling, a sample of items is randomly selected without replacement and the entire batch is rejected if there is at least one defect. The Niko Electronics Company has just manufactured 2500 CDs, and 2% are defective. If 4 of the CDs are selected and tested, what is the probability that the entire batch will be rejected?

11. At Least One Girl Find the probability of getting at least one girl when a couple has five children. Assume that boys and girls are equally likely and that the sex of any child is independent of the others.

12. Selecting Members The board of directors for the Hartford Investment Fund has 10 members.
    a. If 3 members are randomly selected to oversee the auditors, find the probability that the three wealthiest members are selected.
    b. If members are elected to the positions of chairperson, vice chairperson, and treasurer, how many different slates are possible?

13. Roulette When betting on *even* in roulette, there are 38 equally likely outcomes, but only 2, 4, 6, . . . , 36 are winning outcomes.
    a. Find the probability of winning when betting on even.
    b. Find the actual odds against winning with a bet on even.
    c. Casinos pay winning bets according to odds described as 1 : 1. What is your net profit if you bet $5 on even and you win?

14. Is the Pollster Lying? A pollster claims that 12 voters were randomly selected from a population of 200,000 voters (30% of whom are Republicans), and all 12 were Republicans. The pollster claims that this could easily happen by chance. Find the probability of getting 12 Republicans when 12 voters are randomly selected from this population. Based on the result, does it seem that the pollster's claim is correct?

15. Life Insurance The New England Life Insurance Company issues one-year policies to 12 men who are all 27 years of age. Based on data from the Department of Health and Human Services, each of these men has a 99.82% chance of living through the year. What is the probability that they all survive the year?

16. Selecting Acceptable Switches A batch of light switches contains 44 that are good and 4 that are defective. Four different switches are randomly selected from the batch.
    a. Find the probability that they are all good.
    b. Find the probability that exactly three are good.
    c. Find the probability that exactly two are good.

Cumulative Review Exercises

1. Ages of Commercial Jets A recent survey of the U.S. commercial jet fleet showed that there are 1103 Boeing 737s in service. The ages of 12 randomly selected 737s are listed on the next page. The ages are based on data from the Federal Aviation Administration.

13 8 7 15 10 10 13 14 17 29 4 2

a. Find the mean age of the sample group.
b. Find the median age of the sample group.
c. Find the standard deviation of the given sample ages.
d. Find the variance of the given ages.
e. Based on the given ages, would it be unusual to randomly select a Boeing 737 and find that it is new?
f. Which term best identifies the level of measurement of the ages: nominal, ordinal, interval, ratio?
g. If someone skipped Chapter 1 and had to answer part (f) with a random guess, what is the probability of being correct?
h. If one age is randomly selected from this sample, find the probability that it is at least 10 years.
i. If two different ages are randomly selected from this sample, find the probability that they are both less than 10 years.
j. If two ages are randomly selected with replacement, find the probability that they are both less than 10 years.

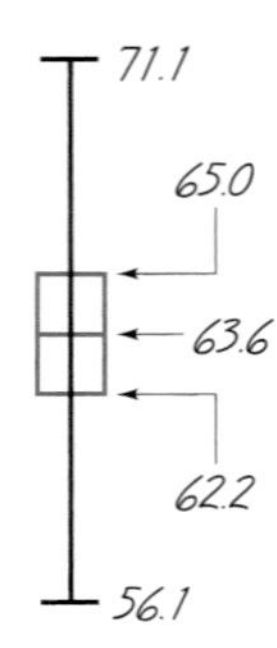

2. Women's Heights The accompanying boxplot depicts heights (in inches) of a large collection of randomly selected adult women.
a. What is the mean height of adult women?
b. If one of these women is randomly selected, find the probability that her height is between 56.1 in. and 62.2 in.
c. If one of these women is randomly selected, find the probability that her height is below 62.2 in. or above 63.6 in.
d. If two women are randomly selected, find the probability that they both have heights between 62.2 in. and 63.6 in.
e. If five women are randomly selected, find the probability that three of them are taller than the mean and the other two are shorter than the mean.

COOPERATIVE GROUP ACTIVITIES

1. *In-class activity:* See Exercise 11 in Section 3-6. Divide into groups of three or four and use coin tossing to develop a simulation that emulates the kingdom that abides by this decree: After a mother gives birth to a son, she will not have any other children. If this decree is followed, does the proportion of girls increase?

2. *In-class activity:* Divide into groups of three or four and use actual thumbtacks to estimate the probability that when dropped, a thumbtack will land with the point up. How many trials are necessary to get a result that seems to be reasonably accurate when rounded to the first decimal place?

3. *Out-of-class activity:* Marine biologists often use the capture-recapture method as a way to estimate the size of a population, such as the number of fish in a lake. This method involves capturing a sample from the population, tagging each member in the sample, then returning them to the population. A second sample is later captured and the tagged members are counted along with the total size of this second sample. As an example, suppose a sample of 50 fish is captured and tagged. Also suppose that a second sample (obtained later) consists of 100 fish with 20 of them tagged, suggesting that when a fish is captured, the probability of it being tagged is estimated to be 0.20. That is, 20% of the fish in the population have been tagged. Because originally 50 sample fish were tagged, we estimate that the population size is 250 fish $(50 \div 20/100 = 250)$.

It's not easy to actually capture and recapture real fish, but we can simulate the procedure using some uniform collection of items such as BB's, colored beads, M&Ms, or Fruit Loop cereal pieces. "Captured" items in the first sample can be "tagged" with a magic marker (or they can be replaced with similar items of a different color). Illustrate the capture-recapture method by designing and conducting such an experiment. Starting with a large package of M&Ms, for example, collect a sample of 50, then use a magic marker to "tag" each one. Replace the tagged items, mix the whole population, then select a second sample and proceed to estimate the population size. Compare the result to the actual population size obtained by counting all of the items.

4. *In-class activity:* Divide into groups of two. In Section 3-6 we discussed the "Monty Hall" problem which, according to *Chance* magazine, has been used to study decision making in business schools at Harvard and Stanford. The problem is based on a television game show that was hosted by Monty Hall. Begin by selecting one of the team members to serve as the host. The other team member is the contestant, and there are three doors numbered 1, 2, and 3. The host should *randomly* select one of the doors, and the selection must not be revealed to the contestant. Pretend that the host has put a prize of a new red Corvette behind the door that was randomly selected, but the other two doors have nothing behind them. The contestant should now select one of the three doors. After the contestant reveals which door has been chosen, the host should select an "empty" door and inform the contestant that this particular door has nothing behind it. The host should now offer the contestant a choice of sticking with the original door or switching to the other door that has not been revealed. After the contestant announces his or her decision to stick or switch, the host should announce that the contestant has (or has not) won the Corvette. Record the result along with the contestant's decision to stick or switch. Repeat the game 20 times with the contestant sticking 10 times and switching 10 times. Then reverse roles and play the game another 20 times. Find the proportion of times the game was won by sticking, and find the proportion of times the game was won by switching. Based on the results, which strategy is better: sticking or switching?

5. *Out-of-class activity:* Divide into groups of two for the purpose of doing an experiment designed to show one approach to dealing with sensitive survey questions, such as those related to drug use, sexual activity, stealing, or cheating. Instead of actually using a controversial question, we will use this innocuous question: "Is your birthday between January 1 and March 31?" About 1/4 of all responses should be "yes," but let's pretend that the question is very sensitive and that survey subjects are reluctant to answer honestly. Survey people by asking them to flip a coin and respond as follows:

- Answer "no" if the coin turns up heads *and* the subject was not born between January 1 and March 31.
- Otherwise, answer "yes" (because the coin turns up tails *or* you were born between January 1 and March 31.

You can see that if someone answers "yes," you don't really know if they were born between January 1 and March 31. Supposedly, respondents tend to be more honest because the coin flip protects their privacy. Survey people and analyze the results to determine the proportion of people born between January 1 and March 31. The accuracy of the results could be checked against their actual birth dates, which can be obtained from a second question. The experiment can be repeated with a question that is more sensitive, but such a question is not given here because the author already receives enough mail.

TECHNOLOGY PROJECT

1. This project will illustrate the law of large numbers described in Section 3-2. Use a computer or TI-83 Plus calculator to simulate 100 births. Accomplish this by randomly generating 100 numbers, each of which is 0 or 1 (where 0 = boy and 1 = girl). Use the results to fill in the following table, then complete the graph by plotting the appropriate points. What happens to the proportion of girls as the sample size increases?

Number of births	10	20	30	40	50	60	70	80	90	100
Proportion of girls										

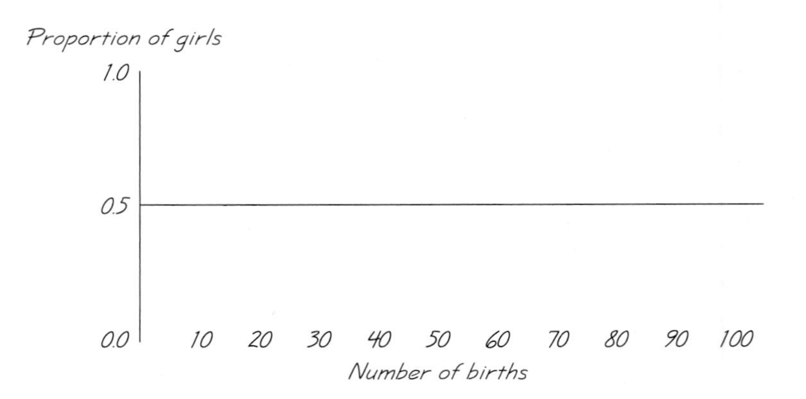

Here are some details for the different applications of technology.

STATDISK: Select **Data** from the main menu bar. Choose **Uniform Generator.** Make the entries for a sample of size 100, a minimum of 0, a maximum of 1, and 0 decimal places (because we want whole numbers). Click on the **Generate** button.

Minitab: Select **Calc** from the main menu bar at the top. Select **Random Data,** then **Integer.** Proceed to enter 100 for the number of rows of data, C1 for the column in which to store the results, 0 for the minimum value, and 1 for the maximum value; then click **OK.**

Excel: Consult the Excel manual that is a supplement to this textbook.

TI-83 Plus: Press the **MATH** key. Select **PRB.** Select the 5th menu item: **randInt(.** Enter 0, 1, 100 and press the **ENTER** key. Press **STO** and **L1** to store the data in list L1. To view the generated births, press **STAT,** then select **Edit.**

For Technology Projects 2–7, refer to the indicated exercise from Section 3–6 and use STATDISK, Minitab, Excel, or a TI-83 Plus calculator.

2. Exercise 5 **3.** Exercise 6 **4.** Exercise 7

5. Exercise 8 **6.** Exercise 10, part (a) **7.** Exercise 10, part (b)

from DATA *to* DECISION

Critical Thinking: When you apply for a job, should you be concerned about drug testing?

According to the American Management Association, most U.S. companies now test at least some employees and job applicants for drug use. The U.S. National Institute on Drug Abuse claims that about 15% of people in the 18–25 age bracket use illegal drugs. Allyn Clark, a 21-year-old college graduate, applied for a job at the Acton Paper Company, took a drug test, and was not offered a job. He suspected that he might have failed the drug test, even though he does not use drugs. In checking with the company's personnel department, he found that the drug test has 99% sensitivity, which means that only 1% of drug users incorrectly test negative. Also, the test has 98% specificity, meaning that only 2% of nonusers are incorrectly identified as drug users. Allyn felt relieved by these figures because he believed that they reflected a very reliable test that usually provides good results—but is this really true?

Analyzing the results

The accompanying table shows data for Allyn and 1999 other job applicants. Based on those results, find *P*(false positive); that is, find the probability of randomly selecting one of the subjects who tested positive and getting someone who does not use drugs. Also find *P*(false negative); that is, find the probability of randomly selecting someone who tested negative and getting someone who does use drugs. Are the probabilities of these wrong results low enough so that job applicants and the Acton Paper Company need not be concerned?

	Drug Users	Nonusers
Positive test result	297	34
Negative test result	3	1666

Internet Project

What's the probability of landing on Boardwalk?

Finding probabilities when rolling dice is easy. With one die, there are six possible outcomes so each outcome, such as rolling a 2, has probability 1/6. For a card game the calculations are more involved, but they are still manageable. But what about a more complicated game, such as the board game Monopoly? What is the probabilty of landing on a particular space on the board? The probability depends on the space your piece currently occupies, the roll of dice, the drawing of cards, as well as other factors. Now consider more true-to-life examples, such as the probability of having an auto accident. The number of factors involved there is too large to even consider, yet such probabilities are nonetheless quoted, for example, by insurance companies.

The Internet Project for this chapter considers methods for computing probabilities in complicated situations. Go to the site

http://www.awlonline.com/triola

where you will be guided in researching probabilities for a board game. Then you will compute such a probability yourself. Finally, you will compute a health-related probability using empirical data.

Statistics *at work*

Based on these statistics on preferred park use, we are now evaluating how we can better service the more diverse population.

Judy Shafer

Deputy Superintendent of Virgin Islands National Park

Judy has worked for the National Park Service for seventeen years and in park management for the past four years.

As Deputy Superintendent of Virgin Islands National Park, do you use statistics in your work?

We use probability and methods of statistics in applications such as analyzing the sustainability of a particular coral species under the environmental stresses of visitor use, pollution, sedimentation, etc. We also use methods of statistics to determine which ethnic and racial populations are using the park, and to determine the manner in which the park is achieving its user-satisfaction goals.

There are so many uses of statistics that it is impossible to describe them all! Our Law Enforcement Rangers collect and use statistics for visitor and resource protection activities; our marine research scientists use statistics on a daily basis to determine the health and viability of various marine species and to determine the level of threat by visitor use and environmental agents, such as pollution and sedimentation. For example, some of the larger fish species, such as the Nassau Grouper, may be so over-fished that this species could be nearing extinction. The statistical data from research studies must guide the park's decisions in how to protect such species as they are inextricably linked to the larger coral reef ecosystem and ultimately to global warming.

Could you give a simple and specific example of how statistics is used?

In the realm of visitor use, the National Park Service has learned that different ethnic and racial groups use national parks in different ways. For example, we have learned that some groups like the traditional backcountry hiking while others prefer the sociability of group picnics. We are considering the incorporation of more group picnic areas to better accommodate visitors of different ethnic and racial backgrounds.

Is your use of statistics increasing or decreasing?

It is definitely increasing and will continue to do so in the future. As in the example above, sometimes we might think we know how a segment of the population enjoys using a park. But then we get the statistical data that may change our entire perception. Based on the statistics, a manager can decide whether a preferred use is compatible with the park's preservation goals and, if it is compatible, attempt to increase visitor satisfaction by accommodating that use.

Does the use of statistics present any major challenges?

As we all know, statistics can be used to prove or disprove almost anything you wish. Ideally, when we establish the parameters to gather data, we try to consider all the variables that might skew the data. The challenge is not so much in the collection of the data but knowing the right questions to ask. In my job, I determine what questions should be asked after I seek the input of many people with diverse racial, ethnic, and cultural backgrounds. That gives me the broadest perspective and stimulates ideas and questions that I may have never thought of alone.

Should job applicants in your area have a statistics course in their background?

Our employees should have a minimum of at least a college-level course in statistics. Without that, you will be a dinosaur in a statistically based 21st century. Without statistics, you will not be as competitive as someone who does have that background.

What other skills are important for employees?

Vision. Leadership. Innovation. Creativity. And the willingness to take some risks to achieve your goals and protect the things that you feel passionate about.

CHAPTER 4

Probability Distributions

CHAPTER PROBLEM

Is the MicroSort gender-selection technique effective?

Advances in medicine and technology are providing us with new capabilities, some of which raise important ethical dilemmas. One such dilemma arises with the potential for parents to select the gender of their children. Although gender-selection techniques appear to be available, they are highly controversial. Many people strongly believe that gender selection is unethical and should be prohibited. Here is an excerpt from a *New York Times Magazine* article ("Getting the Girl", by Lisa Belkin): "If we allow parents to choose the sex of their child today, how long will it be before they order up eye color, hair color, personality traits, and IQ?" Others argue that there is nothing wrong with such techniques, and gender selection should be available to all couples who would like to use them. There are some convincing arguments in favor of at least limited use of gender selection. For example, some couples carry X-linked recessive genes with this result: Any male children have a 50% chance of inheriting a serious disorder, while none of the female children will inherit the disorder. Such couples may want to use gender selection as a way to ensure that they have baby girls.

The Genetics and IVF Institute in Fairfax, Virginia developed a technique called MicroSort, which supposedly increases the chances of a couple having a baby girl. In a preliminary test, 14 couples who wanted baby girls were found. After using the MicroSort technique, 13 of them had girls and one couple had a boy. We will focus on this mathematical issue: Given that 13 out of 14 couples had girls, can we conclude that the MicroSort technique is effective, or might we explain that outcome as just a chance sample result? We will address this issue later in the chapter.

4-1 Overview

In this chapter we combine the methods of *descriptive statistics* presented in Chapter 2 and those of *probability* presented in Chapter 3. Figure 4-1 presents a visual summary of what we will accomplish in this chapter. As the figure shows, using the methods of Chapter 2, we would repeatedly roll the die to collect sample data, which then can be described with graphs (such as a histogram or boxplot), measures of center (such as the mean), and measures of variation (such as the standard deviation). Using the methods of Chapter 3, we could find the probability of each possible outcome. In this chapter we will combine those concepts as we develop probability distributions that describe what will *probably* happen instead of what actually *did* happen. In Chapter 2 we constructed frequency tables and histograms using *observed* sample values that were actually collected, but in this chapter we will construct probability distributions by presenting possible outcomes along with the relative frequencies we *expect*.

A casino "pit boss" knows how a die should behave. The table at the extreme right in Figure 4-1 represents a probability distribution that serves as a model of a theoretically perfect population frequency distribution. In essence, we can describe the relative frequency table for a die rolled an infinite number of times. With this knowledge of the population of outcomes, we are able to find its important characteristics, such as the mean and standard deviation. The remainder of this book and the very core of inferential statistics are based on some knowledge of probability distributions. We begin by examining the concept of a random variable, and then we consider important distributions that have many real applications.

FIGURE 4-1
Combining Descriptive Methods and Probabilities to Form a Theoretical Model of Behavior

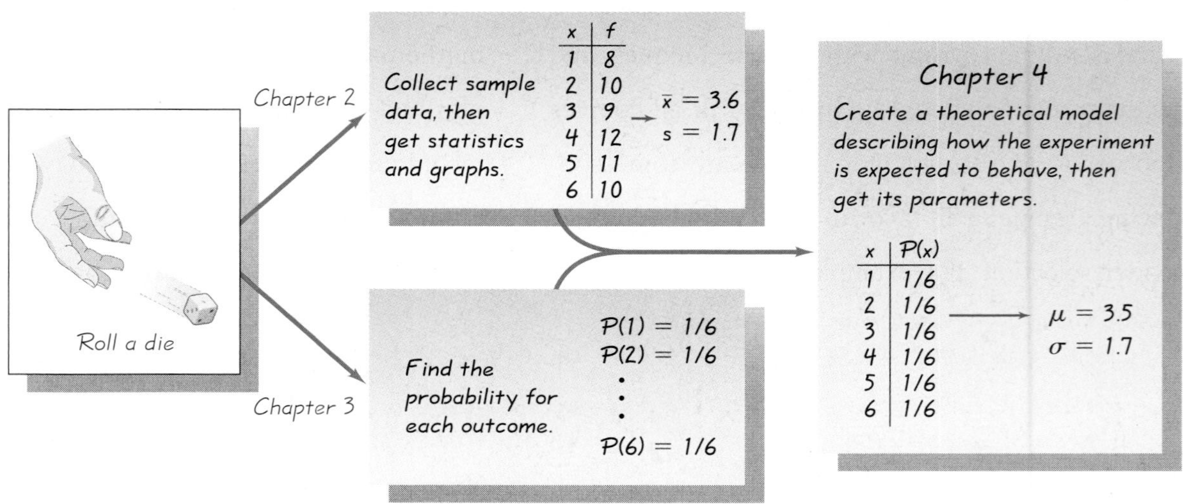

4-2 Random Variables

In this section we discuss random variables, probability distributions, procedures for finding the mean and standard deviation for a probability distribution, and methods for distinguishing between outcomes that are likely to occur by chance and outcomes that are "unusual." We begin with the related concepts of *random variable* and *probability distribution.*

DEFINITIONS

A **random variable** is a variable (typically represented by x) that has a single numerical value, determined by chance, for each outcome of a procedure.

A **probability distribution** is a graph, table, or formula that gives the probability for each value of the random variable.

EXAMPLE Gender of Children A study consists of randomly selecting 14 newborn babies and counting the number of girls (as in the Chapter Problem). If we assume that boys and girls are equally likely and if we let

$$x = \text{number of girls among 14 babies}$$

then x is a random variable because its value depends on chance. The possible values of x are $0, 1, 2, \ldots, 14$. Table 4-1 lists the values of x along with the corresponding probabilities. (In Section 4-3 we will see how to find the probability values, such as those listed in Table 4-1.) Because Table 4-1 gives the probability for each value of the random variable x, that table describes a probability distribution.

TABLE 4-1 Probabilities of Girls

x (girls)	$P(x)$
0	0.000
1	0.001
2	0.006
3	0.022
4	0.061
5	0.122
6	0.183
7	0.209
8	0.183
9	0.122
10	0.061
11	0.022
12	0.006
13	0.001
14	0.000

In Section 1-2 we made a distinction between discrete and continuous data. Random variables may also be discrete or continuous, and the following two definitions are consistent with those given in Section 1-2.

DEFINITIONS

A **discrete random variable** has either a finite number of values or a countable number of values, where "countable" refers to the fact that there might be infinitely many values, but they result from a counting process.

A **continuous random variable** has infinitely many values, and those values can be associated with measurements on a continuous scale in such a way that there are no gaps or interruptions.

This chapter deals exclusively with discrete random variables, but the following chapters will deal with continuous random variables.

(a) Discrete Random Variable: Count of the number of movie patrons.

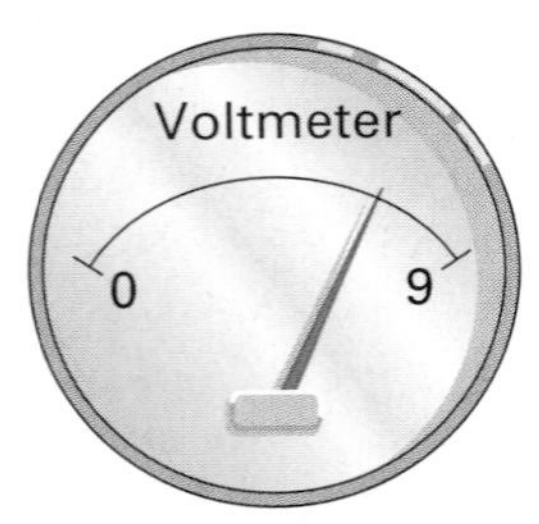

(b) Continuous Random Variable: The measured voltage of a smoke detector battery.

FIGURE 4-2
Devices Used to Count and Measure Discrete and Continuous Random Variables

EXAMPLES The following are examples of discrete and continuous random variables.

1. Let x = the number of eggs that a hen lays in a day. This is a *discrete* random variable because its only possible values are 0, or 1, or 2, and so on. No hen can lay 2.343115 eggs, which would have been possible if the data had come from a continuous scale.
2. The count of the number of patrons attending a rock concert is a whole number and is therefore a discrete random variable. The counting device shown in Figure 4-2(a) is capable of indicating only a finite number of values, so it can be used to obtain values for a *discrete* random variable.
3. Let x = the amount of milk a cow produces in one day. This is a *continuous* random variable because it can have any value over a continuous span. During a single day, a cow might yield an amount of milk that can be any value between 0 gallons and 5 gallons. (The author knows almost nothing about actual cow output per day, but 0 gallons to 5 gallons just seems like it's about right.) It would be possible to get 2.343115 gallons, because the cow is not restricted to the discrete amounts of 0, 1, 2, 3, 4, or 5 gallons.
4. The measure of voltage for a smoke-detector battery can be any value between 0 volts and 9 volts. It is therefore a continuous random variable. The voltmeter shown in Figure 4-2(b) is capable of indicating values on a continuous scale, so it can be used to obtain values for a *continuous* random variable.

Graphs

There are various ways to graph a probability distribution, but we will consider only the **probability histogram.** Figure 4-3 is a probability histogram that is very similar to the relative frequency histogram discussed in Chapter 2, but the vertical scale shows *probabilities* instead of relative frequencies based on actual sample results.

In Figure 4-3, note that along the horizontal axis, the values of 0, 1, 2, . . . , 14 are located at the centers of the rectangles. This implies that the rectangles are each 1 unit wide, so the areas of the rectangles are 0.000, 0.001, 0.006, and so on. The areas of these rectangles are the same as the *probabilities* in Table 4-1. We will see in Chapter 5 and future chapters that this correspondence between area and probability is very useful in statistics.

Every probability distribution must satisfy each of the following two requirements.

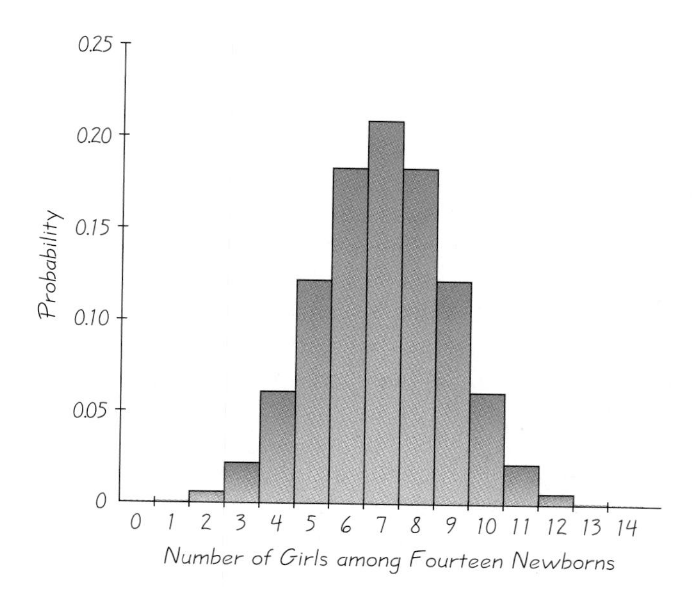

FIGURE 4-3
Probability Histogram for Number of Girls Among 14 Newborn Babies

Requirements for a Probability Distribution

1. $\Sigma P(x) = 1$ where x assumes all possible values
2. $0 \leq P(x) \leq 1$ for every value of x

The first requirement states that the sum of the probabilities for all the possible values of the random variable must equal 1. This makes sense when we realize that the values of the random variable x represent all possible events in the entire sample space, so we are certain (with probability 1) that one of the events will occur. In Table 4-1, the sum of the probabilities is 0.999; it would be 1 if we eliminated the tiny rounding error by carrying more decimal places. Also the probability rule stating $0 \leq P(A) \leq 1$ for any event A (given in Section 3-2) implies that $P(x)$ must be between 0 and 1 for any value of x. Again, refer to Table 4-1 and note that each individual value of $P(x)$ does fall between 0 and 1 for any value of x. Because Table 4-1 does satisfy both of the requirements, it is an example of a probability distribution. A probability distribution may be described by a table, such as Table 4-1, or a graph, such as Figure 4-3, or a formula, as in the following two examples.

EXAMPLE Does $P(x) = x/5$ (where x can take on the values of 0, 1, 2, 3) describe a probability distribution?

SOLUTION To be a probability distribution, $P(x)$ must satisfy the preceding two requirements. But

$$\Sigma P(x) = P(0) + P(1) + P(2) + P(3)$$

Is Parachuting Safe?

About 30 people die each year as more than 100,000 people make about 2.25 million parachute jumps. In comparison, a typical year includes about 200 scuba diving fatalities, 7000 drownings, 900 bicycle deaths, 800 lightning deaths, and 1150 deaths from bee stings. Of course, these figures don't necessarily mean that parachuting is safer than bike riding or swimming. A fair comparison should involve fatality rates, not just the total number of deaths.

The author, with much trepidation, made two parachute jumps but quit after missing the spacious drop zone both times. He has also flown in a hang glider, hot air balloon, and Goodyear blimp.

$$= \frac{0}{5} + \frac{1}{5} + \frac{2}{5} + \frac{3}{5}$$

$$= \frac{6}{5} \qquad \text{(showing that } \Sigma P(x) \neq 1\text{)}$$

Because the first requirement is not satisfied, we conclude that $P(x)$ given in this example is not a probability distribution.

EXAMPLE Does $P(x) = x/3$ (where x can be 0, 1, or 2) determine a probability distribution?

SOLUTION For the given function we find that $P(0) = 0/3$, $P(1) = 1/3$, and $P(2) = 2/3$, so that

1. $\Sigma P(x) = \frac{0}{3} + \frac{1}{3} + \frac{2}{3} = \frac{3}{3} = 1$
2. Each of the $P(x)$ values is between 0 and 1.

Because both requirements are satisfied, the $P(x)$ function given in this example is a probability distribution.

Mean, Variance, and Standard Deviation

Recall that in Chapter 2 we described the following important characteristics of data:

1. *Center:* A representative or average value that indicates where the middle of the data set is located
2. *Variation:* A measure of the amount that the values vary among themselves
3. *Distribution:* The nature or shape of the distribution of the data (such as bell-shaped, uniform, or skewed)
4. *Outliers:* Sample values that lie very far away from the vast majority of the other sample values
5. *Time:* Changing characteristics of the data over time

The probability histogram can give us insight into the nature or shape of the distribution. Also, we can often find the mean, variance, and standard deviation of data, which provide insight into the other characteristics. The mean, variance, and standard deviation for a probability distribution can be found by applying Formulas 4-1, 4-2, 4-3, and 4-4.

Formula 4-1 $\mu = \Sigma[x \cdot P(x)]$ Mean for a probability distribution

Formula 4-2 $\sigma^2 = \Sigma[(x - \mu)^2 \cdot P(x)]$ Variance for a probability distribution

Formula 4-3 $\sigma^2 = [\Sigma x^2 \cdot P(x)] - \mu^2$ Variance for a probability distribution

Formula 4-4 $\sigma = \sqrt{[\Sigma x^2 \cdot P(x)] - \mu^2}$ Standard deviation for a probability distribution

Caution: Evaluate $[\Sigma x^2 \cdot P(x)]$ by first squaring each value of x, then multiplying each square by the corresponding $P(x)$, then adding.

The TI-83 Plus calculator can be used to find the mean and standard deviation. The procedure follows:

Enter the values of the random variable x in list L1. Enter the corresponding probabilities in list L2. Press **STAT,** select **CALC,** then select **1-Var Stats.** Enter L1, L2 (with the comma), then press the **ENTER** key.

Shown here is the TI-83 Plus screen display for the probability distribution described by Table 4-1. In this TI-83 Plus display, the value shown as $\overline{x}$ is actually the value of the mean μ, and the value shown as σx is the value of the standard deviation σ. That is, $\mu = 7$ and $\sigma = 1.876038251$.

TI-83 Plus

```
1-Var Stats
 x̄=7
 Σx=6.993
 Σx²=52.467
 Sx=
 σx=1.876038251
↓n=.999
```

Rationale for Formulas 4-1 through 4-4

Why do Formulas 4-1 through 4-4 work? A probability distribution is a model of a theoretically perfect population frequency distribution. The probability distribution is like a relative frequency distribution based on data that behave perfectly, without the usual imperfections of samples. Because the probability distribution allows us to predict the population of outcomes, we are able to determine the values of the mean, variance, and standard deviation. Formula 4-1 accomplishes the same task as the formula for the mean of a frequency table. (Recall that f represents class frequency and N represents population size.) Rewriting the formula for the mean of a frequency table so that it applies to a population and then changing its form, we get

$$\mu = \frac{\Sigma(f \cdot x)}{N} = \Sigma\frac{f \cdot x}{N} = \Sigma x \cdot \frac{f}{N} = \Sigma x \cdot P(x)$$

In the fraction f/N, the value of f is the frequency with which the value x occurs and N is the population size, so f/N is the probability for the value of x.

Similar reasoning enables us to take the variance formula from Chapter 2 and apply it to a random variable for a probability distribution; the result is Formula 4-2. Formula 4-3 is a shortcut version that will always produce the same result as Formula 4-2. Although Formula 4-3 is usually easier to work with, Formula 4-2 is easier to understand directly. Based on Formula 4-2, we can express the standard deviation as

$$\sigma = \sqrt{\Sigma(x - \mu)^2 \cdot P(x)}$$

or as the equivalent form given in Formula 4-4.

When using Formulas 4-1 through 4-4, use this rule for rounding results.

Autism Cluster Exceeds Expected Number

The Centers for Disease Control and Prevention reports that nationwide the usual rate of autism is about 1 case in 500. But in Brick Township, New Jersey, 40 cases of autism were found among 6000 children in the age bracket of 3-10 years. Where 12 cases of autism would be expected, 40 cases were found. This is an example of a *cluster*—a significantly excessive number of occurrences of some event. In Brick Township, some researchers are looking for possible causes by analyzing water, air, and soil samples. Others argue that families with autistic children are moving to Brick for its model school system. Whatever the cause, methods of statistics are used to determine whether the occurrence of some event, such as autism, is significantly excessive, or whether random chance fluctuations constitute a reasonable explanation.

Round-off Rule for μ, σ^2, and σ

Round results by carrying one more decimal place than the number of decimal places used for the random variable x. If the values of x are integers, round μ, σ^2, and σ to one decimal place.

It is sometimes necessary to use a different rounding rule because of special circumstances, such as results that require more decimal places to be meaningful. For example, with four-engine jets the mean number of jet engines working successfully throughout a flight is 3.999714286, which becomes 4.0 when rounded to one more decimal place than the original data. Here, 4.0 would be misleading because it suggests that all jet engines always work successfully. We need more precision to correctly reflect the true mean, such as the precision in the number 3.999714.

Identifying Unusual Results with the Range Rule of Thumb

The range rule of thumb (discussed in Section 2-5) may also be helpful in interpreting the value of a standard deviation. According to the range rule of thumb, most values should lie within 2 standard deviations of the mean; it is unusual for a score to differ from the mean by more than 2 standard deviations. We can therefore identify "unusual" values by determining that they lie outside of these limits:

$$\text{maximum usual value} = \mu + 2\sigma$$
$$\text{minimum usual value} = \mu - 2\sigma$$

EXAMPLE Table 4-1 describes the probability distribution for the number of girls among 14 randomly selected newborn babies. Assuming that we repeat the study of randomly selecting 14 newborn babies and counting the number of girls each time, find the mean number of girls (among 14), the variance, and the standard deviation. Use those results and the range rule of thumb to find the maximum and minimum usual values.

SOLUTION In Table 4-2, the two columns at the left describe the probability distribution given earlier in Table 4-1, and we create the three columns at the right for the purposes of the calculations required.

Using Formulas 4-1 and 4-3 and the table results, we get

$$\mu = \Sigma[x \cdot P(x)] = 6.993 = 7.0 \quad \text{(rounded)}$$
$$\sigma^2 = [\Sigma x^2 \cdot P(x)] - \mu^2$$
$$= 52.467 - 6.993^2 = 3.564951 = 3.6 \quad \text{(rounded)}$$

The standard deviation is the square root of the variance, so

$$\sigma = \sqrt{3.564951} = 1.9 \quad \text{(rounded)}$$

TABLE 4-2 Calculating μ, σ^2, and σ for a Probability Distribution

x	$P(x)$	$x \cdot P(x)$	x^2	$x^2 \cdot P(x)$
0	0.000	0.000	0	0.000
1	0.001	0.001	1	0.001
2	0.006	0.012	4	0.024
3	0.022	0.066	9	0.198
4	0.061	0.244	16	0.976
5	0.122	0.610	25	3.050
6	0.183	1.098	36	6.588
7	0.209	1.463	49	10.241
8	0.183	1.464	64	11.712
9	0.122	1.098	81	9.882
10	0.061	0.610	100	6.100
11	0.022	0.242	121	2.662
12	0.006	0.072	144	0.864
13	0.001	0.013	169	0.169
14	0.000	0.000	196	0.000
Total		6.993 ↑ $\Sigma[x \cdot P(x)]$		52.467 ↑ $\Sigma x^2 \cdot P(x)$

We now know that among groups of 14 newborn babies, the mean number of girls is 7.0, the variance is 3.6 "girls squared," and the standard deviation is 1.9 girls.

Using the range rule of thumb, we can now find the maximum and minimum usual values as follows:

$$\text{maximum usual value: } \mu + 2\sigma = 7.0 + 2(1.9) = 10.8$$
$$\text{minimum usual value: } \mu - 2\sigma = 7.0 - 2(1.9) = 3.2$$

INTERPRETATION Based on these results, we conclude that in groups of 14 randomly selected babies, the number of girls should usually fall between 3.2 and 10.8.

Identifying Unusual Results with Probabilities

Strong recommendation: The following discussion is recommended for mature audiences. It contains some difficult concepts, but it also contains an extremely important approach used in statistics. You should make every effort to understand this discussion, even if it requires several readings.

In the Chapter Problem, we noted that with the MicroSort technique, there were 13 girls among 14 babies. Is this result unusual? Does this result really suggest that the technique is effective, or could it be that there were 13 girls among 14 babies just by chance? To address this issue, we could use the range rule of thumb to find the minimum and maximum likely outcomes, but here we will consider another approach. We will find the probability of getting *13 or more* girls (not the probability of getting *exactly* 13 girls). It's difficult to see why this probability is the relevant probability, so let's try to clarify it with a more obvious example.

Suppose you were flipping a coin to determine whether it favors heads, and suppose 1000 tosses resulted in 501 heads. This is not evidence that the coin favors heads, because it is very easy to get a result like 501 heads in 1000 tosses just by chance. Yet, the probability of getting *exactly* 501 heads in 1000 tosses is actually quite small: 0.0252. This low probability reflects the fact that with 1000 tosses, any *specific* number of heads will have a very low probability. However, we do not consider 501 heads among 1000 tosses to be *unusual*, because the probability of getting *at least* 501 heads is high: 0.487. This principle can be generalized as follows:

Using Probabilities to Determine When Results Are Unusual

- x successes among n trials is unusually *high* if $P(x \text{ or more})$ is very small (such as less than 0.05)
- x successes among n trials is unusually *low* if $P(x \text{ or fewer})$ is very small (such as less than 0.05)

EXAMPLE **Gender Selection** Using the preceding two criteria based on probabilities, is it unusual to get 13 girls among 14 births? Does the MicroSort technique of gender selection appear to be effective?

SOLUTION Thirteen girls among 14 births is unusually high if P(13 or more girls) is very small. If we refer to Table A-1, we get this result:

$$\begin{aligned} P(13 \text{ or more girls}) &= P(13) + P(14) \\ &= 0.001 + 0.000 \\ &= 0.001 \end{aligned}$$

INTERPRETATION Because the probability 0.001 is so low, we conclude that it is unusual to get 13 girls among 14 births. This suggests that the MicroSort technique of gender selection appears to be effective, because it is highly unlikely that the result of 13 girls among 14 births happened by chance.

Expected Value

The mean of a discrete random variable is the theoretical mean outcome for infinitely many trials. We can think of that mean as the *expected value* in the sense that it is the average value that we would expect to get if the trials could continue indefinitely. The uses of expected value (also called *expectation,* or *mathematical expectation*) are extensive and varied, and they play a very important role in an area of application called *decision theory*.

DEFINITION

The **expected value** of a discrete random variable is denoted by E, and it represents the average value of the outcomes. It is obtained by finding the value of $\Sigma[x \cdot P(x)]$.

$$E = \Sigma[x \cdot P(x)]$$

From Formula 4-1 we see that $E = \mu$. That is, the mean of a discrete random variable is the same as its expected value. Repeat the procedure of flipping a coin five times, and the *mean* number of heads is 2.5; when flipping a coin five times, the *expected value* of the number of heads is also 2.5.

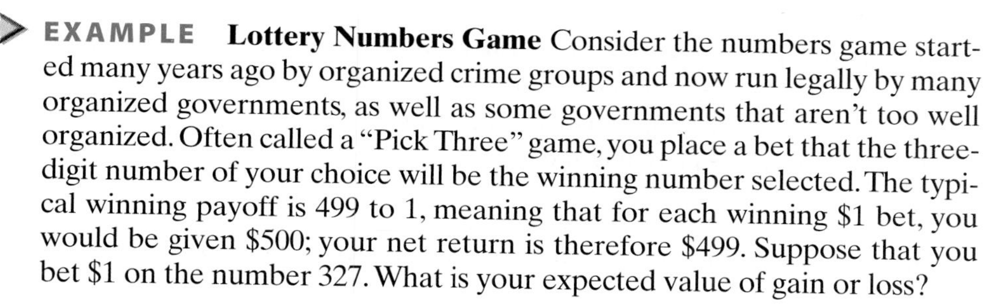

EXAMPLE Lottery Numbers Game Consider the numbers game started many years ago by organized crime groups and now run legally by many organized governments, as well as some governments that aren't too well organized. Often called a "Pick Three" game, you place a bet that the three-digit number of your choice will be the winning number selected. The typical winning payoff is 499 to 1, meaning that for each winning \$1 bet, you would be given \$500; your net return is therefore \$499. Suppose that you bet \$1 on the number 327. What is your expected value of gain or loss?

SOLUTION For this bet there are two simple outcomes: You win or you lose. Because you have selected the number 327 and there are 1000 possibilities (from 000 to 999), your probability of winning is 1/1000 (or 0.001) and your probability of losing is 999/1000 (or 0.999). Table 4-3 summarizes this situation.

From Table 4-3 we can see that when we bet \$1 in the numbers game, our expected value is

$$E = \Sigma[x \cdot P(x)] = -50¢$$

TABLE 4-3 The Numbers Game

Event	x	$P(x)$	$x \cdot P(x)$
Win	\$499	0.001	\$0.499
Lose	−\$1	0.999	−\$0.999
Total			−\$0.50 (or −50¢)

Picking Lottery Numbers

In a typical state lottery, you select six different numbers. After a random drawing, any entries with the correct combination share in the prize. Since the winning numbers are randomly selected, any choice of six numbers will have the same chance as any other choice, but some combinations are better than others. The combination of 1, 2, 3, 4, 5, 6 is a poor choice because many people tend to select it. In a Florida lottery with a \$105 million prize, 52,000 tickets had 1, 2, 3, 4, 5, 6; if that combination had won, the prize would have been only \$1000. It's wise to pick combinations not selected by many others. Avoid combinations that form a pattern on the entry card.

This means that in the long run, for each \$1 bet, we can expect to lose an average of 50¢. This is not a particularly sound investment scheme.

In the preceding example, a player will either lose \$1 or win \$499; there will never be a loss of 50¢, as the expected value of −50¢ might seem to suggest. The expected value of −50¢ is an average over a long run of bets placed. Even if we're thinking of placing only one bet, the expected value of −50¢ is a loss, so this is not a good bet. The potential gain is more than offset by the potential loss.

In this section we learned that a random variable has a numerical value associated with each outcome of some random procedure, and a probability distribution has a probability associated with each value of a random variable. We examined methods for finding the mean, variance, and standard deviation for a probability distribution. We saw that the expected value of a random variable is really the same as the mean. We also learned that lotteries are lousy investments.

4-2 Basic Skills and Concepts

Identifying Discrete and Continuous Random Variables. *In Exercises 1 and 2, identify the given random variable as being discrete or continuous.*

1. **a.** The weight of the cola in a randomly selected can1.
b. The cost of a randomly selected can of Coke
c. The time it takes to fill a can of Pepsi
d. The amount of cola (in ounces) in a can of Pepsi
e. The number of cans of Coke on a beverage delivery truck

2. **a.** The cost of a randomly selected textbook
b. The number of textbooks in a randomly selected bookstore
c. The weight of a randomly selected textbook
d. The number of pages in a randomly selected textbook
e. The time it takes an author to write a textbook (which is a whole lot longer than you might think)

Identifying Probability Distributions. *In Exercises 3–12, determine whether a probability distribution is given. In those cases where a probability distribution is not described, identify the requirements that are not satisfied. In those cases where a probability distribution is described, find its mean and standard deviation.*

3. Gender Selection In a study of the MicroSort gender-selection method, couples in a control group are not given a treatment, and they each have three children. The probability distribution for the number of girls is given in the accompanying table.

x	$P(x)$
0	0.125
1	0.375
2	0.375
3	0.125

4. TV Viewer Surveys When four different households are surveyed on a Monday night, the random variable x is the number of households with televisions tuned to *Monday Night Football* on ABC (based on data from Nielsen Media Research).

x	$P(x)$
0	0.522
1	0.368
2	0.098
3	0.011
4	0.001

5. Gender and Hiring If your college hires the next four employees without regard to gender, and the pool of applicants is large with an equal number of men and women, then the probability distribution for the number x of women hired is described in the accompanying table.

x	$P(x)$
0	0.0625
1	0.2500
2	0.3750
3	0.2500
4	0.0625

6. Videotape Rentals The accompanying table is constructed from data obtained in a study of the number of videotapes rented from Blockbuster.

x	$P(x)$
0	0.04
1	0.26
2	0.36
3	0.20
4	0.08
5	0.04
6	0.02

7. Prior Sentences When randomly selecting a jail inmate convicted of DWI (driving while intoxicated), the probability distribution for the number x of prior DWI sentences is as described in the accompanying table (based on data from the U.S. Department of Justice).

x	$P(x)$
0	0.512
1	0.301
2	0.132
3	0.055

8. Overbooked Flights Air America has a policy of routinely overbooking flights, because past experience shows that some passengers fail to show. The random variable x represents the number of passengers who cannot be boarded because there are more passengers than seats.

x	$P(x)$
0	0.805
1	0.113
2	0.057
3	0.009
4	0.002

9. Paternity Blood Test To settle a paternity suit, two different people are given blood tests. If x is the number having group A blood, then x can be 0, 1, or 2, and the corresponding probabilities are 0.36, 0.48, and 0.16, respectively (based on data from the Greater New York Blood Program).

10. Brand Recognition In a study of brand recognition of Sony, groups of four consumers are interviewed. If x is the number of people in the group who recognize the Sony brand name, then x can be 0, 1, 2, 3, or 4, and the corresponding probabilities are 0.0016, 0.0250, 0.1432, 0.3892, and 0.4096.

11. TV Viewers Five randomly selected households are surveyed on Sunday night, and x represents the number having televisions tuned to *60 Minutes*. The values of x are 0, 1, 2, 3, 4, and 5, and the corresponding probabilities are 0.289, 0.407, 0.229, 0.065, 0.009, and 0.001 (based on data from Nielsen Media Research).

12. Gender Bias in Media A study of gender bias in media coverage involves the selection of people appearing as the subjects in network TV evening news shows. The subjects are randomly selected in groups of four, and the numbers of women are recorded. The probabilities of getting 0, 1, 2, 3, and 4 women are 0.334, 0.421, 0.200, 0.042, and 0.003, respectively (based on data from *USA Today*).

13. Finding Expected Value in Roulette When you give a casino $5 for a bet on the number 7 in roulette, you have a 1/38 probability of winning $175 and a 37/38 probability of losing $5. If you bet $5 that the outcome is an odd number, the probability of winning $5 is 18/38, and the probability of losing $5 is 20/38.

a. If you bet $5 on the number 7, what is your expected value?
b. If you bet $5 that the outcome is an odd number, what is your expected value?
c. Which of these options is best: bet on 7, bet on odd, or don't bet? Why?

14. Finding Expected Value in Craps When you give a casino \$5 for a bet on the "pass line" in the game of craps, there is a 244/495 probability that you will win \$5 and a 251/495 probability that you will lose \$5. What is your expected value? In the long run, how much do you lose for each dollar bet?

15. Finding Expected Value for a Life Insurance Policy The CNA Insurance Company charges Mike \$250 for a one-year \$100,000 life insurance policy. Because Mike is a 21-year-old male, there is a 0.9985 probability that he will live for a year (based on data from U.S. National Center for Health Statistics).

a. From Mike's perspective, what are the values of the two different outcomes?
b. If Mike purchases the policy, what is his expected value?
c. What would be the cost of the insurance policy if the company just breaks even (in the long run with many such policies), instead of making a profit?

16. Finding Expected Value for a Magazine Sweepstakes *Reader's Digest* ran a sweepstakes in which prizes were listed along with the chances of winning: \$5,000,000 (1 chance in 201,000,000), \$150,000 (1 chance in 201,000,000), \$100,000 (1 chance in 201,000,000), \$25,000 (1 chance in 100,500,000), \$10,000 (1 chance in 50,250,000), \$5000 (1 chance in 25,125,000), \$200 (1 chance in 8,040,000), \$125 (1 chance in 1,005,000), and a watch valued at \$89 (1 chance in 3774).

a. Find the expected value of the amount won for one entry.
b. Find the expected value if the cost of entering this sweepstakes is the cost of a postage stamp.

17. Brand Recognition Focus groups are often used for finding detailed information about a product. A focus group of 12 people is randomly selected to discuss products of the Coca Cola Company. In such groups of 12 people, the mean number who recognize the Coca Cola brand name is 11.4, and the standard deviation is 0.75 (based on data from Total Research Corporation). Would it be unusual to randomly select 12 people and find that fewer than eight of them recognize the Coca Cola brand name?

18. Number of Games in a Baseball World Series Based on past results found in the *Information Please Almanac*, there is a 0.120 probability that a baseball World Series contest will last four games, a 0.253 probability that it will last five games, a 0.217 probability that it will last six games, and a 0.410 probability that it will last seven games. Find the mean and standard deviation for the numbers of games that World Series contests lasts. Is it unusual for a team to "sweep" by winning in four games?

x	$P(x)$
0	0.402
1	0.402
2	?
3	0.032
4	0.003
5	0.000

19. Rolling Dice Consider the procedure of rolling a pair of dice 5 times and letting the random variable x represent the number of times that 7 occurs. The accompanying table describes the probability distribution.

a. Find the value of the missing probability.
b. Would it be unusual to roll a pair of dice five times and get at least three 7s? Why or why not?

x	$P(x)$
0	?
1	0.412
2	0.265
3	0.075
4	0.008

20. Car Crash Fatalities An analyst for the National Insurance Company randomly selects deaths of people aged 5–24 and determines the cause of death. The accompanying table describes the probability distribution for four randomly selected people, where x is the number who died as the result of a motor vehicle crash (based on data from the U. S. National Center for Health Statistics). Find the missing probability, then find the mean and standard deviation for the random variable x. Is it unusual to have no motor vehicle fatalities among four randomly selected deaths of people aged 5–24?

21. Determining Whether Gender-Selection Technique Is Effective Refer to Table 4-1 and find the probability of getting 10 or more girls among 14 births. If you are testing a gender-selection technique and you get 10 girls among 14 births, does it appear that the method is effective? Why or why not?

22. Determining Whether Gender-Selection Technique Is Effective Refer to Table 4-1 and find the probability of getting three or fewer girls among 14 births. If you are testing a gender-selection technique and you get three girls among 14 births, does it appear that the method is effective? Why or why not?

4-2 Beyond the Basics

23. Finding Mean and Standard Deviation Let the random variable x represent the number of girls in a family of four children. Construct a table describing the probability distribution, then find the mean and standard deviation. (*Hint:* List the different possible outcomes.)

24. Defective Parts: Finding Mean and Standard Deviation The Sky Ranch is a supplier of aircraft parts. Included in stock are eight altimeters that are correctly calibrated and two that are not. Three altimeters are randomly selected without replacement. Let the random variable x represent the number that are not correctly calibrated. Find the mean and standard deviation for the random variable x.

25. Identifying Probability Distributions In each case, determine whether the given function is a probability distribution.

a. $P(x) = 1/4$, where $x = 1, 2, 3, 4$
b. $P(x) = 1/[2(2 - x)!x!]$, where $x = 0, 1, 2$
c. $P(x) = 1/2x$, where $x = 1, 2, 3, \ldots$
d. $P(x) = 1/2^x$, where $x = 1, 2, 3, \ldots$

26. Determining Effects of Transforming Data Assume that x is a random variable in a probability distribution with mean μ and standard deviation σ. Find expressions for the mean and standard deviation if every value of x is modified as follows.

a. Every value of x is increased by 3.
b. Every value of x is doubled.
c. Every value of x is first tripled, then increased by 5.

27. Computer-Generated Numbers Transformed to z Scores Computers are often used to randomly generate the last digits of telephone numbers of potential survey subjects. The digits are selected so that they are all equally likely. The random variable x is the selected digit.

a. Find the mean and standard deviation of the random variable x.
b. Find the z score for each value of x, then find the mean and standard deviation of these z scores.
c. Will the same mean and standard deviation as you found in part (b) result from every probability distribution?

28. Equally Likely Integers: Mean and Standard Deviation Assume that a probability distribution is described by the discrete random variable x that can assume the values $1, 2, \ldots, n$, and those values are equally likely. This probability distribution has mean and standard deviation described as follows.

$$\mu = \frac{n+1}{2} \quad \text{and} \quad \sigma = \sqrt{\frac{n^2 - 1}{12}}$$

a. Show that $\mu = (n + 1)/2$ for the case of $n = 5$.
b. Show that $\sigma = \sqrt{(n^2 - 1)/12}$ for the case of $n = 5$.
c. In testing someone who claims to have ESP, you randomly select whole numbers between 1 and 20, and the random variable x is the number selected. Find the mean and standard deviation for x.

29. Labeling Dice to Get a Uniform Distribution Assume that you have two blank dice, so that you can label the 12 faces with any numbers. Describe how the dice can be labeled so that, when the two dice are rolled, the totals of the two dice are uniformly distributed so that the outcomes of 1, 2, 3, . . . , 12 each have probability 1/12. (See "Can One Load a Set of Dice So That the Sum Is Uniformly Distributed?" by Chen, Rao, and Shreve, *Mathematics Magazine*, Vol. 70, No. 3.)

4-3 Binomial Probability Distributions

In Section 4-2 we discussed several different discrete probability distributions, but in this section we will focus on one specific type: the binomial probability distribution. Binomial probability distributions are important because they allow us to deal with circumstances in which the outcomes belong to *two* relevant categories, such as pass/fail, or acceptable/defective. The Chapter Problem involves counting the number of girls in 14 births. The problem involves the two categories of boy/girl, so it has the required key element of "twoness." Other requirements are given in the following definition.

DEFINITION

A **binomial probability distribution** results from a procedure that meets all the following requirements:

1. The procedure has a *fixed number of trials.*
2. The trials must be *independent.* (The outcome of any individual trial doesn't affect the probabilities in the other trials.)
3. Each trial must have all outcomes classified into *two categories.*
4. The probabilities must remain *constant* for each trial.

If a procedure satisfies these four requirements, the distribution of the random variable x is called a *binomial probability distribution* (or *binomial distribution*). The following notation is commonly used.

Notation for Binomial Probability Distributions

S and F (success and failure) denote the two possible categories of all outcomes; p and q will denote the probabilities of S and F, respectively, so

$P(\text{S}) = p$ (p = probability of a success)

$P(\text{F}) = 1 - p = q$ (q = probability of a failure)

n	denotes the fixed number of trials.
x	denotes a specific number of successes in n trials, so x can be any whole number between 0 and n, inclusive.
p	denotes the probability of *success* in *one* of the n trials.
q	denotes the probability of *failure* in *one* of the n trials.
$P(x)$	denotes the probability of getting exactly x successes among the n trials.

The word *success* as used here is arbitrary and does not necessarily represent something good. Either of the two possible categories may be called the success S as long as the corresponding probability is identified as p. Once a category has been designated as the success S, be sure that p is the probability of a success and x is the number of successes. That is, be sure that the values of p and x refer to the same category designated as a success. (The value of q can always be found by subtracting p from 1; if $p = 0.95$, then $q = 1 - 0.95 = 0.05$.) Here is an important hint for working with binomial probability problems:

Be sure that x and p both refer to the *same* category being called a success.

One very common application of statistics involves sampling without replacement, as in testing manufactured items or conducting surveys. Strictly speaking, sampling without replacement involves dependent events, which violates the second requirement in the definition. However, the following rule of thumb is based on the fact that if the sample is very small relative to the population size, the difference in results will be negligible if we treat the trials as independent when they are actually dependent.

When sampling without replacement, the events can be treated as if they were independent if the sample size is no more than 5% of the population size. (That is, $n \leq 0.05N$.)

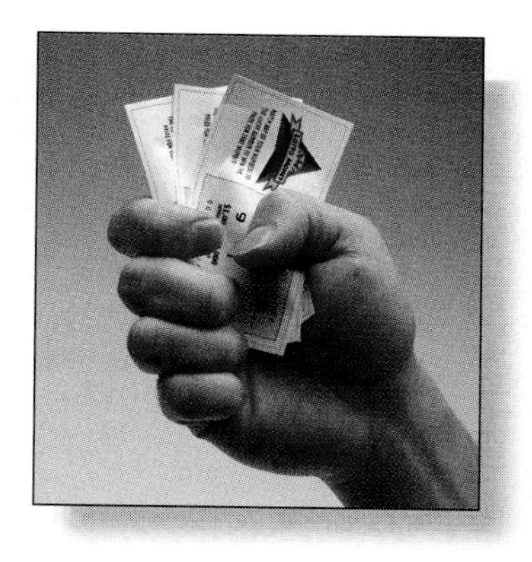

Prophets for Profits

Many books and computer programs claim to be helpful in predicting winning lottery numbers. Some use the theory that particular numbers are "due" (and should be selected) because they haven't been coming up often; others use the theory that some numbers are "cold" (and should be avoided) because they haven't been coming up often; and still others use astrology, numerology, or dreams. Because selections of winning lottery number combinations are independent events, such theories are worthless. A valid approach is to choose numbers that are "rare" in the sense that they are not selected by other people, so that if you win, you will not need to share your jackpot with many others. For this reason, the combination of 1, 2, 3, 4, 5, and 6 is a bad choice because many people use it, whereas 12, 17, 18, 33, 40, 46 is a much better choice, at least until it was published in this book.

EXAMPLE **Using Directory Assistance** AT&T claims that when customers call directory assistance for telephone numbers, the right number is given 90% of the time. Assume a 90% rate of correct responses, and assume that we want to find the probability that among 5 requests, 3 of the responses are correct.

a. Does this procedure result in a binomial distribution?

b. If this procedure does result in a binomial distribution, identify the values of n, x, p, and q.

SOLUTION

a. This procedure does satisfy the requirements for a binomial distribution, as shown below.

1. The number of trials (5) is fixed.

2. The 5 trials are independent because different telephone numbers were used and different operators were reached.
3. Each of the 5 trials has two categories of outcomes: The provided telephone number was either right or wrong.
4. The probability of 0.9 (from 90%) is assumed to remain constant for each of the 5 trials.

b. Having concluded that the given procedure does result in a binomial distribution, we now proceed to identify the values of n, x, p, and q.

1. With 5 requests for telephone numbers, we have $n = 5$.
2. We want the probability for 3 correct responses, so $x = 3$.
3. The probability of a correct response (success) is assumed to be 0.9 (from 90%), so $p = 0.9$.
4. The probability of failure (wrong response) is 0.1, so $q = 0.1$.

Again, it is very important to be sure that x and p both refer to the same concept of "success." In this example, we use x to count the *correct* responses, so p must be the probability of a *correct* response. Therefore, x and p do use the same concept of success (correct response) here.

We will now present three methods for finding the probabilities corresponding to the random variable x in a binomial distribution. The first method involves calculations using the *binomial probability formula* and is the basis for the other two methods. The second method involves the use of Table A-1, and the third method involves the use of statistical software or a calculator. We will describe the three methods, illustrate them, and then provide a rationale for each one. (Note that some instructors prefer to use only one or two of these three methods, so be sure that you know your own instructor's preference. If you are using software or a calculator that automatically produces binomial probabilities, we recommend that you solve one or two exercises using Method 1 to ensure that you understand the basis for the calculations.)

Method 1: Using the Binomial Probability Formula In a binomial distribution, probabilities can be calculated by using the binomial probability formula.

Formula 4-5 $$P(x) = \frac{n!}{(n-x)!x!} \cdot p^x \cdot q^{n-x} \qquad \text{for } x = 0, 1, 2, \ldots, n$$

where
n = number of trials
x = number of successes among n trials
p = probability of success in any one trial
q = probability of failure in any one trial ($q = 1 - p$)

The factorial symbol !, introduced in Section 3-7, denotes the product of decreasing factors. Two examples of factorials are $3! = 3 \cdot 2 \cdot 1 = 6$ and $0! = 1$ (by definition). Many calculators have a factorial key, as well as a key labeled ${}_nC_r$ that can simplify the computations. For calculators with the ${}_nC_r$

key, use this version of the binomial probability formula (where n, x, p, and q are the same as in Formula 4-5):

$$P(x) = {}_nC_x \cdot p^x \cdot q^{n-x}$$

The TI-83 Plus calculator is designed to automatically calculate binomial probabilities using this formula. The use of the TI-83 Plus calculator will be discussed under Method 3.

Special note if you are planning to take The College Board's AP Statistics exam: We denote the probability of a failure by q, but the College Board uses $1 - p$ instead of q, and they give the binomial probability formula in this format:

$$P(X = k) = \binom{n}{k} p^k (1 - p)^{n-k}$$

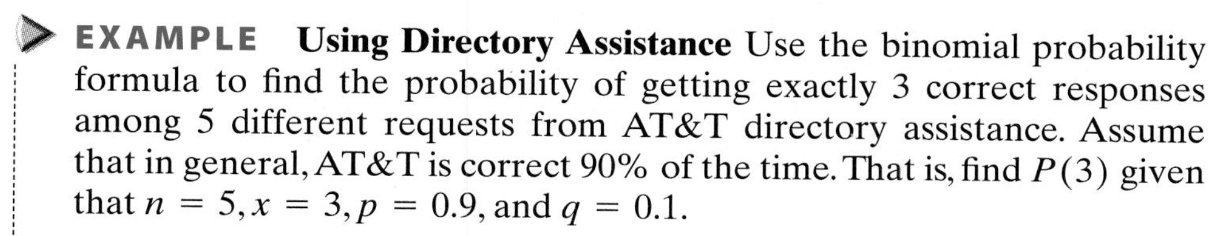

EXAMPLE **Using Directory Assistance** Use the binomial probability formula to find the probability of getting exactly 3 correct responses among 5 different requests from AT&T directory assistance. Assume that in general, AT&T is correct 90% of the time. That is, find $P(3)$ given that $n = 5, x = 3, p = 0.9$, and $q = 0.1$.

SOLUTION Using the given values of n, x, p, and q in the binomial probability formula (Formula 4-5), we get

$$\begin{aligned} P(3) &= \frac{5!}{(5-3)!3!} \cdot 0.9^3 \cdot 0.1^{5-3} \\ &= \frac{5!}{2!3!} \cdot 0.729 \cdot 0.01 \\ &= (10)(0.729)(0.01) = 0.0729 \end{aligned}$$

The probability of getting exactly 3 correct responses for 5 requests is 0.0729.

Calculation hint: When computing a probability with the binomial probability formula, it's helpful to get a single number for $n!/(n-x)!x!$, a single number for p^x, and a single number for q^{n-x}, then simply multiply the three factors together. Don't round too much when you find those three factors; round only at the end.

Method 2: Using Table A-1 in Appendix A In some cases, we can easily find binomial probabilities by simply referring to Table A-1 in Appendix A. First locate n and the corresponding value of x that is desired. At this stage, one row of numbers should be isolated. Now align that row with the proper probability of p by using the column across the top. The isolated number represents the desired probability. A very small probability, such as 0.000000345, is indicated by 0.0+.

Voltaire Beats Lottery

In 1729, the philosopher Voltaire became rich by devising a scheme to beat the Paris lottery. The government ran a lottery to repay municipal bonds that had lost some value. The city added large amounts of money with the net effect that the prize values totaled more than the cost of all tickets. Voltaire formed a group that bought all the tickets in the monthly lottery and won for more than a year. A bettor in the New York State Lottery tried to win a share of an exceptionally large prize that grew from a lack of previous winners. He wanted to write a \$6,135,756 check that would cover all combinations, but the state declined and said that the nature of the lottery would have been changed.

From Table A-1:

n	x	p 0.90
5	0	0.0+
	1	0.0+
	2	0.008
	3	0.073
	4	0.328
	5	0.590

Binomial probability distribution for $n = 5$ and $p = 0.90$

x	$P(x)$
0	0.000
1	0.000
2	0.008
3	0.073
4	0.328
5	0.590

Part of Table A-1 is shown in the margin. When $n = 5$ and $p = 0.90$ in a binomial distribution, the probabilities of 0, 1, 2, 3, 4, and 5 successes are 0.0+, 0.0+, 0.008, 0.073, 0.328, and 0.590, respectively.

EXAMPLE Use the portion of Table A-1 (for $n = 5$ and $p = 0.90$) shown in the margin to find the following.

a. The probability of exactly 3 successes

b. The probability of *at least* 3 successes

SOLUTION

a. The display from Table A-1 shows that when $n = 5$ and $p = 0.90$, the probability of $x = 3$ is given by $P(3) = 0.073$, which is the same value (except for rounding) computed with the binomial probability formula in the preceding example.

b. "At least" 3 successes means that the number of successes is 3 or 4 or 5.

$$\begin{aligned} P(\text{at least } 3) &= P(3 \text{ or } 4 \text{ or } 5) \\ &= P(3) + P(4) + P(5) \\ &= 0.073 + 0.328 + 0.590 \\ &= 0.991 \end{aligned}$$

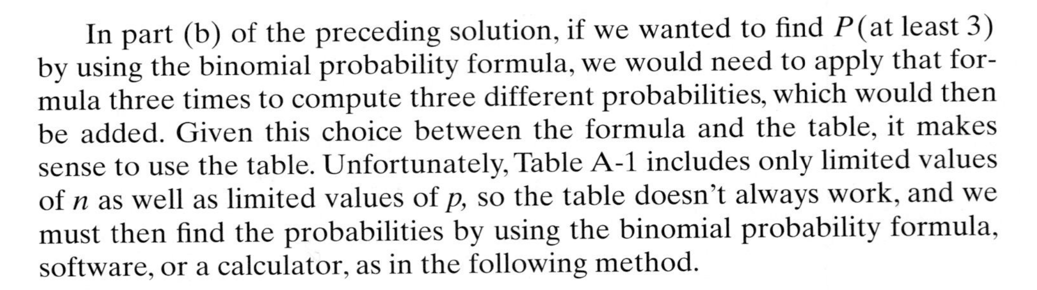

In part (b) of the preceding solution, if we wanted to find $P(\text{at least } 3)$ by using the binomial probability formula, we would need to apply that formula three times to compute three different probabilities, which would then be added. Given this choice between the formula and the table, it makes sense to use the table. Unfortunately, Table A-1 includes only limited values of n as well as limited values of p, so the table doesn't always work, and we must then find the probabilities by using the binomial probability formula, software, or a calculator, as in the following method.

Method 3: Using Technology

STATDISK: Select **Analysis** from the main menu, then select the **Binomial Probabilities** option. Enter the requested values for n and p, and the entire probability distribution will be displayed. Other columns represent cumulative probabilities obtained by adding the values of $P(x)$ as you go down or up the column.

Minitab: First enter a column C1 of the x values for which you want probabilities (such as 0, 1, 2, 3, 4, 5), then select **Calc** from the main menu, and proceed to select the submenu items of **Probability Distributions** and **Binomial.** Enter the number of trials, the probability of success, and C1 for the input column, then click on **OK.**

Excel: List the values of x in column A. Click on cell B1, then click on f_x from the toolbar, and select the function category **Statistical** and then the function name **BINOMDIST**. In the dialog box, enter A1 for the number of successes, enter the number of trials, enter the probability, and enter 0 for the binomial distribution (instead of 1 for the cumulative binomial distribution). A value should appear in cell B1. Click and drag the lower right corner of cell B1 down the column to match the entries in column A, then release the mouse button.

TI-83 Plus: Press **2nd VARS** (to get **DISTR,** which denotes "distributions"), then select the option identified as **binompdf(.** Complete the entry of **binompdf(n, p, x)** with specific values for n, p, and x, then press **ENTER,** and the result will be the probability of getting x successes among n trials. You could also enter **binompdf(n, p)** to get a list of *all* of the probabilities corresponding to $x = 0, 1, 2, \ldots, n$; you could store this list in L2 by pressing **STO → L2.** You could then enter the values of $0, 1, 2, \ldots, n$ in list L1, which would allow you to calculate statistics (by entering **STAT, CALC,** then **L1, L2**) or view the distribution in a table format (by pressing **STAT,** then **EDIT**).

The following are samples of output from STATDISK, Minitab, Excel, and the TI-83 Plus calculator obtained from a binomial distribution in which $n = 5$ and $p = 0.90$.

	A	B
1	0	1E-05
2	1	0.00045
3	2	0.0081
4	3	0.0729
5	4	0.32805
6	5	0.59049

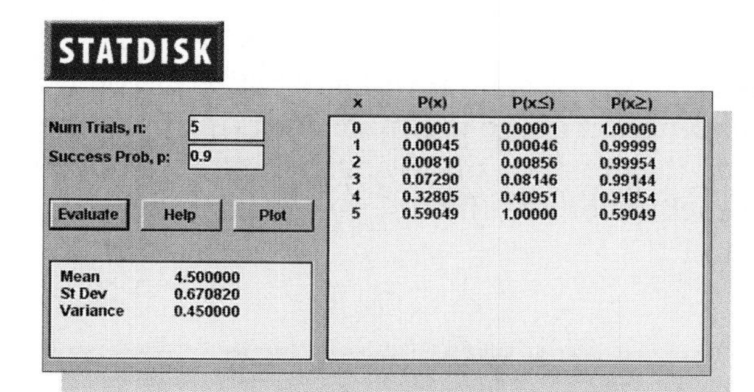

STATDISK

Num Trials, n: 5
Success Prob, p: 0.9

Evaluate Help Plot

Mean 4.500000
St Dev 0.670820
Variance 0.450000

x	P(x)	P(x≤)	P(x≥)
0	0.00001	0.00001	1.00000
1	0.00045	0.00046	0.99999
2	0.00810	0.00856	0.99954
3	0.07290	0.08146	0.99144
4	0.32805	0.40951	0.91854
5	0.59049	1.00000	0.59049

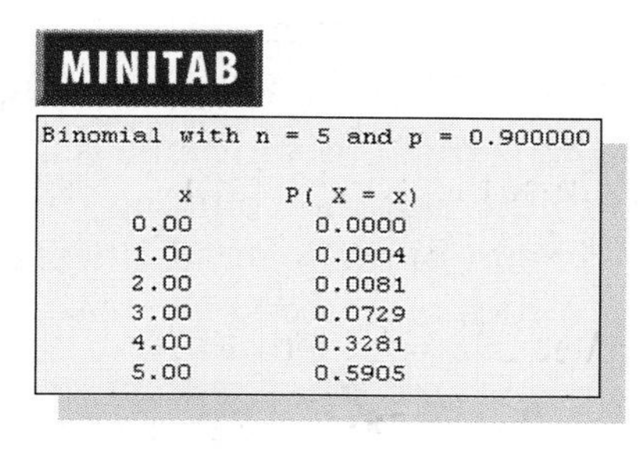

MINITAB

Binomial with n = 5 and p = 0.900000

x	P(X = x)
0.00	0.0000
1.00	0.0004
2.00	0.0081
3.00	0.0729
4.00	0.3281
5.00	0.5905

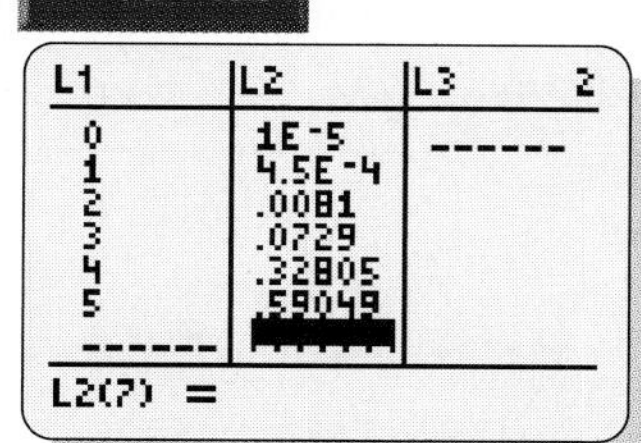

L1	L2	L3
0	1E-5	------
1	4.5E-4	
2	.0081	
3	.0729	
4	.32805	
5	.59049	

L2(7) =

Here is a good strategy for choosing the best method for finding binomial probabilities:

1. Use computer software or a TI-83 Plus calculator, if available.
2. If neither software nor the TI-83 Plus calculator is available, use Table A-1, if possible.
3. If neither software nor the TI-83 Plus calculator is available and the probabilities can't be found using Table A-1, use the binomial probability formula.

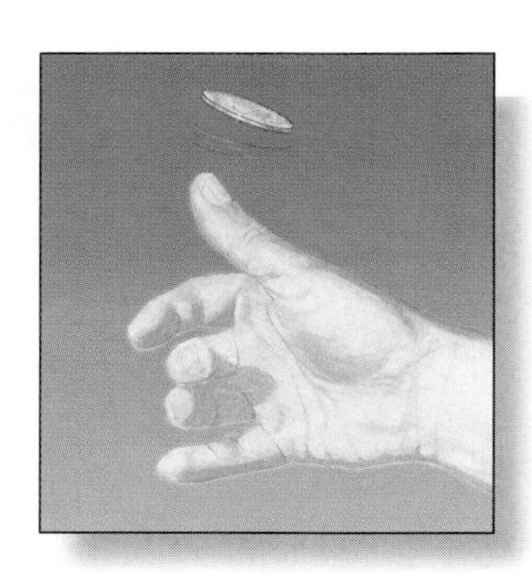

Sensitive Surveys

Survey respondents are sometimes reluctant to honestly answer questions on a sensitive topic, such as employee theft or sex. Stanley Warner (York University, Ontario) devised a scheme that leads to more accurate results in such cases. As an example, ask employees if they stole within the past year and also ask them to flip a coin. The employees are instructed to answer no if they didn't steal and the coin turns up heads. Otherwise, they should answer yes. The employees are more likely to be honest because the coin flip helps protect their privacy. Probability theory can then be used to analyze responses so that more accurate results can be obtained.

Rationale for the Binomial Probability Formula

The binomial probability formula is the basis for all three methods presented in this section. Instead of accepting and using that formula blindly, let's see why it works.

Earlier in this section, we used the binomial probability formula for finding the probability of getting exactly three correct responses among five different requests from AT&T directory assistance. Recall that we used a 90% rate of correct responses. If we use the multiplication rule from Section 3-4, we get the following result:

$$\begin{aligned} &P(\text{3 correct responses followed by 2 wrong responses}) \\ &\quad = 0.9 \cdot 0.9 \cdot 0.9 \cdot 0.1 \cdot 0.1 \\ &\quad = 0.9^3 \cdot 0.1^2 \\ &\quad = 0.00729 \end{aligned}$$

This result does not yield the correct answer because it assumes that the *first* three responses are correct and the *last* two responses are wrong, but there are many other arrangements possible for three correct responses and two wrong responses.

In Section 3-7 we saw that with three items identical to each other (such as *correct* responses) and two other items identical to each other (such as *wrong* responses), the total number of arrangements (permutations) is $5!/(5-3)!\,3!$, or 10. Each of those 10 different arrangements has a probability of $0.9^3 \cdot 0.1^2$, so the total probability is as follows:

$$P(\text{3 correct among 5}) = \frac{5!}{(5-3)!\,3!} \cdot 0.9^3 \cdot 0.1^2$$

Generalize this result as follows: Replace 5 with n, replace x with 3, replace 0.9 with p, replace 0.1 with q, and express 2 as $5 - 3$, which can be replaced with $n - x$. The result is the binomial probability formula. That is, the binomial probability formula is a combination of the multiplication rule of probability and the counting rule for the number of arrangements of n items when x of them are identical to each other and the other $n - x$ are identical to each other. (See Exercises 9 and 10.)

The number of outcomes with exactly x successes among n trials

The probability of x successes among n trials for any one particular order

$$P(x) = \frac{n!}{(n-x)!\,x!} \cdot p^x \cdot q^{n-x}$$

It is important to remember that the binomial probability distribution is one of many probability distributions that can be used for different situations. Because it is designed for situations with two categories of outcomes and a fixed number of independent trials, the binomial distribution is commonly used in applications such as quality control, voter analysis, medical research, advertising, and surveys. Other distributions are considered in Exercises 37–39 and in Section 4-5.

4-3 Basic Skills and Concepts

Identifying Binomial Distributions. *In Exercises 1–8, determine whether the given procedure results in a binomial distribution. For those that are not binomial, identify at least one requirement that is not satisfied.*

1. Guessing the answer to 20 multiple-choice test questions, then determining whether the answers are correct or wrong

2. Rolling a fair die 50 times

3. Rolling a loaded die 50 times

4. Rolling a loaded die 50 times and finding the number of times that 5 occurs

5. Surveying 1000 college students by asking them how many credits they are currently taking

6. Surveying 1000 college students by asking them if they recognize the brand name Microsoft

7. Spinning a roulette wheel 12 times

8. Spinning a roulette wheel 12 times and finding the number of times that the outcome is an odd number.

9. Finding Probabilities When Guessing Answers Multiple-choice questions each have five possible answers, one of which is correct. Assume that you guess the answers to three such questions.

a. Use the multiplication rule to find the probability that the first two guesses are wrong and the third is correct. That is, find $P(\text{WWC})$, where C denotes a correct answer and W denotes a wrong answer.

b. Beginning with WWC, make a complete list of the different possible arrangements of two wrong answers and one correct answer, then find the probability for each entry in the list.

c. Based on the preceding results, what is the probability of getting exactly one correct answer when three guesses are made?

10. Finding Probabilities When Guessing Answers A test consists of multiple-choice questions, each having four possible answers, one of which is correct. Assume that you guess the answers to six such questions.

a. Use the multiplication rule to find the probability that the first two guesses are wrong and the last four guesses are correct. That is, find $P(\text{WWCCCC})$, where C denotes a correct answer and W denotes a wrong answer.

b. Beginning with WWCCCC, make a complete list of the different possible arrangements of two wrong answers and four correct answers, then find the probability for each entry in the list.

c. Based on the preceding results, what is the probability of getting exactly four correct answers when six guesses are made?

Using Table A-1. *In Exercises 11–16, assume that a procedure yields a binomial distribution with a trial repeated n times. Use Table A–1 to find the probability of x successes given the probability p of success on a given trial.*

11. $n = 7, x = 1, p = 0.01$

12. $n = 3, x = 2, p = 0.99$

13. $n = 8, x = 3, p = 0.60$

14. $n = 9, x = 3, p = 0.2$

15. $n = 11, x = 4, p = 0.01$

16. $n = 12, x = 5, p = 0.05$

Using the Binomial Probability Formula. *In Exercises 17–20, assume that a procedure yields a binomial distribution with a trial repeated n times. Use the binomial probability formula to find the probability of x successes given the probability p of success on a single trial.*

17. $n = 5, x = 3, p = 0.25$

18. $n = 8, x = 6, p = 0.75$

19. $n = 10, x = 4, p = 1/3$

20. $n = 12, x = 3, p = 2/3$

Using Computer Results. *In Exercises 21–24, refer to the Minitab display in the margin. The probabilities were obtained by entering the values of $n = 5$ and $p = 0.54$. There is a 0.54 probability that a randomly selected freshman at a two-year college will return the second year (based on data from the College Board). In each case, assume that 5 freshmen at a two-year college are randomly selected and find the indicated probability.*

BINOMIAL WITH

$n = 5$ and $p = 0.54$

x	$P(X = x)$
0.00	0.0206
1.00	0.1209
2.00	0.2838
3.00	0.3332
4.00	0.1956
5.00	0.0459

21. Find the probability that at least four of the freshmen return for the second year.

22. Find the probability that at most two of the freshmen return for the second year.

23. Find the probability that more than one of the freshmen return for the second year.

24. Find the probability that at least one of the freshmen return for the second year.

In Exercises 25–36, find the probability requested.

25. IRS Audits The Hemingway Financial Company prepares tax returns for individuals. (Motto: "We also write great fiction.") According to the Internal Revenue Service, individuals making $25,000–$50,000 are audited at a rate of 1%. The Hemingway Company prepares five tax returns for individuals in that tax bracket, and three of them are audited.
- **a.** Find the probability that when 5 people making $25,000–$50,000 are randomly selected, exactly 3 of them are audited.
- **b.** Find the probability that at least three are audited.
- **c.** Based on the preceding results, what can you conclude about the Hemingway customers? Are they just unlucky, or are they being targeted for audits?

26. On-Time Flights The rates of on-time flights for commercial jets are continuously tracked by the U.S. Department of Transportation. Recently, Southwest Air had the best rate with 80% of its flights arriving on time. A test is conducted by randomly selecting 15 Southwest flights and observing whether they arrive on time.
- **a.** Find the probability that exactly 10 flights arrive on time.
- **b.** Find the probability that at least 10 flights arrive on time.
- **c.** Find the probability that at least 10 flights arrive late.
- **d.** Would it be unusual for Southwest to have 5 flights arrive late? Why or why not?

27. Directory Assistance An article in *USA Today* stated that "Internal surveys paid for by directory assistance providers show that even the most accurate companies give out wrong numbers 15% of the time." Assume that you are testing such a provider by making 10 requests and also assume that the provider gives the wrong number 15% of the time.
- **a.** Find the probability of getting one wrong number.
- **b.** Find the probability of getting at most one wrong number.
- **c.** If you do get at most one wrong number, does it appear that the rate of wrong numbers is not 15%?

28. Taking Courses after Graduation The Market Research Institute found that among employed college graduates aged 30–55 and out of college for at least 10 years, 57% have taken college courses after graduation (as reported in *USA Today*). If a dozen college graduates aged 30–55 and out of college for at least 10 years are randomly selected, find the probability that seven of them have taken college courses after graduation.

29. Color Blindness Nine percent of men and 0.25% of women cannot distinguish between the colors red and green. This is the type of color blindness that causes problems with traffic signals. If six men are randomly selected for a study of traffic signal perceptions, find the probability that exactly two of them cannot distinguish between red and green.

30. TV Viewer Surveys The CBS television show *60 Minutes* has been successful for many years. That show recently had a share of 20, meaning that among the TV sets in use, 20% were tuned to *60 Minutes* (based on data from Nielsen Media Research). Assume that an advertiser wants to verify that 20% share value by conducting its own survey, and a pilot survey begins with 10 households having TV sets in use at the time of a *60 Minutes* broadcast.

a. Find the probability that none of the households are tuned to *60 Minutes.*

b. Find the probability that at least one household is tuned to *60 Minutes.*

c. Find the probability that at most one household is tuned to *60 Minutes.*

d. If at most one household is tuned to *60 Minutes,* does it appear that the 20% share value is wrong? Why or why not?

31. Affirmative Action Programs A study was conducted to determine whether there were significant differences between medical students admitted through special programs (such as affirmative action) and medical students admitted through the regular admissions criteria. It was found that the graduation rate was 94% for the medical students admitted through special programs (based on data from the *Journal of the American Medical Association*).

a. If 10 of the students from the special programs are randomly selected, find the probability that at least 9 of them graduated.

b. Would it be unusual to randomly select 10 students from the special programs and get only 7 that graduate? Why or why not?

32. College Student Housing In a housing study, it was found that 26% of college students live in campus housing (based on data from the Independent Insurance Agents of America). The Providence Insurance Company wants to sell those students special policies insuring their personal property. If they test a marketing strategy by randomly selecting six college students, what is the probability that at least one of them lives in campus housing?

33. Guessing Answers A statistics quiz consists of 10 multiple-choice questions, each with five possible answers. For someone who makes random guesses for all of the answers, find the probability of passing if the minimum passing grade is 60%. Is the probability high enough to make it worth the risk of trying to pass by making random guesses instead of studying?

34. Overbooking Flights Air America has a policy of booking as many as 15 persons on an airplane that can seat only 14. (Past studies have revealed that only 85% of the booked passengers actually arrive for the flight.) Find the probability that if Air America books 15 persons, not enough seats will be available.

35. Acceptance Sampling The Telektronic Company purchases large shipments of fluorescent bulbs and uses this acceptance sampling plan: Randomly select and test 24 bulbs, then accept the whole batch if there is only one or none that doesn't work. If a particular shipment of thousands of bulbs actually has a 4% rate of defects, what is the probability that this whole shipment will be accepted?

36. Identifying Gender Discrimination After being rejected for employment, Kim Kelly learns that the Bellevue Advertising Company has hired only two women among the last 20 new employees. She also learns that the pool of applicants is very large, with an approximately equal number of qualified men and women. Help her address the charge of gender discrimination by finding the probability of getting two or fewer women when 20 people are hired, assuming that there is no discrimination based on gender. Does the resulting probability really support such a charge?

4-3 Beyond the Basics

37. If a procedure meets all the conditions of a binomial distribution except that the number of trials is not fixed, then the **geometric distribution** can be used. The probability of getting the first success on the xth trial is given by $P(x) = p(1 - p)^{x-1}$, where p is the probability of success on any one trial. Assume that the probability of a defective computer component is 0.2. Find the probability that the first defect is found in the seventh component tested.

38. If we sample from a small finite population without replacement, the binomial distribution should not be used because the events are not independent. If sampling is done without replacement and the outcomes belong to one of two types, we can use the **hypergeometric distribution.** If a population has A objects of one type, while the remaining B objects are of the other type, and if n objects are sampled without replacement, then the probability of getting x objects of type A and $n - x$ objects of type B is

$$P(x) = \frac{A!}{(A - x)!x!} \cdot \frac{B!}{(B - n + x)!(n - x)!} \div \frac{(A + B)!}{(A + B - n)!n!}$$

In Lotto 54, a bettor selects six numbers from 1 to 54 (without repetition), and a winning six-number combination is later randomly selected. Find the probability of getting

a. all six winning numbers.
b. exactly five of the winning numbers.
c. exactly three of the winning numbers.
d. no winning numbers.

39. The binomial distribution applies only to cases involving two types of outcomes, whereas the **multinomial distribution** involves more than two categories. Suppose we have three types of mutually exclusive outcomes denoted by A, B, and C. Let $P(\text{A}) = p_1$, $P(\text{B}) = p_2$, and $P(\text{C}) = p_3$. In n independent trials, the probability of x_1 outcomes of type A, x_2 outcomes of type B, and x_3 outcomes of type C is given by

$$\frac{n!}{(x_1!)(x_2!)(x_3!)} \cdot p_1^{x_1} \cdot p_2^{x_2} \cdot p_3^{x_3}$$

(continued)

A genetics experiment involves six mutually exclusive genotypes identified as A, B, C, D, E, and F, and they are all equally likely. If 20 offspring are tested, find the probability of getting exactly five A's, four B's, three C's, two D's, three E's, and three F's by expanding the above expression so that it applies to six types of outcomes instead of only three.

4-4 Mean, Variance, and Standard Deviation for the Binomial Distribution

In Chapter 2 we explored actual collections of real data, and we focused on five important characteristics of a data set: (1) the measure of center, (2) the measure of variation, (3) the nature of the distribution, (4) the presence of outliers, and (5) a pattern over time. A key point of this chapter is that probability distributions describe what will *probably* happen instead of what actually did happen. In Section 4-2, we learned methods for analyzing probability distributions by finding the mean, the standard deviation, and a probability histogram. Because a binomial distribution is a special type of probability distribution, we could use Formulas 4-1, 4-3, and 4-4 (from Section 4-2) for finding the mean, variance, and standard deviation. Fortunately, those formulas can be greatly simplified for binomial distributions, as shown below.

For Any Discrete Probability Distribution

Formula 4-1 $\mu = \Sigma[x \cdot P(x)]$

Formula 4-3 $\sigma^2 = [\Sigma x^2 \cdot P(x)] - \mu^2$

Formula 4-4 $\sigma = \sqrt{[\Sigma x^2 \cdot P(x)] - \mu^2}$

For Binomial Distributions

Formula 4-6 $\mu = n \cdot p$

Formula 4-7 $\sigma^2 = n \cdot p \cdot q$

Formula 4-8 $\sigma = \sqrt{n \cdot p \cdot q}$

EXAMPLE **Gender of Children** In Section 4-2 we included an example illustrating calculations for μ and σ. We used the example of the random variable x representing the number of girls in 14 births. (See Table 4-2 on page 187 for the calculations that illustrate Formulas 4-1 and 4-4.) Use Formulas 4-6 and 4-8 to find the mean and standard deviation for the numbers of girls in groups of 14 births.

SOLUTION Using the values $n = 14$, $p = 0.5$, and $q = 0.5$, Formulas 4-6 and 4-8 can be applied as follows.

$$\mu = n \cdot p = (14)(0.5) = 7.0$$

$$\sigma = \sqrt{n \cdot p \cdot q}$$

$$= \sqrt{(14) \cdot (0.5) \cdot (0.5)} = 1.9 \text{ (rounded)}$$

States Rig Lottery Selections

Many states run a lottery in which players select four digits, such as 1127 (the author's birthday). If a player pays $1 and selects the winning sequence in the correct order, a prize of $5000 is won. States monitor the number selections and, if one particular sequence is selected too often, players are prohibited from placing any more bets on it. The lottery machines are rigged so that once a popular sequence reaches a certain level of sales, the machine will no longer accept that particular sequence. This prevents states from paying out more than they take in. Critics say that this practice is unfair. According to William Thompson, a gambling expert at the University of Nevada in Las Vegas, "they're saying that they (the states) want to be in the gambling business, but they don't want to be gamblers. It just makes a sham out of the whole numbers game."

If you compare these calculations to those required in Table 4-2, it should be obvious that Formulas 4-6 and 4-8 are substantially easier to use.

Formula 4-6 for the mean makes sense intuitively. If we were to ask any statistics student how many girls are expected in 100 births, the usual response would be 50, which can be easily generalized as $\mu = n \cdot p$. The variance and standard deviation are not so easily justified, and we will omit the complicated algebraic manipulations that lead to Formulas 4-7 and 4-8. Instead, refer again to the preceding example and Table 4-2 to verify that for a binomial distribution, Formulas 4-6, 4-7, and 4-8 will produce the same results as Formulas 4-1, 4-3, and 4-4.

EXAMPLE **Gender Selection** The Chapter Problem involved a preliminary trial with 14 couples who wanted to have baby girls. Although the result of 13 girls in 14 births makes it appear that the MicroSort method of gender selection is effective, we would have much more confidence in that conclusion if the sample size had been considerably larger than 14. Suppose the MicroSort method is used with 100 couples, each of whom will have 1 baby. Assume that the result is 68 girls among the 100 babies.

a. Assuming that the MicroSort gender-selection method has no effect, find the mean and standard deviation for the numbers of girls in groups of 100 randomly selected babies.

b. *Interpret* the values from part (a) to determine whether this result (68 girls among 100 babies) supports a claim that the MicroSort method of gender selection is effective.

SOLUTION

a. Assuming that the MicroSort method has no effect and that girls and boys are equally likely, we have $n = 100$, $p = 0.5$, and $q = 0.5$. We can find the mean and standard deviation by using Formulas 4-6 and 4-8 as follows:

$$\mu = n \cdot p = (100)(0.5) = 50$$
$$\sigma = \sqrt{n \cdot p \cdot q} = \sqrt{(100)(0.5)(0.5)} = 5$$

For groups of 100 couples who each have a baby, the mean number of girls is 50 and the standard deviation is 5.

b. We must now interpret the results to determine whether 68 girls among 100 babies is a result that could easily occur by chance, or whether that result is so unlikely that the MicroSort method of gender selection seems to be effective. We will use the range rule of thumb as follows:

$$\text{maximum usual value} = \mu + 2\sigma = 50 + 2(5) = 60$$
$$\text{minimum usual value} = \mu - 2\sigma = 50 - 2(5) = 40$$

INTERPRETATION: According to our range rule of thumb, values are considered to be usual if they are between 40 and 60, so 68 girls is an unusual result. It is very unlikely that we will get 68 girls in 100 births just by chance. If we did get 68 girls in 100 births, we should look for an explanation that is an alternative to chance. If the 100 couples used the MicroSort method of gender selection, it would appear to be effective in increasing the likelihood that a child will be a girl.

You should develop the skills to calculate means and standard deviations using Formulas 4-6 and 4-8, but it is especially important to learn to *interpret* results by using those values. The range rule of thumb, as illustrated in part (b) of the preceding example, suggests that values are unusual if they lie outside of these limits:

$$\textbf{maximum usual value} = \mu + 2\sigma$$

$$\textbf{minimum usual value} = \mu - 2\sigma$$

4-4 Basic Skills and Concepts

Finding μ, σ, and Unusual Values. *In Exercises 1–4, assume that a procedure yields a binomial distribution with n trials and the probability of success for one trial is p. Use the given values of n and p to find the mean μ and standard deviation σ. Also, use the range rule of thumb to find the minimum usual value $\mu - 2\sigma$ and the maximum usual value $\mu + 2\sigma$.*

1. $n = 100, p = 0.25$ **2.** $n = 1068, p = 0.88$

3. $n = 237, p = 2/3$ **4.** $n = 2500, p = 3/5$

5. Guessing Answers Several students are unprepared for a true/false test with 20 questions, and all of their answers are guesses.
- **a.** Find the mean and standard deviation for the number of correct answers for such students.
- **b.** Would it be unusual for a student to pass by guessing and getting at least 12 correct answers? Why or why not?

6. Guessing Answers Several students are unprepared for a multiple-choice quiz with 20 questions, and all of their answers are guesses. Each question has five possible answers, and only one of them is correct.
- **a.** Find the mean and standard deviation for the number of correct answers for such students.
- **b.** Would it be unusual for a student to pass by guessing and getting at least 12 correct answers? Why or why not?

7. Playing Roulette If you bet on any single number in roulette, your probability of winning is 1/38. Assume that you bet on a single number in each of 100 consecutive spins.
- **a.** Find the mean and standard deviation for the number of wins.
- **b.** Would it be unusual to not win once in the 100 trials? Why or why not?

8. Left-Handed People Ten percent of American adults are left-handed. A statistics class has 25 students in attendance. *(continued)*

a. Find the mean and standard deviation for the number of left-handed students in such classes of 25 students.
b. Would it be unusual to survey a class of 25 students and find that 5 of them are left-handed? Why or why not?

9. High School Graduates The Census Bureau reports that 82% of Americans over the age of 25 are high school graduates. A survey of randomly selected Dutchess County residents included 1250 who were over the age of 25, and 1107 of them were high school graduates.
a. Find the mean and standard deviation for the number of high school graduates in groups of 1250 Americans over the age of 25.
b. Is the Dutchess County result of 1107 unusually high? Why or why not? How might such a result be explained?

10. College Student Housing In planning for enrollment, which is expected to grow by 200 students, the College of Newport has found that 9% of college students live in their own off-campus housing (based on data from the Independent Insurance Agents of America).
a. For randomly selected groups of 200 college students, find the mean and standard deviation for the numbers who live in their own off-campus housing.
b. Would it be unusual to find that 50 of the 200 new students live in their own off-campus housing? What is a likely cause of a number that is unusually high?

11. Determining the Effectiveness of an HIV Training Program The New York State Health Department reports a 10% rate of the HIV virus for the "at-risk" population. In one region, an intensive education program is used in an attempt to lower that 10% rate. After running the program, a follow-up study of 150 at-risk individuals is conducted.
a. Assuming that the program has no effect, find the mean and standard deviation for the number of HIV cases in groups of 150 at-risk people.
b. Among the 150 people in the follow-up study, 8% (or 12 people) tested positive for the HIV virus. If the program has no effect, is that rate unusually low? Does this result suggest that the program is effective?

12. Deciphering Messages The Central Intelligence Agency has specialists who analyze the frequencies of letters of the alphabet in an attempt to decipher intercepted messages. In standard English text, the letter r is used at a rate of 7.7%.
a. Find the mean and standard deviation for the number of times the letter r will be found on a typical page of 2600 characters.
b. In an intercepted message sent to Iraq, a page of 2600 characters is found to have the letter r occurring 175 times. Is this unusual?

13. Determining Whether Complaints Are Lower After a Training Program The Newtower Department Store has experienced a 3.2% rate of customer complaints and attempts to lower this rate with an employee training program. After the program, 850 customers are tracked and it is found that only 7 of them filed complaints.
a. Assuming that the training program has no effect, find the mean and standard deviation for the number of complaints in such groups of 850 customers.
b. Based on the results from part (a), is the result of 7 complaints unusual? Does it seem that the training program was effective in lowering the rate of complaints?

14. Are 10% of M&M Candies Blue? Mars, Inc., claims that 10% of its M&M plain candies are blue, and a sample of 100 such candies is randomly selected.

a. Find the mean and standard deviation for the number of blue candies in such groups of 100.

b. Data Set 10 in Appendix B consists of a random sample of 100 M&Ms in which only 5 are blue. Is this result unusual? Does it seem that the claimed rate of 10% is wrong?

15. Are TV Commercials Longer? Are television commercials getting longer, or does it just seem that way? In 1990, 6% of those commercials were one minute long. Recently, 50 commercials were randomly selected, and 16 of them were one minute long.

a. If one-minute commercials continue to be 6% of the total, find the mean and standard deviation for the numbers of one-minute commercials that would be found in groups of 50.

b. Is the recent result of 16 one-minute commercials unusually high, assuming that the same 6% rate applies? Does the recent result suggest that one-minute commercials are now more than 6% of the total?

16. Computer Warranty Repairs The Providence Computer Supply Company knows that 16% of its computers will require warranty repairs within one month of shipment. In a typical month, 279 computers are shipped.

a. If x is the random variable representing the number of computers requiring warranty repairs among the 279 sold in one month, find its mean and standard deviation.

b. For a typical month in which 279 computers are sold, what would be an unusually low figure for the number of computers requiring warranty repair within one month? What would be an unusually high figure? (These values are helpful in determining the number of service technicians the company will need to employ.)

4-4 Beyond the Basics

17. Using the Empirical Rule and Chebyshev's Theorem An experiment is designed to test the effectiveness of the MicroSort method of gender selection, and 100 couples try to have baby girls using the MicroSort method. In an example included in this section, the range rule of thumb was used to conclude that among 100 births, the number of girls should usually fall between 40 and 60.

a. The empirical rule (see Section 2-5) applies to distributions that are bell-shaped. Is the binomial probability distribution for this experiment (approximately) bell-shaped? How do you know?

b. Assuming that the distribution is bell-shaped, how likely is it that the number of girls will fall between 40 and 60 (according to the empirical rule)?

c. Assuming that the distribution is bell-shaped, how likely is it that the number of girls will fall between 35 and 65 (according to the empirical rule)?

d. Using Chebyshev's Theorem, what do we conclude about the likelihood that the number of girls will fall between 40 and 60?

18. Analyzing Results of Experiment in Gender Selection An experiment involving a gender-selection method includes a control group of 15 couples who are not given any treatment intended to influence the genders of the babies.

(continued)

a. Construct a table listing the possible values of the random variable x (which represents the number of girls among the 15 births) and the corresponding probabilities.

b. Find the mean and standard deviation for the numbers of girls in such groups of 15.

c. Find the maximum and minimum usual values for the number of girls. (Use the range rule of thumb.)

d. Using the probabilities found in part (a), find the probability that the number of girls will be outside of the range of usual values found in part (c). That is, find the probability that the number of girls is less than the minimum usual value or greater than the maximum usual value. This result is the specific probability of an unusual outcome for this experiment.

19. Acceptable/Defective Products Mario's Pizza Parlor has just opened. Due to a lack of employee training, there is only a 0.8 probability that a pizza will be edible. An order for five pizzas has just been placed. What is the minimum number of pizzas that must be made in order to be at least 99% sure that there will be five that are edible?

4-5 The Poisson Distribution

This section introduces the *Poisson distribution.* It is particularly important because it is often used as a mathematical model describing arrivals of people in a line, planes arriving at an airport, cars pulling into a gas station, diners arriving at a restaurant, students arriving at a bookstore line, and Internet users logging onto a Web site. We can use the Poisson distribution to find probabilities. For example, suppose your professor has an office hour scheduled every Monday at 11:00 and she finds that during that office hour, the mean number of students who come is 2.3. We can find the probability that for a randomly selected office hour on Monday at 11:00, exactly four students come. We use the Poisson distribution, defined as follows.

DEFINITION

The **Poisson distribution** is a discrete probability distribution that applies to occurrences of some event *over a specified interval.* The random variable x is the number of occurrences of the event in an interval. The interval can be time, distance, area, volume, or some similar unit. The probability of the event occurring x times over an interval is given by Formula 4-9.

Formula 4-9 $$P(x) = \frac{\mu^x \cdot e^{-\mu}}{x!} \quad \text{where } e \approx 2.71828$$

The Poisson distribution has the following requirements:

- The random variable x is the number of occurrences of an event *over some interval.*

- The occurrences must be *random.*
- The occurrences must be *independent* of each other.
- The occurrences must be *uniformly distributed* over the interval being used.

The Poisson distribution has these parameters:

- The mean is μ.
- The standard deviation is $\sigma = \sqrt{\mu}$.

A Poisson distribution differs from a binomial distribution in these fundamental ways:

1. The binomial distribution is affected by the sample size n and the probability p, whereas the Poisson distribution is affected only by the mean μ.
2. In a binomial distribution, the possible values of the random variable x are $0, 1, \ldots, n$, but a Poisson distribution has possible x values of $0, 1, 2, \ldots,$ with no upper limit.

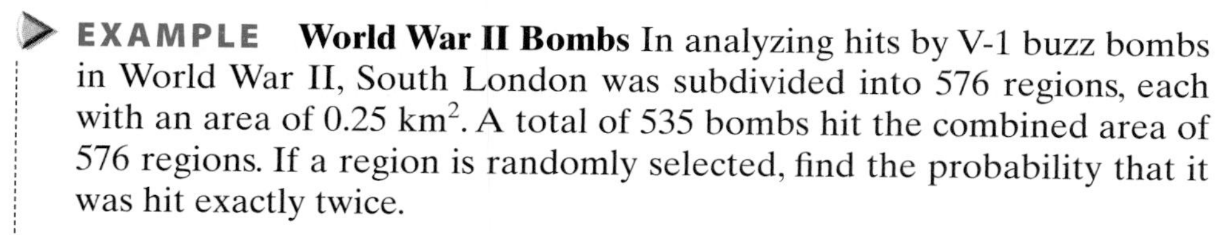

EXAMPLE **World War II Bombs** In analyzing hits by V-1 buzz bombs in World War II, South London was subdivided into 576 regions, each with an area of 0.25 km^2. A total of 535 bombs hit the combined area of 576 regions. If a region is randomly selected, find the probability that it was hit exactly twice.

SOLUTION The Poisson distribution applies because we are dealing with the occurrences of an event (bomb hits) over some interval (a region with area of 0.25 km^2). The mean number of hits per region is

$$\mu = \frac{\text{number of bomb hits}}{\text{number of regions}} = \frac{535}{576} = 0.929$$

Because we want the probability of exactly two hits in a region, we let $x = 2$ and use Formula 4-9 as follows.

$$P(x) = \frac{\mu^x \cdot e^{-\mu}}{x!} = \frac{0.929^2 \cdot 2.71828^{-0.929}}{2!} = \frac{0.863 \cdot 0.395}{2} = 0.170$$

The probability of a particular region being hit exactly twice is $P(2) = 0.170$.

In the preceding example, we can also calculate the probabilities for 0, 1, 3, 4, and 5 hits. (We stop at $x = 5$ because no region was hit more than five times, and the probabilities for $x > 5$ are 0.000 when rounded to three decimal places.) Those probabilities are listed on the next page in Table 4-4, and using them, we are able to find the expected number of regions with 0 hits, 1 hit, and so on. For example, the 576 regions each have a probability of 0.395 of having 0 hits, so we expect that the number of regions with 0 hits is

Probability of an Event That Has Never Occurred

Some events are possible, but are so unlikely that they have never occurred. Here is one such problem of great interest to political scientists: Estimate the probability that your single vote will determine the winner in a U.S. presidential election. Andrew Gelman, Gary King, and John Boscardin write in the *Journal of the American Statistical Association* (Vol. 93, No. 441) that "the exact value of this probability is of only minor interest, but the number has important implications for understanding the optimal allocation of campaign resources, whether states and voter groups receive their fair share of attention from prospective presidents, and how formal 'rational choice' models of voter behavior might be able to explain why people vote at all." The authors show how the probability value of 1 in 10 million is obtained for close elections.

TABLE 4-4 V-1 Buzz Bomb Hits for 576 Regions in South London

Number of Bomb Hits	Probability	Expected Number of Regions	Actual Number of Regions
0	0.395	227.5	229
1	0.367	211.4	211
2	0.170	97.9	93
3	0.053	30.5	35
4	0.012	6.9	7
5	0.002	1.2	1

$576 \cdot 0.395 = 227.5$. (Note that in computing the expected value, we use the number of *regions,* not the number of hits.) This expected value is listed with the others in the third column of Table 4-4. The fourth column describes the results that actually occurred during World War II. There were 229 regions that had no hits, 211 regions that were hit once, and so on. We can now compare the frequencies *predicted* with the Poisson distribution (third column) to the *actual* frequencies (fourth column) to conclude that there is very good agreement. In this case, the Poisson distribution does a good job of predicting the results that actually occurred. (Section 10-2 describes a statistical procedure for determining whether such expected frequencies constitute a good "fit" to the actual frequencies. That procedure does suggest that there is a good fit in this case.)

Poisson as Approximation to Binomial

The Poisson distribution is sometimes used to approximate the binomial distribution when n is large and p is small. One rule of thumb is to use such an approximation when the following two conditions are satisfied:

1. $n \geq 100$

2. $np \leq 10$

If these conditions are satisfied and we want to use the Poisson distribution as an approximation to the binomial distribution, we need a value for μ, and that value can be calculated by using Formula 4-6:

Formula 4-6 $$\mu = n \cdot p$$

EXAMPLE **Roulette** Allyn bets on the number 7 for each of 200 spins of a roulette wheel. Because $P(7) = 1/38$, he expects to win about 5 times. Find the probability that he wins exactly 8 times.

SOLUTION With $n = 200$ and $p = 1/38$, the conditions $n \geq 100$ and $np \leq 10$ are both satisfied, so we can use the Poisson distribution as an approximation to the binomial distribution. We need the value of μ, which is found as follows.

$$\mu = n \cdot p = 200 \cdot \frac{1}{38} = 5.263$$

Having found the value of μ, we can now find $P(8)$:

$$P(8) = \frac{\mu^x \cdot e^{-\mu+}}{x!} = \frac{5.263^8 \cdot 2.71828^{-5.263}}{8!} = \frac{3049.1349}{40320} = 0.0756$$

Using the Poisson distribution as an approximation to the binomial distribution, we find that there is a 0.0756 probability that Allyn will win exactly 8 times in his 200 roulette bets. If we use the binomial distribution, the more accurate probability is 0.0757, so we can see that the Poisson approximation is quite good here.

Using Technology

STATDISK: Select **Analysis** from the main menu bar, then select **Probability Distributions,** then **Poisson.** Click on the OPTIONS button and proceed to enter the value of the mean μ. Use the mouse to move to the right and left for the different values of x.

Minitab: First put the desired value of x in column C1. Now select **Calc** from the main menu bar, then select **Probability Distributions,** then **Poisson.** Enter the value of the mean μ and enter C1 for the input column.

Excel: Click on **fx** on the main menu bar, then select the function category of **Statistical,** then select **POISSON,** then click **OK.** In the dialog box, enter the values for x and the mean, and enter 0 for "Cumulative." (Entering 1 for "Cumulative" results in the probability for values up to and including the entered value of x.)

TI-83 Plus: Press **2nd VARS** (to get **DISTR**), then select option B: **poissonpdf(.** Now press **ENTER,** then proceed to enter μ, x (including the comma). For μ, enter the value of the mean; for x, enter the desired number of occurrences.

4-5 Basic Skills and Concepts

Using a Poisson Distribution to Find Probability. *In Exercises 1–4, assume that the Poisson distribution applies and proceed to use the given mean to find the indicated probability.*

1. If $\mu = 3$, find $P(4)$.

2. If $\mu = 0.555$, find $P(2)$.

3. If $\mu = 50$, find $P(48)$.

4. If $\mu = 0.250$, find $P(20)$.

In Exercises 5–12, use the Poisson distribution to find the indicated probabilities.

5. Faculty Office Hours A statistics professor finds that when she schedules an office hour for student help, an average of 2.3 students arrive. Find the probability that in a randomly selected office hour, the number of student arrivals is
a. 0 **b.** 1 **c.** 2 **d.** 3 **e.** 6

6. Commercial Airplane Crashes According to the National Transportation Safety Board, the United States has the safest aviation system in the world, with commercial planes crashing at a rate of only 0.166 flight per million flights. Find the probability that among the next million flights, the number of crashes is
a. 0 **b.** 1

7. Aircraft Hijacking For the past few years, there has been a yearly average of 29 aircraft hijackings worldwide (based on data from the FAA). The mean number of hijackings per day is estimated as $\mu = 29/365$. If the United Nations is organizing a single international hijacking response team, there is a need to know about the chances of multiple hijackings in one day. Find the probability that the number of hijackings (x) in one day is 0 or 1. Is a single response team sufficient?

8. Deaths from Horse Kicks A classic example of the Poisson distribution involves the number of deaths caused by horse kicks of men in the Prussian Army between 1875 and 1894. Data for 14 corps were combined for the 20-year period, and the 280 corps-years included a total of 196 deaths. After finding the mean number of deaths per corps-year, find the probability that a randomly selected corps-year has the following numbers of deaths.
a. 0 **b.** 1 **c.** 2 **d.** 3 **e.** 4

The actual results consisted of these frequencies: 0 deaths (in 144 corps-years); 1 death (in 91 corps-years); 2 deaths (in 32 corps-years); 3 deaths (in 11 corps-years); 4 deaths (in 2 corps-years). Compare the actual results to those expected from the Poisson probabilities. Does the Poisson distribution serve as a good device for predicting the actual results?

9. Homicide Deaths In one year, there were 116 homicide deaths in Richmond, Virginia (based on "A Classroom Note On the Poisson Distribution: A Model for Homicidal Deaths In Richmond, VA for 1991," *Mathematics and Computer Education,* by Winston A. Richards). For a randomly selected day, find the probability that the number of homicide deaths is
a. 0 **b.** 1 **c.** 2 **d.** 3 **e.** 4

Compare the calculated probabilities to these actual results: 268 days (no homicides); 79 days (1 homicide); 17 days (2 homicides); 1 day (3 homicides); no days with more than 3 homicides.

10. Earthquakes For a recent period of 100 years, there were 93 major earthquakes (at least 6.0 on the Richter scale) in the world (based on data from the *World Almanac and Book of Facts*). Assuming that the Poisson distribution is a suitable model, find the mean number of major earthquakes per year, then find the probability that the number of earthquakes in a randomly selected year is
a. 0 **b.** 1 **c.** 2 **d.** 3 **e.** 4 **f.** 5 **g.** 6 **h.** 7

Here are the actual results: 47 years (0 major earthquakes); 31 years (1 major earthquake); 13 years (2 major earthquakes); 5 years (3 major earthquakes); 2 years (4 major earthquakes); 0 years (5 major earthquakes); 1 year (6 major earthquakes); 1 year (7 major earthquakes). After comparing the calculated probabilities to the actual results, is the Poisson distribution a good model?

4-5 Beyond the Basics

11. Poisson Approximation to Binomial The Poisson distribution can be used to approximate a binomial distribution if $n \geq 100$ and $np \leq 10$. Assume that we have a binomial distribution with $n = 100$ and $p = 0.1$. It is impossible to get 101 successes in such a binomial distribution, but we *can* compute the probability that $x = 101$ with the Poisson approximation. Find that value. How does the result agree with the impossibility of having $x = 101$ with a binomial distribution?

12. Poisson Approximation to Binomial For a binomial distribution with $n = 10$ and $p = 0.5$, we should not use the Poisson approximation because the conditions $n \geq 100$ and $np \leq 10$ are not satisfied. Suppose we go way out on a limb and use the Poisson approximation anyway. Are the resulting probabilities unacceptable approximations? Why or why not?

Food for Thought: Cooking at Home on the Range

BOSTON The eating habits of Americans are continuously monitored by social scientists who are watching for changing trends and their effects on such factors as the economy and family life. The accompanying illustration is based on one from *USA Today* and a survey of adult Americans. It shows that 8% of all adults—roughly 1 in 12—have none of their evening meals cooked at home. At the other extreme, a whopping 19% of adult Americans have all of their evening meals cooked at home. Some social scientists stress the importance of the evening meal as a family meeting with bonding power that rivals the special ingredient mom puts in those dumplings.

USA Today Snapshot. "A look at statistics that shape our lives"

1. Refer to the illustration and construct the probability distribution table represented by the given data.

2. Prove or disprove that the requirements of a probability distribution are satisfied.

3. Find the mean number of days in a week that American adults have evening meals cooked at home.

4. Find the standard deviation of the numbers of days in a week that American adults have evening meals cooked at home.

5. Is it "unusual" for an adult American to have no evening meals cooked at home in a week? Why or why not?

Vocabulary List

random variable
probability distribution
discrete random variable
continuous random variable
probability histogram
expected value
binomial probability distribution
binomial probability formula
Poisson distribution

Review

A major component of this chapter is the concept of a probability distribution, which describes the probability for each value of a random variable. This chapter included only discrete probability distributions, but the following chapters will include continuous probability distributions. The following key points were discussed:

- A *random variable* has values that are determined by chance.
- A *probability distribution* consists of all values of a random variable, along with their corresponding probabilities. A probability distribution must satisfy two requirements: $\Sigma P(x) = 1$ and, for each value of x, $0 \leq P(x) \leq 1$.
- The important characteristics of a *probability distribution* can be explored by computing its mean $(\mu = \Sigma[x \cdot P(x)])$ and standard deviation $\left(\sigma = \sqrt{[\Sigma x^2 \cdot P(x)] - \mu^2}\right)$ and by constructing a probability histogram.
- In a *binomial distribution,* there are two categories of outcomes and a fixed number of independent trials with a constant probability. The probability of x successes among n trials can be found by using the binomial probability formula, or Table A-1, or software (such as STATDISK, Minitab, or Excel), or a TI-83 Plus calculator.
- In a binomial distribution, the mean and standard deviation can be easily found by calculating the values of $\mu = n \cdot p$ and $\sigma = \sqrt{n \cdot p \cdot q}$.
- A *Poisson probability distribution* applies to occurrences of some event over a specific interval, and its probabilities can be computed with Formula 4-9.
- *Unusual outcomes:* This chapter stressed the importance of interpreting results by distinguishing between outcomes that are usual and those that are unusual. We used two different criteria. With the range rule of thumb we have

$$\textbf{maximum usual value} = \mu + 2\sigma$$

$$\textbf{minimum usual value} = \mu - 2\sigma$$

We can also determine whether outcomes are unusual by using probability values.

x successes among n trials is unusually *high* if $P(x$ or more) is very small (such as less than 0.05)

x successes among n trials is unusually *low* if $P(x$ or fewer) is very small (such as less than 0.05)

Review Exercises

x	$P(x)$
0	0.0004
1	0.0094
2	0.0870
3	0.3562
4	0.5470

1. a. What is a random variable?

b. What is a probability distribution?

c. An insurance association's study of home smoke detector use involves homes randomly selected in groups of four. The accompanying table lists values and probabilities for x, the number of homes (in groups of four) that have smoke detectors installed (based on data from the National Fire Protection Association). Does this table describe a probability distribution? Why or why not?

d. Assuming that the table does describe a probability distribution, find its mean.

e. Assuming that the table does describe a probability distribution, find its standard deviation.

f. Is it unusual to randomly select four homes and find that none of them have smoke detectors? Why or why not?

2. TV Ratings The television show *ER* has a 34 share, meaning that while it is being broadcast, 34% of the TV sets in use are tuned to *ER* (based on data from Nielsen Media Research). Assume that during a broadcast of *ER,* 20 TV sets are randomly selected from those that are in use.

a. What is the expected number of sets tuned to *ER?*

b. In such groups of 20, what is the mean number of sets tuned to *ER?*

c. In such groups of 20, what is the standard deviation for the number of sets tuned to *ER?*

d. For such a group of 20, find the probability that exactly 5 TV sets are tuned to *ER.*

e. For such a group of 20, would it be unusual to find that 12 sets are tuned to *ER?* Why or why not?

3. Employee Drug Testing Among companies doing highway or bridge construction, 80% test employees for substance abuse (based on data from the Construction Financial Management Association). A study involves the random selection of 10 such companies.

a. Find the probability that 5 of the 10 companies test for substance abuse.

b. Find the probability that at least half of the companies test for substance abuse.

c. For such groups of 10 companies, find the mean and standard deviation for the number (among 10) that test for substance abuse.

d. Would it be unusual to find that 6 of 10 companies test for substance abuse? Why or why not?

4. Reasons for Being Fired Inability to get along with others is the reason cited in 17% of worker firings (based on data from Robert Half International, Inc.). Concerned about her company's working conditions, the personnel manager at the Boston Finance Company plans to investigate the five employee firings that occurred over the past year.

a. Assuming that the 17% rate applies, find the probability that among those five employees, the number fired because of an inability to get along with others is at least four.

(continued)

b. If the personnel manager actually does find that at least four of the firings are due to an inability to get along with others, does this company appear to be very different from other typical companies? Why or why not?

5. Planning for a New Hospital A new hospital is being considered for the community of Newport, which does not yet have its own hospital. There is a need to know how many babies are born per day. Records show that Newport had 212 births last year.

a. Find the average number of births per day.

b. Find the probability of exactly two births on a given day.

c. Find the probability of exactly three births on a given day. Is this value high enough to warrant the construction of a facility that can accommodate three births on the same day?

CUMULATIVE REVIEW EXERCISES

x	f
0	55
1	2
2	1
3	1
4	0
5	3
6	0
7	2
8	4
9	2

1. Home Run Distances: Analysis of Last Digits The accompanying table lists the last digits of the 70 reported distances (in feet) of the 70 home runs hit by Mark McGwire in 1998 when he broke Roger Maris's home-run record, which had lasted since 1961 (based on data from *USA Today*). The last digits of a data set can sometimes be used to determine whether the data have been measured or simply reported. The presence of disproportionately more 0s and 5s is often a sure indicator that the data have been reported instead of measured.

a. Find the mean and standard deviation of those last digits.

b. Construct the relative frequency table that corresponds to the given frequency table.

c. Construct a table for the probability distribution of randomly selected digits that are all equally likely. List the values of the random variable x $(0, 1, 2, \ldots, 9)$ along with their corresponding probabilities $(0.1, 0.1, 0.1, \ldots, 0.1)$, then find the mean and standard deviation of this probability distribution.

d. Recognizing that sample data naturally deviate from the results we theoretically expect, does it seem that the given last digits roughly agree with the distribution we expect with random selection? Or does it seem that there is something about the sample data (such as disproportionately more 0s and 5s) suggesting that the given last digits are not random? (In Chapter 10, we will present a method for answering such questions much more objectively.)

2. Proving That Dice Are Loaded A casino cheat is caught trying to use a pair of loaded dice. At his court trial, physical evidence reveals that some of the black dots were drilled, filled with lead, then repainted to appear normal. In addition to the physical evidence, the dice are rolled in court with these results:

12	8	9	12	12	9	8	7	12	10
12	3	2	12	10	9	12	11	11	12

A probability expert testifies that when fair dice are rolled, the mean should be 7.0 and the standard deviation should be 2.4.

a. Find the mean and standard deviation of the sample values obtained in court.

b. Based on the outcomes obtained in court, what is the probability of rolling a 12? How does this result compare to the probability of 1/36 (or 0.0278) for fair dice?

c. If the probability of rolling a 12 with fair dice is 1/36, find the probability of getting at least one 12 when fair dice are rolled 20 times.

d. If you are the defense attorney, how would you refute the results obtained in court?

Cooperative Group Activities

1. *In-class activity:* In Chapter 1 we gave several examples of misleading data sets. Suppose we want to identify the probability distribution for the number of children born to randomly selected couples. For each student in the class, find the number of brothers and sisters and record the total number of children (including the student) in each family. Construct the relative frequency table for the result obtained. (The values of the random variable x will be $1, 2, 3, \ldots$.) What is wrong with using this relative frequency table as an estimate of the probability distribution for the number of children born to randomly selected couples?

2. *In-class activity:* Divide into groups of three. Select one person who will be tested for extrasensory perception (ESP) by trying to correctly identify a digit randomly selected by another member of the group. Another group member should record the randomly selected digit, the digit guessed by the subject, and whether the guess was correct or wrong. Construct the table for the probability distribution of randomly generated digits, construct the relative frequency table for the random digits that were actually obtained, and construct a relative frequency table for the guesses. After comparing the three tables, what do you conclude? What proportion of guesses were correct? Does it seem that the subject has the ability to select the correct digit significantly more often than would be expected by chance?

Technology Project

American Air Flight 2705 from New York to San Francisco has seats for 340 passengers. An average of 5% of people with reservations don't show up, so American Air overbooks by accepting 350 reservations for the 340 seats. We can analyze this system by using a binomial distribution with $n = 350$ and $p = 0.95$ (the probability that someone with a reservation does show up).

Find the probability that when 350 reservations are accepted for a particular flight, there are more passengers than seats. That is, find the probability of at least 341 people showing up with reservations, assuming that 350 reservations were accepted. Because of the value of n, Table A-1 cannot be used, and calculations with the binomial probability formula would be extremely time-consuming and painfully tedious. The best approach is to use statistics software or a TI-83 Plus calculator. See Method 3 in Section 4-3 for instructions describing the use of STATDISK, Minitab, Excel, or a TI-83 Plus calculator.

Statistics at work

"Our program is really an education program, but it has wide recognition because the results are released publicly"

Barbara Carvalho:
Director of the Marist College Poll

Lee Miringoff:
Director of the Marist College Institute for Public Opinion

Barbara Carvalho and Lee Miringoff report on their poll results in many interviews for print and electronic media, including news programs for NBC, CBS, ABC, FOX, and public television. Lee Miringoff appears regularly on NBC's "Today" show.

What do you do?

We do public polling. We survey public issues, approval ratings of public officials in New York City, New York State, and nationwide. We don't do partisan polling for political parties, political candidates, or lobby groups. We are independently funded by Marist College and we have no outside funding that in any way might suggest that we are doing research for any particular group on any one issue. Reporters have come to depend on our results for its accuracy and professionalism, but also because they know that it is independent and is not commissioned by any one media source, as many polls are.

How do you select survey respondents?

For a statewide survey we select respondents in proportion to county voter registrations. Different counties have different refusal rates and if we were to select people at random throughout the state, we would get an uneven model of what the state looks like. We stratify by county and use random digit dialing so that we get listed and unlisted numbers.

You mentioned refusal rates. Are they a real problem?

One of the issues that we deal with extensively is the issue of people who don't respond to surveys. That has been increasing over time and there has been much attention from the survey research community. As a research center we do quite well when compared to others. But when you do face-to-face interviews and have refusal rates of 25% to 50%, there's a real concern to find out who is refusing and why they are not responding, and the impact that has on the representativeness of the studies that we're doing.

Is the political process actually influenced by poll results?

Although most polls that people see are public polls, the reality is that the political process is influenced by private polls that the public never sees. No one runs for high office today without using a private poll.

How do you use statistics in your work?

Although we use statistics in a variety of ways, there are two significant areas which are dependent on statistical concepts. First, and perhaps most central, is the concept of probability. The main idea behind doing polls is that by interviewing a few individuals one may generalize about a larger group. It is the method of sampling—determining who is selected to participate in a survey—that allows us to generalize from the opinions of a few to the larger group. Probability theory guides this task. Second, when analyzing survey results, it

is essential to know if the differences we observe in our results or among groups are in fact real or simply owing to chance. Statistics provides the means by which we may address this issue.

What concepts of statistics do you use?

Statistics comes into play in our sampling, even before we get to data analysis. We use statistics to determine our sample size and to develop an estimate of what would be statistically significant. In the data analysis, most of our studies use basic descriptive statistics. Some of the academic studies get into regression analysis.

Would you recommend a statistics course for students?

Absolutely. All numbers are not created equally. Regardless of your field of study or career interests, an ability to critically evaluate research information that is presented to you, to use data to improve services, or to interpret results to develop strategies is a very valuable asset. Surveys, in particular, are everywhere. It is vital that as workers, managers, and citizens we are able to evaluate their accuracy and worth. Numbers are powerful. They are capable of providing support for your ideas or being used against you to undermine your position. Those who can use and understand them will have an advantage over those who can only take them at face value. The study of statistics is important for understanding one aspect of knowledge, and it's a key to opening up other avenues of pursuit. Statistics cuts across disciplines. Students will inevitably find it in their careers at some point.

Do you have any other recommendations for students?

It is important for students to take every opportunity to develop their communication and presentation skills. Sharpen not only your ability to speak and write, but also raise your comfort level with new technologies.

CHAPTER 5

Normal Probability Distributions

CHAPTER PROBLEM

Redesigning Ejection Seats

Ergonomics is the study of problems associated with people adjusting to their environments. The environments are varied and include manufacturing assembly lines, computer stations, cars, elevators, theater seats, and aircraft cockpits. Good ergonomic design results in an environment that is safe, functional, efficient, and comfortable.

Changing times often require ergonomic changes. When visiting homes that are hundreds of years old, you may have noticed that the door openings seem too low. The doorway heights that were comfortable for people a few hundred years ago are not so comfortable for the taller people of today. The United States Air Force recently encountered an ergonomic problem created by its recognition and acceptance of the fact that women make perfectly good pilots of fighter jets. Cockpits of fighter jets were originally designed for men only, so various cockpit changes were required to better accommodate the new women pilots. As this text was being written, the United States Air Force was redesigning its ACES-II ejection seat currently used in fighter jets. The ACES-II model was originally designed for men who weighed between 140 and 211 pounds. Based on data from the National Health and Examination Survey, weights of women have a distribution that is roughly bell-shaped with a mean of 143 pounds and a standard deviation of 29 pounds. Any women pilots weighing less than 140 pounds or more than 211 pounds would have a greater chance of injury if it became necessary to eject. What percent of women weigh between 140 and 211 pounds, so that they are within the design limits of the ACES-II ejection seats? Such questions are important for the ejection seat redesign, and we will answer such questions as we present the important concepts of this chapter.

5-1 Overview

Recall that in Chapter 4, we presented the following concepts:

- A *random variable* is a variable having a single numerical value, determined by chance, for each outcome of some procedure.
- A *probability distribution* describes the probability for each value of the random variable.
- A *discrete* random variable has either a finite number of values or a countable number of values. That is, the *number* of possible values that x can assume is 0, or 1, or 2, and so on.
- A *continuous* random variable has infinitely many values, and those values are often associated with measurements on a continuous scale with no gaps or interruptions.

In Chapter 4 we considered only *discrete* probability distributions, but in this chapter we present *continuous* probability distributions. Although we begin with a uniform distribution, most of the chapter will focus on *normal distributions.* Normal distributions are extremely important because they occur so often in real applications.

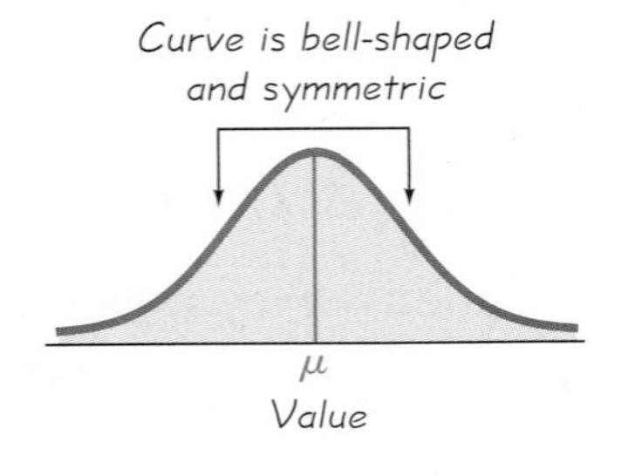

FIGURE 5-1
The Normal Distribution

DEFINITION

If a continuous random variable has a distribution with a graph that is symmetric and bell-shaped, as in Figure 5-1, we say that it has a **normal distribution.** The normal distribution can be described by the equation given as Formula 5-1.

Formula 5-1

$$y = \frac{e^{-\frac{1}{2}\left(\frac{x-\mu}{\sigma}\right)^2}}{\sigma\sqrt{2\pi}}$$

Don't be discouraged by the complexity of Formula 5-1, because it isn't really necessary for us to actually use it. What it shows is that any particular normal distribution is determined by two parameters: the mean, μ, and standard deviation, σ. Once specific values are selected for μ and σ, we can graph Formula 5-1 as we would graph any equation relating x and y; the result is a continuous probability distribution with a bell shape. This is arguably the single most important distribution in statistics, and we will use it often throughout the remainder of the book.

5-2 The Standard Normal Distribution

The focus of this chapter is the concept of a normal probability distribution, but we begin with a *uniform distribution.* The uniform distribution makes it easier for us to see some very important properties, which will be used with normal distributions.

Definition

A continuous random variable has a **uniform distribution** if its values spread evenly over the range of possibilities. The graph of a uniform distribution results in a rectangular shape.

EXAMPLE **Fire Drills** A college dean must select a starting time for a required fire drill. She can select any time between 10:00 A.M. and 3:00 P.M., so we will represent that 5-hour period by the random variable x, which can be any value between 0 and 5. If she selects the time in such a way that every possible time is equally likely, then x is a continuous random variable with a uniform distribution. See Figure 5-2.

When we discussed *discrete* probability distributions in Section 4-2, we identified two requirements: (1) $\Sigma P(x) = 1$, and (2) $0 \leq P(x) \leq 1$ for all values of x. Also in Section 4-2, we stated that the graph of a discrete probability distribution is called a *probability histogram*. The graph of a continuous probability distribution, such as that of Figure 5-2, is called a *density curve*, and it must satisfy two properties similar to the requirements for discrete probability distributions, as listed in the following definition.

Definition

A **density curve** (or **probability density function**) is a graph of a continuous probability distribution. It must satisfy the following properties:

1. The total area under the curve must equal 1.
2. Every point on the curve must have a vertical height that is 0 or greater.

By setting the height of the rectangle in Figure 5-2 to be 0.2, we force the enclosed area to be $5 \times 0.2 = 1$, as required. This property (area = 1) makes it very easy to solve probability problems, so the following statement is important:

Reliability and Validity

The reliability of data refers to the consistency with which results occur, whereas the validity of data refers to how well the data measure what they are supposed to measure. The reliability of an IQ test can be judged by comparing scores for the test given on one date to scores for the same test given at another time. To test the validity of an IQ test, we might compare the test scores to another indicator of intelligence, such as academic performance. Many critics charge that IQ tests are reliable, but not valid; they provide consistent results, but don't really measure intelligence.

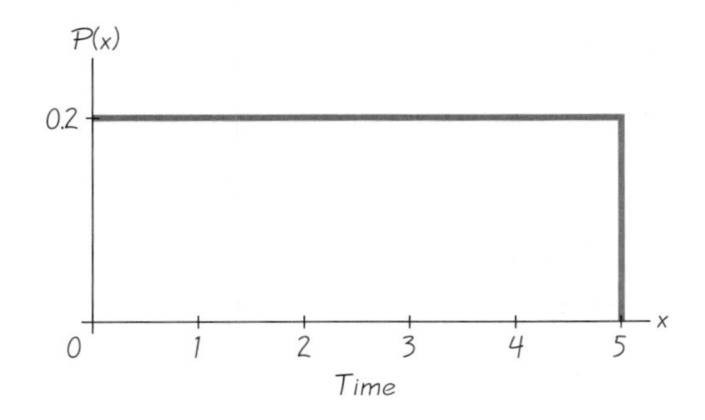

FIGURE 5-2
Uniform Distribution of Times

Because the total area under the density curve is equal to 1, there is a correspondence between area and probability.

EXAMPLE **Fire Drills** A college dean must select a starting time between 0 and 5 hours, and the selection is made in such a way that all possible times are equally likely. If she randomly selects a starting time, what is the probability that it is during the first half hour or the last half hour?

SOLUTION See Figure 5-3, where we shade the regions representing times that are in the first half hour or the last half hour. Because there is a correspondence between area and probability, we can find the probability by using areas as follows:

$$P(\text{first or last half hour}) = P(\text{first half hour}) + P(\text{last half hour})$$
$$= 0.1 + 0.1 = 0.2$$

INTERPRETATION The probability of randomly selecting a starting time during the first half hour or last half hour is 0.2.

FIGURE 5-3
Times in First or Last Half Hour

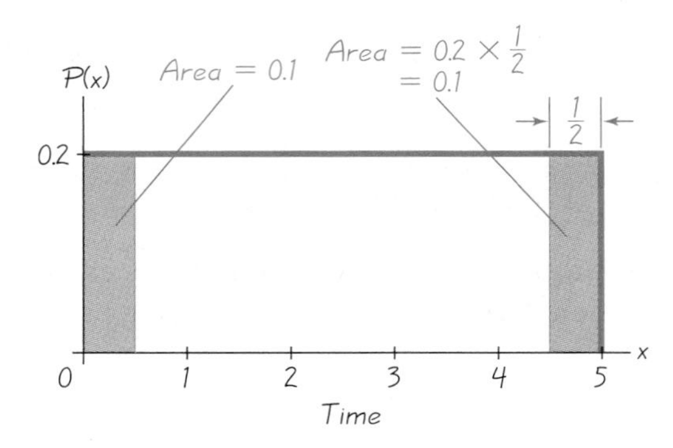

The density curve of a uniform distribution is a horizontal line, so it's easy to find the area of any underlying region by multiplying width and height. The density curve of a normal distribution has the more complicated bell shape shown in Figure 5-1, so it's more difficult to find areas, but the basic principle is the same: There is a correspondence between area and probability.

Just as there are many different uniform distributions (with different ranges of values), there are also many different normal distributions, with each one depending on two parameters: the population mean, μ, and the population standard deviation, σ. (Recall from Chapter 1 that a *parameter* is a numerical measurement describing some characteristic of a *population.*) Figure 5-4 shows density curves for heights of adult women and men. Because men have a larger mean height, the peak of the density curve for men is farther to the right. Because men's heights have a slightly larger standard deviation, the density curve for men is slightly wider. Figure 5-4 shows two different possible normal distributions. There are infinitely many other possibilities, but one is of special interest.

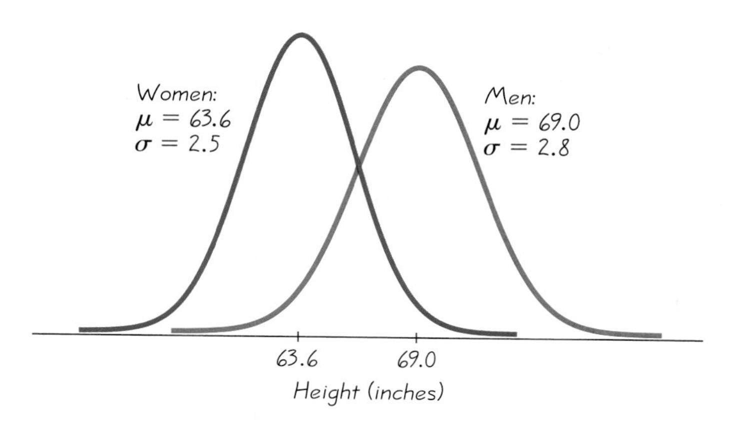

FIGURE 5-4
Heights of Adult Women and Men

DEFINITION

The **standard normal distribution** is a normal probability distribution that has a mean of 0 and a standard deviation of 1. (See Figure 5-5.)

Suppose that somehow we were forced to perform calculations using Formula 5-1. We would quickly see that the most workable values for μ and σ are $\mu = 0$ and $\sigma = 1$. By letting $\mu = 0$ and $\sigma = 1$, mathematicians have calculated areas under the curve. As shown in Figure 5-5, the area under the curve bounded by the mean of 0 and the value of 1 is 0.3413. Remember, the total area under the curve is always 1; this allows us to make the correspondence between area and probability, as we did in the preceding example with the uniform distribution.

Finding Probabilities When Given z Scores

Figure 5-5 shows that the area bounded by the curve, the horizontal axis, and the values of 0 and 1 is an area of 0.3413. Although the figure shows only one area, we can find areas (or probabilities) for many different regions. Such areas can be found by using Table A-2 (in Appendix A and the *Formulas and Tables* insert card), a TI-83 Plus calculator, or software such as STATDISK, Minitab, or Excel. The key features of the different methods are summarized in Table 5-1. It is not necessary to know all five methods; you only need to

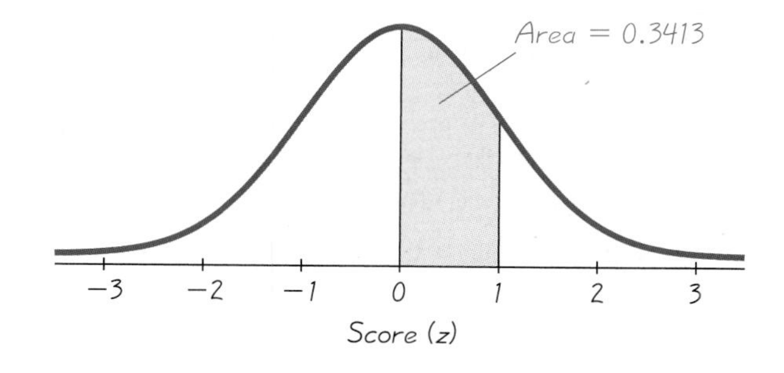

FIGURE 5-5
Standard Normal Distribution, with Mean $\mu = 0$ and Standard Deviation $\sigma = 1$

TABLE 5-1 Methods for Finding Normal Distribution Areas

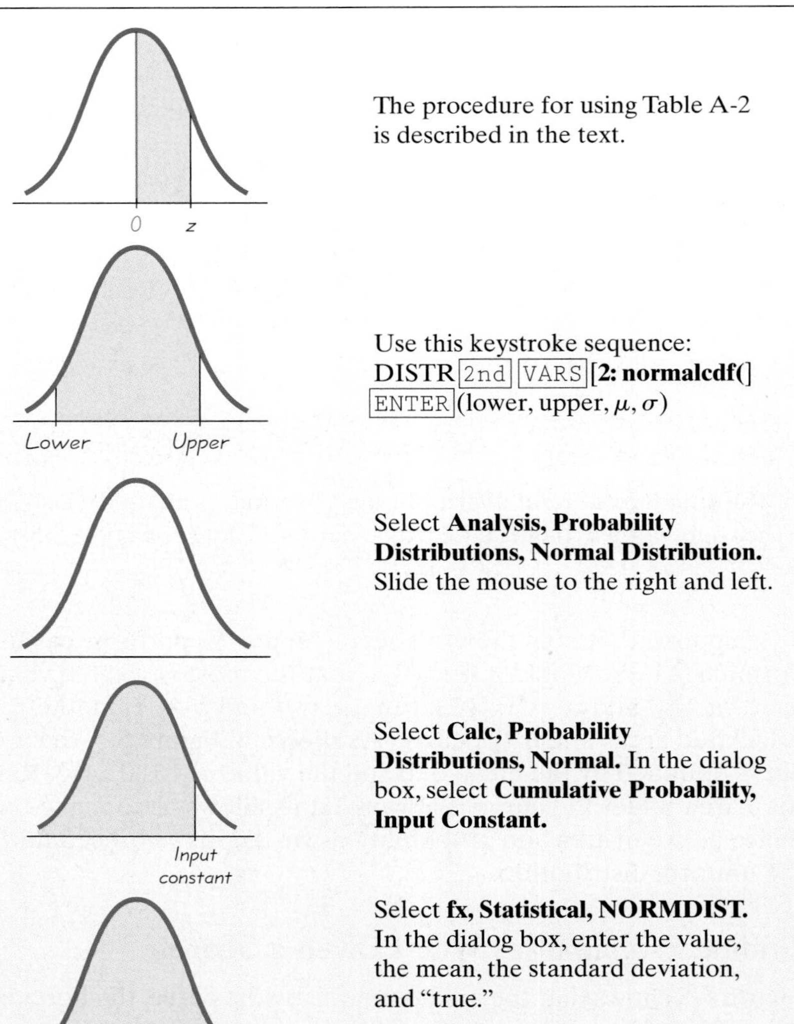

Method	Description	Procedure
Table A-2	Gives area bounded on the left by a vertical line above the mean of 0 and bounded on the right by a vertical line above any specific positive score denoted by z.	The procedure for using Table A-2 is described in the text.
TI-83 Plus Calculator	Gives area bounded on the left and bounded on the right by vertical lines above any specific scores.	Use this keystroke sequence: DISTR [2nd] [VARS] **[2: normalcdf(]** [ENTER] (lower, upper, μ, σ)
STATDISK	Gives a few areas, including the cumulative area from the left, the cumulative area from the right, and the same area as Table A-2.	Select **Analysis, Probability Distributions, Normal Distribution.** Slide the mouse to the right and left.
Minitab	Gives the cumulative area from the left up to a vertical line above a specific value.	Select **Calc, Probability Distributions, Normal.** In the dialog box, select **Cumulative Probability, Input Constant.**
Excel	Gives the cumulative area from the left up to a vertical line above a specific value.	Select **fx, Statistical, NORMDIST.** In the dialog box, enter the value, the mean, the standard deviation, and "true."

know the method you will be using for class and tests. (Because most readers will be using Table A-2 or a TI-83 Plus calculator, the following examples and exercises will discuss their use.)

If you are using Table A-2, it is essential to understand the following points.

1. Table A-2 is designed only for the *standard* normal distribution, which has a mean of 0 and a standard deviation of 1.
2. Each value in the body of the table is an area under the curve bounded on the left by a vertical line above the mean of 0 and bounded on the

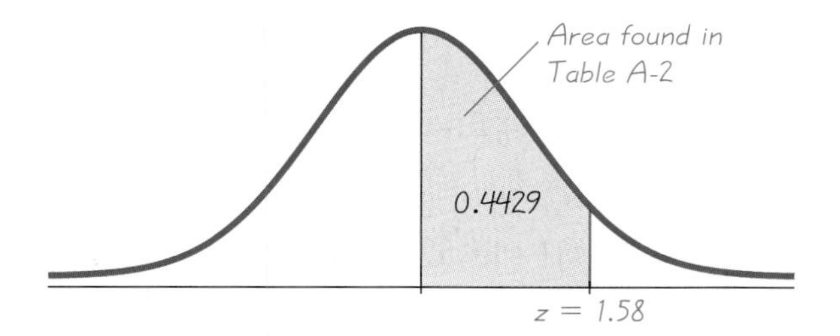

FIGURE 5-6
The Standard Normal Distribution

The area of the shaded region bounded by the mean of 0 and the positive number z can be found in Table A-2.

right by a vertical line above a specific positive score denoted by z, as illustrated in Figure 5-6.

3. When working with a graph, avoid confusing z scores and areas.

   ***z* score: *distance* along the horizontal scale of the standard normal distribution; refer to the leftmost column and top row of Table A-2**

   Area: *region* under the curve; refer to the values in the body of Table A-2

4. The part of the z score denoting hundredths is found across the top row of Table A-2.

The following example requires that we find the probability associated with a value between 0 and 1.58. Begin with the z score of 1.58 by locating 1.5 in the left column; next find the value in the adjoining row of probabilities that is directly below 0.08, as shown in this excerpt from Table A-2.

z	....	0.08
.		.
.		.
.		.
1.5	....	0.4429

The area (or probability) value of 0.4429 indicates that there is a probability of 0.4429 of randomly selecting a z score between 0 and 1.58. (The following sections will consider cases in which the mean is not 0 or the standard deviation is not 1.)

EXAMPLE **Scientific Thermometers** The Precision Scientific Instrument Company manufactures thermometers that are supposed to give readings of 0°C at the freezing point of water. Tests on a large sample of these instruments reveal that at the freezing point of water, some thermometers give readings below 0° (denoted by negative numbers) and some give readings above 0° (denoted by positive numbers). Assume that the mean reading is 0°C and the standard deviation of the readings is 1.00°C. Also assume that the readings are normally distributed. If one thermometer is randomly selected, find the probability that, at the freezing point of water, the reading is between 0° and +1.58°.

SOLUTION The probability distribution of readings is a standard normal distribution, because the readings are normally distributed with $\mu = 0$ and $\sigma = 1$. We need to find the area between 0 and z (the shaded region) in Figure 5-6 with $z = 1.58$. From Table A-2 we find that this area is 0.4429.

If you are using a TI-83 Plus calculator, press the **2nd** key, then press the **VARS** key, then use the down arrow key to select **2:normalcdf** (normal cumulative density function), press the **ENTER** key, then enter 0, 1.58, 0, 1 (for lower value, upper value, mean, standard deviation), and press the **ENTER** key to get a result of .4429465625.

INTERPRETATION The probability of randomly selecting a thermometer with a reading between 0° and +1.58° is therefore 0.4429. Another way to interpret this result is to conclude that 44.29% of the thermometers will have readings between 0° and +1.58°.

EXAMPLE **Scientific Thermometers** Using the thermometers from the preceding example, find the probability of randomly selecting one thermometer that reads (at the freezing point of water) between −2.43° and 0°.

SOLUTION *Using Table A-2:* We are looking for the region that is shaded in Figure 5-7(a), but Table A-2 is designed to apply only to regions to the right of the mean (0), as in Figure 5-7(b). By comparing the shaded area in Figure 5-7(a) to the shaded area in Figure 5-7(b), we can see that those two areas are identical, because the density curve is symmetric. Referring to Table A-2, we can easily determine that the shaded area of Figure 5-7(b) is 0.4925, so the shaded area of Figure 5-7(a) must also be 0.4925.

Using a TI-83 Plus calculator: Press **2nd, VARS, 2,** and proceed to enter −2.43, 0, 0, 1 to get a result of .4924505896.

INTERPRETATION The probability of randomly selecting a thermometer with a reading between −2.43° and 0° is 0.4925. In other words, 49.25% of the thermometers have readings between −2.43° and 0°.

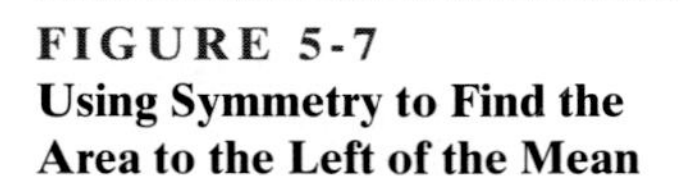

FIGURE 5-7
Using Symmetry to Find the Area to the Left of the Mean

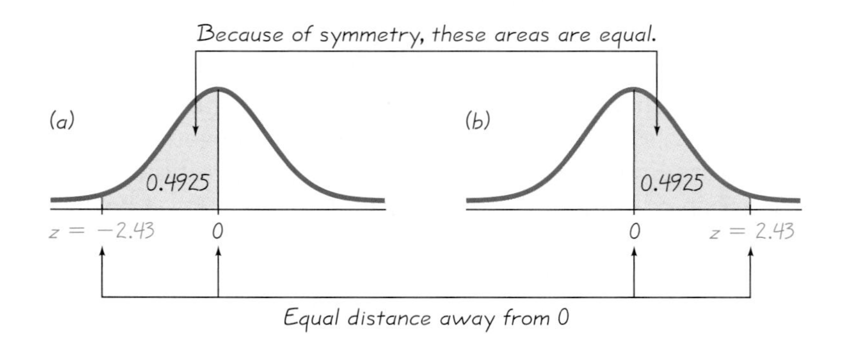

The preceding solution illustrates an important principle:

Although a z score can be negative, the area under the curve (or the corresponding probability) can never be negative.

Now recall the empirical rule (presented in Section 2-5), which stated that for bell-shaped distributions,

- about 68% of all values fall within 1 standard deviation of the mean.
- about 95% of all values fall within 2 standard deviations of the mean.
- about 99.7% of all values fall within 3 standard deviations of the mean.

If we refer to Figure 5-5 with $z = 1$, Table A-2 or a TI-83 Plus calculator will show us that the shaded area is 0.3413. It follows that the proportion of scores between $z = -1$ and $z = 1$ will be $0.3413 + 0.3413 = 0.6826$. That is, about 68% of all scores fall within 1 standard deviation of the mean. A similar calculation with $z = 2$ yields the value of $0.4772 + 0.4772 = 0.9544$ (or about 95%) as the proportion of scores between $z = -2$ and $z = 2$. Similarly, the proportion of scores between $z = -3$ and $z = 3$ is $0.4987 + 0.4987 = 0.9974$ (or about 99.7%). These exact values correspond very closely to those given by the empirical rule. In fact, the values of the empirical rule were found directly from the probabilities in Table A-2 and were slightly rounded for convenience. The empirical rule is sometimes called the *68-95-99.7 rule;* using exact values from Table A-2, it would be called the *68.26-95.44-99.74 rule,* but then it wouldn't sound as snappy.

Because we are dealing with a density curve for a probability distribution, the total area under the curve must be 1. Now refer to Figure 5-8, which shows that a vertical line directly above the mean of 0 divides the area under the curve into two equal parts, each containing an area of 0.5. The following example uses this characteristic.

EXAMPLE **Scientific Thermometers** Once again, make a random selection from the same sample of thermometers. Find the probability that the chosen thermometer reads (at the freezing point of water) more than +1.27°.

SOLUTION We are again dealing with normally distributed values having a mean of 0° and a standard deviation of 1°. The probability of selecting

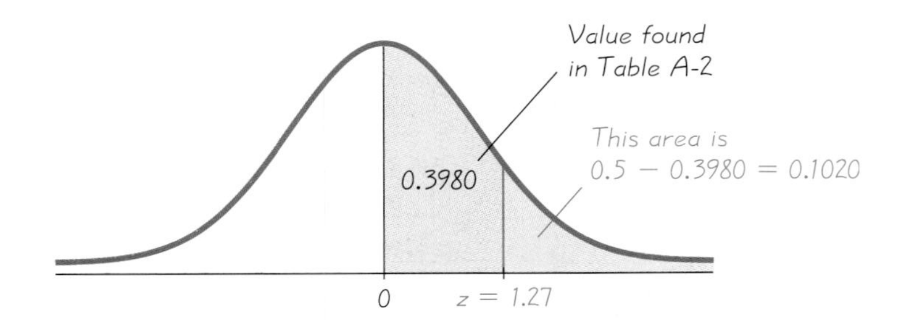

FIGURE 5-8
Finding the Area to the Right of $z = 1.27$

a thermometer that reads greater than $+1.27°$ corresponds to the green-shaded area of Figure 5-8.

Using Table A-2: Table A-2 cannot be used to find that green area directly, but we can use the table to find that $z = 1.27$ corresponds to the orange area of 0.3980, as shown in the figure. We now reason that because the area to the right of zero is one-half of the total area, it has an area of 0.5, and the green-shaded area is $0.5 - 0.3980$, or 0.1020.

Using a TI-83 Plus calculator: Press **2nd, VARS, 2,** and enter the lower value, upper value, mean, and standard deviation. In this example, there is no specific upper value, so we use this trick: For the upper limit, enter a value that is excessively large, such as 999999. The entry of 1.27, 999999, 0, 1 will result in a value of .1020423807.

INTERPRETATION We conclude that there is a probability of 0.1020 of randomly selecting one of the thermometers with a reading greater than $+1.27°$. Another way to interpret this result is to state that if many thermometers are selected and tested, then 0.1020 (or 10.20%) of them will read greater than $+1.27°$.

EXAMPLE **Scientific Thermometers** Assuming that one thermometer in our sample is randomly selected, find the probability that it reads (at the freezing point of water) between 1.20° and 2.30°.

SOLUTION The probability of selecting a thermometer that reads between $+1.20°$ and $+2.30°$ corresponds to the green-shaded area of Figure 5-9.

Using Table A-2: Table A-2 is designed to provide only for regions bounded on the left by the vertical line above 0. We can use the table to find that $z = 1.20$ corresponds to an area of 0.3849 and that $z = 2.30$ corresponds to an area of 0.4893, as shown in the figure. If we denote the area of the green-shaded region by A, we can see from Figure 5-9 that

$$0.3849 + A = 0.4893$$

so

$$A = 0.4893 - 0.3849 = 0.1044$$

FIGURE 5-9
Finding the Area Between $z = 1.20$ and $z = 2.30$

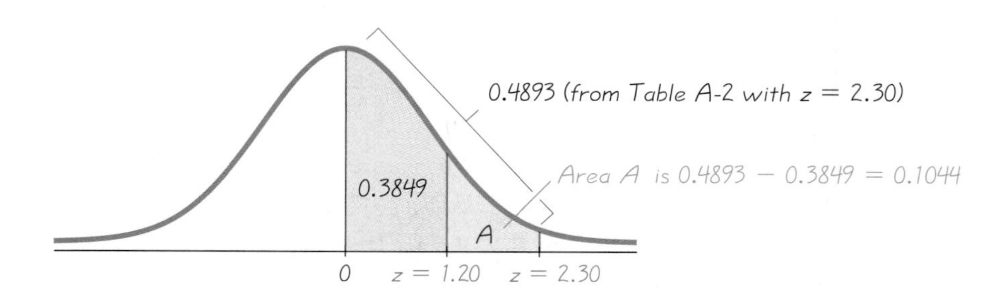

Using a TI-83 Plus calculator: Press **2nd, VARS, 2,** and proceed to enter 1.20, 2.30, 0, 1 to get a result of .1043456505, which differs from the result of 0.1044 by a negligible amount.

INTERPRETATION If one thermometer is randomly selected, the probability that it reads (at the freezing point of water) between 1.20° and 2.30° is therefore 0.1044 (or 0.1043).

The preceding example concluded with the statement that the probability of a reading between 1.20° and 2.30° is 0.1044. Such probabilities can also be expressed with the following notation.

Notation

$P(a < z < b)$	denotes the probability that the z score is between a and b.
$P(z > a)$	denotes the probability that the z score is greater than a.
$P(z < a)$	denotes the probability that the z score is less than a.

Using this notation, we can express the result of the last example as $P(1.20 < z < 2.30) = 0.1044$, which states in symbols that the probability of a z score falling between 1.20 and 2.30 is 0.1044. With a continuous probability distribution such as the normal distribution, the probability of getting any single *exact* value is 0. That is, $P(z = a) = 0$. For example, there is a 0 probability of randomly selecting someone and getting a person whose height is exactly 68.12345678 in. In the normal distribution, any single point on the horizontal scale is represented not by a region under the curve, but by a vertical line above the point. For $P(z = 1.33)$, we have a vertical line above $z = 1.33$, but that vertical line by itself contains no area, so $P(z = 1.33) = 0$. With any continuous random variable, the probability of any one exact value is 0, and it follows that $P(a \le z \le b) = P(a < z < b)$. It also follows that the probability of getting a z score of *at most b* is equal to the probability of getting a z score *less than b.* It is important to correctly interpret key phrases such as *at most, at least, more than, no more than,* and so on. The illustrations in Figure 5-10 on the next page provide an aid to interpreting several of the most common phrases.

Finding *z* Scores When Given Probabilities

So far, the examples of this section involving the standard normal distribution have all followed the same format: Given z scores, we found areas under the curve. These areas represent probabilities. In many other cases, we want a reverse process because we already know the area (or probability), but we need to find the corresponding z score. In such cases, it is very important to

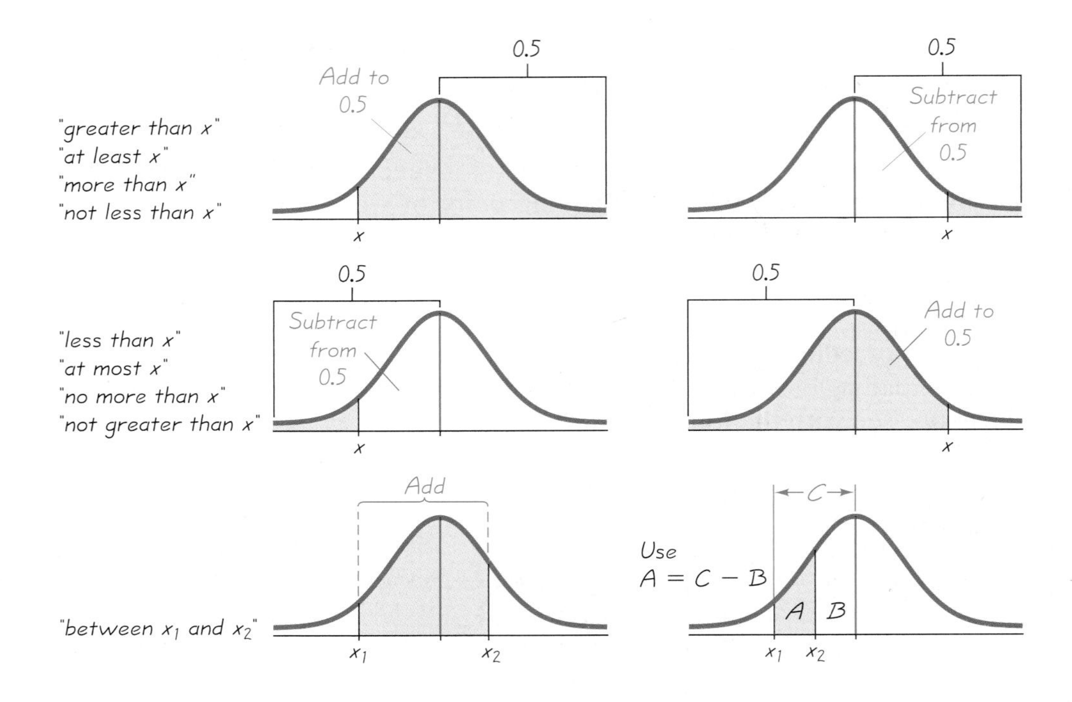

FIGURE 5-10
Interpreting Areas Correctly

avoid confusion between z scores and areas. Remember, z scores are *distances* along the horizontal scale, and they are represented by the numbers in Table A-2 that are in the extreme left column and across the top row. Areas (or probabilities) are regions under the curve, and they are represented by the values in the body of Table A-2. Also, z scores to the left of the centerline are always negative (as in Figure 5-7a). If we already know a probability and want to determine the corresponding z score, we find it as follows.

- **Using Table A-2:**
  1. Draw a bell-shaped curve, draw the centerline, and identify the region under the curve that corresponds to the given probability. If that region is not bounded by the centerline, work with a known region that is bounded by the centerline.
  2. Using the probability representing the area bounded by the centerline, locate the closest probability in the *body* of Table A-2 and identify the corresponding z score.
  3. If the z score is positioned to the left of the centerline, make it negative.
- **Using a TI-83 Plus calculator:** Press **2nd, VARS, 3** (for invNorm), and proceed to enter the total area to the left of the value, the mean, and the standard deviation in the format of

  (total area to the left, mean, standard deviation)

  with the commas included.(See example below.)
- **Using STATDISK:** Select **Analysis, Probability Distributions, Normal Distribution,** and proceed to slide the mouse to the right or left until

you get the desired score. You can get more precision by using the mouse to click and drag part of the curve.

- **Using Minitab:** Select **Calc, Probability Distributions, Normal,** then select **Inverse cumulative probabilities** and the option **Input constant.** For the input constant, enter the total area to the left of the given value.
- **Using Excel:** Select **fx**, **Statistical**, **NORMINV**, and proceed to make the entries in the dialog box. When entering the probability value, enter the total area to the left of the given value.

Because most readers will be using Table A-2 or a TI-83 Plus calculator in class, the following examples illustrate only these two tools for finding z scores.

EXAMPLE **Scientific Thermometers** Use the same thermometers as earlier, with temperature readings that are normally distributed with a mean of 0°C and a standard deviation of 1°C. Find the temperature corresponding to P_{95}, the 95th percentile. That is, find the temperature separating the bottom 95% from the top 5%. See Figure 5-11.

SOLUTION Figure 5-11 shows the z score that is the 95th percentile, separating the top 5% from the bottom 95%.

Using Table A-2: We must refer to Table A-2 to find the unknown z score, and we must use a region bounded by the centerline (where $\mu = 0$) on one side, such as the shaded region of 0.45 in Figure 5-11. (Remember, Table A-2 is designed to directly provide only those areas that are bounded on the left by the centerline and bounded on the right by the z score.) We first search for the area of 0.45 *in the body of the table* and then find the corresponding z score. In Table A-2 the area of 0.45 is between the table values of 0.4495 and 0.4505, but there's an asterisk with a special note indicating that 0.4500 corresponds to a z score of 1.645. We can now conclude that the z score in Figure 5-11 is 1.645, so the 95th percentile is the temperature reading of 1.645°C.

Using a TI-83 Plus calculator: Press **2nd, VARS, 3,** and enter (.95, 0, 1) to get a result of 1.644853626. We entered .95 because the TI-83 Plus calculator requires the total area to the *left* of the value that we are trying to find.

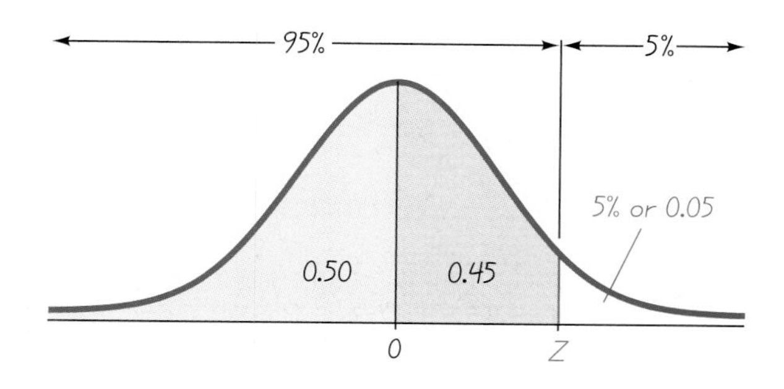

FIGURE 5-11
Finding the 95th Percentile

INTERPRETATION When tested at freezing, 95% of the readings will be less than or equal to 1.645°C, and 5% of them will be greater than or equal to 1.645°C.

z Score	Area
1.645	0.4500
2.575	0.4950

Note that in the preceding solution, Table A-2 led to a z score of 1.645, which is midway between 1.64 and 1.65. When using Table A-2, we can usually avoid interpolation by simply selecting the closest value. There are two special cases listed in the accompanying table that are important because they are used so often in a wide variety of applications. (The value of $z = 2.576$ is slightly closer to the area of 0.4950, but $z = 2.575$ has the advantage of being the value midway between $z = 2.57$ and $z = 2.58$.) Except in these two special cases, we can select the closest value in the table. (If a desired value is midway between two table values, select the larger value.) Also, for z scores above 3.09, we can use 0.4999 as an approximation of the corresponding area between the z score and the centerline.

EXAMPLE **Scientific Thermometers** Using the same thermometers, find P_{10}, the 10th percentile. That is, find the temperature reading separating the bottom 10% of all temperatures from the top 90%.

SOLUTION Refer to Figure 5-12, where the 10th percentile is shown as the z score separating the bottom 10% from the top 90%.

Using Table A-2: Table A-2 is designed for areas bounded by the centerline, so we refer to the shaded area of 0.40 (corresponding to 50% − 10%). In the *body of the table,* we select the closest value of 0.3997 and find that it corresponds to $z = 1.28$. However, because the z score is below the mean of 0, it must be negative. The 10th percentile is therefore −1.28°C.

Using a TI-83 Plus calculator: Enter **2nd, VARS, 3,** and proceed to enter (.10, 0, 1) to get a result of −1.281551567.

INTERPRETATION When tested at freezing, 10% of the thermometer readings will be equal to or less than −1.28°C, and 90% of the readings will be equal to or greater than −1.28°C.

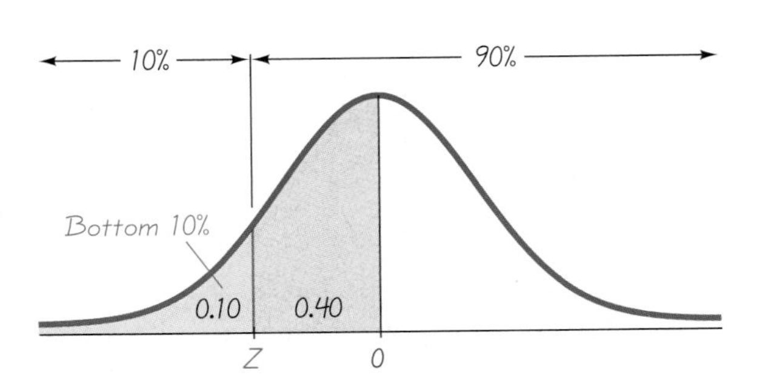

FIGURE 5-12
Finding the 10th Percentile

The examples in this section were contrived so that the mean of 0 and the standard deviation of 1 coincided exactly with the parameters of the standard normal distribution. In reality, it is unusual to find such convenient parameters, because typical normal distributions involve means different from 0 and standard deviations different from 1. In the next section we introduce methods for working with such normal distributions.

5-2 Basic Skills and Concepts

Using a Continuous Uniform Distribution. *In Exercises 1–4, refer to the continuous uniform distribution depicted in Figure 5-2, assume that a time between 0 and 5 hours is randomly selected, and find the probability that the given time is selected.*

1. Less than 4

2. Greater than 2

3. Between 3 and 4

4. Between 1.5 and 3.8

Using a Continuous Uniform Distribution. *In Exercises 5–8, assume that voltages in a circuit vary between 6 volts and 12 volts, and voltages are spread evenly over the range of possibilities, so that there is a uniform distribution. Find the probability of the given range of voltage levels.*

5. Greater than 10 volts

6. Less than 11 volts

7. Between 7 volts and 10 volts

8. Between 6.5 volts and 8 volts

Using the Standard Normal Distribution. *In Exercises 9–28, assume that the readings on the thermometers are normally distributed with a mean of 0° and a standard deviation of 1.00°C. A thermometer is randomly selected and tested. In each case, draw a sketch, and find the probability of each reading in degrees.*

9. Between 0 and 1.50

10. Between 0 and 1.28

11. Between -1.96 and 0

12. Between -0.48 and 0

13. Less than -1.79

14. Greater than 0.37

15. Greater than 2.05

16. Less than -0.92

17. Between 0.50 and 1.50

18. Between 1.50 and 2.50

19. Between -2.00 and -1.00

20. Between 2.00 and 2.34

21. Less than 1.62

22. Less than 2.44

23. Greater than -0.27

24. Greater than -2.09

25. Between -1.08 and 0.33

26. Between -0.90 and 1.95

27. Greater than 0

28. Less than 0

Finding Probability. *In Exercises 29–32, assume that the readings on the thermometers are normally distributed with a mean of 0° and a standard deviation of 1.00°. Find the indicated probability, where z is the reading in degrees.*

29. $P(-1.96 < z < 1.96)$

30. $P(z < 1.645)$

31. $P(z > -2.575)$

32. $P(1.96 < z < 2.33)$

Finding Temperature Values. *In Exercises 33–40, assume that the readings on the thermometers are normally distributed with a mean of 0° and a standard deviation of 1.00°C. A thermometer is randomly selected and tested. In each case, draw a sketch, and find the temperature reading corresponding to the given information.*

33. Find P_{90}, the 90th percentile. This is the temperature reading separating the bottom 90% from the top 10%.

34. Find P_{20}, the 20th percentile.

35. Find Q_1, the temperature reading that is the first quartile.

36. Find D_3, the temperature reading that is the third decile.

37. If 5% of the thermometers are rejected because they have readings that are too low, but all other thermometers are acceptable, find the reading that separates the rejected thermometers from the others.

38. If 12% of the thermometers are rejected because they have readings that are too high, but all other thermometers are acceptable, find the reading that separates the rejected thermometers from the others.

39. A troubleshooter wants to examine thermometers that give readings in the bottom 3%. What reading separates the bottom 3% from the others?

40. If 2.5% of the thermometers are rejected because they have readings that are too high and another 2.5% are rejected because they have readings that are too low, find the two readings that are cutoff values separating the rejected thermometers from the others.

5-2 Beyond the Basics

41. For a standard normal distribution, find the percentage of data that are

a. within 1 standard deviation of the mean.
b. within 1.96 standard deviations of the mean.
c. between $\mu - 3\sigma$ and $\mu + 3\sigma$.
d. between 1 standard deviation below the mean and 2 standard deviations above the mean.
e. more than 2 standard deviations away from the mean.

42. If a continuous uniform distribution has parameters of $\mu = 0$ and $\sigma = 1$, then the minimum is $-\sqrt{3}$ and the maximum is $\sqrt{3}$.

a. For this distribution, find $P(-1 < x < 1)$.
b. Find $P(-1 < x < 1)$ if you incorrectly assume that the distribution is normal instead of uniform.
c. Compare the results from parts (a) and (b). Does the distribution affect the results very much?

43. Assume that z scores are normally distributed with a mean of 0 and a standard deviation of 1.

a. If $P(0 < z < a) = 0.3907$, find a.
b. If $P(-b < z < b) = 0.8664$, find b.
c. If $P(z > c) = 0.0643$, find c.
d. If $P(z > d) = 0.9922$, find d.
e. If $P(z < e) = 0.4500$, find e.

44. In a continuous uniform distribution,

$$\mu = \frac{\text{minimum} + \text{maximum}}{2} \quad \text{and} \quad \sigma = \frac{\text{range}}{\sqrt{12}}$$

Find the mean and standard deviation for the uniform distribution represented in Figure 5-2.

5-3 Normal Distributions: Finding Probabilities

In Section 5-2 we presented the basic methods for working with normal distributions. Because all of the examples and exercises discussed there involved only the *standard* normal distribution (with a mean of 0 and a standard deviation of 1), they were necessarily unrealistic. In this section we include other normal distributions, so we will be able to work with cases that are more realistic and practical.

The basic principle we will be explaining in this section is the following:

If we convert values to standard scores using Formula 5-2, then procedures for working with all normal distributions are the same as those for the standard normal distribution.

Formula 5-2 $$z = \frac{x - \mu}{\sigma}$$

If you used Table A-2 as your primary tool for finding probabilities or z scores in the Section 5-2, you will need to understand and apply the above principle. If you use certain calculators or software programs, the conversion to z scores is not necessary because probabilities can be found directly. Regardless of the method used, however, you need to clearly understand the basic principle, because it is an important foundation for concepts introduced in the following chapters.

We illustrate the principle by showing that the area in Figure 5-13(a) is the same as the area in Figure 5-13(b). That is, the area bounded by values in any normal distribution (as in Figure 5-13a) is the same as the area bounded by the equivalent standardized scores in the standard normal distribution (as in Figure 5-13b). This means that when you are working with a normal distribution that is not the standard normal distribution, you can

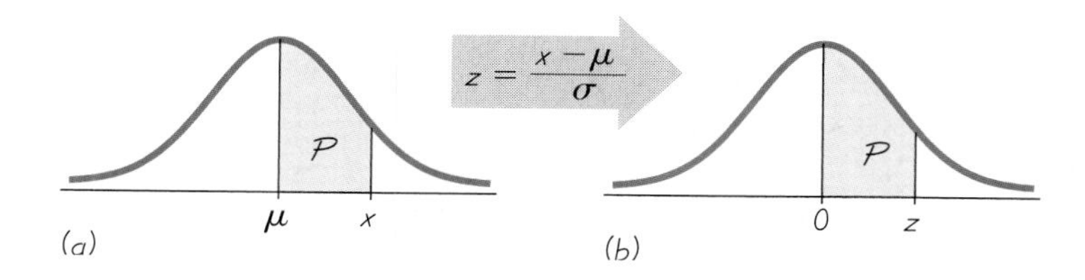

FIGURE 5-13
Converting to the Standard Normal Distribution

Queues

Queuing theory is a branch of mathematics that uses probability and statistics. The study of queues, or waiting lines, is important to businesses such as supermarkets, banks, fast food restaurants, airlines, and amusement parks. Grand Union supermarkets try to keep checkout lines no longer than three shoppers. Wendy's introduced the "Express Pak" to expedite servicing its numerous drive-through customers. Disney conducts extensive studies of lines at its amusement parks so that it can keep patrons happy and plan for expansion. Bell Laboratories uses queuing theory to optimize telephone network usage, and factories use it to design efficient production lines.

use Table A-2 the same way it was used in Section 5-2 as long as you first convert the values to z scores. We therefore recommend the following procedure for finding probabilities for values of a random variable with a normal probability distribution:

1. Sketch a normal curve, label the mean and the specific x values, then *shade* the region representing the desired probability.
2. For each relevant value x that is a boundary for the shaded region, use Formula 5-2 to convert that value to the equivalent z score.
3. Refer to Table A-2 to find the area of the shaded region. This area is the desired probability.

The following example applies these three steps, and it illustrates the relationship between a typical nonnormal distribution and the standard normal distribution.

AS

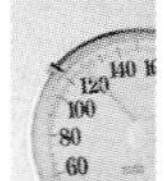

EXAMPLE Jet Ejection Seats In redesigning jet ejection seats to better accommodate women as pilots, it is found that women's weights are normally distributed with a mean of 143 lb and a standard deviation of 29 lb. If a woman is randomly selected, what is the probability that she weighs between 143 lb and 201 lb?

SOLUTION

Step 1: See Figure 5-14, where we label the mean of 143 lb and the value of 201 lb, and we shade the area representing the probability we want.

Step 2: To use Table A-2, we first must use Formula 5-2 to convert the distribution of weights to the standard normal distribution. The weight of 201 lb is converted to a z score as follows.

$$z = \frac{x - \mu}{\sigma} = \frac{201 - 143}{29} = \frac{58}{29} = 2.00$$

This result shows that the weight of 201 lb differs from the mean of 143 lb by 2.00 standard deviations.

FIGURE 5-14
Probability of Weight Between 143 lb and 201 lb

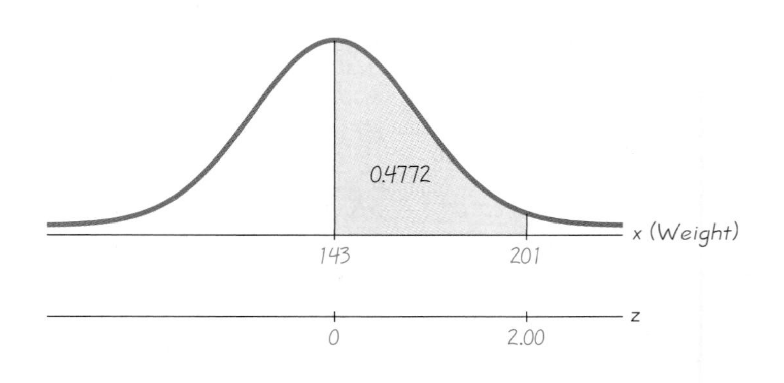

Step 3: Referring to Table A-2, we find that $z = 2.00$ corresponds to an area of 0.4772.

INTERPRETATION There is a probability of 0.4772 of randomly selecting a woman with a weight between 143 lb and 201 lb. This can be expressed in symbols as

$$P(143 < x < 201) = P(0 < z < 2.00) = 0.4772$$

Another way to interpret this result is to conclude that 47.72% of women have weights between 143 lb and 201 lb.

The example is easier to solve with a TI-83 Plus calculator. Press **2nd, VARS, 2** (for **normalcdf**), and enter (143, 201, 143, 29) to get a result of .4772499375. Be sure, however, that you have learned the procedure in the preceding worked-out solution. Know what you are doing before blindly accepting calculator or computer results.

EXAMPLE **Jet Ejection Seats** In the Chapter Problem, we reported that the Air Force designed the current ACES-II ejection seats for men weighing between 140 lb and 211 lb. What percentage of women have weights that are within those limits?

SOLUTION As in the preceding example, women's weights are normally distributed with a mean of 143 lb and a standard deviation of 29 lb. See Figure 5-15, which shows the design limits on the distribution of women's weights. We need to find the area of regions A and B combined. Using Table A-2, we cannot find that combined region directly because Table A-2 isn't designed for such cases, but we can find it indirectly by using the basic procedures presented in Section 5-2. The way to proceed is first to find regions A and B separately, then add them.

For region A:

$$z = \frac{x - \mu}{\sigma} = \frac{140 - 143}{29} = -0.10$$

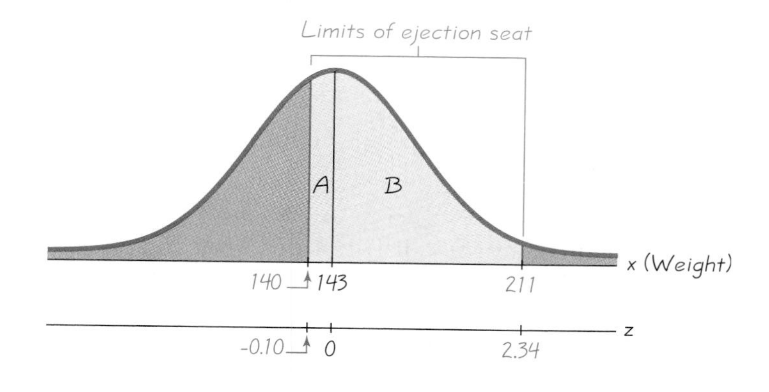

FIGURE 5-15
Weights of Women and Ejection Seat Limits

Using Table A-2, we find that $z = -0.10$ corresponds to an area of 0.0398. Region A has an area of 0.0398.

For region B:

$$z = \frac{x - \mu}{\sigma} = \frac{211 - 143}{29} = 2.34$$

Using Table A-2, we find that $z = 2.34$ corresponds to an area of 0.4904. Region B has an area of 0.4904.

For regions A and B combined:

$$\text{area of } A \text{ and } B \text{ combined} = 0.0398 + 0.4904 = 0.5302$$

INTERPRETATION We found that 53.02% of women have weights between the ejection seat limits of 140 lb and 211 lb. This means that 46.98% of women do *not* have weights between the current limits, so far too many women pilots would risk serious injury if ejection became necessary.

Again, the example is easier to solve if a TI-83 Plus calculator is used. Press **2nd, VARS, 2** (for **normalcdf**), and enter (140, 211, 143, 29) to get a result of 0.5316785365. Because of rounding errors, this differs from the solution of 0.5302 by a small amount.

Test your understanding of the methods of this section by working through the next example on your own. Write your own solution for each part before checking the solution given here.

EXAMPLE **Women's Weights** Again, assume that women's weights are normally distributed with a mean of 143 lb and a standard deviation of 29 lb. Also assume that we want to find the proportion of women weighing between 100 lb and 130 lb.

a. Draw the graph and shade the region representing the proportion of women weighing between 100 lb and 130 lb.

b. Convert the limits of 100 lb and 130 lb to their corresponding z scores.

c. Use Table A-2 to find the two areas corresponding to the two z scores.

d. What is the proportion of women weighing between 100 lb and 130 lb?

SOLUTION

a. Start by sketching a normal distribution curve and entering the value of 143 lb in the middle. The desired limits of 100 lb and 130 lb are both below the mean of 143 lb, so those limits are to the left of the center, as shown in Figure 5-16.

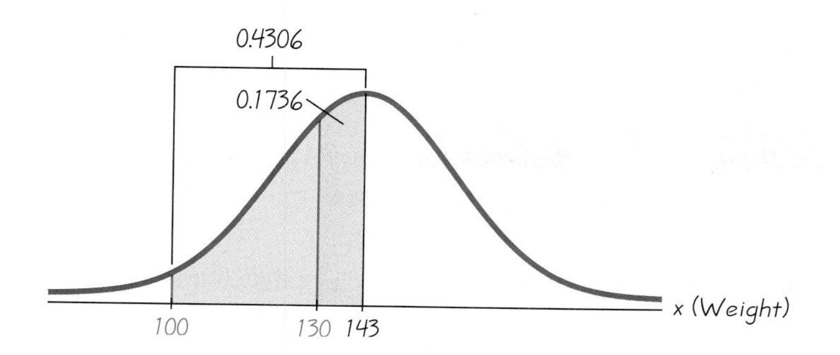

FIGURE 5-16
Weights of Women

b. We convert the values of 100 lb and 130 lb to z scores by using Formula 5-2.

$$z = \frac{x - \mu}{\sigma} = \frac{100 - 143}{29} = -1.48$$

$$z = \frac{x - \mu}{\sigma} = \frac{130 - 143}{29} = -0.45$$

c. Using Table A-2, the z score of 1.48 corresponds to an area of 0.4306. Using Table A-2 again, the z score of 0.45 corresponds to an area of 0.1736. Both of these areas are shown in Figure 5-16.

INTERPRETATION The desired area is the green-shaded region, which is the difference between 0.4306 and 0.1736. Because $0.4306 - 0.1736 = 0.2570$, we conclude that the proportion of women with weights between 100 lb and 130 lb is 0.2570.

In this section we have extended the concepts of Section 5-2 so that they apply to more realistic normal probability distributions. Still, all of the examples and exercises of this section are of the same type: We are given specific limit values and we must find an area (or probability or percentage). In many practical and real cases, the area (or probability or percentage) is known and we must find the relevant value(s). Problems of this type are discussed in the next section.

5-3 Basic Skills and Concepts

Using a Nonstandard Normal Distribution. *In Exercises 1–4, assume that women's weights are normally distributed with a mean given by $\mu = 143$ lb and a standard deviation given by $\sigma = 29$ lb (based on data from the National Health Survey). Also assume that a woman is randomly selected. Draw a graph, and find the indicated probability.*

1. $P(143 \text{ lb} < x < 172 \text{ lb})$

2. $P(150 \text{ lb} < x < 180 \text{ lb})$

3. $P(x > 150 \text{ lb})$

4. $P(x < 186.5 \text{ lb})$

Using a Nonstandard Normal Distribution. *In Exercises 5–8, assume that heights of women are normally distributed with a mean given by $\mu = 63.6$ in. and a* *standard deviation given by $\sigma = 2.5$ in. (based on data from the National Health Survey). Draw a graph, and find the indicated probability or percentage.*

5. Rockette Height Requirement The heights of the Rockette dancers at New York City's Radio City Music Hall must be between 65.5 in. and 68.0 in. If a woman is randomly selected, find the probability that she meets the height requirement to be a Rockette.

6. Beanstalk Club Height Requirement The Beanstalk Club, a social organization for tall people, has a requirement that women must be at least 70 in. (or 5 ft 10 in.) tall. What percentage of women meet that requirement?

7. *Soyuz* Spacecraft Height Requirement In order to fit into a Russian *Soyuz* spacecraft, an astronaut must have a height between 64.5 in. and 72 in. What percentage of women meet that requirement?

8. Height Requirement for Women Soldiers The U.S. Army requires women's heights to be between 58 in. and 80 in. Find the percentage of women meeting that height requirement. Are many women being denied the opportunity to join the Army because they are too short or too tall?

9. SAT Test Required Scores The combined math and verbal scores for females taking the SAT-I test are normally distributed with a mean of 998 and a standard deviation of 202 (based on data from the College Board). If a college includes a minimum score of 900 among its requirements, what percentage of females do *not* satisfy that requirement?

10. SAT Test Required Scores The scores for males on the math portion of the SAT-I test are normally distributed with a mean of 531 and a standard deviation of 114 (based on data from the College Board). If the College of Newport includes a minimum score of 600 among its requirements, what percentage of males do *not* satisfy that requirement?

11. Sugar Packets and Label Requirements Domino sugar packets are labeled as containing 3.5 g. Assume that those packets are actually filled with amounts that are normally distributed with a mean of 3.586 g and a standard deviation of 0.074 g (based on Data Set 4 in Appendix B). What percentage of packets have less than 3.5 g? Are many consumers being cheated?

12. Finding Percentage of Fever Temperatures Based on the sample results in Data Set 6 of Appendix B, assume that human body temperatures are normally distributed with a mean of 98.20°F and a standard deviation of 0.62°F. Bellevue Hospital in New York City uses 100.6°F as the lowest temperature considered to be a fever. What percentage of normal and healthy persons would be considered to have a fever? Does this percentage suggest that a cutoff of 100.6°F is appropriate?

13. Oops One classic use of the normal distribution is inspired by a letter to "Dear Abby" in which a wife claimed to have given birth 308 days after a brief visit from her husband, who was serving in the Navy. The lengths of pregnancies are normally distributed with a mean of 268 days and a standard deviation of 15 days. Given this information, find the probability of a pregnancy lasting 308 days or longer. What does the result suggest?

14. Finding Percentage of Premature Births The lengths of pregnancies are normally distributed with a mean of 268 days and a standard deviation of 15 days. If we stipulate that a baby is *premature* if born at least three weeks early, what percentage of babies are born prematurely? Such information is important so that a hospital can plan to have enough special equipment to handle premature babies.

15. Finding Percentage of Coke Cans Under 12 oz Cans of regular Coke are labeled as containing 12 oz. Assume that the actual contents are normally distributed with a mean of 12.19 oz and a standard deviation of 0.11 oz (based on Data Set 1 in Appendix B). What percentage of cans contain less than the 12 oz printed on the label? Are many consumers being cheated?

16. Finding Percentage of Pepsi Cans Under 12 oz Cans of regular Pepsi are labeled as containing 12 oz. Assume that the actual contents are normally distributed with a mean of 12.29 oz and a standard deviation of 0.09 oz (based on Data Set 1 in Appendix B). What percentage of cans contain less than the 12 oz printed on the label? Are many consumers being cheated?

17. Marinie Corps Height Requirements The U.S. Marine Corps requires that men have heights between 64 in. and 78 in. Find the percentage of men meeting those height requirements. (The National Health Survey shows that heights of men are normally distributed with a mean of 69.0 in. and a standard deviation of 2.8 in.) Are too many men denied the opportunity to join the Marines because they are too short or too tall?

18. Babies at Risk Weights of newborn babies in the United States are normally distributed with a mean of 3420 g and a standard deviation of 495 g (based on data from "Birth Weight and Prenatal Mortality," by Wilcox et al., *Journal of the American Medical Association,* Vol. 273, No. 9). A newborn weighing less than 2200 g is considered to be at risk, because the mortality rate for this group is at least 1%. What percentage of newborn babies are in the "at-risk" category? If the Chicago General Hospital has 900 births in a year, how many of the babies are in the "at-risk" category?

19. Babies at Risk Do Exercise 18 for Norway, where weights of newborn babies are normally distributed with a mean of 3570 g and a standard deviation of 500 g (based on data from "Birth Weight and Prenatal Mortality," by Wilcox et al., *Journal of the American Medical Association,* Vol. 273, No. 9).

20. Computer Chips Meeting Specs An IBM subcontractor was hired to make ceramic substrates, which are used to distribute power and signals to and from computer silicon chips. Specifications require resistance between 1.500 ohms and 2.500 ohms, but the population has normally distributed resistances with a mean of 1.978 ohms and a standard deviation of 0.172 ohms. What percentage of the ceramic substrates will not meet the manufacturer's specifications? Does this manufacturing process appear to be working well?

21. Cholesterol and Coronary Disease The serum cholesterol levels in men aged 18 to 24 are normally distributed with a mean of 178.1 and a standard deviation of 40.7. The units are mg/100 mL, and the data are based on the National Health Survey. One criterion for identifying risk of coronary disease is a cholesterol level above 300. If a man aged 18 to 24 is randomly selected, find the probability that his serum cholesterol level is above 300. Is this probability low enough to warrant serious concern?

22. Dating Skulls Measurements of human skulls from different epochs are analyzed to determine whether they change over time. The maximum breadth is measured for skulls from Egyptian males who lived around 3300 B.C. Results show that those breadths are normally distributed with a mean of 132.6 mm and a standard deviation of 5.4 mm (based on data from *Ancient Races of the Thebaid* by Thomson and Randall-Maciver). An archeologist discovers a male Egyptian skull and a field measurement reveals a maximum breadth of 119 mm. Find the probability of getting a value of 119 mm or less if a skull is randomly selected from the period around 3300 B.C. Is the newly found skull likely to come from that era?

23. Mensa IQ Requirement IQ scores are normally distributed with a mean of 100 and a standard deviation of 15. Mensa is an organization for people with high IQs, and eligibility requires an IQ above 131.5.

a. If someone is randomly selected, find the probability that he or she meets the Mensa requirement.

b. In a typical region of 75,000 people, how many are eligible for Mensa?

24. Counterfeit Coins in Vending Machines Some vending machines are designed so that their owners can adjust the weights of the quarters that are accepted. If many counterfeit coins are found, adjustments are made to reject more coins, with the effect that most of the counterfeit coins are rejected along with many legal coins. Assume that quarters have weights that are normally distributed with a mean of 5.67 g and a standard deviation of 0.070 g. If a vending machine is adjusted to reject quarters weighing less than 5.50 g or more than 5.80 g, what is the percentage of legal quarters that are rejected? Is this result low enough so that not too many customers will be inconvenienced because their coins are rejected?

5-3 Beyond the Basics

25. Units of Measurement Weights of individual women are expressed in units of pounds. What are the units used for the z scores that correspond to individual weights?

26. z Scores Weights of women are normally distributed with a mean of 143 lb and a standard deviation of 29 lb. If weights of all woman are converted to z scores, what are the mean, standard deviation, and distribution of these z scores?

27. Transformations Weights of women are normally distributed with a mean of 143 lb and a standard deviation of 29 lb. What are the distribution, mean, and standard deviation after all of those weights have been converted to kilograms? (1 lb = 0.4536 kg)

28. Using Continuity Correction There are many situations in which a normal distribution can be used as a good approximation to a random variable that has only *discrete* values. In such cases, we can use this *continuity correction:* Represent each whole number by the interval extending from 0.5 below the number to 0.5 above it. Assume that IQ scores are all whole numbers having a distribution that is approximately normal with a mean of 100 and a standard deviation of 15.

a. Without using any correction for continuity, find the probability of randomly selecting someone with an IQ score greater than 105.

b. Using the correction for continuity, find the probability of randomly selecting someone with an IQ score greater than 105.

c. Compare the results from parts (a) and (b).

5-4 Normal Distributions: Finding Values

In this section we consider problems such as this: If women's weights are normally distributed with a mean of 143 lb and a standard deviation of 29 lb, find the weight separating the top 10% from the others. In the examples and exercises in Section 5-3 we used a given value to find a probability or percentage. In this section we follow a reverse procedure—we use a given probability (or percentage) to find a specific value.

In working with problems of finding values when given probabilities, there are three important cautions to keep in mind.

1. *Don't confuse z scores and areas.* Remember, z scores are *distances* along the horizontal scale, but areas are *regions* under the normal curve. Table A-2 lists z scores in the left column and across the top row, but areas are found in the body of the table.
2. *Choose the correct (right/left) side of the graph.* A value separating the top 10% from the others will be located on the right side of the graph, but a value separating the bottom 10% will be located on the left side of the graph.
3. *A z score must be negative whenever it is located to the left of the centerline of 0.*

As in Section 5-3, graphs are extremely helpful, and we strongly urge you to use them. Also as in Section 5-3, we can use Table A-2, or a TI-83 Plus calculator, or software. We again recommend that you begin with Table A-2 and Formula 5-2 so that you can gain a clear understanding of what you are doing before using a calculator or software.

Procedure for Finding Values Using Table A-2 and Formula 5-2

1. Sketch a normal distribution curve, enter the given probability or percentage in the appropriate region of the graph, and identify the x value(s) being sought.
2. Use Table A-2 to find the z score corresponding to the region bounded by x and the centerline of 0. Observe the following cautions:
   - Refer to the *body* of Table A-2 to find the closest area, then identify the corresponding z score.
   - Make the z score *negative* if it is located to the left of the centerline.
3. Using Formula 5-2, enter the values for μ, σ, and the z score found in Step 2, then solve for x. Based on the format of Formula 5-2, we can solve for x as follows.

$$x = \mu + (z \cdot \sigma) \quad \text{(another form of Formula 5-2)}$$

↑

(Make z *negative* if it is to the left of the mean.)

Survey Medium Can Affect Results

In a survey of Catholics in Boston, the subjects were asked if contraceptives should be made available to unmarried women. In personal interviews, 44% of the respondents said yes. But among a similar group contacted by mail or telephone, 75% of the respondents answered yes to the same question.

4. Refer to the sketch of the curve to verify that the solution makes sense in the context of the graph and in the context of the problem.

The following example uses the procedure just outlined. Pay extra attention to Step 2, especially where we make the z score negative because it is to the left of the mean.

EXAMPLE **Women's Weights** In Section 5-3 we used women's weights, which are normally distributed with a mean of 143 lb and a standard deviation of 29 lb (based on data from the National Health Survey). Find the value of P_{10}. That is, find the weight separating the bottom 10% from the top 90%.

SOLUTION

Step 1: We begin with the graph shown in Figure 5-17. We have entered the mean of 143, shaded in green the area representing the bottom 10%, and identified the desired value as x. The orange area between 143 and x must be 40% of the total area (because the left half of the area must combine to be 50% of the total). Because the total area is 1, the orange area constituting 40% of the total must be 0.40.

Step 2: We refer to Table A-2, but we look for an area of 0.4000 in the body of the table. (Remember, Table A-2 is designed to list areas only for those regions bounded on the left by the mean and on the right by some value.) The area closest to 0.4000 is 0.3997, and it corresponds to a z score of 1.28. Because the z score is to the left of the mean, we make it negative and use $z = -1.28$.

Step 3: With $z = -1.28$, $\mu = 143$, and $\sigma = 29$, we solve for x by using Formula 5-2 directly or by using the following version of Formula 5-2:

$$x = \mu + (z \cdot \sigma) = 143 + (-1.28 \cdot 29) = 105.88$$

Step 4: If we let $x = 105.88$ in Figure 5-17, we see that this solution is reasonable because the 10th percentile should be less than the mean of 143.

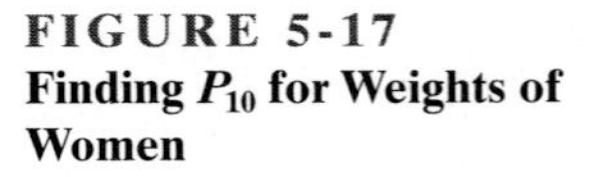

FIGURE 5-17
Finding P_{10} for Weights of Women

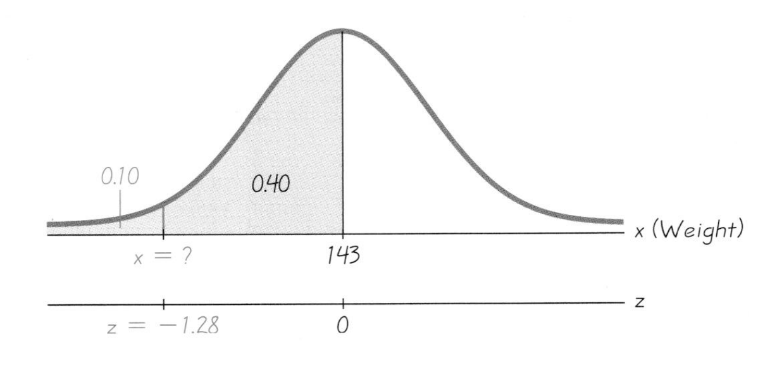

INTERPRETATION The weight of 106 lb (rounded) separates the lowest 10% from the highest 90%.

In this example, it would be very easy to forget the step of making the z score negative because it is below the mean. That error would have resulted in a weight of 180 lb, but in Step 4 we would see that 180 lb is not a reasonable answer. We could then proceed to correct that error.

Procedure for Finding Values Using a TI-83 Plus Calculator To find values given a probability (or percentage) using a TI-83 Plus calculator, press **2nd, VARS, 3** (for **invNorm**) and proceed to enter the following values, including the commas:

$$(\text{total area to the } \textit{left} \text{ of the desired value}, \mu, \sigma)$$

For the preceding example, the total area to the left of the desired value is 0.10. We therefore press **2nd, VARS, 3,** then enter (0.10, 143, 29) to get a result of 105.8350046. Because of rounding errors, this result differs slightly from the one obtained in the solution to the preceding example.

EXAMPLE **Body Temperatures** Assume that body temperatures of healthy adults are normally distributed with a mean of 98.20°F and a standard deviation of 0.62°F (based on Data Set 6 in Appendix B). If a medical researcher wants to study people in the bottom 2.5% and people in the top 2.5%, find the temperatures separating those limits.

SOLUTION

Step 1: We begin with the graph shown in Figure 5-18. We have shaded in green the areas representing the bottom 2.5% and top 2.5% (or 0.025). The areas of 0.475 are found by using the fact that the centerline above the mean divides the total area of 1 into two parts, each with area 0.5. We get $0.5 - 0.025 = 0.475$.

Step 2: We refer to Table A-2, but we look for an area of 0.475 in the body of the table. (Remember, Table A-2 is designed to list areas only for those regions bounded on the left by the mean

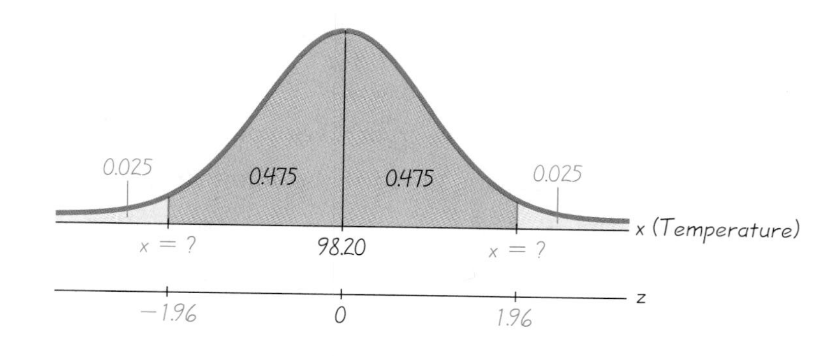

FIGURE 5-18
Body Temperatures

and on the right by some value.) The area of 0.4750 corresponds to $z = 1.96$. For the x value located on the right in Figure 5-18, we use $z = 1.96$, but for the x value located on the left we use $z = -1.96$.

Step 3: With $z = 1.96$, $\mu = 98.20$, and $\sigma = 0.62$, we solve for x using a variation of Formula 5-2:

$$x = \mu + (z \cdot \sigma) = 98.20 + (1.96 \cdot 0.62) = 99.42$$

With $z = -1.96$, $\mu = 98.20$, and $\sigma = 0.62$, we solve for x as follows:

$$x = \mu + (z \cdot \sigma) = 98.20 + (-1.96 \cdot 0.62) = 96.98$$

Step 4: If we let $x = 99.42$ and 96.98 in Figure 5-18, we see that our solutions are reasonable.

INTERPRETATION The researcher should select people with body temperatures below 96.98°F or above 99.42°F.

To solve the preceding problem using a TI-83 Plus calculator, press **2nd, VARS, 3,** and enter (0.025, 98.20, 0.62) to get 96.98482233, then press **2nd, VARS, 3,** and enter (0.975, 98.20, 0.62) to get 99.41517767. Note that in both cases, the first entry is the total area to the left of the desired value, the second entry is the mean, and the third entry is the standard deviation.

The examples and exercises in this section can also be solved using Minitab or Excel. If you are using Minitab, select **Calc, Probability Distributions, Normal,** then select **Inverse cumulative probabilities** and the option **Input constant.** For the input constant, enter the total area to the left of the given value. If using Excel, select **fx, Statistical, NORMINV,** and proceed to make the entries in the dialog box. When entering the probability value, enter the total area to the left of the given value.

5-4 Basic Skills and Concepts

Finding Values in a Normal Distribution. *In Exercises 1–4, assume that a test is designed to measure a person's sense of humor and that scores on this test are normally distributed with a mean of 10 and a standard deviation of 2 (based on the author's fertile imagination). Draw a graph, find the relevant z score, then find the indicated value.*

1. Find the score separating the top 10% from the bottom 90%.

2. Find the score separating the top 25% from the bottom 75%.

3. Find the second decile, D_2, which is the score separating the bottom 20% from the top 80%.

4. Find the first quartile, Q_1, which is the score separating the bottom 25% from the top 75%.

Finding Weights in a Normal Distribution. *In Exercises 5–8, assume that women's weights are normally distributed with a mean given by $\mu = 143$ lb and a standard deviation given by $\sigma = 29$ lb (based on data from the National Health Survey). Find the indicated value.*

5. Find the third quartile, Q_3, which is the weight separating the bottom 75% from the top 25%.

6. Find the sixth decile, D_6, which is the weight separating the bottom 60% from the top 40%.

7. Find P_{15}, which is the weight separating the bottom 15% from the top 85%.

8. Find P_{35}, which is the weight separating the bottom 35% from the top 65%.

In Exercises 9 and 10, assume that women's heights are normally distributed with a mean of 63.6 in. and a standard deviation of 2.5 in.

9. Finding Beanstalk Club Height Requirement The Beanstalk Club, a social organization for tall people, has a requirement that women must be at least 70 in. (or 5 ft 10 in.) tall. Suppose that requirement is changed so that only the tallest 5% of women are eligible. What is the new minimum height requirement for women? Why isn't it practical to state the requirement for women as "a height that is in the top 5% of all women"?

10. Finding Height Requirements for Women Soldiers To be eligible for the U.S. Army, a woman's height must be between 58 in. and 80 in. If that requirement is changed so that only the shortest 1% and tallest 1% are excluded, find the minimum and maximum acceptable heights.

11. Finding SAT Test Requirement Scores for the combined math and verbal portions of the SAT-I test have a mean of 1017 and a standard deviation of 207, and the distribution is approximately normal (based on data from the College Board). If the College of Newport wants to consider only those applicants in the top 42%, what is the minimum required score? What is the problem with the college telling applicants that they must score in the top 42%?

12. Study of Facial Behavior A study compared the facial behavior of nonparanoid schizophrenic persons with that of a control group of normal persons. The control group was timed for eye contact during a period of 5 minutes, or 300 seconds. The eye-contact times were normally distributed with a mean of 184 sec and a standard deviation of 55 sec (based on data from "Ethological Study of Facial Behavior in Nonparanoid and Paranoid Schizophrenic Patients," by Pitman, Kolb, Orr, and Singh, *Psychiatry,* 144:1). Because results showed that nonparanoid schizophrenic patients had much lower eye-contact times than did the control group, you have decided to further analyze people in the control group who are in the bottom 5%. For the control group, find P_5, the 5th percentile. That is, find the eye-contact time separating the bottom 5% from the rest.

13. TV Warranty Replacement times for TV sets are normally distributed with a mean of 8.2 years and a standard deviation of 1.1 years (based on data from "Getting Things Fixed," *Consumer Reports*).

a. Find the probability that a randomly selected TV will have a replacement time less than 5.0 years.

b. If you want to provide a warranty so that only 1% of the TV sets will be replaced before the warranty expires, what is the time length of the warranty?

14. CD Player Warranty Replacement times for CD players are normally distributed with a mean of 7.1 years and a standard deviation of 1.4 years (based on data from "Getting Things Fixed," *Consumer Reports*).

a. Find the probability that a randomly selected CD player will have a replacement time less than 8.0 years.

b. If you want to provide a warranty so that only 2% of the CD players will be replaced before the warranty expires, what is the time length of the warranty?

15. Weights of Discarded Paper Weights of paper discarded by households each week are normally distributed with a mean of 9.4 lb and a standard deviation of 4.2 lb (based on data from the Garbage Project at the University of Arizona).

a. For a randomly selected household, find the probability that the weight of the discarded paper is less than 15 lb.

b. In planning for paper pickup and processing needs, the Orange County Resource Recovery Agency needs to know P_{99}, the weight that is exceeded by only 1% of households. Find that weight.

16. Body Temperatures Healthy people have body temperatures that are normally distributed with a mean of 98.20°F and a standard deviation of 0.62°F (based on Data Set 6 of Appendix B).

a. If a healthy person is randomly selected, what is the probability that he or she has a temperature below 100°F?

b. The Newport Hospital wants to select a minimum temperature for requiring further medical tests. What should that temperature be, if we want only 1.5% of healthy people to exceed it? (Such a result is called a "false positive," meaning that the test result is positive, but the subject is not really sick.)

17. Length of Pregnancies The lengths of pregnancies are normally distributed with a mean of 268 days and a standard deviation of 15 days. If we stipulate that a baby is *premature* if the length of pregnancy is in the lowest 4%, find the length that separates premature babies from those who are not premature. Premature babies often require special care, and this result could be helpful to hospital administrators in planning for that care.

18. High IQ Scores IQ scores are normally distributed with a mean of 100 and a standard deviation of 15.

a. People are considered to be "intellectually very superior" if their score is above 130. What percentage of people fall into that category?

b. If we redefine the category of "intellectually very superior" to be scores in the top 2%, what does the minimum score become?

19. High Cholesterol Levels The serum cholesterol levels in men aged 18 to 24 are normally distributed with a mean of 178.1 and a standard deviation of 40.7. Units are mg/100 mL, and the data are based on the National Health Survey.

a. If a man aged 18 to 24 is randomly selected, find the probability that his serum cholesterol level is greater than 260, a value considered to be "moderately high."

b. The Providence Health Maintenance Organization wants to establish a criterion for recommending dietary changes if cholesterol levels are in the top 3%. What is the cutoff for men aged 18 to 24?

20. Filling Coke Cans Cans of regular Coke are labeled as containing 12 oz. Assume that the actual contents are normally distributed with a mean of 12.19 oz and a standard deviation of 0.11 oz (based on Data Set 1 in Appendix B).

a. What percentage of cans contain less than 11.95 oz?

(*continued*)

b. To save some of the Coke in cans that are filled too much, the quality-control manager is instituting a plan to remove some of the Coke from cans that are in the top 2%. What is the minimum amount of Coke in such cans that will be "reworked"?

5-4 Beyond the Basics

21. Curving Test Scores A teacher informs her psychology class that a test is very difficult, but the grades will be curved. Scores for the test are normally distributed with a mean of 25 and a standard deviation of 5.

a. If she curves by adding 50 to each grade, what is the new mean? What is the new standard deviation?

b. Is it fair to curve by adding 50 to each grade? Why or why not?

c. If the grades are curved according to the following scheme (instead of adding 50), find the numerical limits for each letter grade.

A: Top 10%

B: Scores above the bottom 70% and below the top 10%

C: Scores above the bottom 30% and below the top 30%

D: Scores above the bottom 10% and below the top 70%

F: Bottom 10%

d. Which method of curving the grades is fairer: Adding 50 to each grade or using the scheme given in part (c)? Explain.

22. SAT Scores According to data from the College Entrance Examination Board, scores on the SAT-I test have a mean of 1017, and Q_1 is 880. The scores have a distribution that is approximately normal. Find the standard deviation, then use that result to find P_{99}.

23. Women's Weights Women's weights are normally distributed with a mean of 143 lb and a standard deviation of 29 lb.

a. What weights are within 2 standard deviations of the mean?

b. What percentage of weights are within 2 standard deviations of the mean?

c. Find the two weights having these properties: The mean is midway between them, and 95% of all weights are between them.

24. SAT and ACT Tests Scores by women on the SAT-I test are normally distributed with a mean of 998 and a standard deviation of 202. Scores by women on the ACT test are normally distributed with a mean of 20.9 and a standard deviation of 4.6. Assume that the two tests use different scales to measure the same aptitude.

a. If a woman gets an SAT score that is the 67th percentile, find her actual SAT score and her equivalent ACT score.

b. If a woman gets an SAT score of 1220, find her equivalent ACT score.

5-5 The Central Limit Theorem

This section presents the central limit theorem, one of the most important and useful concepts in statistics. It forms a foundation for estimating population parameters and hypothesis testing—topics discussed at length in the following chapters. We will not present rigorous proofs in this section but will instead focus on the *concepts* and how to apply them. You will need to keep in mind the types of data sets we are considering. Now, instead of using sets of *individual* values, we will work with data sets in which each value is the *mean* of some other sample.

Recall that in Section 4-2 we defined a *random variable* to be a variable that has a single numerical value, determined by chance, for each outcome of a procedure. We also defined a *probability distribution* to be a graph, table, or formula that gives the probability for each value of the random variable. We now define a *sampling distribution* of the mean.

DEFINITION

The **sampling distribution** of the mean is the probability distribution of sample means, with all samples having the same sample size n. (In general, the sampling distribution of any particular statistic is the probability distribution of that statistic.)

AS Confused? Let's try to make sense of our abstract definition by relating it to something more concrete, as in the following example.

EXAMPLE Random Digits Consider the population of digits 0, 1, 2, 3, 4, 5, 6, 7, 8, 9, which are randomly selected with replacement.

a. *Random variable:* If we conduct trials that consist of randomly selecting a single digit, and if we represent the value of the selected digit by x, then x is a random variable (because its value depends on chance).

b. *Probability distribution:* Assuming that the digits are randomly selected, the probability of each digit is 1/10, which can be expressed as the formula $P(x) = 1/10$. This is a probability distribution (because it describes the probability for each value of the random variable x).

c. *Sampling distribution*: Now suppose that we randomly select many different samples, each of size $n = 4$. (Remember, we are sampling with replacement, so any particular sample might have the same digit occurring more than once.) In each sample, we calculate the sample mean $\overline{x}$ (which is itself a random variable because its value depends on chance). The probability distribution of the sample means $\overline{x}$ is a sampling distribution.

Part (c) of the example illustrates a specific sampling distribution of sample means. Our main objective in this section is to present the important properties of such a sampling distribution. Those properties are included in the statement of the *central limit theorem*.

Central Limit Theorem

Given:

1. The random variable x has a distribution (which may or may not be normal) with mean μ and standard deviation σ.
2. Samples all of the same size n are randomly selected from the population of x values. (The samples are selected so that all possible samples of size n have the same chance of being selected.)

Conclusions:

1. The distribution of sample means $\bar{x}$ will, as the sample size increases, approach a *normal* distribution.
2. The mean of the sample means will approach the population mean μ. (That is, the normal distribution from Conclusion 1 has mean μ.)
3. The standard deviation of the sample means will approach $\sigma/\sqrt{n}$. (That is, the normal distribution from Conclusion 1 has standard deviation $\sigma/\sqrt{n}$.)

Practical Rules Commonly Used:

1. For samples of size n larger than 30, the distribution of the sample means can be approximated reasonably well by a normal distribution. The approximation gets better as the sample size n becomes larger.
2. If the original population is itself normally distributed, then the sample means will be normally distributed for *any* sample size n (not just the values of n larger than 30).

The Fuzzy Central Limit Theorem

In *The Cartoon Guide to Statistics* by Gonick and Smith, the authors describe the Fuzzy Central Limit Theorem as follows: "Data that are influenced by many small and unrelated random effects are approximately normally distributed. This explains why the normal is everywhere: stock market fluctuations, student weights, yearly temperature averages, SAT scores: All are the result of many different effects." People's heights, for example, are the results of hereditary factors, environmental factors, nutrition, health care, geographic region, and other influences which, when combined, produce normally distributed values.

The central limit theorem involves two different distributions: the distribution of the original population and the distribution of the sample means. As in previous chapters, we use the symbols μ and σ to denote the mean and standard deviation of the original population, but we now need new notation for the mean and standard deviation of the distribution of sample means.

Notation for the Central Limit Theorem

If all possible random samples of size n are selected from a population with mean μ and standard deviation σ, the mean of the sample means is denoted by $\mu_{\bar{x}}$, so

$$\mu_{\bar{x}} = \mu$$

Also, the standard deviation of the sample means is denoted by $\sigma_{\bar{x}}$, so

$$\sigma_{\bar{x}} = \frac{\sigma}{\sqrt{n}}$$

$\sigma_{\bar{x}}$ is often called the **standard error of the mean.**

TABLE 5-2

SSN digits				$\bar{x}$
1	8	6	4	4.75
5	3	3	6	4.25
9	8	8	8	8.25
5	1	2	5	3.25
9	3	3	5	5.00
4	2	6	2	3.50
7	7	1	6	5.25
9	1	5	4	4.75
5	3	3	9	5.00
7	8	4	1	5.00
0	5	6	1	3.00
9	8	2	2	5.25
6	1	5	7	4.75
8	1	3	0	3.00
5	9	6	9	7.25
6	2	3	4	3.75
7	4	0	7	4.50
5	7	5	6	5.75
4	1	5	7	4.25
1	2	0	6	2.25
4	0	2	8	3.50
3	1	2	5	2.75
0	3	4	0	1.75
1	5	1	0	1.75
9	7	4	0	5.00
7	3	1	1	3.00
9	1	1	3	3.50
8	6	5	9	7.00
5	6	4	1	4.00
9	3	9	5	6.50
6	0	7	3	4.00
8	2	9	6	6.25
0	2	8	6	4.00
2	0	9	7	4.50
5	8	9	0	5.50
6	5	4	9	6.00
4	8	7	6	6.25
7	1	2	0	2.50
2	9	5	0	4.00
8	3	2	2	3.75
2	7	1	6	4.00
6	7	7	1	5.25
2	3	3	9	4.25
2	4	7	5	4.50
5	4	3	7	4.75
0	4	3	8	3.75
2	5	8	6	5.25
7	1	3	4	3.75
8	3	7	0	4.50
5	6	6	7	6.00

EXAMPLE **Random Digits** Again consider the population of digits 0, 1, 2, 3, 4, 5, 6, 7, 8, 9, which are randomly selected with replacement. Assume that we randomly select samples of size $n = 4$. In the original population of digits, all of the values are equally likely. Based on the "Practical Rules Commonly Used" (listed in the central limit theorem box), we *cannot* conclude that the sample means are normally distributed, because the original population does not have a normal distribution and the sample size of 4 is not larger than 30. However, we will explore the sampling distribution to see what can be learned.

Table 5-2 was constructed by recording the last four digits of social security numbers from each of 50 different students. The last four digits of social security numbers are random, unlike the beginning digits, which are used to code particular information. If we combine the four digits from each student into one big collection of 200 numbers, we get a mean of $\bar{x} = 4.5$, a standard deviation of $s = 2.8$, and a distribution with the graph shown in Figure 5-19. Now see what happens when we find the 50 sample means, as shown in Table 5-2. Even though the original collection of data does *not* have a normal distribution, the sample means have a distribution that is approximately *normal.* This can be a confusing concept, so you should stop right here and study this paragraph until its major point becomes clear: The original set of 200 individual numbers does *not* have a normal distribution (because the digits 0–9 occur with approximately equal frequencies), but the 50 sample means *do* have a normal distribution. It's a truly fascinating and intriguing phenomenon in statistics that by sampling from any distribution, we can create a distribution that is normal or at least approximately normal.

Figure 5-20 shows that the distribution of the sample means from the preceding example is approximately normal, even though the original population does not have a normal distribution and the sample size does not

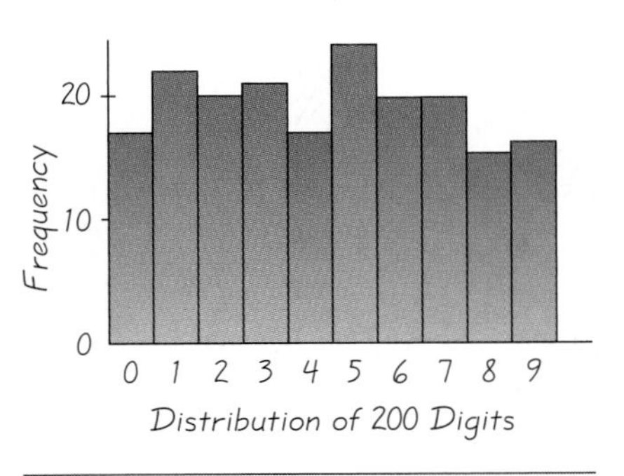

FIGURE 5-19
Distribution of 200 Digits from Social Security Numbers (Last 4 Digits) of 50 Students

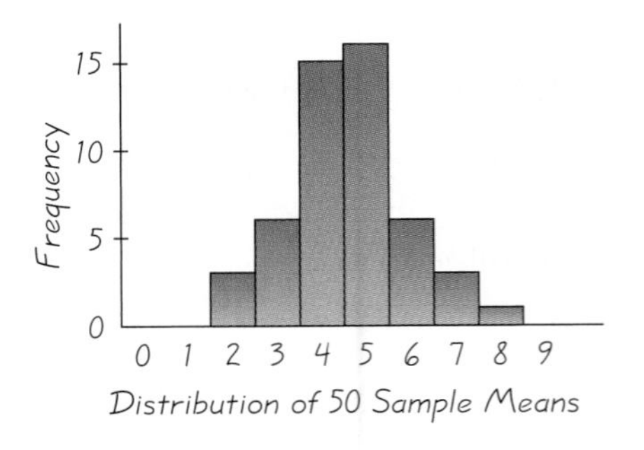

FIGURE 5-20
Distribution of 50 Sample Means for 50 Students

exceed 30. If you closely examine Figure 5-20, you can see that it is not an exact normal distribution, but it would become closer to an exact normal distribution as the sample size increases far beyond 4.

As the sample size increases, the sampling distribution of sample means approaches a *normal* distribution.

AS

Applying the Central Limit Theorem

Many important and practical problems can be solved with the central limit theorem. When working on such problems, remember the following rules.

- ***If you are working with a random sample of size $n > 30$, or if the original population is normally distributed, treat the distribution of sample means as a normal distribution.***
- ***The distribution of the sample means has mean μ.***
- ***The distribution of the sample means has a standard deviation found by computing $\sigma/\sqrt{n}$.***

In the following example, part (a) involves an *individual* value, so we use the same methods presented in Section 5-3; those methods apply to the normal distribution of the random variable x. Part (b), however, involves the mean for a *group* of 36 women, so we must use the central limit theorem in working with the random variable $\bar{x}$. Study this example carefully to understand the significant difference between the procedures used in parts (a) and (b).

EXAMPLE Jet Ejection Seats In the Chapter Problem, we said that engineers were redesigning fighter jet ejection seats to better accommodate women. In human engineering and product design, it is often important to consider people's weights so that airplanes or elevators aren't overloaded, chairs don't break, and other such dangerous or embarrassing mishaps do not occur. Given that the population of women has normally distributed weights with a mean of 143 lb and a standard deviation of 29 lb (based on data from the National Health Survey), find the probability that

a. if 1 woman is randomly selected, her weight is greater than 150 lb.

b. if 36 different women are randomly selected, their mean weight is greater than 150 lb.

SOLUTION

a. *Approach: Use the methods presented in Section 5-3* (because we are dealing with an *individual* value from a normally distributed population). We seek the area of the green-shaded region in Figure 5-21(a). Before using Table A-2, we convert 150 to the corresponding z score:

$$z = \frac{x - \mu}{\sigma} = \frac{150 - 143}{29} = 0.24$$

b. Why can the central limit theorem be used in part (a), even though the sample size does not exceed 30?

6. a. If 4 women are randomly selected, find the probability that they have a mean weight between 135 lb and 140 lb.
b. Why can the central limit theorem be used in part (a), even though the sample size does not exceed 30?

7. Uniform Random-Number Generator The typical computer random-number generator yields numbers from a uniform distribution of values between 0 and 1, with a mean of 0.500 and a standard deviation of 0.289. If 50 random numbers are generated, find the probability that their mean is between 0.6 and 0.7. Would it be *unusual* to generate 50 such numbers and get a mean between 0.6 and 0.7?

8. Mean IQ Score IQ scores are normally distributed with a mean of 100 and a standard deviation of 15. If 25 people are randomly selected for an IQ test, find the probability that their mean IQ score is between 95 and 105.

9. Mean Replacement Times The manager of the Portland Electronics store is concerned that his suppliers have been giving him TV sets with lower than average quality. His research shows that replacement times for TV sets have a mean of 8.2 years and a standard deviation of 1.1 years (based on data from "Getting Things Fixed," *Consumer Reports*). He then randomly selects 50 TV sets sold in the past and finds that the mean replacement time is 7.8 years.
a. Assuming that TV replacement times have a mean of 8.2 years and a standard deviation of 1.1 years, find the probability that 50 randomly selected TV sets will have a mean replacement time of 7.8 years or less.
b. Based on the result from part (a), does it appear that the Portland Electronics store has been given TV sets with lower than average quality?

10. Blood Pressure For women aged 18–24, systolic blood pressures (in mm Hg) are normally distributed with a mean of 114.8 and a standard deviation of 13.1 (based on data from the National Health Survey).
a. If a woman between the ages of 18 and 24 is randomly selected, find the probability that her systolic blood pressure is above 120.
b. If 12 women in that age bracket are randomly selected, find the probability that their mean systolic blood pressure is greater than 120.
c. Given that part (b) involves a sample size that is not larger than 30, why can the central limit theorem be used?

11. Iowa Rain The annual precipitation amounts for Iowa appear to be normally distributed with a mean of 32.473 in. and a standard deviation of 5.601 in. (based on data from the U.S. Department of Agriculture).
a. If 1 year is randomly selected, find the probability that the annual precipitation is less than 29.000 in.
b. If a decade of 10 years is randomly selected, find the probability that the annual precipitation amounts have a mean less than 29.000 in.
c. Given that part (b) involves a sample size that is not larger than 30, why can the central limit theorem be used?

12. Amounts of Coke Assume that cans of Coke are filled so that the actual amounts have a mean of 12.00 oz and a standard deviation of 0.11 oz.
a. Find the probability that a sample of 36 cans will have a mean amount of at least 12.19 oz, as in Data Set 1 in Appendix B

(continued)

b. Based on the result from part (a), is it reasonable to believe that the cans are actually filled with a mean of 12.00 oz? If the mean is not 12.00 oz, are consumers being cheated?

13. Amounts of Pepsi Assume that cans of Pepsi are filled so that the actual amounts have a mean of 12.00 oz and a standard deviation of 0.09 oz.

a. Find the probability that a sample of 36 cans will have a mean amount of at least 12.29 oz, as in Data Set 1 in Appendix B.

b. Based on the result from part (a), is it reasonable to believe that the cans are actually filled with a mean of 12.00 oz? If the mean is not 12.00 oz, are consumers being cheated?

14. Coaching for the SAT Test Scores for women on the verbal portion of the SAT-I test are normally distributed with a mean of 502 and a standard deviation of 109 (based on data from the College Board). Randomly selected women are given the Columbia Review Course before taking the SAT test. Assume that the course has no effect.

a. If 1 of the women students is randomly selected, find the probability that her score is at least 535.

b. If 25 of the women are randomly selected, find the probability that their mean score is at least 535.

c. In finding the probability for part (b), why can the central limit theorem be used even though the sample size does not exceed 30?

d. If the random sample of 25 women does result in a mean score of 535, is there strong evidence to support the claim that the course is actually effective? Why or why not?

15. Coaching for the SAT Test Scores for men on the verbal portion of the SAT-I test are normally distributed with a mean of 509 and a standard deviation of 112 (based on data from the College Board). Randomly selected men are given the Columbia Review Course before taking the SAT test. Assume that the course has no effect.

a. If 1 of the men is randomly selected, find the probability that his score is at least 590.

b. If 16 of the men are randomly selected, find the probability that their mean score is at least 590.

c. In finding the probability for part (b), why can the central limit theorem be used even though the sample size does not exceed 30?

d. If the random sample of 16 men does result in a mean score of 590, is there strong evidence to support the claim that the course is actually effective? Why or why not?

16. Reduced Nicotine in Cigarettes The amounts of nicotine in Dytusoon cigarettes have a mean of 0.941 g and a standard deviation of 0.313 g (based on Data Set 8 in Appendix B). The Huntington Tobacco Company, which produces Dytusoon cigarettes, claims that it has now reduced the amount of nicotine. The supporting evidence consists of a sample of 40 cigarettes with a mean nicotine amount of 0.882 g.

a. Assuming that the given mean and standard deviation have not changed, find the probability of randomly selecting 40 cigarettes with a mean of 0.882 g or less.

b. Based on the result from part (a), is it valid to claim that the amount of nicotine is lower? Why or why not?

17. Overloading of Waste Disposal Facility The town of Newport operates a rubbish waste disposal facility that is overloaded if its 4872 households discard waste with weights having a mean that exceeds 27.88 lb in a week. For many different weeks, it is found that the samples of 4872 households have weights that are normally distributed with a mean of 27.44 lb and a standard deviation of 12.46 lb (based on data from the Garbage Project at the University of Arizona). What is the proportion of weeks in which the waste disposal facility is overloaded? Is this an acceptable level, or should action be taken to correct a problem of an overloaded system?

18. Effect of Diet on Length of Pregnancy The lengths of pregnancies are normally distributed with a mean of 268 days and a standard deviation of 15 days.
- **a.** If 1 pregnant woman is randomly selected, find the probability that her length of pregnancy is less than 260 days.
- **b.** If 25 randomly selected women are put on a special diet just before they become pregnant, find the probability that their lengths of pregnancy have a mean that is less than 260 days (assuming that the diet has no effect).
- **c.** If the 25 women do have a mean of less than 260 days, should the medical supervisors be concerned?

19. Labeling of M&M Packages M&M plain candies have a mean weight of 0.9147 g and a standard deviation of 0.0369 g (based on Data Set 10 in Appendix B). The M&M candies used in Data Set 10 came from a package containing 1498 candies, and the package label stated that the net weight is 1361 g. (If every package has 1498 candies, the mean weight of the candies must exceed $1361/1498 = 0.9085$ g for the net contents to weigh at least 1361 g.)
- **a.** If 1 M&M plain candy is randomly selected, find the probability that it weighs more than 0.9085 g.
- **b.** If 1498 M&M plain candies are randomly selected, find the probability their mean weight is at least 0.9085 g.
- **c.** Given these results, does it seem that the Mars Company is providing M&M consumers with the amount claimed on the label?

20. Elevator Overloading Women's weights are normally distributed with a mean of 143 lb and a standard deviation of 29 lb (based on data from the National Health Survey). An elevator in the Atlanta Women's Club is limited to 16 women, but it will be overloaded if the 16 women have a mean weight in excess of 157.5 lb (yielding a total weight in excess of $16 \cdot 157.5 = 2520$ lb). If 16 women occupy the elevator, find the probability that their mean weight will exceed 157.5 lb, causing the elevator to be overloaded. If the elevator is occupied by 16 women 12 times each day, how many times is it likely to be overloaded in a year? Is that number low enough to be of no concern?

5-5 Beyond the Basics

21. Correcting for a Finite Population Repeat Exercise 20, assuming that the population size is $N = 300$ women. Also, because the women in the elevator must be different, assume the sampling is done without replacement. (*Hint:* See the discussion of the finite population correction factor.)

22. Population Parameters A *population* consists of these values: 2, 3, 6, 8, 11, 18.
- **a.** Find μ and σ.
- **b.** List all samples of size $n = 2$ that are obtained without replacement.

(continued)

c. Find the population of all values of $\bar{x}$ by finding the mean of each sample from part (b).

d. Find the mean $\mu_{\bar{x}}$ and standard deviation $\sigma_{\bar{x}}$ for the population of sample means found in part (c).

e. Verify that

$$\mu_{\bar{x}} = \mu \quad \text{and} \quad \sigma_{\bar{x}} = \frac{\sigma}{\sqrt{n}}\sqrt{\frac{N-n}{N-1}}$$

23. Using Finite Population Correction Factor The finite population correction factor can be ignored when sampling with replacement or when $n \le 0.05N$. When collecting a sample (without replacement) that is 5% of the population N, what do the values of the finite population correction factor have in common for values of $N \ge 600$? (*Hint:* Find the value of the finite population correction factor for a few values of $N \ge 600$ with $n = 0.05N$. What do the results have in common?)

24. Failure of Practical Rule According to a rule commonly used, the distribution of sample means can be considered to be normal provided that the sample size exceeds 30. There are rare cases where this rule does not work. Consider a binomial procedure with 36 trials and a probability of success equal to 0.002. (This is like randomly selecting 36 slips with replacement from a bowl of 500 slips with the number 1 on one slip and the number 0 on the others.) Show that even though the sample size exceeds 30, the means of the numbers of successes have a distribution that is *not* normal. For example, use STATDISK to generate 1000 results of a binomial procedure with $p = 0.002$ and $n = 36$. (Select **Data, Binomial Generator,** then enter 1000, 0.002, and 36.) Use STATDISK's **Sample Transformations** to divide all of those results by 36, then get a histogram to verify that it is not a normal distribution. You could also use Minitab by selecting **Calc, Random Data, Binomial,** or you could use a TI-83 Plus calculator by pressing **MATH,** then selecting **PRB, 7,** and entering **randBin**(36,0.002,500) for 500 results instead of 1000.

5-6 Normal Distribution as Approximation to Binomial Distribution

In Section 4-3 we introduced the *binomial probability distribution,* which has these four requirements:

1. The procedure must have a *fixed number of trials.*
2. The trials must be *independent.*
3. Each trial must have all outcomes classified into *two categories.*
4. The probabilities must remain *constant* for each trial.

In Section 4-3 we presented three methods for finding binomial probabilities: using the binomial probability formula, using Table A-1, and using software (such as STATDISK, Minitab, or Excel) or a TI-83 Plus calculator. In many cases, however, none of those methods are practical, because the calculations require too much time and effort. We now present a new method, which uses a normal distribution as an approximation of a binomial distribution. The following box summarizes the key point of this section.

Normal Distribution as Approximation to Binomial Distribution

If $np \geq 5$ and $nq \geq 5$, then the binomial random variable is approximately normally distributed with the mean and standard deviation given as

$$\mu = np$$
$$\sigma = \sqrt{npq}$$

See Figure 5-23, which shows the probability histogram for a binomial procedure in which there are $n = 100$ trials and the probability of success is $p = 0.5$. The binomial distribution is represented by the *bars* in Figure 5-23, but the approximating normal distribution has been superimposed so that you can clearly see how well it fits. The formal justification that allows us to use the normal distribution as an approximation to the binomial distribution results from more advanced mathematics, but Figure 5-23 is a convincing visual argument supporting that approximation.

When solving binomial probability problems, the normal approximation approach should generally be used after determining that other exact procedures either cannot be used or require too much time and effort. These procedures, taken from Section 4-3, should be tried first because they yield more exact results.

1. Use computer software or a TI-83 Plus calculator, if available.
2. If neither software nor the TI-83 Plus is available, use Table A-1, if possible.

FIGURE 5-23
Probability Histogram for a Binomial Procedure with $n = 100$ and $p = 0.5$

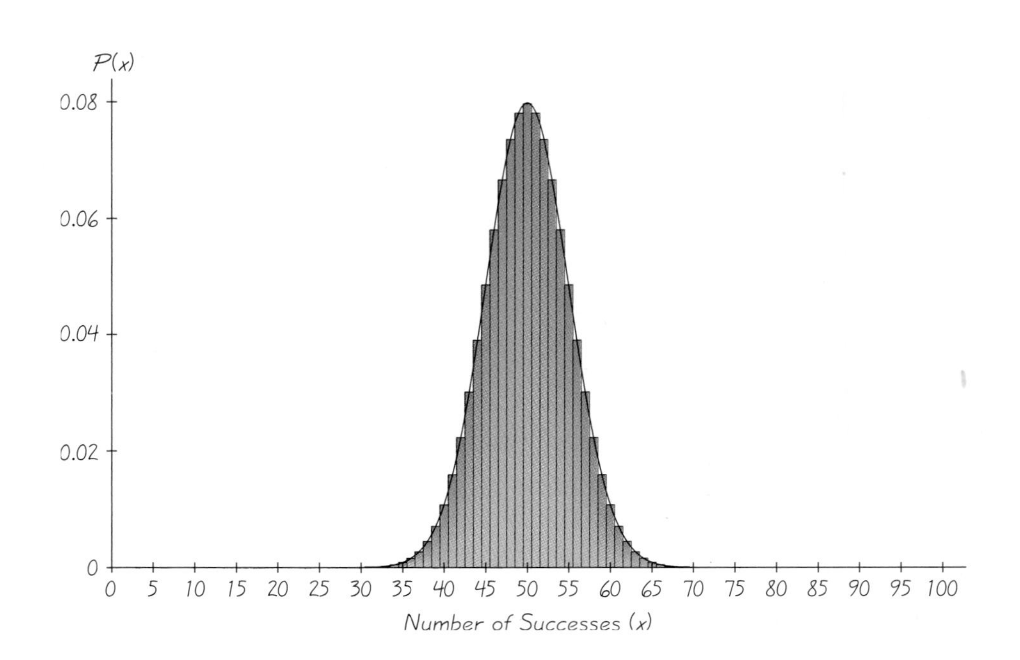

3. If neither software nor the TI-83 Plus is available and the probabilities can't be found using Table A-1, use the binomial probability formula.

If the binomial probability cannot be found using these exact procedures, try the technique of using the normal distribution as an approximation to the binomial distribution. This approach involves the following procedure, which is also shown as a flowchart in Figure 5-24.

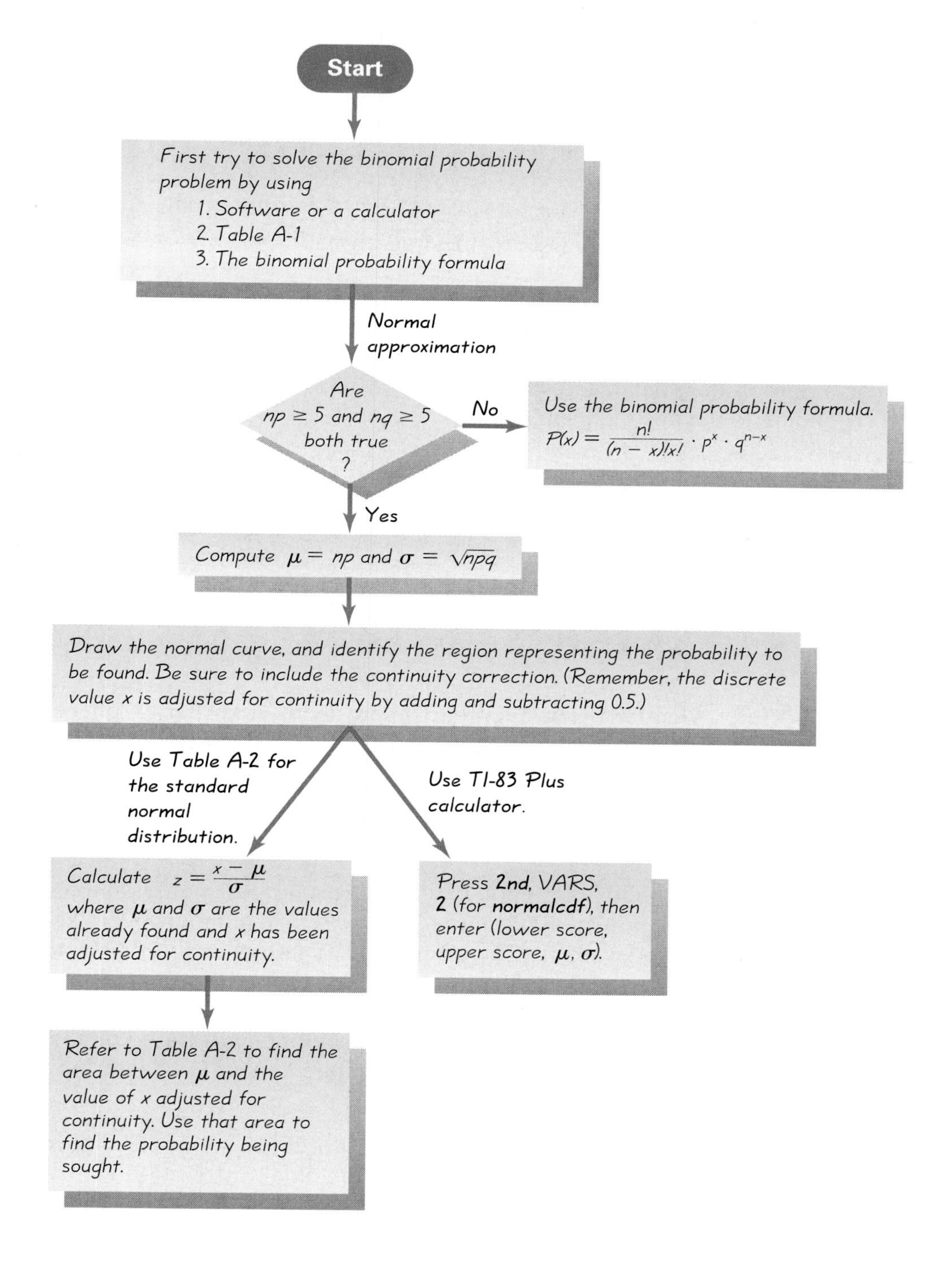

FIGURE 5-24
Solving Binomial Probability Problems Using a Normal Approximation

She Won the Lottery Twice!

Evelyn Marie Adams won the New Jersey Lottery twice in four months. This happy event was reported as an incredible coincidence with a likelihood of only 1 chance in 17 trillion. But Harvard mathematicians Persi Diaconis and Frederick Mosteller note that there is 1 chance in 17 trillion that a particular person with one ticket in each of two New Jersey lotteries will win both times. However, there is about 1 chance in 30 that someone in the United States will win a lottery twice in a four-month period. Diaconis and Mosteller analyzed coincidences and conclude that "with a large enough sample, any outrageous thing is apt to happen."

Procedure for Using a Normal Distribution to Approximate a Binomial Distribution

1. Establish that the normal distribution is a suitable approximation to the binomial distribution by verifying that $np \geq 5$ and $nq \geq 5$. (If these conditions are not both satisfied, then you will have to use software, or a calculator, or calculations with the binomial probability formula.)
2. Find the values of the parameters μ and σ by calculating $\mu = np$ and $\sigma = \sqrt{npq}$.
3. Identify the discrete value x (the number of successes). Change the *discrete* value x by replacing it with the *interval* from $x - 0.5$ to $x + 0.5$. (See the discussion under the subheading "Continuity Corrections" found later in this section.) Draw a normal curve and enter the values of μ, σ, and either $x - 0.5$ or $x + 0.5$, as appropriate.
4. Change x by replacing it with $x - 0.5$ or $x + 0.5$, as appropriate.
5. Find the area corresponding to the desired probability.

   Using Table A-2 for the standard normal distribution: If you have been using Table A-2 in this chapter, find the z score: $z = (x - \mu)/\sigma$. Now use that z score to find the area between μ and either $x - 0.5$ or $x + 0.5$, as appropriate. That area can now be used to identify the area corresponding to the desired probability.

   Using a TI-83 Plus calculator: If you have been using a TI-83 Plus calculator in this chapter, press **2nd, VARS, 2** (for **normalcdf**) and enter the required values in the format of (lower value, upper value, μ, σ) to find the desired probability.

We will illustrate this normal approximation procedure with the following example.

EXAMPLE **Gender Discrimination** The president of Portland College finds that applicants for admission are evenly divided between men and women, and both genders are equally qualified. She concludes that accepted students should be roughly 50% men and 50% women. She checks and finds that among the 1000 students accepted last year, 520 are men.

Find the probability of randomly selecting *at least* 520 men, assuming that the very large pool of applicants consists of equal numbers of equally qualified men and women. (The probability of getting *exactly* 520 men doesn't really tell us anything, because with 1000 trials, the probability of any exact number of men is fairly small. Instead, we need the probability of getting a result *at least* as extreme as the one obtained.) Based on the probability of getting at least 520 men, does there appear to be discrimination based on gender?

SOLUTION Refer to Figure 5-24 for the procedure followed in this solution. The given problem does involve a binomial distribution with a fixed number of trials ($n = 1000$), which are presumably independent, two categories (man, woman) of outcome for each trial, and a probability of 0.5 that presumably remains constant from trial to trial.

We will assume that neither software nor a calculator is available. Table A-1 does not apply, because it stops at $n = 15$. The binomial probability formula is not practical, because we would have to use it 481 times (once for each value of x from 520 to 1000 inclusive), and that would be crazy.

Let's proceed with the five-step approach of using a normal distribution to approximate the binomial distribution.

Step 1: We must first verify that it is reasonable to approximate the binomial distribution by the normal distribution because $np \geq 5$ and $nq \geq 5$, as verified below.

$$np = 1000 \cdot 0.5 = 500 \qquad \text{(Therefore } np \geq 5.)$$
$$nq = 1000 \cdot 0.5 = 500 \qquad \text{(Therefore } nq \geq 5.)$$

Step 2: We now proceed to find the values for μ and σ that are needed for the normal distribution. We get the following:

$$\mu = np = 1000 \cdot 0.5 = 500$$
$$\sigma = \sqrt{npq} = \sqrt{1000 \cdot 0.5 \cdot 0.5} = 15.8113883$$

Step 3: The discrete value of 520 is represented by the vertical strip bounded by 519.5 and 520.5. (See the discussion of continuity corrections, which follows this example.)

Step 4: Because we want the probability of *at least* 520 men, we want the area representing the discrete number of 520 (the region bounded by 519.5 and 520.5), as well as the area to the right, as shown in Figure 5-25.

Step 5: We can now proceed to find the shaded area of Figure 5-25 by using the same methods used in Section 5-3. If we plan to use Table A-2 for the standard normal distribution, we must first convert 519.5 to a z score, then use the table to find the area of

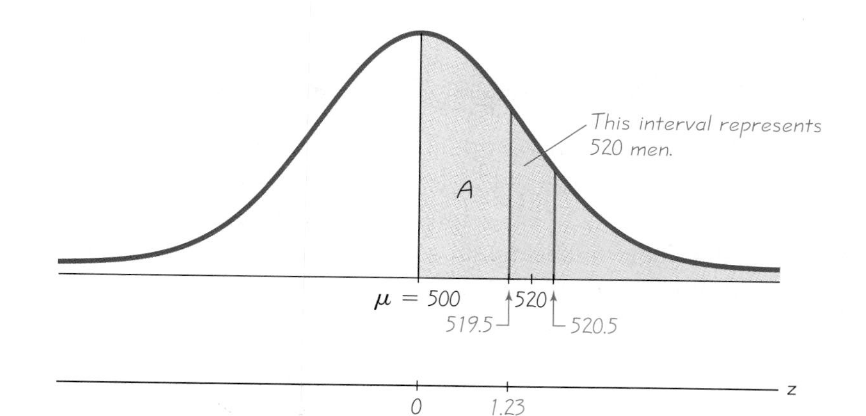

FIGURE 5-25
Finding the Probability of "At Least" 520 Men Among 1000 Accepted Applicants

region A, which is then subtracted from 0.5. The z score is found as follows:

$$z = \frac{x - \mu}{\sigma} = \frac{519.5 - 500}{15.8113883} = 1.23$$

Using Table A-2, we find that $z = 1.23$ corresponds to an area of 0.3907, so the orange region A is 0.3907 and the green-shaded area is $0.5 - 0.3907 = 0.1093$. We can round the probability to three significant digits as 0.109. [If you are using a TI-83 Plus calculator, you can find the green-shaded region of Figure 5-25 by pressing **2nd, VARS, 2.** Then enter (519.5, 999999, 500, 15.8113883) to get a result of 0.1087341775, which is also rounded to 0.109.]

INTERPRETATION There is a 0.109 probability of getting at least 520 men among the 1000 applicants. That is, if applications from men and women are treated equally, then there is a 0.109 probability of getting at least 520 men. Because that probability of 0.109 is not very low, we conclude that the result of 520 men could easily happen by chance. There does not appear to be discrimination based on gender. (We would have concluded that there appears to be discrimination only if the probability had been very small, such as 0.01, indicating that it is very unlikely that we would get at least 520 men by chance.) The reasoning here will be considered in more detail in Chapter 7 when we discuss formal methods of testing hypotheses. For now, we should focus on the method of finding the probability by using the normal approximation technique.

Continuity Corrections

The procedure for using a normal distribution to approximate a binomial distribution includes a step in which we change a discrete number to an interval that is 0.5 below and 0.5 above the discrete number. See the preceding solution, where we changed 520 to the interval between 519.5 and 520.5. This particular step, called a *continuity correction,* is usually difficult to understand, so we will now consider it in more detail.

DEFINITION

When we use the normal distribution (which is continuous) as an approximation to the binomial distribution (which is discrete), a **continuity correction** is made to a discrete whole number x in the binomial distribution by representing the single value x by the *interval* from $x - 0.5$ to $x + 0.5$ (that is, adding and subtracting 0.5).

The following practical suggestions should help you use continuity corrections properly.

Procedure for Continuity Corrections

1. When using the normal distribution as an approximation to the binomial distribution, *always* use the continuity correction. (It is required because we are using the *continuous* normal distribution to approximate the *discrete* binomial distribution.)
2. In using the continuity correction, first identify the discrete whole number x that is relevant to the binomial probability problem. For example, if you're trying to find the probability of getting at least 520 men in 1000 randomly selected people, the discrete whole number of concern is $x = 520$. First focus on the x value itself, and temporarily ignore whether you want at least x, more than x, fewer than x, or whatever.
3. Draw a normal distribution centered about μ, then draw a *vertical strip area* centered over x. Mark the left side of the strip with the number equal to $x - 0.5$, and mark the right side with the number equal to $x + 0.5$. For $x = 520$, for example, draw a strip from 519.5 to 520.5. *Consider the entire area of the strip to represent the probability of the discrete number x itself.*
4. Now determine whether the value of x itself should be included in the probability you want. (For example, "at least x" does include x itself, but "more than x" does not include x itself.) Next, determine whether you want the probability of at least x, at most x, more than x, fewer than x, or exactly x. Shade the area to the right or left of the strip, as appropriate; also shade the interior of the strip itself *if and only if x itself* is to be included. This total shaded region corresponds to the probability being sought.

To see how this procedure results in continuity corrections, see the common cases illustrated in Figure 5-26. Those cases correspond to the statements in the following list.

Statement	Area
At least 520 (includes 520 and above)	To the *right* of 519.5
More than 520 (doesn't include 520)	To the *right* of 520.5
At most 520 (includes 520 and below)	To the *left* of 520.5
Fewer than 520 (doesn't include 520)	To the *left* of 519.5
Exactly 520	Between 519.5 and 520.5

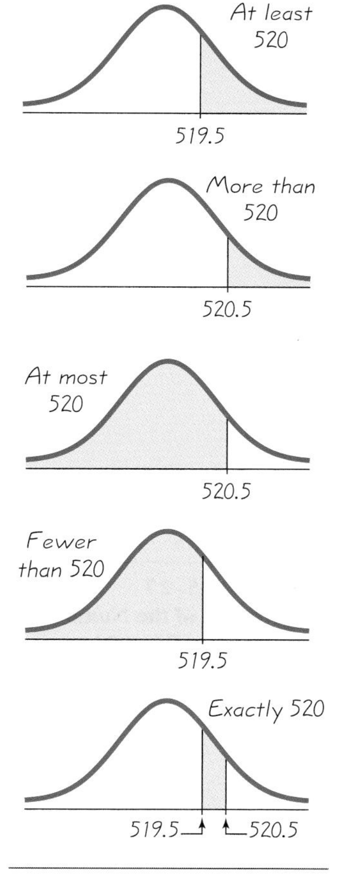

FIGURE 5-26
Identifying the Correct Area

EXAMPLE **Defective Tires** About 4.4% of fatal motor vehicle crashes are caused by defective tires (based on data from the National Safety Council). If a highway safety study begins with the random selection of 750 cases of fatal motor vehicle crashes, estimate the probability that *exactly* 35 of them were caused by defective tires.

SOLUTION The conditions described satisfy the criteria for the binomial distribution with $n = 750$, $p = 0.044$, $q = 0.956$, and $x = 35$. (The value of q is found from $q = 1 - p = 1 - 0.044 = 0.956$.) For the purposes of this example, we assume that neither computer software nor a

4. Probability of exactly 57 defective aircraft emergency locator transmitters

5. Probability of no more than 24 correct answers on an ACT test

6. Probability that the number of students who are absent is between 5 and 8 inclusive

7. Probability that the number of students with jobs is between 16 and 22 inclusive

8. Probability that the number of patients with group A blood is exactly 12

Using Normal Approximation. *In Exercises 9–12, do the following. (a) Find the indicated binomial probability by using Table A-1 in Appendix A. (b) If $np \geq 5$ and $nq \geq 5$, also estimate the indicated probability by using the normal distribution as an approximation to the binomial distribution; if $np < 5$ or $nq < 5$, then state that the normal approximation is not suitable.*

9. With $n = 10$ and $p = 0.8$, find $P(7)$.

10. With $n = 14$ and $p = 0.40$, find $P(8)$.

11. With $n = 15$ and $p = 0.60$, find $P(\text{at least } 9)$.

12. With $n = 12$ and $p = 0.90$, find $P(\text{fewer than } 10)$.

13. Probability of At Least 52 Girls Estimate the probability of getting at least 52 girls in 100 births. Assume that boys and girls are equally likely.

14. Probability of Exactly 53 Girls Estimate the probability of getting exactly 53 girls in 100 births. Assume that boys and girls are equally likely.

15. Probability of At Least Passing Estimate the probability of passing a true/false test of 50 questions if 60% (or 30 correct answers) is the minimum passing grade and all responses are random guesses.

16. Multiple-Choice Test A multiple-choice test consists of 25 questions with possible answers of a, b, c, d, and e. Estimate the probability that with random guessing, the number of correct answers is between 3 and 10 inclusive.

17. Directory Assistance Errors An article in *USA Today* stated that "Internal surveys paid for by directory assistance providers show that even the most accurate companies give out wrong numbers 15% of the time." Assume that you are testing AT&T by making 50 requests and also assume that this provider gives the wrong number 15% of the time. Estimate the probability of getting fewer than three wrong numbers. If you did get fewer than three wrong numbers, what would you conclude about the 15% error rate?

18. Probability of At Least 50 Color-Blind Men Nine percent of men and 0.25% of women cannot distinguish between the colors red and green. This is the type of color blindness that causes problems with traffic signals. Researchers need at least 50 men with this type of color blindness, so they randomly select 600 men for a study of traffic-signal perceptions. Estimate the probability that at least 50 of the men cannot distinguish between red and green. Is the result high enough so that the researchers can be very confident of getting at least 50 men with red and green color blindness?

19. Medical School Admissions: Probability of Fewer Than 456 Graduates A study was conducted to determine whether there were significant differences between medical students admitted through special programs (such as affirmative action) and medical students admitted with the regular admissions criteria. It was found that the graduation rate was 94% for the medical students admitted through special programs (based on data from the *Journal of the American Medical Association*). Would it be unusual to randomly select 500 students from the special programs and get fewer than 456 who graduate? Why or why not?

20. TV Share Value The CBS television show *60 Minutes* recently had a "share" of 20, meaning that among the TV sets in use, 20% were tuned to *60 Minutes* (based on data from Nielsen Media Research). An advertiser wants to verify that value by conducting its own survey of 200 households with TV sets in use at the time of a *60 Minutes* broadcast. Estimate the probability that fewer than 35 of the 200 households are tuned to *60 Minutes.* If fewer than 35 of the 200 households are tuned to *60 Minutes,* is there strong evidence to conclude that the share value of 20% is wrong?

21. Overbooking Flights Air America is considering a new policy of booking as many as 400 persons on an airplane that can seat only 350. (Past studies have revealed that only 85% of the booked passengers actually arrive for the flight.) Estimate the probability that if Air America books 400 persons, not enough seats will be available. Is that probability low enough to be workable, or should the policy be changed?

22. On-Time Flights Recently, Southwest Air had 80% of its flights arriving on time, which was the best rate among U.S. airlines (based on data from the U.S. Department of Transportation). In a check of 150 randomly selected Southwest Air flights, 113 arrived on time. Estimate the probability of getting 113 or fewer on-time flights among 150, assuming that the 80% rate is correct. Is it unusual to get 113 or fewer on-time flights among 150 randomly selected flights?

23. Identifying Gender Discrimination After being rejected for employment, Kim Kelly learns that the Bellevue Advertising Company has hired only 21 women among its last 62 new employees. She also learns that the pool of applicants is very large, with an equal number of qualified men and women. Help her in her charge of gender discrimination by estimating the probability of getting 21 or fewer women when 62 people are hired, assuming no discrimination based on gender. Does the resulting probability really support such a charge?

24. M&M Candies: Are 10% Blue? According to a consumer affairs representative from Mars (the candy company, not the planet), 10% of all M&M plain candies are blue. Data Set 10 in Appendix B shows that among 100 M&Ms chosen, 5 are blue. Estimate the probability of randomly selecting 100 M&Ms and getting 5 or fewer that are blue. Assume that the company's 10% blue rate is correct. Based on the result, is it very unusual to get 5 or fewer blue M&Ms when 100 are randomly selected?

25. Employee Drug Testing Currently, about two-thirds of U.S. companies test newly hired employees for drugs, and 3.8% of those prospective employees test positive (based on data from the American Management Association). The Sigma Electronics Company tests 150 prospective employees and finds that 10 of them test positive for drugs. Estimate the probability of 10 or more positive results among 150 subjects. Based on that value, do the 10 positive test results seem to be unusually high?

26. Blue Genes Some couples have genetic characteristics configured so that one-quarter of all their offspring have blue eyes. A study is conducted of 120 couples believed to have those characteristics, with the result that 20 of their 120 offspring have blue eyes. Estimate the probability that among 120 offspring, 20 or fewer have blue eyes. Based on that probability, does it seem that the one-quarter rate is wrong?

27. Blood Group Providence Memorial Hospital is conducting a blood drive because its supply of group O blood is low, and it needs 177 donors of group O blood. If 400 volunteers donate blood, estimate the probability that the number with group O blood is at least 177. Forty-five percent of us have group O blood, according to data provided by the Greater New York Blood Program.

28. Acceptance Sampling We stated in Section 3-4 that some companies monitor quality by using a method of acceptance sampling, whereby an entire batch of items is rejected if a random sample of a particular size includes more than some specified number of defects. The Dayton Machine Company buys machine bolts in batches of 5000 and rejects a batch if, when 50 of them are sampled, at least 2 defects are found. Estimate the probability of rejecting a batch if the supplier is manufacturing the bolts with a defect rate of 10%. Is this monitoring plan likely to identify the unacceptable rate of defects?

5-6 Beyond the Basics

29. Winning at Roulette Marc Taylor plans to place 200 bets of $1 each on the number 7 at roulette. A win pays off with odds of 35:1 and, on any one spin, there is a probability of 1/38 that 7 will be the winning number. Among the 200 bets, what is the minimum number of wins needed for Marc to make a profit? Estimate the probability that Marc will make a profit.

30. Interpreting Probability Value Twenty-six percent of college students live in campus housing (based on data from the Independent Insurance Agents of America). The Providence Insurance Company randomly selects 1000 college students. Estimate the probability that exactly 250 of them live in campus housing. If 250 of the selected students do live in campus housing, we should *not* conclude that the 26% rate is wrong, even though the probability of exactly 250 is low. Why not?

31. Replacement of TVs Replacement times for TV sets are normally distributed with a mean of 8.2 years and a standard deviation of 1.1 years (based on data from "Getting Things Fixed," *Consumer Reports*). Estimate the probability that for 250 randomly selected TV sets, at least 15 of them have replacement times greater than 10.0 years.

32. Joltin' Joe Assume that a baseball player hits .350, so his probability of a hit is 0.350. (Ignore the complications caused by walks.) Also assume that his hitting attempts are independent of each other.

a. Find the probability of at least 1 hit in 4 tries in 1 game.

b. Assuming that this batter gets up to bat 4 times each game, estimate the probability of getting a total of at least 56 hits in 56 games.

c. Assuming that this batter gets up to bat 4 times each game, find the probability of at least 1 hit in each of 56 consecutive games (Joe DiMaggio's 1941 record).

d. What minimum batting average would be required for the probability in part (c) to be greater than 0.1?

33. Using Continuity Correction In each of the following, find the difference between the answers obtained with and without use of the continuity correction. What do you conclude from the results?

(*continued*)

a. Estimate the probability of getting at least 11 girls in 20 births.
b. Estimate the probability of getting at least 22 girls in 40 births.
c. Estimate the probability of getting at least 220 girls in 400 births.

34. Overbooking Flights Vertigo Airlines works only with advance reservations and experiences a 7% rate of no-shows. How many reservations could be accepted for an airliner with a capacity of 250 if there is at least a 0.95 probability that all reservation holders who show will be accommodated?

5-7 Determining Normality

Some important statistical methods require that the sample data were collected from a population that has a *normal* distribution. It is therefore very important to be able to determine whether we are working with data that are normally distributed. In this section we describe one of many different procedures that could be used for determining whether the population distribution is normal. That procedure follows the definition of the *normal quantile plot.*

DEFINITION

A **normal quantile plot** is a graph of points (x, y), where each x value is from the original set of sample data, and each y value is a z score corresponding to a quantile value of the standard normal distribution. (See Step 3 in the following procedure for details on finding these z scores.)

Procedure for Determining Whether Data Have a Normal Distribution

1. *Histogram:* Construct a histogram. Reject normality if the histogram departs dramatically from a bell shape.
2. *Outliers:* Identify outliers. Reject normality if there is more than one outlier present. (Just one outlier could be an error or the result of chance variation, but be careful, because even a single outlier can have a dramatic effect on results.)
3. *Normal quantile plot:* If the histogram is basically symmetric and there is at most one outlier, construct a *normal quantile plot.* The following steps describe the construction of a normal quantile plot, but the procedure is messy enough so that we usually use software or a calculator to generate the graph.

   a. First sort the data by arranging the values in order from lowest to highest.

   b. With a sample of size n, each value represents a proportion of $1/n$ of the sample, so use the standard normal distribution to find the z score corresponding to these areas under the normal curve and extending from the extreme left: $1/2n, 3/2n, 5/2n, 7/2n$, and so on.

20%	20%	20%	20%	20%
P_{10}	P_{30}	P_{50}	P_{70}	P_{90}
$z = -1.28$	$z = -0.52$	$z = 0$	$z = 0.52$	$z = 1.28$

FIGURE 5-28
Finding Five Quantiles

c. Plot the points (x, y), where each x is an original sample value and y is the corresponding z score found in Step (b).

Examine the normal quantile plot and reject normality if the points do not lie close to a straight line, or if the points exhibit some systematic pattern that is not a straight-line pattern.

Steps 1 and 2 are straightforward, but we will describe the normal quantile plot required in Step 3. Recall that in Section 2-6 we introduced percentiles, quartiles, and deciles, and we noted that they are all types of *quantiles,* which partition data into parts that are approximately equal. For example, suppose we have five sample values, so that each represents 20% of the sample. See Figure 5-28, where we show the five blocks of 20%. See the locations of the percentiles P_{10}, P_{30}, P_{50}, P_{70}, and P_{90}. The z score of -1.28 is found by using the standard normal distribution (from Table A-2 or a TI-83 Plus calculator) to find the value separating an area of 10% in the left tail. The z score of -0.52 separates an area of 30% in the left tail, and so on. If the sample data come from a normal distribution, this plot of the original values against the z scores should result in points that lie close to a straight line.

AS Procedure for Constructing Normal Quantile Plots with a TI-83 Plus Calculator or Software

TI-83 Plus: The TI-83 Plus calculator can be used to generate a normal quantile plot as follows: First enter the sample data in list L1, press **2nd** and the **Y=** key (for **STAT PLOT**), then press **ENTER.** Select **ON,** select the "type" item that is the last item in the second row of options, enter **L1** for the data list. After making all selections, press **ZOOM,** then **9.**

Minitab: Minitab can be used to generate a *normal probability plot,* which can be interpreted the same way as the normal quantile plot. That is, normally distributed data should lie close to a straight line. First enter the values in column C1, then select **Stat, Basic Statistics,** and **Normality Test.** Enter **C1** for the variable, then click on **OK.**

Excel: The Data Desk XL add-in can be used to generate a *normal probability plot,* which can be interpreted the same way as a normal quantile plot. First enter the sample values in column A, then click on **DDXL.** (If DDXL does not appear on the Menu bar, install the Data Desk XL add-in.) Select **Charts and Plots,** then select the function type of **Normal Probability Plot.** Click on the pencil icon for "Quantitative Variable," then enter the range of values, such as A1:A36. Click **OK.**

EXAMPLE **Diet Pepsi** A statistical procedure has a fairly strict requirement that a population must have a normal distribution. Use the sample of 36 weights of diet Pepsi listed in Data Set 1 of Appendix B and determine whether the sample appears to come from a population with a normal distribution.

SOLUTION

Step 1: We begin by constructing a histogram of the 36 weights, with the result shown in the accompanying TI-83 Plus calculator display. An examination of the histogram shows that the data appear to be normally distributed.

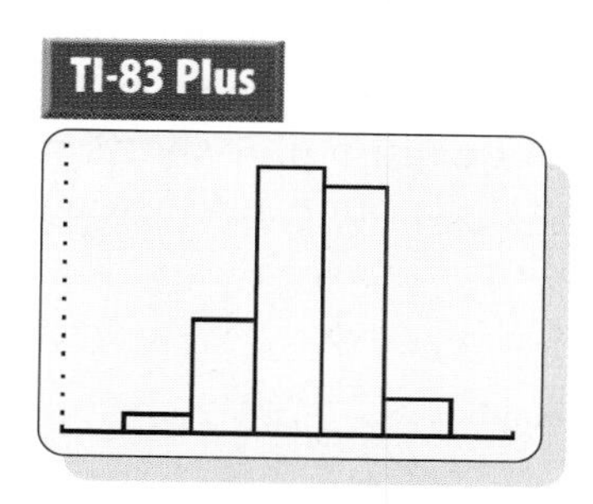

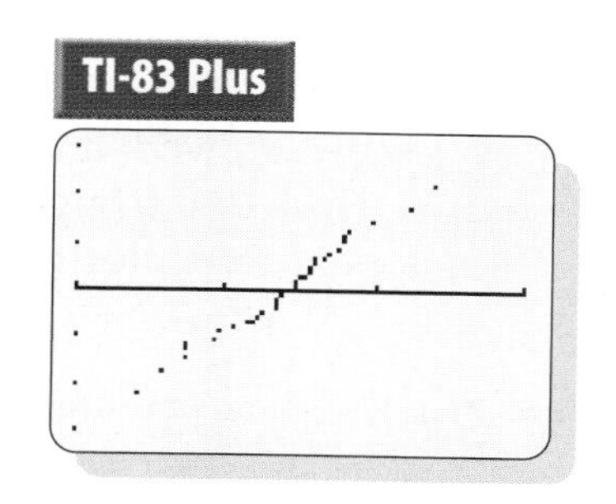

Step 2: There do not appear to be any outliers.

Step 3: The TI-83 Plus screen display of the normal quantile plot is also shown.

INTERPRETATION Examination of the normal quantile plot shows that the pattern of points is close to a straight line, and there is no systematic pattern that is not a straight-line pattern. We conclude that the sample data appear to come from a population with a normal distribution.

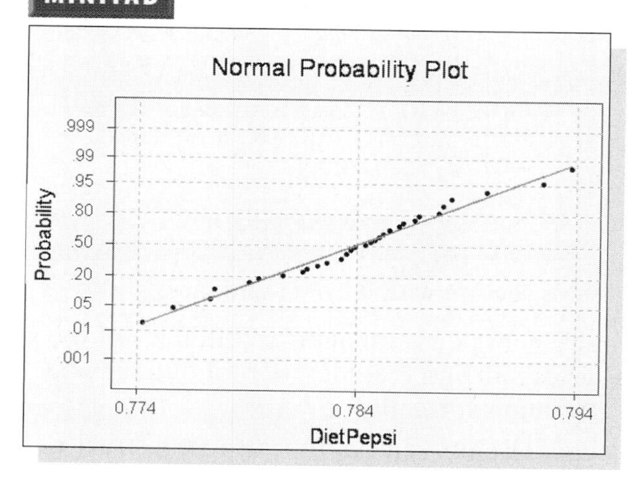

A Professional Speaks About Sampling Error

Daniel Yankelovich, in an essay for *Time*, commented on the sampling error often reported along with poll results. He stated that sampling error refers only to the inaccuracy created by using random sample data to make an inference about a population; the sampling error does not address issues of poorly stated, biased, or emotional questions. He said, "Most important of all, warning labels about sampling error say nothing about whether or not the public is conflict-ridden or has given a subject much thought. This is the most serious source of opinion poll misinterpretation."

The Minitab normal probability plot is also included on the previous page. See the similarity between the arrangements of the points in the two graphs. In both cases, the points appear to be close to a straight line, suggesting that the sample data come from a population with a normal distribution.

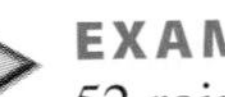

EXAMPLE **Boston Rainfall** In Data Set 17 of Appendix B, use the 52 rainfall amounts listed for Sundays in Boston and test for normality.

SOLUTION The accompanying TI-83 Plus screen displays show the histogram and normal quantile plot. (Without changing the nature of the distribution, we added 0.1 to each value so that the vertical axes would not be superimposed on top of plotted points.) The histogram is extremely skewed, and the normal quantile plot shows points that do not lie close to a straight line. Rainfall amounts in Boston on Sunday do not appear to be normally distributed.

TI-83 Plus: Histogram of Boston Rainfall for Sunday

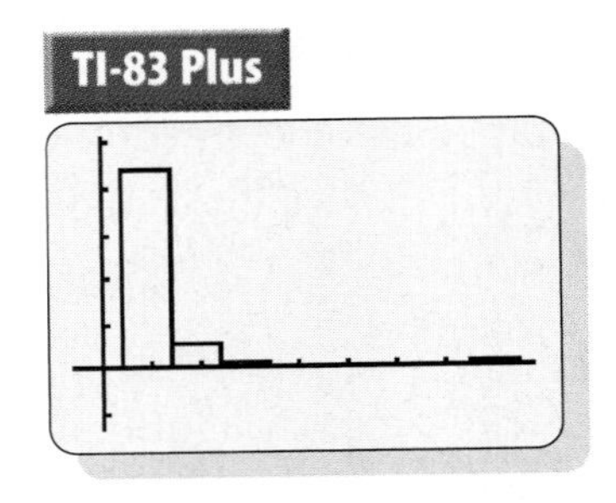

TI-83 Plus: Normal Quantile Plot of Boston Rainfall for Sunday

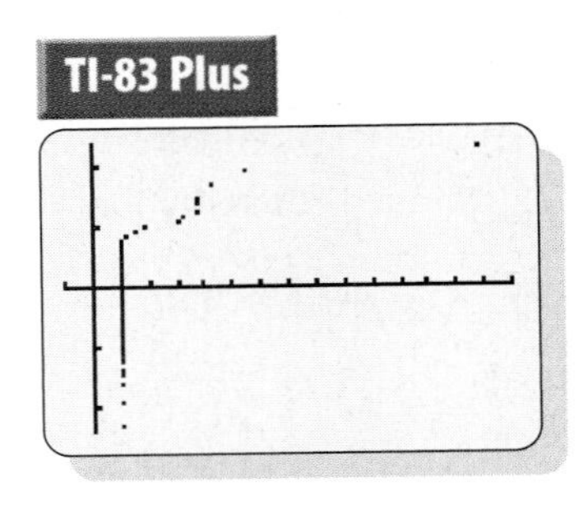

We conclude this section with a brief summary:

- If the requirement of a normal distribution is not too strict, examination of a histogram and consideration of outliers may be all that you need to determine normality.
- Normal quantile plots can be difficult to construct on your own, but they can be generated with a TI-83 Plus calculator or suitable software.

- In addition to the procedures discussed in this section, there are other more advanced procedures, such as the chi-square goodness-of-fit test, the Kolmogorov-Smirnov test, and the Lilliefors test. (See "Beyond Basic Statistics With the Graphing Calculator, Part I: Assessing Goodness-of-Fit," by Calzada and Scariano, *Mathematics and Computer Education*.)

5-7 Basic Skills and Concepts

Interpreting Normal Quantile Plots. *In Exercises 1–4, examine the normal quantile plot and determine whether it depicts data that have a normal distribution.*

1.

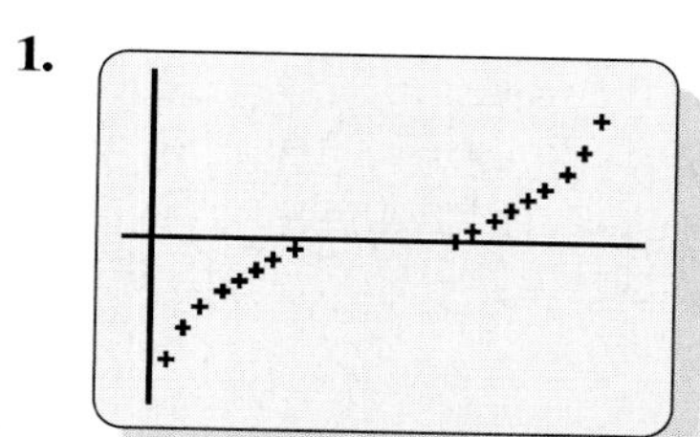

2.

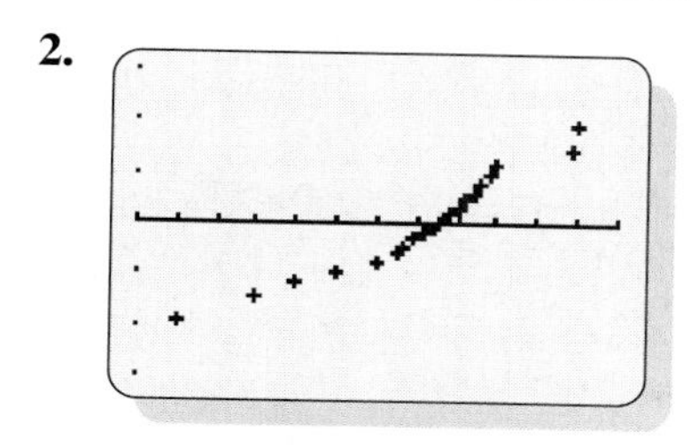

3.

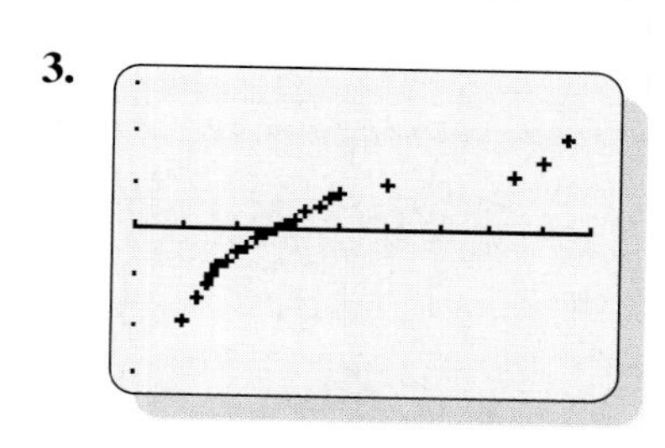

4.

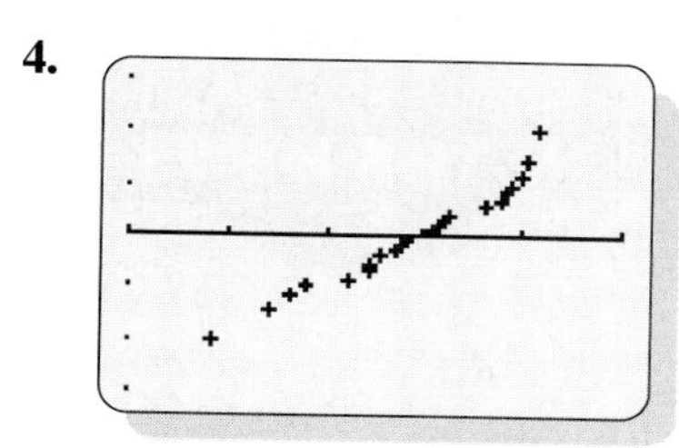

Determining Normality. *In Exercises 5–8, refer to the indicated data set and determine whether the requirement of a normal distribution is satisfied. Assume that this requirement is loose in the sense that the population distribution need not be exactly normal, but it must be a distribution that is basically symmetric with only one mode.*

5. Boston Rainfall The Wednesday rainfall amounts for Boston, as listed in Data Set 17 in Appendix B

6. Weights of M&Ms The weights of the brown M&M candies, as listed in Data Set 10 in Appendix B

7. Old Faithful Geyser The durations of eruptions of the Old Faithful geyser, as listed in Data Set 11 in Appendix B

8. Weights of Sugar Packets The weights of Domino sugar packets, as listed in Data Set 4 in Appendix B

Generating Normal Quantile Plots. *In Exercises 9–12, use the data from the indicated exercise in this section. Use a TI-83 Plus calculator or software (such as Minitab or Excel) capable of generating normal quantile plots or normal probability plots. Generate the graph, then determine whether the data come from a normally distributed population.*

9. Exercise 5

10. Exercise 6

11. Exercise 7

12. Exercise 8

5-7 Beyond the Basics

13. Constructing Normal Quantile Plot Use this sample of weights (in pounds) of randomly selected women: 154, 188, 103, 142, 137, 157, 102.

a. Because there are seven values, the normal quantile z scores can be found by using the areas of $1/2n$, $3/2n$, $5/2n$, $7/2n$, $9/2n$, $11/2n$, and $13/2n$, where $n = 7$. Find those areas.

b. Each area from part (a) is an area of the standard normal distribution extending from the left and bounded on the right by a vertical line above a standard z score. Find the z scores.

c. After first sorting the original set of sample values, pair them with the z scores found in part (b) and construct the normal quantile plot.

14. Instead of using z scores for a normal quantile plot, suppose each value in a sample is converted to its corresponding standard score using $z = (x - \overline{x})/s$. If the (x, z) points are plotted in a graph, can this graph be used to determine whether the sample comes from a normally distributed population? Explain your answer.

IQs Are Rising

NEW YORK In the article "I.Q. Scores Are Up, and Psychologists Wonder Why," *New York Times* reporter Trish Hall noted that people are now scoring substantially higher on IQ tests than they did in the past. "Studies show that if children took that same (1932 IQ) test today, half would score above 120 on the 1932 scale," writes Hall.Currently, IQ scores have a mean of 100, a standard deviation of 15, and a normal distribution. According to Hall's statement, if today's American children took the same IQ test used in 1932, the mean score would be 120, instead of 100 as it was in 1932. The net effect is that today's IQ scores are 20 points higher than in 1932.

1. IQ scores below 70 are often categorized as "intellectually deficient" results. What percentage of people were considered to be intellectually deficient in 1932 (based on the 1932 test)?(Assume that in 1932 the standard deviation was 15.)

2. If the same people who took the 1932 IQ test were to take today's IQ test instead, what percentage of them would be considered to be intellectually deficient (because of scores below 70)?

3. By today's standards, half of the subjects who take IQ tests get scores that are above average (that is, above 100). Based on the 1932 standards, what percentage of us are above average?

4. Identify at least one factor that might partially explain the rise in IQ scores.

Vocabulary List

normal distribution
uniform distribution
density curve
standard normal distribution
sampling distribution
central limit theorem
standard error of the mean
finite population correction factor
continuity correction
normal quantile plot

Review

We introduced the concept of probability distributions in Chapter 4, but included only *discrete* distributions. In this chapter we introduced *continuous* probability distributions and focused on the most important category: normal distributions. Normal distributions will be used extensively in the following chapters.

Normal distributions are bell-shaped when graphed. The total area under the density curve of a normal distribution is 1, so there is a correspondence between areas and probabilities. Specific areas can be found using Table A-2 or a TI-83 Plus calculator or software. (We do not use Formula 5-1, the equation that is used to define the normal distribution.)

In this chapter we presented important methods for working with normal distributions, including those that use the standard score $z = (x - \mu)/\sigma$ for solving problems such as these:

- Given that IQ scores are normally distributed with $\mu = 100$ and $\sigma = 15$, find the probability of randomly selecting someone with an IQ above 90.
- Given that IQ scores are normally distributed with $\mu = 100$ and $\sigma = 15$, find the IQ score separating the bottom 85% from the top 15%.

In Section 5-5 we presented the following important points associated with the central limit theorem:

1. The distribution of sample means will, as the sample size n increases, approach a normal distribution.
2. The mean of the sample means is the population mean μ.
3. The standard deviation of the sample means will be $\sigma/\sqrt{n}$.

In Section 5-6 we noted that we can sometimes approximate a binomial probability distribution by a normal distribution. If both $np \geq 5$ and $nq \geq 5$, the binomial random variable x is approximately normally distributed with the mean and standard deviation given as $\mu = np$ and $\sigma = \sqrt{npq}$. Because the binomial probability distribution deals with discrete data and the normal distribution deals with continuous data, we apply the continuity correction, which should be used in normal approximations to binomial distributions.

Finally, in Section 5-7 we presented a procedure for determining whether sample data appear to come from a population that has a normal distribution. Some of the statistical methods covered later in this book have a loose requirement of a normally distributed population. In such cases, examination of a histogram and outliers might be all that is needed. In other cases, normal quantile plots might be necessary because there is a very strict requirement that the population must have a normal distribution.

Review Exercises

1. Marine Height Requirement for Women To be eligible for the U.S. Marine Corps, a woman must have a height between 58 in. and 73 in. Women have normally distributed heights with a mean of 63.6 in. and a standard deviation of 2.5 in. (based on data from the National Health Survey).
   a. Find the percentage of women who satisfy the Marine Corps height requirement.
   b. If the requirement is changed to exclude the shortest 1% and the tallest 1%, find the heights that are acceptable.
   c. If nine women are randomly selected, find the probability that their mean height is between 63.0 in. and 65.0 in.

2. Men's Heights Men have normally distributed heights with a mean of 69.0 in. and a standard deviation of 2.8 in. (based on data from the National Health Survey).
   a. The Beanstalk Club is a social organization for tall people. Men are eligible for membership if they are at least 74 in. tall. What percentage of men are eligible for membership in the Beanstalk Club?
   b. If a man is randomly selected, what is the probability that he will satisfy the height limitation for the Newport Police Department, which requires men to have heights between 62.0 in. and 76.0 in.?
   c. If the Newport Police Department wants to change its height limitation for men so that only the shortest 2% and the tallest 2% are excluded, what are the minimum and maximum acceptable heights?
   d. If 25 men are randomly selected, what is the probability that their mean height is more than 68.0 in.?

3. Skulls in Anthropology Measurements of human skulls from different epochs are analyzed to determine whether they change over time. The maximum breadth is measured for skulls from Egyptian males who lived around 3300 B.C. Results show that those breadths are normally distributed with a mean of 132.6 mm and a standard deviation of 5.4 mm (based on data from *Ancient Races of the Thebaid* by Thomson and Randall-Maciver).
   a. Find the probability of getting a value greater than 140 mm if a skull is randomly selected from the period of around 3300 B.C.
   b. Find the breadth value that is D_2, the second decile.
   c. If 35 skulls are randomly selected from the 3300 B.C. population, what is the probability that their mean breadth will be between 133.0 mm and 134.0 mm?

4. Magazine Sweepstakes: Probability of New Subscriptions *Entertainment Report* magazine runs a sweepstakes as part of a campaign to acquire new subscribers. In the past, 26% of those who received sweepstakes entry materials have ended up entering the contest and subscribing to the magazine (based on data reported in *USA Today*). Estimate the probability that when sweepstakes entry materials are sent to 500 randomly selected households, the resulting number of new subscriptions is between 125 and 150 inclusive.

5. Measurement Errors The Orange County Bureau of Weights and Measures studies a scale used by the Shop Till U Drop supermarket. It is found that errors on the scale are normally distributed with a mean of 0 oz and a standard deviation of 1 oz. (The errors can be positive or negative.)

(*continued*)

a. If an item is randomly selected and weighed, find the probability that the error is between 0 oz and 1.00 oz.
b. If an item is randomly selected and weighed, find the probability that the error is greater than 2.00 oz.
c. If an item is randomly selected and weighed, find the probability that the error is greater than −1.50 oz.
d. If 25 items are randomly selected and weighed, find the probability that the errors have a mean between 0.25 oz and 0.50 oz.
e. Find the fourth decile, D_4, which is the error value separating the bottom 40% of all errors from the top 60%.

6. Uniform Distribution The Royal Beverage Bottling Company has designed a machine that fills cola cans so that the contents are *uniformly* distributed with a minimum of 11.8 oz and a maximum of 12.2 oz. If one can is randomly selected, find the probability that the amount of cola is
   a. at least 12.0 oz.
   b. between 11.9 oz and 12.1 oz.
   c. less than 11.9 oz.

7. MCAT Scores Scores on the biology portion of the Medical College Admissions Test are normally distributed with a mean of 9.1 and a standard deviation of 2.1.
   a. What percentage of scores are above 8.5, the minimum score required for admission to the Providence School of Medicine?
   b. What percentage of scores are between 7.5 and 9.0?
   c. What is the value of Q_3, the score separating the bottom 75% from the top 25%?
   d. What is the value of Q_1?
   e. If 35 scores are randomly selected, what is the probability that their mean is above 9.0?

8. Telemarketing When telephone solicitors call from the National Magazine Subscription Company, 18% of the people who answer stay on the line for more than one minute. If 800 people are called in one day, find the probability that at least 150 stay on the line for more than one minute.

9. Tire Wear The Chemco Company manufactures car tires that last distances having a normal distribution with a mean of 35,600 mi and a standard deviation of 4275 mi.
   a. If a tire is randomly selected, what is the probability it lasts more than 30,000 mi?
   b. If 40 tires are randomly selected, what is the probability they last distances that have a mean greater than 35,000 mi?
   c. If the manufacturer wants to guarantee the tires so that only 3% will be replaced because of failure before the guaranteed number of miles, for how many miles should the tires be guaranteed?

10. Testing Claimed Psychic Powers The Amazing Bob claims that he has psychic powers and can predict future events. In a test of his ability, he is asked to predict the outcome of a die before it is rolled, and this trial is repeated for a total of 36 rolls. If his predictions are actually random guesses, find the probability that he will be correct at least 13 times. If Bob is correct at least 13 times, is this strong evidence that he has an ability to correctly predict the outcomes? Why or why not?

CUMULATIVE REVIEW EXERCISES

1. Eye Measurement Statistics The listed sample distances (in millimeters) were obtained by using a pupilometer to measure the distances between the pupils of adults (based on data collected by a student of the author).

67 66 59 62 63 66 66 55 63 61 60 56 66 67 59 59 60 62 61 63 62 57

63 60 64 65 63 64 59 60 60 60 56 66 62 60 62 56 60 60 58 52 54

   a. Find the mean $\overline{x}$ of the distances in this sample.
   b. Find the median of the distances in this sample.
   c. Find the mode of the distances in this sample.
   d. Find the standard deviation s of this sample.
   e. Does it seem that the sample values come from a population with a distribution that is approximately normal? Why or why not?
   f. Convert the distance of 66 mm to a z score.
   g. Find the actual percentage of these sample values that exceed 59 mm.
   h. Assuming a normal distribution, find the percentage of *population* distances that exceed 59 mm. Use the sample values of $\overline{x}$ and s as estimates of μ and σ.
   i. What level of measurement (nominal, ordinal, interval, ratio) describes this data set?
   j. The listed measurements appear to be rounded to the nearest millimeter, but are the exact unrounded distances discrete data or continuous data?

2. Left-Handedness According to data from the American Medical Association, 10% of us are left-handed.
   a. If three people are randomly selected, find the probability that they are all left-handed.
   b. If three people are randomly selected, find the probability that at least one of them is left-handed.
   c. Why can't we solve the problem in part (b) by using the normal approximation to the binomial distribution?
   d. If groups of 50 people are randomly selected, what is the mean number of left-handed people in such groups?
   e. If groups of 50 people are randomly selected, what is the standard deviation for the numbers of left-handed people in such groups?
   f. Would it be unusual to get 8 left-handed people in a randomly selected group of 50 people? Why or why not?

COOPERATIVE GROUP ACTIVITIES

1. *Out-of-class activity:* Use the reaction timer on the next page by following the instructions given below. Collect reaction times for a sample of at least 40 different subjects taken from a homogeneous group, such as right-handed college students. For each subject, measure the reaction time for each hand. Construct a histogram for the right-hand times and another histogram for the left-hand results. Based on the histogram shapes, do the two sets of times each appear to be normally distributed? Calculate the values of the mean and standard deviation

for each of the two data sets and compare the two sets of results. Using the right-hand sample mean $\bar{x}$ as an estimate of the population mean μ, and using the right-hand sample standard deviation s as an estimate of the population standard deviation σ, find the quartiles Q_1, Q_2, and Q_3. Repeat that procedure to find Q_1, Q_2, and Q_3 for the left hand. If the right-hand reaction times are to be used to screen job applicants, what time is the cutoff separating the fastest 5% from the slowest 95%?

Instructions for Using the Reaction Timer

a. Cut the reaction timer along the dashed line.

b. Ask the subject to hold his or her thumb and forefinger horizontally; those fingers should be spread apart by a distance that is the same as the width of the reaction timer.

c. Hold the reaction timer so that the bottom edge is just above the subject's thumb and forefinger.

d. Ask the subject to catch the reaction timer as quickly as possible, then release it after a few seconds.

e. Record the reaction time (in seconds) corresponding to the point at which the subject catches it.

2. *Out-of-class activity:* Divide into groups of three or four students. In each group, develop an original procedure to illustrate the central limit theorem. The main objective is to show that when you randomly select samples from a population, the means of those samples tend to be *normally* distributed, regardless of the nature of the population distribution. In Section 5-5, for example, we used the last four digits of social security numbers as a source of samples from a population of digits that are equally likely; we proceeded to show that even though the original population did not have a normal distribution, the sample means tended to be normally distributed.

3. *In-class activity:* Divide into groups of three or four students. Using a coin to simulate births, each individual group member should simulate 25 births and record the number of simulated girls. Combine all the results in the group and record n = total number of births and x = number of girls. Given batches of n births, compute the mean and standard deviation for the number of girls. Is the simulated result usual or unusual? Why?

4. *In-class activity:* Divide into groups of three or four students. Select a set of data from Appendix B (excluding Data Sets 1, 4, 10, 11, and 17, which were used in examples or exercises in Section 5-7). Use the methods of Section 5-7 and construct a histogram and normal quantile plot, then determine whether the data set appears to come from a normally distributed population.

TECHNOLOGY PROJECT

In this chapter we addressed the problem of finding the percentage of women with weights between 140 lb and 211 lb, because those are the safe limits for the ACES-II ejection seats. We found that 53.02% of women have weights between those limits. The solution, given in Section 5-3, involves theoretical calculations based on the assumption that women's weights are normally distributed with a mean of 143 lb and a standard deviation of 29 lb (based on data from the National Health Survey).

REACTION TIMER

−.204
−.202
−.199
−.196
−.194
−.191
−.188
−.185
−.183
−.180
−.177
−.174
−.171
−.168
−.165
−.161
−.158
−.155
−.151
−.148
−.144
−.140
−.137
−.133
−.129
−.125
−.121
−.116
−.112
−.107
−.102
−.097
−.091
−.085
−.079
−.072
−.065
−.056
−.046
−.032

This project describes a different method of solution that is based on a *simulation* technique: We will use a computer or a TI-83 Plus calculator to randomly generate 100 women's weights (from a normally distributed population with $\mu = 143$ and $\sigma = 29$), then we will find the percentage of those simulated weights that fall between 140 lb and 211 lb. The STATDISK, Minitab, Excel, and TI-83 Plus procedures are described as follows.

STATDISK: Select **Data** from the main menu bar, then choose the option of **Normal Generator.** Proceed to generate 100 values with a mean of 143 and a standard deviation of 29. (Use the **Format** option to specify one decimal place.) Next, sort the data by using the options of **Data Sample Editor** and then **Format.** With this sorted list, it becomes quite easy to count the number of weights between 140 and 211. Divide that number by 100 to find the percentage of these simulated weights that are between 140 lb and 211 lb. Compare the result to the theoretical value of 53.02% that was found in Section 5-3.

Minitab: Select the options of **Calc,** then **Random Data,** then **Normal.** Proceed to enter 100 for the number of rows, C1 for the column in which to store the data, 143 for the value of the mean, and 29 for the value of the standard deviation. Now select the option of **Manip,** then **Sort,** and proceed to sort column C1, with the sorted column stored in column C1, and with the sorting to be done by column C1. Examine the values in column C1 and count the number of weights between 140 and 211, then divide that number by 100 to find the percentage of simulated weights that are within the safe limits of the ACES-II ejection seats. Compare the result to the theoretical value of 53.02% that was found in Section 5-3.

Excel: Select **Tools** from the main menu bar, then select **Data Analysis** and **Random Number Generation.** After clicking OK, use the dialog box to enter 1 for the number of variables and 100 for the number of random numbers, then select "normal" for the type of distribution. Enter 143 for the mean and 29 for the standard deviation. Examine the displayed values and count the number of weights between 140 and 211, then divide that number by 100 to find the percentage of simulated weights that are within the safe limits of the ACES-II ejection seats. Compare the result to the theoretical value of 53.02% that was found in Section 5-3.

TI-83 Plus: Press **MATH,** then select **PRB,** then enter **randNorm** (143, 29, 100) to generate 100 values from a normally distributed population with $\mu = 143$ and $\sigma = 29$. Enter **STO → L1** to store the data in list L1. Now press **STAT,** then enter **SortA(L1)** to arrange the data in order. Examine the entries in list L1 to find the number of values between 140 and 211, then divide that number by 100 to find the percentage of simulated weights that are within the safe limits. Compare the result to the theoretical value of 53.02% that was found in Section 5-3.

Critical Thinking: Design a new fighter-jet ejection seat

In the Chapter Problem we stated that fighter jets currently have the ACES-II ejection seats, which were originally designed for men weighing between 140 lb and 211 lb. Now that women have been approved as pilots of fighter jets, those seats must be redesigned. Any redesign must take these population characteristics into account: Women's weights are normally distributed with a mean of 143 lb and a standard deviation of 29 lb, and men's weights are normally distributed with a mean of 172 lb and a standard deviation of 30 lb (based on data from the National Health Survey).

Analyzing the results

1. As of this writing, the goal is to redesign the ejection seats so that any weights between 103 lb and 245 lb can be safely accommodated. What percentage of women's weights fall within those limits? What percentage of men's weights fall within those limits?
2. Plan your own design so that at least 95% of all women's weights will be accommodated and at least 95% of all men's weights will be accommodated. What is the minimum weight and what is the maximum weight that satisfies that 95% requirement simultaneously for both genders? Cost effectiveness requires that the difference between the minimum and maximum be as small as possible.

Internet Project

Exploring the Central Limit Theorem

The Central Limit Theorem is one of the most important results in statistics. It also may be one of the most surprising. Basically, the theorem says that the normal distribution is everywhere. No matter what probability disribution underlies an experiment, there is a corresponding distribution of means that will be aproximately normal in shape. The best way to both understand and appreciate the Central Limit Theorem, however, is to see it in action. The Internet Project for this chapter can be found at this site:

http://www.awlonline.com/triola

You will be asked to view, interpret, and discuss a demonstration of the Central Limit Theorem as part of a dice rolling experiment. In addition, you will be guided in a search through the internet for other such demonstrations.

Statistics *at work*

"It is possible to be a journalist and not be comfortable with statistics, but you're definitely limited in what you can do.

Joel B. Obermayer:

Newspaper reporter for *The News & Observer*

Joel B. Obermayer writes about medical issues and health affairs for *The News & Observer*, a newspaper that covers the eastern half of North Carolina. He reports on managed care, public health, and research at academic medical centers, including Duke University and the University of North Carolina at Chapel Hill.

What concepts of statistics do you use?

I use ideas like statistical significance, error rates, and probability. I don't need to do anything incredibly sophisticated, but I need to be very comfortable with the math and with asking questions about it.

I use statistics to look at medical research to decide whether different studies are significant and to decide how to write about them. Mostly, I need to be able to read statistics and understand them, rather than develop them myself. I use statistics to develop good questions and to bolster the arguments I make in print. I also use statistics to decide if someone is trying to give me a positive spin on something that might be questionable. For example, someone at a local university once sent me a press release about miracle creams that supposedly help slim you down by dissolving fat cells. Well, I doubt that those creams work. The study wasn't too hot either. They were trying to make claims based on a study of only 11 people. The researcher argued that 11 people were enough to make good empirical health claims. Not too impressive. People try to manipulate the media all the time. Good verifiable studies with good verifiable statistical bases make it easier to avoid being manipulated.

Is your use of probability and statistics increasing, decreasing, or remaining stable?

Increasing. People's interest in new therapies that may only be in the clinical trial stage is increasing, partly because of the emphasis on AIDS research and on getting new drugs approved and delivered to patients sooner. It's more important than ever for a medical writer to use statistics to make sure that the studies really prove what the public relations people say they prove.

Should prospective employees have studied some statistics?

It is possible to be a journalist and not be comfortable with statistics, but you're definitely limited in what you can do. If you write about government-sponsored education programs and whether they're effective, or if you write about the dangers of particular contaminants in the environment, you're going to need to use statistics.

In my field, editors often don't think about statistics in the interview process. They worry more about writing skills. A knowledge of statistics is more important for what you can do once you are on the job.

Do you recommend statistics for today's college students? Why?

I think a basic mathematical understanding of the world is a great thing to have, and a knowledge of statistics is a part of that. At the very least, I think it's important to understand some of the basics of statistics. For certain types of jobs, it's critical.

Which other skills are important for today's college students?

As a journalist, I think writing skills are pretty high up there.

6

Estimates and Sample Sizes

CHAPTER PROBLEM

Is the mean body temperature really 98.6°F?

Table 6-1 lists 106 body temperatures (from Data Set 6 in Appendix B) obtained by University of Maryland researchers. Using the methods described in Chapter 2, we can obtain the following important characteristics of the sample data set:

- As revealed by a histogram, the distribution of the data is approximately bell-shaped.
- The mean is $\bar{x} = 98.20$°F.
- The standard deviation is $s = 0.62$°F.
- The sample size is $n = 106$.
- There are no outliers.

It is commonly believed that the mean body temperature is 98.6°F, but the data in Table 6-1 seem to suggest that it is actually 98.20°F. We know that samples tend to vary, so perhaps it is true that the mean body temperature is 98.6°F and the sample mean, $\bar{x} = 98.20$°F, is the result of a chance sample fluctuation. On the other hand, perhaps the sample mean of 98.20°F is correct and the commonly believed value of 98.6°F is wrong. On the basis of an analysis of the sample data in Table 6-1, we will see whether the mean body temperature is or is not 98.6°F.

TABLE 6-1 Body Temperatures (°F) of 106 Healthy Adults

98.6	98.6	98.0	98.0	99.0	98.4	98.4	98.4	98.4	98.6
98.6	98.8	98.6	97.0	97.0	98.8	97.6	97.7	98.8	98.0
98.0	98.3	98.5	97.3	98.7	97.4	98.9	98.6	99.5	97.5
97.3	97.6	98.2	99.6	98.7	99.4	98.2	98.0	98.6	98.6
97.2	98.4	98.6	98.2	98.0	97.8	98.0	98.4	98.6	98.6
97.8	99.0	96.5	97.6	98.0	96.9	97.6	97.1	97.9	98.4
97.3	98.0	97.5	97.6	98.2	98.5	98.8	98.7	97.8	98.0
97.1	97.4	99.4	98.4	98.6	98.4	98.5	98.6	98.3	98.7
98.8	99.1	98.6	97.9	98.8	98.0	98.7	98.5	98.9	98.4
98.6	97.1	97.9	98.8	98.7	97.6	98.2	99.2	97.8	98.0
98.4	97.8	98.4	97.4	98.0	97.0				

Sample temperatures were obtained by Dr. Philip Mackowiak, Dr. Steven Wasserman, and Dr. Myron Levine, University of Maryland researchers.

6-1 Overview

In Chapter 2 we noted that we use *descriptive statistics* to summarize or describe important characteristics of data, but with *inferential statistics* we use sample data to make inferences (or generalizations) about a population. The two major applications of inferential statistics involve the use of sample data to (1) estimate the value of a population parameter, and (2) test some claim (or hypothesis) about a population. In this chapter we introduce methods for estimating values of the following population parameters: population means, proportions, and variances. We also present methods for determining the sample sizes necessary to estimate those parameters. In Chapter 7 we will introduce the basic methods for testing claims (or hypotheses) that have been made about a population.

6-2 Estimating a Population Mean: Large Samples

A Study Strategy: It will become obvious that this section contains a great deal of information and introduces many concepts. The time you devote to this section will be well spent because we introduce the concept of a confidence interval, and that same general concept will be applied to the following sections of this chapter. We suggest that you proceed as follows: First, read this section with the limited objective of simply trying to understand what confidence intervals are, what they accomplish, and why they are needed. Second, try to develop the ability to construct confidence interval estimates of population means. Third, learn how to interpret a confidence interval correctly. Fourth, read the section once again and try to understand the underlying theory. You will always enjoy much greater success if you understand what you are doing, instead of blindly applying mechanical steps in order to obtain an answer.

Here is the main objective of this section: Given a collection of more than 30 sample values, develop an estimate of the value of the population mean μ. For example, the 106 body temperatures given in Table 6-1 in the Chapter Problem constitute a collection of sample values. By using statistics from that sample, we will proceed to estimate the mean μ of *all* body temperatures. This section will consider only samples with more than 30 values. (Section 6-3 will present methods for working with smaller samples.) This sample size condition is listed along with another assumption we make throughout this section:

Assumptions

1. $n > 30$ (The sample has more than 30 values.)
2. The sample is a simple random sample. (All samples of the same size have an equal chance of being selected.)

Recall from Section 1-4 that a simple random sample of n values is obtained if every possible sample of size n has the same chance of being chosen. This requirement of random selection means that the methods of this section cannot be used with some other types of sampling, such as stratified, cluster, and convenience sampling. We should be especially clear about this important point:

Data collected carelessly can be absolutely worthless, even if the sample is quite large.

We know that different samples naturally produce different results. The methods of this section assume that those sample differences are due to chance random fluctuations, not some unsound method of sampling. For example, if you were to collect a sample of IQ test results from 40 of your closest friends, you should not use the results to estimate the mean IQ of all Americans. When samples of friends are used, there are cultural and economic characteristics that cause differences other than the differences found with randomly selected groups of people. Instead of being randomly selected, your sample of friends would be a biased sample, and it would be totally worthless as a basis for estimating the mean IQ of all Americans. That's not what friends are for.

Assuming that we have a simple random sample with more than 30 values, we can now proceed with our major objective: using the sample as a basis for estimating the value of the population mean μ. We begin with some general definitions.

Definitions

An **estimator** is a formula or process for using sample data to estimate a population parameter.

An **estimate** is a specific value or range of values used to approximate a population parameter.

A **point estimate** is a single value (or point) used to approximate a population parameter.

The sample mean $\bar{x}$ is the best point estimate of the population mean μ.

Although we could use another statistic such as the sample median, midrange, or mode as an estimate of the population mean μ, studies have shown that the sample mean $\bar{x}$ usually provides the best estimate, for two reasons.

1. For many populations, the distribution of sample means $\bar{x}$ tends to be more consistent (with *less variation*) than the distributions of other sample statistics. (That is, if you use sample means to estimate the population

Estimating Wildlife Population Sizes

The National Forest Management Act protects endangered species, including the northern spotted owl, with the result that the forestry industry was not allowed to cut vast regions of trees in the Pacific Northwest. Biologists and statisticians were asked to analyze the problem, and they concluded that survival rates and population sizes were decreasing for the female owls, known to play an important role in species survival. Biologists and statisticians also studied salmon in the Snake and Columbia Rivers in Washington, and penguins in New Zealand. In the article "Sampling Wildlife Populations" (*Chance*, Vol. 9, No. 2), authors Bryan Manly and Lyman McDonald comment that in such studies, "biologists gain through the use of modeling skills that are the hallmark of good statistics. Statisticians gain by being introduced to the reality of problems by biologists who know what the crucial issues are."

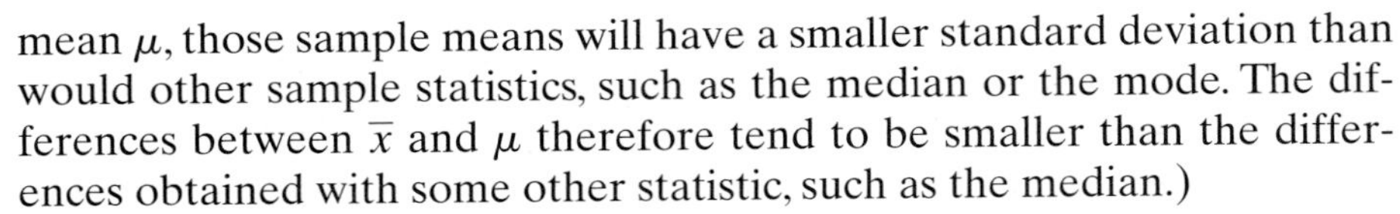

mean μ, those sample means will have a smaller standard deviation than would other sample statistics, such as the median or the mode. The differences between $\bar{x}$ and μ therefore tend to be smaller than the differences obtained with some other statistic, such as the median.)

2. For all populations, the sample mean $\bar{x}$ is an **unbiased estimator** of the population mean μ, meaning that the distribution of sample means tends to center about the value of the population mean μ. (That is, sample means do not systematically tend to overestimate the value of μ, nor do they systematically tend to underestimate μ. Instead, they tend to target the value of μ itself.)

Captured Tank Serial Numbers Reveal Population Size

During World War II, Allied intelligence specialists wanted to determine the number of tanks Germany was producing. Traditional spy techniques provided unreliable results, but statisticians obtained accurate estimates by analyzing serial numbers on captured tanks. As one example, records show that Germany actually produced 271 tanks in June 1941. The estimate based on serial numbers was 244, but traditional intelligence methods resulted in the extreme estimate of 1550. (See "An Empirical Approach to Economic Intelligence in World War II," by Ruggles and Brodie, *Journal of the American Statistical Association*, Vol. 42.)

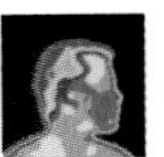

EXAMPLE **Body Temperatures** Use the sample body temperatures given in Table 6-1 to find the best point estimate of the population mean μ of all body temperatures.

SOLUTION For the sample data in Table 6-1, $\bar{x} = 98.20°F$. Because the sample mean $\bar{x}$ is the best point estimate of the population mean μ, we conclude that the best point estimate of the population mean μ of all body temperatures is 98.20°F.

Why Do We Need Confidence Intervals?

In the preceding example we saw that 98.20°F was our *best* point estimate of the population mean μ, but we have no indication of just how *good* our best estimate was. If we knew only the first four temperatures of 98.6, 98.6, 98.0, and 98.0, the best point estimate of μ would be their mean ($\bar{x} = 98.30°F$), but we wouldn't expect this point estimate to be very good because it is based on such a small sample. Because the point estimate has the serious flaw of not revealing anything about how good it is, statisticians have cleverly developed another type of estimate that does reveal how good the estimate is. This estimate, called a *confidence interval* or *interval estimate*, AS consists of a range (or an interval) of values instead of just a single value.

DEFINITION

A **confidence interval** (or **interval estimate**) is a range (or an interval) of values used to estimate the true value of a population parameter.

A confidence interval is associated with a degree of confidence, such as 0.95 (or 95%). The degree of confidence tells us the percentage of times that the confidence interval actually does contain the population parameter, assuming that the estimation process is repeated a large number of times. The definition of degree of confidence uses α (lowercase Greek alpha) to represent a probability or area. The value of α is the complement of the *degree of confidence.* For a 0.95 (or 95%) degree of confidence, $\alpha = 0.05$. For a 0.99 (or 99%) degree of confidence, $\alpha = 0.01$.

DEFINITION

The **degree of confidence** is the probability $1 - \alpha$ (often expressed as the equivalent percentage value) that is the relative frequency of times that the confidence interval actually does contain the population parameter, assuming that the estimation process is repeated a large number of times. (The degree of confidence is also called the **level of confidence,** or the **confidence coefficient.**)

The most common choices for the degree of confidence are 90% (with $\alpha = 0.10$), 95% (with $\alpha = 0.05$), and 99% (with $\alpha = 0.01$). The choice of 95% is most common because it provides a good balance between precision (as reflected in the width of the confidence interval) and reliability (as expressed by the degree of confidence).

Here's an example of a confidence interval based on the sample data of 106 body temperatures given in Table 6-1:

The 0.95 (or 95%) degree of confidence interval estimate of the population mean μ is 98.08°F $< \mu <$ 98.32°F.

Interpreting a Confidence Interval

We must be careful to interpret confidence intervals correctly. There is a correct interpretation and a wrong interpretation of the confidence interval 98.08°F $< \mu <$ 98.32°F.

Correct: We are 95% confident that the interval from 98.08 to 98.32 actually does contain the true value of μ. This means that if we were to select many different samples of size 106 and construct the confidence intervals, 95% of them would actually contain the value of the population mean μ. (Note that in this correct interpretation, the level of 95% refers to the *process* being used to estimate the mean, and it does not refer to the population mean itself.)

Wrong: There is a 95% chance that the true value of μ will fall between 98.08 and 98.32.

At this point in time, there is a fixed and constant value of μ, the mean body temperature of the whole population. If we use sample data to find specific limits, such as 98.08 and 98.32, those limits either enclose the population mean μ or do not, and we cannot determine whether they do or do not without knowing the true value of μ. But it's wrong to say that μ has a 95% chance of falling within the specific limits of 98.08 and 98.32, because μ is a constant, not a random variable. Either μ will fall within these limits or it won't; there's no probability involved. This is a confusing concept, so consider the easier example in which we want to find the probability of a baby being born a girl. If the baby has already been born, but the doctor hasn't yet told us the gender, we can't say that there is a 0.5 probability that the baby is a girl, because the baby is already a girl or is not. There is no chance involved, because the gender has been determined. Similarly, a population mean μ is already determined, and the confidence interval limits either contain μ or do not, so it's wrong to say that there is a 95% chance that μ will fall between 98.08 and 98.32.

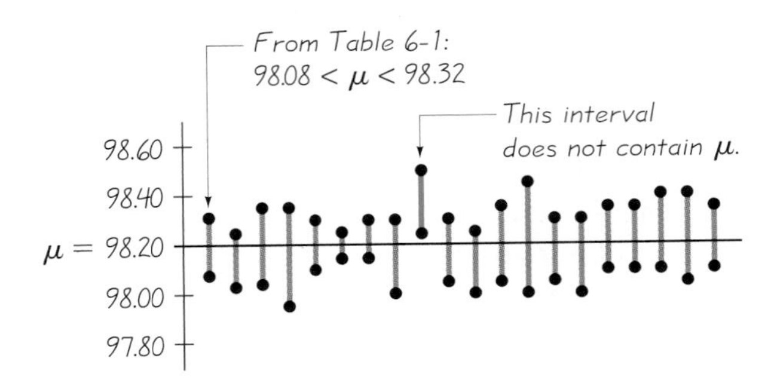

FIGURE 6-1
Confidence Intervals from 20 Different Samples

A confidence level of 95% tells us that the *process* we are using will, in the long run, result in confidence interval limits that contain μ 95% of the time. Suppose that body temperatures really came from a population with a true mean of 98.20° F. Then the confidence interval obtained from the given sample data would contain the population mean, because 98.20 is between 98.08 and 98.32. This is illustrated in Figure 6-1. Figure 6-1 shows the first confidence interval for the real data from Table 6-1, but the other 19 confidence intervals represent hypothetical samples. With 95% confidence, we expect that 19 out of 20 samples should result in confidence intervals that do contain the true value of μ, and Figure 6-1 illustrates this with 19 of the confidence intervals containing μ while one confidence interval does not contain μ.

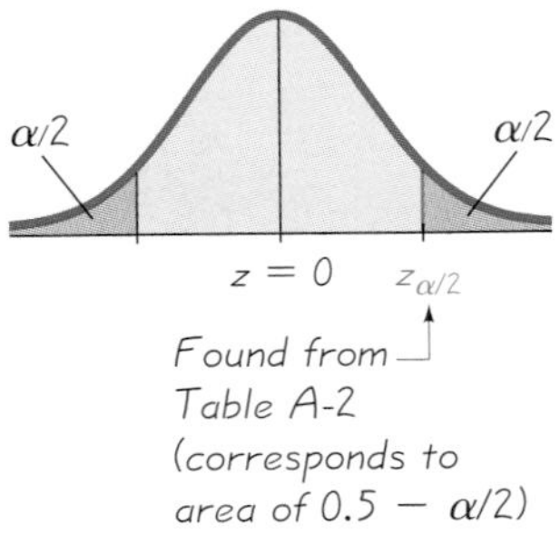

Critical Values

$\alpha/2$ $\alpha/2$

$z = 0$ $z_{\alpha/2}$

Found from Table A-2 (corresponds to area of $0.5 - \alpha/2$)

FIGURE 6-2
The Standard Normal Distribution: The Critical Value $z_{\alpha/2}$

Constructing a confidence interval requires that we find a standard z score that can be used to distinguish between sample statistics that are likely to occur and those that are unlikely. Such a z score is called a *critical value* (defined below), and it is based on the following observations.

1. We know from the central limit theorem that sample means tend to be normally distributed, as in Figure 6-2.
2. Sample means have a relatively small chance (with probability denoted by α) of falling in one of the orange extreme tails of Figure 6-2.
3. Denoting the area of each orange-shaded tail by $\alpha/2$, we see that there is a total probability of α that a sample mean will fall in either of the two tails.

4. By the rule of complements (from Chapter 3), there is a probability of $1 - \alpha$ that a sample mean will fall within the green-shaded region of Figure 6-2.
5. The z score separating the right-tail region is commonly denoted by $z_{\alpha/2}$ and is referred to as a *critical value* because it is on the borderline separating sample means that are likely to occur from those that are unlikely to occur.

These observations can be formalized with the following notation and definition.

Notation for Critical Value

$z_{\alpha/2}$ is the positive z value that is at the vertical boundary separating an area of $\alpha/2$ in the right tail of the standard normal distribution. (The value of $-z_{\alpha/2}$ is at the vertical boundary for the area of $\alpha/2$ in the left tail.) The subscript $\alpha/2$ is simply a reminder that the z score separates an area of $\alpha/2$ in the right tail of the standard normal distribution.

Definition

A **critical value** is the number on the borderline separating sample statistics that are likely to occur from those that are unlikely to occur. The number $z_{\alpha/2}$ is a critical value that is a z score with the property that it separates an area of $\alpha/2$ in the right tail of the standard normal distribution. (See Figure 6-2.)

EXAMPLE Critical Value Find the critical value $z_{\alpha/2}$ corresponding to a 95% degree of confidence.

SOLUTION A 95% degree of confidence corresponds to $\alpha = 0.05$. See Figure 6-3, where we show that the area in each of the orange-shaded

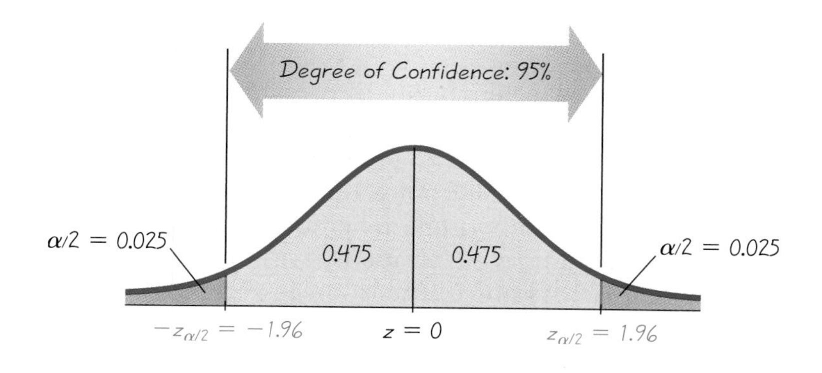

FIGURE 6-3
Finding $z_{\alpha/2}$ for 95% Degree of Confidence

tails is $\alpha/2 = 0.025$. We find $z_{\alpha/2} = 1.96$ by noting that the region to its left (and bounded by the mean of $z = 0$) must be $0.5 - 0.025$, or 0.475. We can refer to Table A-2 and find that the area of 0.4750 (found in the *body* of the table) corresponds exactly to a z score of 1.96. For a 95% degree of confidence, the critical value is therefore $z_{\alpha/2} = 1.96$.

The preceding example showed that a 95% degree of confidence results in a critical value of $z_{\alpha/2} = 1.96$. This is the most common critical value, and it is listed with two other common values in the table that follows.

Degree of Confidence	α	Critical Value, $z_{\alpha/2}$
90%	0.10	1.645
95%	0.05	1.96
99%	0.01	2.575

Margin of Error

When we collect a set of sample data, such as the set of 106 body temperatures listed in Table 6-1, we can calculate the sample mean $\bar{x}$, and that sample mean is typically different from the population mean μ. The difference between the sample mean and the population mean can be thought of as an error. In Section 5-5 we saw that $\sigma/\sqrt{n}$ is the standard deviation of sample means. Using $\sigma/\sqrt{n}$ and the $z_{\alpha/2}$ notation, we now define the *margin of error* E as follows.

DEFINITION

When data from a simple random sample are used to estimate a population mean μ, the **margin of error,** denoted by E, is the maximum likely (with probability $1 - \alpha$) difference between the observed sample mean $\bar{x}$ and the true value of the population mean μ. The margin of error E is also called the *maximum error of the estimate* and can be found by multiplying the critical value and the standard deviation of sample means, as shown in Formula 6-1.

Formula 6-1

$$E = z_{\alpha/2} \cdot \frac{\sigma}{\sqrt{n}}$$

Given the way that the margin of error E is defined, there is a probability of $1 - \alpha$ that a sample mean will be in error (different from the population mean μ) by no more than E, and there is a probability of α that the sample mean will be in error by more than E. The calculation of the margin of error E as given in Formula 6-1 requires that you know the population standard deviation σ, but in reality we seldom know σ when the population mean μ is not known. The following method of calculation is common practice.

Calculating E When σ Is Unknown

If $n > 30$, we can replace σ in Formula 6-1 by the sample standard deviation s.

If $n \leq 30$, the population must have a normal distribution and we must know the value of σ to use Formula 6-1.

[Section 6-3 will present a method for calculating the margin of error E for small $(n \leq 30)$ samples with σ unknown.]

On the basis of the definition of the margin of error E, we can now identify the confidence interval for the population mean μ. The three commonly used formats for expressing the confidence interval are shown in the following box.

Confidence Interval (or Interval Estimate) for the Population Mean μ (Based on Large Samples: $n > 30$)

$$\overline{x} - E < \mu < \overline{x} + E \qquad \text{where} \qquad E = z_{\alpha/2} \cdot \frac{\sigma}{\sqrt{n}}$$

or

$$\mu = \overline{x} \pm E$$

or

$$(\overline{x} - E, \overline{x} + E)$$

DEFINITION

The two values $\overline{x} - E$ and $\overline{x} + E$ are called **confidence interval limits.**

Procedure for Constructing a Confidence Interval for μ (Based on a Large Sample: $n > 30$)

1. Find the critical value $z_{\alpha/2}$ that corresponds to the desired degree of confidence. (For example, if the degree of confidence is 95%, the critical value is $z_{\alpha/2} = 1.96$.)
2. Evaluate the margin of error $E = z_{\alpha/2} \cdot \sigma/\sqrt{n}$. If the population standard deviation σ is unknown, use the value of the sample standard deviation s provided that $n > 30$.
3. Using the value of the calculated margin of error E and the value of the sample mean $\overline{x}$, find the values of $\overline{x} - E$ and $\overline{x} + E$. Substitute those values in the general format for the confidence interval:

$$\overline{x} - E < \mu < \overline{x} + E$$

Estimating Sugar in Oranges

In Florida, members of the citrus industry make extensive use of statistical methods. One particular application involves the way in which growers are paid for oranges used to make orange juice. An arriving truckload of oranges is first weighed at the receiving plant, then a sample of about a dozen oranges is randomly selected. The sample is weighed and then squeezed, and the amount of sugar in the juice is measured. Based on the sample results, an estimate is made of the total amount of sugar in the entire truckload. Payment for the load of oranges is based on the estimate of the amount of sugar because sweeter oranges are more valuable than those less sweet, even though the amounts of juice may be the same.

or

$$\mu = \overline{x} \pm E$$

or

$$(\overline{x} - E, \overline{x} + E)$$

4. Round the resulting values by using the following round-off rule.

Round-Off Rule for Confidence Intervals Used to Estimate μ

1. When using the *original set of data* to construct a confidence interval, round the confidence interval limits to one more decimal place than is used for the original set of data.
2. When the original set of data is unknown and only the *summary statistics* $(n, \overline{x}, s)$ are used, round the confidence interval limits to the same number of decimal places used for the sample mean.

The following example clearly illustrates the relatively simple(!) procedure for actually constructing a confidence interval. The original data from Table 6-1 are listed with one decimal place and the summary statistics use two decimal places, so the confidence interval limits will be rounded to two decimal places.

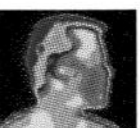

EXAMPLE **Body Temperatures** For the body temperatures in Table 6-1, we have $n = 106$, $\overline{x} = 98.20$°F, and $s = 0.62$°F. Using a 0.95 degree of confidence, find both of the following:

a. The margin of error E

b. The confidence interval for μ

SOLUTION

a. The 0.95 degree of confidence implies that $\alpha = 0.05$, so $z_{\alpha/2} = 1.96$, as shown in the preceding example. The margin of error E is calculated by using Formula 6-1 as follows. (Note that σ is unknown, but we can use $s = 0.62$ for the value of σ because $n > 30$.)

$$E = z_{\alpha/2} \cdot \frac{\sigma}{\sqrt{n}} = 1.96 \cdot \frac{0.62}{\sqrt{106}} = 0.12$$

b. With $\overline{x} = 98.20$ and $E = 0.12$, we construct the confidence interval as follows.

$$\overline{x} - E < \mu < \overline{x} + E$$
$$98.20 - 0.12 < \mu < 98.20 + 0.12$$
$$98.08 < \mu < 98.32$$

INTERPRETATION This result could also be expressed as $\mu = 98.20 \pm 0.12$ or as (98.08, 98.32). Based on the sample of 106 body temperatures listed in Table 6-1, the confidence interval for the population mean μ is $98.08°\text{F} < \mu < 98.32°\text{F}$, and this interval has a 0.95 degree of confidence. This means that if we were to select many different samples of size 106 and construct the confidence intervals as we did here, 95% of them would actually contain the value of the population mean μ.

Note that the confidence interval limits of 98.08°F and 98.32°F do not contain 98.6°F, the value generally believed to be the mean body temperature. Based on the sample data in Table 6-1, it seems very unlikely that 98.6°F is the correct value of μ.

Finding the Point Estimate and *E* from a Confidence Interval

AS

Later in this section we will describe how software and calculators can be used to find a confidence interval. A typical usage requires that you enter a confidence level and sample statistics, and the display shows the confidence interval limits. The sample mean $\bar{x}$ is the value midway between those limits, and the margin of error E is one-half the difference between those limits (because the upper limit is $\bar{x} + E$ and the lower limit is $\bar{x} - E$, the distance separating them is $2E$).

Point estimate of μ:

$$\bar{x} = \frac{(\text{upper confidence interval limit}) + (\text{lower confidence interval limit})}{2}$$

Margin of error:

$$E = \frac{(\text{upper confidence interval limit}) - (\text{lower confidence interval limit})}{2}$$

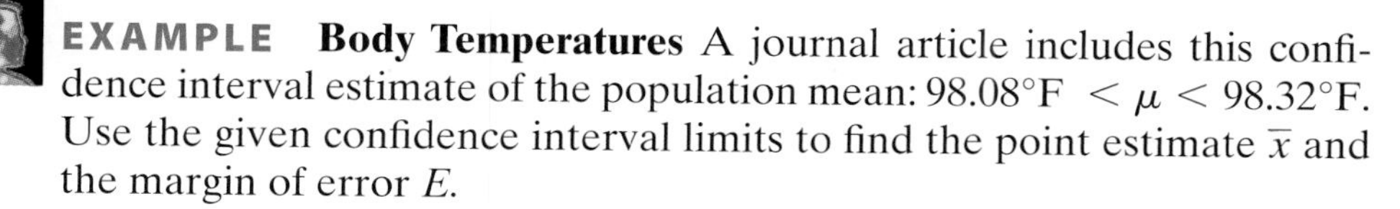

EXAMPLE Body Temperatures A journal article includes this confidence interval estimate of the population mean: $98.08°\text{F} < \mu < 98.32°\text{F}$. Use the given confidence interval limits to find the point estimate $\bar{x}$ and the margin of error E.

SOLUTION

$$\bar{x} = \frac{(\text{upper confidence limit}) + (\text{lower confidence limit})}{2}$$

$$= \frac{98.32°\text{F} + 98.08°\text{F}}{2} = 98.20°\text{F}$$

$$E = \frac{(\text{upper confidence interval limit}) - (\text{lower confidence interval limit})}{2}$$

$$= \frac{98.32°\text{F} - 98.08°\text{F}}{2} = 0.12°\text{F}$$

Using a Confidence Interval to Describe, Explore, or Compare Data

In some cases, we might use a confidence interval to achieve an ultimate goal of estimating the value of a population parameter. For the body temperature data used in this section, an important goal might be to estimate the mean body temperature of healthy adults, and our results strongly suggest that the commonly used value of 98.6°F is incorrect (because we have 95% confidence that the limits of 98.08°F and 98.32°F contain the true value of the population mean). In other cases, a confidence interval might be one of several different tools used to describe, explore, or compare data sets. For example, consider two different data sets consisting of the heights of 100 randomly selected men and the heights of 100 randomly selected women. The accompanying graphs, statistics, and confidence intervals can be used to compare those two data sets.

Men **Women**

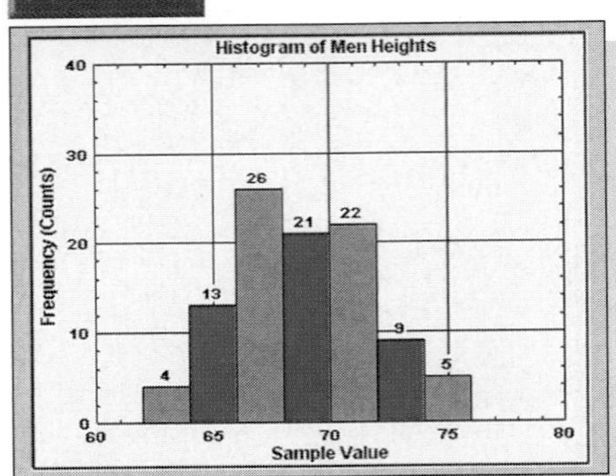

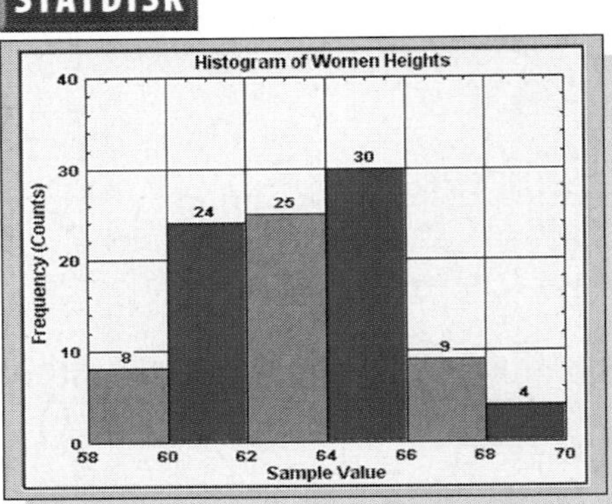

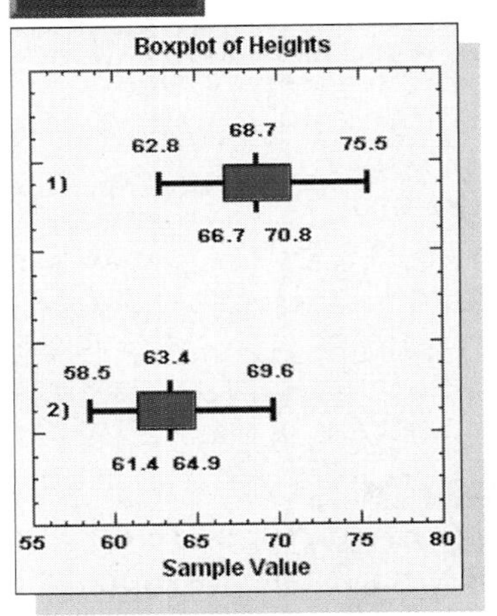

Descriptive Statistics

Men: $n = 100, \bar{x} = 68.76$ in., $s = 2.93$ in.

Women: $n = 100, \bar{x} = 63.39$ in., $s = 2.44$ in.

95% Confidence Intervals

Men: $68.19 \text{ in.} < \mu < 69.33 \text{ in.}$

Women: $62.91 \text{ in.} < \mu < 63.87 \text{ in.}$

The histograms suggest that heights of men and women have distributions that are approximately normal, and the heights of women appear to be generally lower. The boxplots are roughly the same, except for this major difference: The heights of women appear to be considerably lower than those of men. The descriptive statistics also suggest that the mean height of women is less than the mean height of men. Now carefully examine the confidence intervals and see that the upper limit for women is less than the lower limit for men. That is, the confidence intervals do not overlap at all, indicating that women have a mean height that is less than the mean height of men. This is not a big surprise, because most of us already know that women are typically shorter than men, but we now have solid statistical evidence supporting that belief, and we can see how confidence intervals can be used for such comparisons.

What Is the Basis for the Procedure of Finding a Confidence Interval? The basic idea underlying the construction of confidence intervals relates to the central limit theorem, which indicates that with large ($n > 30$) samples, the distribution of sample means is approximately normal with mean μ and standard deviation $\sigma/\sqrt{n}$. The confidence interval format is really a variation of the equation that was already used with the central limit theorem. Express $z = (\bar{x} - \mu_{\bar{x}})/\sigma_{\bar{x}}$ as shown below.

$$z = \frac{\bar{x} - \mu}{\dfrac{\sigma}{\sqrt{n}}}$$

If we solve this equation for μ, we get

$$\mu = \bar{x} - z\frac{\sigma}{\sqrt{n}}$$

Using the positive and negative values for z results in the confidence interval limits we are using.

Let's consider the specific case of a 95% degree of confidence, so $\alpha = 0.05$ and $z_{\alpha/2} = 1.96$. For this case, there is a probability of 0.05 that a sample mean will be more than 1.96 standard deviations (or $z_{\alpha/2}\sigma/\sqrt{n}$, which we denote by E) away from the population mean μ. Conversely, there is a 0.95 probability that a sample mean will be within 1.96 standard deviations (or $z_{\alpha/2}\sigma/\sqrt{n}$) of μ. (See Figure 6-4.) If the sample mean $\bar{x}$ is within

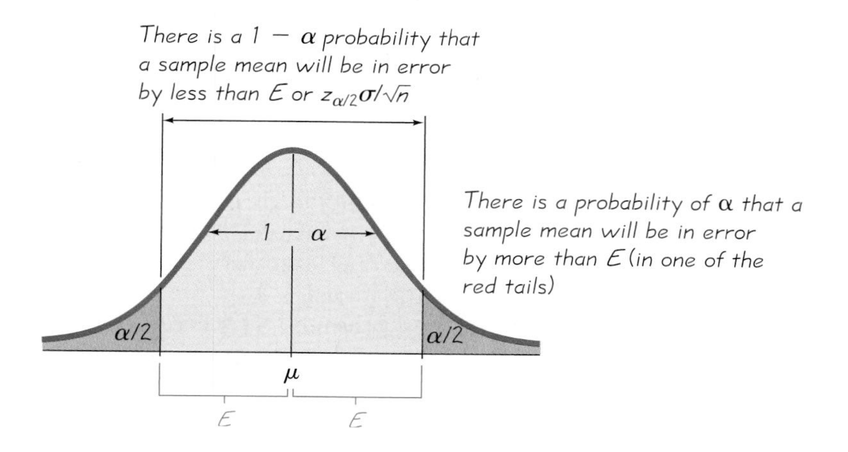

FIGURE 6-4
Distribution of Sample Means

$z_{\alpha/2}\sigma/\sqrt{n}$ of the population mean μ, then μ must be between $\bar{x} - z_{\alpha/2}\sigma/\sqrt{n}$ and $\bar{x} + z_{\alpha/2}\sigma/\sqrt{n}$; this is expressed in the general format of our confidence interval (with $z_{\alpha/2}\sigma/\sqrt{n}$ denoted as E): $\bar{x} - E < \mu < \bar{x} + E$.

Using Technology

STATDISK: You must first find the sample size n, the sample mean $\bar{x}$, and the sample standard deviation s. (See the STATDISK procedure described in Section 2-4.) Select **Analysis** from the main menu bar, select **Confidence Intervals,** then select **Population Mean.** Proceed to enter the items in the dialog box, then click the **Evaluate** button.

Minitab: Minitab requires that you enter a list of the original sample values. Minitab does not perform calculations using only the summary statistics of n, $\bar{x}$, and s. The *Minitab Student Laboratory Manual and Workbook,* which supplements this textbook, describes a trick for working around this Minitab limitation. If you do have a list of the original sample values, enter them in column C1, then select **Stat** and **Basic Statistics.** Next, select **1-sample t** and enter **C1** in the **Variables** box. Click the **OK** button. Again, see the Minitab workbook for more details.

Excel: You must first find the sample size n and the sample standard deviation s (which can be found using **fx, Statistical, STDEV**). Instead of generating the completed confidence interval with specific limits, Excel calculates only the margin of error E. You must then subtract this result from $\bar{x}$ and add it to $\bar{x}$ so that you can identify the actual confidence interval limits. To use Excel, click on **fx,** select the function category of **Statistical,** then select the item of **CONFIDENCE.** In the dialog box, enter the value of α (called the significance level), the standard deviation, and the sample size. The result will be the value of the margin of error E. You can also use the Data Desk XL add-in that is a supplement to this book. Click on **DDXL** and select **Confidence Intervals.** Under the Function Type options, select **1 Var z Interval.** Click on the pencil icon and enter the range of data, such as A1:A52 if you have 52 values listed in column A. Click on **OK.** In the dialog box, set the level of confidence, enter the standard deviation, and click on **Compute Interval.** The confidence interval will be displayed.

TI-83 Plus: The TI-83 Plus calculator can be used to generate confidence intervals for original sample values stored in a list, or you can use the summary statistics n, $\bar{x}$, and s. Either enter the data in list L1 or have the summary statistics available, then press the **STAT** key. Now select **TESTS** and choose **ZInterval** for a confidence interval based on the normal (z) distribution. The TI-83 Plus calculator will require an entry for the population standard deviation, so be prepared to enter the value of s if σ is not known. If you use the data in Table 6-1, the calculator display will include the confidence interval in this format: (98.082, 98.318).

6-2 Basic Skills and Concepts

Interpreting Computer or Calculator Results. *In Exercises 1–4, assume that a computer or calculator was used to generate the given confidence interval limits. Find the mean $\bar{x}$ and margin of error E.*

1. $72.6 < \mu < 82.6$

2. (1007.3, 1009.4)

3. (207, 212)

4. $2.17 < \mu < 2.57$

Finding Critical Values. *In Exercises 5–8, find the critical value $z_{\alpha/2}$ that corresponds to the given degree of confidence.*

5. 99%

6. 96%

7. 97%

8. 92%

Finding Margin of Error and Confidence Interval. *In Exercises 9–12, use the given degree of confidence and sample data to find (a) the margin of error E and (b) the confidence interval for the population mean μ.*

9. Salaries of statistics professors: 95% confidence; $n = 60$, $\bar{x} = \$85{,}678$, $s = \$12{,}345$ (we wish)

10. Distances driven by male students in one year: 90% confidence; $n = 100$, $\bar{x} = 11{,}878$ mi, $s = 3577$ mi

11. Times between uses of a TV remote control by males during commercials: 99% confidence; $n = 150$, $\bar{x} = 7.60$ sec, $s = 2.72$ sec

12. Starting salaries of college graduates who have taken a statistics course: 95% confidence; $n = 100$, $\bar{x} = \$43{,}704$, $s = \$9879$

13. Interpreting Confidence Interval A sample of 35 skulls is obtained for Egyptian males who lived around 1850 B.C. The maximum breadth of each skull is measured with the result that $\bar{x} = 134.5$ mm and $s = 3.48$ mm (based on data from *Ancient Races of the Thebaid* by Thomson and Randall-Maciver). Using these sample results and a confidence level of 95%, the TI-83 Plus calculator display is as shown. Write a statement that interprets the confidence interval.

TI-83 Plus

```
ZInterval
 (133.35,135.65)
 x̄=134.5
 n=35
```

14. Interpreting Confidence Interval The U.S. Department of Health, Education, and Welfare collected sample data for 1525 women, aged 18 to 24. That sample group has a mean serum cholesterol level (measured in mg/100 mL) of 191.7 with a standard deviation of 41.0 (see USDHEW publication 78-1652). Using these sample data and a 90% confidence level, the TI-83 Plus calculator display is as shown. Write a statement that interprets the displayed result. If a doctor claims that the mean serum cholesterol level for women in that age bracket is 200, does that claim seem to be consistent with the confidence interval?

TI-83 Plus

```
ZInterval
 (189.97,193.43)
 x̄=191.7
 n=1525
```

15. Comparing Bowling Lanes Bowling league members bowled (what else?) at two different locations. At the location with oiled lanes, 40 bowling scores had a mean of 205.5 and a standard deviation of 35.4. At the location with dry lanes, 40 bowling scores had a mean of 185.0 and a standard deviation of 28.4 (based on data from a student of the author).

a. Construct a 95% confidence interval estimate of the mean for all scores at the location with oiled lanes.

b. Construct a 95% confidence interval estimate of the mean for all scores at the location with dry lanes.

c. Compare and interpret the results. Does it appear that the lane surface has an effect on the bowling scores?

16. Comparing Service Times

a. The drive-through service times were recorded for 52 randomly selected customers at a Burger King restaurant. Those times had a mean of 181.3 sec and a standard deviation of 82.2 sec. Construct a 95% confidence interval estimate of the population mean.

b. The drive-through service times were recorded for 95 randomly selected customers at a McDonald's restaurant. Those times had a mean of 127.8 sec and a standard deviation of 60.6 sec. Construct a 95% confidence interval estimate of the population mean.

c. Compare and interpret the results. Does it appear that either restaurant has better performance. Why or why not?

17. Shoplifting Listed below are the values (in dollars) of items stolen and recovered for a Sears store. Construct a 99% confidence interval estimate of the mean value of all such items that are stolen and recovered.

144.95	144.00	542.88	346.95	149.99	402.00	34.00	119.96	134.98
16.00	100.97	37.50	209.00	38.00	55.00	23.99	17.99	145.94
99.97	449.99	17.47	24.99	39.99	37.99	529.88	9.99	274.89
379.99	162.97	24.99	229.99	16.96	101.91	389.96	79.99	330.96

18. Lengths of CDs Listed below are the lengths (in minutes) of randomly selected CDs of country, rock, and movie soundtracks. Construct a 97% confidence interval for the mean length of all such CDs.

62.28	56.13	54.03	20.37	56.13	54.03	29.01	58.02	74.14	54.04
50.16	52.15	53.15	47.42	49.19	51.15	49.63	52.00	61.57	57.06
75.12	32.04	32.57	60.41	32.07	37.06	62.06	39.29	33.15	51.37
50.97	52.07	43.16	38.48	73.23	36.26	39.57	69.71	49.22	48.19

19. Bullet Production A Lyman Electronic Digital Caliper was used to accurately measure the lengths of 40 Triton Quik-Shok 380 ACP 90 grain cartridges (bullets). The results (in millimeters) are listed below. The cartridges are supposed to have a mean length of 0.950 mm and they must be between 0.945 mm long and 0.955 mm, otherwise they could be dangerous when fired. Construct a 90% confidence interval estimate of the mean length, then interpret the results.

0.951	0.954	0.954	0.953	0.953	0.953	0.952	0.955	0.951	0.949
0.950	0.950	0.954	0.950	0.948	0.948	0.945	0.947	0.950	0.954
0.951	0.951	0.952	0.943	0.952	0.952	0.949	0.946	0.949	0.949
0.953	0.953	0.953	0.950	0.949	0.950	0.950	0.949	0.950	0.951

20. Times in Voting Booth In studying the time that voters are in the polling booth, a student of the author recorded the given times (in seconds) at a polling station in Marlboro, New York. Construct a 94% confidence interval estimate of the population mean. Interpret the result.

85	60	65	83	45	30	22	43	30	46	115	86	52
100	50	50	35	51	110	15	45	63	40	110	18	34
37	25	55	48	120	45	80	63	44	93	51	57	48
29	83	48	50	110	114	38	60	37	53	70		

21. Comparing Regular and Diet Coke Refer to Data Set 1 in Appendix B.

a. Construct a 95% confidence interval estimate of the population mean weight of the cola in regular Coke cans.

b. Construct a 95% confidence interval estimate of the population mean weight of the cola in diet Coke cans.

c. Compare the confidence intervals from parts (a) and (b) and interpret the results.

22. Comparing Textbook Prices Refer to Data Set 2 in Appendix B. Construct a 98% confidence interval estimate of the mean cost of new textbooks at each of the two colleges. Compare and interpret the results.

23. Comparing Discarded Plastic and Paper Refer to Data Set 5 in Appendix B. Construct a 92% confidence interval estimate of the mean weight of discarded plastic, and construct a 92% confidence interval estimate of the mean weight of discarded paper. Compare and interpret the results. Which is more of an ecological problem: discarded paper or discarded plastic?

24. Second-Hand Smoke Refer to Data Set 9 in Appendix B. Construct a 95% confidence interval estimate of the mean amount of serum cotinine in smokers, then do the same for the ETS group, which consists of nonsmokers exposed to environmental tobacco smoke at home or work. Compare and interpret the results.

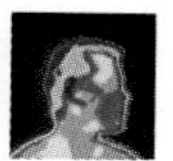

6-2 Beyond the Basics

25. Effect of Outlier Test the effect of an outlier as follows: Use the sample data in Table 6-1 to find a 95% confidence interval estimate of the population mean, but change the first entry from 98.6 to 986. Nobody can really have a body temperature of 986°F, but such an error can easily occur when a decimal point is inadvertently omitted when the sample values are entered.

a. Find the 95% confidence interval.

b. By comparing the result from part (a) to the confidence interval found in this section, describe the effect of an outlier on a confidence interval. Are the confidence interval limits sensitive to outliers?

c. Based on part (b), how should you handle outliers when they are found in sample data sets that will be used for the construction of confidence intervals?

26. Using 5-Number Summary A simple random sample of 200 U.S. Air Force applicants is given standard IQ tests. The IQ scores are normally distributed with this 5-number summary: 80, 100, 108, 116, 136. Construct a 95% confidence interval estimate of the population mean of all such applicants.

27. A 95% confidence interval for the lives (in minutes) of Kodak AA batteries is $430 < \mu < 470$. (See Program 1 of *Against All Odds: Inside Statistics.*) Assume that this result is based on a sample of size 100.

a. What is the value of the sample mean?

b. What is the value of the sample standard deviation?

c. Construct the 99% confidence interval.

d. If the confidence interval $432 < \mu < 468$ is obtained from the same sample data, what is the degree of confidence?

28. Using Finite Population Correction Factor The standard error of the mean is $\sigma/\sqrt{n}$, provided that the population size is infinite. If the population size is finite and is denoted by N, then the correction factor

$$\sqrt{\frac{N-n}{N-1}}$$

should be used whenever $n > 0.05N$. This correction factor multiplies the margin of error E given in Formula 6-1. Find the 95% confidence interval for the mean of 100 IQ scores if a sample of 31 of those scores produces a mean and standard deviation of 132 and 10, respectively.

6-3 Estimating a Population Mean: Small Samples

In Section 6-2 we introduced methods for estimating a population mean μ, but all of the examples and exercises were based on the following assumptions: (1) that sample has more than 30 values ($n > 30$), and (2) that the sample is a simple random sample. Time, cost, and other constraints, however, often require us to work with a small sample. If the sample size n is 30 or fewer, the normal distribution might not be a suitable approximation of the distribution of means from small samples. In this section we describe a procedure for dealing with such cases. Here is the main objective of this section: Given a simple random sample of 30 or fewer values, develop an estimate of the value of the population mean μ.

Assumptions

1. $n \leq 30$
2. The sample is a simple random sample.
3. The sample is from a normally distributed population. (This is a loose requirement, which can be met if the population has only one mode and is basically symmetric.)

If the first assumption is not satisfied, we have a large sample, and we can use the methods described in Section 6-2. If the second assumption is not satisfied, we cannot use the methods described in this book, but it might be possible to use more advanced methods. If the third assumption is not satisfied because the population has a distribution that is very nonnormal, we cannot use the methods of this section, but we might be able to use nonparametric methods (see Chapter 13) or bootstrap resampling methods (see the Computer Project at the end of this chapter).

As in Section 6-2, the sample mean $\bar{x}$ is the best point estimate (or single-valued estimate) of the population mean μ. As in Section 6-2, the distribution of sample means $\bar{x}$ tends to be more consistent (with *less variation*) than the distributions of other sample statistics, and the sample mean $\bar{x}$ is an *unbiased estimator* of the population mean μ, meaning that the distribution of sample means tends to center about the value of the population mean μ.

The sample mean $\bar{x}$ is the best point estimate of the population mean μ.

In Section 6-2 we noted that there is a serious limitation to the usefulness of a point estimate: The single value of a point estimate does not reveal how good that estimate is. Confidence intervals give us much more meaningful information by providing a range of values associated with a degree of likelihood that the range actually does contain the true value of μ. In developing confidence interval estimates of μ in this section, we have two cases: (1) The population standard deviation σ is known; (2) σ is not known.

Case 1 (σ Is Known): The first case is largely unrealistic, because if we don't know the value of the population mean μ and we are trying to estimate that value, it's a pretty safe bet that we also do not know the value of the population standard deviation σ. If it should somehow happen that we satisfy the three assumptions listed near the beginning of this section, and we already know the value of σ, we can construct confidence interval estimates of μ by using the same methods described in Section 6-2. That is, the confidence interval limits are $\bar{x} - E$ and $\bar{x} + E$, where $E = z_{\alpha/2} \cdot \sigma/\sqrt{n}$.

Case 2 (σ Is Not Known): The second case, when σ is not known, is much more realistic and practical, and it is the major focus of this section. Now, instead of using the normal distribution, we use the *Student t distribution* developed by William Gosset (1876–1937). Gosset was a Guinness Brewery employee who needed a distribution that could be used with small samples. The Irish brewery where he worked did not allow the publication of research results, so Gosset published under the pseudonym *Student.*

Student t Distribution

If the distribution of a population is essentially normal (approximately bell-shaped), then the distribution of

$$t = \frac{\bar{x} - \mu}{\frac{s}{\sqrt{n}}}$$

is essentially a **Student t distribution** for all samples of size n. The Student t distribution, often referred to as the **t distribution,** is used to find critical values denoted by $t_{\alpha/2}$.

AS

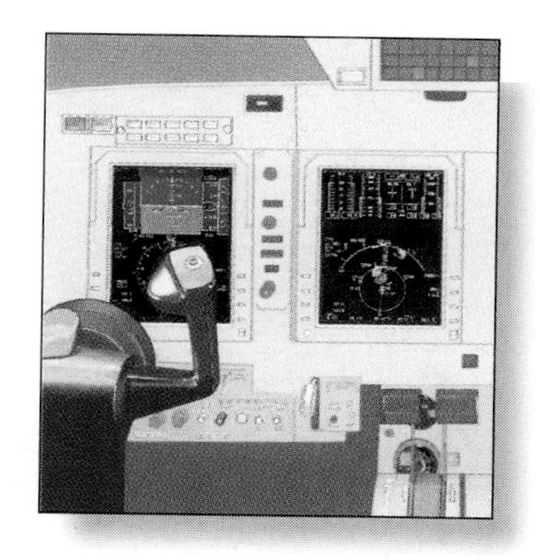

Excerpts from a Department of Transportation Circular

The following excerpts from a Department of Transportation circular concern some of the accuracy requirements for navigation equipment used in aircraft. Note the use of the confidence interval.

"The total of the error contributions of the airborne equipment, when combined with the appropriate flight technical errors listed, should not exceed the following with a 95% confidence (2-sigma) over a period of time equal to the update cycle."

"The system of airways and routes in the United States has widths of route protection used on a VOR system use accuracy of 64.5 degrees on a 95% probability basis."

We will soon discuss some of the important properties of the t distribution, but we will first present the components needed for the construction of confidence intervals. Let's start with the critical value denoted by $t_{\alpha/2}$. A value of $t_{\alpha/2}$ can be found in Table A-3 by locating the appropriate number of *degrees of freedom* in the left column and proceeding across the corresponding row until reaching the number directly below the applicable value of α for two tails.

DEFINITION

The number of **degrees of freedom** for a single data set is the number of sample values that can vary after certain restrictions have been imposed on all data values.

For example, if 10 students have quiz scores with a mean of 80, we can freely assign values to the first 9 scores, but the 10th score is then determined. The sum of the 10 scores must be 800, so the 10th score must equal 800 minus the sum of the first 9 scores. Because those first 9 scores can be freely selected to be any values, we say that there are 9 degrees of freedom available. For the applications of this section, the number of degrees of freedom is simply the sample size minus 1.

$$\textbf{degrees of freedom} = n - 1$$

EXAMPLE **Finding a Critical Value** A sample of size $n = 15$ is a simple random sample selected from a normally distributed population. Find the critical value $t_{\alpha/2}$ corresponding to a 95% degree of confidence.

SOLUTION Because $n = 15$, the number of degrees of freedom is given by $n - 1 = 14$. Using Table A-3, we locate the 14th row by referring to the column at the extreme left. As in Section 6-2, a 95% degree of confidence corresponds to $\alpha = 0.05$, so we find the column with the heading "0.05 (two tails)." The value in the 14th row and the "0.05 (two tails)" column is 2.145, so $t_{\alpha/2} = 2.145$.

Now that we know how to find critical values denoted by $t_{\alpha/2}$, we can go on to describe the margin of error E and the confidence interval.

Margin of Error E for the Estimate of μ [Based on an Unknown σ and a Small Simple Random Sample ($n \leq 30$) from a Normally Distributed Population]

Formula 6-2
$$E = t_{\alpha/2}\frac{s}{\sqrt{n}}$$

where $t_{\alpha/2}$ has $n - 1$ degrees of freedom

Confidence Interval for the Estimate of μ
[Based on an Unknown σ and a Small Simple Random Sample ($n \leq 30$) from a Normally Distributed Population]

$$\bar{x} - E < \mu < \bar{x} + E$$

where
$$E = t_{\alpha/2}\frac{s}{\sqrt{n}}$$

EXAMPLE **Finding a Confidence Interval** Because cardiac deaths appear to increase after heavy snowfalls, a study was designed to compare cardiac demands of snow shoveling to those of using an electric snow thrower. Ten subjects cleared tracts of snow using both methods, and their maximum heart rates (beats per minute) were recorded during both activities. The following results were obtained for the snow shoveling portion of the experiment (based on data from "Cardiac Demands of Heavy Snow Shoveling," by Franklin et al., *Journal of the American Medical Association,* Vol. 273, No. 11):

Maximum Heart Rates During Manual Snow Shoveling:
$n = 10, \bar{x} = 175, s = 15$

Find the 95% confidence interval estimate of the population mean for those who shovel snow.

SOLUTION We will proceed to construct a 95% confidence interval by using the t distribution. We select the t distribution because the following three conditions are met: (1) The sample is small ($n \leq 30$), (2) the population standard deviation σ is unknown, and (3) the population appears to have a normal distribution, because the sample data have a distribution that is approximately bell-shaped.

First we find the critical value $t_{\alpha/2}$, then we use that result to find the margin of error E, and then we use the value of E to find the confidence interval.

1. *Find $t_{\alpha/2}$:* The value of $t_{\alpha/2} = 2.262$ is found in Table A-3 as the critical value corresponding to $n - 1 = 9$ degrees of freedom (left column of Table A-3) and "0.05 (two tails)" (top column heading of Table A-3). Remember that a 95% confidence level corresponds to $\alpha = 0.05$.
2. *Find the margin of error E:* The margin of error $E = 10.7296$ is computed using Formula 6-2 as shown below.

$$E = t_{\alpha/2}\frac{s}{\sqrt{n}} = 2.262 \cdot \frac{15}{\sqrt{10}} = 10.7296$$

3. *Find the confidence interval:* The confidence interval can now be found by using $\overline{x} = 175$ and $E = 10.7296$ as shown below.

$$\overline{x} - E < \mu < \overline{x} + E$$
$$175 - 10.7296 < \mu < 175 + 10.7296$$
$$164 < \mu < 186$$

INTERPRETATION This result could also be expressed in the format of $\mu = 175 \pm 11$ or (164, 186). On the basis of the given sample results, we are 95% confident that the limits of 164 beats per minute and 186 beats per minute actually do contain the value of the population mean μ. As expected, these results suggest that snow shoveling dramatically increases the workload of the heart. (See Data Set 14 in Appendix B for the pulse rates of 100 statistics students. Those at-rest pulse rates have a mean of 70 beats per minute, and the highest is 100.)

We now list the important properties of the t distribution that we are using in this section.

Important Properties of the Student t Distribution

1. The Student t distribution is different for different sample sizes. (See Figure 6-5 for the cases $n = 3$ and $n = 12$.)
2. The Student t distribution has the same general symmetric bell shape as the standard normal distribution, but it reflects the greater variability (with wider distributions) that is expected with small samples.
3. The Student t distribution has a mean of $t = 0$ (just as the standard normal distribution has a mean of $z = 0$).
4. The standard deviation of the Student t distribution varies with the sample size, but it is greater than 1 (unlike the standard normal distribution, which has $\sigma = 1$).
5. As the sample size n gets larger, the Student t distribution gets closer to the standard normal distribution. For values of $(n > 30)$, the differences

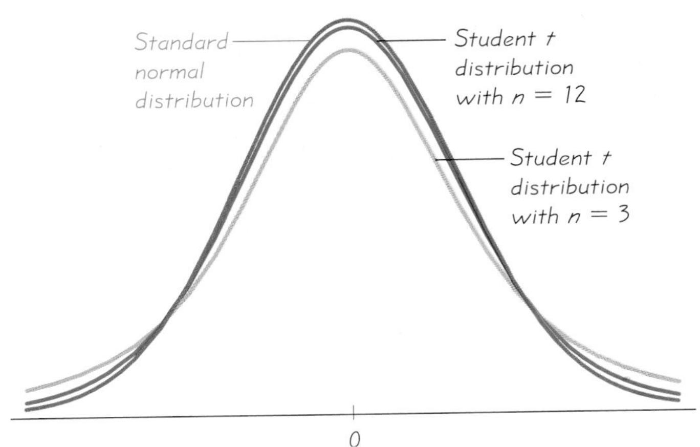

FIGURE 6-5
Student t Distributions for $n = 3$ and $n = 12$

The Student t distribution has the same general shape and symmetry as the standard normal distribution, but it reflects the greater variability that is expected with small samples.

are so small that we can use the critical z values instead of developing a much larger table of critical t values. (The values in the bottom row of Table A-3 are equal to the corresponding critical z values from the standard normal distribution.)

The following is a summary of the conditions indicating use of a t distribution instead of the standard normal distribution. (These same conditions will also apply in Chapter 7.)

Conditions for Using the Student t Distribution

1. The sample is small ($n \leq 30$); and
2. σ is unknown; and
3. The parent population has a distribution that is essentially normal. (Because the distribution of the parent population is often unknown, we often estimate it by constructing a histogram of sample data.)

Choosing the Appropriate Distribution

It is sometimes difficult to decide whether to use the standard normal z distribution or the Student t distribution. The flowchart in Figure 6-6 and Table 6-2 (on the next page) both summarize the key points to be considered when constructing confidence intervals for estimating μ, the population mean. In Figure 6-6 or Table 6-2, note that if we have a small ($n \leq 30$) sample drawn

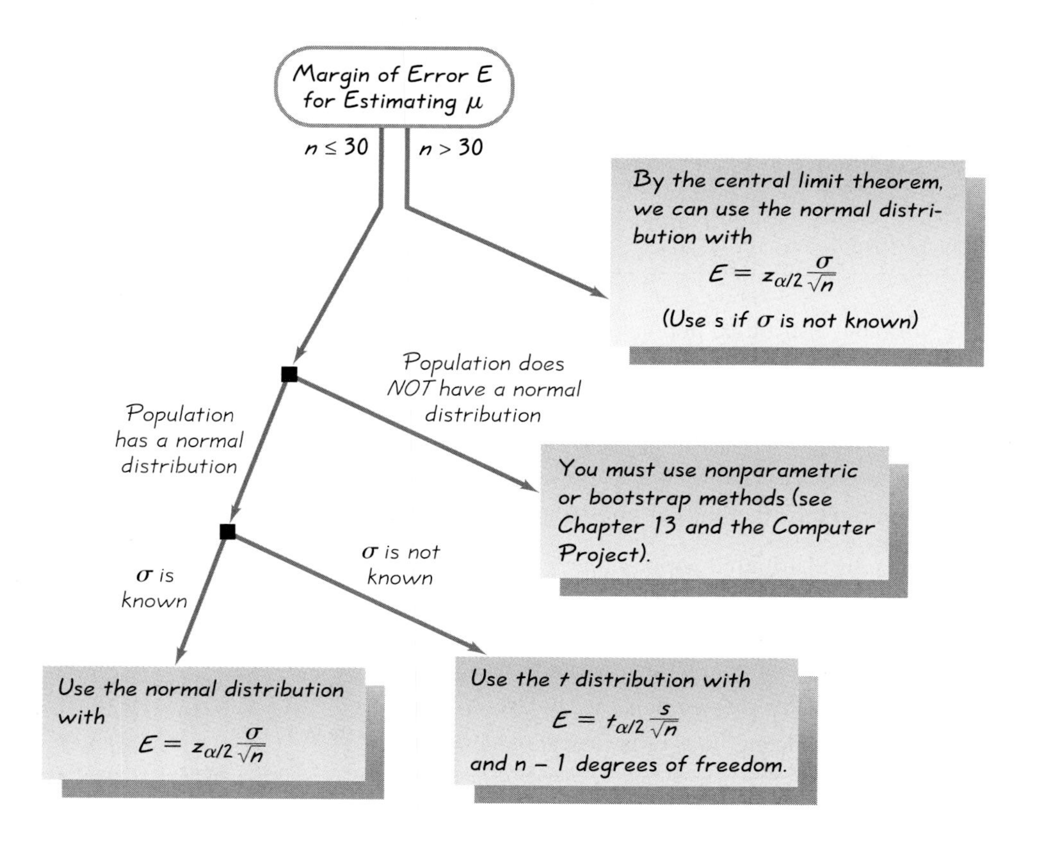

FIGURE 6-6
Using Normal and t Distributions

from a distribution that differs significantly from a normal distribution, we can't use the methods described in this chapter. One alternative is to use non-parametric methods (see Chapter 13), and another alternative is to use the computer bootstrap method, which makes no assumptions about the original population. This method is described in the Computer Project at the end of this chapter.

Another Approach Rather than using *sample size* as the major criterion for choosing between the normal (z) and Student t distributions, some statisticians and software use *knowledge of the population standard deviation* σ as the major criterion, as follows:

- Use the *normal distribution* if the population standard deviation σ is known, and the sample is from a normally distributed population.
- Use the *Student t distribution* if the population standard deviation σ is *not* known, and the sample is from a normally distributed population.

In reality, we seldom know the value of σ, so this procedure suggests that we almost always use the Student t distribution, even if the sample size is large. But using the Student t distribution with a large sample size results in a large number of degrees of freedom, so the value of $t_{\alpha/2}$ is, for all practical purposes, the same as the value of $z_{\alpha/2}$. The confidence interval will be essentially the same with the two approaches. Therefore, we will base our choice on the sample size, as described in Figure 6-6 and Table 6-2.

EXAMPLE **Choosing Distributions** Assuming that you plan to construct a confidence interval for the population mean μ, use the given

TABLE 6-2 Margin of Error E for Confidence Interval Estimate of μ: $\overline{x} - E < \mu < \overline{x} + E$

	Large Sample ($n > 30$)	Small Sample ($n \leq 30$)
σ known	$E = z_{\alpha/2}\frac{\sigma}{\sqrt{n}}$	Normally distributed population: Use $E = z_{\alpha/2}\frac{\sigma}{\sqrt{n}}$ Very nonnormal population distribution: Methods of this chapter don't apply. Use nonparametric methods or bootstrapping.
σ not known	Use s for σ in $E = z_{\alpha/2}\frac{\sigma}{\sqrt{n}}$	Normally distributed population: Use $E = t_{\alpha/2}\frac{s}{\sqrt{n}}$ Very nonnormal population distribution: Methods of this chapter don't apply. Use nonparametric methods or bootstrapping.

data to determine whether the margin of error E should be calculated using a critical value of $z_{\alpha/2}$ (from the normal distribution), a critical value of $t_{\alpha/2}$ (from a t distribution), or neither (so that the methods of this chapter cannot be used).

a. $n = 50$, $\overline{x} = 77.6$, $s = 14.2$, and the shape of the distribution is skewed.

b. $n = 25$, $\overline{x} = 77.6$, $s = 14.2$, and the distribution is bell-shaped.

c. $n = 25$, $\overline{x} = 77.6$, $\sigma = 14.2$, and the distribution is bell-shaped.

d. $n = 25$, $\overline{x} = 77.6$, $\sigma = 14.2$, and the distribution is extremely skewed.

SOLUTION Refer to Figure 6-6 or Table 6-2 to determine the following.

a. Because the sample is large ($n > 30$), the margin of error is calculated using $z_{\alpha/2}$ in Formula 6-1, with the sample standard deviation s used in place of σ.

b. Use $t_{\alpha/2}$ because (1) the sample is small, (2) the population appears to have a normal distribution, and (3) σ is unknown.

c. Use $z_{\alpha/2}$ because the population appears to have a normal distribution and the value of σ is known.

d. Because the sample is small and the population has a distribution that is very nonnormal, the methods of this chapter do not apply. Neither the normal nor the t distribution can be used.

In Section 6-2 we presented the basic concepts of point estimate and confidence interval estimate. In this section we extended those concepts to small sample cases. In the following section we describe a method for determining the sample size needed to estimate a population mean.

AS

Using Technology

STATDISK: Select **Analysis** from the main menu, then select **Confidence Intervals,** followed by **Population Mean,** and enter the items in the dialog box. After you click on **Evaluate,** the confidence interval will be displayed.

Minitab: Minitab requires an original listing of the raw data and will not work with summary statistics. For a way to get around that limitation and for more details on the use of Minitab, see the *Minitab Student Laboratory Manual and Workbook* that supplements this textbook.

Excel: *Caution*: Excel's CONFIDENCE module will generate a value of the margin of error E, *but only for confidence intervals using a critical value of* $z_{\alpha/2}$. To obtain confidence intervals that require critical values of $t_{\alpha/2}$, use the Data Desk XL add-in that is a supplement to this book. Click on **DDXL,** select **Confidence Intervals,** and select **1 Var t**

Interval under from the options listed under Function Type. Click on the pencil icon and enter the range of data, such as A1:A12 if you have 12 values listed in column A. Click on **OK.** In the dialog box, select the level of confidence, then click on **Compute Interval,** and the confidence interval will be displayed.

TI-83 Plus: The TI-83 Plus calculator can be used to generate confidence intervals for original sample values stored in a list or summary statistics for n, $\overline{x}$, and s. Either enter the data in list L1 or have the summary statistics available, then press the **STAT** key. Now select **TESTS** and choose **TInterval** for a confidence interval based on the t distribution.

6-3 Basic Skills and Concepts

Using Correct Distribution *In Exercises 1–8, do one of the following, as appropriate: (a) Find the critical value $t_{\alpha/2}$; (b) find the critical value $z_{\alpha/2}$; (c) state that neither the normal nor the t distribution applies.*

1. 95%; $n = 20$; σ is unknown; population appears to be normally distributed.

2. 95%; $n = 40$; σ is unknown; population appears to be very skewed.

3. 99%; $n = 15$; σ is unknown; population appears to be normally distributed.

4. 90%; $n = 15$; σ is unknown; population appears to be very skewed.

5. 90%; $n = 12$; $\sigma = 24$; population appears to be very skewed.

6. 99%; $n = 100$; $\sigma = 4.2$; population appears to be very skewed.

7. 98%; $n = 9$; σ is unknown; population appears to be normally distributed.

8. 95%; $n = 30$; σ is unknown; population appears to be normally distributed.

Finding Confidence Intervals. *In Exercises 9 and 10, use the given degree of confidence and sample data to find (a) the margin of error and (b) the confidence interval for the population mean μ. Assume that the population has a normal distribution.*

TI-83 Plus Exercise 11

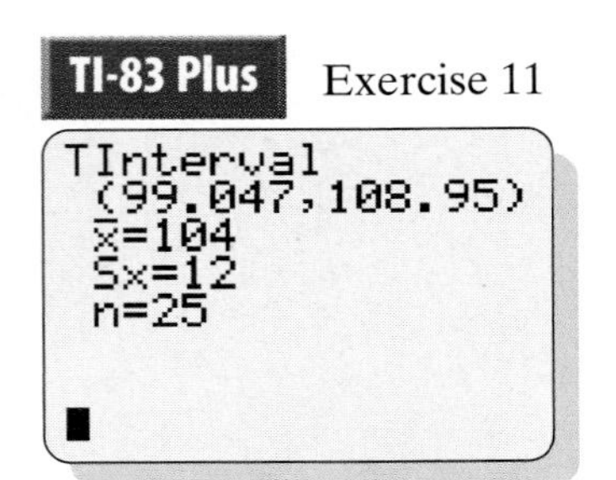

9. Math SAT scores for women: 95% confidence; $n = 15$, $\overline{x} = 496$, $s = 108$

10. Salaries of airline pilots: 99% confidence; $n = 12$, $\overline{x} = \$97{,}334$, $s = \$17{,}747$

Interpreting Calculator Display. *In Exercises 11 and 12, use the given data and the TI-83 Plus calculator display to express the confidence interval in the format of $\overline{x} - E < \mu < \overline{x} + E$. Also write a statement that interprets the confidence interval.*

11. IQ scores of professional athletes: 95% confidence; $n = 25$, $\overline{x} = 104$, $s = 12$

TI-83 Plus Exercise 12

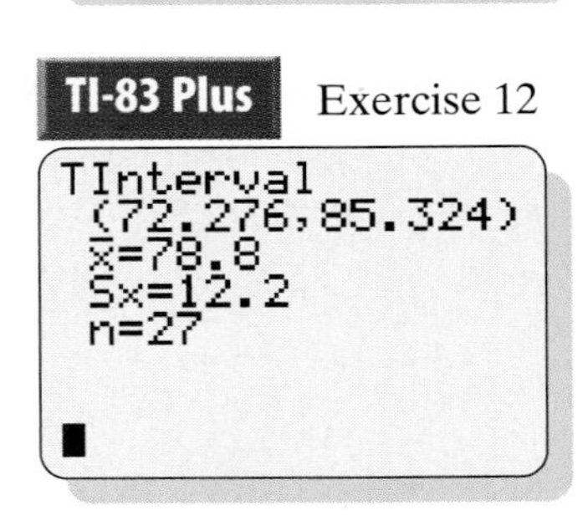

12. Final exam test scores in a statistics course: 99% confidence; $n = 27$, $\overline{x} = 78.8$, $s = 12.2$

Constructing Confidence Intervals. *In Exercises 13–24, construct the confidence interval. Some of these exercises require use of the t distribution, but others require use of the normal distribution.*

13. Snow Shoveling and Heart Attacks Because cardiac deaths appear to increase after heavy snowfalls, a study was designed to compare cardiac demands of snow shoveling to those of using an electric snow thrower. Ten subjects cleared tracts of snow using both methods, and their maximum heart rates (beats per minute) were recorded during both activities. The following results were obtained from the portion of the experiment involving the use of an electric snow thrower (based on data from "Cardiac Demands of Heavy Snow Shoveling," by Franklin et al., *Journal of the American Medical Association,* Vol. 273, No. 11):

Maximum Heart Rates During Automated Snow Removal:
$n = 10, \bar{x} = 124, s = 18$

Find the 95% confidence interval estimate of the population mean for those who use the electric snow thrower. Compare the result to the confidence interval found in the related example of this section and interpret your findings.

14. Weights of Quarters If we refer to the weights (in grams) of quarters listed in Data Set 13 in Appendix B, we will find 50 weights with a mean of 5.622 g and a standard deviation of 0.068 g. Based on this simple random sample of quarters in circulation, construct a 98% confidence interval estimate of the population mean of all quarters in circulation. The U.S. Department of the Treasury claims that it mints quarters to yield a mean weight of 5.670 g. Is this claim consistent with the confidence interval? If not, what is a possible explanation for the discrepancy?

15. Destroying Dodge Vipers With *destructive testing,* sample items are destroyed in the process of testing them. Crash testing of cars is one very expensive example of destructive testing. If you were responsible for such crash tests, there is no way you would want to tell your supervisor that you must crash and destroy more than 30 cars so that you could use the normal distribution. Let's assume that you have crash tested 12 Dodge Viper sports cars (list price: $59,300) under a variety of conditions that simulate typical collisions. Analysis of the 12 damaged cars results in repair costs having a distribution that appears to be bell-shaped, with a mean of $\bar{x}$ = $26,227 and a standard deviation of s = $15,873 (based on data from the Highway Loss Data Institute). Find the 95% interval estimate of μ, the mean repair cost for all Dodge Vipers involved in collisions, and then interpret the result.

16. Costs of Repairing Hondas In crash tests of 15 Honda Odyssey minivans, collision repair costs are found to have a distribution that is roughly bell-shaped, with a mean of $1786 and a standard deviation of $937 (based on data from the Highway Loss Data Institute). Construct the 99% confidence interval for the mean repair cost in all such vehicle collisions.

17. Estimating Car Pollution In a sample of seven cars, each car was tested for nitrogen-oxide emissions (in grams per mile) and the following results were obtained: 0.06, 0.11, 0.16, 0.15, 0.14, 0.08, 0.15 (based on data from the Environmental Protection Agency). Assuming that this sample is representative of the cars in use, construct a 98% confidence interval estimate of the mean amount of nitrogen-oxide emissions for all cars.

18. Crash Hospital Costs A study was conducted to estimate hospital costs for accident victims who wore seat belts. Twenty randomly selected cases have a distribution that appears to be bell-shaped with a mean of $9004 and a standard deviation of $5629 (based on data from the U.S. Department of Transportation). Construct the 99% confidence interval for the mean of all such costs.

19. TV Replacement Times A sample consists of 75 TV sets purchased several years ago. The replacement times of those TV sets have a mean of 8.2 years and a standard deviation of 1.1 years (based on data from "Getting Things Fixed," *Consumer Reports*). Construct a 90% confidence interval for the mean replacement time of all TV sets from that era. Does the result apply to TV sets currently being sold?

20. Interpreting Graphs Statisticians at the College of Newport are conducting research into the ability of people to read and interpret graphs presented by the media. They have developed a comprehensive test of that ability, and they want to estimate the mean score for a simple random sample of people. Test subjects are paid $50 to participate, but the $1000 budget is enough to fund only 20 subjects. Those 20 test scores have a mean of 212 and a standard deviation of 67. Construct a 95% confidence interval for the population mean.

21. Comparing Yellow and Brown M&M Candies Refer to Data Set 10 in Appendix B.
- **a.** Construct a 90% confidence interval for the mean weight of yellow M&M plain candies.
- **b.** Construct a 90% confidence interval for the mean weight of brown M&M plain candies.
- **c.** Compare the methods used to find the two confidence intervals, and interpret the results.

22. Comparing Regular and Diet Pepsi Refer to Data Set 1 in Appendix B.
- **a.** Construct a 95% confidence interval estimate of the mean weight of cola in cans of regular Pepsi.
- **b.** Construct a 95% confidence interval estimate of the mean weight of cola in cans of diet Pepsi.
- **c.** Compare the results from parts (a) and (b) and interpret them.

23. Comparing Textbook Prices Refer to Data Set 2 in Appendix B.
- **a.** Construct a 98% confidence interval estimate of the mean cost of new hardcover textbooks at the University of Massachusetts.
- **b.** Construct a 98% confidence interval estimate of the mean cost of new paperback textbooks at the University of Massachusetts.
- **c.** Compare and interpret the results.

24. Comparing Movies Refer to Data Set 15 in Appendix B.
- **a.** Construct a 99% confidence interval estimate of the mean length (in minutes) of movies with a rating of R.
- **b.** Construct a 99% confidence interval estimate of the mean length (in minutes) of movies with a rating different than R.
- **c.** Compare and interpret the results.

6-3 Beyond the Basics

25. Using the Wrong Distribution Assume that a small ($n \leq 30$) sample is a simple random sample selected from a normally distributed population for which σ is unknown. Construction of a confidence interval should use the t distribution, but how are the confidence interval limits affected if the normal distribution is incorrectly used instead?

26. Effects of Units of Measurement A confidence interval is constructed for a small simple random sample of temperatures (in degrees Fahrenheit) selected from a normally distributed population for which σ is unknown. *(continued)*

a. How is the margin of error E affected if each temperature is converted to the Celsius scale? $\left[C = \frac{5}{9}(F - 32)\right]$

b. If the confidence interval limits are denoted by a and b, find expressions for the confidence interval limits after the original temperatures have been converted to the Celsius scale.

c. Based on the results from part (b), can confidence interval limits for the Celsius temperatures be found by simply converting the confidence interval limits from the Fahrenheit scale to the Celsius scale?

27. Finding Degree of Confidence A simple random sample of Boeing 747 aircraft is selected, and the times (in hours) required to test for structural stress fractures are as follows: 8.1, 9.9, 9.5, 6.9, 9.8. Based on these results, a confidence interval is found to be $\mu = 8.84 \pm 1.24$. Find the degree of confidence.

6-4 Determining Sample Size Required to Estimate μ

In this section we will address the following key question: When we plan to collect a simple random sample of data that will be used to estimate a population mean μ, *how many* sample values must be obtained? In other words, we will find the sample size n that is required to estimate the value of a population mean. For example, suppose we want to estimate the mean weight of airline passengers (an important value for reasons of safety). How many passengers must be randomly selected and weighed? Determining the size of a simple random sample is a very important issue, because samples that are needlessly large waste time and money, and samples that are too small may lead to poor results. In many cases we can find the minimum sample size needed to estimate some parameter, such as the population mean μ.

If we begin with the expression for the margin of error E (Formula 6-1 in Section 6-2) and solve for the sample size n, we get the following.

Sample Size for Estimating Mean μ

Formula 6-3

$$n = \left[\frac{z_{\alpha/2}\sigma}{E}\right]^2$$

where $z_{\alpha/2}$ = critical z score based on the desired degree of confidence

E = desired margin of error

σ = population standard deviation

Formula 6-3 is quite remarkable because it implies that the sample size does not depend on the size (N) of the population; the sample size depends on the desired degree of confidence, the desired margin of error, and the value of the standard deviation σ.

Small Sample

The Children's Defense Fund was organized to promote the welfare of children. The group published *Children Out of School in America,* which reported that in one area, 37.5% of the 16- and 17-year-old children were out of school. This statistic received much press coverage, but it was based on a sample of only 16 children. Another statistic was based on a sample size of only 3 students. (See "Firsthand Report: How Flawed Statistics Can Make an Ugly Picture Look Even Worse," *American School Board Journal,* Vol. 162.)

The sample size must be a whole number, because it represents the number of sample values that must be found. However, when we use Formula 6-3 to calculate the sample size n, we usually get a result that is not a whole number. When this happens, we use the following round-off rule. (It is based on the principle that when rounding is necessary, the required sample size should be rounded *upward* so that it is at least adequately large as opposed to slightly too small.)

Round-Off Rule for Sample Size *n*

When finding the sample size n, if the use of Formula 6-3 does not result in a whole number, always *increase* the value of n to the next *larger* whole number.

The roles of $z_{\alpha/2}$ and E are the same as in Section 6-2. See the following example.

EXAMPLE **IQ Scores** Assume that you want to find the size of the sample needed to estimate the mean IQ score of statistics professors, and you want 95% confidence that your sample mean will be within 2 IQ points of the true population mean μ.

a. Find the value of $z_{\alpha/2}$.

b. Find the value of E.

SOLUTION

a. The 95% degree of confidence corresponds to $\alpha = 0.05$, which results in $z_{\alpha/2} = 1.96$. This result was found in the second example of Section 6-2. (If you are not able to convert a given degree of confidence to the corresponding critical z score, you should return to Section 6-2 and study that procedure.)

b. $E = 2$, because we want the sample mean to be within 2 IQ points of the true population mean μ. That is, the desired error is 2 IQ points. (The only way to have an error of 0 is to give an IQ test to every statistics professor, but that's not practical because they are very busy trying to be effective teachers.)

What If σ Is Not Known?

When applying Formula 6-3, there is one very practical dilemma: The formula requires that we substitute some value for the population standard deviation σ, but in reality, it is usually unknown. There are some ways that we can work around this problem, however.

1. Use the range rule of thumb (see Section 2-5) to estimate the standard deviation as follows: $\sigma \approx \text{range}/4$.
2. Conduct a pilot study by starting the sampling process. Based on the first collection of at least 31 randomly selected sample values, calculate the sample standard deviation s and use it in place of σ. That value can be refined as more sample data are obtained.
3. Estimate the value of σ by using the results of some other study that was done earlier.

In addition, we can sometimes be creative in our use of other known results. For example, IQ tests are typically designed so that the mean is 100 and the standard deviation is 15. Statistics professors have IQ scores with a mean greater than 100 and a standard deviation less than 15 (because they are a more homogeneous group than people randomly selected from the general population). We do not know the specific value of σ for statistics professors, but we can play it safe by using $\sigma = 15$. Using a value for σ that is larger than the correct value will make the sample size larger than necessary, but using a value for σ that is too small would result in a sample size that is inadequate. *When calculating the sample size n, any errors should always be conservative in the sense that they make n too large instead of too small.*

EXAMPLE **IQ Scores of Statistics Professors** Assume that we want to estimate the mean IQ score for the population of statistics professors. How many statistics professors must be randomly selected for IQ tests if we want 95% confidence that the sample mean is within 2 IQ points of the population mean?

SOLUTION The values required for Formula 6-3 are found as follows:

$z_{\alpha/2} = 1.96$ (This is found by converting the 95% degree of confidence to $\alpha = 0.05$, then finding the critical z score as described in Section 6-2.)

$E = 2$ (Because we want the sample mean to be within 2 IQ points of μ, the desired margin of error is 2.)

$\sigma = 15$ (See the discussion that immediately precedes this example.)

With $z_{\alpha/2} = 1.96$, $E = 2$, and $\sigma = 15$, we use Formula 6-3 as follows:

$$n = \left[\frac{z_{\alpha/2}\sigma}{E}\right]^2 = \left[\frac{1.96 \cdot 15}{2}\right]^2 = 216.09 = 217 \text{ (rounded up)}$$

INTERPRETATION Among the thousands of statistics professors, we need to obtain a simple random sample of at least 217 of them, then we need to get their IQ scores. We will be 95% confident that the sample mean $\bar{x}$ will be within 2 IQ points of the true population mean μ.

If we are willing to settle for less accurate results by using a larger margin of error, such as 4, the sample size drops to 54.0225, which is rounded up to 55. Doubling the margin of error causes the required sample size to decrease to one-fourth its original value. Conversely, halving the margin of error quadruples the sample size. Consequently, if you want more accurate results, the sample size must be substantially increased. Because large samples generally require more time and money, there is often a need for a trade-off between the sample size and the margin of error E.

In the next example, the range rule of thumb is used to estimate σ.

EXAMPLE Using the Range Rule of Thumb You plan to estimate the mean selling price of a college textbook. How many textbooks must you sample if you want to be 99% confident that the sample mean is within \$3 of the true population mean μ?

SOLUTION We seek the sample size n given that $\alpha = 0.01$ (from 99% confidence), so $z_{\alpha/2} = 2.575$. We want to be within \$3, so $E = 3$. We don't know the standard deviation σ of all textbook selling prices, but we can estimate σ by using the range rule of thumb. If we reason that typical college textbook prices range from \$10 to \$90, the range is \$80, so that

$$\sigma \approx \frac{\text{range}}{4} = \frac{(90 - 10)}{4} = 20$$

With $z_{\alpha/2} = 2.575$, $E = 3$, and $\sigma \approx 20$, we use Formula 6-3 as follows:

$$n = \left[\frac{z_{\alpha/2}\sigma}{E}\right]^2 = \left[\frac{2.575 \cdot 20}{3}\right]^2 = 294.69444 = 295 \text{ (rounded up)}$$

INTERPRETATION We must randomly select 295 selling prices of college textbooks and then find the value of the sample mean $\bar{x}$. We will be 99% confident that the resulting sample mean is within \$3 of the true mean selling price of all college textbooks.

In the preceding solution, we used the range rule of thumb to estimate σ, but we could also estimate σ by using sample data from another study. For example, Data Set 2 in Appendix B includes textbook prices from two colleges. We could use the sample standard deviation s of one or both of those samples to estimate σ. (See Exercise 10.)

Using Technology

STATDISK: Select **Analysis** from the main menu bar at the top, then select **Sample Size Determination,** followed by **Estimate Mean.** You must now enter the confidence level (such as 0.95), the error E, and the population standard deviation σ. There is also an option that allows you to enter the population size N, assuming that you are sampling without replacement from a finite population. (See Exercise 13.)

Sample size calculations are not included with the TI-83 Plus calculator, or Minitab, or Excel.

6-4 Basic Skills and Concepts

1. Sample Size for Mean IQ of Statistics Students The standard IQ test is designed so that the mean is 100 and the standard deviation is 15 for the population of normal adults. Find the sample size necessary to estimate the mean IQ score of statistics students. We want to be 98% confident that our sample mean is within 1.5 IQ points of the true mean. The mean for this population is clearly greater than 100. The standard deviation for this population is probably less than 15 because it is a group with less variation than a group randomly selected from the general population; therefore, if we use $\sigma = 15$, we are being conservative by using a value that will make the sample size at least as large as necessary. Assume then that $\sigma = 15$ and determine the required sample size.

2. Using Sample Data The Franklin Vending Machine Company must adjust its machines to accept only coins with specified weights. We will obtain a sample of quarters and weigh them to determine the mean. How many quarters must we randomly select and weigh if we want to be 99% confident that the sample mean is within 0.025 g of the true population mean for all quarters? If we use the sample of quarters in Data Set 13 in Appendix B, we can estimate that the population standard deviation is 0.068 g.

3. Estimating Mean Garbage Weight To plan for the proper handling of household garbage, the city of Providence must estimate the mean weight of garbage discarded by households in one week. Find the sample size necessary to estimate that mean if you want to be 96% confident that the sample mean is within 2 lb of the true population mean. For the population standard deviation σ, use the value 12.46 lb, which is the standard deviation of the sample of 62 households included in the Garbage Project study conducted at the University of Arizona.

4. Estimating Mean Weight of Plastic Garbage If we want to estimate the mean weight of plastic discarded by households in one week, how many households must we randomly select if we want to be 99% confident that the sample mean is within 0.250 lb of the true population mean? Data Set 5 in Appendix B includes the weights of plastic discarded for 62 households (based on data from the Garbage Project at the University of Arizona). If we use that sample as a pilot study, we get a standard deviation of $s = 1.065$ lb.

5. Estimating Your Income An economist wants to estimate the mean income for the first year of work for a college graduate who has had the profound wisdom to take a statistics course. How many such incomes must be found if we want to be 95% confident that the sample mean is within \$500 of the true population mean? Assume that a previous study has revealed that for such incomes, $\sigma = \$6250$.

6. Pilot Spatial Perception The College of Newport conducts a study of the spatial perception among commercial aircraft pilots. A new test of spatial perception is developed, and we want to estimate the mean score for all pilots. How many randomly selected pilots must be tested if we want the sample mean to be in error by no more than 5 points, with 96% confidence? A preliminary study suggests that $\sigma = 40$.

7. Students and Television Nielsen Media Research wants to estimate the mean amount of time (in hours) that full-time college students spend watching television each weekday. Find the sample size necessary to estimate that mean with a 0.25 hr (or 15 min) margin of error. Assume that a 96% degree of confidence is desired. Also assume that a pilot study showed that the standard deviation is estimated to be 1.87 hr.

8. Using Range Rule of Thumb You have just been hired by the Boston Marketing Company to conduct a survey to estimate the mean amount of money spent by movie patrons (per movie) in Massachusetts. First use the range rule of thumb to make a rough estimate of the standard deviation of the amounts spent. It is reasonable to assume that typical amounts range from $3 to about $15. Then use that estimated standard deviation to determine the sample size corresponding to 98% confidence and a 25¢ margin of error.

9. Using Range Rule of Thumb Estimate the minimum and maximum ages for typical textbooks currently used in college courses, then use the range rule of thumb to estimate the standard deviation. Next, find the size of the sample required to estimate the mean age (in years) of textbooks currently used in college courses. Assume a 96% degree of confidence that the sample mean will be in error by no more than 0.25 year.

10. Using Sample Data An example in this section used the range rule of thumb to estimate σ, the standard deviation of the prices of college textbooks. That value was then used to find the sample size necessary to estimate the mean price of all such books. Instead of using the range rule of thumb, refer to Data Set 2 in Appendix B.

a. Estimate σ with the value of the standard deviation of the prices of new books at the University of Massachusetts. Use that result to estimate the mean price of new college textbooks. Assume that you want 90% confidence that the sample mean is within $3 of the population mean.

b. Estimate σ with the value of the standard deviation of the prices of new books at Dutchess Community College. Use that result to estimate the mean price of new college textbooks. Assume that you want 90% confidence that the sample mean is within $3 of the population mean.

c. Compare the results from parts (a) and (b) and the result from the example in this section.

11. a. Using Sample Data Find the sample size necessary to estimate the mean daily rainfall amount (in inches) for Boston. Assume that you want 98% confidence that the sample mean is within 0.02 in. of the population mean. Estimate σ by using the sample values for Wednesday rainfall amounts listed in Data Set 17 in Appendix B.

b. Find the sample size necessary to estimate the mean daily rainfall amount for Boston. Assume that you want 98% confidence that the sample mean is within 0.02 in. of the population mean. Estimate σ by using the sample values for Sunday rainfall amounts listed in Data Set 17 in Appendix B.

c. Compare the results from parts (a) and (b).

12. a. Using Sample Data Find the sample size necessary to estimate the mean amount of serum cotinine in *nonsmokers* who have no environmental tobacco smoke exposure at home or at work. Assume that you want to be 90% confident that the sample mean is within 0.3 ng/ml. Estimate σ by using the standard deviation for the sample found in Data Set 9 in Appendix B.

(continued)

b. Find the sample size necessary to estimate the mean amount of serum cotinine in *smokers*. Assume that you want to be 90% confident that the sample mean is within 0.3 ng/ml. Estimate σ by using the standard deviation for the sample found in Data Set 9 in Appendix B.

c. Why are the results from parts (a) and (b) so dramatically different?

6-4 Beyond the Basics

13. Using Finite Population Correction Factor In Formula 6-1 for the margin of error E, we assume that the population is infinite, that we are sampling with replacement, or that the population is very large. If we have a relatively small population and sample without replacement, we should modify E to include a *finite population correction factor* as follows:

$$E = z_{\alpha/2}\frac{\sigma}{\sqrt{n}}\sqrt{\frac{N-n}{N-1}}$$

where N is the population size. Show that the preceding expression can be solved for n to yield

$$n = \frac{N\sigma^2(z_{\alpha/2})^2}{(N-1)E^2 + \sigma^2(z_{\alpha/2})^2}$$

Repeat Exercise 1, assuming that the statistics students are randomly selected without replacement from a population of $N = 200$.

14. Using Finite Population Correction Factor Repeat Exercise 4 assuming that a community consists of 150 households and that all sampling will be done without replacement. Because this sampling is done without replacement from a finite population, use the expression for sample size that is given in Exercise 13.

15. Finding Margin of Error It is found that a sample size of 843 is necessary to estimate the mean weight (in grams) of the sugar in packets supplied by Domino. That sample size is based on a 95% degree of confidence and a population standard deviation that is estimated by the sample standard deviation for Data Set 4 in Appendix B. Find the margin of error E.

6-5 Estimating a Population Proportion

In this chapter we introduce a major function of inferential statistics: estimating the value of a population parameter. In the preceding sections we described (1) point estimates, (2) confidence interval estimates, and (3) a procedure for determining sample size. Our objective in those sections was to estimate a population mean μ. In this section we will take those same three concepts, but will now apply them to the population proportion p. Here are the main objectives of this section:

1. Given sample data with each item belonging to one of two categories, estimate the population proportion p for one of those categories.
2. Determine the sample size required to estimate a population proportion p.

TV Sample Sizes

A *Newsweek* article described the use of people meters as a way of determining how many people are watching different television programs. The article stated, "Statisticians have long argued that the household samples used by the rating services are simply too small to accurately determine what America is watching. In that light, it may be illuminating to note that the 4000 homes reached by the people meters constitute exactly 0.0045% of the wired nation." This implied claim that the sample size of 4000 homes is too small is not valid. Methods of this chapter can be used to show that a random sample size of 4000 can provide results that are quite good, even though the sample might be only 0.0045% of the population.

Television ratings are important because they affect the billions of dollars spent for advertising. *Super Bowl* commercials are among the most expensive, and charges for commercials are based on a Nielsen Media Research estimate of the proportion of households that tune in. Nielsen must address two important questions: (1) How many households must be sampled, and (2) after the sample data have been collected, what is the estimate of the proportion of households tuned to the *Super Bowl*? This section provides the tools that you need to answer such questions.

Estimating the Proportion p

We begin by considering methods for using sample data to estimate the population proportion p. (Later in this section, we will consider the issue of determining sample size.) The following assumptions apply.

Assumptions

1. The sample is a simple random sample.
2. The conditions for the binomial distribution are satisfied. (See Section 4-3.) That is, there is a fixed number of trials, the trials are independent, there are two categories of outcomes, and the probabilities remain constant for each trial.
3. The normal distribution can be used to approximate the distribution of sample proportions because $np \geq 5$ and $nq \geq 5$ are both satisfied. (Because p and q are unknown, we use the sample proportion to estimate their values. Also, there are procedures for dealing with situations in which the normal distribution is not a suitable approximation. See Exercise 40.)

We now introduce the new notation $\hat{p}$ (called "p hat") for the sample proportion.

Notation for Proportions

p = *population* proportion

$\hat{p} = \dfrac{x}{n}$ = *sample* proportion of x successes in a sample of size n

$\hat{q} = 1 - \hat{p}$ = *sample* proportion of x failures in a sample of size n

Probability and Percent Although this chapter focuses on the population proportion p, the procedures discussed here can also be applied to probabilities or percentages, but percentages must be converted to proportions by dropping the percent sign and dividing by 100. For example, the percentage of people who don't buy books, 48.7%, can be expressed in decimal form as 0.487. The symbol p may therefore represent a proportion, a probability, or the decimal equivalent of a percent. For example, if you survey 200 statistics students and find that 80 of them have purchased TI-83 Plus calculators, then the sample proportion is $\hat{p} = x/n = 80/200 = 0.400$, and $\hat{q} = 0.600$ (calculated from $1 - 0.400$). Instead of computing the value of x/n, the value of $\hat{p}$ is sometimes already known because the sample proportion or percentage is given directly. For example, if it is reported that 1068 American television viewers are surveyed and 25% of them are college graduates, then $\hat{p} = 0.25$ and $\hat{q} = 0.75$.

If we want to estimate a population proportion with a single value, the best estimate is $\hat{p}$. As in Sections 6-2 and 6-3, when an estimate of a population parameter is a single value, we call it a *point estimate*.

The sample proportion $\hat{p}$ is the best point estimate of the population proportion p.

We use $\hat{p}$ as the point estimate of p (just as $\bar{x}$ is used as the point estimate of μ) because it is unbiased and is the most consistent of the estimators that could be used. It is unbiased in the sense that the distribution of sample proportions tends to center about the value of p; that is, sample proportions $\hat{p}$ do not systematically tend to underestimate p, nor do they systematically tend to overestimate p. The sample proportion $\hat{p}$ is the most consistent estimator in the sense that the standard deviation of sample proportions tends to be smaller than the standard deviation of any other unbiased estimators.

The point estimate has a major disadvantage: Although the value of $\hat{p}$ is the *best* single value estimate of p, it gives no indication of just how good it is. What we therefore need is an estimate that *does* give us a sense of how accurate it is, and the *confidence interval* meets that objective. We first present the *margin of error* (defined in Section 6-2), which is used for finding a confidence interval, and then we present the format of the confidence interval itself.

Margin of Error of the Estimate of p

Formula 6-4 $$E = z_{\alpha/2}\sqrt{\frac{\hat{p}\hat{q}}{n}}$$

Confidence Interval (or Interval Estimate) for the Population Proportion p

$$\hat{p} - E < p < \hat{p} + E \qquad \text{where } E = z_{\alpha/2}\sqrt{\frac{\hat{p}\hat{q}}{n}}$$

The confidence interval is sometimes expressed in the following formats.

$$p = \hat{p} \pm E$$

or

$$(\hat{p} - E, \hat{p} + E)$$

Push Polling

"Push polling" is the practice of political campaigning under the guise of a poll. Its name is derived from its objective of pushing voters away from opposition candidates by asking loaded questions designed to discredit them. Here's an example of one such question that was used: "Please tell me if you would be more likely or less likely to vote for Roy Romer if you knew that Gov. Romer appoints a parole board which has granted early release to an average of four convicted felons per day every day since Romer took office." The National Council on Public Polls characterizes push polls as unethical, but some professional pollsters do not condemn the practice as long as the questions do not include outright lies.

In Chapter 3, when probabilities were given in decimal form, we rounded to three significant digits. We use that same rounding rule here.

Round-Off Rule for Confidence Interval Estimates of p

Round the confidence interval limits to three significant digits.

EXAMPLE **Misleading Survey Responses** Do people lie about voting? In a survey of 1002 people, 701 people said that they voted in the recent presidential election (based on data from ICR Research Group). Voting records show that 61% of eligible voters actually did vote. Using these survey results,

a. find the point estimate of the proportion of people who say that they voted.

b. find the 95% confidence interval estimate of the proportion of people who say that they voted.

c. determine whether the survey results are consistent with the actual voter turnout of 61%.

SOLUTION

a. The point estimate of p is

$$\hat{p} = \frac{x}{n} = \frac{701}{1002} = 0.6996$$

b. Constructing the confidence interval requires that we first evaluate the margin of error E. The value of E can be found from Formula 6-4. We use $\hat{p} = 0.6996$ (found in part a), $\hat{q} = 0.3004$ (from $\hat{q} = 1 - \hat{p}$), and $z_{\alpha/2} = 1.96$ (from Table A-2, where 95% converts to $\alpha = 0.05$, which is divided equally between the two tails, so $z = 1.96$ is the boundary for the related area of 0.4750).

$$E = z_{\alpha/2}\sqrt{\frac{\hat{p}\hat{q}}{n}} = 1.96\sqrt{\frac{(0.6996)(0.3004)}{1002}} = 0.0283855$$

We can now find the confidence interval by using $\hat{p} = 0.6996$ and $E = 0.0283855$.

$$\hat{p} - E < p < \hat{p} + E$$
$$0.6996 - 0.0283855 < p < 0.6996 + 0.0283855$$
$$0.671 < p < 0.728 \text{ (rounded)}$$

This same result could be expressed in the format of $p = 0.700 \pm 0.028$ or $(0.671, 0.728)$. If we want the 95% confidence interval for the true population *percentage*, we could express the result as $67.1\% < p < 72.8\%$. This confidence interval is often reported with a statement such as this: "Among those eligible to vote, the percentage who say they voted is estimated to be 70%, with a margin of error of plus or minus 2.8 percentage points." That statement is a verbal expression of this format for the confidence interval: $p = 70\% \pm 2.8\%$. (The level of confidence should also be reported, but it rarely is in the media. The media typically use a 95% degree of confidence but omit any reference to it.)

c. Based on the survey results, we are 95% confident that the limits of 67.1% and 72.8% contain the true percentage of eligible voters who *say* that they voted. But we know that 61% of the eligible voters actually did vote. Because 61% does not fall within the 95% confidence interval, we can conclude that the percentage of people who say that they voted appears to be different than the percentage of people who actually did vote.

Interpreting a Confidence Interval As in Sections 6-2 and 6-3, we must be careful to interpret the confidence interval correctly. Examples of correct and wrong interpretations follow.

Correct: We are 95% confident that the interval from 0.671 to 0.728 actually does contain the true value of p. (This means that if we were to select many different samples of size 1002 and construct the confidence intervals as we did in the preceding example, 95% of them would actually contain the value of the population proportion p.)

Wrong: There is a 95% chance that the true value of p will fall between 0.671 and 0.728.

The differences in wording between the correct and the wrong statements might seem subtle, but the real difference is substantial. The "wrong" statement is incorrect because the population proportion p is a fixed value, not a random variable. Either p falls between the confidence interval limits or it does not, and there is no probability or chance involved here.

Rationale for the Margin of Error Because the sampling distribution of proportions is approximately normal (because the conditions $np \geq 5$ and $nq \geq 5$ are both satisfied), we can use results from Section 5-6 to conclude that μ and σ are given by $\mu = np$ and $\sigma = \sqrt{npq}$. Both of these parameters pertain to n trials, but we convert them to a per-trial basis by dividing by n as follows:

$$\text{Mean of sample proportions: } \mu = \frac{np}{n} = p$$

$$\text{Standard deviation of sample proportions: } \sigma = \frac{\sqrt{npq}}{n} = \sqrt{\frac{npq}{n^2}} = \sqrt{\frac{pq}{n}}$$

The first result may seem trivial, because we already stipulated that the true population proportion is p. The second result is nontrivial and is useful in describing the margin of error E, but we replace the product pq by $\hat{p}\hat{q}$ because we don't yet know the value of p (it is the value we are trying to estimate). Formula 6-4 for the margin of error reflects the fact that $\hat{p}$ has a probability of $1 - \alpha$ of being within $z_{\alpha/2}\sqrt{pq/n}$ of p. The confidence interval for p, as given previously, reflects the fact that there is a probability of $1 - \alpha$ that $\hat{p}$ differs from p by less than the margin of error $E = z_{\alpha/2}\sqrt{\hat{p}\hat{q}/n}$.

Determining Sample Size

Suppose we want to collect sample data with the objective of estimating some population proportion. How do we know *how many* sample items must be obtained? If we take the expression for the margin of error E (Formula 6-4), then solve for n, we get Formula 6-5. Formula 6-5 requires $\hat{p}$ as an estimate of the population proportion p, but if no such estimate is known (as is usually the case), we replace $\hat{p}$ by 0.5 and replace $\hat{q}$ by 0.5, with the result given in Formula 6-6.

Sample Size for Estimating Proportion *p*

When an estimate $\hat{p}$ is known: **Formula 6-5** $$n = \frac{[z_{\alpha/2}]^2 \hat{p}\hat{q}}{E^2}$$

When no estimate $\hat{p}$ is known: **Formula 6-6** $$n = \frac{[z_{\alpha/2}]^2 \cdot 0.25}{E^2}$$

Round-Off Rule for Determining Sample Size

If the computed sample size is not a whole number, round it up to the next *higher* whole number.

Use Formula 6-5 when reasonable estimates of $\hat{p}$ can be made by using previous samples, a pilot study, or someone's expert knowledge. When no such estimate can be made, we assign the value of 0.5 to each of $\hat{p}$ and $\hat{q}$, so the resulting sample size will be at least as large as it should be. The underlying reason for the assignment of 0.5 is this: The product $\hat{p} \cdot \hat{q}$ has 0.25 as its largest possible value, which occurs when $\hat{p} = 0.5$ and $\hat{q} = 0.5$. (Try experimenting with different values of $\hat{p}$ to verify that $\hat{p} \cdot \hat{q}$ has 0.25 as the largest possible value.) Note that Formulas 6-5 and 6-6 do not include the population size N, so the size of the population is irrelevant. (*Exception:* When sampling is without replacement from a relatively small finite population. See Exercise 35.)

EXAMPLE **E-Mail** The use of answering machines, fax machines, voice mail, and e-mail is growing rapidly, and they are having a dramatic effect on the way we communicate. Suppose a sociologist wants to determine the current percentage of U.S. households using e-mail. How many households must be surveyed in order to be 90% confident that the sample percentage is in error by no more than four percentage points?

a. Use this result from an earlier study: In 1997, 16.9% of U.S. households used e-mail (based on data from *The World Almanac and Book of Facts*).

b. Assume that we have no prior information suggesting a possible value of $\hat{p}$.

SOLUTION

a. The prior study suggests that $\hat{p} = 0.169$, so $\hat{q} = 0.831$ (found from $\hat{q} = 1 - 0.169$). With a 90% level of confidence, we have $\alpha = 0.10$, so $z_{\alpha/2} = 1.645$. Also, the margin of error is $E = 0.04$ (the decimal equivalent of "four percentage points"). Because we have an estimated value of $\hat{p}$, we use Formula 6-5 as follows.

$$n = \frac{[z_{\alpha/2}]^2 \hat{p}\hat{q}}{E^2} = \frac{[1.645]^2 (0.169)(0.831)}{0.04^2}$$
$$= 237.51965 = 238 \quad \text{(rounded up)}$$

We must survey at least 238 randomly selected households.

b. As in part (a), we again use $z_{\alpha/2} = 1.645$ and $E = 0.04$, but with no prior knowledge of $\hat{p}$ (or $\hat{q}$), we use Formula 6-6 as follows.

$$n = \frac{[z_{\alpha/2}]^2 \cdot 0.25}{E^2} = \frac{[1.645]^2 \cdot (0.25)}{0.04^2}$$
$$= 422.81641 = 423 \quad \text{(rounded up)}$$

INTERPRETATION To be 90% confident that our sample percentage is within four percentage points of the true percentage for all households, we should randomly select and survey 423 households. By comparing this result to the sample size of 238 found in part (a), we can see that if we have no knowledge of a prior study, a larger sample is required to achieve the same results as when the value of $\hat{p}$ can be estimated. But now let's use common sense: We know that the use of e-mail is growing so rapidly that the 1997 estimate is too old to be of much use. Today, substantially more than 16.9% of households use e-mail. Realistically, we need a sample larger than 238 households. Assuming that we don't really know the current rate of e-mail usage, we should randomly select 423 households. With 423 households, we will be 90% confident that we are within four percentage points of the true percentage of households using e-mail.

Common Errors When calculating sample size using Formula 6-5 or 6-6, be sure to substitute the critical z score for $z_{\alpha/2}$. For example, if you are working with 95% confidence, be sure to replace $z_{\alpha/2}$ with 1.96. (Here is the logical sequence: $95\% \Rightarrow \alpha = 0.05 \Rightarrow z_{\alpha/2} = 1.96$ found from Table A-2.) Don't make the mistake of replacing $z_{\alpha/2}$ with 0.95 or 0.05. Also, don't make the mistake of using $E = 3$ as the margin of error corresponding to "three percentage points." When using Formula 6-5 or 6-6, the value of E never exceeds 1, and it's typically a number like 0.03. This error causes the sample size to be 1/10,000th of what it should be, so that you might end up with a sample size of only 1 when the answer is rounded up. You really can't estimate a population proportion by surveying only one person, regardless of how knowledgeable that person claims to be.

Population Size Part (b) of the preceding example involved application of Formula 6-6, the same formula frequently used by Nielsen, Gallup, and other professional pollsters. Many people incorrectly believe that the sample size should be some percentage of the population, but Formula 6-6 shows that the population size is irrelevant. (In reality, the population size is sometimes used, but only in cases in which we sample without replacement from a relatively small population. See Exercise 35.) Most of the polls featured in newspapers, magazines, and broadcast media involve sample sizes in the range of 1000 to 2000. Even though such polls may involve a very small percentage of the total population, they can provide results that are quite good. When Nielsen surveys 4000 TV households from a population of 104 million households, only 0.004% of the households are surveyed; still, we can be 95% confident that the sample percentage will be within one percentage point of the true population percentage.

Polls are now an integral part of American life. They affect the television shows we watch, the leaders we elect, the legislation that governs us, and the products we consume. An understanding of the concepts of this section is truly essential in this new millennium.

Finding the Point Estimate and *E* from a Confidence Interval As in the preceding sections, if we use software or a calculator to find confidence interval limits, the sample proportion $\hat{p}$ and the margin of error E can be found as follows.

Point estimate of p:

$$\hat{p} = \frac{(\text{upper confidence interval limit}) + (\text{lower confidence interval limit})}{2}$$

Margin of error:

$$E = \frac{(\text{upper confidence interval limit}) - (\text{lower confidence interval limit})}{2}$$

EXAMPLE A journal article includes this confidence interval estimate of the population proportion: $0.400 < p < 0.500$. Use the given confidence interval limits to find the point estimate $\hat{p}$ and the margin of error E.

SOLUTION

$$\hat{p} = \frac{(\text{upper confidence interval limit}) + (\text{lower confidence interval limit})}{2}$$

$$= \frac{0.500 + 0.400}{2} = 0.450$$

$$E = \frac{(\text{upper confidence interval limit}) - (\text{lower confidence interval limit})}{2}$$

$$= \frac{0.500 - 0.400}{2} = 0.050$$

AS

Using Technology for Confidence Intervals

STATDISK: Select **Analysis,** then **Confidence Intervals,** then **Population Proportion,** and proceed to enter the requested items.

Minitab: Select **Stat, Basic Statistics,** then **1 Proportion.** In the dialog box, click on the button for **Summarizing Data.** Also click on the **Options** button, enter the desired confidence level (the default is 95%), and click on the box with this statement: "Use test and interval based on normal distribution."

Excel: Use the Data Desk XL add-in that is a supplement to this book. First enter the number of successes in cell A1, then enter the total number of trials in cell B1. Click on **DDXL** and select **Confidence Intervals,** then select **Summ 1 Var Prop Interval** (which is an abbreviated form of "confidence interval for a proportion using summary data for one variable"). Click on the pencil icon for "Num successes"and enter A1. Click on the pencil icon for "Num trials"and enter B1. Click on **OK**. In the dialog box, select the level of confidence, then click on **Compute Interval**.

TI-83 Plus: Press **STAT,** select **TESTS,** then select **1-PropZInt** and proceed to enter the required items.

Using Technology for Sample Size Determination

STATDISK: Select **Analysis,** then **Sample Size Determination,** then **Estimate Proportion.** Proceed to enter the required items in the dialog box.

Sample size determination is not available as a built-in function with Minitab, Excel, or TI-83 Plus.

6-5 Basic Skills and Concepts

Interpreting Confidence Interval Limits. *In Exercises 1–4, use the given confidence interval limits to find the point estimate $\hat{p}$ and the margin of error E.*

1. $0.800 < p < 0.840$

2. $(0.444, 0.474)$

3. $(0.432, 0.455)$

4. $0.560 < p < 0.598$

Finding Margin of Error. *In Exercises 5–10, assume that a sample is used to estimate a population proportion p. Find the margin of error E that corresponds to the given statistics and the degree of confidence.*

5. $n = 500, x = 400, 95\%$

6. $n = 1000, x = 250, 99\%$

7. $n = 1068, x = 237, 98\%$

8. $n = 777, x = 543, 90\%$

9. 95% confidence; the sample size is 200, of which 35% are successes

10. 99% confidence; the sample size is 550, of which 90% are successes

Constructing Confidence Intervals. *In Exercises 11–14, use the given sample data and degree of confidence to construct the confidence interval estimate of the population proportion p.*

11. $n = 750, x = 150$, 99% confidence

12. $n = 1200, x = 800$, 90% confidence

13. $n = 1357, x = 222$, 95% confidence

14. $n = 3622, x = 412$, 98% confidence

Determining Sample Size. *In Exercises 15–18, use the given data to find the minimum sample size required to estimate a population proportion or percentage.*

15. Margin of error: 0.025; confidence level: 99%; $\hat{p}$ and $\hat{q}$ unknown

16. Margin of error: 0.04; confidence level: 98%; $\hat{p}$ and $\hat{q}$ unknown

17. Margin of error: three percentage points; confidence level: 95%; $\hat{p}$ is estimated to be 0.15 from a prior study

18. Margin of error: four percentage points; confidence level: 97%; $\hat{p}$ is estimated to be 0.33 from a prior study

TI-83 Plus

```
1-PropZInt
 (.93053,.94926)
 p̂=.9398971001
 n=4276
```

19. Interpreting Calculator Display In 1920 only 35% of U.S. households had telephones, but that rate is now much higher. A recent survey of 4276 randomly selected households showed that 4019 of them had telephones (based on data from the U.S. Census Bureau). Using those survey results and a 99% confidence level, the TI-83 Plus calculator display is as shown.

a. Write a statement that interprets the confidence interval.

b. Based on the preceding result, should pollsters be concerned about results from surveys conducted by telephone?

TI-83 Plus

```
1-PropZInt
 (.61292,.66708)
 p̂=.64
 n=850
```

20. Interpreting Calculator Display The Hartford Insurance Company wants to estimate the percentage of drivers who change tapes or CDs while driving. A random sample of 850 drivers results in 544 who change tapes or CDs while driving (based on data from *Prevention* magazine). Using the sample data and a 90% confidence level, the TI-83 Plus calculator display is as shown.

a. Write a statement that interprets the confidence interval.

b. Based on the preceding result, does the practice of changing tapes or CDs while driving appear to be a threat to safety?

21. Confidence Interval for Theme Park Attendance Each year, billions of dollars are spent at theme parks owned by Disney, Universal Studios, Sea World, Busch Gardens, and others. A survey of 1233 people who took trips revealed that 111 of them included a visit to a theme park (based on data from the Travel Industry Association of America).

a. Find the point estimate of the *percentage* of people who visit a theme park when they take a trip.

b. Find a 95% confidence interval estimate of the *percentage* of all people who visit a theme park when they take a trip.

22. Confidence Interval for Voter Surveys In a recent presidential election, 611 voters were surveyed and 308 of them said that they voted for the candidate who won (based on data from the ICR Survey Research Group).

a. Find the point estimate of the *percentage* of voters who said that they voted for the candidate who won.

b. Find a 98% confidence interval estimate of the *percentage* of voters who said that they voted for the candidate who won.

c. Of those who voted, 43% actually voted for the candidate who won. Is this result consistent with the survey results? How might a discrepancy be explained?

23. Pilot Fatalities Researchers studied crashes of general aviation (noncommercial and nonmilitary) airplanes and found that pilots died in 5.2% of 8411 crash landings (based on data from "Risk Factors for Pilot Fatalities in General Aviation Airplane Crash Landings," by Rostykus, Cummings, and Mueller, *Journal of the American Medical Association*, Vol. 280, No. 11).

a. Construct a 95% confidence interval estimate of the percentage of pilot deaths in all general aviation crashes.

b. Among crashes with an explosion or fire on the ground, the pilot fatality rate is estimated by the 95% confidence interval of (15.5%, 26.9%). Is this result substantially different from the result from part (a)? What can you conclude about an explosion or fire as a risk factor?

24. Effectiveness of Gender Selection In a study of the Clark method of gender selection, 40 couples tried to have baby girls. Among the 40 babies, 62.5% were girls.

a. Construct a 95% confidence interval estimate for the proportion of girls from all couples who try to have baby girls with the Clark method of gender selection.

b. Based on the result from part (a), can we conclude that the Clark method is effective, with a rate of girls that is greater than 50%?

25. Drug Testing The drug Ziac is used to treat hypertension. In a clinical test, 3.2% of 221 Ziac users experienced dizziness (based on data from Lederle Laboratories).

a. Construct a 99% confidence interval estimate of the percentage of all Ziac users who experience dizziness.

b. In the same clinical test, people in the placebo group didn't take Ziac, but 1.8% of them reported dizziness. Based on the result from part (a), what can we conclude about dizziness as an adverse reaction to Ziac?

26. Smoking and College Education The tobacco industry closely monitors all surveys that involve smoking. One survey showed that among 785 randomly selected subjects who completed four years of college, 18.3% smoke (based on data from the American Medical Association). *(continued)*

a. Construct the 98% confidence interval for the true percentage of smokers among all people who completed four years of college.

b. Based on the result from part (a), does the smoking rate for those with four years of college appear to be substantially different than the 27% rate for the general population?

27. Sample Size for School Computer Usage A researcher wants to estimate the percentage of students aged 12–18 who use computers in school. How many randomly selected students must be surveyed if she wants to be 98% confident that the margin of error is five percentage points?

a. Assume that we have an estimate of $\hat{p}$ found from a prior study that revealed a percentage of 82% (based on data from the Consumer Electronics Manufacturers Association).

b. Assume that we have no prior information suggesting a possible value of $\hat{p}$.

28. Sample Size for Left-Handed Golfers As a manufacturer of golf equipment, the Spalding Corporation wants to estimate the proportion of golfers who are left-handed. (The company can use this information in planning for the number of right-handed and left-handed sets of golf clubs to make.) How many golfers must be surveyed if we want 99% confidence that the sample proportion has a margin of error of 0.025?

a. Assume that there is no available information that could be used as an estimate of $\hat{p}$.

b. Assume that we have an estimate of $\hat{p}$ found from a previous study that suggests that 15% of golfers are left-handed (based on a *USA Today* report).

29. Sample Size for Motor Vehicle Ownership You have been hired by the Ford Motor Company to do market research, and you must estimate the percentage of households in which a vehicle is owned. How many households must you survey if you want to be 94% confident that your sample percentage has a margin of error of three percentage points?

a. Assume that a previous study suggested that vehicles are owned in 86% of households.

b. Assume that there is no available information that can be used to estimate the percentage of households in which a vehicle is owned.

30. Sample Size for Weapons on Campus Concerned about campus safety, college officials want to estimate the percentage of students who carry a gun, knife, or other such weapon. How many randomly selected students must be surveyed in order to be 95% confident that the sample percentage has a margin of error of three percentage points?

a. Assume that another study indicated that 7% of college students carry weapons (based on a study by Cornell University).

b. Assume that there is no available information that can be used to estimate the percentage of college students carrying weapons.

31. Color Blindness In a study of perception, 80 men are tested and 7 are found to have red/green color blindness (based on data from *USA Today*).

a. Construct a 90% confidence interval estimate of the proportion of all men with this type of color blindness.

b. What sample size would be needed to estimate the proportion of male red/green color blindness if we wanted 96% confidence that the sample proportion is in error by no more than 0.03? Use the sample proportion as a known estimate.

(continued)

c. Women have a 0.25% rate of red/green color blindness. Based on the result from part (a), can we safely conclude that women have a lower rate of red/green color blindness than men?

32. TV Ratings Nielsen ratings showed that *60 Minutes* received a 22 market share, meaning that among the TV sets in use at the time, 22% were tuned to *60 Minutes*. Assume that this is based on a sample size of 4000 (typical for Nielsen surveys).

a. Construct a 97% confidence interval estimate of the proportion of all sets in use that were tuned to *60 Minutes* at the time of the broadcast.

b. What sample size would be required to estimate the percentage of sets tuned to *60 Minutes* if we wanted 99% confidence that the sample percentage is in error by no more than one-half of one percentage point? (Assume that we have no estimate of the proportion.)

c. At the time of this particular *60 Minutes* broadcast, ABC ran "Exposed: Pro Wrestling," and that show received a share of 11. Based on the result from part (a), can we conclude that *60 Minutes* had a greater proportion of viewers? Did professional wrestling really need to be exposed?

33. Red M&M Candies Refer to Data Set 10 in Appendix B and find the sample proportion of M&Ms that are red. Use that result to construct a 95% confidence interval estimate of the population percentage of M&Ms that are red. Is the result consistent with the 20% rate that is reported by the candy maker Mars?

34. Credit Cards Refer to Data Set 14 in Appendix B.

a. Find the sample proportion of statistics students who have at least one credit card.

b. Use the sample proportion from part (a) to construct a 95% confidence interval estimate of the proportion of all statistics students who have at least one credit card.

c. Find the sample size necessary to estimate the proportion of all statistics students who have at least one credit card. Use 95% confidence and a 0.03 margin of error. Also use the sample proportion from part (a) as an estimate of the proportion of statistics students who have at least one credit card.

6-5 Beyond the Basics

35. Using Finite Population Correction Factor This section presented Formulas 6-5 and 6-6, which are used for determining sample size. In both cases we assumed that the population is infinite or very large and that we are sampling with replacement. When we have a relatively small population with size N and sample without replacement, we modify E to include the *finite population correction factor* shown here, and we can solve for n to obtain the result given here. Use this result to repeat part (b) of Exercise 32, assuming that we limit our population to a town with 5000 television sets in use.

$$E = z_{\alpha/2}\sqrt{\frac{\hat{p}\hat{q}}{n}}\sqrt{\frac{N-n}{N-1}} \qquad n = \frac{N\hat{p}\hat{q}[z_{\alpha/2}]^2}{\hat{p}\hat{q}[z_{\alpha/2}]^2 + (N-1)E^2}$$

36. Poll Accuracy A *New York Times* article about poll results states, "In theory, in 19 cases out of 20, the results from such a poll should differ by no more than one percentage point in either direction from what would have been obtained by interviewing all voters in the United States." Find the sample size suggested by this statement.

37. Fax Machine Ownership Because a proposed survey is time-consuming, an enterprising pollster posts it on the Internet and promises free software to everyone who responds by completing the survey. Results include 2250 responses, and 80% of them indicate that a fax machine is owned. Construct a 95% confidence interval for the percentage of all people who have a fax machine. Interpret the result.

38. Women's Heights Women's heights are normally distributed with a mean of 63.6 in. and a standard deviation of 2.5 in. How many women must be surveyed if we want to estimate the percentage who are taller than 5 ft? Assume that we want 98% confidence that the error is no more than 2.5 percentage points. (*Hint:* The answer is substantially smaller than 2172.)

39. One-Sided Confidence Interval A *one-sided confidence interval* for p can be written as $p < \hat{p} + E$ or $p > \hat{p} - E$, where the margin of error E is modified by replacing $z_{\alpha/2}$ with z_{α}. If Air America wants to report an on-time performance of at least x percent with 95% confidence, construct the appropriate one-sided confidence interval and then find the percent in question. Assume that a simple random sample of 750 flights results in 630 that are on time.

40. Confidence Interval from Small Sample Special tables are available for finding confidence intervals for proportions involving small numbers of cases, where the normal distribution approximation cannot be used. For example, given three successes among eight trials, the 95% confidence interval found in *Standard Probability and Statistics Tables and Formulae* (CRC Press) is $0.085 < p < 0.755$. Find the confidence interval that would result if you were to use the normal distribution incorrectly as an approximation to the binomial distribution. Are the results reasonably close?

41. Interpreting Confidence Interval Limits Assume that a coin is modified so that it favors heads, and 100 tosses result in 95 heads. Find the 99% confidence interval estimate of the proportion of heads that will occur with this coin. What is unusual about the results obtained by the methods of this section? Does common sense suggest a modification of the resulting confidence interval?

42. Rule of Three Suppose n trials of a binomial experiment result in no successes. According to the *Rule of Three*, we have 95% confidence that the true population proportion has an upper bound of $3/n$. (See "A Look at the Rule of Three," by Jovanovic and Levy, *American Statistician*, Vol. 51, No. 2.)

a. If n independent trials result in no successes, why can't we find confidence interval limits by using the methods described in this section?

b. If 20 patients are treated with a drug and there are no adverse reactions, what is the 95% upper bound for p, the proportion of all patients who experience adverse reactions to this drug?

6-6 Estimating a Population Variance

In this section we consider the same three concepts introduced earlier in this chapter: (1) point estimate, (2) confidence interval, and (3) determining the required sample size. Whereas the preceding sections applied these concepts to estimates of means and proportions, this section applies them to the population variance σ^2 or standard deviation σ. Here are the main objectives of this section:

1. Given sample values, estimate the population standard deviation σ or the population variance σ^2.
2. Determine the sample size required to estimate a population standard deviation or variance.

Many real situations, such as quality control in a manufacturing process, require that we estimate values of population variances or standard deviations. In addition to making products with measurements yielding a desired mean, the manufacturer must make products of *consistent* quality that do not run the gamut from extremely good to extremely poor. As this consistency can often be measured by the variance or standard deviation, these become vital statistics in maintaining the quality of products and services.

Assumptions

1. The sample is a simple random sample.
2. The population must have normally distributed values (even if the sample is large).

The assumption of a normally distributed population was made in earlier sections, but that requirement is more critical here. For the methods of this section, the use of populations with very nonnormal distributions can lead to gross errors. Consequently, the requirement to have a reasonably normal distribution is much more strict, and we should check the distribution of data by constructing histograms and quantile plots, as described in Section 5-7.

When we considered estimates of means and proportions, we used the normal and Student t distributions. When developing estimates of variances or standard deviations, we use another distribution, referred to as the chi-square distribution. We will examine important features of that distribution before proceeding with the development of confidence intervals.

Chi-Square Distribution

In a normally distributed population with variance σ^2, we randomly select independent samples of size n and compute the sample variance s^2 (see Formula 2-5) for each sample. The sample statistic $\chi^2 = (n-1)s^2/\sigma^2$ has a distribution called the **chi-square distribution.**

Chi-Square Distribution

Formula 6-7 $$\chi^2 = \frac{(n-1)s^2}{\sigma^2}$$

where

n = sample size

s^2 = sample variance

σ^2 = population variance

We denote chi-square by χ^2, pronounced "kigh square." (The specific mathematical equations used to define this distribution are not given here because they are beyond the scope of this text.) To find critical values of the chi-square distribution, refer to Table A-4. The chi-square distribution is determined by the number of degrees of freedom, and in this chapter we use $n - 1$ degrees of freedom.

$$\textbf{degrees of freedom} = n - 1$$

In later chapters we will encounter situations in which the degrees of freedom are not $n - 1$, so we should not make the incorrect generalization that the number of degrees of freedom is always $n - 1$.

Properties of the Distribution of the Chi-Square Statistic

1. The chi-square distribution is not symmetric, unlike the normal and Student t distributions (see Figure 6-7). (As the number of degrees of freedom increases, the distribution becomes more symmetric, as Figure 6-8 illustrates.)
2. The values of chi-square can be zero or positive, but they cannot be negative (see Figure 6-7).
3. The chi-square distribution is different for each number of degrees of freedom (see Figure 6-8), and the number of degrees of freedom is given

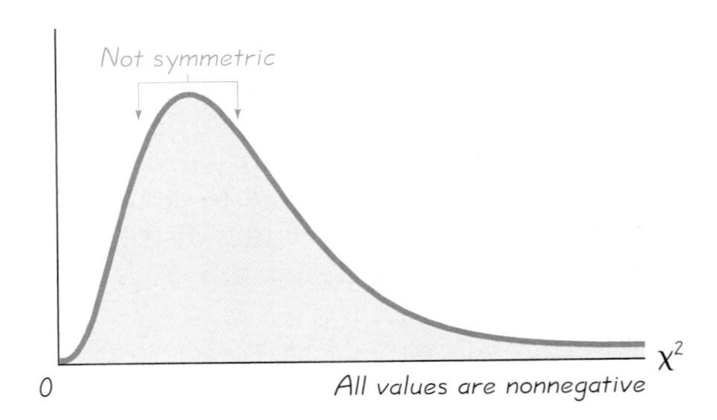

FIGURE 6-7
Chi-Square Distribution

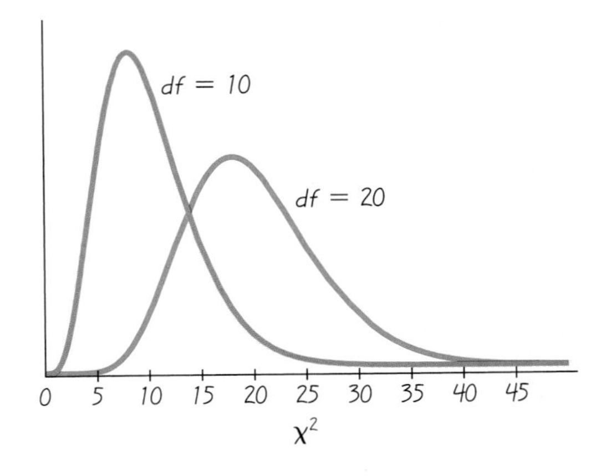

FIGURE 6-8
Chi-Square Distribution for df = 10 and df = 20

by df $= n - 1$ in this section. As the number of degrees of freedom increases, the chi-square distribution approaches a normal distribution.

Because the chi-square distribution is skewed instead of symmetric, the confidence interval does not fit a format of $s^2 \pm E$, and we must do separate calculations for the upper and lower confidence interval limits. There is a different procedure for finding critical values, illustrated in the following example. Note the following essential feature of Table A-4:

In Table A-4, each critical value of χ^2 corresponds to an area given in the top row of the table, and that area represents the *total region located to the right* of the critical value.

EXAMPLE **Critical Values** Find the critical values of χ^2 that determine critical regions containing an area of 0.025 in each tail. Assume that the relevant sample size is 10 so that the number of degrees of freedom is $10 - 1$, or 9.

SOLUTION See Figure 6-9 and refer to Table A-4. The critical value to the right ($\chi^2 = 19.023$) is obtained in a straightforward manner by locating 9 in the degrees-of-freedom column at the left and 0.025 across the top. The critical value of $\chi^2 = 2.700$ to the left once again corresponds to 9 in the degrees-of-freedom column, but we must locate 0.975 (found by subtracting 0.025 from 1) across the top because the values in the top row are always *areas to the right* of the critical value. Refer to Figure 6-9 and see that the total area to the right of $\chi^2 = 2.700$ is 0.975. Figure 6-9 shows that, for a sample of 10 values taken from a normally distributed population, the chi-square statistic $(n - 1)s^2/\sigma^2$ has a 0.95

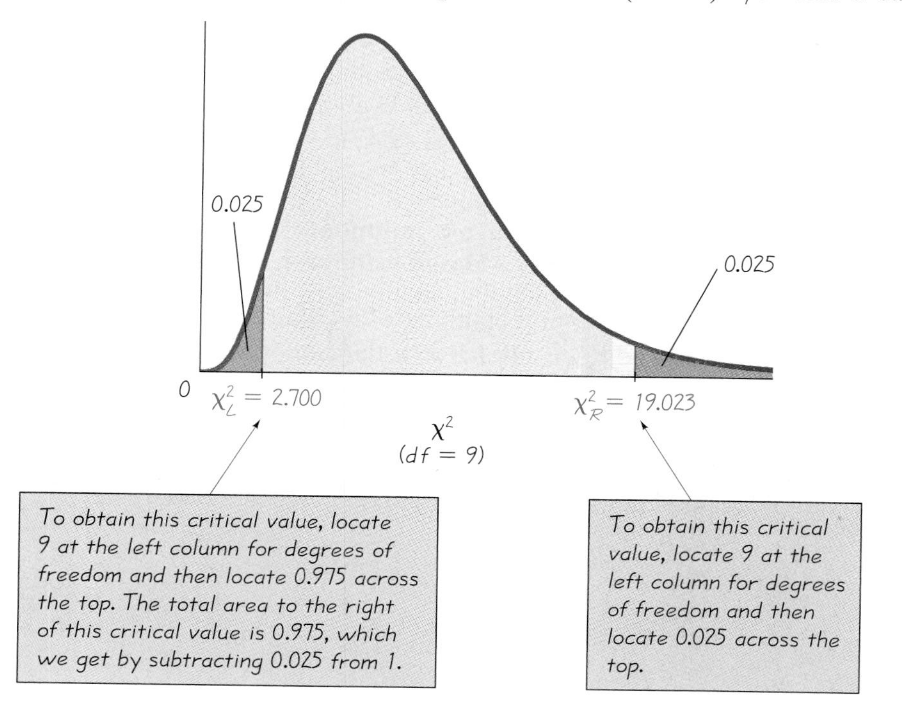

FIGURE 6-9
Critical Values of the Chi-Square Distribution

probability of falling between the chi-square critical values of 2.700 and 19.023.

When obtaining critical values of χ^2 from Table A-4, note that the numbers of degrees of freedom are consecutive integers from 1 to 30, followed by 40, 50, 60, 70, 80, 90, and 100. When a number of degrees of freedom (such as 52) is not found on the table, you can usually use the closest critical value. For example, if the number of degrees of freedom is 52, refer to Table A-4 and use 50 degrees of freedom. (If the number of degrees of freedom is exactly midway between table values, such as 55, simply find the mean of the two χ^2 values.) For numbers of degrees of freedom greater than 100, use the equation given in Exercise 22, or a more detailed table, or a statistical software package.

Estimators of σ^2

Because sample variances s^2 (found by using Formula 2-5) tend to center on the value of the population variance σ^2, we say that s^2 is an *unbiased estimator* of σ^2. That is, sample variances s^2 do not systematically tend to overestimate the value of σ^2, nor do they systematically tend to underestimate σ^2. Instead, they tend to target the value of σ^2 itself. Also, the values of s^2 tend to produce smaller errors by being closer to σ^2 than do other measures of variation. For these reasons, the value of s^2 is generally the best single value (or point estimate) of the various possible statistics we could use to estimate σ^2.

The sample variance s^2 is the best point estimate of the population variance σ^2.

Because s^2 is an unbiased estimator of σ^2, we might expect that s would be an unbiased estimator of σ, but this is not the case. (See Exercise 35 in Section 2-5.) If the sample size is large, however, the bias is so small that we can use s as a reasonably good estimate of σ. Even though it is a biased estimate, s is often used as a point estimate of σ.

The sample standard deviation s is commonly used as a point estimate of σ (even though it is a biased estimate).

Although s^2 is the best point estimate of σ^2, there is no indication of how good it actually is. To compensate for that deficiency, we develop an interval estimate (or confidence interval) that is more informative.

Confidence Interval (or Interval Estimate) for the Population Variance σ^2

$$\frac{(n-1)s^2}{\chi_R^2} < \sigma^2 < \frac{(n-1)s^2}{\chi_L^2}$$

This expression is used to find a confidence interval for the variance σ^2, but the confidence interval (or interval estimate) for the standard deviation σ is found by taking the square root of each component, as shown below.

$$\sqrt{\frac{(n-1)s^2}{\chi_R^2}} < \sigma < \sqrt{\frac{(n-1)s^2}{\chi_L^2}}$$

The notations χ_R^2 and χ_L^2 in the preceding expressions are described as follows. (Note that some other texts use $\chi_{\alpha/2}^2$ in place of χ_R^2, and they use $\chi_{1-\alpha/2}^2$ in place of χ_L^2.)

Notation

With a total area of α divided equally between the two tails of a chi-square distribution, χ_L^2 denotes the left-tailed critical value and χ_R^2 denotes the right-tailed critical value. (See Figure 6-10.)

Confidence interval limits for σ^2 and σ should be rounded by using the following round-off rule, which is really the same basic rule given in Section 6-2.

Round-Off Rule for Confidence Interval Estimates of σ or σ^2

1. When using the original set of data to construct a confidence interval, round the confidence interval limits to one more decimal place than is used for the original set of data.
2. When the original set of data is unknown and only the summary statistics (n, s) are used, round the confidence interval limits to the same number of decimal places used for the sample standard deviation or variance.

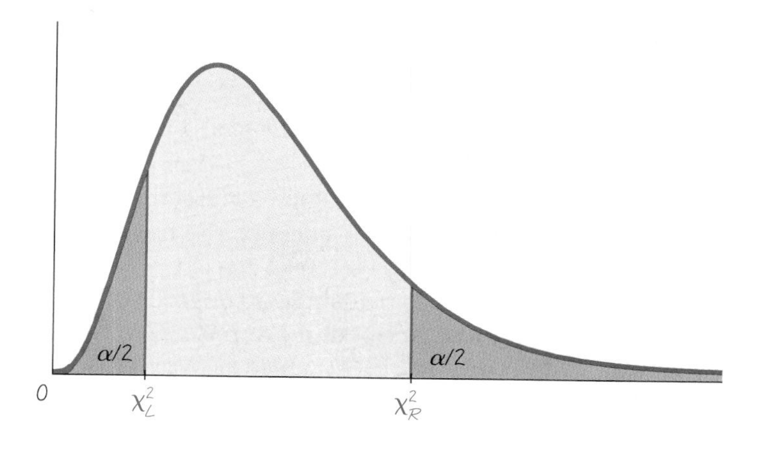

FIGURE 6-10
Chi-Square Distribution with Critical Values χ_R^2 and χ_L^2

The critical values χ_R^2 and χ_L^2 separate the extreme areas corresponding to sample variances that are unlikely (with probability α).

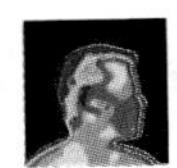

EXAMPLE **Body Temperatures** Table 6-1 in the Chapter Problem lists 106 body temperatures (from Data Set 6 in Appendix B) obtained by University of Maryland researchers. Use the following characteristics of the data set to construct a 95% confidence interval estimate of σ, the standard deviation of the body temperatures of the whole population:

a. As revealed by a histogram, the distribution of the data is approximately bell-shaped.

b. The mean is 98.20° F.

c. The standard deviation is $s = 0.62°$ F.

d. The sample size is $n = 106$.

e. There are no outliers.

SOLUTION We begin by finding the critical values of χ^2. With a sample of 106 values, we have 105 degrees of freedom. This isn't too far away from the 100 degrees of freedom found in Table A-4, so we will go with that. (See Exercise 22 for a method that will yield more accurate critical values.) For a 95% degree of confidence, we divide $\alpha = 0.05$ equally between the two tails of the chi-square distribution, and we refer to the values of 0.975 and 0.025 across the top row of Table A-4. The critical values of χ^2 are $\chi_L^2 = 74.222$ and $\chi_R^2 = 129.561$. Using these critical values, the sample standard deviation of $s = 0.62$, and the sample size of 106, we construct the 95% confidence interval by evaluating the following.

$$\frac{(106 - 1)(0.62)^2}{129.561} < \sigma^2 < \frac{(106 - 1)(0.62)^2}{74.222}$$

This becomes $0.31 < \sigma^2 < 0.54$. Taking the square root of each part (before rounding) yields $0.56°\text{F} < \sigma < 0.74°\text{F}$.

INTERPRETATION Based on this result, we have 95% confidence that the limits of 0.56°F and 0.74°F contain the true value of σ.

The confidence interval $0.56 < \sigma < 0.74$ can also be expressed as (0.56, 0.74), but the format of $\sigma = s \pm E$ *cannot* be used because the confidence interval does not have s at its center.

Instead of approximating the critical values by using 100 degrees of freedom, we could use software or the method described in Exercise 22, and the confidence interval becomes $0.55°\text{F} < \sigma < 0.72°\text{F}$, which is very close to the result obtained here.

Rationale We now explain why the confidence intervals for σ^2 and σ have the forms just given. If we obtain samples of size n from a population with variance σ^2, the distribution of the $(n-1)s^2/\sigma^2$ values will be as shown in Figure 6-10. For a simple random sample, there is a probability of $1-\alpha$ that the statistic $(n-1)s^2/\sigma^2$ will fall between the critical values of χ_L^2 and χ_R^2. In other words (and symbols), there is a $1-\alpha$ probability that both of the following are true:

$$\frac{(n-1)s^2}{\sigma^2} < \chi_R^2 \quad \text{and} \quad \frac{(n-1)s^2}{\sigma^2} > \chi_L^2$$

If we multiply both of the preceding inequalities by σ^2 and divide each inequality by the appropriate critical value of χ^2, we see that the two inequalities can be expressed in the equivalent forms

$$\frac{(n-1)s^2}{\chi_R^2} < \sigma^2 \quad \text{and} \quad \frac{(n-1)s^2}{\chi_L^2} > \sigma^2$$

These last two inequalities can be combined into one inequality:

$$\frac{(n-1)s^2}{\chi_R^2} < \sigma^2 < \frac{(n-1)s^2}{\chi_L^2}$$

There is a probability of $1-\alpha$ that these confidence interval limits contain the population variance σ^2.

Determining Sample Size

The procedures for finding the sample size necessary to estimate σ^2 are much more complex than the procedures given earlier for means and proportions. Instead of using very complicated procedures, we will use Table 6-3 on page 350. STATDISK also provides sample sizes. With STATDISK, select **Analysis, Sample Size Determination,** and then **Estimate St Dev.** Minitab, Excel, and the TI-83 Plus calculator do not provide such sample sizes.

EXAMPLE With 95% confidence, you wish to estimate σ to within 10%. How large should your sample be? Assume that the population is normally distributed.

SOLUTION From Table 6-3, we can see that 95% confidence and an error of 10% for σ correspond to a sample of size 191. You should randomly select 191 values from the population.

be underweight (cheating consumers) and others will be overweight (lowering profit). A consumer would not be happy with a doughnut so small that it can be seen only with an electron microscope, nor would a consumer be happy with a doughnut so large that it resembles a tractor tire. The quality-control supervisor has found that he can stay out of trouble if the doughnuts have a mean of 3.50 oz and a standard deviation of 0.06 oz or less. Twelve doughnuts are randomly selected from the production line and weighed, with the results given here (in ounces). Construct a 95% confidence interval for σ^2 and a 95% confidence interval for σ, then determine whether the quality- control supervisor is in trouble.

3.43 3.37 3.58 3.50 3.68 3.61 3.42 3.52 3.66 3.50 3.36 3.42

15. Estimating Standard Deviation of Weights of M&Ms Refer to Data Set 10 in Appendix B.

a. Use the range rule of thumb (see Section 2-5) to estimate σ, the standard deviation of the weights of brown M&M plain candies.

b. Use the listed weights to construct a 98% confidence interval for σ.

c. Does the confidence interval contain your estimated value of σ?

16. Snow Shoveling and Heart Attacks: Comparing Variation Because cardiac deaths appear to increase after heavy snowfalls, a study was designed to compare cardiac demands of snow shoveling to those of using an electric snow thrower. Ten subjects cleared tracts of snow using both methods, and their maximum heart rates (beats per minute) were recorded during both activities. The following results were obtained (based on data from "Cardiac Demands of Heavy Snow Shoveling," by Franklin et al., *Journal of the American Medical Association*, Vol. 273, No. 11):

Maximum Heart Rates During Manual Snow Shoveling:
$n = 10, \bar{x} = 175, s = 15$

Maximum Heart Rates During Automated Snow Removal:
$n = 10, \bar{x} = 124, s = 18$

a. Construct a 95% confidence interval estimate of the population standard deviation σ for those who did manual snow shoveling.

b. Construct a 95% confidence interval estimate of the population standard deviation σ for those who used the automated electric snow thrower.

c. Compare and interpret the results. Does the variation appear to be different for the two groups?

17. a. Comparing Waiting Lines The listed values are waiting times (in minutes) of customers at the Jefferson Valley Bank, where customers enter a single waiting line that feeds three teller windows. Construct a 95% confidence interval for the population standard deviation σ.

6.5 6.6 6.7 6.8 7.1 7.3 7.4 7.7 7.7 7.7

b. The listed values are waiting times (in minutes) of customers at the Bank of Providence, where customers may enter any one of three different lines that have formed at three teller windows. Construct a 95% confidence interval for the population standard deviation σ.

4.2 5.4 5.8 6.2 6.7 7.7 7.7 8.5 9.3 10.0

c. Interpret the results found in parts (a) and (b). Do the confidence intervals suggest a difference in the variation among waiting times? Which arrangement seems better: the single-line system or the multiple-line system?

18. Comparing Weights of Regular and Diet Pepsi Refer to Data Set 1 in Appendix B.

a. Construct a 95% confidence interval estimate of the standard deviation of weights of cola in cans of regular Pepsi.

b. Construct a 95% confidence interval estimate of the standard deviation of weights of cola in cans of diet Pepsi.

c. Compare and interpret the results.

19. Comparing Hardcover and Paperback Textbooks Refer to Data Set 2 in Appendix B.

a. Construct a 98% confidence interval estimate of the standard deviation of the costs of new hardcover textbooks at the University of Massachusetts.

b. Construct a 98% confidence interval estimate of the standard deviation of the costs of new paperback textbooks at the University of Massachusetts.

c. Compare and interpret the results.

20. Movie Lengths and Ratings Refer to Data Set 15 in Appendix B.

a. Construct a 99% confidence interval estimate of the standard deviation of the lengths (in minutes) of movies with a rating of R.

b. Construct a 99% confidence interval estimate of the standard deviation of the lengths (in minutes) of movies with a rating different than R.

c. Compare and interpret the results.

6-6 Beyond the Basics

21. Finding Missing Data A journal article includes a graph showing that sample data are normally distributed.

a. The degree of confidence is inadvertently omitted when this confidence interval is given: $2.8 < \sigma < 6.0$. Find the degree of confidence for these given sample statistics: $n = 20$, $\bar{x} = 45.2$, and $s = 3.8$.

b. This 95% confidence interval is given: $19.1 < \sigma < 45.8$. Given $n = 12$, find the value of the standard deviation s, which was omitted from the article.

22. Finding Critical Values In constructing confidence intervals for σ or σ^2, we use Table A-4 to find the critical values χ^2_L and χ^2_R, but that table applies only to cases in which $n \leq 101$, so the number of degrees of freedom is 100 or fewer. For larger numbers of degrees of freedom, we can approximate χ^2_L and χ^2_R by using

$$\chi^2 = \frac{1}{2}\left[\pm z_{\alpha/2} + \sqrt{2k - 1}\,\right]^2$$

where k is the number of degrees of freedom and $z_{\alpha/2}$ is the critical z score first described in Section 6-2. Construct the 95% confidence interval for σ by using the following sample data: The measured heights of 772 men between the ages of 18 and 24 have a standard deviation of 2.8 in. (based on data from the National Health Survey).

Discrepancy Between Actual Church Attendance and Poll Results

ASHTABULA COUNTY For about 60 years, the Gallup Organization has been polling people about their attendance at a religious service. Results have been fairly consistent, with about 40% stating that they attend a religious service at least once a week. Some survey experts charge that when people are polled about this issue, they tend to be less than totally honest. These experts claim that poll respondents have a tendency to provide answers that they think the pollster wants to hear. Researchers C. Kirk Hadaway and Penny Long Marler contacted pastors and did actual head counts at various religious services. In a Knight-Ridder News Service Report, it was stated that "a telephone poll of Ashtabula County's (Ohio) Roman Catholics showed 51 percent said they attended church in the week before the poll. Yet only 24 percent actually showed up during the time covered."

1. The article refers to a telephone poll conducted in Ashtabula County, Ohio. The population of that county is 99,821 (based on the last census report). How many of those people would have to be surveyed in order to be 95% confident that the sample percentage is within two percentage points of the population proportion?

2. The article refers to a discrepancy between what people say in a survey and what they actually do. Can that discrepancy be reduced or eliminated by increasing the number of people included in the survey?

3. Describe a procedure for determining whether the Ashtabula County results also apply to *your* county.

4. Researchers Hadaway and Marler commented that if the survey results are accurate, then "three decades of otherwise corrosive social and cultural changes has left American church attendance virtually untouched." Identify some of those corrosive changes and how they might affect the actual rate of church attendance.

VOCABULARY LIST

estimator
estimate
point estimate
unbiased estimator
confidence interval
interval estimate
degree of confidence
level of confidence
confidence coefficient
critical value
margin of error (E)
confidence interval limits
Student t distribution
t distribution
degrees of freedom
chi-square distribution

Review

In this chapter and the following chapter, we present the fundamental and important concepts of inferential statistics. In this chapter we focused on methods for finding *estimates* of population means, proportions, and variances and developed procedures for finding each of the following.

- point estimate
- confidence interval
- required sample size

We discussed point estimate (or single-valued estimate) and formed these conclusions:

- The best point of estimate of μ is $\overline{x}$.
- The best point estimate of p is $\hat{p}$.
- The best point estimate of p is $\hat{p}$
- The value of s is commonly used as a point estimate of σ, even though it is a biased estimate.

As single values, the point estimates don't convey any real sense of how reliable they are, so we introduced confidence intervals (or interval estimates) as more informative estimates. We also considered ways of determining the sample sizes necessary to estimate parameters to within given margins of error. This chapter also introduced the Student t and chi-square distributions. We must be careful to use the correct probability distribution for each set of circumstances.

It is important to know that all of the confidence interval and sample size procedures in this chapter require that we have a population with a distribution that is approximately normal. If the distribution is very nonnormal, we must use other methods, such as the bootstrap method described in the Computer Project at the end of this chapter.

Review Exercises

1. Qwerty/Dvorak Keyboards The Chapter Problem for Chapter 2 included word ratings for two keyboard configurations: the QWERTY keyboard and the Dvorak keyboard. Here are the sample statistics expressed with extra digits for more precision:

QWERTY word ratings: $n = 52, \overline{x} = 4.4038, s = 2.8440$

Dvorak word ratings: $n = 52, \overline{x} = 1.7308, s = 1.7502$

a. Construct a 98% confidence interval estimate of the population mean of all word ratings based on the QWERTY keyboard.
b. Construct a 98% confidence interval estimate of the population mean of all word ratings based on the Dvorak keyboard.
c. Compare and interpret the results from parts (a) and (b).
d. Why can't we use the methods of Section 6-6 to construct confidence interval estimates of the population standard deviations?

2. Determining Sample Size You want to estimate the percentage of U.S. statistics students who get grades of B or higher. How many such students must you survey if you want 97% confidence that the sample percentage is off by no more than two percentage points?

3. Estimating a Mean
   a. You want to estimate the mean final exam score of all U.S. statistics students. (Assume that the final exam scores are numerical grades, not letter grades.) How many such students must you survey if you want 97% confidence that the sample mean is off by no more than two points? Assume that a pilot study suggests that the population standard deviation is given by $\sigma = 15.7$.
   b. In surveying the number of students found in part (a), what would be wrong with selecting the students at the colleges closest to you?

4. Length of Car Ownership A NAPA Auto Parts supplier wants information about how long car owners plan to keep their cars. A simple random sample of 25 car owners results in $\bar{x} = 7.01$ years and $s = 3.74$ years, respectively (based on data from a Roper poll). Assume that the sample is drawn from a normally distributed population.
   a. Find a 95% confidence interval estimate of the population mean.
   b. Find a 95% confidence interval estimate of the population standard deviation.

5. Air Bag Fatalities In the first two months of a recent year, 94 car occupants were killed by air bags, and 61 of them were "improperly belted" (based on data from the National Highway Traffic Safety Administration). Construct a 95% confidence interval estimate of the percentage of car occupants who were killed by air bags while being improperly belted. Also, write a statement that interprets the confidence interval.

6. Alcohol Service Policy: Determining Sample Size In a Gallup poll of 1004 adults, 93% indicated that restaurants and bars should refuse service to patrons who have had too much to drink. If you plan to conduct a new poll to confirm that the percentage continues to be correct, how many randomly selected adults must you survey if you want 98% confidence that the margin of error is four percentage points?

7. Confidence Interval for Mean Arm Length In designing a new machine to be used on an assembly line at a General Motors plant, an engineer obtains measurements of arm lengths of a simple random sample of male machine operators. The following values (in centimeters) are obtained. Construct the 95% confidence interval for the mean arm length of all such employees.

76.8	75.6	69.3	75.7	75.5	71.2	72.5	71.9
70.9	69.4	71.7	72.5	72.2	68.5	75.9	73.0

8. Confidence Interval for Mean GRE Score Among those who take the Graduate Record Examination (GRE), 67 people are randomly selected. This sample group has a mean score of 558 and a standard deviation of 139 on the quantitative portion of the GRE (based on data from the Educational Testing Service). Find a 99% confidence interval estimate of the population mean.

9. Determining Sample Size A medical researcher wishes to estimate the serum cholesterol level (in mg/100 mL) of all women aged 18 to 24. There is strong evidence suggesting that $\sigma = 41.0$ mg/100 mL (based on data from a survey of 1524 women aged 18 to 24, as part of the National Health Survey). If the researcher wants to be 95% confident of obtaining a sample mean that is off by no more than four units (mg/100 mL), how large must the sample be?

10. TV Commercials In a Roper survey of 1,998 randomly selected adults, 24% included loud commercials among the annoying aspects of television. Construct the 99% confidence interval for the percentage of all adults who are annoyed by loud commercials.

Cumulative Review Exercises

1. Analyzing Weights of Supermodels Supermodels are sometimes criticized on the grounds that their low weights encourage unhealthy eating habits among young women. Listed below are the weights (in pounds) of nine randomly selected supermodels.

125 (Taylor)	119 (Auermann)	128 (Schiffer)	128 (MacPherson)
119 (Turlington)	127 (Hall)	105 (Moss)	123 (Mazza)
115 (Hume)			

Find each of the following.

a. mean
b. median
c. mode
d. midrange
e. range
f. variance
g. standard deviation
h. Q_1
i. Q_2
j. Q_3
k. What is the level of measurement of these data (nominal, ordinal, interval, ratio)?
l. Construct a boxplot for the data. l.
m. Construct a 99% confidence interval for the population mean.
n. Construct a 99% confidence interval for the standard deviation σ.
o. Find the sample size necessary to estimate the mean weight of all supermodels so that there is 99% confidence that the sample mean is in error by no more than 2 lb. Use the sample standard deviation s from part (g) as an estimate of the population standard deviation σ.
p. When women are randomly selected from the general population, their weights are normally distributed with a mean of 143 lb and a standard deviation of 29 lb (based on data from the National Health and Examination Survey). Based on the given sample values, do the weights of supermodels appear to be substantially less than the weights of randomly selected women? Explain.

2. X-Linked Recessive Disorders A genetics expert has determined that for certain couples, there is a 0.25 probability that any child will have an X-linked recessive disorder.

a. Find the probability that among 200 such children, at least 65 have the X-linked recessive disorder.
b. A subsequent study of 200 actual births reveals that 65 of the children have the X-linked recessive disorder. Based on these sample results, construct a 95% confidence interval for the proportion of all such children having the disorder.

(continued)

c. Based on parts (a) and (b), does it appear that the expert's determination of a 0.25 probability is correct? Explain.

COOPERATIVE GROUP ACTIVITIES

1. *Out-of-class activity:* Collect sample data, and use the methods of Section 6-2 or 6-3 to construct confidence interval estimates of population means. Here are some suggestions for sample data:

- Ages of cars driven by statistics students and/or ages of cars driven by faculty
- Ages of math books and ages of science books in your college library (based on the copyright dates)
- Heights of male statistics students and heights of female statistics students
- Lengths of words in *New York Times* editorials and lengths of words in editorials found in your local newspaper
- Lengths of words in *Time* magazine, *Newsweek* magazine, and *People* magazine
- Proportion of students at your college who can correctly identify the president, and vice president and the governor of their home state

2. *In-class activity:* Divide into groups with approximately 10 students in each group. Get the reaction timer from the first Cooperative Group Activity given in Chapter 5, and measure the reaction time of each group member. (Right-handed students should use their right hand, and left-handed students should use their left hand.) Use the methods of this chapter to estimate the mean reaction time for all college students. Construct a 90% confidence interval estimate of that mean. Compare the results to those found in other groups.

3. *In-class activity:* Divide into groups of three or four. Examine a current magazine such as *Time* or *Newsweek,* and find the proportion of pages that include advertising. Based on the results, construct a 95% confidence interval estimate of the percentage of all such pages that have advertising. Compare results with other groups.

4. *In-class activity:* Divide into groups with approximately 10 students in each group. First, each group member should write an estimate of the mean amount of cash being carried by students in the group. Next, each group member should report the actual amount of cash being held (including paper money and coins). (The amounts should be written anonymously on separate sheets of paper which are mixed, so that nobody's privacy is compromised.) Use the reported values to find $\bar{x}$ and s, then construct a 95% confidence interval estimate of the mean μ. Describe the precise population that is being estimated. Which group member came closest to the value of $\bar{x}$? Compare the results with other groups. Are the results consistent, or are there large variations among the group results?

TECHNOLOGY PROJECT

The *bootstrap method* can be used to construct confidence intervals for situations in which traditional methods cannot (or should not) be used. For example, the following sample of 10 scores was randomly selected from a very nonnormal distribution, so the methods previously discussed cannot be used.

2.9 564.2 1.4 4.7 67.6 4.8 51.3 3.6 18.0 3.6

The methods of this chapter require that the population have a distribution that is at least approximately normal. The bootstrap method, which makes no assumptions about the original population, typically requires a computer to build a bootstrap population by replicating (duplicating) a sample many times. We can draw from the sample with replacement, thereby creating an approximation of the original population. In this way, we pull the sample up "by its own bootstraps" to simulate the original population. Using the sample data given above, construct a 95% confidence interval estimate of the population mean μ by using the bootstrap method as described in the following Minitab steps.

a. Create 500 new samples, each of size 10, by selecting 10 scores with replacement from the 10 sample scores given above. With Minitab, first enter the sample scores in column C1, then enter probabilities of 0.1, 0.1, . . . , 0.1 (ten times) in column C2. Now select **Calc** from the main menu bar, then select **Random Data,** followed by **Discrete.** Proceed to generate 500 rows of data, stored in columns C11–C20, with the values in C1 and probabilities in C2, and then click OK.

b. Find the means of the 500 bootstrap samples generated in part (a). Select **Calc, Row Statistics,** and **Mean,** enter input variables of C11–C20 with results to be stored in C21, and then click OK.

c. Sort the 500 means (arrange them in order). Select **Manip** from the main menu bar, choose the option of **Sort,** and proceed to sort column C21. Store the sorted column in C21, and sort by column C21. Click OK.

d. Find the percentiles $P_{2.5}$ and $P_{97.5}$ for the sorted means that result from the preceding step. ($P_{2.5}$ is the mean of the 12th and 13th scores in the ranked list of column C21; $P_{97.5}$ is the mean of the 487th and 488th scores in column C21.) Identify the resulting confidence interval by substituting the values for $P_{2.5}$ and $P_{97.5}$ in $P_{2.5} < \mu < P_{97.5}$. Does this confidence interval contain the true value of μ, which is 148?

Now use the bootstrap method to find a 95% confidence interval for the population standard deviation σ. (Use the same steps listed above, but specify *standard deviation* instead of mean in part b.) Compare your result to the interval $318.4 < \sigma < 1079.6$, which was obtained by incorrectly using the methods described in Section 6-6. (Their use is incorrect because the population distribution is very nonnormal.) This incorrect confidence interval for σ does not contain the true value of σ, which is 232.1. Does the bootstrap procedure yield a confidence interval for σ that contains 232.1, verifying that the bootstrap method is effective?

An alternative to using Minitab is to use special software designed specifically for bootstrap resampling methods. The author recommends Resampling Stats, available from Resampling Stats, Inc., 612 N. Jackson St., Arlington, VA, 22201.

Critical Thinking
He's angry, but is he right?

The following excerpt is taken from a letter written by a corporation president and sent to the Associated Press.

> *When you or anyone else attempts to tell me and my associates that 1223 persons account for our opinions and tastes here in America, I get mad as hell! How dare you! When you or anyone else tells me that 1223 people represent America, it is astounding and unfair and should be outlawed.*

The writer then goes on to claim that because the sample size of 1223 people represents 120 million people, his letter represents 98,000 people (120 million divided by 1223) who share the same views.

Analyzing the results

a. Given that the sample size is 1223 and the degree of confidence is 95%, find the margin of error for the proportion. Assume that there is no prior knowledge about the value of that proportion.

b. The writer of the letter is taking the position that a sample size of 1223 taken from a population of 120 million people is too small to be meaningful. Do you agree or disagree? Write a response that either supports or refutes the writer's position that the sample is too small.

c. The writer also makes the claim that because the poll of 1223 people was projected to reflect the opinions of 120 million, any 1 person actually represents 98,000 other people. As the writer is 1 person, he claims to represent 98,000 other people. Is this claim correct? Explain why or why not.

Confidence Intervals

The confidence intervals in this chapter illustrate an important point in the science of statistical estimation. Namely, estimations based on sample data are made with certain degrees of confidence. In the Internet Project for Chapter 6, you will use confidence intervals to make a statement about the temperature where you live. Go to this site:

http://awlonline.com/triola

Here you will find instructions on how to use the internet to find temperature data collected by the weather station nearest your home. With this data in hand, you will construct confidence intervals for temperatures during different time periods and attempt to draw some conclusions about temperature change in your area.

Statistics *at work*

"Statistics is the language in which we communicate about data, and as such, it is an essential tool for conducting our business."

Barry Cook:
Senior Vice President and Cheif Research Officer, Nielsen Media Research

What is your job and what do you do?

I am responsible for research methods and statistical operations at Nielsen Media Research (Our estimates are used as audience currency for over $40 Billion in television advertising expenditures each year.) There are 80 people in the Statistical Research Department (statisticians, analysts, computer support staff, and field surveyors) and 10 people in the Methodology Research Department (responsible for research method experimentation and analysis).

What concepts of statistics do you use?

Probability samples are the core asset of our business. Means and standard errors are the products of our business. We use categorical and interval statistics to analyze data (including regression, chi-square, analysis of variance). Statistical modeling is playing an increasing role in refining our estimation procedures.

How do you use statistics on the job?

We look for relationships and anomalies in our data. Since most of the data are time-series data, we look at trends and deviations from trends. We conduct experiments on methodological changes, and perform significance tests to evaluate the experimental data.

Please describe one specific example illustrating how the use of statistics was successful in improving a product or service.

We monitor the demographic characteristics of our television measurement panels, and compare these characteristics to Census data. If, through statistical analysis, we conclude that the differences are unlikely to be a result of sampling error, we examine our procedures and processes to find possible causes. When our analyses confirmed that our samples tended to underrepresent households with younger adults, we traced the cause to mobility (and our response time). When new occupants moved into a housing unit that is part of our sample, it used to take us up to six months to discover it. Younger households move more frequently, and this lag time caused them to be out of the sample more than households that didn't move as often. We changed our procedures to reduce this six-month-maximum lag to two weeks, and the proportion of young households in our samples no longer averages below the Census estimates.

How do you ensure objectivity?

Corporate business conduct guidelines are a condition of employment. The leadership of the organization is outspoken on the principles of independence and research objectivity, and there is an ongoing effort to communicate throughout the organization (and with our research respondents) the importance of fairness in audience measurement procedures. Our customers often raise questions about potential biases—and across all the segments of the industry (sellers as well as buyers of advertising time), a certain balance is achieved.

How beneficial do you find your knowledge of statistics for performing your responsibilities?

Statistics is the language in which we communicate about data, and as such, it is an essential tool for conducting our business. Statistical testing is the input to many major decisions, and the ability to collect structured data and analyze it appropriately is central to good decision making.

Please cite an example of how your data are used.

Nielsen Media Research's television audience estimates are used by advertisers, ad agencies and television programmers (e.g., networks, stations and syndicators) to establish the relative value of what they buy and sell (advertising exposure). They negotiate a cost per thousand persons reached, and then multiply that cost by the Nielsen Media Research audience estimate. For example, if a 30 second commercial in a particular program is purchased at a rate of $15 per thousand adults between the ages of 18 and 49 and if the estimated average audience for that program is 7.2 million adults 18-49, then the cost of the commercial is $15*(7,200,000/1000) = $108,000.

In terms of statistics, what would you recommend for prospective employees?

There are many different jobs within this company—and a knowledge of basic statistics is helpful to all members of our organization. For those who provide data analysis services, two semesters of statistics are a minimum. In our statistical research department, we look for people who have majored in Statistics, with degrees ranging from B.S. to Ph.D. (Master's degrees are typical).

Do you feel job applicants are viewed more favorably if they have studied some statistics?

Formal training in statistics is a plus, but the standout candidates also ask good questions about data.

Which other skills are important for today's college students?

Critical thinking, clear and persuasive communication skills, the ability to work collaboratively, and a core sense of ethics.

CHAPTER 7

Hypothesis Testing

CHAPTER PROBLEM

The mean body temperature is 98.6°F, right?

When asked, most of us will identify the mean body temperature for healthy adults as 98.6°F. Table 7-1 lists 106 measured body temperatures found in Data Set 2 of Appendix B (for 12 A.M. on Day 2). Those 106 temperatures, found by University of Maryland researchers, have a mean of $\bar{x} = 98.20$°F and a standard deviation of $s = 0.62$°F. In Chapter 6 we used the same set of temperatures to estimate μ, the mean body temperature, and we found this 95% confidence interval: $98.08°\text{F} < \mu < 98.32°\text{F}$ which does *not* contain 98.6°F, the value generally believed to be the mean body temperature. The confidence interval is an *estimate* of the mean body temperature, but the researchers went further by making a claim that 98.6°F "should be abandoned as a concept having any particular significance for the normal body temperature." Should we reject the common belief that the mean body temperature of healthy adults is 98.6° F? There is a standard procedure for testing such claims, and this chapter will describe that procedure.

TABLE 7-1 Body Temperature (F°)of 106 Healthy Adults

98.6	98.6	98.0	98.0	99.0	98.4	98.4	98.4	98.4	98.6	98.6
98.8	98.6	97.0	97.0	98.8	97.6	97.7	98.8	98.0	98.0	98.3
98.5	97.3	98.7	97.4	98.9	98.6	99.5	97.5	97.3	97.6	98.2
99.6	98.7	99.4	98.2	98.0	98.6	98.6	97.2	98.4	98.6	98.2
98.0	97.8	98.0	98.4	98.6	98.6	97.8	99.0	96.5	97.6	98.0
96.9	97.6	97.1	97.9	98.4	97.3	98.0	97.5	97.6	98.2	98.5
98.8	98.7	97.8	98.0	97.1	97.4	99.4	98.4	98.6	98.4	98.5
98.6	98.3	98.7	98.8	99.1	98.6	97.9	98.8	98.0	98.7	98.5
98.9	98.4	98.6	97.1	97.9	98.8	98.7	97.6	98.2	99.2	97.8
98.0	98.4	97.8	98.4	97.4	98.0	97.0				

7-1 Overview

In Chapter 6 we introduced methods for using sample data to estimate values of population parameters. This chapter introduces another major topic of inferential statistics: testing claims (or *hypotheses*) made about population parameters.

DEFINITION

In statistics, a **hypothesis** is a claim or statement about a property of a population.

The following statements are examples of hypotheses that will be tested by the procedures we develop in this chapter.

- Medical researchers claim that the mean body temperature of healthy adults is not equal to 98.6° F(a claim about a population *mean*).
- Drivers who use cell phones have a car crash rate that is greater than the 13% rate for those who do not use cell phones (a claim about a population *proportion*).
- When new equipment is used to manufacture aircraft altimeters, the variation in the errors is reduced so that the readings are more consistent (a claim about a population *variance*).

Before beginning to study this chapter, you should recall—and understand clearly—this basic rule, first introduced in Section 3-1.

Rare Event Rule for Inferential Statistics

If, under a given assumption, the probability of a particular observed event is exceptionally small, we conclude that the assumption is probably not correct.

Following this rule, we analyze a sample in an attempt to distinguish between results that can *easily* occur and results that are *highly unlikely*. We can explain the occurrence of highly unlikely results by saying either that a rare event has indeed occurred or that things aren't as they are assumed to be. Let's apply this reasoning in the following example.

EXAMPLE **Gender Selection** ProCare Industries, Ltd., once provided a product called "Gender Choice," which, according to advertising claims, allowed couples to "increase your chances of having a boy up to 85%, a girl up to 80%." Gender Choice was available in blue packages for couples wanting a baby boy and (you guessed it) pink packages for couples wanting a baby girl. Suppose we conduct an experiment with 100 couples who want to have baby girls, and they all follow the Gender Choice "easy-to-use in-home system" described in the pink package.

Using common sense and no formal statistical methods, what should we conclude about the effectiveness of Gender Choice if the 100 babies include

a. 52 girls?

b. 97 girls?

SOLUTION

a. We normally expect around 50 girls in 100 births. The result of 52 girls is close to 50, so we should not conclude that the Gender Choice product is effective. If the 100 couples used no special methods of gender selection, the result of 52 girls could easily occur by chance.

b. The result of 97 girls in 100 births is extremely unlikely to occur by chance. We could explain the occurrence of 97 girls in one of two ways: Either an *extremely* rare event has occurred by chance, or Gender Choice is effective. Because of the extremely low probability of getting 97 girls, the more likely explanation is that the product is effective.

The key point of the preceding example is that we should conclude that the product is effective only if we get *significantly* more girls than we would expect under normal circumstances. Although the outcomes of 52 girls and 97 girls are both "above average," the result of 52 girls is not significant, whereas 97 girls does constitute a significant result.

This brief example illustrates the basic approach used in testing hypotheses. The formal method involves a variety of standard terms and conditions incorporated into an organized procedure. We suggest that you begin the study of this chapter by first reading Sections 7-2 and 7-3 casually to obtain a general idea of their concepts and then rereading Section 7-2 more carefully to become familiar with the terminology.

7-2 Fundamentals of Hypothesis Testing

We begin this section with an informal example, then we describe the formal components of the standard method of hypothesis testing: null hypothesis, alternative hypothesis, test statistic, critical region, significance level, critical value, type I error, and type II error. After studying this chapter, you should be able to do the following:

- Given a claim, identify the null hypothesis and the alternative hypothesis, and express them both in symbolic form.
- Given a significance level, identify the critical value(s).
- Given a claim and sample data, calculate the value of the test statistic.
- State the conclusion of a hypothesis test in simple, nontechnical terms.

- Identify the type I and type II errors that could be made when testing a given claim.

You should study the following example until you thoroughly understand it. Once you do, you will have captured a major concept of statistics.

EXAMPLE **Body Temperatures** In the Chapter Problem, we noted that researchers claimed that the mean body temperature of healthy adults is not equal to 98.6°F. The University of Maryland researchers collected sample data with these characteristics: $n = 106, \bar{x} = 98.20°\text{F}, s = 0.62°\text{F}$, and the shape of the distribution is approximately normal. The key question is this: *Do the sample data (with $\bar{x} = 98.20°F$) provide enough evidence to warrant rejection of the common belief that $\mu = 98.6°F$?*

We conclude that there is sufficient evidence to warrant rejection of the belief that $\mu = 98.6$ because, if the mean is really 98.6, the probability of getting the sample mean of 98.20 is approximately 0.0002, which is too small. (Later, we will show how that probability value of 0.0002 is determined.)

If we assume that $\mu = 98.6$ and use the central limit theorem (Section 5-5), we know that sample means tend to be normally distributed with these parameters:

$$\mu_{\bar{x}} = \mu = 98.6 \quad \text{(by assumption)}$$

$$\sigma_{\bar{x}} = \frac{\sigma}{\sqrt{n}} \approx \frac{s}{\sqrt{n}} = \frac{0.62}{\sqrt{106}} = 0.06$$

We construct Figure 7-1 by assuming that $\mu = 98.6$ and by using the parameters just shown. Figure 7-1 also shows that if μ is really 98.6, then 95% of all sample means should fall between 98.48 and 98.72. (The values of 98.48 and 98.72 were found by using the methods of Section 5-5. Specifically, 95% of sample means should fall within 1.96 standard deviations of μ. With $\sigma_{\bar{x}} = 0.06$, 95% of sample means should fall within $1.96 \times 0.06 \approx 0.12$ of 98.6. Falling within 0.12 of 98.6 is equivalent to falling between 98.48 and 98.72.)

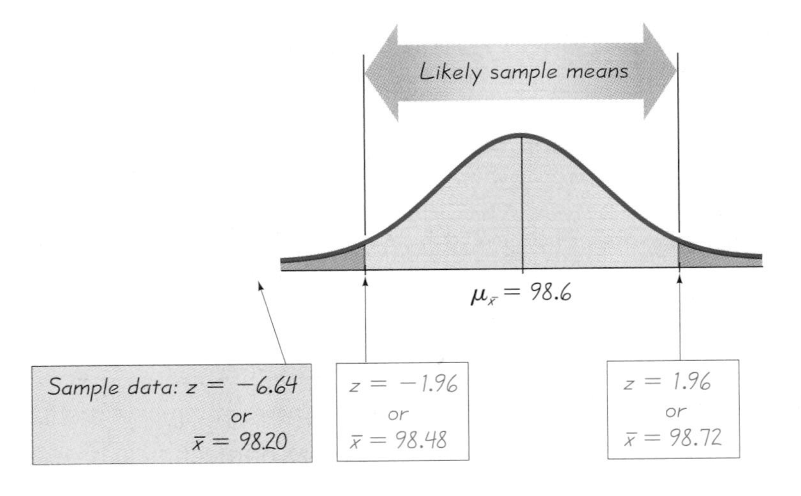

FIGURE 7-1
Central Limit Theorem

The expected distribution of sample means assuming that $\mu = 98.6$

Here are the key points:

- The common belief is that $\mu = 98.6$.
- The sample resulted in $\bar{x} = 98.20$.
- Considering the distribution of sample means, the sample size, and the magnitude of the discrepancy between 98.6 and 98.20, we find that a sample mean of 98.20 is *unlikely* (with less than a 5% chance) to occur if μ is really 98.6.
- There are two reasonable explanations for the sample mean of 98.20: Either a very rare event has occurred, or μ is not really 98.6. Because the probability of getting a sample mean of 98.20 (when $\mu = 98.6$) is so low, we go with the more reasonable explanation: The value of μ is not 98.6 as is commonly believed.

The preceding example illustrates well the basic method of reasoning we will use throughout this chapter. Read it carefully several times until you understand it. Try not to dwell on the details of the calculations. Instead, focus on the key idea that although there is a common belief that $\mu = 98.6$, the sample mean is $\bar{x} = 98.20$. By using the central limit theorem, we find that if the mean is really 98.6, then the probability of getting a sample with a mean of 98.20 is very small, which suggests that the belief that $\mu = 98.6$ should be rejected. In Section 7-3 we will describe the specific steps used in hypothesis testing, but let's first describe the components of a formal **hypothesis test,** or **test of significance**.

Components of a Formal Hypothesis Test

Null and Alternative Hypotheses

- The **null hypothesis** (denoted by H_0) is a statement about the value of a population parameter (such as the mean), and it must contain the condition of equality and must be written with the symbol $=$, $\leq$, or $\geq$. (When actually conducting the test, we operate under the assumption that the parameter *equals* some specific value.) For the mean, the null hypothesis will be stated in one of the following three possible forms:

  H_0: $\mu =$ some value $\quad$ H_0: $\mu \leq$ some value $\quad$ H_0: $\mu \geq$ some value

  For example, the null hypothesis corresponding to the common belief that the mean body temperature is 98.6°F is expressed as H_0: $\mu = 98.6$. We test the null hypothesis directly in the sense that we assume it is true and reach a conclusion to either reject H_0 or fail to reject H_0.
- The **alternative hypothesis** (denoted by H_1) is the statement that must be true if the null hypothesis is false. For the mean, the alternative hypothesis will be stated in only one of three possible forms:

  H_1: $\mu \neq$ some value $\quad$ H_1: $\mu >$ some value $\quad$ H_1: $\mu <$ some value

Basically, H_1 is the opposite of H_0. For example, if H_0 is given as $\mu = 98.6$ then it follows that the alternative hypothesis is H_1: $\mu \neq 98.6$.

Note about Using $\leq$ or $\geq$ in H_0: Even though we sometimes express H_0 with the symbol $\leq$ or $\geq$, as in H_0: $\mu \leq 98.6$ or H_0: $\mu \geq 98.6$, we conduct the test by assuming that $\mu = 98.6$ is true. This is because we must have a single fixed value for μ so that we can work with a single distribution having a specific mean. (Some textbooks and some software packages use notation in which H_0 *always* contains only the equals symbol. Where this and many other textbooks might use $\mu \leq 98.6$ and $\mu > 98.6$ for H_0 and H_1, respectively, some others might use $\mu = 98.6$ and $\mu > 98.6$ instead.)

Note about Forming Your Own Claims (Hypotheses): If you are conducting a study and want to use a hypothesis test to *support* your claim, the claim must be worded so that it becomes the alternative hypothesis. In other words, your claim must be expressed in one of these formats:

$\mu \neq$ some value $\qquad \mu >$ some value $\qquad \mu <$ some value

For example, suppose you have developed a magic potion that raises IQ scores so that the mean becomes greater than 100. If you want to provide evidence of the potion's effectiveness, you must state the claim as $\mu > 100$. (In this context of trying to support the goal of the research, the alternative hypothesis is sometimes referred to as the *research hypothesis*. Also in this context, the null hypothesis of $\mu \leq 100$ is assumed to be true for the purpose of conducting the hypothesis test, but it is hoped that the conclusion will be rejection of the null hypothesis so that the claim of $\mu > 100$ is supported.)

Note about Testing the Validity of Someone Else's Claim: Sometimes we test the validity of someone else's claim, such as the claim of the Coca Cola Bottling Company that "the mean amount of Coke in cans is at least 12 oz," which becomes the null hypothesis of H_0: $\mu \geq 12$. In this context of testing the validity of someone else's claim, their original claim sometimes becomes the null hypothesis (because it contains equality), and it sometimes becomes the alternative hypothesis (because it does not contain equality).

Figure 7-2 summarizes the procedures for identifying the null and alternative hypotheses. Pay special attention to the fact that the original claim can sometimes become a null hypothesis and can sometimes be an alternative hypothesis, depending on the wording of the original claim.

EXAMPLE **Identifying the Null and Alternative Hypotheses** As an employee of the Orange County Bureau of Weights and Measures, you have been assigned the project of testing this claim: When meters on the gas pumps at the Jefferson Road Auto Service Center indicate 1 gal, the mean amount of gas supplied is actually *less* than 1 gal. Identify the null hypothesis and the alternative hypothesis, and express them both in symbolic form.

SOLUTION Refer to Figure 7-2, which shows the three-step procedure.

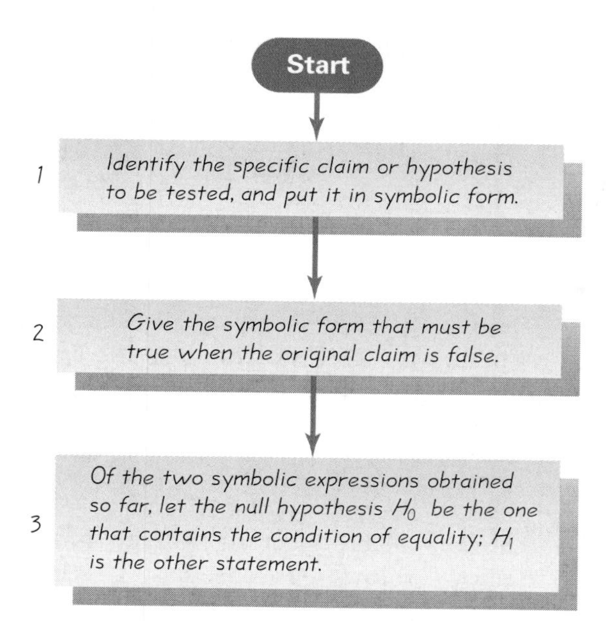

FIGURE 7-2
Identifying the Null and Alternative Hypotheses

Step 1: The claim that the mean is less than 1 gal can be expressed symbolically as $\mu < 1$ gal.

Step 2: If the original claim of $\mu < 1$ gal is false, then $\mu \geq 1$ gal.

Step 3: Of the two symbolic expressions found so far, the expression $\mu \geq 1$ contains the condition of equality, so it becomes the null hypothesis. We have

null hypothesis	H_0: $\mu \geq 1$	
alternative hypothesis	H_1: $\mu < 1$	(original claim)

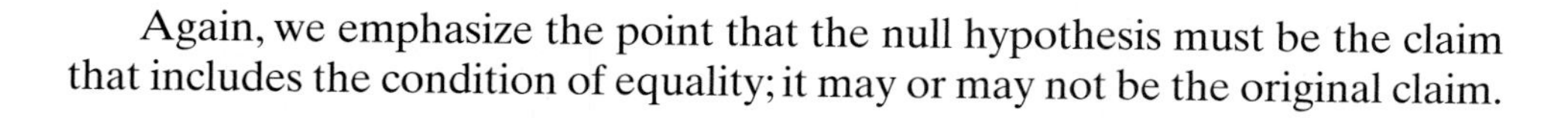

Again, we emphasize the point that the null hypothesis must be the claim that includes the condition of equality; it may or may not be the original claim.

Test Statistic

- The **test statistic** is a value computed from the sample data that is used in making the decision about the rejection of the null hypothesis. The test statistic converts the sample statistic (such as the sample mean $\overline{x}$) to a score (such as the z score) with the assumption that the null hypothesis is true. The test statistic can therefore be used to gauge whether the discrepancy between the sample and the claim is significant. In this section and the following section, we use only large samples in testing claims made about population means. In such cases, we can use the central limit theorem (Section 5-5) with the test statistic based on the discrepancy between the sample mean $\overline{x}$ and the claimed

Lie Detectors

Why not require all criminal suspects to take lie detector tests and dispense with trials by jury? The Council of Scientific Affairs of the American Medical Association states, "It is established that classification of guilty can be made with 75% to 97% accuracy, but the rate of false positives is often sufficiently high to preclude use of this (polygraph) test as the sole arbiter of guilt or innocence." A "false positive" is an indication of guilt when the subject is actually innocent. Even with accuracy as high as 97%, the percentage of false positive results can be 50%, so half of the innocent subjects incorrectly appear to be guilty.

population mean μ, calculated by using Formula 7-1, which is another form of $z = (\bar{x} - \mu_{\bar{x}})/\sigma_{\bar{x}}$.

Formula 7-1
$$z = \frac{\bar{x} - \mu_{\bar{x}}}{\frac{\sigma}{\sqrt{n}}} \quad \text{Test statistic}$$

EXAMPLE Finding the Test Statistic Use the sample data ($n = 106$, $\bar{x} = 98.20$, and $s = 0.62$) in the preceding informal example, and find the value of the test statistic for the claim that the population mean is given by $\mu = 98.6$.

SOLUTION With $n = 106$, $\bar{x} = 98.20$, and $s = 0.62$, and with the assumption that $\mu = 98.6$, the test statistic is calculated by using Formula 7-1 as shown below (with s used as an estimate of σ).

$$z = \frac{\bar{x} - \mu_{\bar{x}}}{\frac{\sigma}{\sqrt{n}}} = \frac{98.20 - 98.6}{\frac{0.62}{\sqrt{106}}} = -6.64$$

Critical Region, Significance Level, and Critical Value

- The **critical region** is the set of all values of the test statistic that cause us to reject the null hypothesis. For the preceding informal example, the critical region is represented by the red-shaded part of Figure 7-1 and consists of values of the test statistic less than $z = -1.96$ or greater than $z = 1.96$.
- The **significance level** (denoted by α) is the probability that the test statistic will fall in the critical region when the null hypothesis is actually true. If the test statistic falls in the critical region, we will reject the null hypothesis, so α is the probability of making the mistake of rejecting the null hypothesis when it is true. This is the same α introduced in Section 6-2, where we defined the degree of confidence for a confidence interval to be the probability $1 - \alpha$. Common choices for α are 0.05, 0.01, and 0.10. In Figure 7-1, the red-shaded regions in the two tails combine to be $\alpha = 0.05$. The significance level is described by $\alpha = 0.05$.
- A **critical value** is any value that separates the critical region (where we reject the null hypothesis) from the values of the test statistic that do not lead to rejection of the null hypothesis. The critical values depend on the nature of the null hypothesis, the relevant sampling distribution, and the significance level α. For the preceding example, the critical values of $z = -1.96$ and $z = 1.96$ separate the red-shaded critical regions, as shown in Figure 7-1.

EXAMPLE **Finding Critical Values** Suppose that we have the mean from a large ($n > 30$) sample of the amounts of Coke in cans labeled 12 oz. The normal distribution applies (because such sample means are approximately normally distributed according to the central limit theorem). Using a significance level of $\alpha = 0.05$, find the critical z values for each of the following null hypotheses:

a. $\mu = 12$ oz (so the critical region is in *both* tails of the normal distribution, where we reject the given null hypothesis)

b. $\mu \geq 12$ oz (so the critical region is in the *left* tail of the normal distribution, where we reject the given null hypothesis)

c. $\mu \leq 12$ oz (so the critical region is in the *right* tail of the normal distribution, where we reject the given null hypothesis)

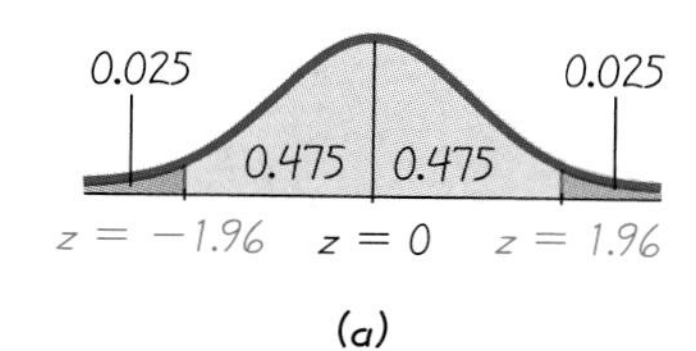

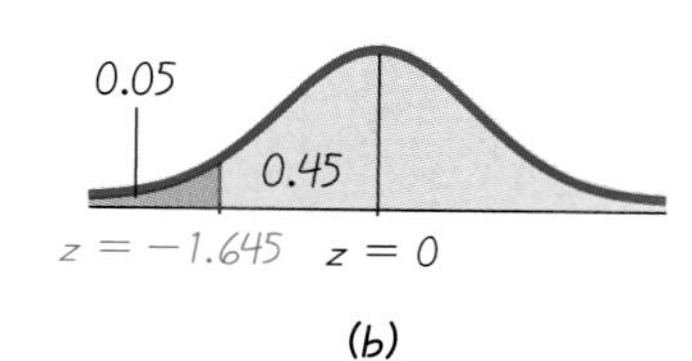

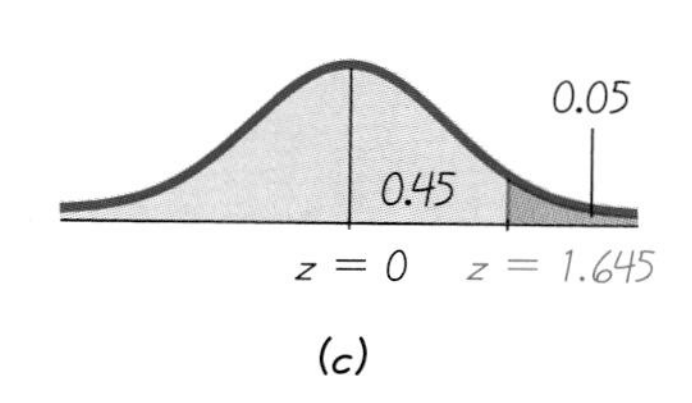

FIGURE 7-3
Finding Critical Values

SOLUTION

a. See Figure 7-3(a). The shaded tails contain a total area of $\alpha = 0.05$, so each tail contains an area of 0.025. Using the methods of Section 5-2 (see the subsection "Finding z Scores When Given Probabilities"), the values of $z = 1.96$ and $z = -1.96$ separate the right and left tail regions. The critical values are therefore $z = 1.96$ and $z = -1.96$.

b. See Figure 7-3(b). With a null hypothesis of $\mu \geq 12$ oz, the critical region is in the left tail, where we *reject* the null hypothesis. (More about that later, too.) With a left-tail area of 0.05, the critical value is found to be $z = -1.645$ (by using the methods of Section 5-2).

c. See Figure 7-3(c). With a null hypothesis of $\mu \leq 12$ oz, the critical region is in the right tail, where we *reject* the null hypothesis. (More about that later.) With a right-tail area of 0.05, the critical value is found to be $z = 1.645$ (by using the methods of Section 5-2).

Two-Tailed, Left-Tailed, Right-Tailed The *tails* in a distribution are the extreme regions bounded by critical values. Some hypothesis tests are two-tailed, some are right-tailed, and some are left-tailed.

- **Two-tailed test:** The critical region is in the two extreme regions (tails) under the curve.
- **Right-tailed test:** The critical region is in the extreme right region (tail) under the curve.
- **Left-tailed test:** The critical region is in the extreme left region (tail) under the curve.

In two-tailed tests, the significance level α is divided equally between the two tails that constitute the critical region. For example, in a two-tailed test with a significance level of $\alpha = 0.05$, there is an area of 0.025 in each of the two tails. In tests that are right- or left-tailed, the area of the critical region is α.

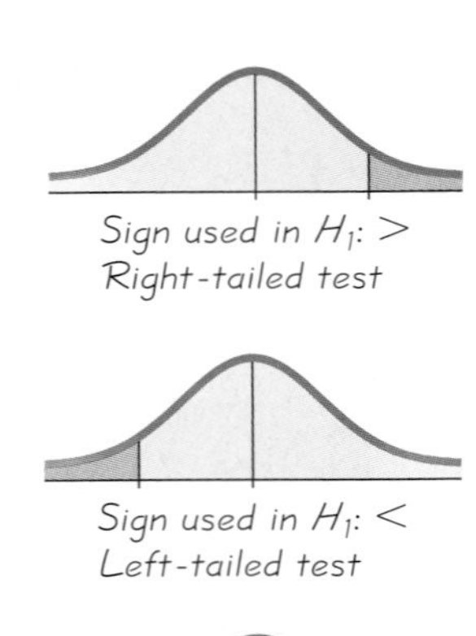

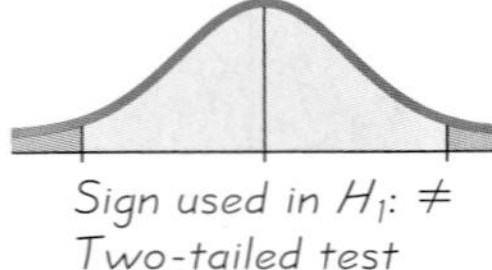

By examining the null hypothesis, we should be able to deduce whether a test is right-tailed, left-tailed, or two-tailed. The tail will correspond to the critical region containing the values that would conflict significantly with the null hypothesis. A useful check is summarized in the margin figures, which show that the inequality sign in H_1 points in the direction of the critical region. The symbol ≠ is often expressed in programming languages as <>, and this reminds us that an alternative hypothesis such as $\mu \neq 98.6$ corresponds to a two-tailed test.

Conclusions in Hypothesis Testing

We have seen that the original claim sometimes becomes the null hypothesis and at other times becomes the alternative hypothesis. However, our standard procedure of hypothesis testing requires that we always test the null hypothesis, and we will see in the following section that our initial conclusion will always be one of the following:

1. Reject the null hypothesis.
2. Fail to reject the null hypothesis.

Wording the Final Conclusion: The conclusion of rejecting the null hypothesis or failing to reject it is fine for those of us with the wisdom to take a statistics course, but we should use simple, nontechnical terms in stating what the conclusion really means. Figure 7-4 shows how to formulate the correct wording of the final conclusion. Note that only one case leads to wording indicating that the sample data actually *support* the conclusion. If you want to support some claim, state it in such a way that it becomes the alternative hypothesis, and then hope that the null hypothesis gets rejected. For example, to support the claim that the mean body temperature is different from 98.6°F, make the claim that $\mu \neq 98.6$. This claim will be an alternative hypothesis that will be supported if you reject the null hypothesis, H_0: $\mu = 98.6$. If, on the other hand, you claim that $\mu = 98.6$, you will either reject or fail to reject the claim; in either case, you will not *support* the claim that $\mu = 98.6$.

Accept/Fail to Reject: Some texts say "accept the null hypothesis" instead of "fail to reject the null hypothesis." Whether we use the term *accept* or *fail to reject*, we should recognize that *we are not proving the null hypothesis;* we are merely saying that the sample evidence is not strong enough to warrant rejection of the null hypothesis. It's like a jury's saying that there is not enough evidence to convict a suspect. The term *accept* is somewhat misleading, because it seems to imply incorrectly that the null hypothesis has been proved. The phrase *fail to reject* says more correctly that the available evidence isn't strong enough to warrant rejection of the null hypothesis. In this text we will use the terminology *fail to reject the null hypothesis*, instead of *accept the null hypothesis.*

EXAMPLE Stating the Final Conclusion Suppose researchers claim that the mean body temperature is different from 98.6°F. This claim of

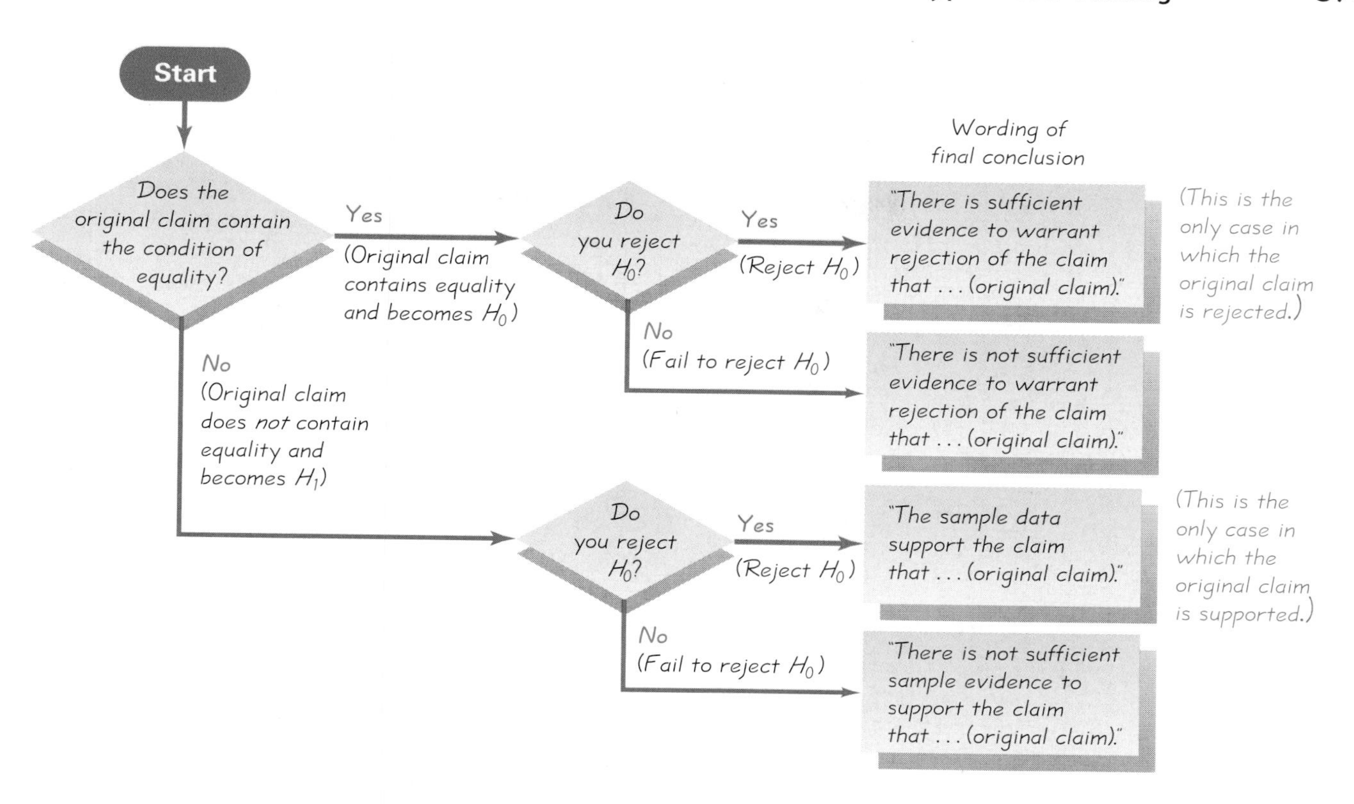

FIGURE 7-4
Wording of Final Conclusion

$\mu \neq 98.6$ becomes the alternative hypothesis (because it does not contain equality), while the null hypothesis becomes $\mu = 98.6$. Further suppose that the sample evidence causes us to reject the null hypothesis of $\mu = 98.6$. State the conclusion in simple, nontechnical terms.

SOLUTION Refer to Figure 7-4. The original claim does not contain the condition of equality, and we do reject the null hypothesis. The wording of the final conclusion should therefore be as follows: "The sample data support the claim that the mean body temperature is different from 98.6°F."

Type I and Type II Errors When testing a null hypothesis, we arrive at a conclusion of rejecting it or failing to reject it. Such conclusions are sometimes correct and sometimes wrong (even if we do everything correctly). Table 7-2 on the next page summarizes the two different types of errors that can be made, along with the two different types of correct decisions. We distinguish between the two types of errors by calling them *type I and type II errors.*

- **Type I error:** The mistake of rejecting the null hypothesis when it is actually true. The symbol $\boldsymbol{\alpha}$ (alpha) is used to represent the probability of a type I error.
- **Type II error:** The mistake of failing to reject the null hypothesis when it is actually false. The symbol $\boldsymbol{\beta}$ (beta) is used to represent the probability of a type II error.

TABLE 7-2 Type I and Type II Errors

		True State of Nature	
		The null hypothesis is true.	The null hypothesis is false.
Decision	We decide to reject the null hypothesis.	**Type I error** (rejecting a true null hypothesis) α	Correct decision
	We fail to reject the null hypothesis.	Correct decision	**Type II error** (failing to reject a false null hypothesis) β

Notation

α (alpha) = probability of a type I error (the probability of rejecting the null hypothesis when it is true)

β (beta) = probability of a type II error (failing to reject a false null hypothesis)

EXAMPLE **Identifying Type I and Type II Errors** Assume that we are conducting a hypothesis test with the following null and alternative hypotheses:

$$H_0: \mu = 98.6$$
$$H_1: \mu \neq 98.6$$

Give statements identifying

a. a type I error.

b. a type II error.

SOLUTION

a. A type I error is the mistake of rejecting a true null hypothesis, so this is a type I error: Rejecting the claim that the mean is 98.6 when the mean really does equal 98.6.

b. A type II error is the mistake of failing to reject the null hypothesis when it is false, so this is a type II error: Failing to reject the claim that the mean equals 98.6 when the mean is really different from 98.6.

Controlling Type I and Type II Errors: One step in our standard procedure for testing hypotheses involves the selection of the significance level α, which is the probability of a type I error. However, we don't select β [P(type II error)]. It would be great if we could always have $\alpha = 0$ and $\beta = 0$, but in reality that is not possible, so we must attempt to manage the α and β error probabilities. Mathematically, it can be shown that α, β, and the sample size n are all related, so when you choose or determine any two of them, the third is automatically determined. The usual practice in research and industry is to select the values of α and n, so the value of β is determined. Depending on the seriousness of a type I error, try to use the largest α that you can tolerate. For type I errors with more serious consequences, select smaller values of α. Then choose a sample size n as large as is reasonable, based on considerations of time, cost, and other relevant factors. (Sample size determinations were discussed in Section 6-4.) The following practical considerations may be relevant:

1. For any fixed α, an increase in the sample size n will cause a decrease in β. That is, a larger sample will lessen the chance that you make the error of not rejecting the null hypothesis when it's actually false.
2. For any fixed sample size n, a decrease in α will cause an increase in β. Conversely, an increase in α will cause a decrease in β.
3. To decrease both α and β, increase the sample size.

To make sense of these abstract ideas, let's consider M&Ms (produced by Mars, Inc.) and Bufferin brand aspirin tablets (produced by Bristol-Myers Products).

- The mean weight of the M&M candies is supposed to be at least 0.9085 g (in order to conform to the weight printed on the package label).
- The Bufferin tablets are supposed to have a mean weight of 325 mg of aspirin.

Because M&Ms are candies used for enjoyment, whereas Bufferin tablets are drugs used for treatment of health problems, we are dealing with two very different levels of seriousness. If the M&Ms don't have a mean weight of 0.9085 g, the consequences are not very serious, but if the Bufferin tablets don't contain a mean of 325 mg of aspirin, the consequences could be very serious, possibly including consumer lawsuits and actions on the part of the Federal Drug Administration. Consequently, in testing the claim that $\mu = 0.9085$ g for M&Ms, we might choose $\alpha = 0.05$ and a sample size of $n = 100$; in testing the claim that $\mu = 325$ mg for Bufferin tablets, we might choose $\alpha = 0.01$ and a sample size of $n = 500$. The smaller significance level α and larger sample size n are chosen because of the more serious consequences associated with testing a commercial drug.

Power of a Test: We have presented notation indicating that β is the probability of failing to reject a false null hypothesis (type II error). It follows that $1 - \beta$ is the probability of rejecting a false null hypothesis. Statisticians refer to this probability as the *power* of a test, and they often use it to gauge the test's effectiveness in recognizing that a null hypothesis is false.

Large Sample Size Isn't Good Enough

Biased sample data should not be used for inferences, no matter how large the sample is. For example, in Women and Love: A Cultural Revolution in Progress, Shere Hite bases her conclusions on 4500 replies that she received after mailing 100,000 questionnaires to various women's groups. A random sample of 4500 subjects would usually provide good results, but Hite's sample is biased. It is criticized for overrepresenting women who join groups and women who feel strongly about the issues addressed. Because Hite's sample is biased, her inferences are not valid, even though the sample size of 4500 might seem to be sufficiently large.

DEFINITION

The **power** of a hypothesis test is the probability $(1 - \beta)$ of rejecting a false null hypothesis, which is computed by using a particular significance level α and a particular value of the mean that is an alternative to the value assumed true in the null hypothesis.

Suppose we are using a 0.05 significance level to test the null hypothesis that the mean height of men is 6 ft (or 72 in.). Given sample data and given the alternative height of 69 in., we can compute the power of the test to reject $\mu = 72$ in. If our sample consists of only a few observations, the power will be low, but if it consists of hundreds of observations, the power will be much higher. (In addition to increasing the sample size, there are other ways to increase the power, such as increasing the significance level, using a more extreme value for the population mean, or decreasing the standard deviation.) Just as 0.05 is a common choice for a significance level, a power of at least 0.80 is a common requirement for determining that a hypothesis test is effective. Because the calculations of power are really tough, only Exercise 40 deals with power.

In Section 7-3 we will show how the components presented in this section are used in formal procedures for testing hypotheses.

7-2 Basic Skills and Concepts

Stating Conclusions about Claims. *In Exercises 1–4, what do you conclude? (Don't use formal procedures and exact calculations. Use only the rare event rule described in Section 7-1, and make subjective estimates to determine whether events are likely.)*

1. Claim: A coin is fair, and heads turns up 27 times in 30 tosses.

2. Claim: A die is fair, and the outcome of 1 occurs 9 times in 60 rolls.

3. Claim: Women who eat blue M&M candies have a better chance of having a baby boy, and 50 such women gave birth to 27 boys and 23 girls.

4. Claim: A roulette wheel is fair, and the number 7 occurs each time in seven consecutive trials. (A roulette wheel has 38 equally likely slots, one of which is 7.)

Formulating Claims and Conclusions. *In Exercises 5–8, write the claim that is suggested by the given statement, then write a conclusion about the claim. Don't use symbolic expressions or formal procedures; use common sense.*

5. As the owner of the Travelsafe Auto Insurance Company, you observe that one customer, Lisa Kerr, has filed a claim for a different stolen car in each of the past five years.

6. Yuri is a "mentalist" who claims that he has powers of mental telepathy. He is asked to identify the month on a calendar in an adjacent room, and he is correct six times in 60 such trials.

7. The rate of Lyme disease cases in Clinton County is 2%. When 1000 people from that county are given a placebo, it is found that 22 of them contract Lyme disease. When 1000 people from that county are given a new experimental vaccine, it is found that 18 of them contract Lyme disease.

8. In a test of the polio vaccine, 200,000 children are given a placebo and 200,000 other children are given the Salk vaccine. Among those in the placebo group, 115 later developed polio. Among those given the Salk vaccine, 33 later developed polio.

Identifying Hypotheses. *In Exercises 9–16, examine the given statement, then express the null hypothesis H_0 and alternative hypothesis H_1 in symbolic form. Be sure to use the correct symbol (μ, p, σ) for the indicated parameter.*

9. The mean age of gamblers is greater than 30 years.

10. The mean salary of statistics professors is at least $60,000.

11. Fewer than one-half of all Internet users make on-line purchases.

12. The proportion of defective computers is less than 0.05.

13. Women's heights have a standard deviation less than 2.8 in., which is the standard deviation for men's heights.

14. The percentage of viewers tuned to *60 Minutes* is equal to 24%.

15. The mean amount of Coke in cans is at least 12 oz.

16. Salaries among women business analysts have a standard deviation greater than $3000.

Finding Critical Values. *In Exercises 17–24, find the critical z values. In each case, assume that the normal distribution applies.*

17. Two-tailed test; $\alpha = 0.05$

18. Two-tailed test; $\alpha = 0.01$

19. Right-tailed test; $\alpha = 0.02$

20. Left-tailed test; $\alpha = 0.05$

21. $\alpha = 0.04$; H_1 is $\mu \neq 250$.

22. $\alpha = 0.10$; H_1 is $\mu < 98.6$.

23. $\alpha = 0.06$; H_0 is $\mu \leq 980$.

24. $\alpha = 0.02$; H_0 is $\mu \geq 56.7$.

Finding Test Statistics. *In Exercises 25–28, use Formula 7-1 to find the value of the test statistic z.*

25. From a study of consumer buying: The claim is $\mu = 0.21$, and the sample statistics include $n = 32$, $\overline{x} = 0.83$, and $s = 0.24$.

26. From a *New York Times* article on ages: The claim is $\mu = 73.4$, and the sample statistics include $n = 35$, $\overline{x} = 69.5$, and $s = 8.7$.

27. From a study of seat belt use: The claim is $\mu < 1.39$, and the sample statistics include $n = 123$, $\overline{x} = 0.83$, and $s = 0.16$.

28. From a study of the amounts of Coke in cans: The claim is $\mu = 12.00$, and the sample statistics include $n = 36$, $\overline{x} = 12.19$, and $s = 0.11$.

Stating Conclusions. *In Exercises 29–32, identify the null hypothesis, then state the final conclusion in simple, nontechnical terms. Be sure to address the original claim. (See Figure 7-4.)*

29. Original claim: Women have a mean height equal to 60.0 in.

Initial conclusion: Reject the null hypothesis.

30. Original claim: The mean IQ score of statistics professors is greater than 100.

Initial conclusion: Reject the null hypothesis.

31. Original claim: The mean amount of Coke in cans labeled 12 oz is actually less than 12 oz.

Initial conclusion: Fail to reject the null hypothesis.

32. Original claim: The mean body temperature of healthy adults is equal to 98.2°F.

Initial conclusion: Fail to reject the null hypothesis.

Identifying Type I and Type II Errors. *In Exercises 33–36, identify the type I error and the type II error that correspond to the given hypothesis. (Hint: Begin by identifying the null hypothesis.)*

33. The mean amount of Coke in cans is equal to 12 oz.

34. The mean annual income of college graduates is greater than $20,000.

35. The mean IQ of convicted felons is less than 100.

36. The mean weight loss of Slimnow diet pill users is at least 5 lb.

7-2 Beyond the Basics

37. Why Must Equality Be Assumed? When testing a claim that a mean is at least 100, the null hypothesis becomes $\mu \geq 100$. When actually conducting the hypothesis test, why is it necessary to assume that $\mu = 100$ instead of $\mu \geq 100$?

38. Proving Claims Under standard conditions, the Ford Taurus travels 21 mi while consuming exactly 1 gal of gas. You have developed a modification to the engine that you believe increases that distance. If you want to justify your claim of increased mileage by using a hypothesis test, is it possible to state your claim so that it becomes the null hypothesis? Why or why not?

39. Using $\alpha = 0$ Someone suggests that in testing hypotheses, you can eliminate a type I error by making $\alpha = 0$. In a two-tailed test, what critical values correspond to $\alpha = 0$? If $\alpha = 0$, will the null hypothesis ever be rejected?

40. Power of a Test Assume that you are using a significance level of $\alpha = 0.05$ to test the claim that $\mu < 2$ and that your sample is a simple random sample of 50 values.

a. Find β, the probability of making a type II error, given that the population actually has a normal distribution with $\mu = 1.5$ and $\sigma = 1$. (*Hint:* With H_0: $\mu \geq 2$, begin by finding the values of the sample means that do not lead to rejection of H_0, then find the probability of getting a sample mean with one of those values.)

b. Find $1 - \beta$, which is the *power* of the test. If β is the probability of *failing* to reject a false null hypothesis, describe the probability of $1 - \beta$.

7-3 Testing a Claim about a Mean: Large Samples

In this section we combine the components introduced in Section 7-2 into a unified and systematic procedure for testing claims made about population means. We will present three methods of testing hypotheses that might appear to be different but are equivalent in the sense that they always lead to similar conclusions. The first procedure is the traditional method; the second procedure is the P-value method; and the third procedure is the confidence interval method. Because confidence intervals were thoroughly discussed in Chapter 6, our emphasis now will be placed on the traditional and P-value methods.

We begin by identifying the three assumptions that apply to the methods of this section.

Assumptions

1. The sample is a simple random sample. (Remember this very important point made in Chapter 1: *Data carelessly collected may be so completely useless that no amount of statistical torturing can salvage them.*)
2. The sample is large ($n > 30$). (This allows us to apply the central limit theorem so that we can use the normal distribution. *There is no requirement that the population data must have a normal distribution.*)
3. If the value of the population standard deviation σ is unknown, we can use the sample standard deviation s as an estimate of σ. (We can do this only if the sample size is large, with more than 30 sample values.)

Our first step is to *explore* the data set. Using the methods introduced in Chapter 2, we investigate center, variation, and distribution by drawing a graph; finding the mean, standard deviation, and 5-number summary; and identifying any outliers. In this section we will use the 106 body temperatures given in the Chapter Problem. Almost everyone assumes that the mean body temperature is 98.6°F, but we will test that belief. For now, we might note that the 106 body temperatures in Table 7-1 have a mean of 98.20°F. Those sample values also have the following 5-number summary: minimum = 96.5, $Q_1 = 97.8$, median = 98.4, $Q_3 = 98.6$, and maximum = 99.6. It's worth noting that 98.6 is the third quartile instead of being at the center. There do not appear to be any outliers that are far away from all of the other sample values. (Outliers can dramatically affect some results, such as those discussed in this chapter.) The accompanying STATDISK display of a histogram on the next page shows that the 106 body temperatures have a distribution that is roughly bell-shaped, and the histogram verifies that there are no outliers very far away from the other sample values. The histogram also shows that

98.6 is not a perfect balance point for the histogram; if a fulcrum is placed at 98.6, the histogram would tilt to the left, indicating that the mean appears to be lower than 98.6.

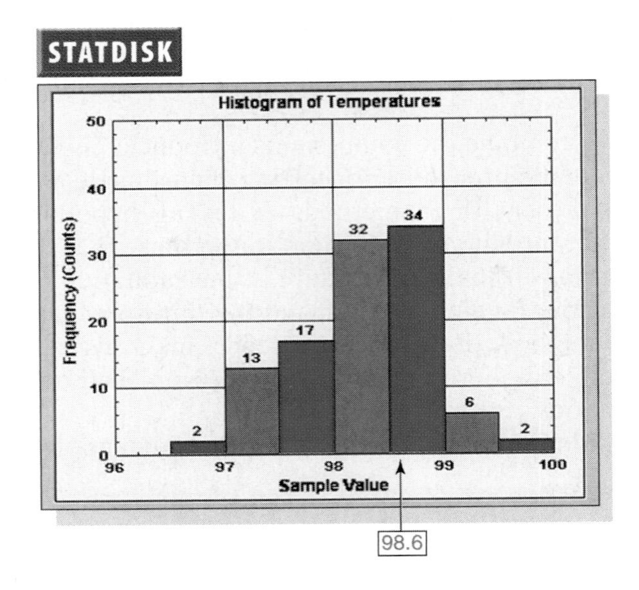

The Traditional Method of Testing Hypotheses

Figure 7-5 summarizes the steps used in the **traditional (or classical) method** of testing hypotheses. This procedure uses the components described in Section 7-2 as part of a system to identify a sample result that is *significantly* different from the claimed value. The relevant sample statistic is converted to a test statistic, which we compare to a critical value. The following expression can be used for the examples and exercises of this section.

Test Statistic for Claims about μ When $n > 30$

$$z = \frac{\bar{x} - \mu_{\bar{x}}}{\frac{\sigma}{\sqrt{n}}}$$

Note that Step 7 in Figure 7-5 uses this decision criterion:

Reject the null hypothesis if the test statistic is in the critical region.

Fail to reject the null hypothesis if the test statistic is not in the critical region.

Section 7-2 presented an informal example of a hypothesis test; the following example formalizes that test.

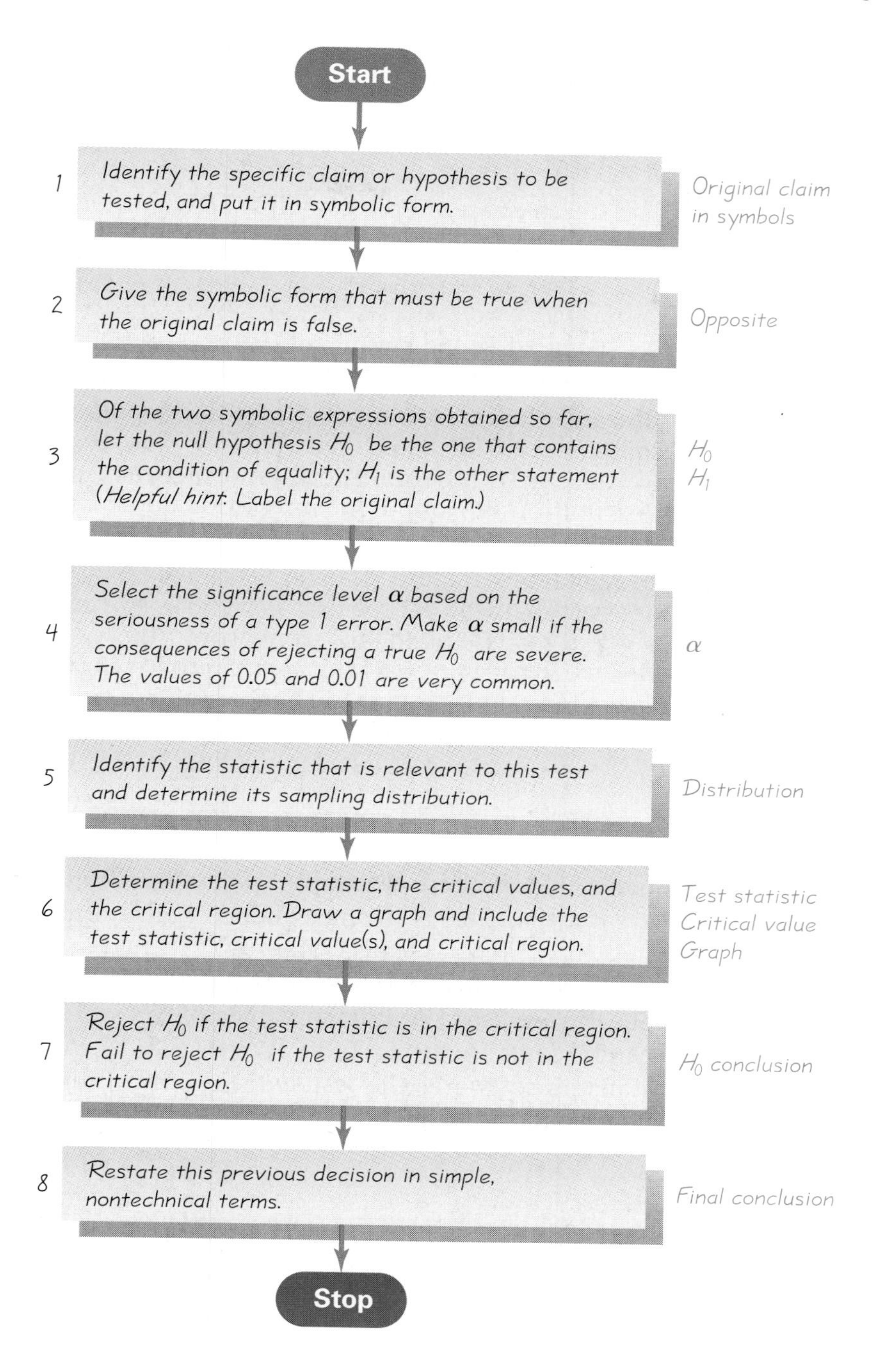

FIGURE 7-5
Traditional Method of Hypothesis Testing

EXAMPLE **Body Temperatures** Using the sample data given in the Chapter Problem ($n = 106$, $\overline{x} = 98.20$, $s = 0.62$) and a 0.05 significance level, test the claim that the mean body temperature of healthy adults is equal to 98.6°F. Use the traditional method by following the procedure outlined in Figure 7-5.

SOLUTION Refer to Figure 7-5 and follow these steps.

Step 1: The claim that the mean is equal to 98.6 is expressed in symbolic form as $\mu = 98.6$.

Step 2: The alternative (in symbolic form) to the original claim is $\mu \neq 98.6$.

Step 3: The statement $\mu = 98.6$ contains the condition of equality, and so it becomes the null hypothesis. We have

$$H_0: \mu = 98.6 \text{ (original claim)} \qquad H_1: \mu \neq 98.6$$

Step 4: As specified in the statement of the problem, the significance level is $\alpha = 0.05$.

Step 5: Because the claim is made about the population mean μ, the sample statistic most relevant to this test is $\bar{x} = 98.20$. Because $n > 30$, the central limit theorem indicates that the distribution of sample means can be approximated by a *normal* distribution.

Step 6: In calculating the test statistic, we can use $s = 0.62$ as a reasonable estimate of σ (because $n > 30$), so the test statistic of $z = -6.64$ is found by converting the sample mean of $\bar{x} = 98.20$ to $z = -6.64$ through the following computation.

$$z = \frac{\bar{x} - \mu_{\bar{x}}}{\frac{\sigma}{\sqrt{n}}} = \frac{98.20 - 98.6}{\frac{0.62}{\sqrt{106}}} = -6.64$$

The critical z values are found by first noting that the test is two-tailed because a sample mean significantly less than or greater than 98.6 is strong evidence against the null hypothesis that $\mu = 98.6$. Because the test is two-tailed, we divide $\alpha = 0.05$ equally between the two tails to get 0.025 in each tail. See Figure 7-6. The critical values of $z = -1.96$ and $z = 1.96$ can be found by using Table A-2, or a TI-83 Plus calculator, or statistics software. The test statistic, critical region, and critical values are shown in Figure 7-6.

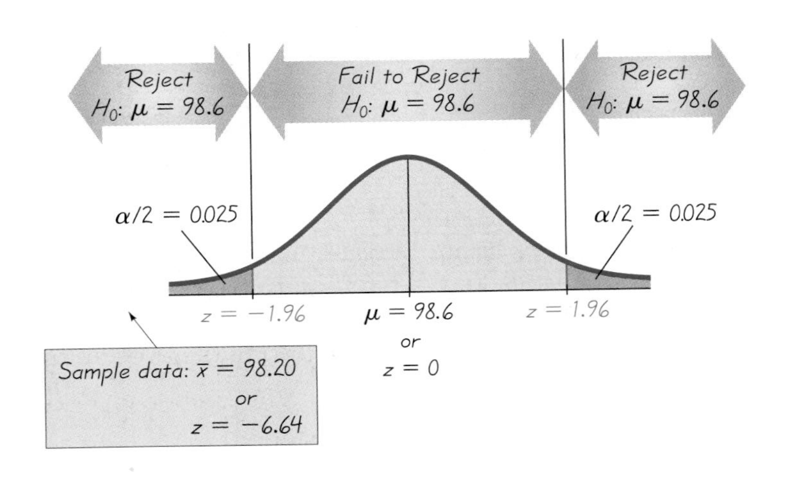

FIGURE 7-6 Distribution of Means of Body Temperatures Assuming $\mu = 98.6$

Step 7: Because the test statistic of $z = -6.64$ falls within the critical region, we reject the null hypothesis.

INTERPRETATION To restate the Step 7 conclusion in nontechnical terms, we might refer to Figure 7-4 in the preceding section. We are rejecting the null hypothesis, which is the original claim. We conclude that there is sufficient evidence to warrant rejection of the claim that the mean body temperature of healthy adults is 98.6°F.

In the preceding example, the sample mean of 98.20 was converted to a test statistic of $z = -6.64$. That is, the sample mean is 6.64 standard deviations away from the claimed mean. In Chapter 2 we saw that a z score of 6.64 is so large that it represents an unusual result (if, in fact, $\mu = 98.6$). This evidence is strong enough to make us believe that the mean is actually different from 98.6. The critical values and critical region clearly identify the range of unusual values that cause us to reject the claimed value of μ.

The preceding example illustrated a two-tailed hypothesis test; the following example illustrates a right-tailed test. The following example also illustrates the difference between *statistical significance* and *practical significance.*

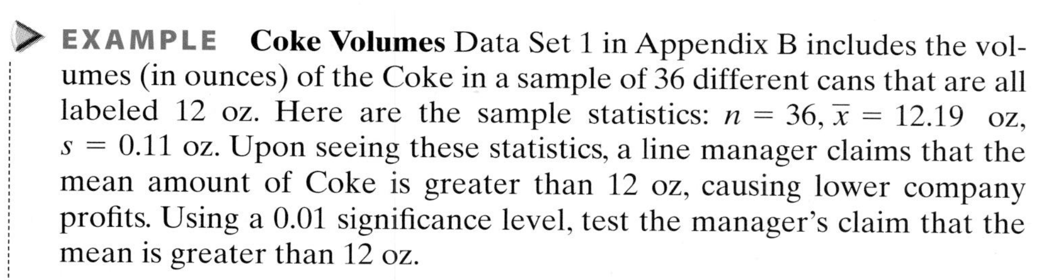

EXAMPLE **Coke Volumes** Data Set 1 in Appendix B includes the volumes (in ounces) of the Coke in a sample of 36 different cans that are all labeled 12 oz. Here are the sample statistics: $n = 36$, $\bar{x} = 12.19$ oz, $s = 0.11$ oz. Upon seeing these statistics, a line manager claims that the mean amount of Coke is greater than 12 oz, causing lower company profits. Using a 0.01 significance level, test the manager's claim that the mean is greater than 12 oz.

SOLUTION We know that measured amounts are subject to random chance variation, so we need to determine whether the sample mean of 12.19 oz is *significantly* greater than 12 oz. Refer to Figure 7-5 and follow these steps.

Step 1: Express the manager's claim as $\mu > 12$.

Step 2: The opposite (in symbolic form) of the original claim is $\mu \leq 12$.

Step 3: The statement $\mu \leq 12$ contains the condition of equality, so it becomes the null hypothesis and we have

$$H_0: \mu \leq 12 \qquad H_1: \mu > 12 \text{ (original claim)}$$

Step 4: With a 0.01 significance level, we have $\alpha = 0.01$.

Step 5: We will use a *normal* distribution because the claim is directed at the population mean, so we are working with the distribution of sample means. Because $n > 30$, the central limit theorem indicates that the distribution of sample means can be approximated by a normal distribution.

Statistics: Jobs and Employers

Here is a small sample of advertised jobs in the field of statistics: forecaster, database analyst, marketing scientist, credit-risk manager, cancer researcher and evaluator, insurance-risk analyst, educational testing researcher, biostatistician, statistician for pharmaceutical products, cryptologist, statistical programmer.

Here is a small sample of firms offering jobs in the field of statistics: Centers for Disease Control and Prevention, Cardiac Pacemakers, Inc., National Institutes of Health, National Cancer Institute, CNA Insurance Companies, Educational Testing Service, Roswell Park Cancer Institute, Cleveland Clinic Foundation, National Security Agency, Quantiles, 3M, IBM, Nielsen Media Research, AT&T Labs, Bell Labs, Hewlett Packard, Johnson & Johnson, Smith Hanley.

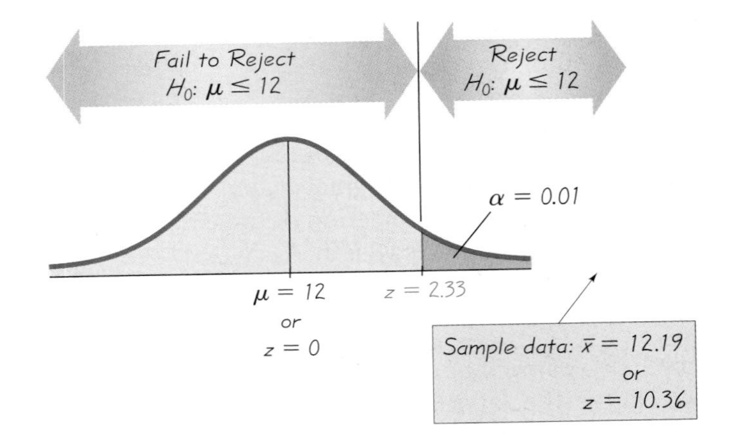

FIGURE 7-7
Distribution of Means of Amounts of Coke Assuming $\mu = 12$ oz

Step 6: In calculating the test statistic, we can use $s = 0.11$ as a reasonable estimate of σ (because $n > 30$), so the test statistic of $z = 10.36$ is computed as follows:

$$z = \frac{\bar{x} - \mu_{\bar{x}}}{\frac{\sigma}{\sqrt{n}}} = \frac{12.19 - 12}{\frac{0.11}{\sqrt{36}}} = 10.36$$

The critical value of $z = 2.33$ is found in Table A-2 as the z score corresponding to an area of 0.49. (The test is right-tailed because the null hypothesis of $\mu \le 12$ is rejected if the sample mean $\bar{x}$ is *greater than* 12 by a significant amount.) The test statistic, critical region, and critical value are shown in Figure 7-7.

Step 7: From Figure 7-7 we see that the sample mean of 12.19 oz does fall within the critical region, so we reject the null hypothesis H_0.

INTERPRETATION Our final conclusion is that the sample data support the claim that the mean amount of Coke is greater than 12 oz. (Refer to Figure 7-4 for help with wording this final conclusion.) The Coca Cola Company could increase profits by filling cans with slightly less Coke. As consumers, we know that we are not being cheated by the Coca Cola Company, but the extra Coke we are given is so small that there would not be a perceptible difference. From the perspective of consumers, even though the result is *statistically significant,* it does not appear to have *practical significance* because the mean appears to exceed 12 oz by a very small amount.

In presenting the results of a hypothesis test, it is not always necessary to show all of the steps included in Figure 7-5. However, the results should include the null hypothesis, the alternative hypothesis, the calculation of the test statistic, a graph such as Figure 7-7, the initial conclusion (reject H_0 or fail to reject H_0), and the final conclusion stated in nontechnical terms. The graph should show the test statistic, critical value(s), critical region, and significance level.

The *P*-Value Method of Testing Hypotheses

The *P*-value method of testing hypotheses is very similar to the traditional method, but the key difference is the way in which we decide whether to reject the null hypothesis. The approach used by the *P*-value method is to find the *probability* (*P*-value) of getting a result such as the one obtained, and reject the null hypothesis if that probability is very low. Let's start with a formal definition of the *P*-value.

DEFINITION

A ***P*-value** (or **probability value)** is the probability of getting a value of the sample test statistic that is *at least as extreme* as the one found from the sample data, assuming that the null hypothesis is true.

It is important to understand why the preceding definition uses the probability of a value *at least as extreme* as the one found. Why not simply find the probability of the particular result that occurred? Suppose you toss a coin 1000 times to determine whether it favors heads, and suppose you get 501 heads. Intuition should suggest that this is not evidence that the coin favors heads, because it is quite easy to get 501 heads in 1000 tosses. Yet, the probability of getting exactly 501 heads in 1000 tosses is actually quite small: 0.0252. With 1000 tosses, any specific number of heads will have a very low probability. However, the result of 501 heads among 1000 tosses is not *unusual* because the probability of getting *at least* 501 heads is high: 0.487. The point to keep in mind is that when you are deciding whether a sample result is unusual, you need to find the probability of getting a result *at least as extreme* as the result obtained, not the probability of getting the particular result by itself. This is the concept underlying *P*-values.

The *P*-value is the likelihood that a sample, such as the one obtained, will occur when the null hypothesis is actually true. A very small *P*-value (such as 0.05 or lower) suggests that the sample results are very unlikely under the assumption of the null hypothesis, so that such a small *P*-value is evidence against the null hypothesis. It would be helpful to remember these associations:

P-Value	Interpretation
Small *P*-values (such as 0.05 or lower)	*Unusual* sample results. *Significant* difference from the null hypothesis
Large *P*-values (such as above 0.05)	Sample results are *not unusual.* *Not a significant* difference from the null hypothesis

AS

If you are a medical researcher and you believe that you have developed a new drug to cure baldness, you want your hypothesis test to result in a *small* *P*-value, indicating that your results are highly significant.

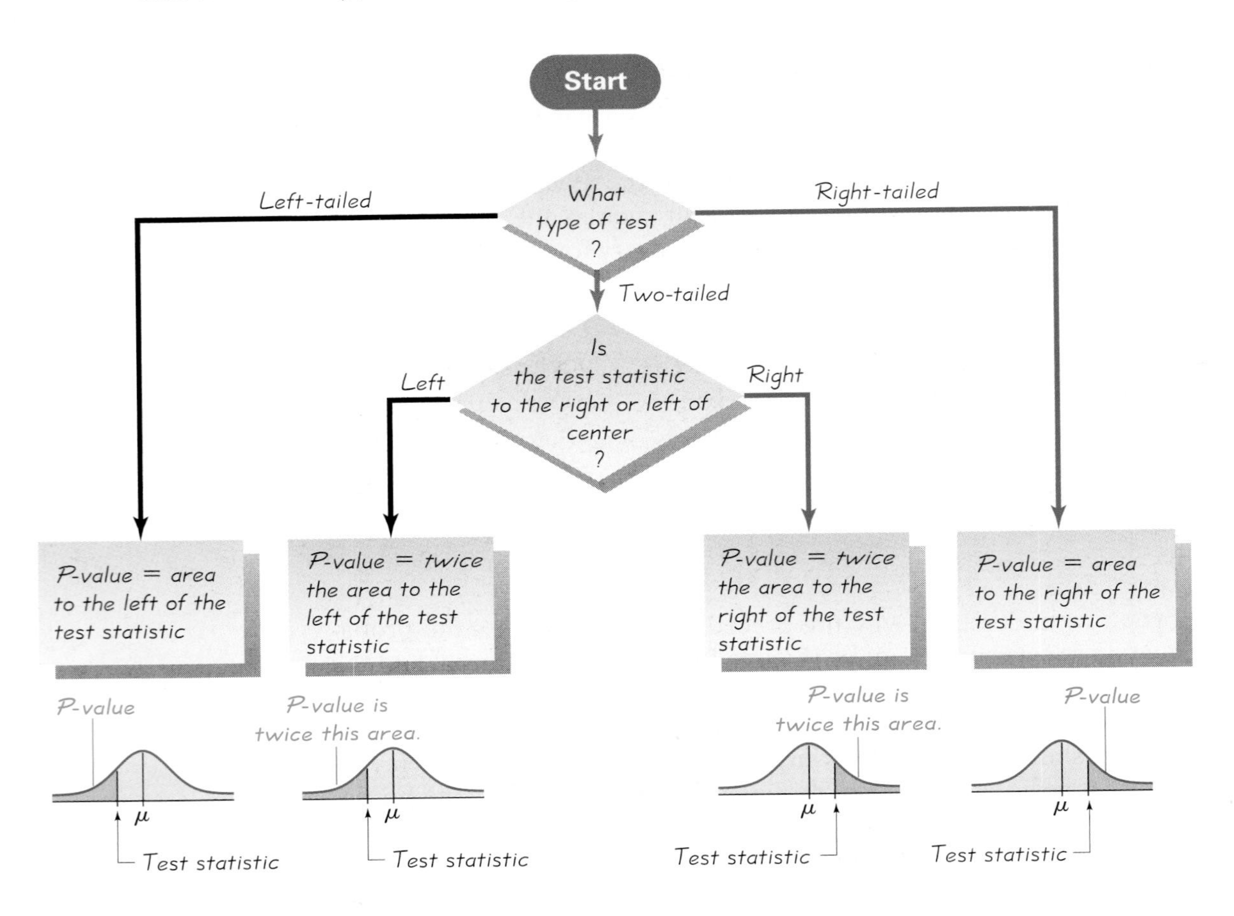

FIGURE 7-8
Finding *P*-Values

The procedure for finding *P*-values is summarized in Figure 7-8. The figure shows the following:

Right-tailed test: The *P*-value is the area to the right of the test statistic.

Left-tailed test: The *P*-value is the area to the left of the test statistic.

Two-tailed test: The *P*-value is *twice* the area of the extreme region bounded by the test statistic.

Figure 7-9 summarizes the *P*-value method for testing hypotheses. A comparison of the traditional method (summarized in Figure 7-5) and the *P*-value method (Figure 7-9) shows that they are essentially the same, but they differ in the decision criterion. The traditional method compares the test statistic to the critical values, whereas the *P*-value method compares the *P*-value to the significance level. However, the traditional and *P*-value methods are equivalent in the sense that they will always result in the same conclusion. They are simply different paths leading to the same final destination.

The *P*-value approach uses most of the same basic procedures as the traditional approach, but Steps 6 and 7 are different.

Step 6: Find the *P*-value (as shown in Figure 7-8).

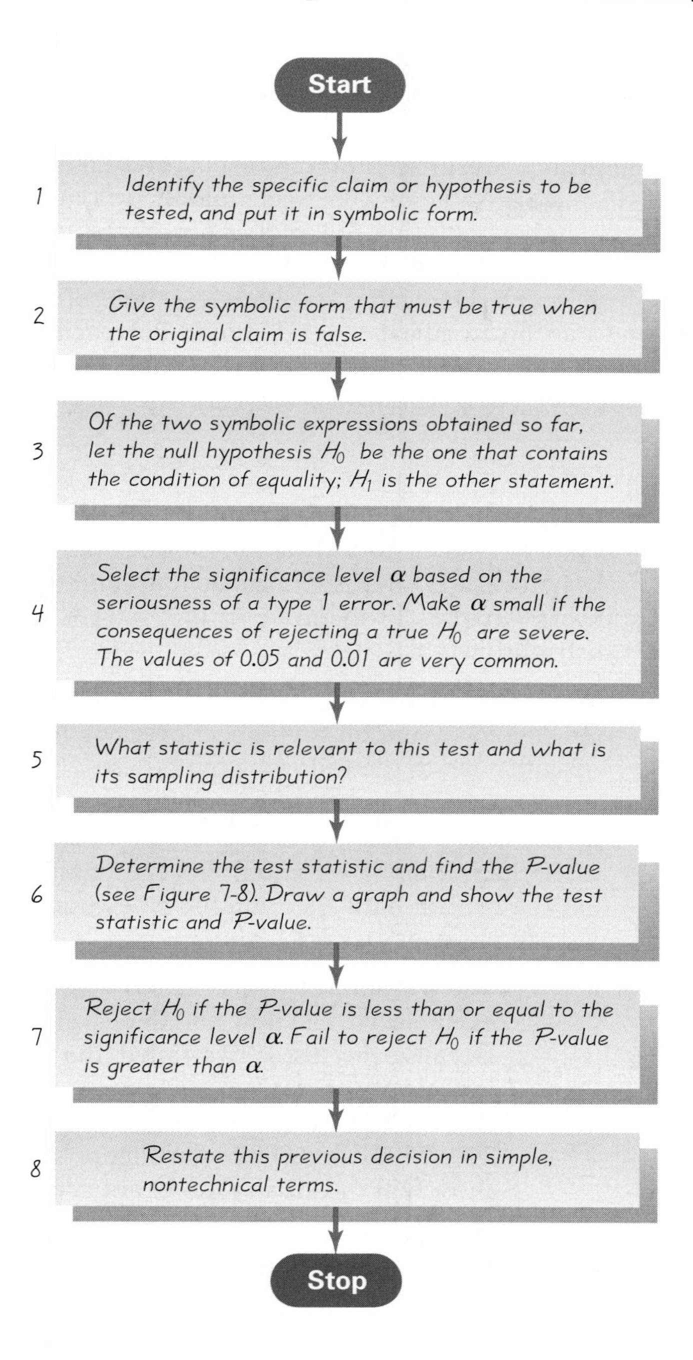

FIGURE 7-9
***P*-Value Method of Testing Hypotheses**

Step 7: Report the *P*-value. Some statisticians prefer to simply report the *P*-value and leave the conclusion to the reader. Others prefer to use the following decision criterion:

- *Reject the null hypothesis* if the *P*-value is less than or equal to the significance level α.
- *Fail to reject the null hypothesis* if the *P*-value is greater than the significance level α.

Many statisticians consider it good practice to always select a significance level *before* doing a hypothesis test. This is a particularly good procedure when using *P*-values because we may be tempted to adjust the significance level based on the results. For example, with a 0.05 significance level and a *P*-value of 0.06, we should fail to reject the null hypothesis, but it is sometimes tempting to say that a probability of 0.06 is small enough to warrant rejection of the null hypothesis. Other statisticians argue that prior selection of a significance level reduces the usefulness of *P*-values. They contend that no significance level should be specified and that the conclusion should be left to the reader. We will use the decision criterion that involves a comparison of a significance level and the *P*-value.

Ethics in Reporting

The American Association for Public Opinion Research developed a voluntary code of ethics to be used in news reports of survey results. This code requires that the following be included: (1) identification of the survey sponsor, (2) date the survey was conducted, (3) size of the sample, (4) nature of the population sampled, (5) type of survey used, (6) exact wording of survey questions. Surveys funded by the U.S. government are subject to a prescreening that assesses the risk to those surveyed, the scientific merit of the survey, and the guarantee of the subject's consent to participate.

EXAMPLE **Body Temperatures** Use the *P*-value method to test the claim that the mean body temperature of healthy adults is equal to 98.6°F. As before, use a 0.05 significance level and the sample data summarized in the Chapter Problem ($n = 106$, $\bar{x} = 98.20$, $s = 0.62$, a bell-shaped distribution).

SOLUTION Except for Steps 6 and 7, this solution is the same as the one developed earlier in this section using the traditional method. Steps 1, 2, and 3 led to the following hypotheses:

$$H_0: \mu = 98.6 \text{ (original claim)} \qquad H_1: \mu \neq 98.6$$

In Steps 4 and 5 we noted that the significance level is $\alpha = 0.05$ and that the central limit theorem indicates use of the normal distribution. We now proceed to Steps 6 and 7.

Step 6: The test statistic of $z = -6.64$ was already found in a preceding example. We can now find the *P*-value by referring to Figure 7-8. Because the test is two-tailed, the *P*-value is twice the area to the left of the test statistic $z = -6.64$. Using Table A-2, we find that the area to the left of $z = -6.64$ is 0.0001, so the *P*-value is $2 \times 0.0001 = 0.0002$ (see Figure 7-10). (More precise results show that the area to the left of $z = -6.64$ is actually much less than 0.0001.)

Step 7: Because the *P*-value of 0.0002 is less than the significance level of $\alpha = 0.05$, we reject the null hypothesis.

INTERPRETATION As with the traditional method, we conclude in Step 8 that there is sufficient evidence to warrant rejection of the claim that the mean body temperature is 98.6°F.

The following example applies the *P*-value approach to the Coke data discussed in an earlier example in this section.

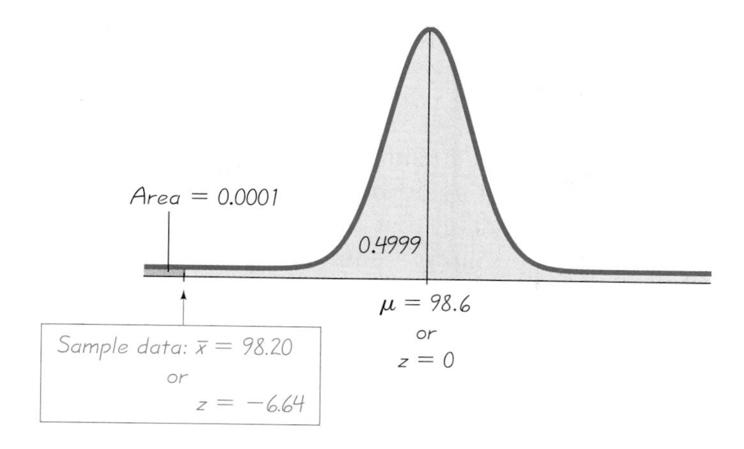

FIGURE 7-10
***P*-Value Method of Testing H_0: $\mu = 98.6$**

Because the test is two-tailed, the *P*-value is *twice* the red-shaded area.

EXAMPLE **Coke Volumes** Use the *P*-value approach and a 0.01 significance level to test the claim that the amounts of Coke in cans have a mean that is greater than 12 oz. Recall that the sample data consist of 36 cans with a mean of 12.19 oz and a standard deviation of 0.11 oz.

SOLUTION Again, refer to the steps used earlier to solve this problem by the traditional method. Steps 1, 2, and 3 resulted in the following hypotheses:

$$H_0\text{: } \mu \leq 12 \text{ oz} \qquad H_1\text{: } \mu > 12 \text{ oz (original claim)}$$

Steps 4 and 5 led to a significance level of $\alpha = 0.01$ and the decision that the normal distribution is relevant to this test of a claim about a sample mean. We now proceed to Steps 6 and 7.

Step 6: The test statistic of $z = 10.36$ was computed in the earlier solution to this same problem. To find the *P*-value, we refer to Figure 7-8, which indicates that for this right-tailed test, the *P*-value is the area to the right of the test statistic. Referring to Table A-2, we find that the area to the right of $z = 10.36$ is $0.5 - 0.4999$, or 0.0001. The *P*-value is therefore 0.0001. (Using a reference more precise than Table A-2 shows that the *P*-value is actually much smaller than 0.0001.)

Step 7: Because the *P*-value of 0.0001 is less than the significance level of $\alpha = 0.01$, we reject the null hypothesis that $\mu \leq 12$.

INTERPRETATION As in the previous solution of this same problem, we conclude that there is sufficient evidence to support the claim that the mean is greater than 12 oz. The *P*-value of 0.0001 shows that it is very unlikely that we would get a sample mean of 12.19 oz if the population mean were really 12 oz.

The next procedure for testing hypotheses is based on confidence intervals and therefore requires the concepts discussed in Section 6-2.

Testing Claims with Confidence Intervals

Confidence intervals can be used to identify results that are highly unlikely, so we can determine whether there is a *significant* difference between sample results and a claimed value of a parameter. For example, let's again consider the hypothesis-testing problem described at the beginning of the chapter. We want to test the claim that the mean body temperature of healthy adults is equal to 98.6°F. Sample data consist of $n = 106$ temperatures with mean $\overline{x} = 98.20$ and standard deviation $s = 0.62$. Chapter 6 described methods of constructing confidence intervals. In particular, we used the body temperature sample data to construct the following 95% confidence interval:

$$98.08 < \mu < 98.32$$

We are 95% confident that the limits of 98.08 and 98.32 contain the population mean μ. (This means that if we were to repeat the experiment of collecting a sample of 106 body temperatures, 95% of the samples would result in confidence interval limits that actually do contain the value of the population mean μ.) This confidence interval suggests that it is very unlikely that the population mean is equal to 98.6. That is, on the basis of the confidence interval given here, we reject the common belief that the mean body temperature of healthy adults is 98.6°F. We can generalize this procedure as follows: First use the sample data to construct a confidence interval, and then apply the following decision criterion.

> **A confidence interval estimate of a population parameter contains the likely values of that parameter. We should therefore reject a claim that the population parameter has a value that is not included in the confidence interval.**

Using this criterion, we see that the confidence interval of $98.08 < \mu < 98.32$ does not contain the claimed value of 98.6, and we therefore reject the claim that the population mean equals 98.6.

We can make a direct correspondence between a confidence interval and a hypothesis test only when the test is two-tailed. A one-tailed hypothesis test with significance level α corresponds to a confidence interval with degree of confidence $1 - 2\alpha$.

Hypothesis Test	Confidence Interval
Two-tailed; significance level α	Use degree of confidence $1 - \alpha$.
One-tailed; significance level α	Use degree of confidence $1 - 2\alpha$.

For example, if we want to conduct a two-tailed hypothesis test with a 0.05 significance level, we can use a 95% confidence interval. But if we want to conduct a right-tailed hypothesis test with a 0.05 significance level, we should use a 90% confidence interval.

In the remainder of the text, we will apply methods of hypothesis testing to other circumstances, such as those involving claims about proportions or standard deviations or those involving more than one population. It is easy to become entangled in a complex web of steps without ever understanding the underlying rationale of hypothesis testing. The key to that understanding lies in the rare event rule for inferential statistics: **If, under a given observed assumption, the probability of getting the observed sample is exceptionally small, we conclude that the assumption is probably not correct.** When testing a claim, we make an assumption (null hypothesis) that contains equality. We then compare the assumption and the sample results to form one of the following conclusions.

- If the sample results can easily occur when the assumption (null hypothesis) is true, we attribute the relatively small discrepancy between the assumption and the sample results to chance.
- If the sample results cannot easily occur when the assumption (null hypothesis) is true, we explain the relatively large discrepancy between the assumption and the sample results by concluding that the assumption is not true, so we reject the assumption.

Using Technology

STATDISK: If working with a list of the original sample values, first find the sample size, sample mean, and sample standard deviation by using the STATDISK procedure described in Section 2-4. After finding the values of n, $\overline{x}$, and s, proceed to select the main menu bar item **Analysis,** then select **Hypothesis Testing,** followed by **Mean-One Sample.** Enter the data in the dialog box. See the accompanying STATDISK display showing the results from the body temperature example of this section.

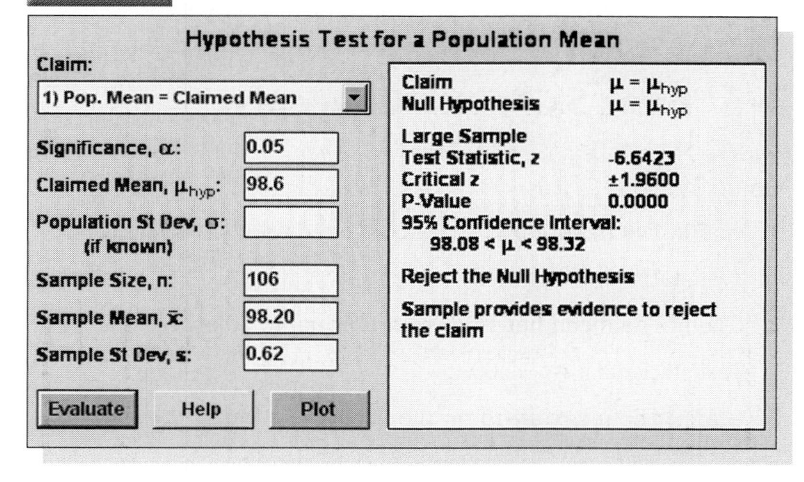

Minitab: Minitab works only with the list of the original data. (For a way to circumvent this restriction, see the *Minitab Student*

Laboratory Manual and Workbook that is a supplement to this text.) First enter the data in column C1, then select the menu items **Stat, Basic Statistics,** and **1-Sample z,** and enter the required data. The box identified as **alternative** is used to select the form of the alternative hypothesis, and it can include either **not equal, less than,** or **greater than.** See Exercises 25 and 26 for samples of Minitab displays.

Excel: Excel's built-in ZTEST function is extremely tricky to use, because the generated *P*-value is not always the same standard *P*-value described in this section. Instead, use the Data Desk XL add-in that is a supplement to this book. First enter the sample data in column A. Select **DDXL,** then **Hypothesis Tests.** Under the function type options, select **1 Var Z Test.** Click on the pencil icon and enter the range of data values, such as A1 : A106 if you have 106 values listed in column A. Click on **OK.** Follow the four steps listed in the dialog box. After clicking on **Compute** in step 4, you will get the *P*-value, test statistic, and conclusion.

TI-83 Plus: If using a TI-83 Plus calculator, press **STAT,** then select **TESTS** and choose the first option, **Z-Test.** You can use the original data or the summary statistics **(Stats)** by providing the entries indicated in the window display. If you choose to enter the summary statistics, you must first find the value of the sample standard deviation s, which can be entered for the value of σ (provided that the sample size is greater than 30). The first three items of the TI-83 Plus results will include the alternative hypothesis, the test statistic, and the *P*-value. See Exercises 27 and 28 for samples of TI-83 Plus calculator displays.

7-3 Basic Skills and Concepts

Finding P-Values. *In Exercises 1–4, use the given claim and test statistic to find the P-value.*

1. Claim: The mean IQ score of convicted felons is equal to 100.

Test statistic: $z = -1.03$

2. Claim: The mean height of men is equal to 70 in.

Test statistic: $z = -2.00$

3. Claim: The mean score on the verbal portion of the Graduate Record Exam is at least 500.

Test statistic: $z = -12.67$

4. Claim: The mean price of college textbooks is greater than \$60.

Test statistic: $z = 2.34$

In Exercises 5–8, identify the following components of the hypothesis test: (a)test statistic, (b)critical value(s), (c)P-Value, (d)final conclusion.

5. Exam Scores Test the claim that for the population of statistics final exams, the mean score is given by $\mu = 75$. Sample data are summarized as $n = 500$, $\bar{x} = 77$, and $s = 15$. Test at the $\alpha = 0.05$ significance level.

6. Heights of Males Test the claim that for the population of male college students, the mean height is given by $\mu = 69.0$ in. Sample statistics include $n = 50$, $\bar{x} = 69.6$ in., and $s = 3.0$ in. Use a significance level of $\alpha = 0.01$.

7. Grade-Point Average Test the claim that the population of sophomore college students has a mean grade point average greater than 2.25. Sample statistics include $n = 150$, $\bar{x} = 2.35$, and $s = 0.80$. Use a significance level of $\alpha = 0.02$.

8. Ages of Jets Test the claim that for the population of U.S. commercial jets, the mean age is less than 18 years. Sample statistics include $n = 85$, $\bar{x} = 15.8$ years, and $s = 8.3$ years. Use a significance level of $\alpha = 0.04$.

Traditional Method. *In Exercises 9–14, test the given claim by using the traditional method of testing hypotheses.*

9. Body Temperatures In this section we presented an example about testing the claim that the mean body temperature equals 98.6°F. Use the same sample data ($n = 106$, $\bar{x} = 98.20$, $s = 0.62$ oz) to test the claim that the mean body temperature is *less than* 98.6°F. Use a significance level of $\alpha = 0.01$.

10. Coke Volumes In this section we presented an example about testing the claim that the mean volume of Coke in cans is greater than 12 oz. Use the same sample data ($n = 36$, $\bar{x} = 12.19$ oz, $s = 0.11$ oz) to test the claim that the mean *equals* 12 oz. Use a significance level of $\alpha = 0.02$.

11. Baseballs In previous tests, baseballs were dropped 24 ft onto a concrete surface, and they bounced an average of 92.84 in. In a test of a sample of 40 new balls, the bounce heights had a mean of 92.67 in. and a standard deviation of 1.79 in. (based on data from Brookhaven National Laboratory and *USA Today*). Use a 0.05 significance level to determine whether there is sufficient evidence to support the claim that the new balls have bounce heights with a mean different from 92.84 in. Does it appear that the new baseballs are different?

12. Compulsive Buyers Researchers designed a questionnaire to identify compulsive buyers. For a sample of consumers who identified themselves as compulsive buyers, questionnaire scores have a mean of 0.83 and a standard deviation of 0.24 (based on data from "A Clinical Screener for Compulsive Buying," by Faber and Guinn, *Journal of Consumer Research,* Vol. 19). Assume that the subjects were randomly selected and that the sample size was 32. Using a 0.01 significance level, test the claim that the self-identified compulsive-buyer population has a mean greater than 0.21, the mean for the general population. Does the questionnaire seem to be effective in identifying compulsive buyers?

13. Drug Production The nighttime cold medicine Dozenol bears a label indicating the presence of 600 mg of acetaminophen in each fluid ounce of the drug. The Food and Drug Administration randomly selected 40 one-ounce samples and found that the mean acetaminophen content is 593 mg, whereas the standard deviation is 21 mg. Use $\alpha = 0.01$ and test the claim of the Medassist Pharmaceutical Company that the population mean is equal to 600 mg. Would you buy this cold medicine?

14. Effectiveness of Seat Belts A study included 123 children who were wearing seat belts when injured in motor vehicle crashes. The amounts of time spent in an intensive care unit have a mean of 0.83 day and a standard deviation of 0.16 day (based on data from "Morbidity Among Pediatric Motor Vehicle Crash Victims: The Effectiveness of Seat Belts," by Osberg and Di Scala, *American Journal of Public Health,* Vol. 82, No. 3). Using a 0.01 significance level, test the claim that the seat belt sample comes from a population with a mean of less than 1.39 days, which is the mean for the population who were not wearing seat belts when injured in motor vehicle crashes. Do the seat belts seem to help?

P-Value Method. *In Exercises 15–20, test the given claim by using the P-value method of testing hypotheses.*

15. Testing Effectiveness of SAT Course The effectiveness of a test preparation course was studied for a simple random sample of 75 subjects who took the SAT before and after coaching. The differences between the scores resulted in a mean increase of 0.6 and a standard deviation of 3.8. (See "An Analysis of the Impact of Commercial Test Preparation Courses on SAT Scores," by Sesnowitz, Bernhardt, and Kwain, *American Education Research Journal,* Vol. 19, No. 3.) Using a 0.05 significance level, test the claim that the population mean increase is greater than 0, indicating that the course is effective in raising scores. Should people take this course?

16. Prison Time for Convicted Embezzlers When 70 convicted embezzlers were randomly selected, the mean length of prison sentence was found to be 22.1 months and the standard deviation was found to be 8.6 months (based on data from the U.S. Department of Justice). Kim Patterson is running for political office on a platform of tougher treatment of convicted criminals. Test her claim that prison terms for convicted embezzlers have a mean of less than 2 years. Use a 0.01 significance level.

17. Conductor Life Span A *New York Times* article noted that the mean life span for 35 male symphony conductors was 73.4 years, in contrast to the mean of 69.5 years for males in the general population. Assuming that the 35 males have life spans with a standard deviation of 8.7 years, use a 0.05 significance level to test the claim that male symphony conductors have a mean life span that is different from 69.5 years.

18. Weights of Garbage Data Set 5 in Appendix B includes the *total* weights of garbage discarded by households in one week (based on data collected as part of the Garbage Project at the University of Arizona). For that data set, the mean is 27.44 lb and the standard deviation is 12.46 lb. Using a 0.01 significance level, test the claim of the city of Providence supervisor that the mean weight of all garbage discarded by households each week is less than 29 lb, the amount that can be handled by the town. Based on the result, is there any cause for concern that there might be too much garbage to handle?

19. Weights of Quarters If we refer to the weights (in grams) of quarters listed in Data Set 13 in Appendix B, we find 50 weights with a mean of 5.622 g and a standard deviation of 0.068 g. The U.S. Department of the Treasury claims that the procedure it uses to mint quarters yields a mean weight of 5.670 g. Use a 0.01 significance level to test the claim that the mean weight of quarters in circulation is 5.670 g. If the claim is rejected, what is a possible explanation for the discrepancy?

20. Testing Wristwatch Accuracy Forty people are selected and the accuracy of their wristwatches is checked, with positive errors representing watches that are ahead of the correct time and negative errors representing watches that are behind the correct time. The 40 values have a mean of 117.3 sec and a standard deviation of 185.0 sec. Use a 0.01 significance level to test the claim that the population of all watches has a mean equal to 0 sec. What do you conclude about the accuracy of people's wristwatches?

Testing Claims Using Sample Data. *In Exercises 21–24, use the given data set to test the stated claim.*

21. Weights of Sugar Packets Refer to Data Set 4 in Appendix B and test the claim that the weights of the sugar packets have a mean equal to 3.5 g, as indicated on the label. If the mean does not appear to equal 3.5 g, what do you conclude?

22. Second-Hand Smoke Refer to Data Set 9 in Appendix B and use the measured levels of serum cotinine (in ng/ml) in people who are nonsmokers and have no exposure to environmental tobacco smoke at home or at work (NOETS). Test the claim of tobacco company spokesmen that these people have "positive" levels of cotinine, suggesting that this measure is not a good indicator of exposure to tobacco smoke. Use a significance level of 0.005.

23. Pepsi Volumes Refer to Data Set 1 in Appendix B and use the volumes (in ounces) of regular Pepsi. Test the claim that the mean is greater than 12 oz, as indicated on the can labels. Can Pepsi be supplied in a way that is more profitable to the company?

24. Textbook Prices Refer to Data Set 2 in Appendix B and use the prices of the 40 *new* textbooks at the University of Massachusetts. Test the claim that such textbooks have a mean price less than $70. Use a 0.05 significance level.

Interpreting Computer and Calculator Displays. *In Exercises 25–28, use the computer or calculator display to form a conclusion.*

25. Mean Weight of M&M Candies A package of M&M plain candies is labeled as containing 1361 g, and there are 1498 candies, so the mean weight of the individual candies should be 1361/1498, or 0.9085 g. In a test to determine whether consumers are being cheated, a sample of 33 brown M&Ms is randomly selected. (See Data Set 10 in Appendix B.) When the 33 weights are used with Minitab, the display is as shown here. Interpret those results. Are consumers being cheated?

```
TEST OF MU = 0.90850 VS MU < 0.90850

        N   MEAN     STDEV    SE MEAN   T      P VALUE

C1     33   .91282   .03952   .00688    0.63   0.73
```

26. Analysis of Last Digits Analysis of the last digits of sample data values sometimes reveals whether the data have been accurately measured and reported. If the last digits are uniformly distributed from 0 through 9, their mean should be 4.5. Reported data (such as weights or heights) are often rounded so that the last digits include disproportionately more 0s and 5s. The last digits in the reported lengths (in feet) of the 70 home runs hit by Mark McGwire in 1998 are used to test the claim that they come from a population with a mean of 4.5 (based on data from *USA Today*). When Minitab is used to test that claim, the display is as shown here. Using a 0.05 significance level, interpret the Minitab results. Does it appear that the distances were accurately measured?

```
Test of mu = 4.50000 vs mu not = 4.50000
The assumed sigma = 2.693

Variable     N     Mean    StDev   SE Mean        Z         P
C1          70    1.229    2.693     0.322   -10.16    0.0000
```

TI-83 Plus

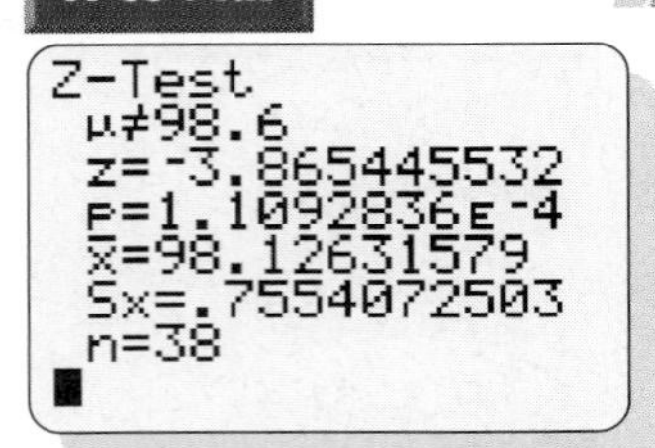

27. Testing Mean Body Temperature In an example in this section we tested the claim that the mean body temperature is equal to 98.6°F, but we used 106 readings taken at 12:00 A.M. on the second day of observations. If we use the 38 body temperatures taken at 8:00 A.M. on the first day, the TI-83 Plus calculator will display the results as shown. Interpret those results. Are these results consistent with the results found in the example in this section? For these results, could there be a potential problem due to the large number of missing values?

TI-83 Plus

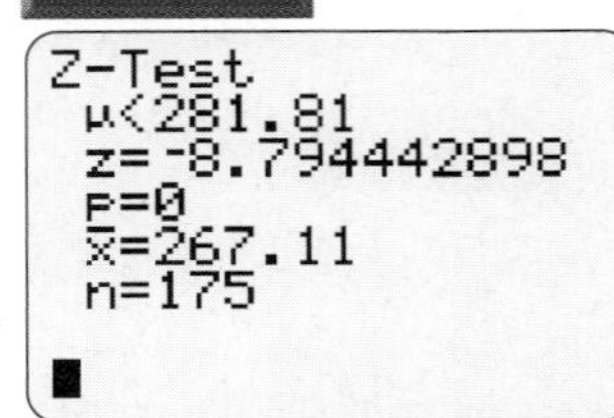

28. Are Thinner Aluminum Cans Weaker? Data Set 12 in Appendix B includes the measured axial loads (in pounds) of 175 cola cans that use aluminum 0.0109 in. thick. Before obtaining these sample results, the standard cans had a thickness of 0.0111 in. and the mean axial load was 281.81 lb. When using the axial loads of the thinner cans in a test of the claim that the mean axial load is less than 281.81 lb, the TI-83 Plus calculator provides the accompanying display. Assume that we are using a 0.01 significance level. Interpret the results. Do the thinner cans appear to have a mean axial load less than 281.81 lb?

7-3 Beyond the Basics

29. Finding Standard Deviation A journal article reported that a null hypothesis of $\mu = 100$ was rejected because the P-value was less than 0.01. The sample size was given as 62, and the sample mean was given as 103.6. Find the largest possible standard deviation.

30. Axial Loads of Aluminum Cans In Exercise 28, find the *largest* sample mean below 281.81 lb that will support the claim that the mean axial load of the 0.0109-in. cans is less than 281.81 lb. (Use the same sample size and sample standard deviation.)

31. Finding Probability of Type II Error For a given hypothesis test, the probability α of a type I error is fixed, whereas the probability β of a type II error depends on the particular value of μ that is used as an alternative to the null hypothesis. For hypothesis tests of the type found in this section, we can find β as follows:

Step 1: Find the value(s) of $\bar{x}$ that correspond to the critical value(s). In

$$z = \frac{\bar{x} - \mu_{\bar{x}}}{\sigma_{\bar{x}}}$$

substitute the critical value(s) for z, enter the values for $\mu_{\bar{x}}$ and $\sigma_{\bar{x}}$, and solve for $\bar{x}$.

Step 2: Given a particular value of μ that is an alternative to the null hypothesis H_0, draw the normal curve with this new value of μ at the center. Also plot the value(s) of $\bar{x}$ found in Step 1.

Step 3: Refer to graph in Step 2, and find the area of the new critical region bounded by the value(s) of $\overline{x}$ found in step 1. This is the probability of rejecting the null hypothesis, given that the new value of μ is correct

Step 4: The value of β is 1 minus the area from Step 3. This is the probability of failing to reject the null hypothesis, given that the new value of μ is correct.

The preceding steps allow you to find the probability of failing to reject H_0 when it is false. You are determining the area under the curve that excludes the critical region in which you reject H_0; this area corresponds to a failure to reject a false H_0, because we use a particular value of μ that goes against H_0. Refer to the body-temperature example discussed in this section and find β (the probability of a type II error) corresponding to the following:

a. $\mu = 98.7$

b. $\mu = 98.4$

32. Power of Test The *power* of a test, expressed as $1 - \beta$, is the probability of rejecting a false null hypothesis. Refer to Exercise 28. If the test of that claim has a power of 0.2, find the mean μ that is being used as an alternative to H_0 (see Exercise 31).

7-4 Testing a Claim about a Mean: Small Samples

One great advantage of learning the methods of hypothesis testing described in Sections 7-2 and 7-3 is that those same methods can be easily modified for use in many other circumstances, such as those discussed in this section. The examples and exercises in Sections 7-2 and 7-3 all use large samples in tests of claims about means, but in this section we will deal with small sample cases. The main objective of this section is to develop the ability to test claims made about population means when the number of sample values is 30 or fewer. The following assumptions apply to the methods described in this section.

Assumptions

1. The samples are simple random samples. (We stress again that if data are carelessly collected, they may be so completely useless that no amount of statistical torturing can salvage them.)
2. The sample is small ($n \leq 30$).
3. The value of the population standard deviation σ is unknown. (We seldom know the population standard deviation when we do not know the population mean.)
4. The sample values come from a population with a distribution that is approximately normal.

Death Penalty as Deterrent

A common argument supporting the death penalty is that it discourages others from committing murder. Jeffrey Grogger of the University of California analyzed daily homicide data in California for a four-year period during which executions were frequent. Among his conclusions published in the Journal of the American Statistical Association (Vol. 85, No. 410): "The analyses conducted consistently indicate that these data provide no support for the hypothesis that executions deter murder in the short term." This is a major social policy issue, and the efforts of people such as Professor Grogger help to dispel misconceptions so that we have accurate information with which to address such issues.

P-Values

The preceding example followed the traditional approach to hypothesis testing, but STATDISK, Minitab, the TI-83 Plus calculator, and much of the literature will display *P*-values. Because the t distribution table (Table A-3) includes only selected values of the significance level α, we cannot usually find an exact *P*-value from Table A-3. Instead, we can use that table to identify limits that contain the *P*-value. In the last example we found the test statistic to be $t = 6.440$, and we know that the test is one-tailed with 19 degrees of freedom. By examining the row of Table A-3 corresponding to 19 degrees of freedom, we see that the test statistic of 6.440 exceeds the largest critical value in that row. Although we cannot pinpoint an exact *P*-value from Table A-3, we can conclude that it must be less than 0.005. That is, we conclude that *P*-value < 0.005. (Some calculators and computer programs allow us to find exact *P*-values. For the preceding example, STATDISK displays a *P*-value of 0.0000, Minitab displays a *P*-value of 0.0000, and the TI-83 Plus calculator displays a *P*-value of 0.00000197.) With a significance level of 0.01 and a *P*-value less than 0.005, we reject the null hypothesis (because the *P*-value is less than the significance level), as we did using the traditional method in the preceding example.

EXAMPLE **Finding a *P*-Value** Use Table A-3 to find the *P*-value corresponding to these results: The Student t distribution is used in a two-tailed test with a sample of $n = 10$ scores, and the test statistic is $t = 2.567$.

SOLUTION Refer to the row of Table A-3 with 9 degrees of freedom and note that the test statistic of 2.567 falls between the critical values of 2.821 and 2.262. Because the test is two-tailed, we consider the values of α at the top that are identified with "two tails." The critical values of 2.821 and 2.262 correspond to 0.02 (two tails) and 0.05 (two tails), so we express the *P*-value as follows:

$$0.02 < P\text{-value} < 0.05$$

With a true null hypothesis, the chance of getting a sample mean (from 10 sample values) that converts to a test statistic of $t = 2.567$ is somewhere between 0.02 and 0.05.

So far, we have discussed tests of hypotheses made about population means only. In the next section we will test hypotheses made about population proportions or percentages, and in the last section we will consider claims made about standard deviations or variances.

Using Technology

When conducting hypothesis tests using the Student t distribution, the procedures for using STATDISK, Minitab, Excel, and the TI-83 Plus calculator are essentially the same as described in Section 7-3, except for these selections:

Minitab: Select **1-Sample t** instead of **1-Sample z.**

Excel: Select **1 Var t Test** instead of **1 Var z Test.**

TI-83 Plus: Select **T-Test** instead of **Z-Test.**

7-4 Basic Skills and Concepts

Traditional Method. *In Exercises 1–4, assume that the sample is a simple random sample selected from a normally distributed population. Test the given claim by using the traditional method of testing hypotheses. Include the test statistic, critical value(s), and conclusions.*

1. Exam Scores Test the claim that for the population of statistics final exams, the mean score is given by $\mu = 75$. Sample data are summarized as $n = 16$, $\bar{x} = 77$, and $s = 15$. Test at the $\alpha = 0.05$ significance level.

2. Heights of Males Test the claim that for the population of male college students, the mean height is given by $\mu = 68.0$ in. Sample statistics include $n = 20$, $\bar{x} = 70.0$ in., and $s = 2.6$ in. Use a significance level of $\alpha = 0.01$.

3. Grade-Point Average Test the claim that the population of sophomore college students has a mean grade-point average greater than 2.00. Sample statistics include $n = 24$, $\bar{x} = 2.35$, and $s = 0.70$. Use a significance level of $\alpha = 0.01$.

4. Ages of Jets Test the claim that for the population of U.S. commercial jets, the mean age is less than 18 years. Sample statistics include $n = 8$, $\bar{x} = 15.8$ years, and $s = 8.3$ years. Use a significance level of $\alpha = 0.025$.

P-Value Method. *In Exercises 5–8, refer to the indicated exercise and Table A-3 to describe the P-value. (See the example in this section.)*

5. Exercise 1

6. Exercise 2

7. Exercise 3

8. Exercise 4

Traditional Method. *For hypothesis tests in Exercises 9–24, use the traditional approach summarized in Figure 7-5. Draw a graph showing the test statistic and critical values. In each case, assume that the population has a distribution that is approximately normal and that the sample is a simple random sample. Caution: Some of the exercises require the use of the normal distribution (as described in the preceding section) instead of the Student t distribution (as described in this section); be sure to check the conditions to determine which distribution is appropriate.*

9. Body Temperatures Refer to Data Set 6 in Appendix B. Using only the first 25 body temperatures listed for 12 A.M. of Day 2, test the claim that the mean body temperature of all healthy adults is equal to 98.6°F. For the significance level, use $\alpha = 0.05$. For the first 25 scores, $\bar{x} = 98.24$ and $s = 0.56$.

10. Body Temperatures Refer to Data Set 6 in Appendix B. Using only the first 35 body temperatures listed for 12 A.M. of Day 2, test the claim that the mean body temperature of all healthy adults is equal to 98.6°F. Use a 0.05 significance level. For the first 35 scores, $\bar{x} = 98.27$ and $s = 0.65$.

11. BMW Crash Tests Because of the expense involved, car crash tests often use small samples. When five BMW cars are crashed under standard conditions, the repair costs (in dollars) are as shown in the accompanying list. Use a 0.05 significance level to test the claim that the mean for all BMW cars is less than \$1000. Would BMW be justified in advertising that under the standard conditions, the repair costs are less than \$1000?

797 571 904 1147 418

12. Effectiveness of SAT Training Course Tom Wolfson is absolutely convinced that his Tom Wolfson School for Standard Tests is effective in helping test subjects raise their SAT scores. Tom has never been accused of being too humble. Test the claim that the Wolfson students have a mean score greater than 1017, which is the mean for the general population. Use these results from 20 subjects, who took the Wolfson course prior to taking the SAT: $\bar{x} = 1040$, $s = 207$. Is there a better way to design the experiment for testing the effectiveness of the Wolfson course?

13. Testing via Video A study was conducted to determine whether a standard clerical test would need revision for use on video display terminals (VDTs). The VDT scores of 22 subjects have a mean of 170.2 and a standard deviation of 35.3 (based on data from "Modification of the Minnesota Clerical Test to Predict Performance on Video Display Terminals," by Silver and Bennett, *Journal of Applied Psychology,* Vol. 72, No. 1). At the 0.05 significance level, test the claim that the mean for all subjects taking the VDT test differs from the mean of 243.5 for the standard printed version of the test. Based on the result, should the VDT test be revised?

14. Effects of Hypnotism In a study of factors affected by hypnotism, visual analogue scale (VAS) sensory ratings were obtained for 16 subjects. For these sample ratings, the mean is 8.33 and the standard deviation is 1.96 (based on data from "An Analysis of Factors That Contribute to the Efficacy of Hypnotic Analgesia," by Price and Barber, *Journal of Abnormal Psychology,* Vol. 96, No. 1). At the 0.01 significance level, test the claim that this sample comes from a population with a mean rating of less than 10.00.

15. Reading Test Scores The following reading test results were obtained for a sample of 15 third-grade students: $\bar{x} = 31.0$, $s = 10.5$. (The data are based on "A Longitudinal Study of the Effects of Retention/Promotion on Academic Achievement," by Peterson et al., *American Educational Research Journal,* Vol. 24, No. 1.) Does this third-grade sample mean differ significantly from a first-grade population mean of 41.9? Assume a 0.01 significance level.

16. Mean Time for Four-Year Degree Kim Greco is a high school senior who is concerned about attending college because she knows that many college students require more than four years to earn a bachelor's degree. At the 0.10 significance level, test the claim of her guidance counselor, who states that the mean time is greater than 5 years. Sample data consist of a mean of 5.15 years and a standard deviation of 1.68 years for 80 randomly selected college graduates (based on data from the National Center for Education Statistics).

17. Weights of Brown M&Ms Using the weights of only the *brown* M&Ms listed in Data Set 10 in Appendix B, test the claim that the mean is greater than 0.9085 g, the mean value necessary for the 1498 M&Ms to produce a total of 1361 g as the package indicates. Use a 0.05 significance level. For the brown M&Ms, $\bar{x} = 0.9128$ g and $s = 0.0395$ g. Based on the result, can we conclude that the packages contain more than the claimed weight printed on the label?

18. Weights of Blue M&Ms Using the weights of only the *blue* M&Ms listed in Data Set 10 in Appendix B, test the claim that the mean is at least 0.9085 g, the mean value necessary for the 1498 M&Ms to produce a total of 1361 g as the package indicates. Use a 0.05 significance level. For the blue M&Ms, $\overline{x} = 0.9014$ g and $s = 0.0573$ g. Based on the result, can we conclude that the package contents do not agree with the claimed weight printed on the label?

19. Effectiveness of New Chicken Feed When a poultry farmer uses her regular feed, the newborn chickens have normally distributed weights with a mean of 62.2 oz. In an experiment with an enriched feed mixture, nine chickens are born with the weights (in ounces) given below. Use a 0.01 significance level to test the claim that the mean weight is higher with the enriched feed.

61.4 62.2 66.9 63.3 66.2 66.0 63.1 63.7 66.6

20. Olympic Winners The accompanying list shows the winning times (in seconds) of men in the 100-meter dash for consecutive summer Olympic games, listed in order by row. Assuming that these results are sample data randomly selected from the population of all past and future Olympic games, test the claim that the mean time is equal to 10.5 sec. What do you observe about the precision of the numbers? What extremely important characteristic of the data set is not considered in this hypothesis test? Do the results from the hypothesis test suggest that future winning times should be around 10.5 sec, and is such a conclusion valid?

12.0 11.0 11.0 11.2 10.8 10.8 10.8 10.6 10.8 10.3 10.3 10.3

10.4 10.5 10.2 10.0 9.95 10.14 10.06 10.25 9.99 9.92 9.96

21. Mean Pepsi Volumes Refer to Data Set 1 in Appendix B for the volumes (in ounces) of cans of diet Pepsi. Is there sufficient evidence to conclude that the mean is greater than the 12 oz that is printed on the label?

22. Nicotine in Cigarettes Refer to the nicotine contents of cigarettes listed in Data Set 8 in Appendix B. Use the sample data to test the claim that the population mean is less than 1.0 mg. Does the conclusion apply to the mean nicotine level of all 100 mm cigarettes (not menthol or light) smoked by consumers? Why or why not?

23. Effect of Vitamin Supplement on Birth Weight The accompanying list shows the birth weights (in kilograms) of male babies born to mothers taking a special vitamin supplement (based on data from the New York State Department of Health). At the 0.05 significance level, test the claim that the mean birth weight for all male babies of mothers given vitamins is equal to 3.39 kg, which is the mean for the population of all males. Based on the result, does the vitamin supplement appear to have an effect on birth weight?

3.73 4.37 3.73 4.33 3.39 3.68 4.68 3.52

3.02 4.09 2.47 4.13 4.47 3.22 3.43 2.54

24. Nicotine in Cigarettes The Carolina Tobacco Company advertised that its best-selling nonfiltered cigarettes contain at most 40 mg of nicotine, but *Consumer Advocate* magazine ran tests of 10 randomly selected cigarettes and found the amounts shown in the accompanying list. The sample is small because the laboratory work required to extract the nicotine is time-consuming and expensive.

(*continued*)

It's a serious matter to charge that the company advertising is wrong, so the magazine editor chooses a significance level of $\alpha = 0.01$ in testing her belief that the mean nicotine content is greater than 40 mg. Using a 0.01 significance level, test the editor's belief that the mean is greater than 40 mg.

47.3 39.3 40.3 38.3 46.3 43.3 42.3 49.3 40.3 46.3

TI-83 Plus

```
T-Test
 μ<1000
 t=-1.412000693
 p=.0866601222
 x̄=985.6666667
 Sx=46.51809684
 n=21
```

TI-83 Plus

```
T-Test
 μ>63.2
 t=1.27082077
 p=.1122518742
 x̄=69.53333333
 Sx=19.30161602
 n=15
```

25. Calcium Carbonate in Antacid Tablets Bottles of antacid tables include a printed label claiming that they contain 1000 mg of calcium carbonate. In a simple random sample of tablets made by the Medassist Pharmaceutical Company, the amounts of calcium carbonate are measured. When testing the claim that consumers are being cheated because the mean is less than 1000 mg, the TI-83 Plus calculator displays the results shown. Interpret those results and determine whether there is sufficient evidence to support the claim that consumers are being cheated.

26. Time for Laughing Rita Gibbons is a stand-up comedian who videotapes her performances and records the total of the times she must wait for audience laughter to subside. She recorded the times (in seconds) for 15 different shows in which she used a new routine. When testing the claim that her new routine is better than the old one, she obtains the TI-83 Plus calculator results shown here. Interpret those results. Does her new routine seem to be better than the old one?

27. Acid Rain The results shown below result from using Minitab to test the claim that the amounts of acid rain sulfate deposits in Pennsylvania come from a population with a mean greater than 12.00 kg/hectare (based on data from the U.S. Department of Agriculture). Use a 0.01 significance level and interpret the Minitab results from the hypothesis test.

```
Test of mu = 12.000 vs mu > 12.000
Variable    N     Mean    StDev   SE Mean      T   P-value
C1         11   13.286    1.910     0.576   2.23     0.025
```

28. Reliability of Aircraft Radios The mean time between failures (in hours) for a Telektronic Company radio used in light aircraft is 420 h. After 15 new radios were modified in an attempt to improve reliability, tests were conducted to measure the times between failures. When Minitab is used to test the claim that the modified radios have a mean greater than 420 h, the results are as shown here. Does it appear that the modifications improved reliability?

```
Test of mu = 420.00 vs mu > 420.00
Variable    N     Mean    StDev   SE Mean      T        P
C1         15   434.73    18.01      4.65   3.17   0.0034
```

7-4 Beyond the Basics

29. Using Computer Results Refer to the Minitab display in Exercise 28. If the claim is changed from "greater than 420 h" to "not equal to 420 h," how are the values in the bottom row affected?

30. Using the Wrong Distribution Because of certain conditions, a hypothesis test requires the Student t distribution, as described in this section. Assume that the standard normal distribution was incorrectly used instead. Does using the standard normal distribution make you more or less likely to reject the null hypothesis, or does it not make a difference? Explain.

31. Finding Critical t Values When finding critical values, we sometimes need significance levels other than those available in Table A-3. Some computer programs approximate critical t values by calculating

$$t = \sqrt{\text{df} \cdot (e^{A^2/\text{df}} - 1)}$$

where

$$\text{df} = n - 1$$

$$e = 2.718$$

$$A = z\left(\frac{8 \cdot \text{df} + 3}{8 \cdot \text{df} + 1}\right)$$

and z is the critical z score. Use this approximation to find the critical t score corresponding to $n = 10$ and a significance level of 0.05 in a right-tailed case. Compare the results to the critical t value found in Table A-3.

32. Probability of Type II Error Refer to Exercise 24 and assume that you're testing the claim that $\mu > 40$ mg. Find β (the probability of a type II error), given that the actual value of the population mean is $\mu = 45.0518$ mg. (See Exercise 31 in Section 7-3.)

7-5 Testing a Claim about a Proportion

In the preceding sections we introduced the basic methods of hypothesis testing, but they were used to address claims made about population *means* only. In this section we see how to apply those same basic methods to claims made about population *proportions.* The proportions can also represent probabilities or the decimal equivalents of percents. The following are examples of the types of claims we will be able to test.

- Fewer than 1/4 of all college graduates smoke.
- The percentage of late-night television viewers who watch *The Late Show with David Letterman* is equal to 18%.
- If a fatal car crash occurs, there is a 0.44 probability that it involves a driver who had been drinking.

Assumptions

1. The sample observations are a simple random sample. (Let us never forget the critical importance of sound sampling methods.)
2. The conditions for a *binomial distribution* are satisfied. (There are a fixed number of independent trials having constant probabilities, and each trial has two outcome categories, called "success" and "failure.")
3. The conditions $np \geq 5$ and $nq \geq 5$ are both satisfied, so **the binomial distribution of sample proportions can be approximated by a normal distribution with $\mu = np$ and $\sigma = \sqrt{npq}$** (as described in Section 5-6).

Polls and Psychologists

Poll results can be dramatically affected by the wording of questions. A phrase such as "over the last few years" is interpreted differently by different people. Over the last few years (actually, since 1980), survey researchers and psychologists have been working together to improve surveys by decreasing bias and increasing accuracy. In one case, psychologists studied the finding that 10 to 15 percent of those surveyed say they voted in the last election when they did not. They experimented with theories of faulty memory, a desire to be viewed as responsible, and a tendency of those who usually vote to say that they voted in the most recent election, even if they did not. Only the last theory was actually found to be part of the problem.

If these assumptions are not all satisfied, we may be able to use other methods not described in this section. In this section, all examples and exercises involve cases in which the assumptions are satisfied, so the sampling distribution of sample proportions can be approximated by the normal distribution. We use the following notation and test statistic.

Notation

n = number of trials

$\hat{p} = \frac{x}{n}$ (*sample* proportion)

p = population proportion (used in the null hypothesis)

$q = 1 - p$

Test Statistic for Testing a Claim about a Proportion

$$z = \frac{\hat{p} - p}{\sqrt{\frac{pq}{n}}}$$

The Traditional Method

If you are using the traditional method of testing hypotheses as outlined in Figure 7-5, find critical z values from Table A-2 (standard normal distribution) by using the procedures described in Section 7-2, or find critical z values from STATDISK or a TI-83 Plus calculator. For example, in a two-tailed test with significance level $\alpha = 0.05$, divide α equally between the two tails, then refer to Table A-2 for the z score corresponding to an area of $0.5 - 0.025 = 0.475$; the result is $z = 1.96$, so the critical values are $z = -1.96$ and $z = 1.96$.

The *P*-Value Method

To find the *P*-value corresponding to a z test statistic, use the following procedure, which was shown in Figure 7-8 and described in Section 7-3.

Right-tailed test: *P*-value = area to right of test statistic z

Left-tailed test: *P*-value = area to left of test statistic z

Two-tailed test: *P*-value = *twice* the area of the extreme region bounded by the test statistic z

As in Section 7-3, we use this decision criterion:

Reject the null hypothesis if the *P*-value is less than or equal to the significance level α.

When testing a claim about a population proportion p, be careful to identify correctly the sample proportion $\hat{p}$. The sample proportion $\hat{p}$ is sometimes given directly, but in other cases it must be calculated. See the examples below.

Given Statement	Finding $\hat{p}$
10% of the observed sports cars are red.	$\hat{p}$ is given directly: $\hat{p} = 0.10$
96 surveyed households have cable TV and 54 do not.	$\hat{p}$ must be calculated using $\hat{p} = x/n$: $\hat{p} = \frac{x}{n} = \frac{96}{(96 + 54)} = 0.64$

Caution: When a calculator or computer display of $\hat{p}$ results in many decimal places, use all of those decimal places when evaluating the z test statistic. Large errors can result from rounding $\hat{p}$ too much.

EXAMPLE **Survey of Voters** In a survey of 1002 people, 701 said that they voted in the recent presidential election (based on data from the ICR Research Group). Test the claim that when surveyed, the proportion of people who say that they voted is equal to 0.61, which is the proportion of people who actually did vote. Use a 0.05 significance level with the traditional method of testing hypotheses.

SOLUTION We will use the traditional method of testing hypotheses as outlined in Figure 7-5. We use the sample proportion of $\hat{p} = 701/1002 = 0.6996007984$ and the claimed population proportion of $p = 0.61$.

Step 1: The original claim is that the proportion of people who say that they voted is 0.61. We express this in symbolic form as $p = 0.61$.

Step 2: The opposite of the original claim is $p \neq 0.61$.

Step 3: Because $p = 0.61$ contains equality, we have

$$H_0: p = 0.61 \text{ (null hypothesis and original claim)}$$
$$H_1: p \neq 0.61 \text{ (alternative hypothesis)}$$

Step 4: The significance level is α 5 0.05.

Step 5: The statistic relevant to this test is $\hat{p} = 701/1002 = 0.6996007984$. The sampling distribution of sample proportions is approximated by the normal distribution. (The requirements $np \geq 5$ and $nq \geq 5$ are both satisfied with $n = 1002$, $p = 0.61$, and $q = 0.39$.)

Step 6: The test statistic of $z = 5.81$ is found as follows:

$$z = \frac{\hat{p} - p}{\sqrt{\frac{pq}{n}}} = \frac{0.6996007984 - 0.61}{\sqrt{\frac{(0.61)(0.39)}{1002}}} = 5.81$$

The critical values of $z = -1.96$ and $z = 1.96$ are found from Table A-2. The test statistic and critical values are shown in the STATDISK graph.

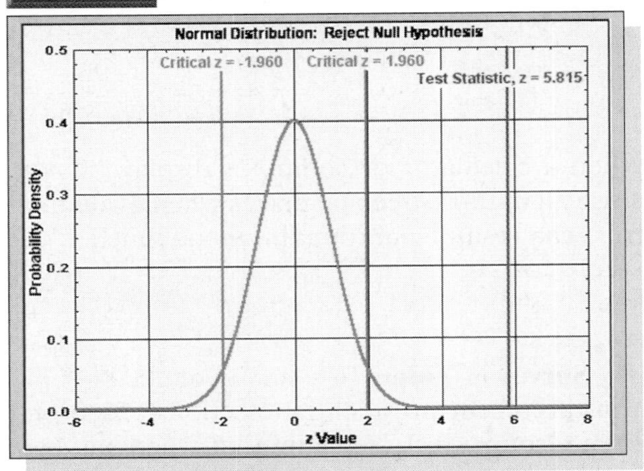

Step 7: Because the test statistic does fall within the critical region, reject the null hypothesis.

INTERPRETATION There is sufficient evidence to warrant rejection of the claim that when surveyed, the proportion of people who say they voted is equal to 0.61. It appears that people do not accurately report whether they voted.

EXAMPLE ***P*-Value Approach** Repeat the preceding example, but use the *P*-value approach.

SOLUTION Steps 1–4 will be the same as in the preceding solution. In Step 6, the test statistic $z = 5.81$ is computed as shown in the preceding example. Instead of finding the critical values, we proceed to find the *P*-value as follows (see Figure 7-8).

The test is two-tailed and the test statistic is to the right of center, so the *P*-value is twice the area to the right of the test statistic. Table A-2 indicates that the area between $z = 0$ and $z = 5.81$ is 0.4999, so the area to the right of the test statistic is $0.5 - 0.4999 = 0.0001$. The *P*-value is $2 \times 0.0001 = 0.0002$. (More precise tables, STATDISK, or a TI-83 Plus calculator will produce a smaller *P*-value.)

INTERPRETATION Because the *P*-value of 0.0002 is less than or equal to the significance level of 0.05, we reject the null hypothesis. We again conclude that there is sufficient evidence to warrant rejection of the claim that when surveyed, the proportion of people who say they voted is equal to 0.61. Again, the *P*-value method is simply another way of arriving at the same conclusion reached using the traditional approach.

Rationale for Test Statistic: The test statistic used in this section is justified by noting that when using the normal distribution to approximate a binomial distribution, we substitute $\mu = np$ and $\sigma = \sqrt{npq}$ to get

$$z = \frac{x - \mu}{\sigma} = \frac{x - np}{\sqrt{npq}}$$

In this expression, x is the number of successes among n trials. Divide the numerator and denominator of this expression by n, then replace x/n by the symbol $\hat{p}$, and you get the test statistic we are using. In other words, the test statistic is simply the same standard score (from Section 2-5) of $z = (x - \mu)/\sigma$, but modified for the binomial notation.

Using Technology

STATDISK: Select **Analysis, Hypothesis Testing, Proportion-One Sample,** then proceed to enter the data in the dialog box.

Minitab: Select **Stat, Basic Statistics, 1 Proportion,** then click on the button for "Summarized data." Enter the sample size and number of successes, then click on **Options** and proceed to enter the data in the dialog box.

Excel: First enter the number of successes in cell A1, and enter the total number of trials in cell B1. Use the Data Disk XL add-in by clicking on **DDXL,** then select **Hypothesis Tests.** Under the function type options, select **Summ 1 Var Prop Test** (for testing a claimed proportion using summary data for one variable). Click on the pencil icon for "Num successes" and enter A1. Click on the pencil icon for "Num trials" and enter B1. Click **OK.** Follow the four steps listed in the dialog box. After clicking on **Compute** in step 4, you will get the *P*-value, test statistic, and conclusion.

TI-83 Plus: Press **STAT,** select **TESTS,** and then select **1-PropZTest.** Enter the claimed value of the population proportion for p0, then enter the values for x and n, and then select the type of test. Highlight **Calculate,** then press the **ENTER** key.

15. Percentage of Weapons on Campus Concerned about campus safety, the president of the College of Little Falls commissions a study. Among 400 students surveyed at this college, it is found that 7% carry a gun, knife, or other such weapon. Use a 0.01 significance level to test the claim that less than 10% of all U.S. college students carry such weapons. What is wrong with this conclusion?

16. TV Ratings A simple random sample of 1200 households with TV sets in use shows that 22% of them were tuned to *60 Minutes.* Use a 0.025 significance level to test the claim of a CBS executive that "*60 Minutes* gets more than a 20 share," which means that more than 20% of the sets in use are tuned to *60 Minutes.* If you are a commercial advertiser and you are trying to negotiate lower costs, what would you argue?

TI-83 Plus

```
1-PropZTest
 prop>.75
 z=8.262364472
 p=7.221833E-17
 p̂=.91
 n=500
```

17. Interpreting Calculator Display The Federal Aviation Administration will fund research on spatial disorientation of pilots if there is sufficient sample evidence (at the 0.01 significance level) to conclude that among aircraft accidents involving such disorientation, more than three-fourths result in fatalities. A study of 500 aircraft accidents involving spatial disorientation of the pilot found that 91% of those accidents resulted in fatalities (based on data from the U. S. Department of Transportation). The accompanying TI-83 Plus calculator display is obtained. Interpret that display. Based on these sample results, will the funding be approved?

TI-83 Plus

```
1-PropZTest
 prop<.5
 z=-1.789749481
 p=.0367470453
 p̂=.47997998
 n=1998
```

18. Interpreting Calculator Display A television executive claims that "fewer than half of all adults are annoyed by the violence shown on television." Sample data from a Roper poll showed that 48% of 1,998 surveyed adults indicated their annoyance with television violence. The accompanying TI-83 Plus calculator display is obtained. Use a 0.05 significance level and interpret that display. Is the executive's claim supported by the sample data?

19. Using M&M Data Refer to Data Set 10 in Appendix B and find the sample proportion of M&Ms that are blue. Use that result to test the claim of Mars, Inc., that 10% of its plain M&M candies are blue.

20. Using Survey Data Refer to Data Set 14 in Appendix B and consider only those statistics students who are 21 or older. Find the sample percentage of smokers in that age group, then test the claim that statistics students aged 21 or over smoke at a rate that is less than 32%, which is the smoking rate for the general population of persons aged 21 and over (based on data from the U.S. National Institute on Drug Abuse). Is there a reason that statistics students 21 and over would smoke at a rate lower than the rate for the general population in that age group?

7-5 Beyond the Basics

21. Using a Different Method In a study of perception, 80 men are tested and 7 are found to have red/green color blindness (based on data from *USA Today*). We want to use a 0.01 significance level to test the claim that men have a red/green color-blindness rate that is greater than the 0.25% rate for women.

a. Why can't we use the methods of this section?

b. Assuming that the red/green color-blindness rate for men is equal to the 0.25% rate for women, find the probability that among 80 randomly selected men, at least 7 will have that type of color blindness. Describe the method used to find that probability.

c. Based on the result from part (b), what do you conclude?

22. Dealing with No Successes In a simple random sample of 50 plain M&M candies, it is found that none of them are blue. We want to use a 0.01 significance level to test the claim of Mars, Inc., that the proportion of M&M candies that are blue is equal to 0.10. Can the methods of this section be used? If so, test the claim. If not, explain why not.

23. Misleading with Statistics Chemco, a supplier of chemical-waste containers, finds that 3% of a sample of 500 units are defective. Being fundamentally dishonest, the Chemco production manager wants to make a claim that the rate of defective units is no more than some specified percentage, and he doesn't want that claim rejected at the 0.05 significance level if the sample data are used. What is the *lowest* defective rate he can claim under these conditions?

24. Probability of Type II Error Refer to Exercise 18. If the true value of p is 0.45, find β, the probability of a type II error (see Exercise 31 in Section 7-3). [*Hint:* In Step 3, use the values $p = 0.45$ and $pq/n = (0.45)(0.55)/1998$.]

25. False Claim A researcher claimed that when 20 mice were treated, the success rate was equal to 47%. What is the basis for rejecting that claim?

7-6 Testing a Claim about a Standard Deviation or Variance

Quality-control engineers want to ensure that a product is, on the average, acceptable, but they also want to produce items of *consistent* quality so that there will be few defective products. For example, the consistency of aircraft altimeters is governed by Federal Aviation Regulation 91.36, which requires that aircraft altimeters be tested and calibrated to give a reading "within 125 feet (on a 95-percent probability basis)." Even if the mean altitude reading is exactly correct, an excessively large standard deviation will result in individual readings that are dangerously low or high. Consistency is improved by reducing the standard deviation or variance. In the preceding sections of this chapter we described methods for testing claims made about population means and proportions. This section focuses on variation, which is critically important in many applications, including quality control. The main objective of this section is to present methods to test claims made about a population standard deviation σ or variance σ^2. We begin by identifying the assumptions that apply to such tests.

Assumptions

1. The samples are simple random samples. (Remember the importance of good sampling methods.)
2. The population has values that are normally distributed.

Other methods of testing hypotheses also require a normally distributed population, but tests of claims about standard deviations or variances are not as *robust,* meaning that the inferences can be very misleading if the population does not have a normal distribution. Therefore, the condition of a normally distributed population is a much stricter requirement in this section.

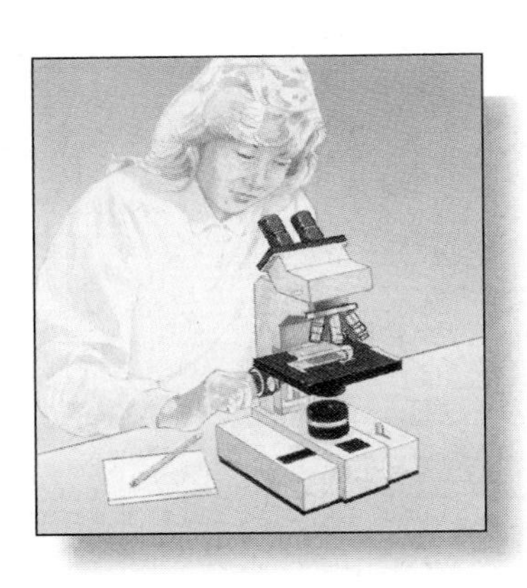

Ethics in Experiments

Sample data can often be obtained by simply observing or surveying members selected from the population. Many other situations require that we somehow manipulate circumstances to obtain sample data. In both cases ethical questions may arise. Researchers in Tuskegee, Alabama, withheld the effective penicillin treatment to syphilis victims so that the disease could be studied. This experiment continued for a period of 27 years!

When testing a claim made about a population standard deviation σ or variance σ^2, we use the chi-square test statistic and critical values described as follows.

Test Statistic for Testing Hypotheses about σ or σ^2

$$\chi^2 = \frac{(n-1)s^2}{\sigma^2}$$

where n = sample size
s^2 = sample variance
σ^2 = population variance (given in the null hypothesis)

Critical Values

1. Critical values are found in Table A-4.
2. Degrees of freedom = $n - 1$.

Don't be confused by the use of both the normal and the chi-square distributions. After verifying that the sample data appear to come from a normally distributed population, we should then shift gears and think "chi-square." The chi-square distribution was introduced in Section 6-6, where we noted the following important properties.

Properties of the Chi-Square Distribution

1. All values of χ^2 are nonnegative, and the distribution is not symmetric (see Figure 7-12).
2. There is a different distribution for each number of degrees of freedom (see Figure 7-13).
3. The critical values are found in Table A-4 using

degrees of freedom = $n - 1$

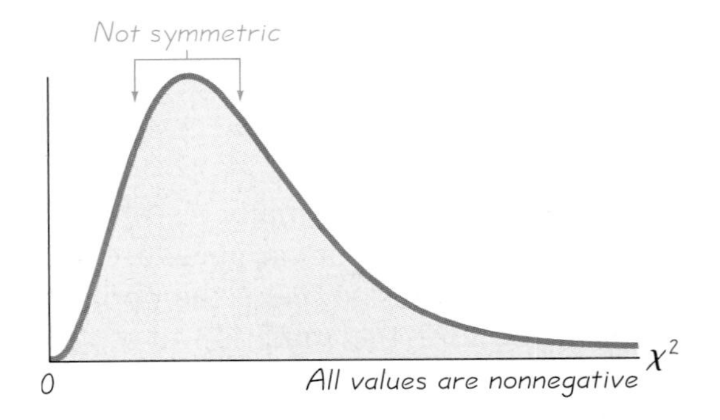

FIGURE 7-12
Properties of the Chi-Square Distribution

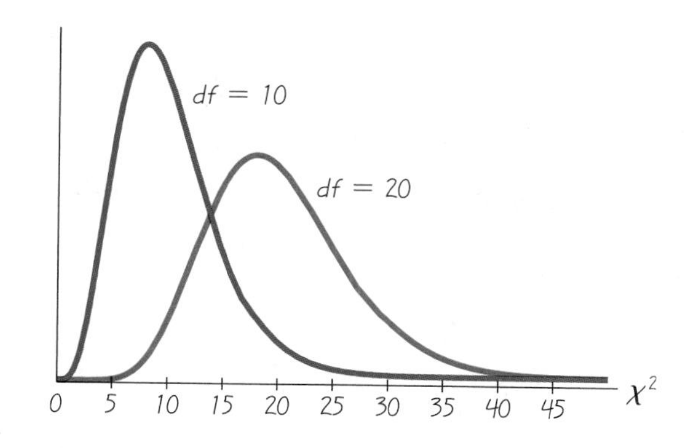

FIGURE 7-13
Chi-Square Distributions for 10 and 20 Degrees of Freedom

There is a different distribution for each number of degrees of freedom.

Critical values are found in Table A-4 by first locating the row corresponding to the appropriate number of degrees of freedom (where df $= n - 1$). Next, the significance level α is used to determine the correct column as described in the following list.

Right-tailed test: Locate the area at the top of Table A-4 that is equal to the significance level α.

Left-tailed test: Calculate $1 - \alpha$, then locate the area at the top of Table A-4 that is equal to $1 - \alpha$.

Two-tailed test: Calculate $1 - \alpha/2$ and $\alpha/2$, then locate the area at the top of Table A-4 that equals $1 - \alpha/2$ (for the left critical value) and locate the area at the top equal to $\alpha/2$ (for the right critical value). (See Figure 6-9 and the example on pages 345-346.)

EXAMPLE **IQ Scores of Statistics Professors** For a simple random sample of adults, IQ scores are normally distributed with a mean of 100 and a standard deviation of 15. A simple random sample of 13 statistics professors yields a standard deviation of $s = 7.2$. A psychologist is quite sure that statistics professors have IQ scores that have a mean greater than 100. He doesn't understand the concept of standard deviation very well and does not realize that the standard deviation should be lower than 15 (because statistics professors have less variation than the general population). Instead, he claims that statistics professors have IQ scores with a standard deviation equal to 15, the same standard deviation for the general population. Use a 0.05 significance level to test the claim that $\sigma = 15$. Based on the result, what do you conclude about the standard deviation of IQ scores for statistics professors?

SOLUTION We will use the traditional method of testing hypotheses as outlined in Figure 7-5.

Step 1: The claim is expressed in symbolic form as $\sigma = 15$.

Step 2: If the original claim is false, then $\sigma \neq 15$.

Step 3: The null hypothesis must contain equality, so we have

$$H_0: \sigma = 15 \text{ (original claim)} \qquad H_1: \sigma \neq 15$$

Step 4: The significance level is $\alpha = 0.05$.

Step 5: Because the claim is made about σ, we use the chi-square distribution.

Step 6: The test statistic is

$$\chi^2 = \frac{(n-1)s^2}{\sigma^2} = \frac{(13-1)(7.2)^2}{15^2} = 2.765$$

The critical values of 4.404 and 23.337 are found in Table A-4, in the 12th row (degrees of freedom $= n - 1 = 12$) in the columns corresponding to 0.975 and 0.025. See the test statistic and critical values shown in Figure 7-14.

Step 7: Because the test statistic is in the critical region, we reject the null hypothesis.

INTERPRETATION There is sufficient evidence to warrant rejection of the claim that the standard deviation is equal to 15. It appears that statistics professors have IQ scores with a standard deviation that is significantly different than the standard deviation of 15 for the general population.

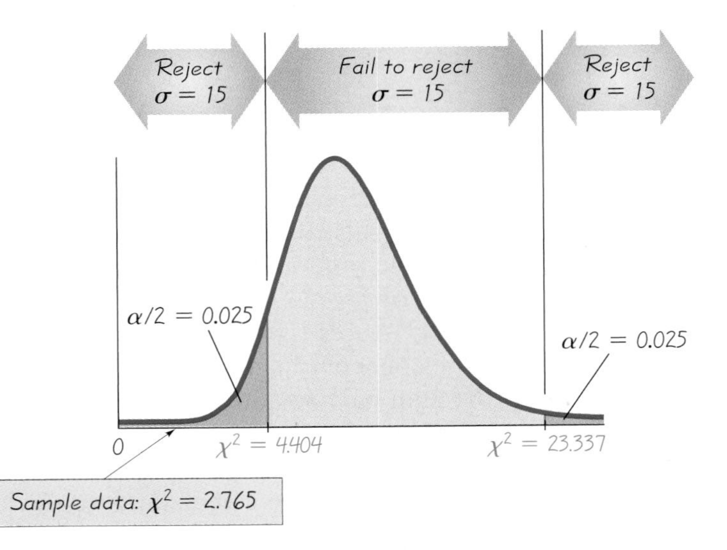

FIGURE 7-14
Hypothesis Test of Claim that $\sigma = 15$

The *P*-Value Method

Instead of the traditional approach to hypothesis testing, we can also use the *P*-value approach summarized in Figures 7-8 and 7-9. If STATDISK is used for the preceding example, the *P*-value of 0.0060 will be found. If we use Table A-4, we usually cannot find *exact* *P*-values because that chi-square distribution table includes only selected values of α. If using Table A-4, we can identify limits that contain the P-value. The test statistic from the last example is $\chi^2 = 2.765$, and we know that the test is two-tailed with 12 degrees of freedom. Refer to the 12th row of Table A-4 and see that the test statistic of 2.765 is less than every entry in that row, which means that the area to the left of the test statistic is less than 0.005. The P-value is therefore less than 0.01. (Figure 7-8 shows that in a two-tailed test with a test statistic to the left of the center, the P-value is *twice* the tail area beyond the test statistic.) Because the P-value is less than the significance level of $\alpha = 0.05$, we reject the null hypothesis. Again, the traditional method and the P-value method are equivalent in the sense that they always lead to the same conclusion.

Using Technology

STATDISK: Select **Analysis,** then **Hypothesis Testing,** then **StDev-One Sample.** Proceed to provide the required entries in the dialog box, then click on **Evaluate.** STATDISK will display the test statistic, critical value, P-value, conclusion, and confidence intervals.

Minitab, Excel, TI-83 Plus: These technologies are not yet designed to test claims made about σ or σ^2.

7-6 Basic Skills and Concepts

Finding Critical Values. *In Exercises 1 and 2, use Table A-4 to find the critical values of χ^2 based on the given information.*

1. a. H_0: $\sigma = 15$, $n = 10$, $\alpha = 0.01$
b. H_1: $\sigma > 0.62$, $n = 27$, $\alpha = 0.01$
c. H_1: $\sigma < 14.4$, $n = 21$, $\alpha = 0.05$

2. a. H_1: $\sigma < 1.22$, $n = 23$, $\alpha = 0.025$
b. H_1: $\sigma > 92.5$, $n = 12$, $\alpha = 0.10$
c. H_0: $\sigma = 0.237$, $n = 16$, $\alpha = 0.05$

Testing Claims about Variation. *In Exercises 3–14, use the traditional method to test the given hypotheses. Follow the steps outlined in Figure 7-5, and draw the appropriate graph. In all cases, assume that the population is normally distributed and that the sample is a simple random sample.*

3. Use a 0.05 significance level to test the claim that $\sigma = 15$. The sample statistics are as follows: $n = 10$, $\bar{x} = 103.5$, $s = 18.0$.

4. Use a 0.01 significance level to test the claim that $\sigma > 5$. The sample statistics are as follows: $n = 16$, $\bar{x} = 47.3$, $s = 8.0$.

5. Variation in Peanut M&Ms Use a 0.01 significance level to test the claim that peanut M&M candies have weights that vary more than the weights of plain M&M candies. The standard deviation for the weights of plain M&M candies is 0.04 g. A sample of 40 peanut M&Ms has weights with a standard deviation of

0.31 g. Why should peanut M&M candies have weights that vary more than the weights of plain M&M candies?

6. Random Generation of Data The TI-83 Plus calculator can be used to generate random data from a normally distributed population. The command **randNorm(100,15,50)** generates 50 values from a normally distributed population with $\mu = 100$ and $\sigma = 15$. One such generated sample of 50 values has a mean of 98.4 and a standard deviation of 16.3. Use a 0.10 significance level to test the claim that this sample actually does come from a population with a standard deviation equal to 15. What does the result say about the variation among the generated sample values?

7. Manufacturing Aircraft Altimeters The Stewart Aviation Products Company uses a new production method to manufacture aircraft altimeters. A simple random sample of 81 altimeters is tested in a pressure chamber, and the errors in altitude are recorded as positive values (for readings that are too high) or negative values (for readings that are too low). The sample has a standard deviation of $s = 52.3$ ft. At the 0.05 significance level, test the claim that the new production line has errors with a standard deviation different from 43.7 ft, which was the standard deviation for the old production method. If it appears that the standard deviation has changed, does the new production method appear to be better or worse than the old method?

8. Is the New Machine Better? The Medassist Pharmaceutical Company uses a machine to pour cold medicine into bottles in such a way that the standard deviation of the weights is 0.15 oz. A new machine is tested on 71 bottles, and the standard deviation for this sample is 0.12 oz. The Dayton Machine Company, which manufactures the new machine, claims that it fills bottles with a lower variation. At the 0.05 significance level, test the claim made by the Dayton Machine Company. If Dayton's machine is being used on a trial basis, should its purchase be considered?

9. Statistics Test Scores Tests in the author's past statistics classes have scores with a standard deviation equal to 14.1. One of his current classes now has 27 test scores with a standard deviation of 9.3. Use a 0.01 significance level to test the claim that this current class has less variation than past classes. Does a lower standard deviation suggest that the current class is doing better?

10. Bank Customer Waiting Times With individual lines at its various windows, the Jefferson Valley Bank found that the standard deviation for normally distributed waiting times on Friday afternoons was 6.2 min. The bank experimented with a single main waiting line and found that for a simple random sample of 25 customers, the waiting times have a standard deviation of 3.8 min. On the basis of previous studies, we can assume that the waiting times are normally distributed. Use a 0.05 significance level to test the claim that a single line causes lower variation among the waiting times. Why would customers prefer waiting times with less variation? Does the use of a single line result in a shorter wait?

11. Blood Pressure Measurements Systolic blood pressure results from contraction of the heart. In comparing systolic blood pressure levels of men and women, Dr. Jane Taylor obtained readings for a simple random sample of 50 women. The sample mean and standard deviation were found to be 130.7 and 23.4, respectively. If systolic blood pressure levels for men are known to have a mean and standard deviation of 133.4 and 19.7, respectively, test the claim that women have more variation. Use a 0.05 significance level. (All readings are in millimeters of mercury, and data are based on the National Health Survey.)

12. Body Temperatures In Section 7-3, we used the sample data given at the beginning of the chapter ($n = 106$, $\bar{x} = 98.20$, $s = 0.62$) to test the claim that the mean body temperature of healthy adults is equal to 98.6°F, and we rejected that claim. The test statistic will cause rejection of $\mu = 98.6$°F as long as the standard deviation is less than 2.11°F. Use the sample statistics and a 0.005 significance level to test the claim that $\sigma < 2.11$°F.

13. Using Data of Heights of Men Based on data from the National Health Survey, men aged 25–34 have heights with a standard deviation of 2.9 in. Test the claim that men aged 45–54 have heights with a different standard deviation. The heights of 25 randomly selected men in the 45–54 age bracket are listed below.

66.80 71.22 65.80 66.24 69.62 70.49 70.00 71.46 65.72

68.10 72.14 71.58 66.85 69.88 68.69 72.77 67.34 68.40

68.96 68.70 72.69 68.67 67.79 63.97 67.19

14. Using Birth Weight Data Shown below are birth weights (in kilograms) of male babies born to mothers taking a special vitamin supplement (based on data from the New York State Department of Health). Test the claim that this sample comes from a population with a standard deviation equal to 0.470 kg, which is the standard deviation for male birth weights in general. Does the vitamin supplement appear to affect the variation among birth weights?

3.73 4.37 3.73 4.33 3.39 3.68 4.68 3.52

3.02 4.09 2.47 4.13 4.47 3.22 3.43 2.54

15. Supermodel Weights Use a 0.01 significance level to test the claim that weights of female supermodels vary less than the weights of women in general. The standard deviation of weights of the population of women is 29 lb. Listed below are the weights (in pounds) of nine randomly selected supermodels.

125 (Taylor) 119 (Auermann) 128 (Schiffer) 128 (MacPherson)
119 (Turlington) 127 (Hall) 105 (Moss) 123 (Mazza)
115 (Hume)

16. Supermodel Heights Use a 0.05 significance level to test the claim that heights of female supermodels vary less than the heights of women in general. The standard deviation of heights of the population of women is 2.5 in. Listed below are the heights (in inches) of randomly selected supermodels (Taylor, Harlow, Mulder, Goff, Evangelista, Avermann, Schiffer, MacPherson, Turlington, Hall, Crawford, Campbell, Herzigova, Seymour, Banks, Moss, Mazza, Hume).

71 71 70 69 69.5 70.5 71 72 70 70 69 69.5 69 70 70 66.5 70 71

7-6 Beyond the Basics

17. Finding *P*-Values Use Table A-4 to find the range of possible *P*-values in the given exercises.

a. Exercise 3 **b.** Exercise 5 **c.** Exercise 15

18. Finding Critical Values of χ^2 For large numbers of degrees of freedom, we can approximate critical values of χ^2 as follows:

$$\chi^2 = \frac{1}{2}(z + \sqrt{2k - 1})^2$$

Here k is the number of degrees of freedom and z is the critical value, found in Table A-2. For example, if we want to approximate the two critical values of χ^2 in a two-tailed hypothesis test with $\alpha = 0.05$ and a sample size of 150, we let $k = 149$ with $z = -1.96$, followed by $k = 149$ and $z = 1.96$.

a. Use this approximation to estimate the critical values of χ^2 in a two-tailed hypothesis test with $n = 101$ and $\alpha = 0.05$. Compare the results to those found in Table A-4.

b. Use this approximation to estimate the critical values of χ^2 in a two-tailed hypothesis test with $n = 150$ and $\alpha = 0.05$.

19. Finding Critical Values of χ^2 Repeat Exercise 18 using this approximation (with k and z as described in Exercise 18):

$$\chi^2 = k\left(1 - \frac{2}{9k} + z\sqrt{\frac{2}{9k}}\right)^3$$

20. Effect of Outlier When using the hypothesis testing procedure of this section, will the result be dramatically affected by the presence of an outlier? Describe how you arrived at your response.

21. Last-Digit Analysis The last digits of sample data are sometimes used in an attempt to determine whether the data have been measured or simply reported by the subject. Reported data often have last digits with disproportionately more 0s and 5s. Measured data tend to have last digits with a mean of 4.5, a standard deviation of about 3, and the digits should occur with roughly the same frequency.

a. How is the standard deviation of the data affected if there are disproportionately more 0s and 5s?

b. Why can't we use the methods of this section to test that the last digits of the sample data have a standard deviation equal to 3?

22. Probability of a Type II Error Refer to Exercise 10. Assuming that σ is actually 4.0, find β (the probability of a type II error). See Exercise 31 in Section 7-3, and modify the procedure so that it applies to a hypothesis test involving σ instead of μ.

"Freshman 15": Fact or Fantasy?

BOSTON Along with all of the typical "back-to-school" hype about lunch boxes and school buses, each September is typically greeted with media reports and advice about the "freshman 15," which is the popular name given to the phenomenon of first-year college students gaining 15 lb during their freshman year. But does this 15-lb weight gain actually occur, or is it simply a myth? Carole Nhu'y Hodge, Linda Jackson, and Linda Sullivan are Michigan State University researchers who conducted their own investigation. They studied 61 Michigan State female students who took an introductory psychology course. The volunteers, who were given extra credit for participation in the experiment, were weighed at the beginning of their freshman year and at a point in time six months later. Among their findings reported in *Psychology of Women Quarterly*: "Body weight at the beginning of the first college year (Time 1) was compared with weight approximately 6 months later (Time 2). Average weight at Time 2, 131.45 lb (59.62 kg), was no different from average weight at Time 1, 130.57 lb (59.23 kg)." They also state that "Our findings suggest it (the 15-lb weight gain) is fantasy, although additional research is needed before drawing firm conclusions."

1. What do the researchers infer when they say that there is "no difference" between the mean weight at Time 1 (130.57 lb) and the mean weight at Time 2 (131.45 lb), when there is an apparent difference of 0.88 lb?
2. What are the limitations of this particular study? That is, if the sample data are used to make inferences about a population, identify the specific population in question.
3. Identify any aspects of the experiment that could potentially threaten the validity of the results.
4. Identify a possible null hypothesis and alternative hypothesis for this experiment.

Vocabulary List

hypothesis
hypothesis test
test of significance
null hypothesis
alternative hypothesis
test statistic
critical region
significance level
critical value
two-tailed test
right-tailed test
left-tailed test
type I error
type II error
alpha (α)
beta (β)
power
traditional method
classical method
P-value
probability value

5. Testing Consistency of Filled Bags The Kansas Farm Products Company uses a machine that fills 50-lb corn-seed bags. In the past, the machine has had a standard deviation of 0.75 lb. In an attempt to get more consistent weights, mechanics have replaced some worn machine parts. A simple random sample of 61 bags taken from the repaired machinery produced a sample mean of 50.13 lb and a sample standard deviation of 0.48 lb. At the 0.05 significance level, test the claim that the weights are "more consistent" with the repaired machinery than in the past. If the weights are more consistent, how is the standard deviation affected?

6. Percentage Believing That Elvis Is Alive *USA Today* ran a report about a University of North Carolina poll of 1248 adults from the southern United States. It was reported that 8% of those surveyed believe that Elvis Presley still lives. The article began with the claim that "almost 1 out of 10" Southerners still thinks Elvis is alive. At the 0.01 significance level, test the claim that the true percentage is less than 10%. Based on the result, determine whether the 8% sample result justifies the phrase "almost 1 out of 10."

7. Driving Distances of Women The New England Insurance Company is reviewing the driving habits of women aged 16–24 to determine whether they should continue to pay higher premiums than women in a higher age bracket. In a study of 750 randomly selected women drivers aged 16–24, the mean driving distance for one year is 6047 mi and the standard deviation is 2944 mi (based on data from the Federal Highway Administration). Use a 0.01 significance level to test the claim that the population mean for women in the 16–24 age bracket is less than 7124 mi, which is the known mean for women in the higher age bracket. If women in the 16–24 age bracket drive less, should they be charged lower insurance premiums?

8. Uses of Fruitcakes A Bruskin-Goldring Research poll of 1012 adults showed that among those who used fruitcakes, 28% ate them and 72% used them for other purposes, such as doorstops, birdfeed, and landfill. This surprised fruitcake producers, who believe that a fruitcake is an appealing food. The president of the Kansas Food Products Company claims that the poll results are a fluke and, in reality, at least half of all adults eat their fruitcakes. Use a 0.01 significance level to test that claim. Based on the result, does it appear that fruitcake producers should consider changes to make their product more appealing as a food or better suited for its uses as a doorstop, birdfeed, and so on?

9. Seizure Reduction Drug The Medassist Pharmaceutical Company makes a pill intended to reduce the chance of seizures among children. The pill is supposed to contain 20.0 mg of phenobarbital. A simple random sample of 20 pills yielded the amounts (in mg) listed here. Are these pills acceptable at the $\alpha = 0.01$ significance level?

27.5	26.0	22.9	23.4	23.0
23.9	32.6	20.9	22.9	24.3
24.8	16.1	24.3	17.3	18.9
20.7	33.0	15.6	24.3	23.3

10. Claim about Mean Annual Income The Orange County Supervisor boasts that full-time salaried employees in her county have a mean annual income that is greater than $30,000. If she based that claim on a sample of 150 such employees who had a mean of $30,122 and a standard deviation of $14,276, is her claim justified? Use a 0.01 significance level.

Cumulative Review Exercises

1. Analyzing SAT Scores The College of Newport randomly selects female applicants for admission and records their scores on the math portion of the SAT. Here are the results: 490, 570, 630, 440, 460, 720, 410, 780.
   a. Find the mean of this sample.
   b. Find the median.
   c. Find the standard deviation.
   d. Find the variance.
   e. Find the range.
   f. Construct a 95% confidence interval estimate of the population mean.
   g. Use a 0.05 significance level to test the claim that this sample comes from a population with a mean equal to 496, which is the reported mean of the SAT math scores for all women.
   h. Based on the preceding results, what do you conclude about the female applicants to the College of Newport? Are their SAT math scores about the same as the general population, or are they better or worse? (The population of all women who take the math portion of the SAT has a mean score of 496.)

2. SAT Math Scores of Women The math SAT scores for women are normally distributed with a mean of 496 and a standard deviation of 108.
   a. If a woman who takes the math portion of the SAT is randomly selected, find the probability that her score is above 500.
   b. If five math SAT scores are randomly selected from the population of women who take the test, find the probability that all five of the scores are above 500.
   c. If five women who take the math portion of the SAT are randomly selected, find the probability that their mean is above 500.
   d. Find P_{90}, the score separating the bottom 90% from the top 10%.

3. Blood Pressure Readings A medical researcher obtains the systolic blood pressure readings (in mm Hg) in the accompanying list from a sample of women aged 18–24 who have a new strain of viral infection. (Healthy women in that age group have readings that are normally distributed with a mean of 114.8 and a standard deviation of 13.1.)

134.9	78.7	108.9	133.0	123.7	96.1	126.9	89.8
132.0	134.7	132.1	121.7	112.3	150.2	158.3	154.4

   a. Find the sample mean $\bar{x}$ and standard deviation s.
   b. Use a 0.05 significance level to test the claim that the sample comes from a population with a mean blood pressure equal to 114.8.
   c. Use the sample data to construct a 95% confidence interval for the population mean μ. Do the confidence interval limits contain the value of 114.8, which is the mean for healthy women aged 18–24?
   d. Use a 0.05 significance level to test the claim that the sample comes from a population with a standard deviation equal to 13.1, which is the standard deviation for healthy women aged 18–24.
   e. Based on the preceding results, does it seem that the new strain of viral infection affects systolic blood pressure?

4. ESP A student majoring in psychology designs an experiment to test for extrasensory perception (ESP). In this experiment, a card is randomly selected

from a shuffled deck, and the blindfolded subject must guess the suit (clubs, diamonds, hearts, spades) of the card selected. The experiment is repeated 25 times, with the card replaced and the deck reshuffled each time.

a. For subjects who make random guesses with no ESP, find the mean number of correct responses.

b. For subjects who make random guesses with no ESP, find the standard deviation for the numbers of correct responses.

c. For subjects who make random guesses with no ESP, find the probability of getting more than 12 correct responses.

d. If a subject gets more than 12 correct responses, test the claim that they made random guesses. Use a 0.05 significance level.

e. You want to conduct a survey to estimate the percentage of adult Americans who believe that some people have ESP. How many people must you survey if you want 90% confidence that your sample percentage is in error by no more than four percentage points?

Cooperative Group Activities

1. *In-class activity:* Each student should estimate the length of the classroom. The values should be based on visual estimates, with no actual measurements being taken. After the estimates have been collected, measure the length of the room, then test the claim that the sample mean is equal to the actual length of the classroom. Is there a "collective wisdom," whereby the class mean is approximately equal to the actual room length?

2. *Out-of-class activity:* Using a wristwatch that is reasonably accurate, set the time to be exact. Use a radio station or telephone time report which states that "at the tone, the time is . . . . " If you cannot set the time to the nearest second, record the error for the watch you are using. Now compare the time on your watch to the time on others. Record the errors with positive signs for watches that are ahead of the actual time and negative signs for those watches that are behind the actual time. Use the data to test the claim that the mean error of all wristwatches is equal to 0. Do we collectively run on time, or are we early or late? Also test the claim that the standard deviation of errors is less than 1 min. What are the practical implications of a standard deviation that is excessively large?

3. *In-class activity:* In a group of three or four people, conduct an ESP experiment by selecting one of the group members as the subject. Draw a circle on one small piece of paper and draw a square on another sheet of the same size. Repeat this experiment 20 times: Randomly select the circle or the square and place it in the subject's hand behind his or her back so that it cannot be seen, then ask the subject to identify the shape (without looking at it); record whether the response is correct. Test the claim that the subject has ESP because the proportion of correct responses is greater than 0.5.

4. *In-class activity:* After dividing into groups with sizes between 10 and 20 people, each group member should record the number of heartbeats in a minute. After calculating $\bar{x}$ and s, each group should proceed to test the claim that the mean is greater than 60, which is the author's result. (When people exercise, they tend to have lower pulse rates, and the author runs five miles a few times each week. What a guy.)

TECHNOLOGY PROJECT

STATDISK, Minitab, Excel, the TI-83 Plus calculator, and many other tools can be used to randomly generate data from a normally distributed population with a given mean and standard deviation.

a. Use such a tool to randomly generate five scores from a normally distributed population with a mean of 100 and a standard deviation of 15 (the parameters for typical IQ tests).

b. Using the five sample values generated in part (a), test the claim that the sample is from a population with a mean equal to 100. Use a 0.10 significance level.

c. Repeat parts (a) and (b) nine more times, so that a total of 10 different samples have been generated, and 10 different hypothesis tests have been executed.

d. With a 0.10 significance level, there is 0.10 probability of making a type I error (rejecting the null hypothesis when it is true). Because of the way that the sample data are generated, we know that 100 is the true population mean, so we do make a type I error in this experiment anytime that we reject the null hypothesis of $\mu = 100$. For the 10 trials of this experiment, how often was the null hypothesis actually rejected? When we conduct 10 such trials, how many times do we expect to reject the null hypothesis? Are the actual results consistent with the theoretical results? Explain.

Here are specific instructions for steps (a) and (b) using STATDISK, Minitab, Excel, and the TI-83 Plus calculator.

STATDISK:

a. Click on **Data,** then click on **Normal Generator.** In the dialog box, enter a sample size of 5, a mean of 100, a standard deviation of 15, and enter 0 for the number of decimal places. Click on **Generate.**

Now proceed to find the values of the sample mean and standard deviation. With the five generated values displayed, click on **Copy,** then then click on the main menu bar item of **Data.** Click on the menu item of **Descriptive Statistics,** then click on **Paste.** The generated sample data should appear. Click on **Evaluate.** Record the values of the sample mean and standard deviation.

b. Click on the main menu item of **Analysis,** then click on **Hypothesis Testing.** Select **Mean - One Sample.** In the dialog box, enter a significance level of 0.10, a claimed mean of 100, a sample size of 5, and enter the values of the sample mean and standard deviation that were recorded in step (a). Click on **Evaluate** and record the result (reject null hypothesis or fail to reject the null hypothesis).

Minitab

a. Click on the main menu item of **Calc,** then click on **Random Data,** then **Normal.** In the dialog box, enter 5 for the number of rows to be generated, enter C1 for the column in which data will be stored, enter a mean of 100, and enter a standard deviation of 15. Click **OK.**

b. Click on the main menu item of **Stat,** select **Basic Statistics,** then select **1-Sample t.** In the dialog box, enter C1 for the variable name, click on the button for **Test Mean,** enter 100 in the adjacent box, then click **OK.** Interpret the displayed results and record the conclusion (reject the null hypothesis or fail to reject the null hypothesis).

Excel

a. Click on **Tools,** select **Data Analysis,** then select **Random Number Generation** and click **OK.** In the dialog box, enter 1 for the number of variables, enter 5 for the number of random numbers, select the distribution option of **Normal,** enter a mean of 100, enter a standard deviation of 15, then click **OK.**

b. Click **DDXL** and select **Hypothesis Tests.** Select the item of **1 Var t Test.** Click on the pencil icon and enter the range of cells containing the generated sample data. For example, enter A1:A5 for five values in rows 1 through 5 of column A. Click **OK.** In the dialog box, click on the bar in step 1 and proceed to enter a claimed mean of 100. Proceed to click on the bars in the remaining steps. After clicking **Compute,** record the conclusion (reject the null hypothesis or fail to reject the null hypothesis).

TI-83 Plus

a. First clear list L1 by pressing **STAT,** then **4:ClrList,** then **L1.** Now press **MATH,** then select **PRB** and select the menu item of **6:randNorm(**. Press **ENTER,** then proceed to enter 100,15,5 and press **ENTER.** Store the generated sample data by pressing **STO L1** followed by the **ENTER** key.

b. Press **STAT,** select **TESTS,** then select **2:T-Test** and press **ENTER.** Select **Data** (because we have the generated data in list L1), enter 100 for the claimed mean, and proceed to obtain the results of the hypothesis test. Interpret the displayed results and record the conclusion (reject the null hypothesis or fail to reject the null hypothesis).

Critical Thinking: Should this drug be approved?

Standard methods of hypothesis testing are routinely used to decide whether newly developed drugs are actually effective. Pharmaceutical companies routinely research and develop drugs that accomplish objectives such as lowering blood pressure, lowering cholesterol levels, and growing hair on the bald pates of men. Men aged 18–24 have a mean cholesterol level of 178.1 (based on data from the National Health Survey), and one criterion for identifying risk of coronary disease is a cholesterol level above 300.

Analyzing the results

The accompanying list shows the serum cholesterol levels (in mg/100 mL) of randomly selected men aged 18–24 who took the drug Sinatrastatin.

a. Use the given data to test the claim that this treatment group has a mean level less than 178.1.

b. Based on the result, does it appear that the drug is effective in lowering cholesterol?

c. This experiment involves treating randomly selected subjects, then comparing the sample mean to the population mean. Is there a more effective way that the experiment could be designed? If so, describe the better design.

131	159	174	227	232	204
111	194	159	148	233	112
169	111	168	173	180	257
138	192	227	147	153	178
164	150	147	151	206	213
225	189	158	123	198	215

Internet Project

Hypothesis Testing

This chapter discussed the important method of inferential statistics known as hypothesis testing. This Internet Project will have you conduct tests using a variety of data sets. For each data set, you will be asked to

- Collect data available on the Internet.
- Formulate a hypothesis based on a given question.
- Conduct the test at a specified level of significance.
- Summarize your conclusions

Go to the web site for this text:

http://www.awlonline.com/triola

There you will find instructions outlining tasks using data from business and economics, education, and sports, as well as a classic example from physical sciences.

Statistics *at work*

"If I didn't have any background in statistics, I would not be able to fully understand the data my company produces... help protect our workers and customers."

Jeffrey Foy

Jeffrey Foy is a toxicologist working for the Cabot Corporation, a chemical company.

Jeffrey Foy is also responsible for the hazard evaluation of the chemicals Cabot Corporation produces. It is his job to understand how the company's products may affect humans or the environment and help decide on the best ways to protect both.

What do you do in your job?

My responsibilities include arranging and evaluating toxicological studies, writing material safety data sheets, and helping our research and development groups produce materials that are safe for both people and the environment or to understand what potential hazards the materials might have.

What concepts of statistics do you use?

The primary concept I use is hypothesis testing (probability testing).

How do you use statistics on the job?

I use statistics daily. Statistical methods have been and are used in two ways in my work. First, statistics is used to help determine how I design my experiments. Second, statistics is used to determine if the data generated are significant or sometimes if they're even good enough to use. If you start an experiment without thinking of how you are going to use or evaluate the data, you may not sufficiently account for all the endpoints you need to examine. If you start your work and realize that you didn't include enough subjects, time points, or dose ranges, then you may have just wasted time. If you're lucky, then that's all you've wasted. Studies that I am involved in can cost from as little as \$1,000, to as much as \$500,000 or more, and if you don't properly determine how you are going to evaluate the data, you could cost your company a great deal of time and money. If the experiment is done properly, then we move on to analyze the data. The data from the studies we perform are used to assess any potential health effects our products may have on our workers, customers or the environment. The results are used to determine how chemicals can be sold or handled. When performing experiments at a testing laboratory or drug company, you want to determine if your materials have an effect, whether desired (a drug curing a disease) or undesired (that same drug being toxic). Statistics plays an enormous role in our evaluation of the significance of the effects.

Please describe one specific example illustrating how the use of statistics was successful in improving a product or service.

A toxicology study costing about \$300,000 dollars was recently conducted. The data from the study were to be used to help determine if a particular chemical caused any effects in the subjects studied. After the study was performed, flaws were found in both the data and statistics used. It took an additional 2 years to properly review the data and finish the health evaluation. If the proper methods and endpoints had

been chosen, then the additional time and money may not have been necessary. It was the understanding of the data and correct statistical evaluation that helped prevent the failure and potential repeat of the study.

How do you ensure objectivity in your studies?

Objectivity is attained by having an external source do our studies and perform the initial statistical analysis of the data.

How beneficial is your knowledge of statistics in your job?

If I didn't have any background in statistics, I would not be able to fully understand the data my company produces. I would not be able to use the data to help protect our workers and customers.

In terms of statistics, what would you recommend for prospective employees?

If you are doing scientific research on a daily basis, then you must thoroughly understand which methods of statistics are needed to evaluate your data. Not all data are evaluated the same way, and a researcher needs to know the correct ways of testing his or her hypothesis. It could make the difference between the commercial success or failure of a product. If you are not doing as much with data evaluation, then you should be able to look at data and statistics compiled by others and be able to evaluate their results.

Do you recommend statistics for today's college student? Why?

Statistics is essential.

Which other skills are important for today's college students?

Public speaking and a working knowledge of a second language (2 years of study), since so many companies are now global enterprises.

CHAPTER 8

Inferences from Two Samples

CHAPTER PROBLEM

Do male statistics students exaggerate their heights?

Male statistics students were given a survey that included a question asking them to report their height in inches. They weren't told that their height would be measured, but heights were accurately measured after the survey was completed. Anonymity was maintained, with code numbers used instead of names, so that no personal information would be publicly announced and nobody would be embarrassed by the results. Sample data are listed in Table 8-1. Is there a significant difference between the reported heights and the measured heights?

TABLE 8-1 Reported and Measured Heights in inches of Male Statistics Students

Student	A	B	C	D	E	F	G	H	I	J	K	L
Reported height	68	74	82.25	66.5	69	68	71	70	70	67	68	70
Measured height	66.8	73.9	74.3	66.1	67.2	67.9	69.4	69.9	68.6	67.9	67.6	68.8
Difference	1.2	0.1	**7.95**	0.4	1.8	0.1	1.6	0.1	1.4	**−0.9**	0.4	1.2

8-1 Overview

Chapter 6 introduced an important activity of inferential statistics: Sample data were used to estimate values of population parameters. Chapter 7 introduced a second important activity of inferential statistics: Sample data were used to test hypotheses about population parameters. In both of those chapters, all examples and exercises involved the use of *one* sample to form an inference about *one* population. In reality, however, there are many important and meaningful situations in which it becomes necessary to compare *two* sets of sample data. The following are examples typical of those found in this chapter, which presents methods for using data from two samples so that inferences can be made about the populations from which they came.

- When testing the effectiveness of a drug, determine whether there is a significant difference between a group of subjects who were treated with the drug and a second group of subjects who were given a placebo.
- When surveying car drivers, determine whether the percentage of women who use cell phones is different from the percentage of men who use cell phones.

8-2 Inferences about Two Means: Independent and Large Samples

In this section we consider methods for using sample data from two independent and large ($n > 30$) samples to test hypotheses made about two population means or to construct confidence interval estimates of the difference between two population means. We begin by formally defining *independent* and *dependent*.

DEFINITIONS

Two samples are **independent** if the sample values selected from one population are not related to or somehow paired with the sample values selected from the other population. If the values in one sample are related to the values in the other sample, the samples are **dependent.** Such samples are often referred to as **matched pairs,** or **paired samples.** (We will use *matched pairs.*)

EXAMPLE Drug Testing

Independent samples: One group of subjects is treated with a drug while a separate group of subjects is given a placebo. These two sample groups are independent because there is no direct relationship between the treatment group and the placebo group.

Matched pairs: Blood pressure levels of subjects are measured before and after taking a drug. Each "before" value is matched with the "after" value because each before/after pair of measurements comes from the same person.

For the hypothesis tests and confidence intervals described in this section, we make the following assumptions.

Assumptions

1. The two samples are *independent.*
2. The two sample sizes are *large.* That is, $n_1 > 30$ and $n_2 > 30$.
3. Both samples are *simple random samples.*

Exploring the Data Sets

Let's suppose that you want to investigate the difference between two simple random samples selected from two populations. Instead of immediately running a hypothesis test or constructing a confidence interval, you might first consider *exploring* the two samples using the methods described in Chapter 2. It could be very helpful to do the following:

- Find descriptive statistics for both data sets, including n, $\bar{x}$, and s.
- Create boxplots of both data sets, drawn on the same scale so that they can be compared.
- Create histograms of both data sets, so that their distributions can be compared.
- Identify any outliers.

Hypothesis Tests

In testing a claim about the difference $\mu_1 - \mu_2$, we use the following test statistic.

Test Statistic for Two Means: Independent and Large Samples

$$z = \frac{(\bar{x}_1 - \bar{x}_2) - (\mu_1 - \mu_2)}{\sqrt{\dfrac{\sigma_1^2}{n_1} + \dfrac{\sigma_2^2}{n_2}}}$$

σ_1 and σ_2: If σ_1 and σ_2 are not known, use s_1 and s_2 in their places, provided that both samples are large.

***P*-value:** Use the computed value of the test statistic z, and find the P-value by following the procedure summarized in Figure 7-8.

Critical values: Based on the significance level α, find critical values by using the procedures introduced in Section 7-2.

Except for the form of this test statistic, the method of hypothesis testing is essentially the same as that described in Chapter 7. (Later in this section, we will discuss the rationale for the form of the given test statistic.) As in Chapter 7, if the values of σ_1 and σ_2 are not known, we can use s_1 and s_2 in their places, provided that both samples are large. If σ_1 and σ_2 are known, we use those values in calculating the test statistic, but in reality they are rarely known.

EXAMPLE **Coke versus Pepsi** Data Set 1 in Appendix B includes the weights (in pounds) of samples of regular Coke and regular Pepsi. Sample statistics and boxplots are shown below. Use a 0.01 significance level to test the claim that the mean weight of regular Coke is different from the mean weight of regular Pepsi.

	Regular Coke	Regular Pepsi
n	36	36
$\bar{x}$	0.81682	0.82410
s	0.007507	0.005701

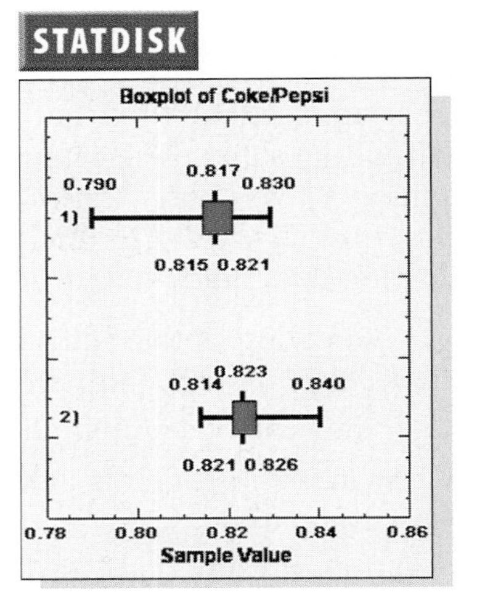

SOLUTION In the spirit of exploring the two data sets, we see that the means do not appear to be dramatically different, but the boxplots seem to display a difference. Let's proceed with a formal hypothesis test to determine whether the difference between the two sample means is really significant.

Step 1: The claim of different means can be expressed symbolically as $\mu_1 \neq \mu_2$.

Step 2: If the original claim is false, then $\mu_1 = \mu_2$.

Step 3: The null hypothesis must contain equality, so we have

$$H_0: \mu_1 = \mu_2 \qquad H_1: \mu_1 \neq \mu_2 \text{ (original claim)}$$

We now proceed with the assumption that $\mu_1 = \mu_2$, or $\mu_1 - \mu_2 = 0$.

Step 4: The significance level is $\alpha = 0.01$.

Step 5: Because we have two independent and large samples and we are testing a claim about the two population means, we use a normal distribution with the test statistic given earlier in this section.

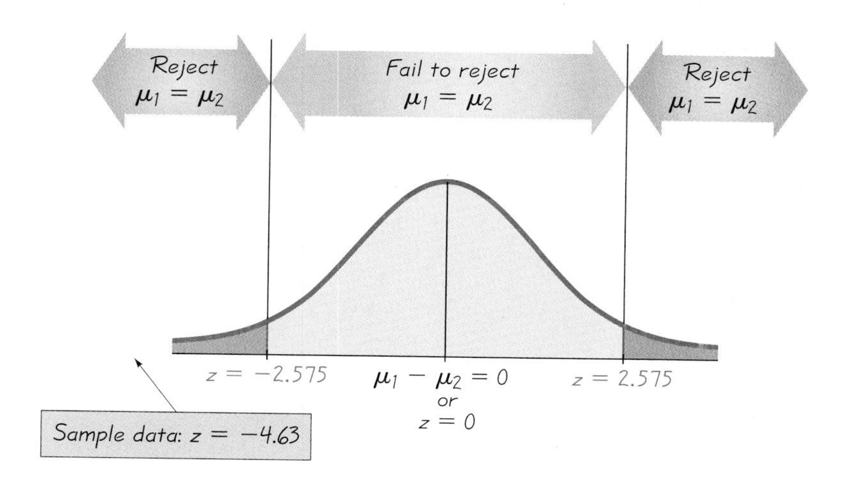

FIGURE 8-1
Distribution of Differences Between Means of Weights of Regular Coke and Regular Pepsi

Step 6: The values of σ_1 and σ_2 are unknown, but the samples are both large, so we can use the sample standard deviations as estimates of the population standard deviations, and the test statistic is calculated as follows:

$$z = \frac{(\bar{x}_1 - \bar{x}_2) - (\mu_1 - \mu_2)}{\sqrt{\frac{\sigma_1^2}{n_1} + \frac{\sigma_2^2}{n_2}}} = \frac{(0.81682 - 0.82410) - 0}{\sqrt{\frac{0.007507^2}{36} + \frac{0.005701^2}{36}}} = -4.63$$

Because we are using a normal distribution, the critical values of $z = \pm 2.575$ are found from Table A-2. (With $\alpha = 0.01$ divided equally between the two tails, find the z score corresponding to an area of $0.5 - 0.01 \div 2$, or 0.4950.) The P-value is twice the area to the left of the test statistic (see Figure 7-8); using Table A-2, the P-value is 0.0002. (Using more accurate software, the P-value is smaller.) The test statistic, critical value, and critical region are shown in Figure 8-1.

Step 7: Because the test statistic falls within the critical region, reject the null hypothesis $\mu_1 = \mu_2$ (or $\mu_1 - \mu_2 = 0$).

INTERPRETATION There is sufficient evidence to support the claim that there is a difference between the mean weight of Coke and the mean weight of Pepsi. The *practical* significance of that difference is another issue not addressed by this hypothesis test, but the magnitude of the difference does not appear to be anything that a consumer would ever notice. Also, note that we are dealing with weights, not volumes, so the apparent lower mean for Coke simply indicates that the ingredients weigh less; this does not suggest that Coke is providing less of its drink than Pepsi.

Confidence Intervals

We can construct confidence interval estimates of the difference between two population means, which we denote as $\mu_1 - \mu_2$.

Confidence Interval Estimate of $\mu_1 - \mu_2$ (for Independent and Large Samples)

The confidence interval estimate of the difference $\mu_1 - \mu_2$ is

$$(\bar{x}_1 - \bar{x}_2) - E < (\mu_1 - \mu_2) < (\bar{x}_1 - \bar{x}_2) + E$$

where
$$E = z_{\alpha/2}\sqrt{\frac{\sigma_1^2}{n_1} + \frac{\sigma_2^2}{n_2}}$$

As with hypothesis testing, if σ_1 and σ_2 are not known, use s_1 and s_2 in their places, provided that both samples are large.

Commercials

Television networks have their own clearance departments for screening commercials and verifying claims. The National Advertising Division, a branch of the Council of Better Business Bureaus, investigates advertising claims. The Federal Trade Commission and local district attorneys also become involved. In the past, Firestone had to drop a claim that its tires resulted in 25% faster stops, and Warner Lambert had to spend $10 million informing customers that Listerine doesn't prevent or cure colds. Many deceptive ads are voluntarily dropped, and many others escape scrutiny simply because the regulatory mechanisms can't keep up with the flood of commercials.

EXAMPLE **Coke and Pepsi** Using the sample data given in the preceding example, construct a 99% confidence interval estimate of the difference between the mean weights of regular Coke and regular Pepsi.

SOLUTION We first find the value of the margin of error E.

$$E = z_{\alpha/2}\sqrt{\frac{\sigma_1^2}{n_1} + \frac{\sigma_2^2}{n_2}} = 2.575\sqrt{\frac{0.007507^2}{36} + \frac{0.005701^2}{36}} = 0.004045$$

We now find the desired confidence interval as follows:

$$(\bar{x}_1 - \bar{x}_2) - E < (\mu_1 - \mu_2) < (\bar{x}_1 - \bar{x}_2) + E$$
$$(0.81682 - 0.82410) - 0.004045 < (\mu_1 - \mu_2) < (0.81682 - 0.82410) + 0.004045$$
$$-0.01133 < (\mu_1 - \mu_2) < -0.00324$$

INTERPRETATION We are 99% confident that the limits of -0.01133 and -0.00324 actually do contain the difference between the two population means. This result could be more clearly presented by stating that μ_2 exceeds μ_1 by an amount that is between 0.00324 lb and 0.01133 lb. Because those limits do not contain 0, this confidence interval suggests that it is very unlikely that the two population means are equal.

Rationale: Why Do the Test Statistic and Confidence Interval Have the Particular Forms We Have Presented? Recall from the central limit theorem that sample means tend to be normally distributed. When applied to differences between sample means, the central limit theorem leads to the following key properties of the values that are the differences $\bar{x}_1 - \bar{x}_2$:

- The *differences* between sample means $(\bar{x}_1 - \bar{x}_2)$ tend to be normally distributed.
- The population mean of the sample $\bar{x}_1 - \bar{x}_2$ values is equal to $\mu_1 - \mu_2$.
- The population standard deviation of the $\bar{x}_1 - \bar{x}_2$ values is $\sqrt{\sigma_1^2/n_1 + \sigma_2^2/n_2}$.

To clarify the third statement above: The variance of the *differences* between two independent random variables equals the variance of the first random variable *plus* the variance of the second random variable. That is, the variance of sample values $\bar{x}_1 - \bar{x}_2$ will tend to equal $\sigma^2_{\bar{x}_1} + \sigma^2_{\bar{x}_2}$, provided that $\bar{x}_1$ and $\bar{x}_2$ are independent. (See Exercise 23.) For large simple random samples, the standard deviation of sample means is $\sigma/\sqrt{n}$, so the variance of sample means is σ^2/n. If we use this expression, σ^2/n, for the variance of sample means and use the property that the variance of the *differences* between two independent random variables equals the variance of the first random variable *plus* the variance of the second random variable, we get the following:

$$\sigma^2_{\bar{x}_1 - \bar{x}_2} = \sigma^2_{\bar{x}_1} + \sigma^2_{\bar{x}_2} = \frac{\sigma_1^2}{n_1} + \frac{\sigma_2^2}{n_2}$$

Taking the square root, we see that the standard deviation of the differences $\bar{x}_1 - \bar{x}_2$ can be expressed as follows:

$$\text{standard deviation of the } \bar{x}_1 - \bar{x}_2 \text{ sample values} = \sqrt{\frac{\sigma_1^2}{n_1} + \frac{\sigma_2^2}{n_2}}$$

Using the three properties listed above, and assuming that the samples are independent and large, the test statistic used in hypothesis tests has the same basic format we have already used:

$$\frac{(\text{sample statistic}) - (\text{claimed population parameter})}{(\text{standard deviation of sample statistics})} = \frac{(\bar{x}_1 - \bar{x}_2) - (\mu_1 - \mu_2)}{\sqrt{\dfrac{\sigma_1^2}{n_1} + \dfrac{\sigma_2^2}{n_2}}}$$

The format of the confidence interval given in this section is really the same format introduced in Chapter 6, except that the sample mean $\bar{x}$ is replaced by $\bar{x}_1 - \bar{x}_2$, the population mean μ is replaced by $\mu_1 - \mu_2$, and the standard deviation $\sigma/\sqrt{n}$ is replaced by $\sqrt{\sigma_1^2/n_1 + \sigma_2^2/n_2}$.

Using Technology

STATDISK: Select **Analysis,** then **Hypothesis Testing,** then **Mean-Two Independent Samples.** Enter the required values in the dialog box. Confidence interval limits are included with hypothesis test results.

Minitab: Minitab requires the original lists of sample data and does not work with summary statistics. If the original sample values are known, enter them in columns C1 and C2. Select **Stat** from the main menu, then select **Basic Statistics,** followed by **2-Sample t.** Select the option **Samples in different columns,** and enter C1 for the first sample and C2 for the second sample. Now select the appropriate form of the alternative hypothesis, and enter the confidence level. If you don't know the original sample values, there is a way to use Minitab, but it's tricky; see the *Minitab Student Laboratory Manual and Workbook.*

The Placebo Effect

It has been a common belief that when patients are given a placebo (a treatment with no medicinal value), about one-third of them show some improvement. However, a more recent study of 6000 patients showed that for those with mild medical problems, the placebos seemed to result in improvement in about two-thirds of the cases. The placebo effect seems to be strongest when patients are very anxious and they like their physicians. Because it could cloud studies of new treatments, the placebo effect is minimized by using a double-blind experiment in which neither the patient nor the physician knows whether the treatment is a placebo or a real medicine.

Excel: Enter the values for the first sample in column A, then enter the values for the second sample in column B.

To use the Data Desk XL add-in, click on **DDXL.** Select **Hypothesis Tests** and **2 Var t Test,** or select **Confidence Interval** and **2 Var t Interval.** In the dialog box, click on the pencil icon for the first quantitative column, and enter the range of values for the first sample, such as A1:A50. Click on the pencil icon for the second quantitative column, and enter the range of values for the second sample. Click on **OK.** Now complete the new dialog box by following the indicated steps. In step 1, be sure to select **2 Sample.**

To use Excel's Data Analysis add-in, select **Tools** from the main menu, then **Data Analysis,** then **z-Test: 2 Samples for Means.** Proceed to make the required entries in the dialog box. Enter the "Variable 1 Range" with something like A1:A36 (for sample values in the first 36 cells of column A). Also enter the "Variable 2 Range" to identify the location of the values from the second sample. Enter the hypothesized mean difference; enter 0 to test the claim that $\mu_1 = \mu_2$. Enter the significance level in the "alpha" box, then click **OK.** Excel does not generate a confidence interval for the difference between two sample means.

TI-83 Plus: The TI-83 Plus calculator can also be used for the methods of this section. Begin by entering the data in lists L1 and L2, then press **STAT** and select **TESTS.** Choose the option **2-SampZTest,** and proceed to enter the data requested. (If summary statistics are known, you can use the **Stats** option. If both sample sizes are larger than 30, enter the sample standard deviations when prompted for σ_1 and σ_2.) To get a confidence interval, press **STAT,** then select **TESTS** and choose the option **2-SampZInt.**

8-2 Basic Skills and Concepts

Testing Equality of Means. *In Exercises 1 and 2, use a 0.05 significance level to test the claim that the two samples come from populations with the same mean. In each case, assume that the two samples are independent simple random samples.*

1.

Treatment Group	Placebo Group
$n_1 = 50$	$n_2 = 100$
$\bar{x}_1 = 7.00$	$\bar{x}_2 = 6.00$
$s_1 = 1.00$	$s_2 = 2.00$

2.

Math Majors	English Majors
$n_1 = 40$	$n_2 = 60$
$\bar{x}_1 = 75.0$	$\bar{x}_2 = 70.0$
$s_1 = 15.0$	$s_2 = 14.0$

3. Constructing Confidence Interval Using the sample data given in Exercise 1, construct a 95% confidence interval estimate of the difference between the two population means. Do the confidence interval limits contain 0? What do you conclude about the difference between the treatment group and the placebo group?

4. Constructing Confidence Interval Using the sample data given in Exercise 2, construct a 95% confidence interval estimate of the difference between the two population means. Do the confidence interval limits contain 0? What do you conclude about the difference between the math majors and the English majors?

5. Inferences from Samples of Regular Coke and Diet Coke Using Data Set 1 in Appendix B, we find the sample statistics for the weights (in pounds) of regular Coke and diet Coke as listed in the margin.

a. Use a 0.01 significance level to test the claim that cans of regular Coke and diet Coke have the same mean weight.

b. Construct a 99% confidence interval estimate of $\mu_1 - \mu_2$, the difference between the mean weight of regular Coke and the mean weight of diet Coke.

Regular Coke	Diet Coke
$n_1 = 36$	$n_2 = 36$
$\overline{x}_1 = 0.81682$	$\overline{x}_2 = 0.78479$
$s_1 = 0.007507$	$s_2 = 0.004391$

6. Testing Equality of Means for M&Ms Data Set 10 in Appendix B contains weights of 100 randomly selected M&M plain candies. Those weights have a mean of 0.9147 g and a standard deviation of 0.0369 g. A previous edition of this book used a different sample of 100 M&M plain candies (before blue was introduced), with a mean and standard deviation of 0.9160 g and 0.0433 g, respectively. Is there sufficient evidence to conclude that the two population means are different?

7. Police as Eyewitnesses Does stress affect the recall ability of police eyewitnesses? This issue was studied in an experiment that tested eyewitness memory a week after a nonstressful interrogation of a cooperative suspect and a stressful interrogation of an uncooperative and belligerent suspect. The numbers of details recalled a week after the incident are summarized in the margin (based on data from "Eyewitness Memory of Police Trainees for Realistic Role Plays," by Yuille et al., *Journal of Applied Psychology,* Vol. 79, No. 6). Use a 0.01 significance level to test the claim in the article that "stress decreased the amount recalled."

Nonstress	Stress
$n_1 = 40$	$n_2 = 40$
$\overline{x}_1 = 53.3$	$\overline{x}_2 = 45.3$
$s_1 = 11.6$	$s_2 = 13.2$

8. Ages of Student and Faculty Cars Students at the author's college randomly selected 217 student cars and found that they had ages with a mean of 7.89 years and a standard deviation of 3.67 years. They also randomly selected 152 faculty cars and found that they had ages with a mean of 5.99 years and a standard deviation of 3.65 years.

a. Using a 0.05 significance level, is there sufficient evidence to support the claim that student cars are older than faculty cars?

b. Construct a 95% confidence interval estimate of the difference $\mu_1 - \mu_2$, where μ_1 is the mean age of student cars.

9. Weights of Men in Different Age Brackets As part of the National Health Survey, data were collected on the weights of men in two different age brackets. For 804 men aged 25–34, the mean is 176 lb and the standard deviation is 35.0 lb. For 1657 men aged 65–74, the mean and standard deviation are 164 lb and 27.0 lb, respectively.

a. Test the claim that the older men come from a population with a mean that is less than the mean for men in the 25–34 age bracket. Use a 0.01 significance level.

b. Construct a 99% confidence interval for the difference between the means of the men in the two age brackets. Do the confidence interval limits contain 0?

(continued)

Research in Twins

Identical twins occur when a single fertilized egg splits in two, so that both twins share the same genetic makeup. There is now an explosion in research focused on those twins. Speaking for the Center for Study of Multiple Birth, Louis Keith notes that now "we have far more ability to analyze the data on twins using computers with new, built-in statistical packages." A common goal of such studies is to explore the classic issue of "nature versus nurture." For example, Thomas Bouchard, who runs the Minnesota Study of Twins Reared Apart, has found that IQ is 50%-60% inherited, while the remainder is the result of external forces.

Identical twins are matched pairs that provide better results by allowing us to reduce the genetic variation that is inevitable with unrelated pairs of people.

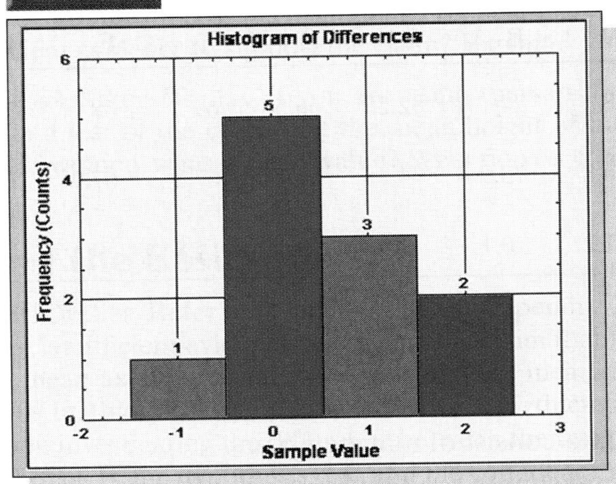

Hypothesis Tests

In dealing with sample data consisting of matched pairs, we base our calculations on the differences (d) between the pairs of data, because comparing two individual sample means would waste important information about the paired data. The following notation is based on the differences.

Notation for Matched Pairs

μ_d = mean value of the differences d for the *population* of paired data

$\bar{d}$ = mean value of the differences d for the paired *sample* data (equal to the mean of the $x - y$ values)

s_d = standard deviation of the differences d for the paired *sample* data

n = number of *pairs* of data

We now use this notation to describe the test statistic to be used in hypothesis tests of claims made about the difference between the means of two populations, given that the sample data consist of matched pairs. If we have simple random paired sample data from populations in which the paired differences have a normal distribution, we use the sample differences and proceed with a single test. We use the following test statistic with a Student t distribution.

Test Statistic for Matched Pairs of Sample Data

$$t = \frac{\bar{d} - \mu_d}{\frac{s_d}{\sqrt{n}}}$$

(continued)

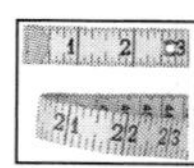

EXAMPLE How Much Do Male Statistics Students Exaggerate Their Heights? Using the sample data from Table 8-1 with the outlier excluded, construct a 95% confidence interval estimate of μ_d, which is the mean of the differences between reported heights and measured heights of male statistics students.

SOLUTION We use the values of $\bar{d} = 0.672727$, $s_d = 0.825943$, $n = 11$, and $t_{\alpha/2} = 2.228$ (found from Table A-3 with 10 degrees of freedom and 0.05 in two tails). We first find the value of the margin of error E.

$$E = t_{\alpha/2}\frac{s_d}{\sqrt{n}} = (2.228)\left(\frac{0.825943}{\sqrt{11}}\right) = 0.554841$$

The confidence interval can now be found.

$$\bar{d} - E < \mu_d < \bar{d} + E$$
$$0.672727 - 0.554841 < \mu_d < 0.672727 + 0.554841$$
$$0.12 < \mu_d < 1.23$$

INTERPRETATION The result is sometimes expressed as $\mu_d = 0.67 \pm 0.55$ or as (0.12, 1.23). In the long run, 95% of such samples will lead to confidence interval limits that actually do contain the true population mean of the differences. Note that the confidence interval limits do not contain 0, indicating that the true value of μ_d is significantly different from 0. That is, the mean value of the "reported − measured" differences is different from 0. On the basis of the confidence interval, we conclude that there is sufficient evidence to support the claim that there is a difference between the reported heights and the measured heights of male statistics students.

Using Technology

AS

STATDISK: Select **Analysis,** then **Hypothesis Testing,** then **Matched Pairs.** In the dialog box, choose the format of the claim, enter a significance level, enter the sample data, and then click on **Evaluate.** STATDISK automatically provides confidence interval limits.

Minitab: Enter the paired sample data in columns C1 and C2. Click on **Stat,** select **Basic Statistics,** then select **Paired t.** Enter C1 for the first sample, enter C2 for the second sample, then click on the **Options** box to change the confidence level or form of the alternative hypothesis.

Excel: Enter the paired sample data in columns A and B.

To use the Data Desk XL add-in, click on **DDXL.** Select **Hypotheses Tests** and **Paired t Test** or select **Confidence Intervals** and **2 Var t Interval.** In the dialog box, click on the pencil icon for the first quantitative column and enter the range of values for the first sample, such as A1:A14. Click on the pencil icon for the second quantitative column and enter the range of values for the second sample. Click on **OK.** Now complete the new dialog box by following the indicated steps.

To use Excel's Data Analysis add-in, click on **Tools,** found on the main menu bar, then select **Data Analysis,** and proceed to select **t-test Paired Two Sample for Means.** In the dialog box, enter the range of values for each of the two samples, enter the desired population mean difference, and enter the significance level. The displayed results will include the test statistic, the *P*-values for a one-tailed test and a two-tailed test, and the critical values for a one-tailed test and a two-tailed test.

TI-83 Plus: *Caution:* Do not use the menu item **2-SampTTest** because it applies to *independent* samples. Instead, enter the data for the first variable in list L1, enter the data for the second variable in list L2, then clear the screen and enter **L1 − L2 → L3.** Next press **STAT,** then select **TESTS,** and choose the option of **T-Test.** Using the input option of **Data,** enter the indicated data, including list L3, and press **ENTER** when done. A confidence interval can also be found by pressing **STAT,** then selecting **TESTS,** then **TInterval.**

8-3 Basic Skills and Concepts

Calculations for Matched Pairs. *In Exercises 1 and 2, assume that you want to test the claim that the paired sample data come from a population for which the mean difference is $\mu_d = 0$. Assuming a 0.05 significance level, find (a) $\bar{d}$, (b) s_d, (c) the t test statistic, and (d) the critical values.*

1.

x	8	8	6	9	7
y	3	2	6	4	9

2.

x	20	25	27	27	23	29	30	26
y	20	24	25	29	20	29	32	29

3. Using the sample paired data in Exercise 1, construct a 95% confidence interval for the population mean of all differences $x - y$.

4. Using the sample paired data in Exercise 2, construct a 99% confidence interval for the population mean of all differences $x - y$.

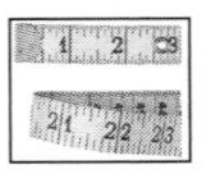

5. Reported and Measured Female Heights This section used reported heights and measured heights for a sample of *male* statistics students. Listed here are data collected from a sample of *female* statistics students.

a. Is there sufficient evidence to support the claim that female statistics students exaggerate by reporting heights that are greater than their actual measured heights? Use a 0.05 significance level.

b. Construct a 95% confidence interval estimate of the mean difference between reported heights and measured heights. Interpret the resulting confidence interval, and comment on the implications of whether the confidence interval limits contain 0.

Reported height	64	63	64	65	64	64	63	59	66	64
Measured height	63.5	63.1	63.8	63.4	62.1	64.4	62.7	59.3	65.4	62.2

6. Crash Test Repair Costs In low-speed crash tests of five BMW cars, the repair costs were computed for a factory-authorized repair center and an independent repair facility. The results are listed in the accompanying table.

a. Is there sufficient evidence to support the claim that the independent center has lower repair costs? Use a 0.01 significance level.

b. Construct a 99% confidence interval estimate of the mean difference between the repair costs of the factory-authorized repair center and the independent repair center. Do the confidence interval limits contain 0?

c. Apart from any monetary difference in repair costs, what other key factor is critical in choosing between the authorized repair center and the independent repair center?

Authorized repair center	$797	$571	$904	$1147	$418
Independent repair center	$523	$488	$875	$911	$297

7. Effectiveness of SAT Course Refer to the data in the table that lists SAT scores before and after the sample of 10 students took a preparatory course.

a. Is there sufficient evidence to conclude that the preparatory course is effective in raising scores? Use a 0.05 significance level.

b. Construct a 95% confidence interval estimate of the mean difference between the before and after scores. Write a statement that interprets the resulting confidence interval.

Student	A	B	C	D	E	F	G	H	I	J
SAT score before course (x)	700	840	830	860	840	690	830	1180	930	1070
SAT score after course (y)	720	840	820	900	870	700	800	1200	950	1080

Based on data from the College Board and "An Analysis of the Impact of Commercial Test Preparation Courses on SAT Scores," by Sesnowitz, Bernhardt, and Knain, *American Educational Research Journal,* Vol. 19, No. 3.

8. Effectiveness of Drug Captopril is a drug designed to lower systolic blood pressure. When subjects were tested with this drug, their systolic blood pressure readings (in mm of mercury) were measured before and after the drug was taken, with the results given in the accompanying table.

a. Use the sample data to construct a 99% confidence interval for the mean difference between the before and after readings.

b. Is there sufficient evidence to support the claim that captopril is effective in lowering systolic blood pressure?

Subject	A	B	C	D	E	F	G	H	I	J	K	L
Before	200	174	198	170	179	182	193	209	185	155	169	210
After	191	170	177	167	159	151	176	183	159	145	146	177

Based on data from "Essential Hypertension: Effect of an Oral Inhibitor of Angiotensin-Converting Enzyme," by MacGregor et al., *British Medical Journal,* Vol. 2.

9. Effectiveness of Hypnotism in Reducing Pain A study was conducted to investigate the effectiveness of hypnotism in reducing pain. Results for randomly selected subjects are given in the accompanying table. The values are before and after hypnosis; the measurements are in centimeters on a pain scale.

a. Construct a 95% confidence interval for the mean of the "before − after" differences.

b. At the 0.05 significance level, test the claim that the sensory measurements are lower after hypnotism.

c. Does hypnotism appear to be effective in reducing pain?

Subject	A	B	C	D	E	F	G	H
Before	6.6	6.5	9.0	10.3	11.3	8.1	6.3	11.6
After	6.8	2.4	7.4	8.5	8.1	6.1	3.4	2.0

Based on "An Analysis of Factors That Contribute to the Efficacy of Hypnotic Analgesia," by Price and Barber, *Journal of Abnormal Psychology,* Vol. 96, No. 1.

10. Measuring Intelligence in Children Mental measurements of young children are often made by giving them blocks and telling them to build a tower as tall as possible. One experiment of block building was repeated a month later, with the times (in seconds) listed in the accompanying table.

a. Is there sufficient evidence to support the claim that there is a difference between the two times? Use a 0.01 significance level.

b. Construct a 99% confidence interval for the mean of the differences. Do the confidence interval limits contain 0, indicating that there is not a significant difference between the times of the first and second trials?

Child	A	B	C	D	E	F	G	H	I	J	K	L	M	N	O
First trial	30	19	19	23	29	178	42	20	12	39	14	81	17	31	52
Second trial	30	6	14	8	14	52	14	22	17	8	11	30	14	17	15

Based on data from "Tower Building," by Johnson and Courtney, *Child Development,* Vol. 3.

11. Morning and Night Body Temperatures Refer to Data Set 6 in Appendix B. Use the paired data consisting of body temperatures of women at 8:00 A.M. and at 12:00 A.M. on Day 2.

a. Construct a 95% confidence interval for the mean difference of 8 A.M. temperatures minus 12 A.M. temperatures.

b. Using a 0.05 significance level, test the claim that for those temperatures, the mean difference is 0. Based on the result, do morning and night body temperatures appear to be about the same?

12. Qwerty/Dvorak Keyboards The following sentence, which contains every letter of the alphabet, was used for testing typewriters: "The quick brown fox jumped over the lazy red dog." If we use the word-rating procedure described in the Chapter Problem for Chapter 2, we find that the 10 words in that sentence have Qwerty ratings of 2, 5, 7, 3, 5, 5, 2, 3, 2, 1. The corresponding Dvorak ratings are 0, 5, 5, 3, 5, 3, 0, 4, 1, 1.

a. Is there sufficient evidence to support the claim that the Dvorak keyboard has lower ratings?

b. Construct a 95% confidence interval estimate of the difference. Does it contain 0, suggesting that there is not a significant difference between the Qwerty and Dvorak word ratings?

13. Treating Motion Sickness The following Minitab display resulted from an experiment in which 10 subjects were tested for motion sickness before and after taking the drug astemizole. The Minitab data column C3 consists of differences in the number of head movements that the subjects could endure without becoming nauseous. (The differences were obtained by subtracting the "after" values from the "before" values.)

a. Use a 0.05 significance level to test the claim that astemizole has an effect (for better or worse) on vulnerability to motion sickness. Based on the result, would you use astemizole if you were concerned about motion sickness while on a cruise ship?

b. Instead of testing for some effect (for better or worse), suppose we want to test the claim that astemizole is effective in *preventing* motion sickness? What is the P-value, and what do you conclude?

c. Use the Minitab results to construct a 95% confidence interval for the mean difference between the numbers of head movements that the subject could endure without becoming nauseous. Do the confidence interval limits contain 0, indicating that astemizole does not have a significant effect on vulnerability to motion sickness?

```
TEST OF MU = 0.0 VS MU N.E. 0.0
      N    MEAN      STDEV      SE MEAN        T          P VALUE
C3   10    -7.5       57.7         18.2    -0.41             0.69
```

14. Dieting: Interpreting Calculator Display Researchers obtained weight loss data from a sample of dieters using the New World Athletic Club facilities. The before/after weights are recorded, then the differences (before − after) are computed. The TI-83 Plus calculator results are shown for a test of the claim that the diet is effective.

a. Is there sufficient evidence to support the claim that the diet is effective? Explain.

b. What is the mean weight loss? Is it large enough to make the diet practical for someone wanting to lose weight?

c. Use the displayed results to construct a 95% confidence interval for the mean weight loss.

TI-83 Plus

```
T-Test
 μ>0
 t=6.431306409
 p=3.340322E-4
 x̄=5.142857143
 Sx=2.115700942
 n=7
```

8-3 Beyond the Basics

15. Effects of an Outlier In the examples in this section, we excluded the outlier in Table 8-1 and used the remaining 11 pairs of data. How are the results affected if that outlier is included? In general, can an outlier have a dramatic effect on the hypothesis test and confidence interval?

16. Effects of Units of Measurement The examples in this section used heights measured in inches. Suppose we convert all reported and measured heights from inches to centimeters. Is the hypothesis test affected by such a change in units? Is the confidence interval affected by such a change in units? How?

17. Confidence Intervals and One-Sided Tests The 95% confidence interval for a collection of paired sample data is $0.0 < \mu_d < 1.2$. Based on this confidence interval, the traditional method of hypothesis testing leads to the conclusion that the claim of $\mu_d > 0$ is supported. What is the smallest possible value of the significance level of the hypothesis test?

18. Using the Wrong Procedure Section 8-2 included an example of a hypothesis test that used the weights of 36 cans of regular Coke and 36 cans of regular Pepsi. The weights are listed in Data Set 1 in Appendix B. In Section 8-2 we tested the hypothesis of no difference between the two population means. Suppose we incorrectly believe that those weights are paired, so that the first regular Coke weight is paired with the first regular Pepsi weight, the second weights are paired, and so on. Compare the results obtained by using the methods of this section to the results found in Section 8-2. Are the results seriously affected if we choose the wrong procedure?

8-4 Inferences about Two Proportions

A strong argument could be made that this section is one of the most important sections in the book because this is where we describe methods for dealing with two sample proportions—a situation that is very common in real applications. Although this section is based on proportions, we can deal with percentages by using the corresponding decimal equivalents. For example, we might want to determine whether there is a difference between the percentage of adverse reactions in a placebo group and the percentage of adverse reactions in a drug treatment group. We can convert the percentages to their corresponding decimal values and proceed to use the methods of this section.

When testing a hypothesis made about two population proportions—such as the proportions of cured patients in a population given some treatment and a second population given a placebo—or when constructing a confidence interval for the difference between two population proportions, we make the following assumptions and use the following notation.

Assumptions

1. We have proportions from two *independent* simple random samples.
2. For both samples, the conditions $np \geq 5$ and $nq \geq 5$ are satisfied. That is, there are at least five successes and five failures in each of the two samples. (In many cases, we will test the claim that two populations have equal proportions so that $p_1 - p_2 = 0$. Because we assume that $p_1 - p_2 = 0$, it is not necessary to specify the particular value that p_1 and p_2 have in common. In such cases, the conditions $np \geq 5$ and $nq \geq 5$ can be checked by replacing p with the estimated pooled proportion $\overline{p}$, which will be described later.)

Notation for Two Proportions

For population 1 we let

p_1 = *population* proportion
n_1 = size of the sample
x_1 = number of successes in the sample

(continued)

$\hat{p}_1 = \frac{x_1}{n_1}$ (the *sample* proportion)

$\hat{q}_1 = 1 - \hat{p}_1$

The corresponding meanings are attached to p_2, n_2, x_2, $\hat{p}_2$, and $\hat{q}_2$, which come from population 2.

Finding the Numbers of Successes x_1 and x_2: The calculations for hypothesis tests and confidence intervals require that we have specific values for x_1, n_1, x_2, and n_2. Sometimes the available sample data include those specific numbers, but sometimes it is necessary to calculate the values of x_1 and x_2. The first example in this section reports that "when 734 men were treated with Viagra, 16% of them experienced headaches." From that statement we can see that $n_1 = 734$ and $\hat{p}_1 = 0.16$, but the actual number of successes x_1 is not given. However, from $\hat{p}_1 = x_1/n_1$, we know that

$$x_1 = n_1 \cdot \hat{p}_1$$

so that $x_1 = 734 \cdot 0.16 = 117.44$. But you cannot have 117.44 men who experienced headaches, because everyone either experiences a headache or does not, and the number of successes x_1 must therefore be a whole number. We can round 117.44 to 117. We can now use $x_1 = 117$ in the calculations that require its value. It's really quite simple: 16% of 734 means 0.16×734, which results in 117.44, which we round to 117.

Hypothesis Tests

In Section 7-5 we discussed tests of hypotheses made about a single population proportion. We will now consider tests of hypotheses made about two population proportions, but *we will be testing only claims that* $p_1 = p_2$, and we will use the following pooled (or combined) estimate of the value that p_1 and p_2 have in common. (For claims that the difference between p_1 and p_2 is equal to a nonzero constant, see Exercise 21 of this section.) You can see from the form of the pooled estimate $\overline{p}$ that it basically combines the two different samples into one big sample.

Pooled Estimate of p_1 and p_2

The **pooled estimate of p_1 and p_2** is denoted by $\overline{p}$ and is given by

$$\overline{p} = \frac{x_1 + x_2}{n_1 + n_2}$$

We denote the complement of $\overline{p}$ by $\overline{q}$, so $\overline{q} = 1 - \overline{p}$.

Polio Experiment

In 1954 an experiment was conducted to test the effectiveness of the Salk vaccine as protection against the devastating effects of polio. Approximately 200,000 children were injected with an ineffective salt solution, and 200,000 other children were injected with the vaccine. The experiment was "double blind" because the children being injected didn't know whether they were given the real vaccine or the placebo, and the doctors giving the injections and evaluating the results didn't know either. Only 33 of the 200,000 vaccinated children later developed paralytic polio, whereas 115 of the 200,000 injected with the salt solution later developed paralytic polio. Statistical analysis of these and other results led to the conclusion that the Salk vaccine was indeed effective against paralytic polio.

Test Statistic for Two Proportions

The following test statistic applies to null and alternative hypotheses that fit one of these three formats:

$$H_0: p_1 = p_2 \qquad H_0: p_1 \geq p_2 \qquad H_0: p_1 \leq p_2$$
$$H_1: p_1 \neq p_2 \qquad H_1: p_1 < p_2 \qquad H_1: p_1 > p_2$$

$$z = \frac{(\hat{p}_1 - \hat{p}_2) - (p_1 - p_2)}{\sqrt{\dfrac{\overline{p}\,\overline{q}}{n_1} + \dfrac{\overline{p}\,\overline{q}}{n_2}}}$$

where $\quad p_1 - p_2 = 0$ (assumed in the null hypothesis)

$$\hat{p}_1 = \frac{x_1}{n_1} \quad \text{and} \quad \hat{p}_2 = \frac{x_2}{n_2}$$

$$\overline{p} = \frac{x_1 + x_2}{n_1 + n_2}$$

$$\overline{q} = 1 - \overline{p}$$

***P*-value:** Use the computed value of the test statistic z and find the P-value by following the procedure summarized in Figure 7-8.

Critical values: Based on the significance level α, find critical values by using the procedures introduced in Section 7-2.

Once again, the test statistic fits the common format of

$$\frac{(\text{sample statistic}) - (\text{claimed population parameter})}{(\text{standard deviation of sample statistics})}$$

The following example will illustrate one method used to compare a placebo group and a drug treatment group. It will also help clarify the roles of x_1, n_1, $\hat{p}_1$, $\overline{p}$, and so on. In particular, you should recognize that under the assumption of equal proportions, the best estimate of the common proportion is obtained by pooling both samples into one big sample, so that $\overline{p}$ becomes a more obvious estimate of the common population proportion.

EXAMPLE **Viagra Treatment and Placebo** The drug Viagra has become quite well known, and it has had a substantial economic impact on its producer, Pfizer Pharmaceuticals. In preliminary tests for adverse reactions, it was found that when 734 men were treated with Viagra, 16% of them experienced headaches. (There's some real irony there.) Among 725 men in a placebo group, 4% experienced headaches. These results were provided by Pfizer Pharmaceuticals. Is there sufficient evidence to support the claim that among those men who take Viagra, headaches occur at a rate that is greater than the rate for those who do not take Viagra? Use a 0.01 significance level.

SOLUTION For notation purposes, we stipulate that sample 1 is the treatment group of men who took Viagra, and sample 2 is the group of men who took a placebo instead. Because 16% of the 734 men in the Viagra treatment group experienced headaches, the actual number of those men who experienced headaches is

$$\begin{aligned} 16\% \text{ of } 734 &= 0.16 \times 734 \\ &= 117.44 = 117 \text{ (rounded to a whole number)} \end{aligned}$$

A similar calculation shows that if 4% of the 725 men in the placebo group experienced headaches, then the actual number of those men who experienced headaches is 29. We can now summarize the sample data as follows.

Viagra Treatment Group	Placebo Group
$n_1 = 734$	$n_2 = 725$
$x_1 = 117$	$x_2 = 29$
$\hat{p}_1 = \dfrac{117}{734} = 0.159401$	$\hat{p}_2 = \dfrac{29}{725} = 0.04$

We will now use the traditional method of hypothesis testing, as summarized in Figure 7-5.

Step 1: The claim of a greater headache rate for Viagra users can be represented by $p_1 > p_2$.

Step 2: If $p_1 > p_2$ is false, then $p_1 \leq p_2$.

Step 3: Because our claim of $p_1 > p_2$ does not contain equality, it becomes the alternative hypothesis, and we have

$$H_0\text{: } p_1 \leq p_2 \qquad H_1\text{: } p_1 > p_2 \text{ (original claim)}$$

Step 4: The significance level is $\alpha = 0.01$.

Step 5: We will use the normal distribution (with the test statistic previously given) as an approximation to the binomial distribution. We have two independent samples, and the conditions $np \geq 5$ and $nq \geq 5$ are satisfied for each of the two samples. To check this, we note that in conducting this test, we assume that $p_1 = p_2$, where their common value is the pooled estimate $\overline{p}$, calculated as

$$\overline{p} = \frac{x_1 + x_2}{n_1 + n_2} = \frac{117 + 29}{734 + 725} = 0.100069$$

With $\overline{p} = 0.100069$, it follows that $\overline{q} = 1 - 0.100069 = 0.899931$. We verify that $np \geq 5$ and $nq \geq 5$ for both samples as follows:

The Lead Margin of Error

Authors Stephen Ansolabehere and Thomas Belin wrote in their article "Poll Faulting" (*Chance* magazine) that "our greatest criticism of the reporting of poll results is with the margin of error of a single proportion (usually $\pm 3\%$) when media attention is clearly drawn to the *lead* of one candidate." They point out that the lead is really the *difference* between two proportions ($p_1 - p_2$) and go on to explain how they developed the following rule of thumb: The lead is approximately $\sqrt{3}$ times larger than the margin of error for any one proportion. For a typical pre-election poll, a reported $\pm 3\%$ margin of error translates to about $\pm 5\%$ for the lead of one candidate over the other. They write that the margin of error for the lead should be reported.

Sample 1	Sample 2
$n_1 p = (734)(0.100069) = 73.45 \geq 5$	$n_2 p = (725)(0.100069) = 72.55 \geq 5$
$n_1 q = (734)(0.899931) = 660.55 \geq 5$	$n_2 q = (725)(0.899931) = 652.45 \geq 5$

Step 6: We can now find the value of the test statistic.

$$z = \frac{(\hat{p}_1 - \hat{p}_2) - (p_1 - p_2)}{\sqrt{\dfrac{\bar{p}\,\bar{q}}{n_1} + \dfrac{\bar{p}\,\bar{q}}{n_2}}}$$

$$= \frac{\left(\dfrac{117}{734} - \dfrac{29}{725}\right) - 0}{\sqrt{\dfrac{(0.100069)(0.899931)}{734} + \dfrac{(0.100069)(0.899931)}{725}}} = 7.60$$

The critical value of $z = 2.33$ is found by observing that we have a right-tailed test with $\alpha = 0.01$. The value of $z = 2.33$ is found from Table A-2 as the z score corresponding to an area of $0.5 - 0.01 = 0.4900$.

Step 7: In Figure 8-3 we see that the test statistic falls within the critical region, so we reject the null hypothesis of $p_1 \leq p_2$.

INTERPRETATION We conclude that there is sufficient evidence to support the claim that the headache rate is greater for the men treated with Viagra than for the men who took a placebo.

These steps use the traditional approach to hypothesis testing, but it would be quite easy to use the *P*-value approach. In Step 6, instead of finding the critical value of *z,* we would find the *P*-value by using the procedure summarized in Figures 7-8 and 7-9. With a test statistic of $z = 7.60$ and a right-tailed test, we get

$$P\text{-value} = (\text{area to the right of } z = 7.60\,) = 0.0001$$

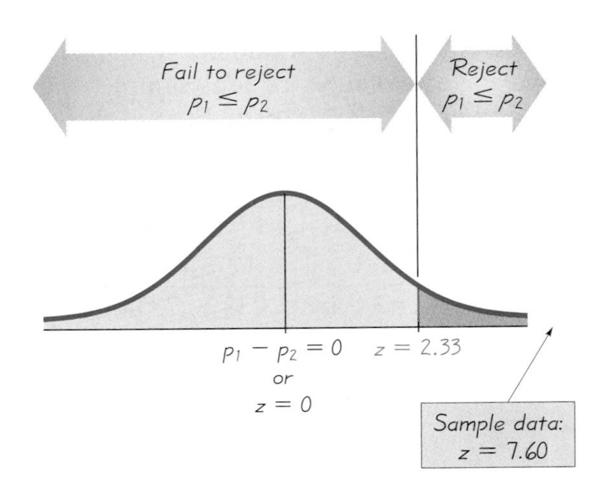

FIGURE 8-3
Distribution of Differences Between Proportions of Headaches in the Viagra Treatment Group and the Placebo Group

The P-value of 0.0001 comes from Table A-2, where all z scores above 3.09 yield a right-tailed area of 0.0001. If we use STATDISK, Minitab, Excel, or a TI-83 Plus calculator, we get P-values such as 0.0000 or 1.507947E-14 (which can be expressed as 0.00000000000000151). Again, we reject the null hypothesis because the P-value is less than the significance level of $\alpha = 0.01$. Again, we conclude that there is sufficient evidence to support the claim that the headache rate is greater for the men treated with Viagra than for the men who took a placebo

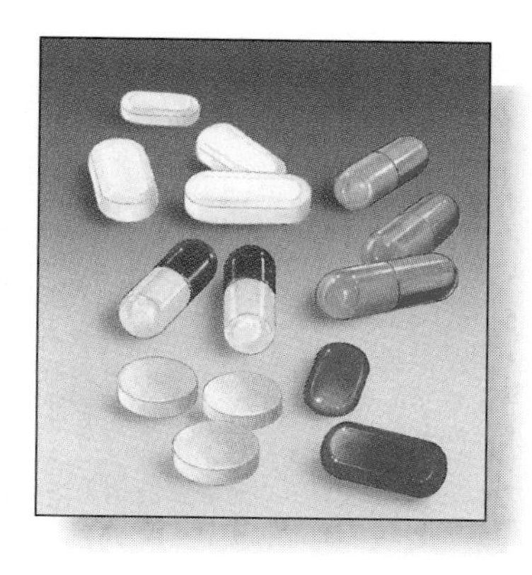

Drug Screening: False Positives

The American Management Association reports that more than half of all companies test for drugs. For a job applicant undergoing drug screening, a false positive is an indication of drug use when he or she does not use drugs. A false negative is an indication of no drug use when he or she is a user. The test sensitivity is the probability of a positive indication for a drug user; the test specificity is the probability of a negative indication for a nonuser. Suppose that 3% of job applicants use drugs, and a test has a sensitivity of 0.99 and a specificity of 0.98. The high probabilities make the test seem reliable, but 40% of the positive indications will be false positives.

Confidence Intervals

In the preceding example we found that the 16% rate of headaches among men who take Viagra is significantly greater than the 4% rate for men who do not. If we compute the difference between the two rates, we get 12%, which is a point estimate of the difference between the two population rates. But, with the same assumptions given at the beginning of this section, we can construct a confidence interval estimate of the difference between population proportions ($p_1 - p_2$) by evaluating the following.

Confidence Interval Estimate of $p_1 - p_2$

The confidence interval estimate of the difference $p_1 - p_2$ is:

$$(\hat{p}_1 - \hat{p}_2) - E < (p_1 - p_2) < (\hat{p}_1 - \hat{p}_2) + E$$

where
$$E = z_{\alpha/2}\sqrt{\frac{\hat{p}_1\hat{q}_1}{n_1} + \frac{\hat{p}_2\hat{q}_2}{n_2}}$$

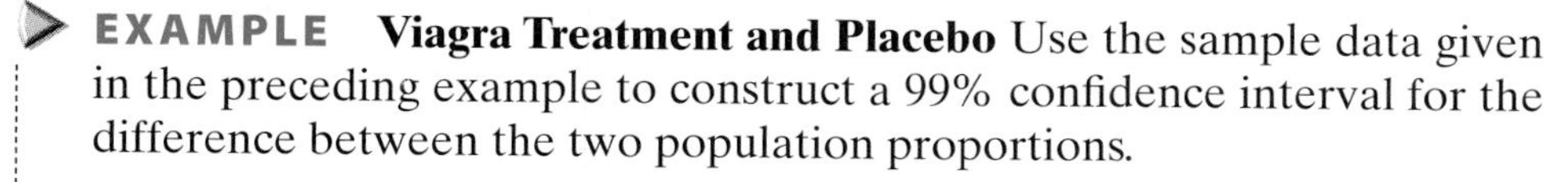

EXAMPLE **Viagra Treatment and Placebo** Use the sample data given in the preceding example to construct a 99% confidence interval for the difference between the two population proportions.

SOLUTION With a 99% degree of confidence, $z_{\alpha/2} = 2.575$ (from Table A-2). We first calculate the value of the margin of error E as shown.

$$E = z_{\alpha/2}\sqrt{\frac{\hat{p}_1\hat{q}_1}{n_1} + \frac{\hat{p}_2\hat{q}_2}{n_2}} = 2.575\sqrt{\frac{\left(\frac{117}{734}\right)\left(\frac{617}{734}\right)}{734} + \frac{\left(\frac{29}{725}\right)\left(\frac{696}{725}\right)}{725}}$$
$$= 0.0395$$

With $\hat{p}_1 = 117/734 = 0.1594$, $\hat{p}_2 = 29/725 = 0.0400$, and $E = 0.0395$, the confidence interval is evaluated as follows.

$$(\hat{p}_1 - \hat{p}_2) - E < (p_1 - p_2) < (\hat{p}_1 - \hat{p}_2) + E$$
$$(0.1594 - 0.0400) - 0.0395 < (p_1 - p_2) < (0.1594 - 0.0400) + 0.0395$$
$$0.0799 < (p_1 - p_2) < 0.1589$$

Does Aspirin Help Prevent Heart Attacks?

In a recent study of 22,000 male physicians, half were given regular doses of aspirin while the other half were given placebos. The study ran for six years at a cost of $4.4 million. Among those who took the aspirin, 104 suffered heart attacks. Among those who took the placebos, 189 suffered heart attacks. (The figures are based on data from Time and the New England Journal of Medicine, Vol. 318, No. 4.) This is a classic experiment involving a treatment group (those who took the aspirin) and a placebo group (those who took pills that looked and tasted like the aspirin pills, but no aspirin was present). We can use methods presented in this chapter to address the issue of whether the results show a statistically significant lower rate of heart attacks among the sample group who took aspirin. The issue is clearly important because it can affect many lives. This particular study was criticized because only male physicians were included, and the results might not apply to women.

INTERPRETATION The confidence interval limits do not contain 0, suggesting that there is a significant difference between the two proportions. As always, we should be careful when interpreting confidence intervals. Because p_1 and p_2 have fixed values and are not variables, it is wrong to state that there is a 99% chance that the value of $p_1 - p_2$ falls between 0.0799 and 0.1589. Instead, we should say that if we were to repeat the same sampling process and construct 99% confidence intervals, in the long run 99% of the intervals will actually contain the value of $p_1 - p_2$.

Rationale: Why Do the Procedures of This Section Work? The test statistic given for hypothesis tests is justified by the following:

1. With $n_1 p_1 \geq 5$ and $n_1 q_1 \geq 5$, the distribution of $\hat{p}_1$ can be approximated by a normal distribution with mean p_1, standard deviation $\sqrt{p_1 q_1 / n_1}$, and variance $p_1 q_1 / n_1$. These conclusions are based on Sections 5-6 and 6-5, and they also apply to the second sample.
2. Because $\hat{p}_1$ and $\hat{p}_2$ are each approximated by a normal distribution, $\hat{p}_1 - \hat{p}_2$ will also be approximated by a normal distribution with mean $p_1 - p_2$ and variance

$$\sigma^2_{(\hat{p}_1 - \hat{p}_2)} = \sigma^2_{\hat{p}_1} + \sigma^2_{\hat{p}_2} = \frac{p_1 q_1}{n_1} + \frac{p_2 q_2}{n_2}$$

(In Section 8-2 we established that the variance of the *difference* between two independent random variables is the *sum* of their individual variances.)

3. Because the values of p_1, p_2, q_1, and q_2 are typically unknown and from the null hypothesis we assume that $p_1 = p_2$, we can pool (or combine) the sample data. The pooled estimate of the common value of p_1 and p_2 is $\bar{p} = (x_1 + x_2)/(n_1 + n_2)$. If we replace p_1 and p_2 by $\bar{p}$ and replace q_1 and q_2 by $\bar{q} = 1 - \bar{p}$, the variance from Step 2 leads to the following standard deviation.

$$\sigma_{(\hat{p}_1 - \hat{p}_2)} = \sqrt{\frac{\bar{p}\bar{q}}{n_1} + \frac{\bar{p}\bar{q}}{n_2}}$$

4. We now know that the distribution of $p_1 - p_2$ is approximately normal, with mean $p_1 - p_2$ and standard deviation as shown in Step 3, so that the z test statistic has the form given earlier.

The form of the confidence interval requires an expression for the variance different from the one given in Step 3. In Step 3 we are assuming that $p_1 = p_2$, but if we don't make that assumption (as in the construction of a confidence interval), we estimate the variance of $\hat{p}_1 - \hat{p}_2$ as

$$\sigma^2_{(\hat{p}_1 - \hat{p}_2)} = \sigma^2_{\hat{p}_1} + \sigma^2_{\hat{p}_2} = \frac{\hat{p}_1 \hat{q}_1}{n_1} + \frac{\hat{p}_2 \hat{q}_2}{n_2}$$

and the standard deviation becomes

$$\sqrt{\frac{\hat{p}_1\hat{q}_1}{n_1} + \frac{\hat{p}_2\hat{q}_2}{n_2}}$$

In the test statistic

$$z = \frac{(\hat{p}_1 - \hat{p}_2) - (p_1 - p_2)}{\sqrt{\frac{\hat{p}_1\hat{q}_1}{n_1} + \frac{\hat{p}_2\hat{q}_2}{n_2}}}$$

let z be positive and negative (for two tails) and solve for $p_1 - p_2$. The results are the limits of the confidence interval given earlier.

Using Technology

STATDISK: Select **Analysis** from the main menu bar, then select **Hypothesis Testing,** then **Proportion-Two Samples.** Enter the required items in the dialog box. Confidence interval limits are included with the hypothesis test results.

Minitab: Minitab can now handle summary statistics for two samples. Select **Stat** from the main menu bar, then select **Basic Statistics,** then **2 Proportions.** Click on the button for **Summarize data.** Click on the **Options** bar. Enter the desired confidence level, enter the claimed value of $p_1 - p_2$, and click on the box to use the pooled estimate of p for the test. Click **OK** twice.

Excel: You must use the Data Desk XL add-in, which is a supplement to this book. First make these entries: In cell A1 enter the number of successes for sample 1, in cell B1 enter the number of trials for sample 1, in cell C1 enter the number of successes for sample 2, and in cell D1 enter the number of trials for sample 2. Click on **DDXL.** Select **Hypothesis Tests** and **Summ 2 Var Prop Test** or select **Confidence Intervals** and **Summ 2 Var Prop Interval.** In the dialog box, click on the four pencil icons and enter A1, B1, C1, and D1 in the four input boxes. Click **OK.** Proceed to complete the new dialog box.

TI-83 Plus: The TI-83 Plus calculator can be used for hypothesis tests and confidence intervals. Press **STAT** and select **TESTS.** Then choose the option of **2-PropZTest** (for a hypothesis test) or **2-PropZInt** (for a confidence interval). When testing hypotheses, the TI-83 Plus calculator will display a P-value instead of critical values, so the P-value method of testing hypotheses is used.

8-4 Basic Skills and Concepts

Finding Number of Successes. *In Exercises 1–4, find the number of successes x suggested by the given statement.*

1. From the *Uniform Crime Reports*: Of 2750 randomly selected arrests of criminals younger than 21 years of age, 4.25% involve violent crimes.

2. From the *New York Times*: Among 240 vinyl gloves subjected to stress tests, 63% leaked viruses.

3. From *Sociological Methods and Research*: When 294 central-city residents were surveyed, 28.9% refused to respond.

4. From *Science* magazine: Among 850 teens surveyed with an anonymous computer program, 12.4% said that they carried a gun within the last 30 days.

Calculations for Testing Claims. *In Exercises 5 and 6, assume that you plan to use a significance level of $\alpha = 0.05$ to test the claim that $p_1 = p_2$. Use the given sample sizes and numbers of successes to find (a) the pooled estimate $\overline{p}$, (b) the z test statistic, (c) the critical z values, and (d) the P-value.*

5.

Treatment Group	Placebo Group
$n_1 = 20$	$n_2 = 25$
$x_1 = 10$	$x_2 = 15$

6.

Students	Faculty
$n_1 = 150$	$n_2 = 50$
$x_1 = 30$	$x_2 = 20$

7. Testing Effectiveness of Vaccine In a *USA Today* article about an experimental nasal spray vaccine for children, the following statement was presented: "In a trial involving 1602 children only 14 (1%) of the 1070 who received the vaccine developed the flu, compared with 95 (18%) of the 532 who got a placebo." The article also referred to a study claiming that the experimental nasal spray "cuts children's chances of getting the flu." Is there sufficient sample evidence to support the stated claim?

8. Color Blindness in Men and Women In a study of red/green color blindness, 500 men and 2100 women are randomly selected and tested. Among the men, 45 have red/green color blindness. Among the women, 6 have red/green color blindness (based on data from *USA Today*).

a. Is there sufficient evidence to support the claim that men have a higher rate of red/green color blindness than women?

b. Construct the 99% confidence interval for the difference between the color blindness rates of men and women. Do the confidence interval limits contain 0?

c. Why would the sample size for women be so much larger than the sample size for men?

9. Seat Belts and Hospital Time A study was made of 413 children who were hospitalized as a result of motor vehicle crashes. Among 290 children who were not using seat belts, 50 were injured severely. Among 123 children using seat belts, 16 were injured severely (based on data from "Morbidity Among Pediatric Motor Vehicle Crash Victims: The Effectiveness of Seat Belts," by Osberg and Di Scala, *American Journal of Public Health,* Vol. 82, No. 3). Is there sufficient sample evidence to conclude, at the 0.05 significance level, that the rate of severe injuries is lower for children wearing seat belts? Based on these results, what action should be taken?

10. Interpreting a Computer Display A U.S. Department of Justice report (NCJ-156831) included the claim that "in spouse murder cases, wife defendants were less likely to be convicted than husband defendants." Sample data consisted of 277 convictions among 318 husband defendants, and 155 convictions among 222 wife defendants. Test the stated claim and identify one possible explanation for the result. The Minitab results are shown here.

```
Sample     X        N      Sample p
1          277      318    0.871069
2          155      222    0.698198
Estimate for p(1)- p(2): 0.172871
95% CI for p(1)- p(2): (0.102140, 0.243602)
Test for p(1) - p(2) = 0 (vs > 0): z = 4.94 P-Value = 0.000
```

11. Effectiveness of Salk Vaccine for Polio In initial tests of the Salk vaccine, 33 of 200,000 vaccinated children later developed polio. Of 200,000 children vaccinated with a placebo, 115 later developed polio. The TI-83 Plus calculator display is shown here. At the 0.01 significance level, test the claim that the Salk vaccine is effective in lowering the polio rate. Does it appear that the vaccine is effective?

TI-83 Plus

```
2-PropZTest
 p1<p2
 z=-6.741605792
 p=7.88E-12
 p̂1=1.65E-4
 p̂2=5.75E-4
↓p̂=3.7E-4
```

12. Airline Load Factor In a recent year, Southwest Airlines had 3,131,727 aircraft seats available on all of its flights, and 2,181,604 of them were occupied by passengers. America West had 2,091,859 seats available, and 1,448,255 of them were occupied. The percentage of seats occupied is called the *load factor*, so these results show that the load factor is 69.7% (rounded) for Southwest Airlines and 69.2% (rounded) for America West. (The data are from the U. S. Department of Transportation.) Answer the following by assuming that the results are from randomly selected samples.

a. Test the claim that both airlines have the same load factor.

b. Given that 69.7% and 69.2% appear to be so obviously close, how do you explain the results from part (a)?

c. Generalize the key point of this example by completing the following statement: "If two sample sizes are extremely large, even seemingly small differences in sample proportions . . . "

13. Violent Crime and Age Group The newly appointed head of the state mental health agency claims that a smaller proportion of the crimes committed by persons younger than 21 years of age are violent crimes (when compared to the crimes committed by persons 21 years of age or older). Of 2750 randomly selected arrests of criminals younger than 21 years of age, 4.25% involve violent crimes. Of 2200 randomly selected arrests of criminals 21 years of age or older, 4.55% involve violent crimes (based on data from the Uniform Crime Reports). Construct a 95% confidence interval for the difference between the two proportions of violent crimes. Do the confidence interval limits contain 0? Does this indicate that there isn't a significant difference between the two rates of violent crimes?

14. Testing Laboratory Gloves The *New York Times* ran an article about a study in which Professor Denise Korniewicz and other Johns Hopkins researchers subjected laboratory gloves to stress. Among 240 vinyl gloves, 63% leaked viruses. Among 240 latex gloves, 7% leaked viruses. At the 0.005 significance level, test the claim that vinyl gloves have a larger virus leak rate than latex gloves.

15. Written Survey and Computer Survey In a study of 1700 teens aged 15–19, half were given written surveys and half were given surveys using an anonymous computer program. Among those given the written surveys, 7.9% say that they carried a gun within the last 30 days. Among those given the computer surveys, 12.4% say that they carried a gun within the last 30 days (based on data from the Urban Institute).

a. The sample percentages of 7.9% and 12.4% are obviously not equal, but is the difference significant? Explain.

b. Construct a 99% confidence interval estimate of the difference between the two population percentages, and interpret the result.

16. Adverse Drug Reactions In clinical tests of adverse reactions to the drug Viagra, 4.0% of the 734 subjects in the treatment group experienced nasal congestion, but 2.1% of the 725 subjects in the placebo group experienced nasal congestion (based on data from Pfizer Pharmaceuticals).

a. Using a 0.01 significance level, is there sufficient evidence to support the claim that nasal congestion occurs at a higher rate among Viagra users than those who do not use Viagra?

b. Construct the 99% confidence interval estimate of the difference between the rate of nasal congestion among Viagra users and the nasal congestion rate for those who use a placebo. Do the confidence interval limits contain 0, and what does this suggest about the two rates?

17. Drinking and Crime Karl Pearson, who developed many important concepts in statistics, collected crime data in 1909. Of those convicted of arson, 50 were drinkers and 43 abstained. Of those convicted of fraud, 63 were drinkers and 144 abstained. Use a 0.01 significance level to test the claim that the proportion of drinkers among convicted arsonists is greater than the proportion of drinkers convicted of fraud. Does it seem reasonable that drinking might have had an effect on the type of crime? Why?

18. Poll Refusal Rate Professional pollsters are becoming concerned about the growing rate of refusals among potential survey subjects. In analyzing the problem, there is a need to know if the refusal rate is universal or if there is a difference between the rates for central-city residents and those not living in central cities. Specifically, it was found that when 294 central-city residents were surveyed, 28.9% refused to respond. A survey of 1015 residents not in a central city resulted in a 17.1% refusal rate (based on data from "I Hear You Knocking But You Can't Come In," by Fitzgerald and Fuller, *Sociological Methods and Research,* Vol. 11, No. 1). At the 0.01 significance level, test the claim that the central-city refusal rate is the same as the refusal rate in other areas.

19. Warmer Surgical Patients Recover Better? An article published in *USA Today* stated that "in a study of 200 colorectal surgery patients, 104 were kept warm with blankets and intravenous fluids; 96 were kept cool. The results show: Only 6 of those warmed developed wound infections vs. 18 who were kept cool." Use a 0.05 significance level to test the claim of the article's headline: "Warmer surgical patients recover better." If these results are verified, should surgical patients be routinely warmed?

20. Home Field Advantage When games were sampled from throughout a season, it was found that the home team won 127 of 198 professional *basketball* games, and the home team won 57 of 99 professional *football* games (based on data from "Predicting Professional Sports Game Outcomes from Intermediate Game Scores," by Cooper et al., *Chance,* Vol. 5, No. 3–4). Construct a 95% confidence interval for the difference between the proportions of home wins. Do the confidence interval limits contain 0? Based on the results, is there a significant difference between the proportions of home wins? What do you conclude about the home field advantage?

8-4 Beyond the Basics

21. Testing for Constant Difference To test the null hypothesis that the difference between two population proportions is equal to a nonzero constant c, use the test statistic

$$z = \frac{(\hat{p}_1 - \hat{p}_2) - c}{\sqrt{\dfrac{\hat{p}_1(1 - \hat{p}_1)}{n_1} + \dfrac{\hat{p}_2(1 - \hat{p}_2)}{n_2}}}$$

As long as n_1 and n_2 are both large, the sampling distribution of the test statistic z will be approximately the standard normal distribution. Refer to the sample data included with the example presented in this section, and use a 0.05 significance level to test the claim that the headache rate of Viagra users is 10 percentage points more than the percentage for those who use a placebo.

22. Transitivity of Hypothesis Tests Sample data are randomly drawn from three independent populations, each of size 100. The sample proportions are $\hat{p}_1 = 40/100$, $\hat{p}_2 = 30/100$, and $\hat{p}_3 = 20/100$.

a. At the 0.05 significance level, test H_0: $p_1 = p_2$.
b. At the 0.05 significance level, test H_0: $p_2 = p_3$.
c. At the 0.05 significance level, test H_0: $p_1 = p_3$.
d. In general, if hypothesis tests lead to the conclusions that $p_1 = p_2$ and $p_2 = p_3$ are reasonable, does it follow that $p_1 = p_3$ is also reasonable? Why or why not?

23. Determining Sample Size The sample size needed to estimate the difference between two population proportions to within a margin of error E with a confidence level of $1 - \alpha$ can be found as follows. In the expression

$$E = z_{\alpha/2}\sqrt{\frac{p_1 q_1}{n_1} + \frac{p_2 q_2}{n_2}}$$

replace n_1 and n_2 by n (assuming that both samples have the same size) and replace each of p_1, q_1, p_2, and q_2 by 0.5 (because their values are not known). Then solve for n.

Use this approach to find the size of each sample if you want to estimate the difference between the proportions of men and women who own cars. Assume that you want 95% confidence that your error is no more than 0.03.

24. Interpreting Drug Test Results Ziac is a Lederle Laboratories drug developed to treat hypertension. Lederle Laboratories reported that when 221 people were treated with Ziac, 3.2% of them experienced dizziness. It was also reported that among the 144 people in the placebo group, 1.8% experienced dizziness.

a. Can you use the methods of this section to test the claim that there is a significant difference between the two rates of dizziness? Why or why not?
b. Can the given information be correct? Why or why not?

8-5 Comparing Variation in Two Samples

Because the characteristic of variation among data is extremely important, this section presents a method for using two samples to compare the variances of the populations from which the samples are drawn. In Section 2-5 we saw that variation in a sample could be described by measures that include standard deviation and variance. Because standard deviation is a very effective measure of variation, and because it is easier to understand than variance, we have tended to stress the use of standard deviation instead of variance. Let's briefly review the relationship between standard deviation and variance by noting that the variance is the square of the standard deviation.

Measures of Variation

s = standard deviation of *sample* $\quad$ s^2 = variance of *sample* (sample standard deviation squared)

σ = standard deviation of *population* $\quad$ σ^2 = variance of *population* (population standard deviation squared)

Do Air Bags Save Lives?

The National Highway Transportation Safety Administration reported that for a recent year, 3,448 lives were saved because of air bags. It was reported that for car drivers involved in frontal crashes, the fatality rate was reduced 31%; for passengers, there was a 27% reduction. It was noted that "calculating lives saved is done with a mathematical analysis of the real-world fatality experience of vehicles with air bags compared with vehicles without air bags. These are called double-pair comparison studies, and are widely accepted methods of statistical analysis."

Although the basic procedure of this section is designed for variances, we can also use it for standard deviations. The method we use requires the following assumptions.

Assumptions

1. The two populations are *independent* of each other. (Recall from Section 8-2 that two samples are independent if the sample selected from one population is not related to the sample selected from the other population.)
2. The two populations are each *normally distributed.* (This assumption is important because the methods of this section are extremely sensitive to departures from normality.)

Exploring the Data

Because the requirement of normal distributions is so important, we should begin by comparing the two sets of sample data by using tools such as histograms, boxplots, and normal quantile plots (see Section 5-7). We should search for outliers, and we should find the values of the sample statistics, especially the standard deviations. For example, consider the 36 weights of regular Coke in 36 different cans. Section 8-2 included a boxplot of that data set. Shown here are a histogram from a TI-83 Plus calculator and a normal probability plot from Minitab. The histogram shows that the data have a dis-

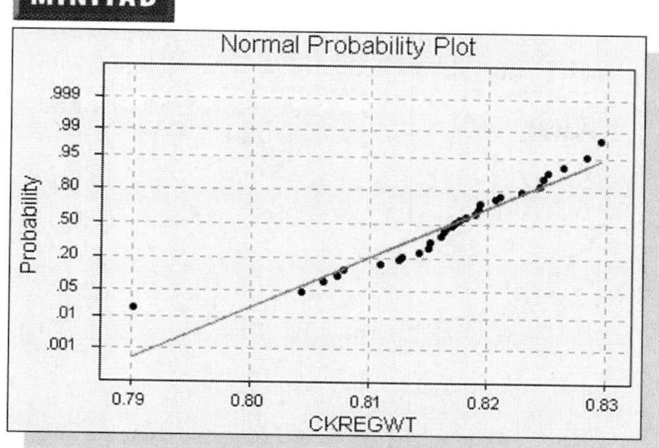

TI-83 Plus

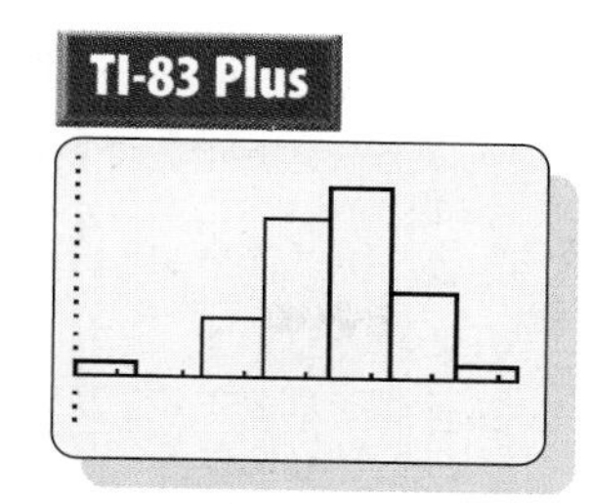

tribution that is approximately normal and that there is one value that is a potential outlier. The normal probability plot, which can be interpreted as if it were a normal quantile plot, shows that the points are reasonably close to a straight line, but they don't fit the straight line perfectly. This data set clearly satisfies a requirement of a distribution that is *approximately* normal, but it isn't so clear that this data set satisfies the stricter requirements of normality that apply to the methods of this section.

Hypothesis Tests

The computations of this section will be greatly simplified if we stipulate that s_1^2 represents the *larger* of the two sample variances. It doesn't really matter which sample is designated as sample 1, so we choose to use the following notation.

Notation for Hypothesis Tests with Two Variances or Standard Deviations

s_1^2 = *larger* of the two sample variances

n_1 = size of the sample with the *larger* variance

σ_1^2 = variance of the population from which the sample with the *larger* variance was drawn

The symbols s_2^2, n_2, and σ_2^2 are used for the other sample and population.

For two normally distributed populations with equal variances (that is, $\sigma_1^2 = \sigma_2^2$), the sampling distribution of the following test statistic is the ***F* distribution** shown in Figure 8-4 on the next page with critical values listed in Table A-5. If you continue to repeat an experiment of randomly selecting samples from two normally distributed populations with equal variances, the distribution of the ratio s_1^2/s_2^2 of the sample variances is the F distribution.

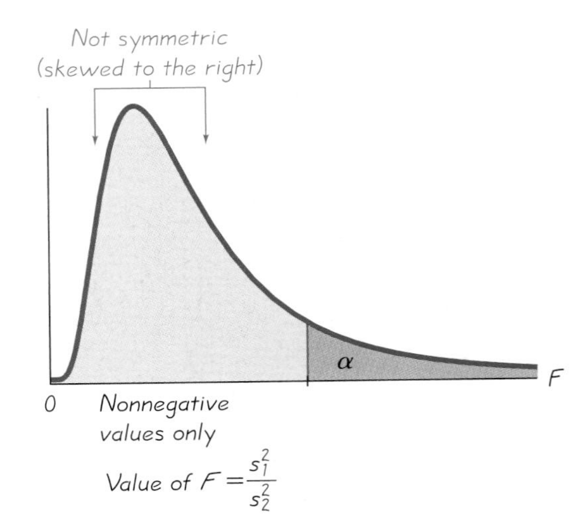

FIGURE 8-4
***F* Distribution**

There is a different F distribution for each different pair of degrees of freedom for the numerator and the denominator.

In Figure 8-4, note these properties of the *F* distribution:

- The *F* distribution is not symmetric.
- Values of the *F* distribution cannot be negative.
- The exact shape of the *F* distribution depends on two different degrees of freedom.

Test Statistic for Hypothesis Tests with Two Variances

$$F = \frac{s_1^2}{s_2^2}$$

Critical values: Using Table A-5, we obtain critical *F* values that are determined by the following three values:

1. The significance level α (Table A-5 has six pages of critical values for $\alpha = 0.01, 0.025$, and 0.05.)
2. **Numerator degrees of freedom** $= n_1 - 1$
3. **Denominator degrees of freedom** $= n_2 - 1$

To find a critical value, first refer to the part of Table A-5 corresponding to α (for a one-tailed test) or $\alpha/2$ (for a two-tailed test), then intersect the column representing the degrees of freedom for s_1^2 with the row representing the degrees of freedom for s_2^2. Because we are stipulating that the larger sample variance is s_1^2, all one-tailed tests will be right-tailed and all two-tailed tests will require that we find only the critical value located to the right. Good news: We have no need to find a critical value separating a left-tailed critical region. (Because the *F* distribution is not symmetric and has only nonnegative values, a left-tailed critical value cannot be found by using the negative of the right-tailed critical value; instead, a left-tailed critical value is found by using the reciprocal of the right-tailed value with the numbers of degrees of freedom reversed. See Exercise 14.)

We often have numbers of degrees of freedom that are not included in Table A-5. We could use linear interpolation to approximate the missing values, but in most cases that's not necessary because the F test statistic is either less than the lowest possible critical value or greater than the largest possible critical value. For example, Table A-5 shows that for $\alpha = 0.025$ in the right tail, 20 degrees of freedom for the numerator, and 34 degrees of freedom for the denominator, the critical F value is between 2.0677 and 2.1952. Any F test statistic below 2.0677 will result in failure to reject the null hypothesis, any F test statistic above 2.1952 will result in rejection of the null hypothesis, and interpolation is necessary only if the F test statistic happens to fall between 2.0677 and 2.1952. The use of a statistical software package such as STATDISK or Minitab eliminates this problem by providing critical values or P-values.

Interpreting the F Test Statistic: If the two populations really do have equal variances, then s_1^2/s_2^2 tends to be close to 1 because s_1^2 and s_2^2 tend to be close in value. But if the two populations have radically different variances, s_1^2 and s_2^2 tend to be very different numbers. Denoting the larger of the sample variances by s_1^2, we see that the ratio s_1^2/s_2^2 will be a large number whenever s_1^2 and s_2^2 are far apart in value. Consequently, a value of F near 1 will be evidence in favor of the conclusion that $\sigma_1^2 = \sigma_2^2$, but a large value of F will be evidence against the conclusion of equality of the population variances.

***Large* values of F are evidence against $\sigma_1^2 = \sigma_2^2$.**

Claims about Standard Deviations: The F test statistic applies to a claim made about two variances, but we can also use it for claims about two population standard deviations. Any claim about two population standard deviations can be restated in terms of the corresponding variances.

EXAMPLE **Coke versus Pepsi** Data Set 1 in Appendix B includes the weights (in pounds) of samples of regular Coke and regular Pepsi. The sample statistics are summarized in the accompanying table. Use a 0.05 significance level to test the claim that the weights of regular Coke and the weights of regular Pepsi have the same standard deviation.

	Regular Coke	Regular Pepsi
n	36	36
$\bar{x}$	0.81682	0.82410
s	0.007507	0.005701

SOLUTION Instead of using the sample standard deviations to test the claim of equal standard deviations, we will use the sample variances to test the claim of equal variances. Because we stipulate in this section that the larger variance is denoted by s_1^2, we let $s_1^2 = 0.007507^2$, $n_1 = 36$, $s_2^2 = 0.005701^2$, and $n_2 = 36$. We now proceed to use the traditional method of testing hypotheses as outlined in Figure 7-5.

Step 1: The claim of equal standard deviations is equivalent to a claim of equal variances, which we express symbolically as $\sigma_1^2 = \sigma_2^2$.

Step 2: If the original claim is false, then $\sigma_1^2 \neq \sigma_2^2$.

Step 3: Because the null hypothesis must contain equality, we have

$$H_0: \sigma_1^2 = \sigma_2^2 \text{ (original claim)} \qquad H_1: \sigma_1^2 \neq \sigma_2^2$$

Step 4: The significance level is $\alpha = 0.05$.

Step 5: Because this test involves two population variances, we use the F distribution.

Step 6: The test statistic is

$$F = \frac{s_1^2}{s_2^2} = \frac{0.007507^2}{0.005701^2} = 1.7339$$

For the critical values, first note that this is a two-tailed test with 0.025 in each tail. As long as we are stipulating that the larger variance is placed in the numerator of the F test statistic, we need to find only the right-tailed critical value. From Table A-5 we see that the critical value of F is between 1.8752 and 2.0739, which we find by referring to 0.025 in the right tail, with 35 degrees of freedom for the numerator and 35 degrees of freedom for the denominator. (STATDISK and Excel provide a critical value of 1.9611.)

Step 7: Figure 8-5 shows that the test statistic $F = 1.7339$ does not fall within the critical region, so we fail to reject the null hypothesis of equal variances.

INTERPRETATION There is not sufficient evidence to warrant rejection of the claim that the two variances are equal. However, we should recog-

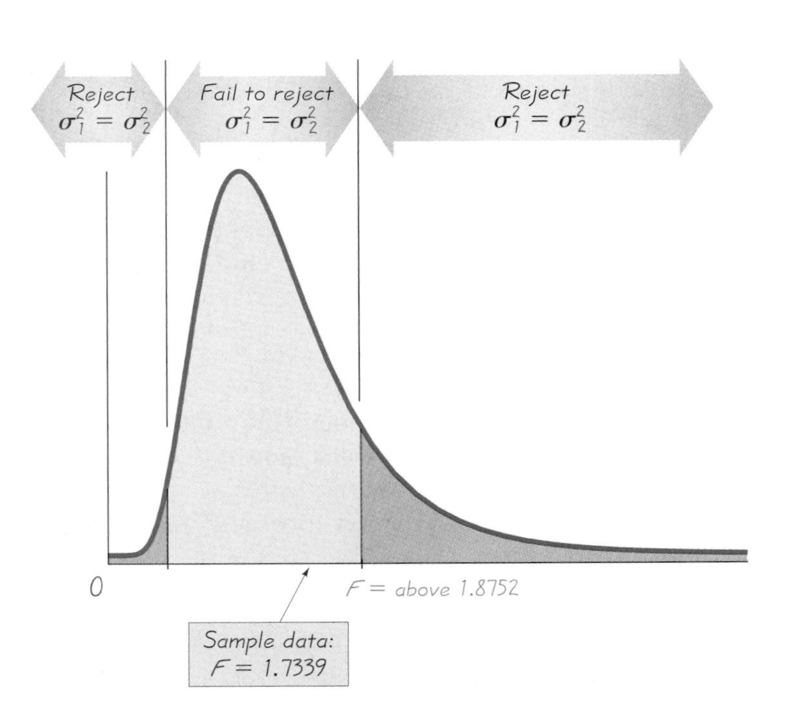

FIGURE 8-5
Distribution of s_1^2/s_2^2 for Weights of Regular Coke and Regular Pepsi.

nize that the F test is extremely sensitive to distributions that are not normally distributed, so this conclusion might make it appear that there is no significant difference between the population variances when there really is a difference that was hidden by nonnormal distributions.

In the preceding example we used a two-tailed test for the claim of equal variances. A right-tailed test would yield the same test statistic of $F = 1.7339$, but a different critical value of F.

We have described the traditional method of testing hypotheses made about two population variances. Exercise 13 deals with the P-value approach, and Exercise 15 deals with the construction of confidence intervals.

Lower Variation, Higher Quality

Ford and Mazda were producing similar transmissions that were supposed to be made with the same specifications. But the American-made transmissions required more warranty repairs than the Japanese-made transmissions. When investigators inspected samples of the Japanese transmission gearboxes, they first thought that their measuring instruments were defective because they weren't detecting any variability among the Mazda transmission gearboxes. They realized that although the American transmissions were within the specifications, the Mazda transmissions were not only within the specifications, but consistently close to the desired value. By reducing variability among transmission gearboxes, Mazda reduced the costs of inspection, scrap, rework, and warranty repair.

Using Technology

STATDISK: Select **Analysis** from the main menu, then select **Hypothesis Testing,** then **StDev-Two Samples.** Enter the required items in the dialog box.

Minitab: First enter all of the data from the two samples in column C1, with the values of the first sample stacked above the values of the second sample. In column C2 enter the identifying "subscripts," consisting of a 1 next to every value from the first sample and a 2 next to every value from the second sample. Now select **Stat,** then **ANOVA,** then **Homogeneity of Variance.** Enter C1 for the response variable, and enter C2 for the factors. Enter the confidence level, with 0.95 corresponding to a 0.05 significance level. Click **OK.** Among the various results displayed, find the F test statistic and the corresponding P-value. If the P-value is less than or equal to the significance level, reject the null hypothesis of equal variances.

Excel: First enter the data from the first sample in the first column A, then enter the values of the second sample in column B. Select **Tools, Data Analysis,** and then **F-Test Two-Sample for Variances.** In the dialog box, enter the range of values for the first sample (such as A1:A36) and the range of values for the second sample. Enter the value of the significance level in the "Alpha" box. Excel will provide the F test statistic, the P-value for the one-tailed case, and the critical F value for the one-tailed case. For a two-tailed test, double the P-value given by Excel.

TI-83 Plus: Press the **STAT** key, then select **TESTS,** then **2-SampFTEST.** You can use the summary statistics or you can use the data that are entered as lists.

8-5 Basic Skills and Concepts

Hypothesis Test of Equal Variances. *In Exercises 1 and 2, test the given claim. Use a significance level of* $\alpha = 0.05$, *and assume that all populations are normally distributed. Use the traditional method of testing hypotheses outlined in Figure 7-5, and draw the appropriate graphs.*

1. Claim: The treatment population and the placebo population have different variances ($\sigma_1^2 \neq \sigma_2^2$).

Treatment group: $n = 16, \bar{x} = 482.8, s = 150.0$

Placebo group: $n = 30, \bar{x} = 503.7, s = 75.0$

2. Claim: Women have a larger variance than men.

Men: $n = 15, \bar{x} = 107, s = 12.4$

Women: $n = 25, \bar{x} = 75.7, s = 16.8$

3. Weights of Regular Coke and Diet Coke This section included an example about a hypothesis test of the claim that weights of regular Coke and regular Pepsi have the same standard deviation. Test the claim that regular Coke and diet Coke have weights with different standard deviations. Sample weights are found in Data Set 1 in Appendix B, but here are the summary statistics: The sample of 36 weights of regular Coke have a standard deviation of 0.007507 lb, and the sample of 36 weights of diet Coke have a standard deviation of 0.004391 lb. Use a 0.05 significance level. If the results were to show that the standard deviations are significantly different, what would be an important factor that might explain the difference?

4. Axial Loads of Aluminum Cans Data Set 12 in Appendix B includes axial loads (in pounds) of a sample of 175 aluminum cans that are 0.0109 in. thick and another sample of 175 aluminum cans that are 0.0111 in. thick. (An axial load is the maximum weight that the sides can support. It is measured by using a plate to apply increasing pressure to the top of the can until it collapses.) The sample of 0.0109 in. cans has axial loads with a mean of 267.1 lb and a standard deviation of 22.1 lb. The sample of 0.0111 in. cans has axial loads with a mean of 281.8 lb and a standard deviation of 27.8 lb. Use a 0.05 significance level to test the claim that the samples come from populations with the same standard deviation.

Nonstress	Stress
$n_1 = 40$	$n_2 = 40$
$\bar{x}_1 = 53.3$	$\bar{x}_2 = 45.3$
$s_1 = 11.6$	$s_2 = 13.2$

5. Police as Eyewitnesses An experiment was conducted to investigate the effect of stress on the recall ability of police eyewitnesses. The experiment involved a nonstressful interrogation of a cooperative suspect and a stressful interrogation of an uncooperative and belligerent suspect. The numbers of details recalled a week after the incident are summarized in the margin (based on data from "Eyewitness Memory of Police Trainees for Realistic Role Plays," by Yuille et al., *Journal of Applied Psychology,* Vol. 79, No. 6). Use a 0.10 significance level to test the claim that the samples come from populations with different standard deviations.

6. Weights of Men in Different Age Groups As part of the National Health Survey, data were collected on the weights of men. For 804 men aged 25 to 34, the mean is 176 lb and the standard deviation is 35.0 lb. For 1657 men aged 65 to 74, the mean and standard deviation are 164 lb and 27.0 lb, respectively. At the 0.01 significance level, test the claim that the older men come from a population with a standard deviation less than that for men in the 25 to 34 age bracket.

7. Ages of Faculty and Student Cars Students at the author's college randomly selected 217 student cars and found that they had ages with a mean of 7.89 years and a standard deviation of 3.67 years. They also randomly selected 152 faculty cars and found that they had ages with a mean of 5.99 years and a standard deviation of 3.65 years. Is there sufficient evidence to support the claim that the ages of faculty cars vary less than the ages of student cars?

8. Testing Effects of Zinc A study of zinc-deficient mothers was conducted to determine effects of zinc supplementation during pregnancy. Sample data are listed in the margin (based on data from "The Effect of Zinc Supplementation on Pregnancy Outcome," by Goldenberg et al., *Journal of the American Medical Association*, Vol. 274, No. 6). The weights were measured in grams. Using a 0.05 significance level, is there sufficient evidence to support the claim that the variation of birth weights for the placebo population is greater than the variation for the population treated with zinc supplements?

Zinc Supplement Group	Placebo Group
$n_1 = 294$	$n_2 = 286$
$\bar{x}_1 = 3214$	$\bar{x}_2 = 3088$
$s_1 = 669$	$s_2 = 728$

9. Blanking Out on Tests Many students have had the unpleasant experience of panicking on a test because the first question was exceptionally difficult. The arrangement of test items was studied for its effect on anxiety. The following scores are measures of "debilitating test anxiety" (which most of us call panic or blanking out). The Excel display for the given values is also shown. At the 0.05 significance level, test the claim that the two given samples come from populations with the same variance.

Questions Arranged from Easy to Difficult

24.64	39.29	16.32	32.83	28.02
33.31	20.60	21.13	26.69	28.90
26.43	24.23	7.10	32.86	21.06
28.89	28.71	31.73	30.02	21.96
25.49	38.81	27.85	30.29	30.72

Questions Arranged from Difficult to Easy

33.62	34.02	26.63	30.26
35.91	26.68	29.49	35.32
27.24	32.34	29.34	33.53
27.62	42.91	30.20	32.54

Based on data from "Item Arrangement, Cognitive Entry Characteristics, Sex and Test Anxiety as Predictors of Achievement in Examination Performance," by Klimko, *Journal of Experimental Education,* Vol. 52, No. 4.

EXCEL

	A	B	C
1	F-Test Two-Sample for Variances		
2		*Variable 1*	*Variable 2*
3	Mean	27.1152	31.72813
4	Variance	47.01983	18.1489
5	Observations	25	16
6	df	24	15
7	F	2.590782	
8	P(F<=f) one-tail	0.029928	
9	F Critical one-tail	2.287827	

TI-83 Plus

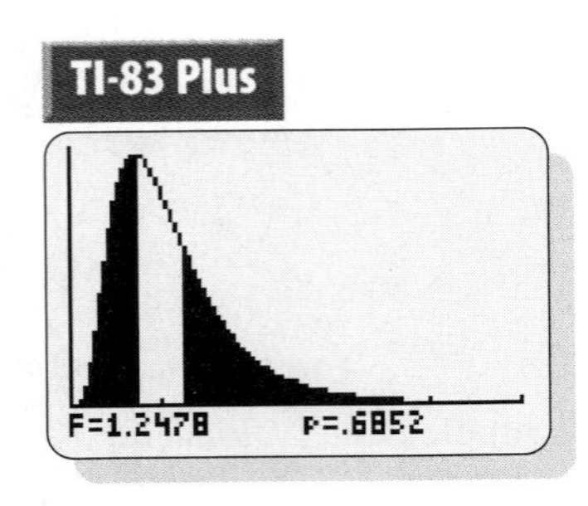

10. Calcium and Blood Pressure Sample data were collected in a study of calcium supplements and their effects on blood pressure. A placebo group and a calcium group began the study with blood pressure measurements (based on data from "Blood Pressure and Metabolic Effects of Calcium Supplementation in Normotensive White and Black Men," by Lyle et al., *Journal of the American Medical Association,* Vol. 257, No. 13). Sample values are listed and a TI-83 Plus display is shown. At the 0.10 significance level, test the claim that the two sample groups come from populations with the same standard deviation. If the experiment requires groups with equal standard deviations, are these two groups acceptable?

Placebo:	124.6	104.8	96.5	116.3	106.1	128.8	107.2	123.1
	118.1	108.5	120.4	122.5	113.6			
Calcium:	129.1	123.4	102.7	118.1	114.7	120.9	104.4	116.3
	109.6	127.7	108.0	124.3	106.6	121.4	113.2	

11. Comparing Diet Coke and Diet Pepsi Refer to Data Set 1 in Appendix B, and use a 0.05 significance level to test the claim that diet Coke and diet Pepsi have weights with different standard deviations.

12. Comparing Textbook Prices Refer to Data Set 2 in Appendix B. Is there sufficient evidence to support the claim that the standard deviations of new textbook prices at the two colleges are different? Do the sample values appear to satisfy the requirement of coming from populations with normal distributions?

8-5 Beyond the Basics

13. Determining *P*-Values To test a claim about two population variances by using the *P*-value approach, first find the *F* test statistic, then refer to Table A-5 to determine how it compares to the critical values listed for $\alpha = 0.01$, $\alpha = 0.025$, and $\alpha = 0.05$. Referring to Exercise 3, what can be concluded about the *P*-value?

14. Finding Lower Critical *F* Values In this section, for hypothesis tests that were two-tailed, we found only the upper critical value. Let's denote that value by F_R, where the subscript suggests the critical value for the right tail. The lower critical value F_L (for the left tail) can be found as follows: First interchange the degrees of freedom, and then take the reciprocal of the resulting F value found in Table A-5. (F_R is often denoted by $F_{\alpha/2}$, and F_L is often denoted by $F_{1-\alpha/2}$.) Find the critical values F_L and F_R for two-tailed hypothesis tests based on the following values.
   a. $n_1 = 10, n_2 = 10, \alpha = 0.05$
   b. $n_1 = 10, n_2 = 7, \alpha = 0.05$
   c. $n_1 = 7, n_2 = 10, \alpha = 0.05$
   d. $n_1 = 25, n_2 = 10, \alpha = 0.02$
   e. $n_1 = 10, n_2 = 25, \alpha = 0.02$

15. Contructing Confidence Intervals In addition to testing claims involving σ_1^2 and σ_2^2, we can also construct confidence interval estimates of the ratio σ_1^2/σ_2^2 using the following expression.

$$\left(\frac{s_1^2}{s_2^2} \cdot \frac{1}{F_{\text{R}}}\right) < \frac{\sigma_1^2}{\sigma_2^2} < \left(\frac{s_1^2}{s_2^2} \cdot \frac{1}{F_{\text{L}}}\right)$$

(continued)

Here F_L and F_R are as described in Exercise 14. Refer to the data in Exercise 10 and construct a 95% confidence interval estimate for the ratio of the placebo group variance to the calcium-supplement group variance.

16. Effects of Transformations on Data Sample data consist of temperatures recorded for two different groups of items that were produced by two different production techniques. A quality-control specialist plans to analyze the results. She begins by testing for equality of the two population standard deviations.

a. If she adds the same constant to every temperature from both groups, how is the value of the test statistic F affected?

b. If she multiplies every value from both groups by the same constant, how is the value of the test statistic F affected?

c. If she converts all temperatures from the Fahrenheit scale to the Celsius scale, how is the value of the test statistic F affected?

8-6 Inferences about Two Means: Independent and Small Samples

In this section we present methods of inferential statistics (hypothesis tests and confidence intervals) for situations involving the means of two independent populations, when (unlike Section 8-2) at least one of the two samples is small (with $n \leq 30$). We describe methods for testing hypotheses and constructing confidence intervals for situations in which the following assumptions apply.

Assumptions

1. The two samples are *independent.*
2. The two samples are simple random samples *selected from normally distributed* populations.
3. At least one of the two samples is small $(n \leq 30)$.

When these conditions are satisfied, we use one of three different procedures corresponding to the following cases:

Case 1: The values of both population variances are known. (In reality, this case seldom occurs.)

Case 2: The two populations have equal variances. (That is, $\sigma_1^2 = \sigma_2^2$.)

Case 3: The two populations have unequal variances. (That is, $\sigma_1^2 \neq \sigma_1^2$.)

Case 1: Both Population Variances Are Known

In reality, Case 1 almost never occurs. Finding σ_1^2 and σ_2^2 typically requires that we know all of the values of both populations, and we can therefore find the values of μ_1 and μ_2, so there is no need to make inferences about μ_1 and μ_2 by testing claims or constructing confidence intervals. If some strange set of circumstances allows us to know the values of σ_1^2 and σ_2^2 but not μ_1 and μ_2, then we use the methods described in Section 8-2 for large samples. Because this case of known population variances is so unlikely to occur, we will not pursue it.

Choosing Between Cases 2 and 3: How Do We Decide That Two Populations Have Equal or Unequal Variances?

If the three assumptions listed at the beginning of this section are satisfied, inferences about the two population means are made with the procedures we will describe in Cases 2 and 3. Case 2 applies when the two populations have *equal* variances, and Case 3 applies when the two populations have *different* variances. But how do we determine whether the two population variances are equal? In Section 8-5 we described a procedure for testing the claim that the two populations have equal variances, and this approach, which we will now call the *preliminary F test approach*, is used by some statisticians and is designed into some software. It is somewhat controversial, however. Some argue that if we apply the F test with a certain significance level and then do a t test at the same level, the overall result will not be at the same level of significance. Also, in addition to being sensitive to differences in population variances, the F statistic is so sensitive to departures from normal distributions that we might choose a method for the wrong reason. (For example, the F test might make us think that the two populations have unequal variances, when the variances are actually equal but the test was misleading because the two population distributions were not both normal.)

Some textbooks completely dodge the issue by presenting only those methods that assume equal population variances. Other textbooks describe both methods but ignore any discussion of criteria for choosing between them. (The exercises include notes, such as "assume equal population variances.") Moser and Stevens, in "Homogeneity of Variance in the Two-Sample Means Test" (*The American Statistician*, Vol. 46, No. 1), argue against the preliminary F test and recommend instead that we always assume that the two population variances are *not* equal (as in Case 3 below). Here's another strategy: Make the inferences using *both* methods (for equal variances and for unequal variances). If the two sets of results agree, your conclusion is probably OK. Here is a summary of two different schools of thought:

1. *Preliminary F test approach*: Apply the F test described in Section 8-5 to test the null hypothesis that $\sigma_1^2 = \sigma_2^2$. Use the conclusion of that test as follows.
   - *Fail to reject* $\sigma_1^2 = \sigma_2^2$: Assume *equal* population variances. (Case 2)
   - *Reject* $\sigma_1^2 = \sigma_2^2$: Assume *unequal* population variances. (Case 3)
2. *Assume unequal variances*: Assume that the two populations have *unequal* variances ($\sigma_1^2 \neq \sigma_2^2$) and use the methods of Case 3. (Remember, this assumption applies only to cases in which we have two independent samples, at least one of which is small, and the population variances are not known.)

Case 2: Equal Population Variances: Pool the Two Sample Variances

If the two populations have unknown but equal variances, we calculate a **pooled estimate of σ^2** that is common to both populations. The pooled estimate is denoted by s_p^2 and is a weighted average of s_1^2 and s_2^2, as shown in the box. (Minitab and the TI-83 Plus calculator are two of many systems using the "pooled" option of this case.) Hypotheses about μ_1 and μ_2 can be tested and confidence interval estimates of $\mu_1 - \mu_2$ can be constructed by using the following methods.

Test Statistic (Small Independent Samples and Equal Variances)

$$t = \frac{(\bar{x}_1 - \bar{x}_2) - (\mu_1 - \mu_2)}{\sqrt{\dfrac{s_p^2}{n_1} + \dfrac{s_p^2}{n_2}}}$$

where
$$s_p^2 = \frac{(n_1 - 1)s_1^2 + (n_2 - 1)s_2^2}{(n_1 - 1) + (n_2 - 1)}$$

and the number of degrees of freedom is given by df $= n_1 + n_2 - 2$.

Confidence Interval (Small Independent Samples and Equal Variances)

$$(\bar{x}_1 - \bar{x}_2) - E < (\mu_1 - \mu_2) < (\bar{x}_1 - \bar{x}_2) + E$$

where
$$E = t_{\alpha/2}\sqrt{\frac{s_p^2}{n_1} + \frac{s_p^2}{n_2}}$$

and s_p^2 is as given in the test statistic.

EXAMPLE **Cigarette Filters and Nicotine** Refer to Table 8-2 for the measured nicotine contents of randomly selected filtered and nonfiltered king-size cigarettes. All measurements are in milligrams, and the data are from the Federal Trade Commission. Use a 0.05 significance level to test the claim that the mean amount of nicotine in filtered king-size cigarettes is equal to the mean amount of nicotine in nonfiltered king-size cigarettes.

TABLE 8-2 Nicotine Contents (in mg) of Cigarettes

Filtered	1.2	1.3	1.1	1.1	1.0	0.1	1.1	1.0	0.8	1.0	0.9	0.8	1.0	1.0	1.0	0.9	1.2	1.1	0.1	1.2	0.9
Nonfiltered	1.6	1.9	1.6	1.8	1.7	1.7	1.4	1.5													

Nicotine (mg)

Filtered Kings	Nonfiltered Kings
$n_1 = 21$	$n_2 = 8$
$\bar{x}_1 = 0.94$	$\bar{x}_2 = 1.65$
$s_1 = 0.31$	$s_2 = 0.16$

SOLUTION Before performing the hypothesis test, first explore the data sets. The sample statistics are listed in the margin, and the STATDISK-generated boxplots are also shown.

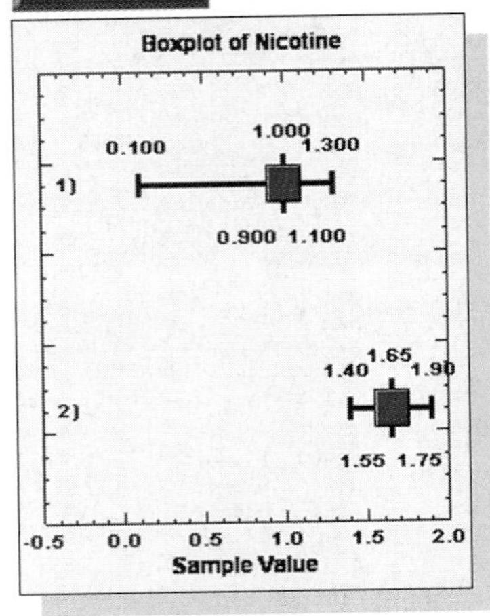

The sample statistics and boxplots seem to indicate a clear difference between the two means, but those graphs are based on small samples, so are the differences really significant? The two samples are independent (because they are separate and are not matched), the sample sizes are small, and we don't know the values of σ_1 and σ_2, so we are dealing with either Case 2 or Case 3. Using a preliminary F test (described in Section 8-5) to choose between Case 2 and Case 3, we test H_0: $\sigma_1^2 = \sigma_2^2$ with the test statistic

$$F = \frac{s_1^2}{s_2^2} = \frac{0.31^2}{0.16^2} = 3.7539$$

With $\alpha = 0.05$ in a two-tailed F test and with 20 and 7 degrees of freedom for the numerator and denominator, respectively, we use Table A-5 to find the critical F value of 4.4667. Because the computed test statistic of $F = 3.7539$ does *not* fall within the critical region, we fail to reject the null hypothesis of equal variances and proceed by using the approach outlined in Case 2.

In using the Case 2 approach, we test the claim that $\mu_1 = \mu_2$, with the following null and alternative hypotheses:

$$H_0: \mu_1 = \mu_2 \text{ (or } \mu_1 - \mu_2 = 0) \qquad H_1: \mu_1 \neq \mu_2$$

The test statistic for this case of equal variances requires the value of the pooled variance s_p^2 so we find that value first.

$$s_p^2 = \frac{(n_1 - 1)s_1^2 + (n_2 - 1)s_2^2}{(n_1 - 1) + (n_2 - 1)}$$

$$= \frac{(21 - 1)\cdot 0.31^2 + (8 - 1)\cdot 0.16^2}{(21 - 1) + (8 - 1)}$$

$$= 0.078$$

We can now find the value of the test statistic:

$$t = \frac{(\bar{x}_1 - \bar{x}_2) - (\mu_1 - \mu_2)}{\sqrt{\dfrac{s_p^2}{n_1} + \dfrac{s_p^2}{n_2}}} = \frac{(0.94 - 1.65) - 0}{\sqrt{\dfrac{0.078}{21} + \dfrac{0.078}{8}}} = -6.119$$

The critical values of $t = -2.052$ and $t = 2.052$ are found from Table A-3 by referring to the column for $\alpha = 0.05$ (two tails) and to the row for df = 27 (the value of 21 + 8 − 2). Figure 8-6 shows the test statistic, critical values, and critical region. We see that the test statistic does fall within the critical region, so we reject the null hypothesis that $\mu_1 = \mu_2$.

INTERPRETATION There is sufficient evidence to warrant rejection of the claim that filtered king cigarettes and nonfiltered kings have the same mean amounts of nicotine. The subjective conclusion we formed by intuitively analyzing the boxplots and the sample statistics is now upheld by a formal hypothesis test.

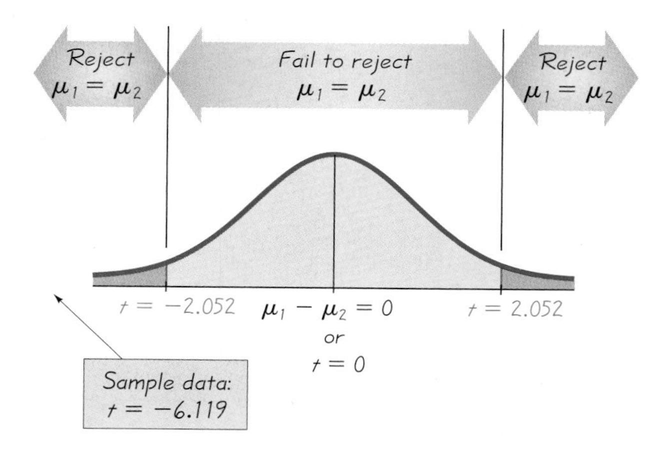

FIGURE 8-6
Distribution of Differences Between Means of Nicotine Amounts in Filtered and Nonfiltered Cigarettes

Instead of using the preliminary F test approach, we could use the same data with the methods to be described in Case 3, and we would again reject the null hypothesis that $\mu_1 = \mu_2$.

EXAMPLE **Cigarette Filters and Nicotine** Using the data in the preceding example, construct a 95% confidence interval estimate of $\mu_1 - \mu_2$.

SOLUTION It is easy to find that $\bar{x}_1 - \bar{x}_2 = 0.94 - 1.65 = -0.71$. Next, we find the value of the margin of error E.

$$E = t_{\alpha/2}\sqrt{\frac{s_p^2}{n_1} + \frac{s_p^2}{n_2}} = 2.052\sqrt{\frac{0.078}{21} + \frac{0.078}{8}} = 0.238$$

With $\overline{x}_1 - \overline{x}_2 = -0.71$ and with $E = 0.238$, we proceed to construct the confidence interval as follows.

$$(\overline{x}_1 - \overline{x}_2) - E < (\mu_1 - \mu_2) < (\overline{x}_1 - \overline{x}_2) + E$$
$$-0.71 - 0.238 < (\mu_1 - \mu_2) < -0.71 + 0.238$$
$$-0.95 < (\mu_1 - \mu_2) < -0.47$$

INTERPRETATION This final format is somewhat awkward with its negative signs, so we might interpret it by saying that we are 95% confident that nonfiltered king-size cigarettes have a mean nicotine content level that exceeds the mean for filtered kings by an amount that is between 0.47 mg and 0.95 mg.

Case 3: Unequal Population Variances

If the two populations have unequal variances, there is no exact method for testing equality of means and constructing confidence intervals. An *approximate* method is to use the following test statistic and confidence interval.

Test Statistic (Small Independent Samples and Unequal Variances)

$$t = \frac{(\overline{x}_1 - \overline{x}_2) - (\mu_1 - \mu_2)}{\sqrt{\frac{s_1^2}{n_1} + \frac{s_2^2}{n_2}}}$$

where df = smaller of $n_1 - 1$ and $n_2 - 1$

Confidence Interval (Small Independent Samples and Unequal Variances)

$$(\overline{x}_1 - \overline{x}_2) - E < (\mu_1 - \mu_2) < (\overline{x}_1 - \overline{x}_2) + E$$

where $E = t_{\alpha/2}\sqrt{\frac{s_1^2}{n_1} + \frac{s_2^2}{n_2}}$

and df = smaller of $n_1 - 1$ and $n_2 - 1$

This test statistic and confidence interval give the number of degrees of freedom as the smaller of $n_1 - 1$ and $n_2 - 1$, but this is a more conservative and simpler alternative to computing the number of degrees of freedom by using Formula 8-1. (Statdisk uses Formula 8-1.)

Formula 8-1

$$\text{df} = \frac{(A + B)^2}{\frac{A^2}{n_1 - 1} + \frac{B^2}{n_2 - 1}}$$

where $A = \frac{s_1^2}{n_1}$ and $B = \frac{s_2^2}{n_2}$

More exact results are obtained by using Formula 8-1, but they continue to be only approximate.

AS

EXAMPLE **Cigarette Filters and Tar** Refer to the data listed in Table 8-3 and use a 0.05 significance level to test the claim that the mean amount of tar in filtered king-size cigarettes is *less than* the mean amount of tar in nonfiltered king-size cigarettes. (All measurements are in milligrams, and the data are from the Federal Trade Commission.)

SOLUTION The sample statistics are listed in the margin, and they appear to suggest a dramatic difference between the two means. But are those differences really significant?

Tar (mg)	
Filtered Kings	Nonfiltered Kings
$n_1 = 21$	$n_2 = 8$
$\bar{x}_1 = 13.3$	$\bar{x}_2 = 24.0$
$s_1 = 3.7$	$s_2 = 1.7$

The two samples are independent (because they are separate and are not matched), the sample sizes are small, and we don't know the values of σ_1 and σ_2, so we are dealing with either Case 2 or Case 3. If we use a preliminary F test to choose between Case 2 and Case 3, we test H_0: $\sigma_1^2 = \sigma_2^2$ with the test statistic

$$F = \frac{s_1^2}{s_2^2} = \frac{3.7^2}{1.7^2} = 4.7370$$

With $\alpha = 0.05$ in a two-tailed F test and with 20 and 7 degrees of freedom for the numerator and denominator, respectively, we use Table A-5 to find the critical F value of 4.4667. Because the computed test statistic of $F = 4.7370$ does fall within the critical region, we reject the null hypothesis of equal variances and proceed by using the approach outlined in Case 3.

In using the Case 3 approach, we test the claim that $\mu_1 < \mu_2$ with the following null and alternative hypotheses:

$$H_0: \mu_1 \geq \mu_2 \text{ (or } \mu_1 - \mu_2 \geq 0)$$
$$H_1: \mu_1 < \mu_2 \quad \text{(original claim)}$$

The test statistic for this case of unequal variances is

$$t = \frac{(\bar{x}_1 - \bar{x}_2) - (\mu_1 - \mu_2)}{\sqrt{\dfrac{s_1^2}{n_1} + \dfrac{s_2^2}{n_2}}} = \frac{(13.3 - 24.0) - 0}{\sqrt{\dfrac{3.7^2}{21} + \dfrac{1.7^2}{8}}} = -10.630$$

INTERPRETATION The critical value of $t = -1.895$ is found from Table A-3 by referring to the column for $\alpha = 0.05$ (one tail) and to the row for df $= 7$ (the smaller of $21 - 1$ and $8 - 1$). (If Formula 8-1 is used to compute the

TABLE 8-3 Tar Contents (in mg) of Cigarettes

Filtered	16	15	16	14	16	1	16	18	10	14	12	11	14	13	13	13	16	16	8	16	11
Nonfiltered	23	23	24	26	25	26	21	24													

number of degrees of freedom, you get df = 26, so the critical value is $t = -1.706$.) Because this left-tailed test has a test statistic of $t = -10.630$ and a critical value of $t = -1.895$, the test statistic does fall within the critical region, so we reject the null hypothesis that $\mu_1 \geq \mu_2$. There is sufficient evidence to support the claim that filtered king-size cigarettes have a mean tar level that is less than the mean for nonfiltered kings. (If we use the same data with the methods of Case 2 for equal variances, we again reject the null hypothesis that $\mu_1 \geq \mu_2$, so the conclusion is the same.)

EXAMPLE **Cigarette Filters and Tar** Using the data in the preceding example, construct a 95% confidence interval estimate of $\mu_1 - \mu_2$.

SOLUTION With $\bar{x}_1 - \bar{x}_2 = 13.3 - 24.0 = -10.7$, we proceed to find the value of the margin of error E.

$$E = t_{\alpha/2}\sqrt{\frac{s_1^2}{n_1} + \frac{s_2^2}{n_2}} = 2.365\sqrt{\frac{3.7^2}{21} + \frac{1.7^2}{8}} = 2.381$$

With $\bar{x}_1 - \bar{x}_2 = -10.7$ and with $E = 2.381$, we construct the confidence interval as follows.

$$(\bar{x}_1 - \bar{x}_2) - E < (\mu_1 - \mu_2) < (\bar{x}_1 - \bar{x}_2) + E$$
$$-10.7 - 2.381 < (\mu_1 - \mu_2) < -10.7 + 2.381$$
$$-13.1 < (\mu_1 - \mu_2) < -8.3$$

INTERPRETATION Again, this final format is somewhat awkward with its negative signs, so we might interpret it by saying that we are 95% confident that the mean tar content for nonfiltered kings exceeds the mean tar content for filtered kings by an amount that is between 8.3 mg and 13.1 mg.

Using Statistics to Identify Thieves

Methods of statistics can be used to determine that an employee is stealing, and they can also be used to estimate the amount stolen. The following are some of the indicators that have been used. For comparable time periods, samples of sales have means that are significantly different. The mean sale amount decreases significantly. There is a significant increase in the proportion of "no sale" register openings. There is a significant decrease in the ratio of cash receipts to checks. Methods of hypothesis testing can be used to identify such indicators. (See "How To Catch a Thief" by Manly and Thomson, Chance, Vol. 11, No. 4.)

This section, Section 8-2, and Section 8-3 all deal with inferences about the means of two populations. Determining the correct procedure can be difficult because we must consider issues such as the independence of the samples, the sizes of the samples, and whether the two populations appear to have equal variances. Figure 8-7 is designed to simplify the decisions that lead to the correct procedure.

Using Technology

STATDISK: In choosing between Case 2 and Case 3, STATDISK automatically uses a preliminary F test. Begin by selecting the menu items **Analysis, Hypothesis Testing,** and **Mean-Two Independent Samples.** When relevant, the conclusion of the preliminary F test is included in the STATDISK display. Also, the STATDISK display will include a confidence interval.

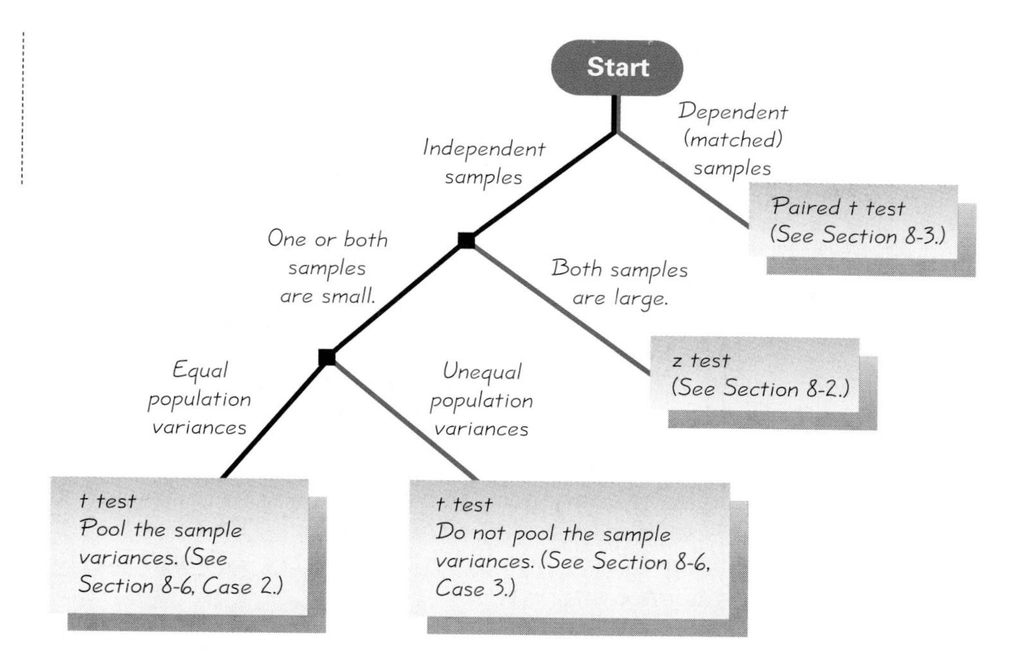

FIGURE 8-7
Inferences about the Means of Two Populations

Minitab: If the two population variances appear to be equal, Minitab does allow use of a pooled estimate of the common variance. After entering the sample data in columns C1 and C2, select the options **Stat, Basic Statistics,** and **2-Sample t,** then click on **Samples in different columns** and proceed to enter C1 for the first sample and C2 for the second sample. In the box identified as **alternative,** select the wording for the alternative hypothesis (*not equal* or *less than* or *greater than*), and enter the confidence level appropriate for the test (such as 0.95 for $\alpha = 0.05$). There will be a box next to **Assume equal variances,** and click on that box only if you want to assume that the two populations have equal variances (Case 2 of this section).The Minitab display also includes the confidence interval limits.

Excel: Enter the data for the two samples in columns A and B.

To use the Data Desk XL add-in, click on **DDXL.** Select **Hypothesis Tests** and **2 Var t Test** or select **Confidence Intervals** and **2 Var t Interval.** In the dialog box, click on the pencil icon for the first quantitative column and enter the range of values for the first sample, such as A1:A14. Click on the pencil icon for the second quantitative column and enter the range of values for the second sample. Click on **OK.** Now complete the new dialog box by following the indicated steps. In step 1, select **2-sample** for the assumption of unequal population variances, or select **Pooled** for the assumption of equal population variances.

To use Excel's Data Analysis add-in, click on **Tools** and select **Data Analysis.** Select one of the following two items:

t-test: Two-Sample Assuming Equal Variances
t-test: Two-Sample Assuming Unequal Variances

Proceed to enter the range for the values of the first sample (such as A1:A21) and then the range of values for the second sample. Enter a value for the claimed difference between the two population means, which will often be 0. Enter the significance level in the Alpha box and click on **OK.** (Excel does not provide a confidence interval.)

TI-83 Plus: The TI-83 Plus calculator does give you the option of using "pooled" variances (if you believe that $\sigma_1^2 = \sigma_2^2$) or not pooling the variances. To conduct tests of the type found in this section, press **STAT,** then select **TESTS** and choose **2-SampTTest** (for a hypothesis test) or **2-SampTInt** (for a confidence interval).

8-6 Basic Skills and Concepts

Testing Equality of Means. *In Exercises 1 and 2, test the given claim. Use a significance level of $\alpha = 0.05$, and assume that all populations are normally distributed. Use the traditional method of testing hypotheses outlined in Figure 7-5. Instead of using a preliminary F test to determine whether the two populations have equal variances, simply examine the standard deviations and make a judgment based on their obvious values.*

1. Claim: The treatment and placebo groups have the same mean ($\mu_1 = \mu_2$).

Treatment Group: $n_1 = 5, \bar{x}_1 = 98.20, s_1 = 0.62$

Placebo Group: $n_2 = 25, \bar{x}_2 = 104.60, s_2 = 0.61$

2. Claim: The treatment and placebo groups have the same mean ($\mu_1 = \mu_2$,).

Treatment Group: $n_1 = 20, \bar{x}_1 = 235.3, s_1 = 5.6$

Placebo Group: $n_2 = 10, \bar{x}_2 = 177.7, s_2 = 31.4$

3. Using the sample data in Exercise 1, construct a 95% confidence interval estimate of the difference between the means of the two populations.

4. Using the sample data in Exercise 2, construct a 95% confidence interval estimate of the difference between the means of the two populations.

5. Testing Equality of Heights of Men and Women Samples of men and women are randomly selected, and their heights are accurately measured with the results (in inches) listed below (based on data from the National Health Survey).

a. Assuming that the heights of men and women have *equal* variances, use a 0.01 significance level to test the claim that men and women have the same mean height.

b. Assuming that the heights of men and women have *unequal* variances, use a 0.01 significance level to test the claim that men and women have the same mean height.

c. Compare the conclusions to parts (a) and (b). Do both procedures lead to the same result?

d. Based on the given sample data, what do you conclude about the heights of men and women? Do the sample data suggest that men and women have the same mean height? Do the results agree with your knowledge of the heights of men and women?

Men:	73.0	71.6	68.7	65.9	70.4	$\bar{x} = 69.920, s = 2.747$
Women:	63.9	67.1	63.2	64.7		$\bar{x} = 64.725, s = 1.698$

6. Heights of Men and Women: Confidence Intervals Refer to the sample data used in Exercise 5.

a. Assuming that heights of men and women have *equal* variances, construct a 99% confidence interval estimate of the difference $\mu_1 - \mu_2$. Do the confidence interval limits contain 0, and what does this suggest about equality of the two population means?

b. Assuming that heights of men and a women have *unequal* variances, construct a 99% confidence interval estimate of the difference $\mu_1 - \mu_2$. Do the confidence interval limits contain 0, and what does this suggest about equality of the two population means?

c. Compare the results from parts (a) and (b). Does the assumption of equal variances or unequal variances have much of an effect on the confidence interval limits?

7. Brain Right Cordate and Psychiatric Disorders: Hypothesis Test Are severe psychiatric disorders related to biological factors that can be physically observed? One study used X-ray computed tomography (CT) to collect data on brain volumes for a group of patients with obsessive-compulsive disorders and a control group of healthy persons. Sample results for volumes (in mL) follow for the right cordate (based on data from "Neuroanatomical Abnormalities in Obsessive-Compulsive Disorder Detected with Quantitative X-Ray Computed Tomography," by Luxenberg et al., *American Journal of Psychiatry,* Vol. 145, No. 9). At the 0.01 significance level, test the claim that obsessive-compulsive patients and healthy persons have the same mean brain volume. Based on this result, does it seem that obsessive-compulsive disorders have a biological basis?

Obsessive-compulsive patients: $n = 10, \bar{x} = 0.34, s = 0.08$

Control group: $n = 10, \bar{x} = 0.45, s = 0.08$

8. Brain Right Cordate and Psychiatric Disorders: Confidence Interval Using the sample data from Exercise 7, construct a 99% confidence interval estimate of the difference between the mean brain volume for the disorder patient group and the mean brain volume for the healthy control group. What does the confidence interval suggest about the difference between the two population means?

9. Brain Volume and Psychiatric Disorders: Confidence Intervals The study cited in Exercise 7 resulted in the values summarized here for total brain volumes (in mL).

a. Using these sample statistics, construct a 95% confidence interval for the difference between the mean brain volume of obsessive-compulsive patients and the mean brain volume of healthy persons. Assume that the two populations have *equal* variances.

b. Construct a 95% confidence interval for the difference between the mean brain volume of obsessive-compulsive patients and the mean brain volume of healthy persons. Assume that the two populations have *unequal* variances.

(continued)

c. Compare the results from parts (a) and (b). Are they very different? Which results are better?

Obsessive-compulsive patients: $n = 10$, $\bar{x} = 1390.03$, $s = 156.84$

Control group: $n = 10$, $\bar{x} = 1268.41$, $s = 137.97$

10. Brain Volume and Psychiatric Disorders: Hypothesis Test Refer to the sample data in Exercise 9.

a. Assuming that the population variances are *equal*, use a 0.05 significance level to test the claim that there is no difference between the mean for obsessive-compulsive patients and the mean for healthy persons.

b. Assuming that the population variances are *unequal*, use a 0.05 significance level to test the claim that there is no difference between the mean for obsessive-compulsive patients and the mean for healthy persons.

c. Compare the results from parts (a) and (b). Which results are better? Based on the results, does it seem that obsessive-compulsive disorders have a biological basis?

11. Red/Orange M&Ms: Hypothesis Test Data Set 10 in Appendix B includes a sample of 21 red M&Ms with weights having a mean of 0.9097 g and a standard deviation of 0.0275 g. The data set also includes a sample of 8 orange M&Ms with weights having a mean of 0.9251 g and a standard deviation of 0.0472 g. Use a preliminary F test to determine whether the populations have equal variances, then use a 0.05 significance level to test for equality of the two population means. If you were responsible for controlling production so that red and orange M&Ms have weights with the same mean, would you take corrective action?

12. Testing the Anchoring Effect Randomly selected statistics students were given five seconds to estimate the value of a product of numbers with the results given in the accompanying table. Use a preliminary F test to determine whether the two populations have equal variances, then use a 0.05 significance level to test the claim that the two samples come from populations with the same mean. Does the order of the numbers appear to have an effect on the estimate? (See the Cooperative Group Activities at the end of Chapter 2.)

Estimates from Students Given $1 \times 2 \times 3 \times 4 \times 5 \times 6 \times 7 \times 8$					**Estimates from Students Given $8 \times 7 \times 6 \times 5 \times 4 \times 3 \times 2 \times 1$**				
1560	169	5635	25	842	100,000	2000	42,000	1500	52,836
40,320	5000	500	1110	10,000	2050	428	372	300	225
200	1252	4000	2040	175	500	1200	400	49,000	4000
856	42,200	49,654	560	800	1876	3600	354	750	640
					64,582	23,410			

13. Carbon Monoxide and Cigarettes Refer to the given data for the measured amounts of carbon monoxide (CO) from samples of filtered and nonfiltered king-size cigarettes. All measurements are in milligrams, and the data are from the Federal Trade Commission. Use a 0.05 significance level to test the claim that the mean amount of carbon monoxide in filtered king-size cigarettes is equal to the mean amount of carbon monoxide for nonfiltered king-size cigarettes. Based on this result and the examples of this section, are cigarette filters effective, or are they just a sales gimmick with no real effect?

Filtered:	14 12 14 16 15 2 14 16 11 13 13 12 13 12 13 14 14 14 9 17 12
Nonfiltered:	14 15 17 17 16 16 14 16

14. Effects of Alcohol An experiment was conducted to test the effects of alcohol. The errors were recorded in a test of visual and motor skills for a treatment group of people who drank ethanol and another group given a placebo. The results are shown in the accompanying table. Use a 0.05 significance level to test the claim that the two groups come from populations with the same mean. Do these results support the common belief that drinking is hazardous for drivers, pilots, ship captains, and so on?

Treatment Group	Placebo Group
$n_1 = 22$	$n_2 = 22$
$\bar{x}_1 = 4.20$	$\bar{x}_2 = 1.71$
$s_1 = 2.20$	$s_2 = 0.72$

Based on data from "Effects of Alcohol Intoxication on Risk Taking, Strategy, and Error Rate in Visuomotor Performance," by Streufert et al., *Journal of Applied Psychology,* Vol. 77, No. 4.

15. Peanut and Plain M&Ms A simple random sample of 21 peanut M&M candies is selected, and each candy is weighed. The mean is 2.4658 g and the standard deviation is 0.3127 g. The 100 M&M plain candies listed in Data Set 10 in Appendix B have a mean of 0.9147 g and a standard deviation of 0.0369 g. Is there sufficient evidence to support the claim that there is a difference between the mean weight of peanut M&Ms and the mean weight of plain M&Ms?

16. Blanking Out on Tests Many students have had the unpleasant experience of panicking on a test because the first question was exceptionally difficult. The arrangement of test items was studied for its effect on anxiety. The following scores are measures of "debilitating test anxiety" (which most of us call panic or blanking out). Is there sufficient evidence to support the claim that the two populations of scores have the same mean? Is there sufficient evidence to support the claim that the arrangement of the test items has an effect on the score?

Questions Arranged from Easy to Difficult					**Questions Arranged from Difficult to Easy**			
24.64	39.29	16.32	32.83	28.02	33.62	34.02	26.63	30.26
33.31	20.60	21.13	26.69	28.90	35.91	26.68	29.49	35.32
26.43	24.23	7.10	32.86	21.06	27.24	32.34	29.34	33.53
28.89	28.71	31.73	30.02	21.96	27.62	42.91	30.20	32.54
25.49	38.81	27.85	30.29	30.72				

Based on data from "Item Arrangement, Cognitive Entry Characteristics, Sex and Test Anxiety as Predictors of Achievement in Examination Performance," by Klimko, *Journal of Experimental Education,* Vol. 52, No. 4.

8-6 Beyond the Basics

17. Effect of No Variation in Sample An experiment was conducted to test the effects of alcohol. The breath alcohol levels were measured for a treatment group of people who drank ethanol and another group given a placebo. The results are given in the accompanying table. Use a 0.05 significance level to test the claim that the two groups come from populations with the same mean.

Treatment Group	Placebo Group
$n_1 = 22$	$n_2 = 22$
$\bar{x}_1 = 0.049$	$\bar{x}_2 = 0.000$
$s_1 = 0.015$	$s_2 = 0.000$

Based on data from "Effects of Alcohol Intoxication on Risk Taking, Strategy, and Error Rate in Visuomotor Performance," by Streufert et al., *Journal of Applied Psychology,* Vol. 77, No. 4.

18. Equal Sample Standard Deviations Assume that two samples have the same standard deviation and that both are independent, small, and randomly selected from normally distributed populations. Also assume that we want to test the claim that the samples come from populations with the same mean.

a. Is it necessary to conduct a preliminary F test?

b. If both samples have standard deviation s, what is the value of s_p^2 expressed in terms of s?

19. a. Refer to the two examples of this section that used the tar amounts in cigarettes. (See the Case 3 subsection.) How are the hypothesis test and confidence interval affected if the number of degrees of freedom is calculated with Formula 8-1, instead of by using the smaller of $n_1 - 1$ and $n_2 - 1$?

b. Refer to the two examples in this section that used the tar amounts in cigarettes. (See the Case 3 subsection.) How are the hypothesis test and confidence interval affected if the one extreme outlier is deleted?

20. Consequences of Wrong Test Refer to Table 8-1 at the beginning of this chapter for the reported/measured heights of male statistics students. Delete the outlier values for the third student, as in Section 8-3. Because the reported/measured heights are matched pairs, we used the paired t test described in Section 8-3, where we obtained a test statistic of $t = 2.701$ when we tested the claim that reported heights are greater than measured heights. How are the results affected if we treat the sample data as two *independent* sets of data, instead of *matched pairs*? In general, if paired data are incorrectly treated as being independent, what is the effect on the results?

The Fear of Flying

POUGHKEEPSIE The lives of many people are affected by a fear that prevents them from flying. Sports announcer John Madden gained notoriety as he crossed the country by rail or motorhome, traveling from one football stadium to another. The Marist Institute for Public Opinion conducted a poll of 1014 adults, 48% of whom were men. The results are depicted in the accompanying illustration from *USA Today*. The poll results show that 12% of the men and 33% of the women fear flying.

Just plane scared

While 47% of adults believe flying is the safest way to travel (vs. 39% who say cars and 14% for trains), about 41 million are afraid. Percent of Americans who fear flying:

All adults 22%

Men 12%

Women 33%

Source: Marist Institute for Public Opinion By Scott Boeck and Genevieve Lynn, USA TODAY

USA Today Snapshot. "A look at statistics that shape the nation."

1. Is there sufficient evidence to conclude that there is a significant difference between the percentage of men and the percentage of women who fear flying?
2. Construct a 95% confidence interval estimate of the difference between the percentage of men and the percentage of women who fear flying. Do the confidence interval limits contain 0, and what is the significance of whether they do or do not?
3. Construct a 95% confidence interval for the percentage of men who fear flying.
4. Based on the result from Exercise 3, complete the following statement, which is typical of the statement that would be reported in a newspaper or magazine: "Based on the Marist Institute for Public Opinion poll, the percentage of men who fear flying is 22% with a margin of error of _____."
5. Examine the completed statement in Exercise 4. What important piece of information should be included, but is not included?

VOCABULARY LIST

independent samples
dependent samples
matched pairs
paired samples
pooled estimate of p_1 and p_2
F distribution
numerator degrees of freedom
denominator degrees of freedom
pooled estimate of σ^2

REVIEW

In Chapters 6 and 7 we introduced two major concepts of inferential statistics: the estimation of population parameters and the methods of testing hypotheses made about population parameters. Whereas those chapters considered only cases involving a single population, this chapter considered two samples drawn from two populations.

- Section 8-2 considered inferences made about the means of two independent populations, but the involved samples were both large ($n > 30$).
- Section 8-3 considered inferences made about the means of population data consisting of matched pairs.
- Section 8-4 discussed the procedures for making inferences about two population proportions.
- Section 8-5 presented methods for testing claims about the means of two population standard deviations or variances.
- Section 8-6 described procedures for making inferences about the means of two independent populations, with at least one of the samples being small ($n \leq 30$). Sections 8-2, 8-3, and 8-6 dealt with a variety of different cases for making inferences about two population means; see Figure 8-7 for help in selecting the correct procedure.

REVIEW EXERCISES

1. Testing Adverse Drug Reactions In clinical tests of adverse reactions to the drug Viagra, 7% of the 734 subjects in the treatment group experienced dyspepsia (indigestion), but 2% of the 725 subjects in the placebo group experienced dyspepsia (based on data from Pfizer Pharmaceuticals).

a. Is there sufficient evidence to support the claim that dyspepsia occurs at a higher rate among Viagra users than those who do not use Viagra?

b. Construct a 95% confidence interval estimate of the difference between the dyspepsia rate for Viagra users and the dyspepsia rate for those who use a placebo. Do the confidence interval limits contain 0, and what does this suggest about the two dyspepsia rates?

2. Testing Effectiveness of Training Program The effectiveness of a mental training program was tested in the military. In an antiaircraft artillery examination, scores for an experimental group and a control group were recorded.

(continued)

a. Use the given data to test the claim that both groups come from populations with the same variance. Use a 0.05 significance level.

b. Use the given data to test the claim that both groups come from populations with the same mean. (Based on the results from part a, assume that the two populations have the same variance.)

c. Use the given data to construct a 95% confidence interval estimate of the difference between the two population means.

Experimental Group				**Control Group**			
60.83	117.80	44.71	75.38	122.80	70.02	119.89	138.27
73.46	34.26	82.25	59.77	118.43	54.22	118.58	74.61
69.95	21.37	59.78	92.72	121.70	70.70	99.08	120.76
72.14	57.29	64.05	44.09	104.06	94.23	111.26	121.67
80.03	76.59	74.27	66.87				

Based on "Routinization of Mental Training in Organizations: Effects on Performance and Well-Being," by Larsson, *Journal of Applied Psychology,* Vol. 72, No. 1.

3. People Helping People In a study of people who stop to help drivers with disabled cars, researchers hypothesized that more people would stop to help someone if they first saw another driver with a disabled car getting help. In one experiment, 2000 drivers first saw a woman being helped with a flat tire and then saw a second woman who was alone, farther down the road, with a flat tire; 2.90% of those 2000 drivers stopped to help the second woman. Among 2000 other drivers who did not see the first woman being helped, only 1.75% stopped to help (based on data from "Help on the Highway," by McCarthy, *Psychology Today*). At the 0.05 significance level, test the claim that the percentage of people who stop after first seeing a driver with a disabled car being helped is greater than the percentage of people who stop without first seeing someone else being helped.

4. Couch Patrol: Confidence Interval Nielsen Media Research reported that adult women watch TV an average of 5 hr and 1 min per day, compared to an average of 4 hr and 17 min for adult men. Assume that those results are found from a sample of 100 men and 100 women and that the two groups have the same standard deviation of 57 min. Construct a 99% confidence interval for the difference $\mu_1 - \mu_2$, where μ_1 is the mean for adult women. Do the confidence interval limits contain 0? Does this suggest that there is or is not a significant difference between the two means?

5. Testing Effects of Physical Training A study was conducted to investigate some effects of physical training. Sample data are listed below, with all weights given in kilograms. (See "Effect of Endurance Training on Possible Determinants of VO_2 During Heavy Exercise," by Casaburi et al., *Journal of Applied Physiology,* Vol. 62, No. 1.) Is there sufficient evidence to conclude that there is a difference between the pretraining and posttraining weights? What do you conclude about the effect of training on weight?

Pretraining:	99	57	62	69	74	77	59	92	70	85
Posttraining:	94	57	62	69	66	76	58	88	70	84

6. Physical Training: Confidence Interval Use the data from Exercise 5 to construct a 95% confidence interval for the mean of the differences between pretraining and posttraining weights.

7. Power of Education A study of the economic effects of education included annual incomes of women with a high school diploma and women with a college degree. The sample statistics are shown for randomly selected women from each group (based on data from the U.S. Census Bureau).
   a. Using a 0.01 significance level, is there enough evidence to support the claim that women with a college degree have incomes with a higher mean?
   b. Construct a 99% confidence interval estimate of the difference between the mean income of women high school graduates and women college graduates. Do the confidence interval limits contain 0, which would suggest that there is not a significant difference between the two means?

High School	College
$n_1 = 85$	$n_2 = 120$
$\bar{x}_1 = \$26,588$	$\bar{x}_2 = \$44,765$
$s_1 = \$8441$	$s_2 = \$12,469$

8. Customer Waiting Times Customer waiting times are studied at the Jefferson Valley Bank. When 25 randomly selected customers enter any one of several waiting lines, their times have a mean of 6.896 min and a standard deviation of 3.619 min. When 20 randomly selected customers enter a single main waiting line that feeds the individual teller stations, their waiting times have a mean of 7.460 min and a standard deviation of 1.841 min.
   a. Use a 0.01 significance level to test the claim that waiting times for the single line have a lower standard deviation.
   b. Use a 0.01 significance level to test the claim that the two populations have the same mean.

9. Drug Solubility Twelve Dozenol tablets are tested for solubility before and after being stored for one year. The indexes of solubility are given in the table below.

Before	472	487	506	512	489	503	511	501	495	504	494	462
After	562	512	523	528	554	513	516	510	524	510	524	508

   a. At the 0.05 significance level, test the claim that the Dozenol tablets are more soluble after the storage period.
   b. Construct a 95% confidence interval estimate of the mean difference (after − before).

10. Yellow and Green M&Ms Refer to Data Set 10 in Appendix B and test the claim that yellow M&Ms and green M&Ms have the same mean weight. Use a preliminary *F* test to determine whether the two populations have equal variances. If you were part of a quality-control team with responsibility for ensuring that yellow M&Ms and green M&Ms have the same mean weight, would you take corrective action?

Cumulative Review Exercises

1. Speeding Tickets for Men and Women The data in the accompanying table were obtained through a survey of randomly selected subjects.
   a. If one of the survey subjects is randomly selected, find the probability of getting someone ticketed for speeding.
   b. If one of the survey subjects is randomly selected, find the probability of getting a man or someone ticketed for speeding.
   c. Find the probability of getting someone ticketed for speeding, given that the selected person is a man.
   d. Find the probability of getting someone ticketed for speeding, given that the selected person is a woman.
   e. Use a 0.05 significance level to test the claim that the percentage of women ticketed for speeding is less than the percentage of men. Can we conclude that men generally speed more than women?

	Ticketed for Speeding Within the Last Year?	
	Yes	No
Men	26	224
Women	27	473

Based on data from R.H. Bruskin Associates.

2. Cell Phones and Crashes: Analizing Newspaper Report In an article from the Associated Press, it was reported that researchers "randomly selected 100 New York motorists who had been in an accident and 100 who had not. Of those in accidents, 13.7 percent owned a cellular phone, while just 10.6 percent of the accident-free drivers had a phone in the car." Analyze these results.

3. Good Test Question A test question is considered good if it discriminates between prepared and unprepared students. The first question on a test was answered correctly by 62 of 80 prepared students and by 23 of 50 unprepared students.
   a. At the 0.05 significance level, test the claim that the majority of the prepared students answer this question correctly.
   b. At the 0.05 significance level, test the claim that the proportion of prepared students who answer this question correctly is greater than the proportion of unprepared students who answer it correctly.

4. Gender and GRE Test Scores A study was conducted to investigate relationships between different types of standard test scores. On the Graduate Record Examination verbal test, 68 women had a mean of 538.82 and a standard deviation of 114.16, and 86 men had a mean of 525.23 and a standard deviation of 97.23. (See "Equivalencing MAT and GRE Scores Using Simple Linear Transformation and Regression Methods," by Kagan and Stock, *Journal of Experimental Education,* Vol. 49, No. 1.)
   a. At the 0.02 significance level, test the claim that the two groups come from populations with the same standard deviation.
   b. At the 0.02 significance level, test the claim that the two groups come from populations with the same mean.

(*continued*)

c. Construct a 98% confidence interval estimate of the mean score for all women who take the test.

Cooperative Group Activities

1. *Out-of-class activity:* Are estimates influenced by anchoring numbers? Refer to the related Chapter 2 Cooperative Group Activity. In Chapter 2 we noted that, according to author John Rubin, when people must estimate a value, their estimate is often "anchored" to (or influenced by) a preceding number. In that Chapter 2 activity, some subjects were asked to quickly estimate the value of $8 \times 7 \times 6 \times 5 \times 4 \times 3 \times 2 \times 1$, and others were asked to quickly estimate the value of $1 \times 2 \times 3 \times 4 \times 5 \times 6 \times 7 \times 8$. In Chapter 2, we could compare the two sets of results by using statistics (such as the mean) and graphs (such as boxplots). The methods of Chapter 8 now allow us to compare the results with a formal hypothesis test. (See Exercise 12 in Section 8-6.) Specifically, collect your own sample data and test the claim that when we begin with larger numbers (as in $8 \times 7 \times 6$), our estimates tend to be larger.

2. *In-class activity:* Collect your own data to test the claim that male and female statistics students have the same mean pulse rate. Divide into groups according to gender, with about 10 or 12 students in each group. Each group member should record his or her pulse rate by counting the number of heartbeats in one minute, and the group statistics (n, $\bar{x}$, s) should be calculated. The groups should exchange their results and test the null hypothesis of no difference between the two population means. Is there a difference? Explain.

3. *In-class activity:* Divide into groups of about 10 or 12 students and use the reaction timer included with the Chapter 5 Cooperative Group Activities. Each group member should be tested for right-hand reaction time and left-hand reaction time. Using the group results, test the claim that there is no difference between the right-hand and left-hand reaction times. Compare the conclusion to the conclusions reached by other groups. Is there a difference?

4. *In-class activity:* Divide into groups of three or four to test the claim that the Dvorak keyboard is easier to use than the Qwerty keyboard. See the Chapter Problem for Chapter 2, and also see the exercise in this chapter in the feature titled "From Data to Decision: Critical Thinking" (page 500). Instead of using the given data sets, collect your own data. Determine whether it is better to select a sample of text that will be used to rate the words by both methods, or whether it is better to select two separate samples of text so that the Qwerty rating scheme can be applied to one while the Dvorak rating scheme can be applied to the other. Collect your own data, then use the appropriate method to test the hypothesis that the Dvorak keyboard configuration is better.

Technology Project

STATDISK, Minitab, Excel, and the TI-83 Plus calculator are all capable of generating normally distributed data drawn from a population with a specified mean and standard deviation. Generate two sets of sample data that represent simulated IQ scores, as shown below.

IQ Scores of Treatment Group:	Generate 10 sample values from a normally distributed population with mean 100 and standard deviation 15.
IQ Scores of Placebo Group:	Generate 12 sample values from a normally distributed population with mean 100 and standard deviation 15.

STATDISK: Select **Data,** then select **Normal Generator.**

Minitab: Select **Calc, Random Data, Normal.**

Excel: Select **Tools, Data Analysis, Random Number Generator,** and be sure to select **Normal** for the distribution.

TI-83 Plus: Press **MATH,** select **PRB,** then select **6:randNorm(** and proceed to enter the mean, the standard deviation, and the number of scores (such as 100, 15, 10).

You can see from the way the data are generated that both data sets really come from the same population, so there should be no difference between the two sample means.

a. After generating the two data sets, use a 0.10 significance level to test the claim that the two samples come from populations with the same mean.

b. If this experiment is repeated, what percentage of trials will lead to the conclusion that the two population means are different? How does this relate to a type I error?

c. If your generated data should lead to the conclusion that the two population means are different, would this conclusion be correct or incorrect in reality? How do you know?

d. Repeat the experiment until you get two samples that cause you to reject the null hypothesis of equal means. Based on that number, how would you describe the likelihood of rejecting equality of the population means?

from DATA to DECISION

Critical Thinking: Is there a real difference between typing on a Qwerty keyboard and typing on a Dvorak keyboard?

Is there a real difference in ease of typing between using the Qwerty keyboard configuration and the Dvorak keyboard configuration? (See the Chapter Problem for Chapter 2 for illustrations of the actual keyboard configurations.) The Chapter 2 introductory problem presented two data sets obtained by using the 52 words in the Preamble to the Constitution. The first data set consisted of the point rating score for each word typed on the Qwerty keyboard, and the second data set listed the word ratings for the Dvorak keyboard. Using the methods of Chapter 2, we can compare the data sets using tools such as the sample statistics $(n, \bar{x}, s)$ and boxplots. The sample statistics are summarized in the accompanying table. Comparing the sample means in that table, we see that the Qwerty mean of 4.4 appears to be considerably higher than the Dvorak mean of 1.7, suggesting that the Qwerty keyboard is more difficult to use (because higher word ratings result from words that are more difficult to type). But is this apparent difference really *significant*?

Analyzing the results

Refer to the original data listed in Tables 2-1 and 2-2 at the beginning of Chapter 2, and conduct a hypothesis test to determine whether the Qwerty population mean is different from the Dvorak population mean. Identify the method you use and be sure to justify the requirements that must be satisfied for that method. Write a brief report summarizing your procedure and conclusions.

Qwerty and Dvorak Sample Statistics

	Qwerty	Dvorak
n	52	52
$\bar{x}$	4.4	1.7
s	2.8	1.8

TABLE 2-1 Qwerty Keyboard Word Ratings

2	2	5	1	2	6	3	3	4	2
4	0	5	7	7	5	6	6	8	10
7	2	2	10	5	8	2	5	4	2
6	2	6	1	7	2	7	2	3	8
1	5	2	5	2	14	2	2	6	3
1	7								

TABLE 2-2 Dvorak Keyboard Word Ratings

2	0	3	1	0	0	0	0	2	0
4	0	3	4	0	3	3	1	3	5
4	2	0	5	1	4	0	3	5	0
2	0	4	1	5	0	4	0	1	3
0	1	0	3	0	1	2	0	0	0
1	4								

Internet Project

Hypothesis Testing

Whereas Chapter 7 presented methods for testing hypotheses about a single population, the methods of this chapter extend those methods to two populations. Similarly, the Internet Project for this chapter difers from that of Chapter 7 through the use of two data sets in each of the hypothesis tests. The project can be found at

http://awlonline.com/triola

There you will find several hypothesis-testing problems that examine population demographics, salary fairness, and a traditional superstition. In each case you will formulate the problem as a hypothesis test using the notation of the chapter, seek out appropriate data, conduct the test, and then interpret the results with a summarizing conclusion.

Statistics at work

"The basic role of statistics here at Consumers Union is to ensure that the studies we do are valid, reliable, and understandable."

David Burn

Director, Technical Information Management for Consumers Union, which publishes Consumer Reports.

When they're not designing experiments or analyzing the results, statisticians (and other Consumers Union staff) have the opportunity to participate in product tests. As an avid runner and triathlete, David often volunteers for tests of fitness equipment. In this case, his assessment of various treadmills is data used to rate their quality.

Author's note: The author met with David Burn and the other statisticians at Consumers Union: M. Edna Derderian, Keith Newsom-Stewart, Martin Romm, and Jed Schneider. The author toured the testing facility and saw that Consumers Union meticulously designs, executes, and analyzes objective experiments.

What is the role of statistics in your product evaluation?

The basic role of statistics here at Consumers Union is to ensure that the studies we do are valid, reliable, and understandable. We look for significance in results in the sense that if we conclude that product A is better than product B, that conclusion is based on good, solid statistical theory. But we always remind ourselves that statistical significance does not outweigh practical significance. If there is a statistical significance of 0.01 in the difference between two speakers, but there is a practical significance of 0.1 in the way the human ear hears those speakers, then the statistical significance is irrelevant. That's the way the article would be written. We might say that the tested speakers are of such a high quality that there is really no difference, so you could buy based on price.

Who works on a typical evaluation project?

We work with different departments having people from a variety of backgrounds. Typically, a project leader is from the department where the product belongs. An electronics engineer might be the project leader for a study of CD players. Other team members typically include a statistician, a market information analyst, a writer, and an editor.

What does the statistician do?

For the statistician, the work really begins when the market survey report comes in and we can see what consumers want to know about. The decision on what to test is made by the project team with input from all members, including the statistician. We design an experiment and method of sampling. For a CD player, we might want to test sound quality and resistance to skipping when the unit is jarred. We work with the project leader on the test protocol, sampling, and other experimental design issues. The statistician begins analyzing the test data by using general exploratory methods with graphs and numerical measurements. We explore the data because we don't necessarily know what we are going to get. At this stage we are using tools like scatterplots, boxplots, and histograms to look for features such as outliers. Then we do analyses that might make some products stand out or appear to be different from the others. Ultimately, we use statistical methods to develop overall scores we call ratings. A statistical report is written and checked by one of the other statisticians. The type of analysis is questioned to be sure that it is the best choice, and we check that everything was done correctly and that the results are valid.

How do you ensure objectivity?

We don't accept any type of outside advertising. We don't represent any organizations or companies: We represent the consumer. We buy all of our products on the open market, just the way any consumer would. We have anonymous shoppers throughout the United States in all major areas, and the statistical design for the sampling plan might require shopping in San Francisco, Florida, New York, and Texas—all at the same time for the same product. That way we get a representative sample of the product that the manufacturer is making.

Do you use computers in your work?

We all use state-of-the-art computers, and a variety of statistical and other software packages.

Would you recommend a statistics course to college students?

I would recommend several statistics courses. Certainly, elementary statistics courses are important for an overview of parametric and nonparametric methods. In addition, linear models, sampling, and design of experiment provide the depth and appreciation needed for critical review of many issues involving statistical inquiries. Beyond this, the ability to communicate your ideas is essential. Because we work on several projects simultaneously, personal organization and project management are critical.

CHAPTER 9

Correlation and Regression

CHAPTER PROBLEM

How much of a tip should I leave?

All of us are confronted with the problem of determining how much of a tip we should leave a waiter or waitress in a restaurant. Although unofficial, many of us have heard that the tip should be 15% of the bill. Some of us try to walk a fine line between giving away too much money and looking too cheap. Others take genuine pleasure in rewarding service with a very generous tip, and there are those who have a firm policy of giving either nothing or as little as possible. Almost all of us believe that there is a relationship between the amount of the bill and the amount of the tip: Larger bills result in larger tips. Let's consider Table 9-1, which lists some sample data collected from the author's students. (Much better results could be obtained by using a sample that is larger than the six pairs of data included in Table 9-1, but we are using a sample of six pairs so that it will be easy to follow the calculations in the sections that follow.)

In Table 9-1, there appears to be a pattern of larger tips for larger bills. We will analyze the data in Table 9-1 as we address two questions:

1. Is there sufficient evidence to conclude that there is a relationship between the amount of the bill and the amount of the tip?
2. If there is a relationship, how do we use it to determine how much of a tip should be left?

TABLE 9-1 Paired Data for Six Dining Parties

Bill ($)	33.46	50.68	87.92	98.84	63.60	107.34
Tip ($)	5.50	5.00	8.08	17.00	12.00	16.00

When we examine such a scatterplot, we should study the overall pattern of the plotted points. If there is a pattern, we should note its direction. That is, as one variable increases, does the other seem to increase or decrease? We should observe whether there are any outliers, which are points that lie very far away from all of the other points. The Excel-generated scatterplot does seem to reveal a pattern showing that larger bills tend to go along with larger tips. There do not appear to be any outliers.

Other examples of scatterplots are shown in Figure 9-1. The graphs in Figure 9-1(a), (b), and (c) depict a pattern of increasing values of y that correspond to increasing values of x. As you proceed from (a) to (c), the

FIGURE 9-1
Scatterplots

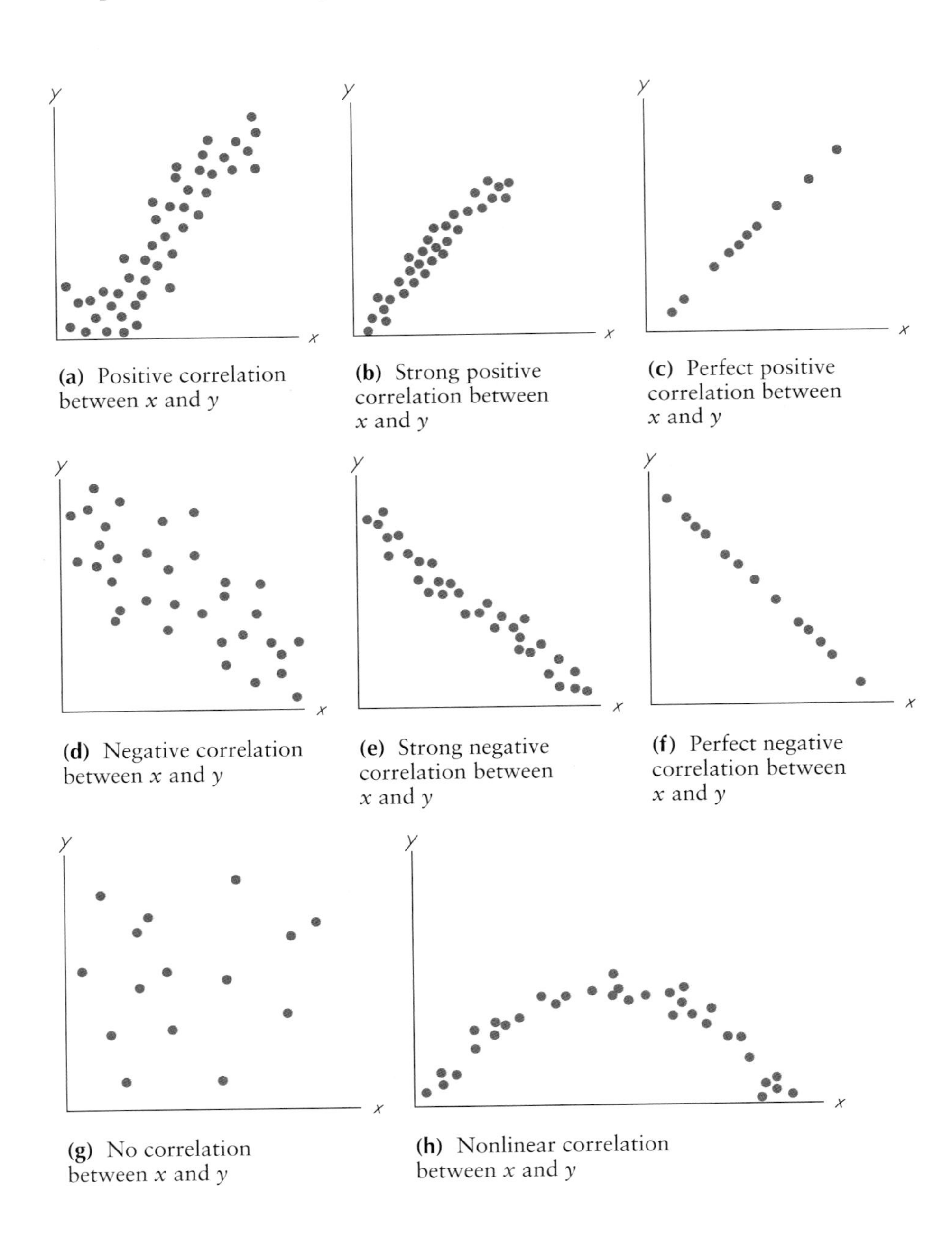

(a) Positive correlation between x and y

(b) Strong positive correlation between x and y

(c) Perfect positive correlation between x and y

(d) Negative correlation between x and y

(e) Strong negative correlation between x and y

(f) Perfect negative correlation between x and y

(g) No correlation between x and y

(h) Nonlinear correlation between x and y

dot pattern becomes closer to a straight line, suggesting that the relationship between x and y becomes stronger. The scatterplots in (d), (e), and (f) depict patterns in which the y-values decrease as the x-values increase. Again, as you proceed from (d) to (f), the relationship becomes stronger. In contrast to the first six graphs, the scatterplot of (g) shows no pattern and suggests that there is no correlation (or relationship) between x and y. Finally, the scatterplot of (h) shows a pattern, but it is not a straight-line pattern.

Linear Correlation Coefficient

Because visual examinations of scatterplots are largely subjective, we need more precise and objective measures. We will use the linear correlation coefficient r, which is useful for detecting straight-line patterns.

DEFINITION

The **linear correlation coefficient** r measures the strength of the linear relationship between the paired x- and y-values in a *sample.* Its value is computed by using Formula 9-1, which follows. [The linear correlation coefficient is sometimes referred to as the **Pearson product moment correlation coefficient** in honor of Karl Pearson (1857–1936), who originally developed it.]

Formula 9-1

$$r = \frac{n\Sigma xy - (\Sigma x)(\Sigma y)}{\sqrt{n(\Sigma x^2) - (\Sigma x)^2}\sqrt{n(\Sigma y^2) - (\Sigma y)^2}}$$

Because r is calculated using sample data, it is a sample statistic used to measure the strength of the linear correlation between x and y. If we had every pair of population values for x and y, the result of Formula 9-1 would be a population parameter, represented by ρ (Greek rho).

We will describe how to compute and interpret the linear correlation coefficient r given a list of paired data, but let's first identify the notation relevant to Formula 9-1. Later in this section we will present the underlying theory that led to the development of Formula 9-1.

Notation for the Linear Correlation Coefficient

n	represents the number of pairs of data present.
Σ	denotes the addition of the items indicated.
Σx	denotes the sum of all x-values.
Σx^2	indicates that each x-value should be squared and then those squares added.
$(\Sigma x)^2$	indicates that the x-values should be added and the total then squared. It is extremely important to avoid confusing Σx^2 and $(\Sigma x)^2$.

(continued)

Σxy	indicates that each x-value should first be multiplied by its corresponding y-value. After obtaining all such products, find their sum.
r	represents the linear correlation coefficient for a *sample.*
ρ	represents the linear correlation coefficient for a *population.*

Rounding the Linear Correlation Coefficient

Round the linear correlation coefficient r to three decimal places (so that its value can be directly compared to critical values in Table A-6). When calculating r and other statistics in this chapter, rounding in the middle of a calculation often creates substantial errors, so try using your calculator's memory to store intermediate results and round off only at the end. Many inexpensive calculators have Formula 9-1 built in so that you can automatically evaluate r after entering the sample data.

EXAMPLE **Tipping** Using the data in Table 9-1, find the value of the linear correlation coefficient r. (A later example will use this value to determine whether there is a relationship between the amounts of bills and the amounts of tips.)

SOLUTION For the sample paired data in Table 9-1, $n = 6$ because there are six pairs of data. The other components required in Formula 9-1 are found from the calculations in Table 9-2. Note how this vertical format makes the calculations easier.

TABLE 9-2 Finding Statistics Used to Calculate r

	Bill ($)	Tip ($)			
	x	y	$x \cdot y$	x^2	y^2
	33.46	5.50	184.0300	1119.5716	30.2500
	50.68	5.00	253.4000	2568.4624	25.0000
	87.92	8.08	710.3936	7729.9264	65.2864
	98.84	17.00	1680.2800	9769.3456	289.0000
	63.60	12.00	763.2000	4044.9600	144.0000
	107.34	16.00	1717.4400	11,521.8756	256.0000
Total	441.84	63.58	5308.7436	36,754.1416	809.5364
	↑	↑	↑	↑	↑
	Σx	Σy	Σxy	Σx^2	Σy^2

Using the calculated values and Formula 9-1, we can now evaluate r as follows:

$$r = \frac{n\Sigma xy - (\Sigma x)(\Sigma y)}{\sqrt{n(\Sigma x^2) - (\Sigma x)^2}\sqrt{n(\Sigma y^2) - (\Sigma y)^2}}$$

$$= \frac{6(5308.7436) - (441.84)(63.58)}{\sqrt{6(36{,}754.1416) - (441.84)^2}\sqrt{6(809.5364) - (63.58)^2}}$$

$$= \frac{3760.2744}{\sqrt{25{,}302.264}\sqrt{814.802}} = 0.828$$

These calculations get quite messy with large data sets, so it's fortunate that the linear correlation coefficient can be found automatically with many different calculators and computer programs. See "Using Technology" at the end of this section for comments about STATDISK, Minitab, Excel, and the TI-83 Plus calculator.

Student Ratings of Teachers

Many colleges equate high student ratings with good teaching—an equation often fostered by the fact that student evaluations are easy to administer and measure.

However, one study that compared student evaluations of teachers with the amount of material learned found a strong negative correlation between the two factors. Teachers rated highly by students seemed to induce less learning.

In a related study, an audience gave a high rating to a lecturer who conveyed very little information but was interesting and entertaining.

Interpreting the Linear Correlation Coefficient

We need to interpret a calculated value of r, such as the value of 0.828 found in the preceding example. Given the way that Formula 9-1 is constructed, the value of r must always fall between -1 and $+1$ inclusive. If r is close to 0, we conclude that there is no significant linear correlation between x and y, but if r is close to -1 or $+1$, we conclude that there is a significant linear correlation between x and y. Interpretations of "close to" 0 or 1 or -1 are vague, so we use the following very specific decision criterion:

If the absolute value of the computed value of r exceeds the value in Table A-6, conclude that there is a significant linear correlation. Otherwise, there is not sufficient evidence to support the conclusion of a significant linear correlation.

When there really is no linear correlation between x and y, Table A-6 lists values that are "critical" in this sense: They separate *usual* values of r from those that are *unusual.* For example, Table A-6 shows us that with $n = 6$ pairs of sample data, the critical values are 0.811 (for $\alpha = 0.05$) and 0.917 (for $\alpha = 0.01$). Critical values and the role of α are carefully described in Chapters 6 and 7. Here's how we interpret those numbers: With six pairs of data and no linear correlation between x and y, there is a 5% chance that the absolute value of the computed linear correlation coefficient r will exceed 0.811. With $n = 6$ and no linear correlation, there is a 1% chance that $|r|$ will exceed 0.917.

EXAMPLE **Tipping** Given the sample data in Table 9-1, for which $r = 0.828$, refer to Table A-6 to determine whether there is a significant linear correlation between the amounts of bills and the amounts of tips. In Table A-6, use the critical value for $\alpha = 0.05$. (With $\alpha = 0.05$, we will

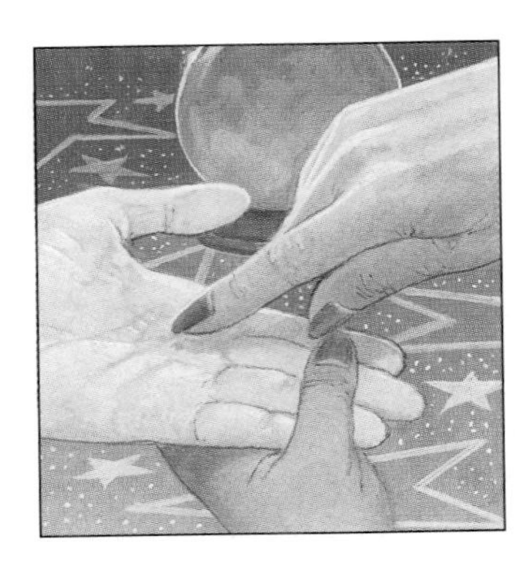

conclude that there is a significant linear correlation only if the sample is unlikely in the sense that such a value of r occurs 5% of the time or less.)

SOLUTION Referring to Table A-6, we locate the row for which $n = 6$ (because there are six pairs of data). That row contains the critical values of 0.811 (for $\alpha = 0.05$) and 0.917 (for $\alpha = 0.01$). Using the critical value for $\alpha = 0.05$, we see that there is less than a 5% chance that with no linear correlation, the absolute value of the computed r will exceed 0.811. Because $r = 0.828$, its absolute value does exceed 0.811, so we conclude that there is a significant linear correlation between bills and tips.

Palm Reading

Some people believe that the length of their palm's lifeline can be used to predict longevity. In a letter published in the Journal of the American Medical Association, authors M. E. Wilson and L. E. Mather refuted that belief with a study of cadavers. Ages at death were recorded, along with the lengths of palm lifelines. The authors concluded that there is no significant correlation between age at death and length of lifeline. Palmistry lost, hands down.

We have already noted that the format of Formula 9-1 requires that the calculated value of r always fall between -1 and $+1$ inclusive. We list that property along with other important properties.

Properties of the Linear Correlation Coefficient r

1. The value of r is always between -1 and 1 inclusive. That is,

$$-1 \leq r \leq 1$$

2. *The value of r does not change if all values of either variable are converted to a different scale.* For example, if the U.S. dollar amounts in Table 9-1 are converted to French francs, the value of r will not change.
3. *The value of r is not affected by the choice of x or y.* Interchange all x- and y-values and the value of r will not change.
4. *r measures the strength of a linear relationship.* It is not designed to measure the strength of a relationship that is not linear.

Interpreting r: Explained Variation

If we conclude that there is a significant linear correlation between x and y, we can find a linear equation that expresses y in terms of x, and that equation can be used to predict values of y for given values of x. In Section 9-3 we will describe a procedure for finding such equations and show how to predict values of y when given values of x. But a predicted value of y will not necessarily be the exact result, because in addition to x, there are other factors affecting y, such as random variation and other characteristics not included in the study. In Section 9-4 we will present a rationale and more details about this important principle:

The value of r^2 is the proportion of the variation in y that is explained by the linear relationship between x and y.

EXAMPLE **Tipping** Using the bill/tip data in Table 9-1, we have found that the linear correlation coefficient is $r = 0.828$. What proportion of the variation in the tip can be explained by the variation in the bill?

SOLUTION With $r = 0.828$, we get $r^2 = 0.686$, and we conclude that 0.686 (or about 69%) of the variation in tips can be explained by the linear relationship between the bill and the tip. This implies that about 31% of the variation in tips can be explained by factors other than the amount of the bill. Such other factors might include the friendliness of the waiter or waitress, the state of the economy, the wealth of the person giving the tip, the quality of the food, the ambience, and so on.

Common Errors Involving Correlation

We now identify three of the most common sources of errors made in interpreting results involving correlation:

1. *A common source of error involves concluding that correlation implies causality.* One study showed a strong correlation between the salaries of statistics professors and per capita beer consumption, but those two variables are affected by the state of the economy, a third variable lurking in the background. (A **lurking variable** is one that affects the variables being studied, but is not included in the study.)
2. *Another source of error arises with data based on averages.* Averages suppress individual variation and may inflate the correlation coefficient. One study produced a 0.4 linear correlation coefficient for paired data relating income and education among individuals, but the linear correlation coefficient became 0.7 when regional averages were used. AS
3. *A third source of error involves the property of linearity.* A relationship may exist between x and y even when there is no significant linear correlation. The data depicted in Figure 9-2 result in a value of $r = 0$, which is an indication of no *linear* correlation between the two variables. However, we can easily see from looking at the figure that there is a pattern reflecting a very strong *nonlinear* relationship. (Figure 9-2 is a scatterplot that depicts the relationship between distance above ground and time elapsed for an object thrown upward.)

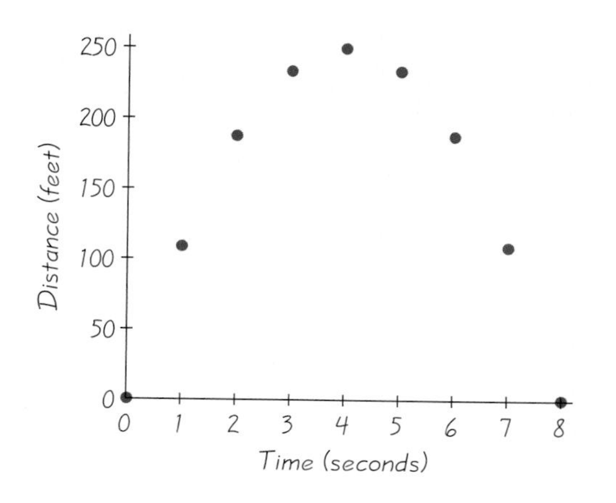

FIGURE 9-2
Scatterplot of Distance above Ground and Time for Object Thrown Upward

Formal Hypothesis Test (Requires Coverage of Chapter 7)

We present two methods (summarized in Figure 9-3) for using a formal hypothesis test to determine whether there is a significant linear correlation between two variables. Some instructors prefer Method 1 because it reinforces concepts introduced in earlier chapters. Others prefer Method 2 because it involves easier calculations.

Figure 9-3 shows that the null and alternative hypotheses will be expressed as follows:

$$H_0: \rho = 0 \qquad \text{(No linear correlation)}$$

$$H_1: \rho \neq 0 \qquad \text{(Linear correlation)}$$

For the test statistic, we use one of the following methods.

Method 1: Test Statistic Is t This method follows the format presented in earlier chapters. It uses the Student t distribution with a test statistic having the form $t = (r - \mu_r)/s_r$, where μ_r and s_r denote the claimed value of the mean and the sample standard deviation of r values. Because we assume that $\rho = 0$, it follows that $\mu_r = 0$. Also, it can be shown that s_r, the standard deviation of linear correlation coefficients, can be expressed as $\sqrt{(1 - r^2)/(n - 2)}$. We can therefore use the following test statistic.

Test Statistic *t* for Linear Correlation

$$t = \frac{r}{\sqrt{\dfrac{1 - r^2}{n - 2}}}$$

Critical values: Use Table A-3 with degrees of freedom $= n - 2$.

Method 2: Test Statistic Is r This method requires fewer calculations. Instead of calculating the test statistic just given, we use the computed value of r as the test statistic. Critical values are found in Table A-6.

Test statistic *r* for Linear Correlation

Test statistic: r
Critical values: Refer to Table A-6.

Figure 9-3 shows that the decision criterion is to reject the null hypothesis of $\rho = 0$ if the absolute value of the test statistic exceeds the critical values; rejection of $\rho = 0$ means that there is sufficient evidence to support a claim of a linear correlation between the two variables. If the absolute value of the test statistic does not exceed the critical values, then we fail to reject

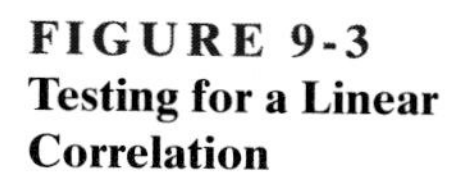
FIGURE 9-3
Testing for a Linear Correlation

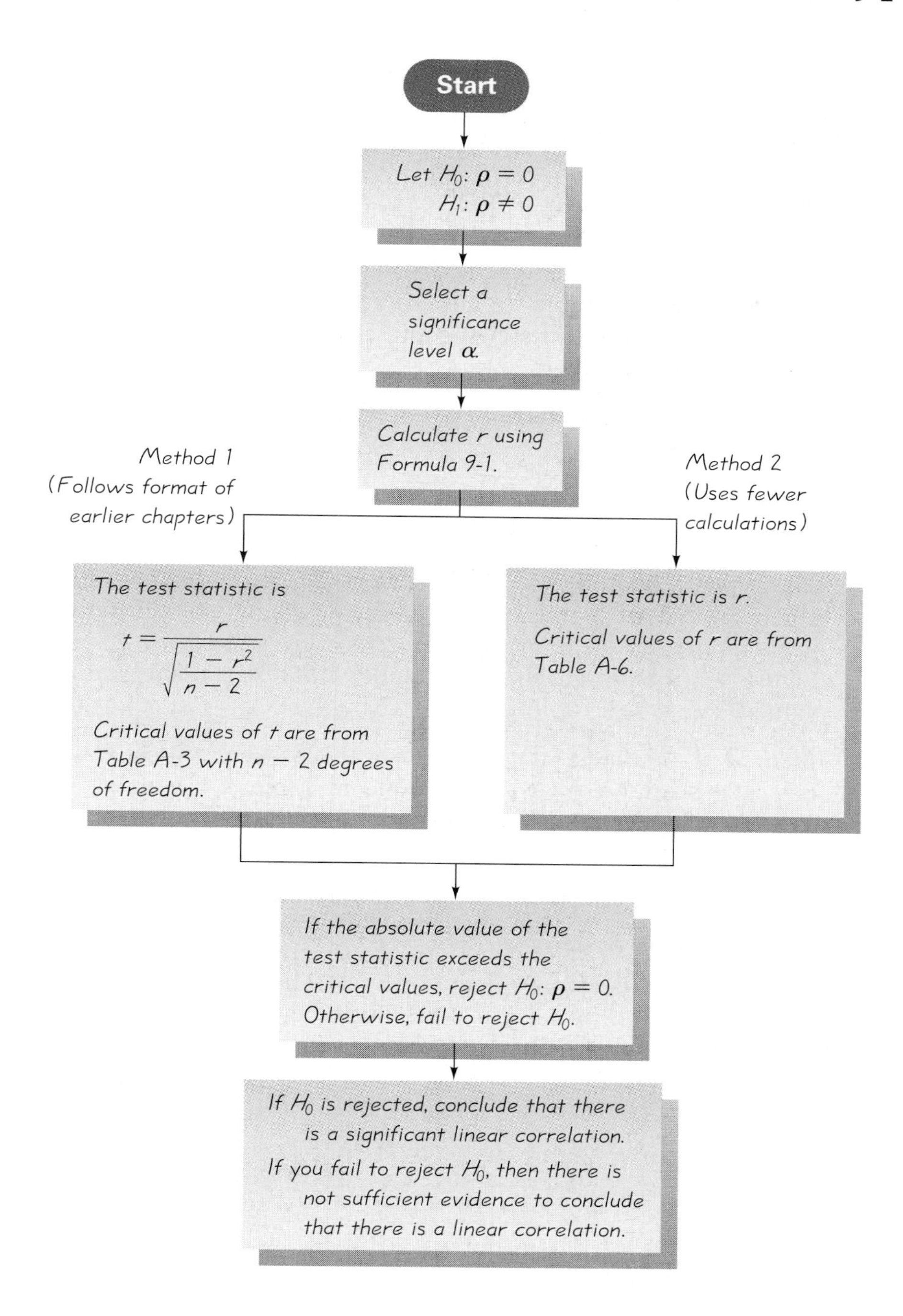

$\rho = 0$; that is, there is not sufficient evidence to conclude that there is a linear correlation between the two variables.

EXAMPLE **Tipping** Using the sample data in Table 9-1, test the claim that there is a linear correlation between amounts of bills and amounts of tips. For the test statistic, use both (a) Method 1 and (b) Method 2.

SOLUTION Refer to Figure 9-3. To claim that there is a significant linear correlation is to claim that the population linear correlation coefficient ρ is different from 0. We therefore have the following hypotheses.

$$H_0: \rho = 0 \quad \text{(No linear correlation)}$$
$$H_1: \rho \neq 0 \quad \text{(Linear correlation)}$$

No significance level α was specified, so use $\alpha = 0.05$.

In a preceding example we already found that $r = 0.828$. With that value, we now find the test statistic and critical value, using each of the two methods just described.

a. *Method 1:* The test statistic is

$$t = \frac{r}{\sqrt{\dfrac{1 - r^2}{n - 2}}} = \frac{0.828}{\sqrt{\dfrac{1 - 0.828^2}{6 - 2}}} = 2.953$$

The critical values of $t = -2.776$ and $t = 2.776$ are found in Table A-3, where 2.776 corresponds to 0.05 divided between two tails (with 0.025 in each tail) and the number of degrees of freedom is $n - 2 = 4$. See Figure 9-4 for the graph that includes the test statistic and critical values.

b. *Method 2:* The test statistic is $r = 0.828$. The critical values of $r = -0.811$ and $r = 0.811$ are found in Table A-6 with $n = 6$ and $\alpha = 0.05$. See Figure 9-5 for a graph that includes this test statistic and critical values.

Using either of the two methods, we find that the absolute value of the test statistic does exceed the critical value (Method 1: $2.953 > 2.776$; Method 2: $0.828 > 0.811$); that is, the test statistic does fall within the critical region. We therefore reject H_0: $\rho = 0$. There is sufficient evidence to support the claim of a linear correlation between bills and tips. The

FIGURE 9-4
Testing H_0: $\rho = 0$ with Method 1

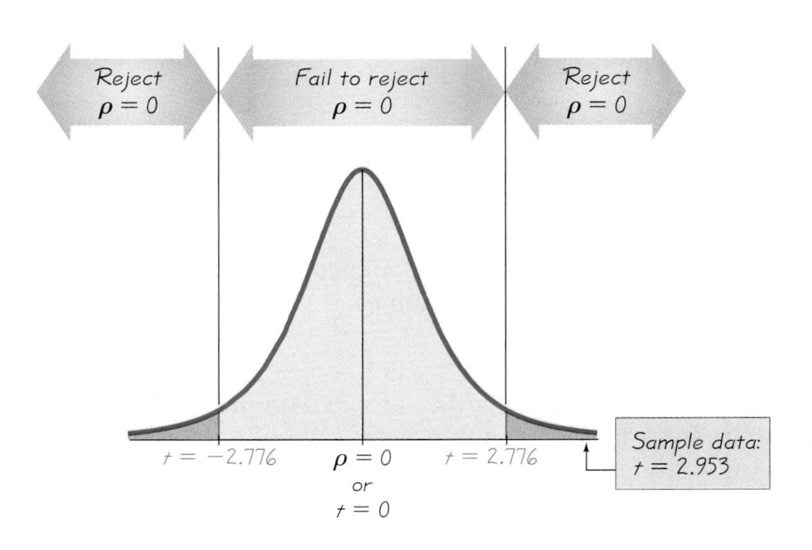

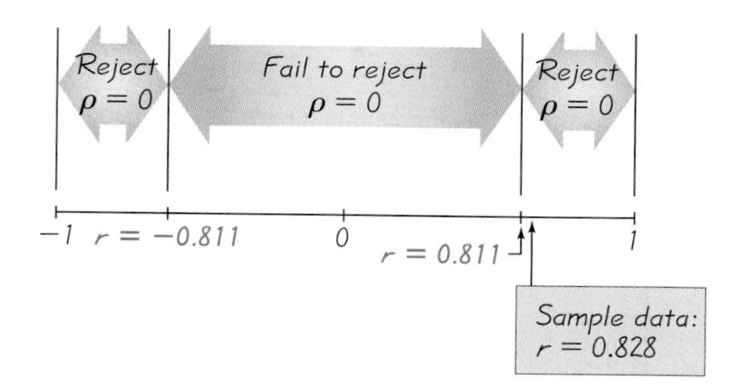

FIGURE 9-5
Testing H_0: $\rho = 0$ with Method 2

amount of the tip does seem to correspond to the amount of the bill, as we would expect.

One-tailed Tests: The preceding example and Figures 9-4 and 9-5 illustrate a two-tailed hypothesis test. The examples and exercises in this section will generally involve only two-tailed tests, but one-tailed tests can occur with a claim of a positive linear correlation or a claim of a negative linear correlation. In such cases, the hypotheses will be as shown here.

Claim of Negative Correlation (Left-tailed test)	Claim of Positive Correlation (Right-tailed test)
H_0: $\rho \geq 0$	H_0: $\rho \leq 0$
H_1: $\rho < 0$	H_1: $\rho > 0$

For these one-tailed tests, Method 1 can be handled as in earlier chapters. For Method 2, either calculate the critical value as described in Exercise 23 or modify Table A-6 by replacing the column headings of $\alpha = 0.05$ and $\alpha = 0.01$ by the one-sided critical values of $\alpha = 0.025$ and $\alpha = 0.005$, respectively.

Rationale: We have presented Formula 9-1 for calculating r and have illustrated its use; we will now give a justification for it. Formula 9-1 simplifies the calculations used in this equivalent formula:

$$r = \frac{\Sigma(x - \bar{x})(y - \bar{y})}{(n - 1)s_x s_y}$$

We will temporarily use this latter version of Formula 9-1 because its form relates more directly to the underlying theory. We will consider the following paired data, which are depicted in the scatterplot shown in Figure 9-6.

x	1	1	2	4	7
y	4	5	8	15	23

Figure 9-6 on the next page includes the point $(\bar{x}, \bar{y}) = (3, 11)$, which is called the *centroid* of the sample points.

DEFINITION

Given a collection of paired (x, y) data, the point $(\bar{x}, \bar{y})$ is called the **centroid.**

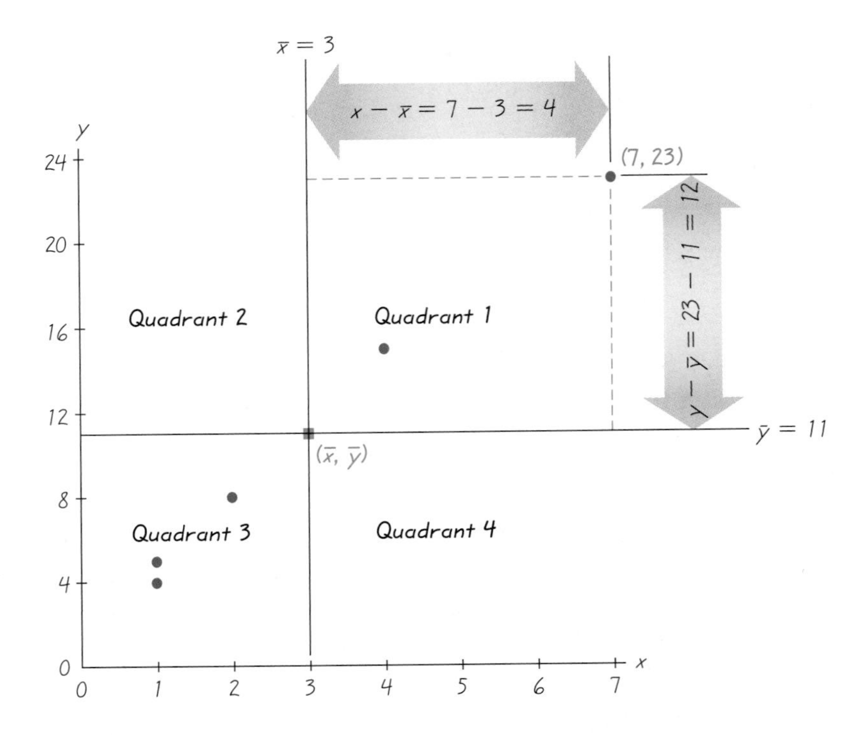

FIGURE 9-6
Scatterplot Partitioned into Quadrants

The statistic r, sometimes called the *Pearson product moment,* was first developed by Karl Pearson. It is based on the sum of the product of the moments $(x - \overline{x})$ and $(y - \overline{y})$; that is, on the statistic $\Sigma(x - \overline{x})(y - \overline{y})$. In any scatterplot, vertical and horizontal lines through the centroid $(\overline{x}, \overline{y})$ divide the diagram into four quadrants, as in Figure 9-6. If the points of the scatterplot tend to approximate an uphill line (as in the figure), individual values of the product $(x - \overline{x})(y - \overline{y})$ tend to be positive because most of the points are found in the first and third quadrants, where the products of $(x - \overline{x})$ and $(y - \overline{y})$ are positive. If the points of the scatterplot approximate a downhill line, most of the points are in the second and fourth quadrants, where $(x - \overline{x})$ and $(y - \overline{y})$ are opposite in sign, so $\Sigma(x - \overline{x})(y - \overline{y})$ is negative. Points that follow no linear pattern tend to be scattered among the four quadrants, so the value of $\Sigma(x - \overline{x})(y - \overline{y})$ tends to be close to 0.

The sum $\Sigma(x - \overline{x})(y - \overline{y})$ depends on the magnitude of the numbers used. For example, if you change x from inches to feet, that sum will change. To make r independent of the particular scale used, we include the sample standard deviations as follows:

$$r = \frac{\Sigma(x - \overline{x})(y - \overline{y})}{(n - 1)s_x s_y}$$

This expression can be algebraically manipulated into the equivalent form of Formula 9-1.

In preceding chapters we discussed methods of inferential statistics by addressing methods of hypothesis testing, as well as methods for constructing confidence interval estimates. A similar procedure may be used to find confidence intervals for ρ. However, because the construction of such confidence intervals involves somewhat complicated transformations, that process is presented in Exercise 24 (Beyond the Basics).

We can use the linear correlation coefficient to determine whether there is a linear relationship between two variables. Using the data in Table 9-1, we have concluded that there is a linear correlation between amounts of restaurant bills and amounts of tips left by the customers. Having concluded that a relationship exists, we would like to determine what that relationship is so that we can calculate the amount of the tip once we know the amount of the bill. This next stage of analysis is addressed in the following section.

Using Technology

STATDISK: Select **Analysis** from the main menu bar, then use the option **Correlation and Regression.** Enter the paired data, or use Copy/Paste to copy the data. Enter a value for the significance level. Click on the **Evaluate** button. The STATDISK display will include the value of the linear correlation coefficient along with the critical value of *r*, the conclusion, and other results to be discussed in later sections. Graphs, including a scatterplot, can also be obtained by clicking on the Plot 1 and Plot 2 buttons.

Minitab: Enter the paired data in columns C1 and C2, then select **Stat** from the main menu bar, choose **Basic Statistics,** followed by **Correlation,** and proceed to enter C1 and C2 for the columns to be used. Minitab will provide the value of the linear correlation coefficient *r* as well as a *P*-value. To obtain a scatterplot, select **Graph,** followed by **Plot,** then enter C1 and C2 for *X* and *Y*, and click **OK.**

Excel: Excel has a function that calculates the value of the linear correlation coefficient. First enter the paired sample data in columns A and B. Click on the **fx** function key located on the main menu bar. Select the function category **Statistical** and the function name **CORREL,** then click **OK.** In the dialog box, enter the cell range of values for *x*, such as A1:A6. Also enter the cell range of values for *y*, such as B1:B6. To obtain a scatterplot, click on the Chart Wizard on the main menu, then select the chart type identified as **XY(Scatter).** In the dialog box, enter the input range of the data, such as A1:B6. Click **Next** and proceed to use the dialog boxes to modify the graph as desired.

The Data Desk XL add-in can also be used. Click on **DDXL** and select **Regression,** then click on the Function Type box and select **Correlation.** In the dialog box, click on the pencil icon for the X-Axis Variable and enter the range of values for the first sample, such as A1:A14. Click on the pencil icon for the Y-Axis Variable and enter the range of values for the second sample. Click on **OK.** A scatter diagram and the correlation coefficient will be displayed.

TI-83 Plus: Enter the paired data in lists L1 and L2, then press **STAT** and select **TESTS.** Using the option of **LinRegTTest** will result in several displayed values, including the value of the linear correlation coefficient r. To obtain a scatterplot, press **2nd,** then **Y=** (for STAT PLOT). Press **Enter** twice to turn Plot 1 on, then select the first graph type, which resembles a scatterplot. Set the X list and Y list labels to L1 and L2 and press the **ZOOM** key, then select **ZoomStat** and press the **Enter** key.

9-2 Basic Skills and Concepts

In Exercises 1 and 2, use a significance level of $\alpha = 0.05$.

1. Testing for Correlation and Describing Variation Assume that 40 pairs of data result in a linear correlation coefficient of $r = 0.401$.

a. Is there a significant linear correlation between x and y? Explain.

b. What proportion of the variation in y can be explained by the variation in x?

2. Testing for Correlation and Describing Variation Assume that 60 pairs of data result in a linear correlation coefficient of $r = -0.222$.

a. Is there a significant linear correlation between x and y? Explain.

b. What proportion of the variation in y can be explained by the variation in x?

Testing for a Linear Correlation. *In Exercises 3 and 4, use a scatterplot and the linear correlation coefficient r to determine whether there is a correlation between the two variables.*

3.

x	1	2	3	5	5
y	19	12	4	20	18

4.

x	2	3	3	5	8
y	5	9	10	12	25

Testing for a Linear Correlation. *In Exercises 5–16, construct a scatterplot, find the value of the linear correlation coefficient r and use a significance level of $\alpha = 0.05$ to determine whether there is a significant linear correlation between the two variables. Save your work because the same data sets will be used in the next section.*

5. Supermodel Heights and Weights Listed below are heights (in inches) and weights (in pounds) for supermodels Niki Taylor, Nadia Avermann, Claudia Schiffer, Elle MacPherson, Christy Turlington, Bridget Hall, Kate Moss, Valerie Mazza, and Kristy Hume.

Height (in.)	71	70.5	71	72	70	70	66.5	70	71
Weight (lb.)	125	119	128	128	119	127	105	123	115

6. Bear Chest Sizes and Weights When bears were anesthetized, researchers measured the distances (in inches) around the bears' chests and weighed the bears (in pounds). The results are given below for eight male bears. Based on the results, does a bear's weight seem to be related to its chest size? Do the results change if the chest measurements are converted to feet, with each of those values divided by 12?

x Chest (in.)	26	45	54	49	41	49	44	19
y Weight (lb)	90	344	416	348	262	360	332	34

Based on data from Minitab and Gary Alt.

7. Discarded Plastic and Household Size The accompanying table lists weights (in pounds) of plastic discarded by a sample of households, along with the sizes of the households. Is there a significant linear correlation? This issue is important to the Census Bureau, which provided project funding, because the presence of a correlation implies that we can predict population size by analyzing discarded garbage.

Plastic (lb)	0.27	1.41	2.19	2.83	2.19	1.81	0.85	3.05
Household size	2	3	3	6	4	2	1	5

Based on data provided by Masakazu Tani and the Garbage Project at the University of Arizona.

8. Discarded Paper and Household Size The paired data below consist of weights (in pounds) of discarded paper and sizes of households.

Paper (lb)	2.41	7.57	9.55	8.82	8.72	6.96	6.83	11.42
Household size	2	3	3	6	4	2	1	5

Based on data obtained from the Garbage Project at the University of Arizona.

9. DWI and Jail A study was conducted to investigate the relationship between age (in years) and BAC (blood alcohol concentration) measured when convicted DWI (driving while intoxicated) jail inmates were first arrested. Sample data are given below for randomly selected subjects. Based on the result, does the BAC level seem to be related to the age of the person tested?

Age	17.2	43.5	30.7	53.1	37.2	21.0	27.6	46.3
BAC	0.19	0.20	0.26	0.16	0.24	0.20	0.18	0.23

Based on data from the Dutchess County STOP-DWI Program.

10. Weapons and Murder Rate The accompanying table lists the number of registered automatic weapons (in thousands), along with the murder rate (in murders per 100,000 of population), for randomly selected states. Automatic weapons are guns that continue to fire repeatedly while the trigger is held back. Are firearm murders often committed with automatic weapons? Does a significant linear correlation imply that increased numbers of automatic weapons result in more murders?

Automatic weapons	11.6	8.3	3.6	0.6	6.9	2.5	2.4	2.6
Murder rate	13.1	10.6	10.1	4.4	11.5	6.6	3.6	5.3

Data provided by the FBI and the Bureau of Alcohol, Tobacco, and Firearms.

11. Old Faithful Geyser Refer to Data Set 11 in Appendix B.

a. Use the paired data for durations and intervals after eruptions of the geyser. Is there a significant linear correlation, suggesting that the interval after an eruption is related to the duration of the eruption?

b. Use the paired data for intervals after eruptions and heights of eruptions of the Old Faithful geyser. Is there a significant linear correlation, suggesting that the interval after an eruption is related to the height of the eruption?

c. Assume that you want to develop a method for predicting the time interval to the next eruption. Based on the results from parts (a) and (b), which factor would be more relevant: eruption duration or eruption height? Why?

12. Cigarette Nicotine, Tar, and Carbon Monoxide Refer to Data Set 8 in Appendix B.

a. Use the paired data consisting of tar and nicotine. Based on the result, does there appear to be a significant linear correlation between cigarette tar and nicotine? If so, can researchers reduce their laboratory expenses by measuring only one of these two variables?

b. Use the paired data consisting of carbon monoxide and nicotine. Based on the result, does there appear to be a significant linear correlation between cigarette carbon monoxide and nicotine? If so, can researchers reduce their laboratory expenses by measuring only one of these two variables?

c. Assume that researchers want to develop a method for predicting the amount of nicotine, and they want to measure only one other item. In choosing between tar and carbon monoxide, which is the better choice? Why?

13. Diamond Prices, Carats, and Color Refer to Data Set 3 in Appendix B.

a. Use the paired data consisting of price and carat (weight). Is there a significant linear correlation between the price of a diamond and its weight in carats?

b. Use the paired price/color data. Is there a significant linear correlation between the price of a diamond and its color?

c. Assume that you are planning to buy a diamond engagement ring. In considering the value of a diamond, which characteristic should you consider to be more important: the carat weight or the color? Why?

14. Car Fuel Consumption, Weight, and Displacement Refer to Data Set 18 in Appendix B.

a. Use the paired data consisting of highway fuel consumption rate (mi/gal) and weight (pounds). Is there a significant linear correlation between those two variables?

b. Use the paired data consisting of highway fuel consumption rate (mi/gal) and the engine displacement (in liters). Is there a significant linear correlation between those two variables?

c. Assume that you want to develop a way to predict a car's fuel consumption rate without actually measuring it. Which factor would be more relevant: the car's weight or its engine displacement? Why?

15. Movie Grosses, Budgets, and Viewer Ratings Refer to Data Set 15 in Appendix B.

a. Use the paired data consisting of the amount of money grossed and the amount of money budgeted. Is there a significant linear correlation between those two variables?

b. Use the paired data consisting of the amount of money grossed and the viewer rating. Is there a significant linear correlation between those two variables?

(continued)

c. Based on the preceding results, if you plan to invest in a company that makes movies, which is the better choice: a company that makes big budget movies or a company that makes movies that viewers rate highly? Why?

16. Home Selling Prices, List Prices, and Taxes Refer to Data Set 16 in Appendix B.

a. Use the paired data consisting of home selling price and list price. We might expect that these variables would be related, but is there sufficient evidence to support that expectation?

b. Use the paired data consisting of home selling price and the amount of taxes. The tax bill is supposed to be based on the value of the house. Is it? Explain.

Identifying Correlation Errors. *In Exercises 17–20, describe the error in the stated conclusion. (See the list of common sources of errors included in this section.)*

17. *Given:* The paired sample data of the ages of subjects and their scores on a test of reasoning result in a linear correlation coefficient very close to 0.

Conclusion: Younger people tend to get higher scores.

18. *Given:* There is a significant linear correlation between personal income and years of education.

Conclusion: More education causes a person's income to rise.

19. *Given:* Subjects take a test of verbal skills and a test of manual dexterity, and those pairs of scores result in a linear correlation coefficient very close to 0.

Conclusion: Scores on the two tests are not related in any way.

20. *Given:* There is a significant linear correlation between state average tax burdens and state average incomes.

Conclusion: There is a significant linear correlation between individual tax burdens and individual incomes.

9-2 Beyond the Basics

21. Using Data from Scatterplot Sometimes, instead of having numerical data, we have only graphical data. The accompanying Excel scatterplot on the next page is similar to one that was included in "The Prevalence of Nosocomial Infection in Intensive Care Units in Europe," by Jean-Louis Vincent et al., *Journal of the American Medical Association,* Vol. 274, No. 8. Each point represents a different European country. Estimate the value of the linear correlation coefficient, and determine whether there is a significant linear correlation between the mortality rate and the rate of infections acquired in intensive care units.

22. Correlations with Transformed Data In addition to testing for a linear correlation between x and y, we can often use transformations of data to explore for other relationships. For example, we might replace each x-value by x^2 and use the methods of this section to determine whether there is a linear correlation between y and x^2. Given the paired data in the accompanying table, construct the scatterplot and then test for a linear correlation between y and each of the following. Which case results in the largest value of r?

a. x **b.** x^2 **c.** $\log x$ **d.** $\sqrt{x}$ **e.** $1/x$

x	1.3	2.4	2.6	2.8	2.4	3.0	4.1
y	0.11	0.38	0.41	0.45	0.39	0.48	0.61

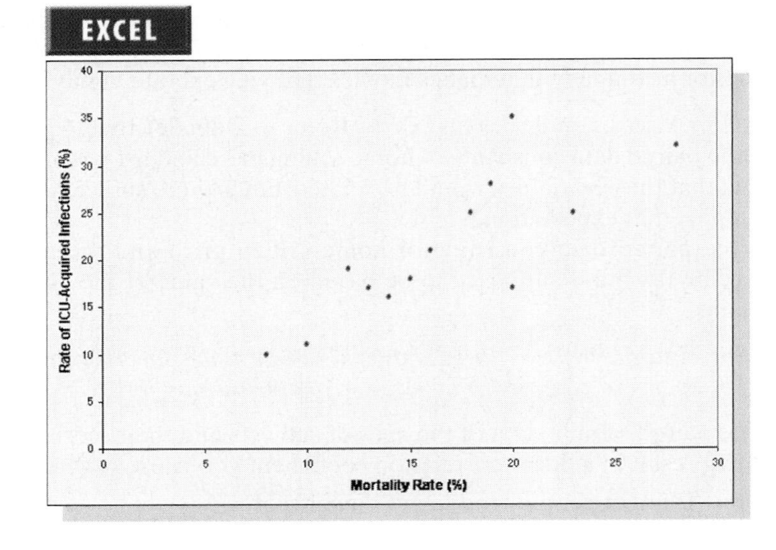

23. Finding Critical r-Values The critical values of r in Table A-6 are found by solving

$$t = \frac{r}{\sqrt{\dfrac{1 - r^2}{n - 2}}}$$

for r to get

$$r = \frac{t}{\sqrt{t^2 + n - 2}}$$

where the t-value is found from Table A-3 by assuming a two-tailed case with $n - 2$ degrees of freedom. Table A-6 lists the results for selected values of n and α. Use the formula for r given here and Table A-3 (with $n - 2$ degrees of freedom) to find the critical values of r for the given cases.

a. H_0: $\rho = 0$, $n = 50$, $\alpha = 0.05$
b. H_1: $\rho \neq 0$, $n = 75$, $\alpha = 0.10$
c. H_0: $\rho \geq 0$, $n = 20$, $\alpha = 0.05$
d. H_0: $\rho \leq 0$, $n = 10$, $\alpha = 0.05$
e. H_1: $\rho > 0$, $n = 12$, $\alpha = 0.01$

24. Constructing Confidence Intervals for ρ Given n pairs of data from which the linear correlation coefficient r can be found, use the following procedure to construct a confidence interval about the population parameter ρ.

Step a. Use Table A-2 to find the $z_{\alpha/2}$ value that corresponds to the desired degree of confidence.

Step b. Evaluate the interval limits w_L and w_R:

$$w_L = \frac{1}{2}\ln\left(\frac{1 + r}{1 - r}\right) - z_{\alpha/2} \cdot \frac{1}{\sqrt{n - 3}}$$

$$w_R = \frac{1}{2}\ln\left(\frac{1 + r}{1 - r}\right) + z_{\alpha/2} \cdot \frac{1}{\sqrt{n - 3}}$$

(continued)

Step c. Now evaluate the confidence interval limits in the expression below.

$$\frac{e^{2w_L} - 1}{e^{2w_L} + 1} < \rho < \frac{e^{2w_R} - 1}{e^{2w_R} + 1}$$

Use this procedure to construct a 95% confidence interval for ρ, given 50 pairs of data for which $r = 0.600$.

25. Is the Real Issue Correlation? Refer to Table 8-1, located with the Chapter Problem at the beginning of Chapter 8. Table 8-1 includes paired data representing the reported heights and the measured heights of 12 male statistics students. Can correlation be used to address the issue of whether male statistics students exaggerate their heights? Why or why not?

9-3 Regression

In Section 9-2 we analyzed paired data with the goal of determining whether there is a significant linear correlation between two variables. The main objective of this section is to describe the relationship between two variables by finding the graph and equation of the straight line that represents the relationship. This straight line is called the *regression line,* and its equation is called the *regression equation.* Sir Francis Galton (1822–1911) studied the phenomenon of heredity and showed that when tall or short couples have children, the heights of those children tend to *regress,* or revert to the more typical mean height for people of the same gender. We continue to use Galton's "regression" terminology, even though our data might not involve the same height phenomena.

DEFINITIONS

Given a collection of paired sample data, the **regression equation**

$$\hat{y} = b_0 + b_1 x$$

algebraically describes the relationship between the two variables. The graph of the regression equation is called the **regression line** (or *line of best fit,* or *least-squares line*).

This definition expresses a relationship between x (called the **independent variable,** or **predictor variable**) and $\hat{y}$ (called the **dependent variable,** or **response variable**). In the preceding definition, the typical equation of a straight line $y = mx + b$ is expressed in the form $\hat{y} = b_0 + b_1 x$, where b_0 is the y-intercept and b_1 is the slope. The following notation box shows that b_0 and b_1 are sample statistics used to estimate the population parameters β_0 and β_1.

Notation for Regression Equation

	Population Parameter	Sample Statistic
y-intercept of regression equation	β_0	b_0
Slope of regression equation	β_1	b_1
Equation of the regression line	$y = \beta_0 + \beta_1 x$	$\hat{y} = b_0 + b_1 x$

For the regression methods given in this section, we make the following assumptions.

Assumptions

1. We are investigating only *linear* relationships.
2. For each x-value, y is a random variable having a normal (bell-shaped) distribution. All of these y distributions have the same variance. Also, for a given value of x, the distribution of y-values has a mean that lies on the regression line. (Results are not seriously affected if departures from normal distributions and equal variances are not too extreme.)

An important goal of this section is to use paired sample data to estimate the regression equation. Using only sample data, we can't find the exact values of the population parameters β_0 and β_1, but we can use the sample data to estimate them with b_0 and b_1, which are found by using Formulas 9-2 and 9-3.

Formula 9-2 $$b_0 = \frac{(\Sigma y)(\Sigma x^2) - (\Sigma x)(\Sigma xy)}{n(\Sigma x^2) - (\Sigma x)^2}$$ y-intercept

Formula 9-3 $$b_1 = \frac{n(\Sigma xy) - (\Sigma x)(\Sigma y)}{n(\Sigma x^2) - (\Sigma x)^2}$$ (slope)

These formulas might look intimidating, but they are programmed into many calculators and computer programs, so the values of b_0 and b_1 can be easily found. (See "Using Technology" at the end of this section.) In those cases when we must use formulas instead of a calculator or computer, the required computations will be much easier if we keep in mind the following facts:

1. If the linear correlation coefficient r has been computed using Formula 9-1, the values of Σx, Σy, Σx^2, and Σxy have already been found, and they can be used again in Formula 9-3. (Also, the numerator for r in Formula 9-1 is the same numerator for b_1 in Formula 9-3; the denominator for r includes the denominator for b_1. If the calculation for r is set up carefully, the calculation for b_1 requires the simple division of one known number by another.)
2. If you find the slope b_1 first, you can use Formula 9-4 to find the y-intercept b_0. [The regression line always passes through the centroid $(\overline{x}, \overline{y})$ so that $\overline{y} = b_0 + b_1\overline{x}$ must be true, and this equation can be expressed as

Formula 9-4. It is usually easier to find the y-intercept b_0 by using Formula 9-4 than by using Formula 9-2.]

Formula 9-4 $$b_0 = \bar{y} - b_1\bar{x}$$

Once we have evaluated b_0 and b_1, we can identify the estimated regression equation, which has the following special property: *The regression line fits the sample points best.* (The specific criterion used to determine which line fits "best" is the least-squares property, which will be described later.) We will now briefly discuss rounding and then illustrate the procedure for finding and applying the regression equation.

Cell Phones and Crashes

Because some countries have banned the use of cell phones in cars while other countries are considering such a ban, researchers studied the issue of whether the use of cell phones while driving increases the chance of a crash. A sample of 699 drivers was obtained. Members of the sample group used cell phones and were involved in crashes. Subjects completed questionnaires and their telephone records were checked. Telephone usage was compared to the time interval immediately preceding a crash to a comparable time period the day before. Conclusion: Use of a cell phone was associated with a crash risk that was about four times as high as the risk when a cell phone was not used. (See "Association between Cellular-Telephone Calls and Motor Vehicle Collisions," by Redelmeier and Tibshirani, New England Journal of Medicine, Vol. 336, No. 7.)

Rounding the y-Intercept b_0 and the Slope b_1

It's difficult to provide a simple universal rule for rounding values of b_0 and b_1, but we usually try to round each of these values to *three significant digits* or use the values provided by STATDISK, Minitab, Excel, or a TI-83 Plus calculator. Because these values are very sensitive to rounding at intermediate steps of calculations, try to carry at least six significant digits (or use exact values) in the intermediate steps. Depending on how you round, this book's answers to examples and exercises may be slightly different from your answers.

EXAMPLE **Tipping** In Section 9-2 we used the Table 9-1 data (x = amount of bill; y = amount of tip) to find that the linear correlation coefficient of $r = 0.828$ indicates that there is a significant linear correlation. Now find the regression equation of the straight line that relates x and y.

SOLUTION We will find the regression equation by using Formulas 9-3 and 9-4 and these values already found in Table 9-2 in Section 9-2:

$$n = 6 \qquad \Sigma x = 441.84 \qquad \Sigma y = 63.58$$
$$\Sigma x^2 = 36{,}754.1416 \qquad \Sigma y^2 = 809.5364 \qquad \Sigma xy = 5308.7436$$

First find the slope b_1 by using Formula 9-3:

$$\begin{aligned} b_1 &= \frac{n(\Sigma xy) - (\Sigma x)(\Sigma y)}{n(\Sigma x^2) - (\Sigma x)^2} \\ &= \frac{6(5308.7436) - (441.84)(63.58)}{6(36{,}754.1416) - (441.84)^2} = \frac{3760.2744}{25{,}302.264} = 0.148614 \\ &= 0.149 \quad \text{(rounded)} \end{aligned}$$

Next, find the y-intercept b_0 by using Formula 9-4:

$$\begin{aligned} b_0 &= \bar{y} - b_1\bar{x} \\ &= \frac{63.58}{6} - (0.148614)\frac{441.84}{6} = -0.347 \quad \text{(rounded)} \end{aligned}$$

Pizza Correlates with Crisis

When former President Clinton was threatened with impeachment by Congress, government employees worked late and ordered record numbers of pizzas. Frank Meeks, owner of 59 Domino's Pizza outlets in Washington, D.C., reported that on the Saturday during the height of the impeachment crisis, Capitol Hill pizza deliveries exceeded $10,000 while White House pizza deliveries totaled $3,000. Meeks noted that pizza sales also peaked during the Persian Gulf War, and they peak annually during budget debates.

Knowing the slope b_1 and y-intercept b_0, we can now express the estimated equation of the regression line as

$$\hat{y} = -0.347 + 0.149x$$

We should realize that this equation is an *estimate* of the true regression equation $y = \beta_0 + \beta_1 x$. This estimate is based on one particular set of sample data listed in Table 9-1, but another sample drawn from the same population would probably lead to a slightly different equation.

The Minitab display shows the regression line plotted on the scatterplot.

MINITAB

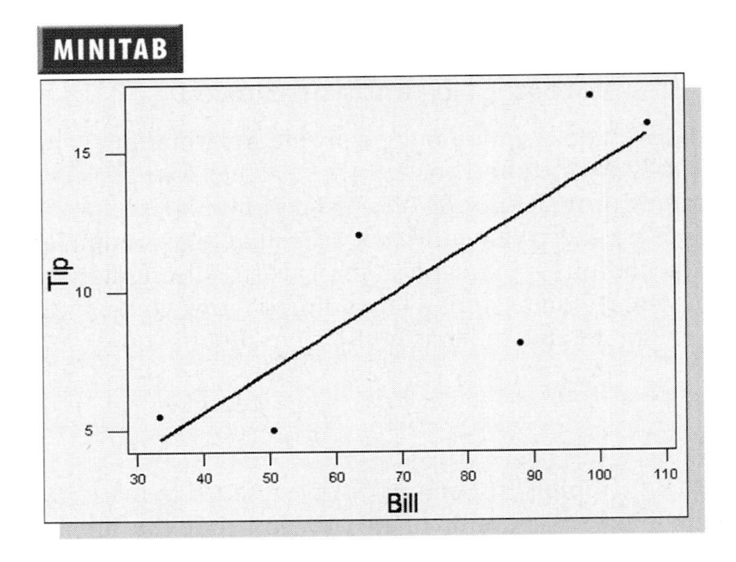

We can see that this line fits the data well. Note that even though the regression line fits the data points well, it does not actually pass through any of the original data points. In some other cases, the regression line might pass through one or more of the sample data points, but it is common for the regression line to miss all of them.

Using the Regression Equation for Predictions

Regression equations can be helpful when used in *predicting* the value of one variable, given some particular value for the other variable. If the regression line fits the data quite well, then it makes sense to use its equation for predictions, provided that we don't go beyond the scope of the available values. However, *we should use the equation of the regression line only if r indicates that there is a significant linear correlation. In the absence of a significant linear correlation, we should not use the regression equation for projecting or predicting; instead, our best estimate of the second variable is simply its sample mean.*

In predicting a value of y based on some given value of x...

1. **If there is *not* a significant linear correlation, the best predicted y-value is $\bar{y}$.**

2. If there is a significant linear correlation, the best predicted y-value is found by substituting the x-value into the regression equation.

Figure 9-7 summarizes this process, which is easier to understand if we think of r as a measure of how well the regression line fits the sample data. If r is near -1 or $+1$, then the regression line fits the data well, but if r is near 0, then the regression line fits poorly (and should not be used for predictions).

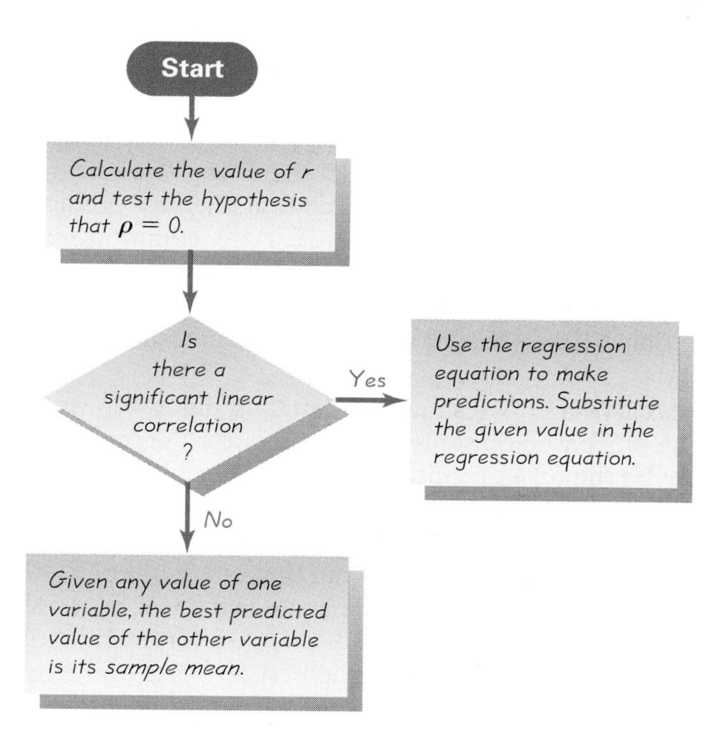

FIGURE 9-7
Predicting the Value of a Variable

EXAMPLE **Predicting Tip** Using the sample data in Table 9-1, we found that there is a significant linear correlation between amounts of bills and amounts of tips, and we also found that the regression equation is $\hat{y} = -0.347 + 0.149x$. If the amount of a bill is \$100, predict the amount of the tip that will be left.

SOLUTION There's a strong temptation to jump in and substitute 100 for x in the regression equation, but we should first consider whether there is a significant linear correlation that justifies the use of that equation. In this example, we do have a significant linear correlation (with $r = 0.828$), so our predicted value is found as follows:

$$\begin{aligned}\hat{y} &= -0.347 + 0.149x \\ &= -0.347 + 0.149(100) = \$14.55\end{aligned}$$

The predicted tip for a \$100 bill is \$14.55. (If there had not been a significant linear correlation, our best predicted tip would have been $\bar{y} = \$63.58/6 = \$10.60$.)

EXAMPLE **Shoe Sizes and Tips** There is obviously no significant linear correlation between shoe sizes of adults and the amounts that they tip waiters and waitresses. Given that an adult has a shoe size of 9, what is the best prediction of the amount of tip this person would leave?

SOLUTION Because there is no significant linear correlation, we do not use a regression equation. Instead, the best predicted tip is simply the *mean* tip that is left. Based on the sample data in Table 9-1, the mean tip is $\bar{y} = \$63.58/6 = \$10.60$, so that is the best predicted value.

Carefully compare the solutions to the preceding two examples and note that we used the regression equation when there was a significant linear correlation, but in the absence of such a correlation, the best predicted value of y is simply the value of the sample mean $\bar{y}$. Also, common sense suggests that if someone is given a restaurant bill of $100 and that person has a shoe size of 9, the tip value of $14.55 (based on the regression equation) is a better prediction than $10.60 (based on the mean tip amount for everyone).

Interpreting Predicted Values We should understand that when we determine a predicted value of y based on some given value of x, the predicted value will not necessarily be the exact result. There are other factors, such as random variation and characteristics not included in the study, that can also affect y. Based on the sample data in Table 9-1, we see that for a bill of $100, the best predicted tip is $14.55. This does not mean that every bill of $100 will result in a tip of exactly $14.55; some will be less and some will be more. Based on the regression equation we have found, many different bills of $100 will result in many different tips, but the single best estimate of the tip is $14.55.

A common error is to use the regression equation for making a prediction when there is no significant linear correlation. That error violates the first of the following guidelines.

Guidelines for Using the Regression Equation

1. *If there is no significant linear correlation, don't use the regression equation to make predictions.*
2. *When using the regression equation for predictions, stay within the scope of the available sample data.* If you find a regression equation that relates women's heights and shoe sizes, it's absurd to predict the shoe size of a woman who is 10 ft tall. (The regression equation is like a function whose domain is the scope of the sample x-values.)
3. *A regression equation based on old data is not necessarily valid now.* The regression equation relating used-car prices and ages of cars is no longer usable if it's based on data from the 1970s.
4. *Don't make predictions about a population that is different from the population from which the sample data were drawn.* If we collect sample data from men and develop a regression equation relating age and TV remote-control usage, the results don't necessarily apply to women. If we use state

averages to develop a regression equation relating SAT math scores and SAT verbal scores, the results don't necessarily apply to *individuals.*

Interpreting the Regression Equation: Marginal Change

We can use the regression equation to see the effect on one variable when the other variable changes by some specific amount.

DEFINITION

In working with two variables related by a regression equation, the **marginal change** in a variable is the amount that it changes when the other variable changes by exactly one unit.

The slope b_1 in the regression equation represents the marginal change in y that occurs when x changes by one unit. For the bill/tip data of Table 9-1, the regression line has a slope of 0.149, which shows that if we increase x (the amount of the bill) by 1 dollar, the tip will increase by 0.149 dollars, or about 15 cents. Marginal change in this case can be simply described by stating that for every dollar the bill goes up, the tip goes up by about 15 cents.

Outliers and Influential Points

A correlation/regression analysis of bivariate (paired) data should include an investigation of *outliers* and *influential points,* defined as follows.

DEFINITIONS

In a scatterplot, an **outlier** is a point lying far away from the other data points.

Paired sample data may include one or more **influential points,** which are points that strongly affect the graph of the regression line.

An outlier is easy to identify: Examine the scatterplot and identify a point that is far away from the others. Here's how to determine whether a point is an influential point: Graph the regression line resulting from the data with the point included, then graph the regression line resulting from the data with the point excluded. If the graph changes by a considerable amount, the point is influential. Influential points are often found by identifying those outliers that are *horizontally* far away from the other points.

For example, refer to the preceding Minitab display. Suppose that we include one more pair of data coming from a really hungry couple who ran up a dinner bill of $500, then left a tip of $75. This additional point would be an outlier because it would be far away from the other points, but it would not be an influential point because it would not change the position of the regression line very much. Now suppose the same couple ran up a $500 bill but left no tip ($0). This additional point would now become an influential point because the graph of the regression line would change considerably, as shown by the Minitab display here. Compare this regression line to the one shown in the last Minitab display, and you will see clearly that the addition of that one hungry and stingy couple has a dramatic effect on the regression line.

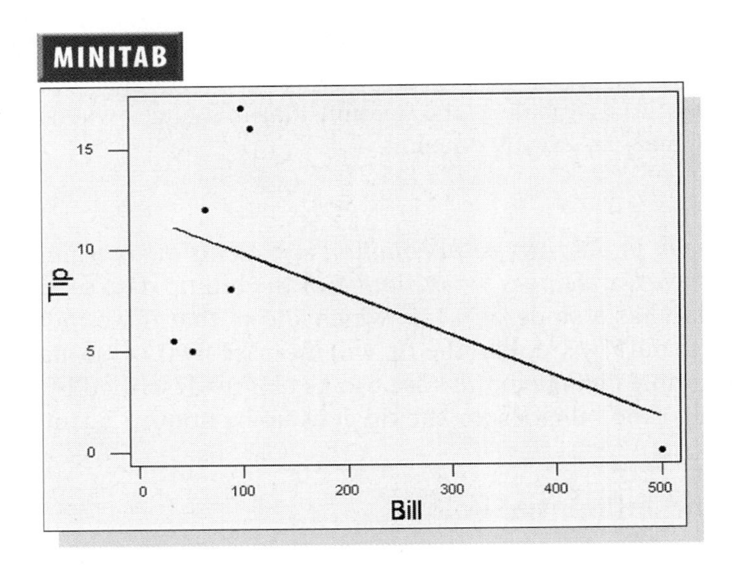

Residuals and the Least-Squares Property

We have stated that the regression equation represents the straight line that fits the data "best," and we will now describe the criterion used in determining the line that is better than all others. This criterion is based on the vertical distances between the original data points and the regression line. Such distances are called *residuals*.

DEFINITION

For a sample of paired (x, y) data, a **residual** is the difference $(y - \hat{y})$ between an observed sample y-value and the value of $\hat{y}$, which is the value of y that is predicted by using the regression equation.

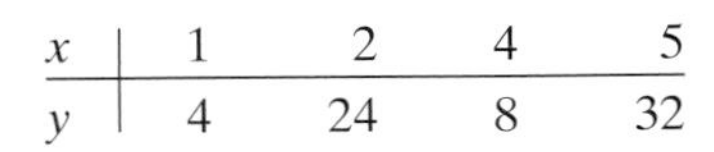

x	1	2	4	5
y	4	24	8	32

This definition might seem as clear as tax-form instructions, but you can easily understand residuals by referring to Figure 9-8, which corresponds to the paired sample data in the margin. In Figure 9-8, the residuals are represented by the dashed lines. For a specific example, see the residual indicated as 7, which is directly above $x = 5$. If we substitute $x = 5$ into the regression

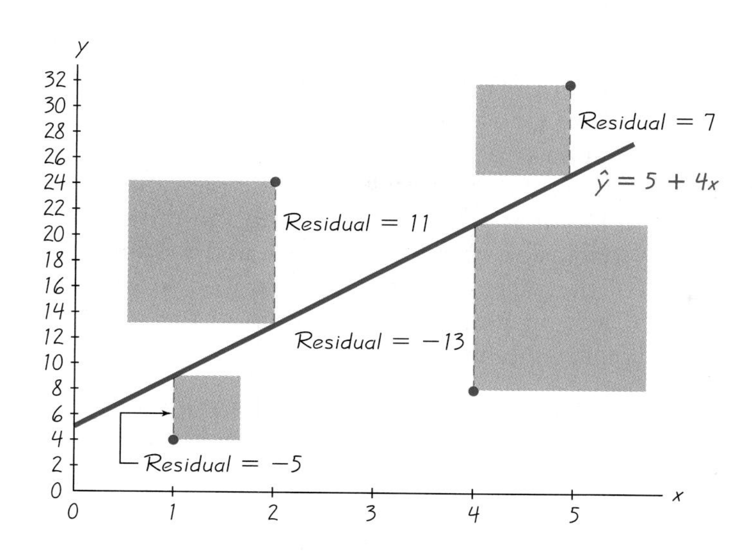

FIGURE 9-8
Scatter Diagram with Regression Line, Residuals, and Squares of Residuals

equation $\hat{y} = 5 + 4x$, we get a predicted value of $\hat{y} = 25$. When $x = 5$, the *predicted* value of y is $\hat{y} = 25$, but the actual *observed* sample value is $y = 32$. The difference $y - \hat{y} = 32 - 25 = 7$ is a residual.

The regression equation represents the line that fits the points "best" according to the *least-squares property.*

DEFINITION

A straight line satisfies the **least-squares property** if the sum of the squares of the residuals is the smallest sum possible.

From Figure 9-8, we see that the residuals are −5, 11, −13, and 7, so the sum of their squares is

AS

$$(-5)^2 + 11^2 + (-13)^2 + 7^2 = 364$$

We can visualize the least-squares property by referring to Figure 9-8, where the squares of the residuals are represented by the red square areas. The sum of the red square areas is 364, which is the smallest sum possible. Use any other straight line, and the red squares will combine to produce an area larger than the combined red area of 364.

Fortunately, we need not deal directly with the least-squares property when we want to find the equation of the regression line. Calculus has been used to build the least-squares property into Formulas 9-2 and 9-3. Because the derivations of these formulas require calculus, we don't include them in this text.

Using Technology

Because of the messy calculations involved, the linear correlation coefficient r and the slope and y-intercept of the regression line are usually found by using a calculator or computer software.

STATDISK: Select **Analysis** from the main menu bar, then use the option **Correlation and Regression.** Enter the paired data, or use Copy/Paste to copy the data. Enter a value for the significance level. Click on the **Evaluate** button. The STATDISK display will include the value of the linear correlation coefficient along with the critical value of r, the conclusion about correlation, and the intercept and slope of the regression equation, as well as some other results. Click on **Plot 1** to get a graph of the scatterplot with the regression line included.

Minitab: First enter the paired data in columns C1 and C2. In Section 9-2 we saw that we could find the value of the linear correlation coefficient r by selecting **Stat/Basic Statistics/Correlation.** To get the equation of the regression line, select **Stat/Regression/Regression,** and enter C2 for "response" and C1 for "predictor." To get the graph of the scatterplot with the regression line, select **Stat/Regression/Fitted Line Plot,** then enter C2 for the response variable and C1 for the predictor variable. Select the "linear" model.

Excel: Enter the paired data in columns A and B. Use Excel's Data Analysis add-in by selecting **Tools** from the main menu, then selecting **Data Analysis** and **Regression,** then clicking **OK.** Enter the range for the y values, such as B1:B6. Enter the range for the x values, such as A1:A6. Click on the box adjacent to Line Fit Plots, then click **OK.** Among all of the information provided by Excel, the slope and intercept of the regression equation can be found under the table heading "Coefficient." The displayed graph will include a scatterplot of the original sample points along with the points that would be predicted by the regression equation. You can easily get the regression line by connecting the "predicted y" points.

To use the Data Desk XL add-in, click on **DDXL** and select **Regression,** then click on the Function Type box and select **Simple Regression.** Click on the pencil icon for the response variable and enter the range of values for the y (or dependent) variable. Click on the pencil icon for the explanatory variable and enter the range of values for the x (or independent) variable. Click on **OK.** The slope and intercept of the regression equation can be found under the table heading "Coefficient."

TI-83 Plus: Enter the paired data in lists L1 and L2, then press **STAT** and select **TESTS,** then choose the option **LinRegTTest.** The displayed results will include the y-intercept and slope of the regression equation. Instead of b_0 and b_1, the TI-83 display represents these values as a and b.

9-3 Basic Skills and Concepts

Finding the Equation of the Regression Line. *In Exercises 1–4, use the given data to find the equation of the regression line.*

1.

x	1	3	3	5
y	6	8	8	10

2.

x	2	4	5	6	8
y	3	1	0	−1	−3

3.

x	1	2	3	5	5
y	19	12	4	20	18

4.

x	2	3	3	5	8
y	5	9	10	12	25

Finding the Equation of the Regression Line and Making Predictions. *Exercises 5–16 use the same data sets as the exercises in Section 9-2. In each case, find the regression equation, letting the first variable be the independent (x) variable. Find predicted values where they are requested. Caution: When finding predicted values, be sure to follow the prediction procedure described in this section.*

5. Supermodel Heights and Weights

Height (in.)	71	70.5	71	72	70	70	66.5	70	71
Weight (lb)	125	119	128	128	119	127	105	123	115

Find the best predicted weight of a supermodel who is 69 in. tall.

6. Bear Chest Sizes and Weights

x Chest (in.)	26	45	54	49	41	49	44	19
y Weight (lb)	90	344	416	348	262	360	332	34

Find the best predicted weight of a bear with a chest size of 52 in.

7. Discarded Plastic and Household Size

Plastic (lb)	0.27	1.41	2.19	2.83	2.19	1.81	0.85	3.05
Household size	2	3	3	6	4	2	1	5

What is the best predicted size of a household that discards 0.50 lb of plastic?

8. Discarded Paper and Household Size

Paper (lb)	2.41	7.57	9.55	8.82	8.72	6.96	6.83	11.42
Household size	2	3	3	6	4	2	1	5

What is the best predicted size of a household that discards 10.00 lb of paper?

9. DWI and Jail

Age	17.2	43.5	30.7	53.1	37.2	21.0	27.6	46.3
BAC	0.19	0.20	0.26	0.16	0.24	0.20	0.18	0.23

What is the best predicted blood alcohol level of a person 21.0 years of age who has been convicted and jailed for DWI? (The BAC level is measured at the time of arrest.)

10. Weapons and Murder Rate

Automatic weapons	11.6	8.3	3.6	0.6	6.9	2.5	2.4	2.6
Murder rate	13.1	10.6	10.1	4.4	11.5	6.6	3.6	5.3

What is the best predicted murder rate for a state with 10,000 registered automatic weapons? (In this table, the numbers of automatic weapons are in thousands, and the murder rate is number of murders per 100,000 people.)

11. Old Faithful Geyser Refer to Data Set 11 in Appendix B.

a. Use the paired data for durations and intervals after eruptions of the Old Faithful geyser. What is the best predicted time before the next eruption if the previous eruption lasted for 210 sec?

b. Use the paired height and interval data for eruptions of the Old Faithful geyser. What is the best predicted time before the next eruption if the previous eruption had a height of 275 ft?

c. Which predicted time is better: the result from part (a) or the result from part (b)? Why?

12. Cigarette Nicotine, Tar, and Carbon Monoxide Refer to Data Set 8 in Appendix B.

a. Use the paired data consisting of tar and nicotine. What is the best predicted amount of nicotine for a cigarette with 3 mg of tar?

b. Use the paired data consisting of carbon monoxide and nicotine. What is the best predicted amount of nicotine for a cigarette with 18 mg of carbon monoxide?

c. Which prediction is better: the result from part (a) or the result from part (b)? Why?

13. Diamond Prices, Carats, and Color Refer to Data Set 3 in Appendix B.

a. Use the paired data consisting of price and carat (weight). What is the best predicted price of a diamond with a weight of 1.5 carats?

b. Use the paired price/color data. What is the best predicted price of a diamond with a color rating of 3?

c. Which prediction is better: the result from part (a) or the result from part (b)? Why?

14. Car Fuel Consumption, Weight, and Displacement Refer to Data Set 18 in Appendix B.

a. Use the paired data consisting of highway fuel consumption rate (mi/gal) and weight (pounds). What is the best predicted fuel consumption rate for a car that weighs 3000 lb?

b. Use the paired data consisting of highway fuel consumption rate (mi/gal) and the engine displacement (in liters). What is the best predicted fuel consumption rate for a car with an engine displacement of 4.0 liters?

c. Which prediction is better: the result from part (a) or the result from part (b)? Why?

15. Movie Grosses, Budgets, and Viewer Ratings Refer to Data Set 15 in Appendix B.

a. Use the paired data consisting of the amount of money grossed and the amount of money budgeted. What is the best predicted gross amount if a movie has a budgeted amount of $15 million?

b. Use the paired data consisting of the amount of money grossed and the viewer rating. What is the best predicted gross amount if a movie has a viewer rating of 7?

c. Is the prediction from part (a) better than the prediction from part (b)? Why or why not?

16. Home Selling Prices, List Prices, and Taxes Refer to Data Set 16 in Appendix B.

a. Use the paired data consisting of home selling price and list price. What is the best predicted selling price of a home with a list price of $200,000?

b. Use the paired data consisting of home selling price and the amount of taxes. The tax bill is supposed to be based on the value of the house. Is it?

17. Predicting Values In each of the following cases, find the best predicted value of y given that $x = 3.00$. The given statistics are summarized from paired sample data.

a. $r = 0.931$, $\bar{y} = 7.00$, $n = 10$, and the equation of the regression line is $\hat{y} = 4.00 + 2.00x$.

b. $r = -0.033$, $\bar{y} = 2.50$, $n = 80$, and the equation of the regression line is $\hat{y} = 5.00 - 2.00x$.

18. Predicting Values In each of the following cases, find the best predicted value of y given that $x = 2.00$. The given statistics are summarized from paired sample data.

a. $r = -0.882$, $\bar{y} = 3.57$, $n = 15$, and the equation of the regression line is $\hat{y} = 23.00 - 8.00x$.

b. $r = 0.187$, $\bar{y} = 9.33$, $n = 60$, and the equation of the regression line is $\hat{y} = 4.00 + 8.00x$.

19. Identifying Outliers and Influential Points Refer to the six pairs of data listed in Table 9-1. If we include another dining party with a bill of $400 and a tip of $50, is the new point an outlier? Is it an influential point?

20. Identifying Outliers and Influential Points Refer to the six pairs of data listed in Table 9-1. If we include another dining party with a bill of $600 and a tip of $5, is the new point an outlier? Is it an influential point?

9-3 Beyond the Basics

21. How Is Regression Equation Affected by Change in Scale? Large numbers, such as those in the accompanying table, often cause computational problems. First use the given data to find the equation of the regression line, then find the equation of the regression line after each x-value has been divided by 1000. How are the results affected by the change in x? How would the results be affected if each y-value were divided by 1000?

x	924,736	832,985	825,664	793,427	857,366
y	142	111	109	95	119

22. Testing Least-Squares Property According to the least-squares property, the regression line minimizes the sum of the squares of the residuals. We noted that with the paired data in the margin, the regression equation is $\hat{y} = 5 + 4x$ and the sum of the squares of the residuals is 364. Show that the equation $\hat{y} = 8 + 3x$ results in a sum of squares of residuals that is greater than 364.

x	1	2	4	5
y	4	24	8	32

23. Using Logarithms to Transform Data If the scatterplot reveals a nonlinear (not a straight line) pattern that you recognize as another type of curve, you may be able to apply the methods of this section. For the data given in the margin, find the linear equation ($y = b_0 + b_1x$) that best fits the sample data, and find the logarithmic equation ($y = a + b \ln x$) that best fits the sample data. (*Hint:* Begin by replacing each x-value with $\ln x$.) Which of these two equations fits the data better? Why?

x	2.0	2.5	4.2	10.0
y	12.0	18.7	53.0	225.0

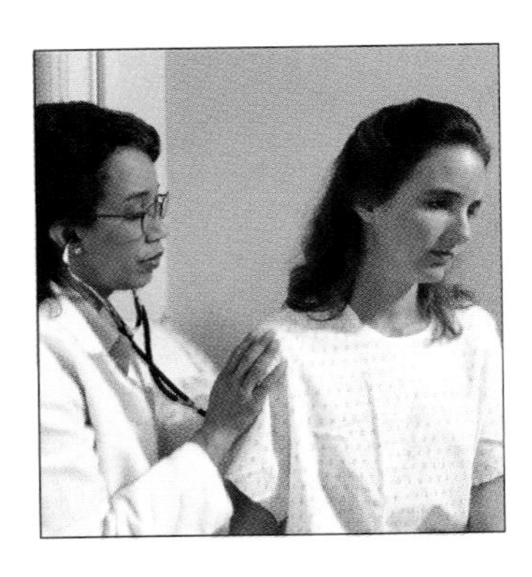

Wage Gender Gap

Many articles note that, on average, full-time female workers earn about 70¢ for each $1 earned by full-time male workers. Researchers at the Institute for Social Research at the University of Michigan analyzed the effects of various key factors and found that about one-third of the discrepancy can be explained by differences in education, seniority, work interruptions, and job choices. The other two-thirds remains unexplained by such labor factors.

whereas y should be 13, *it is* 19. The discrepancy between 13 and 19 cannot be explained by the regression line, and it is called an *unexplained deviation,* or a *residual.* The specific case illustrated in Figure 9-9 can be generalized as follows:

$$\begin{array}{lccccc} & \text{(total deviation)} & = & \text{(explained deviation)} & + & \text{(unexplained deviation)} \\ \text{or} & (y - \bar{y}) & = & (\hat{y} - \bar{y}) & + & (y - \hat{y}) \end{array}$$

This last expression applies to a particular point (x, y), but it can be further generalized and modified to include all of the pairs of sample data, as shown in Formula 9-5. In that formula, the **total variation** is expressed as the sum of the squares of the total deviation values, the **explained variation** is the sum of the squares of the explained deviation values, and the **unexplained variation** is the sum of the squares of the unexplained deviation values.

Formula 9-5

$$\begin{array}{lccccc} & \text{(total variation)} & = & \text{(explained variation)} & + & \text{(unexplained variation)} \\ \text{or} & \Sigma(y - \bar{y})^2 & = & \Sigma(\hat{y} - \bar{y})^2 & + & \Sigma(y - \hat{y})^2 \end{array}$$

Coefficient of Determination

The components of Formula 9-5 are used in the following important definition.

DEFINITION

The **coefficient of determination** is the amount of the variation in y that is explained by the regression line. It is computed as

$$r^2 = \frac{\text{explained variation}}{\text{total variation}}$$

We can compute r^2 by using the definition just given with Formula 9-5, or we can simply square the linear correlation coefficient r, which is found by using the methods described in Section 9-2. For example, in Section 9-2 we noted that if $r = 0.828$, then $r^2 = 0.686$, which means that *68.6% of the total variation in y can be explained by the linear relationship between x and y* (*as described by the regression equation*). *It follows that 31.4% of the total variation in y remains unexplained.*

EXAMPLE **Diamonds** In Exercise 13(a) in Section 9-2, we found that for the paired data consisting of weights (in carats) of diamonds and the prices of the diamonds, the linear correlation coefficient is given by $r = 0.767$. Find the percentage of the variation in y (price) that can be explained by the linear relationship between the weight and the price.

SOLUTION The coefficient of determination is $r^2 = 0.767^2 = 0.588$, indicating that the ratio of explained variation in y to total variation in y is 0.588. We can now state that 58.8% of the total variation in y can be explained by the regression equation. We interpret this to mean that 58.8% of the total variation in diamond prices can be explained by the variation in their weights; the other 41.2% is attributable to other factors, such as color, clarity, and random chance. But remember that these results are estimates based on the given sample data. Other sample data will likely result in different estimates.

Prediction Intervals

In Section 9-3 we used the Table 9-1 sample data to find the regression equation $\hat{y} = -0.347 + 0.149x$, where $\hat{y}$ represents the predicted amount of tip and x represents the amount of the bill. We then used that equation to predict the y-value, given that $x = \$100$. We found that the best predicted tip amount for a \$100 restaurant bill is \$14.55. If we use the unrounded values for slope and intercept, we get the more accurate result of \$14.51. Because \$14.51 is a single value, it is referred to as a *point estimate.* In Chapter 6 we saw that point estimates have the serious disadvantage of not giving us any information about how accurate they might be. Here, we know that \$14.51 is the best predicted value, but we don't know how accurate that value is. In Chapter 6 we developed confidence interval estimates to overcome that disadvantage, and in this section we follow that precedent. We will use a **prediction interval,** which is a confidence interval estimate of a predicted value of y.

The development of a prediction interval requires a measure of the spread of sample points about the regression line. Recall that the unexplained deviation (or residual) is the vertical distance between a sample point and the regression line, as illustrated in Figure 9-9. The *standard error of estimate* is a collective measure of the spread of the sample points about the regression line; it is formally defined as follows.

DEFINITION

The **standard error of estimate,** denoted by s_e, is a measure of the differences (or distances) between the observed sample y-values and the predicted values $\hat{y}$ that are obtained using the regression equation. It is given as

$$s_e = \sqrt{\frac{\Sigma(y - \hat{y})^2}{n - 2}}$$

where $\hat{y}$ is the predicted y-value.

STATDISK, Minitab, Excel, and the TI-83 Plus calculator are all designed to automatically compute the value of s_e. See "Using Technology" at the end of this section.

The development of the standard error of estimate s_e closely parallels that of the ordinary standard deviation introduced in Chapter 2. Just as the standard deviation is a measure of how values deviate from their mean, the standard error of estimate s_e is a measure of how sample data points deviate from their regression line. The reasoning behind dividing by $n - 2$ is similar to the reasoning that led to division by $n - 1$ for the ordinary standard deviation. It is important to note that smaller values of s_e reflect points that stay close to the regression line, and larger values occur with points farther away from the regression line.

Formula 9-6 can also be used to compute the standard error of estimate s_e. It is algebraically equivalent to the expression in the definition, but this form is generally easier to work with because it doesn't require that we compute each of the predicted values $\hat{y}$ by substitution in the regression equation. However, Formula 9-6 does require that we find the y-intercept b_0 and the slope b_1 of the estimated regression line.

Formula 9-6 $$s_e = \sqrt{\frac{\Sigma y^2 - b_0 \Sigma y - b_1 \Sigma xy}{n - 2}}$$ (standard error of estimate)

EXAMPLE Use Formula 9-6 to find the standard error of estimate s_e for the bill/tip sample data listed in Table 9-1.

SOLUTION In Section 9-2 we used the Table 9-1 data to find:

$$n = 6 \qquad \Sigma y^2 = 809.5364 \qquad \Sigma y = 63.58 \qquad \Sigma xy = 5308.7436$$

In Section 9-3 we used the Table 9-1 data to find the y-intercept and the slope of the regression line. Those values are given here with extra decimal places for greater precision.

$$b_0 = -0.347279 \qquad b_1 = 0.148614$$

We can now use these values in Formula 9-6 to find the standard error of estimate s_e.

$$s_e = \sqrt{\frac{\Sigma y^2 - b_0 \Sigma y - b_1 \Sigma xy}{n - 2}}$$

$$= \sqrt{\frac{809.5364 - (-0.347279)(63.58) - (0.148614)(5308.7436)}{6 - 2}}$$

$$= 3.26584 = 3.27 \text{ (rounded)}$$

We can measure the spread of the sample points about the regression line with the standard error of estimate $s_e = 3.27$.

We can use the standard error of estimate s_e to construct interval estimates that will help us see how dependable our point estimates of y really are. Assume that for each fixed value of x, the corresponding sample values of y are normally distributed about the regression line, and those normal distributions have the same variance. The following interval estimate applies to an *individual* y-value. (For a confidence interval used to predict the *mean* of all y-values for some given x-value, see Exercise 24.)

Prediction Interval for an Individual y

Given the fixed value x_0, the prediction interval for an individual y is

$$\hat{y} - E < y < \hat{y} + E$$

where the margin of error E is

$$E = t_{\alpha/2} s_e \sqrt{1 + \frac{1}{n} + \frac{n(x_0 - \bar{x})^2}{n(\Sigma x^2) - (\Sigma x)^2}}$$

and x_0 represents the given value of x, $t_{\alpha/2}$ has $n - 2$ degrees of freedom, and s_e is found from Formula 9-6.

EXAMPLE **Tipping** Refer to the Table 9-1 sample data listing amounts of restaurant bills (x) along with the corresponding amounts of tips (y). In previous sections we have shown the following:

- There is a significant linear correlation (at the 0.05 significance level).
- The regression equation is $\hat{y} = -0.347 + 0.149x$.
- When $x = \$100$, the predicted y-value is \$14.55, but a more accurate value of \$14.51 is obtained if we calculate the predicted value using unrounded values of slope and intercept.

Construct a 95% prediction interval for the tip, given that the bill is \$100. This will provide a sense of how accurate the predicted tip of \$14.51 really is.

SOLUTION We have already used the Table 9-1 sample data to find the following values:

$n = 6 \qquad \bar{x} = 73.64 \qquad \Sigma x = 441.84 \qquad \Sigma x^2 = 36{,}754.1416 \qquad s_e = 3.26584$

From Table A-3 we find $t_{\alpha/2} = 2.776$. (We used $6 - 2 = 4$ degrees of freedom with $\alpha = 0.05$ in two tails.) We can now calculate the margin of error E by letting $x_0 = 100$, because we want the prediction interval of the amount of tip y given that the amount of the bill is $x = 100$.

$$E = t_{\alpha/2} s_e \sqrt{1 + \frac{1}{n} + \frac{n(x_0 - \bar{x})^2}{n(\Sigma x^2) - (\Sigma x)^2}}$$

$$= (2.776)(3.26584)\sqrt{1 + \frac{1}{6} + \frac{6(100 - 73.64)^2}{6(36{,}754.1416) - (441.84)^2}}$$

$$= (2.776)(3.26584)(1.15388) = 10.46$$

With $\hat{y} = \$14.51$ and $E = \$10.46$, we get the prediction interval as follows:

$$\hat{y} - E < y < \hat{y} + E$$
$$\$14.51 - \$10.46 < y < \$14.51 + \$10.46$$
$$\$4.05 < y < \$24.97$$

That is, for a \$100 restaurant bill, we have 95% confidence that the tip is between \$4.05 and \$24.97. We might also say that for a \$100 restaurant bill, we have 95% confidence that the tip is between 4% and 25%. That's a relatively large range. (One factor contributing to the large range is that the sample size is very small because we are using only six pairs of sample data.)

Minitab can be used to find the prediction interval limits. If Minitab is used here, it will provide the result of (4.05, 24.98) below the heading "95.0% P.I." This corresponds to the same prediction interval found above, with the upper limit being off by a penny because of rounding.

In addition to knowing that the predicted tip amount is \$14.51, we now have a sense of how reliable that estimate really is. The 95% prediction interval found in this example shows that the \$14.51 predicted tip can vary substantially.

Using Technology

STATDISK: STATDISK can be used to find the linear correlation coefficient r, the equation of the regression line, the standard error of estimate s_e, the total variation, the explained variation, the unexplained variation, and the coefficient of determination. Select Analysis from the main menu bar, then use the option **Correlation and Regression.** Enter the paired data, or use Copy/Paste to copy the data. Enter a value for the significance level. Click on the **Evaluate** button. The STATDISK display will include the linear correlation coefficient, the coefficient of determination, the regression equation, and the value of the standard error of estimate s_e.

Minitab: Minitab can be used to find the regression equation, the standard error of estimate s_e (labeled S), the value of the coefficient of determination (labeled R-sq), and the limits of a prediction interval. Enter the x-data in column C1 and the y-data in column C2, then select the options **Stat, Regression,** and **Regression.** Enter C2 in the box labeled "Response" and enter C1 in the box labeled "Predictors."

If you want a prediction interval for some given value of x, click on the **Options** box and enter 100 (or whatever value of x_0 is desired) in the box labeled "Prediction intervals for new observations."

Excel: Excel can be used to find the regression equation, the standard error of estimate s_e, and the coefficient of determination (labeled as R square). Enter the paired data in columns A and B.

To use Excel's Data Analysis add-in, select **Tools** from the main menu, then select **Data Analysis,** followed by **Regression,** and then click **OK.** Enter the range for the y values, such as B1:B6. Enter the range for the x values, such as A1:A6. Click on **OK.**

To use the Data Desk XL add-in, click **DDXL** and select **Regression,** then click on the Function Type box and select **Simple Regression.** Click on the pencil icon for the response variable and enter the range of values for the y (or dependent) variable. Click on the pencil icon for the explanatory variable and enter the range of values for the x (or independent) variable. Click on **OK.**

TI-83 Plus: The TI-83 Plus calculator can be used to find the linear correlation coefficient r, the equation of the regression line, the standard error of estimate s_e, and the coefficient of determination (labeled as r^2). Enter the paired data in lists L1 and L2, then press **STAT** and select **TESTS,** and then choose the option **LinRegTTest.**

9-4 Basic Skills and Concepts

Interpreting the Coefficient of Determination. *In Exercises 1–4, use the value of the linear correlation coefficient r to find the coefficient of determination and the percentage of the total variation that can be explained by the linear relationship between the two variables.*

1. $r = 0.3$

2. $r = -0.05$

3. $r = -0.327$

4. $r = 0.777$

Interpreting a Computer Display *In Exercises 5–8, refer to the Minitab display on the next page that was obtained by using the paired data consisting of bear chest size and bear weight as listed in Data Set 7 in Appendix B. Along with the paired sample data, Minitab was also given a chest size of 50 (in.) to be used for predicting weight.*

5. Testing for Correlation Using the information provided in the display determine the value of the linear correlation coefficient. Given that there are 54 pairs of data, is there a significant linear correlation between the chest sizes of bears and their weights?

6. Identifying Total Variation What percentage of the total variation in weights of bears can be explained by the linear relationship between chest size and weight?

MINITAB

```
The regression equation is
Weight = -264 + 12.5 CHEST

Predictor        Coef      StDev         T        P
Constant      -264.48      17.90    -14.77    0.000
CHEST         12.5444     0.4859     25.82    0.000

S = 33.08       R-Sq = 92.8%      R-Sq(adj) = 92.6%


Predicted Values

    Fit   StDev Fit        95.0% CI              95.0% PI
 362.74        8.29  (346.10, 379.38)   (294.31, 431.17)
```

7. Predicting Bear Weight If a bear has a chest size of 50 in., what is the single value that is the best predicted weight? (Assume that there is a significant linear correlation between chest size and weight.)

8. Finding Prediction Interval For a given chest size of 50 in., identify the 95% prediction interval and write a statement interpreting that interval.

Finding Measures of Variation. *In Exercises 9–12, find the (a) explained variation, (b) unexplained variation, (c) total variation, (d) coefficient of determination, and (e) standard error of estimate s_e.*

9. Tile Quantity and Cost The accompanying table lists numbers x of patio tiles and costs y (in dollars) of having them manually cut to fit. (The equation of the regression line is $\hat{y} = 2 + 3x$.)

x	1	2	3	5	6
y	5	8	11	17	20

10. Bear Chest Size and Weight The paired data below consist of the chest sizes (in inches) and weights (in pounds) of a sample of male bears. (The equation of the regression line is $\hat{y} = -187.462 + 11.2713x$.)

x Chest (in.)	26	45	54	49	41	49	44	19
y Weight (lb)	90	344	416	348	262	360	332	34

11. Discarded Plastic and Household Size The paired data below consist of the weights (in pounds) of discarded plastic and sizes of households. (The equation of the regression line is $\hat{y} = 0.549270 + 1.47985x$.)

Plastic (lb)	0.27	1.41	2.19	2.83	2.19	1.81	0.85	3.05
Household size	2	3	3	6	4	2	1	5

12. Discarded Garbage and Household Size The paired data below consist of the total weights (in pounds) of discarded garbage and sizes of households. (The equation of the regression line is $\hat{y} = 0.182817 + 0.119423x$.)

Total (lb)	10.76	19.96	27.60	38.11	27.90	21.90	21.83	49.27	33.27	35.54
Household size	2	3	3	6	4	2	1	5	6	4

13. Effect of Variation on Prediction Interval Refer to the data given in Exercise 9 and assume that the necessary conditions of normality and variance are met.
a. For $x = 4$, find $\hat{y}$, the predicted value of y.
b. How does the value of s_e affect the construction of the 95% prediction interval of y for $x = 4$?

14. Finding Predicted Value and Prediction Interval Refer to Exercise 10 and assume that the necessary conditions of normality and variance are met.
a. For a bear with a measured chest size of 52 in., find $\hat{y}$, the predicted weight.
b. Find the 99% prediction interval of y for $x = 52$.

15. Finding Predicted Value and Prediction Interval Refer to the data given in Exercise 11 and assume that the necessary conditions of normality and variance are met.
a. Find the predicted size of a household that discards 2.50 lb of plastic.
b. Find the 95% prediction interval for the size of a household that discards 2.50 lb of plastic.

16. Finding Predicted Value and Prediction Interval Refer to the data given in Exercise 12 and assume that the necessary conditions of normality and variance are met.
a. For a household discarding 20.0 lb of garbage, find the predicted household size.
b. Find the 99% prediction interval for the size of a household that discards 20.0 lb of garbage.

Finding a Prediction Interval. *In Exercises 17–20, refer to the Table 9-1 sample data. Let x represent the amount of a restaurant bill, and let y represent the amount of the tip. Use the given amount of the bill and the given degree of confidence to construct a prediction interval estimate of the tip amount. (See the example in this section.)*

17. \$50 bill; 95% confidence

18. \$40 bill; 99% confidence

19. \$80 bill; 98% confidence

20. \$75.84 bill; 90% confidence

9-4 Beyond the Basics

21. Confidence Intervals for β_0 and β_1 Confidence intervals for the y-intercept β_0 and slope β_1 for a regression line ($y = \beta_0 + \beta_1 x$) can be found by evaluating the limits in the intervals below.

$$b_0 - E < \beta_0 < b_0 + E$$

where

$$E = t_{\alpha/2} s_e \sqrt{\frac{1}{n} + \frac{\bar{x}^2}{\Sigma x^2 - \frac{(\Sigma x)^2}{n}}}$$

$$b_1 - E < \beta_1 < b_1 + E$$

(continued)

where $$E = t_{\alpha/2} \cdot \frac{s_e}{\sqrt{\Sigma x^2 - \frac{(\Sigma x)^2}{n}}}$$

In these expressions, the y-intercept b_0 and the slope b_1 are found from the sample data, and $t_{\alpha/2}$ is found from Table A-3 by using $n - 2$ degrees of freedom. Using the bill/tip data in Table 9-1, find the 95% confidence interval estimates of β_0 and β_1. From those results, what can you conclude about the predicted amount of a tip?

22. Understanding Variation

a. If a collection of paired data includes at least three pairs of values, what do you know about the linear correlation coefficient if $s_e = 0$?

b. If a collection of paired data is such that the total explained variation is 0, what do you know about the slope of the regression line?

c. Using the coefficient of determination with the bear measurements in Data Set 7 in Appendix B, we find that $r^2 = 0.747$ for the paired bear length and bear weight data, and $r^2 = 0.928$ for the paired bear chest size and bear weight data. From these results we conclude that 74.7% of the variation in the weight of a bear can be explained by the linear relationship between length and weight, and 92.8% of the variation in weight can be explained by the linear relationship between chest size and weight. Are these results inconsistent because the sum of 74.7% and 92.8% exceeds 100%? How is it possible to have two such percentages that exceed 100%?

23. Understanding Variation

a. Find an expression for the unexplained variation in terms of the sample size n and the standard error of estimate s_e.

b. Find an expression for the explained variation in terms of the coefficient of determination r^2 and the unexplained variation.

c. Suppose we have a collection of paired data for which $r^2 = 0.900$ and the regression equation is $\hat{y} = 3 - 2x$. Find the linear correlation coefficient.

24. Finding Confidence Interval for Mean Predicted Value From the expression given in this section for the margin of error corresponding to a prediction interval for y, we can get the expression

$$s_{\hat{y}} = s_e\sqrt{1 + \frac{1}{n} + \frac{n(x_0 - \bar{x})^2}{n(\Sigma x^2) - (\Sigma x)^2}}$$

which is the *standard error of the prediction* when predicting for a *single y*, given that $x = x_0$. When predicting for the *mean* of all values of y for which $x = x_0$, the point estimate $\hat{y}$ is the same, but $s_{\hat{y}}$ is as follows:

$$s_{\hat{y}} = s_e\sqrt{\frac{1}{n} + \frac{n(x_0 - \bar{x})^2}{n(\Sigma x^2) - (\Sigma x)^2}}$$

Use the data from Table 9-1 and extend the last example of this section to find a point estimate and a 95% confidence interval estimate of the mean amount of tip of all restaurant bills that are $100.

9-5 Multiple Regression

The preceding sections of this chapter have all involved relationships between exactly *two* variables. The main objective of this section is to develop a method for analyzing relationships involving *more than two* variables. For example, in addition to considering the relationship between the amount of a restaurant bill and the amount of the tip, we might also include such variables as the age of the waiter or waitress, the annual income of the person giving the tip, and so on. We will focus on three key elements: the multiple regression equation, the value of adjusted R^2, and the P-value. As in the previous sections of this chapter, we will work with *linear* relationships only. We begin with the *multiple regression equation*.

Multiple Regression Equation

DEFINITION

A **multiple regression equation** expresses a linear relationship between a dependent variable y and two or more independent variables $(x_1, x_2, \ldots, x_k)$. The general form of a multiple regression equation is $\hat{y} = b_0 + b_1x_1 + b_2x_2 + \cdots + b_kx_k$.

We will use the following notation, which follows naturally from the notation used in Section 9-3.

Notation

$\hat{y} = b_0 + b_1x_1 + b_2x_2 + \cdots + b_kx_k$ (General form of the estimated multiple regression equation)

n = sample size

k = number of *independent* variables (The independent variables are also called **predictor variables,** or x variables.)

$\hat{y}$ = predicted value of the dependent variable y (computed by using the multiple regression equation)

$x_1, x_2 \ldots, x_k$ are the independent variables.

β_0 = the y-intercept, or the value of y when all of the predictor variables are 0 (This value is a population parameter.)

b_0 = estimate of β_0 based on the sample data (b_0 is a sample statistic.)

$\beta_1, \beta_2, \ldots, \beta_k$ are the coefficients of the independent variables $x_1, x_2, \ldots, x_k$.

$b_1, b_2, \ldots, b_k$ are the sample estimates of the coefficients $\beta_1, \beta_2, \ldots, \beta_k$.

Although it is not easy, the calculations of the preceding sections of this chapter can all be done with a scientific calculator, even if the calculator does not have any special statistical features. The computations required for multiple regression are so complicated that a statistical software package must be used. Later in this section we will describe procedures for using STATDISK, Minitab, and Excel, then we will focus on *interpreting* computer displays. (The TI-83 Plus calculator did not have multiple regression features when this text was written, but a supplement is likely to be available as an application that can be stored.)

EXAMPLE **Bears** Data Set 7 in Appendix B has measurements taken from 54 bears that were anesthetized, but we will consider the data from only eight of those bears, as listed in Table 9-3. Using the data in Table 9-3, find the multiple regression equation in which the dependent (y) variable is weight and the independent variables are head length (HEADLEN) and total overall length (LENGTH).

SOLUTION Using Minitab, we obtain the results shown in the display on the next page. The multiple regression equation is shown as

WEIGHT = −374 + 18.8 HEADLEN + 5.87 LENGTH

Using the notation presented earlier in this section, we could write this equation as

$$\hat{y} = -374 + 18.8x_3 + 5.87x_6$$

This equation best fits the given data (for weight, head length, and overall length) according to the least-squares criterion described in Section 9-3. Remember, this equation is based on sample data, so another sample will likely result in a slightly different equation. If the equation fits the data well,

TABLE 9-3 Data from Anesthetized Male Bears

Variable	Minitab Column	Name	Sample Data							
y	C1	WEIGHT	80	344	416	348	262	360	332	34
x_2	C2	AGE	19	55	81	115	56	51	68	8
x_3	C3	HEADLEN	11.0	16.5	15.5	17.0	15.0	13.5	16.0	9.0
x_4	C4	HEADWDTH	5.5	9.0	8.0	10.0	7.5	8.0	9.0	4.5
x_5	C5	NECK	16.0	28.0	31.0	31.5	26.5	27.0	29.0	13.0
x_6	C6	LENGTH	53.0	67.5	72.0	72.0	73.5	68.5	73.0	37.0
x_7	C7	CHEST	26	45	54	49	41	49	44	19

it can be used for predictions. For example, if we determine that the equation is suitable for predictions, and if we have a bear with a 14.0-in. head length and a 71.0-in. overall length, we can predict its weight by substituting those values into the regression equation to get a predicted weight of 306 lb. Also, the coefficients $b_3 = 18.8$ and $b_6 = 5.87$ can be used to determine marginal change, as described in Section 9-3. For example, the coefficient $b_3 = 18.8$ shows that when the overall length of a bear remains constant, the predicted weight increases by 18.8 lb for each 1-in. increase in the length of the head.

Predictors for Success

When a college accepts a new student, it would like to have some positive indication that the student will be successful in his or her studies. College admissions deans consider SAT scores, standard achievement tests, rank in class, difficulty of high school courses, high school grades, and extracurricular activities. In a study of characteristics that make good predictors of success in college, it was found that class rank and scores on standard achievement tests are better predictors than SAT scores. A multiple regression equation with college grade-point average predicted by class rank and achievement test score was not improved by including another variable for SAT score. This particular study suggests that SAT scores should not be included among the admissions criteria, but supporters argue that SAT scores are useful for comparing students from different geographic locations and high school backgrounds.

Adjusted R^2

R^2 denotes the **multiple coefficient of determination,** which is a measure of how well the multiple regression equation fits the sample data. A perfect fit would result in $R^2 = 1$. A very good fit results in a value near 1. A very poor fit results in a value of R^2 close to 0. The value of $R^2 = 0.828$ in the Minitab display indicates that 82.8% of the variation in bear weight can be explained by the head length x_3 and the overall length x_6. *The multiple coefficient of determination* R^2 is a measure of how well the regression equation fits the sample data, but it has a serious flaw: As more variables are included, R^2 increases. (Actually, R^2 could remain the same, but it usually increases.) Although the largest R^2 is thus achieved by simply including all of the available variables, the best multiple regression equation does not necessarily use all of the available variables. Consequently, it is better to use the adjusted coefficient of determination when comparing different multiple regression

MINITAB

```
The regression equation is
WEIGHT = -374 + 18.8 HEADLEN + 5.87 LENGTH

Predictor       Coef   Stdev  t-ratio       p
Constant      -374.3   134.1    -2.79   0.038
HEADLEN        18.82   23.15     0.81   0.453
LENGTH         5.875   5.065     1.16   0.299

s = 68.56   R-sq = 82.8%   R-sq(adj) = 75.9%

Analysis of Variance

SOURCE       DF       SS      MS       F      p
Regression    2   113142   56571   12.03  0.012
Error         5    23506    4701
Total         7   136648

SOURCE       DF   SEQ SS
HEADLEN       1   106819
LENGTH        1     6323
```

(1) Multiple regression equation

$R^2 = 0.828$

(2) Adjusted $R^2 = 0.759$

(3) Overall significance of multiple regression equation

Model for Alumni Contributions

One study developed a multiple regression equation that is a good predictor of alumni donations at liberal arts colleges. The college hoped to use the results of the study to improve its fund-raising strategies. The dependent variable was the amount of money donated in a year. Independent variables included income, age, whether the donor was single, whether the donor belonged to a fraternity or sorority, whether the donor was active in alumni activities, the donor's major, distance to the college, and the nation's unemployment rate (used as a measure of the economy). Other independent variables (such as whether the donor had children) were excluded for lack of statistical significance. (See "An Econometric Model of Alumni Giving: A Case Study for a Liberal Arts College," by Bruggink and Siddiqui, The American Economist, Vol. 39, No. 2.)

equations, because doing so adjusts the R^2 value based on the number of variables and the sample size.

DEFINITION

The **adjusted coefficient of determination** is the multiple coefficient of determination R^2 modified to account for the number of variables and the sample size. It is calculated by using Formula 9-7.

Formula 9-7
$$\text{adjusted } R^2 = 1 - \frac{(n-1)}{[n-(k+1)]}(1-R^2)$$

where n = sample size
k = number of independent (x) variables

The Minitab display shows the adjusted coefficient of determination as `R-sq(adj) = 75.9%`. If we use Formula 9-7 with the R^2 value of 0.828, $n = 8$, and $k = 2$, we find that the adjusted R^2 value is 0.759, confirming Minitab's displayed value of 75.9%. For the weight, head length, and length data in Table 9-3, the R^2 value of 82.8% indicates that 82.8% of the variation in weight can be explained by the head length x_3 and overall length x_6, but when we compare this multiple regression equation to others, it is better to use the adjusted R^2 of 75.9% (or 0.759).

P-Value

The *P*-value is a measure of the overall significance of the multiple regression equation. The Minitab display shows a *P*-value of 0.012. In this case, the small value of 0.012 suggests that the multiple regression equation has good overall significance and is usable for predictions. That is, it makes sense to predict weights of bears based on their head lengths and overall lengths. Like the adjusted R^2, this *P*-value is a good measure of how well the equation fits the sample data. The value of 0.012 results from a test of the null hypothesis that $\beta_3 = \beta_6 = 0$. Rejection of $\beta_3 = \beta_6 = 0$ implies that at least one of β_3 and β_6 is not 0, suggesting that this regression equation is effective in determining bear weights.

A complete analysis of the Minitab results might include other important elements, such as the significance of the individual coefficients, but we will limit our discussion to the three key components—multiple regression equation, adjusted R^2, and *P*-value.

Finding the Best Multiple Regression Equation

If you refer to Table 9-3, you can see that there are seven different variables of measurement for the eight different bears. The Minitab display is based on the selection of weight as the dependent variable and the selection of head length and overall length as the independent variables. But if we want to predict the weight of a bear, is there some other combination of variables that might be better than head length and overall length? Table 9-4 lists a few of the combinations of variables, and we are now confronted with the important objective of finding the *best* multiple regression equation. *Although determination of the best multiple regression equation is often quite difficult and beyond the scope of this book,* the following guidelines should provide some help.

Guidelines for Finding the Best Multiple Regression Equation

1. *Use common sense and practical considerations to include or exclude variables.* For example, we might exclude the variable of age because inexperienced researchers might not know how to determine the age of a bear and, when questioned, bears are reluctant to reveal their ages.
2. *Include only a few variables.* Instead of including almost every available variable, include relatively few independent (x) variables. In weeding out independent variables that don't have an effect on the dependent variable, it might be helpful to find the linear correlation coefficient r for each pair of variables being considered. For example, using the data in Table 9-3, we will find that there is a 0.955 linear correlation for the paired `NECK/HEADLEN` data. Because there is such a high correlation between neck size and head length, there is no need to include both of those variables. In choosing between `NECK` and `HEADLEN`, we should include `NECK` for this reason: `NECK` is a better predictor of `WEIGHT` because the `NECK/WEIGHT` paired data have a linear correlation coefficient of $r = 0.971$, which is higher than $r = 0.884$ for the paired `HEADLEN/WEIGHT` data.
3. *Use adjusted R^2.* Select an equation having a value of adjusted R^2 with this property: If an additional independent variable is included, the value of adjusted R^2 does not increase by a substantial amount. For example, Table 9-4 shows that if we use only the independent variable `CHEST`, the adjusted R^2 is 0.980, but when we include all six variables, the adjusted R^2 increases to 0.996. Using six variables instead of only one is too high a price to pay for such a small increase in the adjusted R^2. We're better off using the single independent variable `CHEST` than using all six independent variables.

TABLE 9-4 Searching for the Best Multiple Regression Equation

	LENGTH	CHEST	HEADLEN/ LENGTH	AGE/NECK/ LENGTH/CHEST	AGE/HEADLEN/HEADWDTH/ NECK/LENGTH/CHEST
R^2	0.805	0.983	0.828	0.999	0.999
Adjusted R^2	0.773	0.980	0.759	0.997	0.996
Overall significance	0.002	0.000	0.012	0.000	0.046

4. *For a given number of independent (x) variables, select the equation with the largest value of adjusted R^2.* That is, choose those variables with the property that no other combination of the same number of independent variables will yield a larger value of adjusted R^2. For example, if we decide to use a single independent variable, we should use CHEST, because its adjusted R^2 of 0.980 is larger than the adjusted R^2 for any other single variable. Consequently, CHEST is the best *single* predictor of WEIGHT.
5. *Consider the P-value.* Select an equation having overall significance, as determined by the P-value in the computer display. For example, see the values of overall significance in Table 9-4. The use of all six independent variables results in an overall significance of 0.046, which is just barely significant at the $\alpha = 0.05$ level; we're better off with the single variable CHEST, which has overall significance of 0.000.

Making Music with Multiple Regression

Sony manufactures millions of compact discs in Terre Haute, Indiana. At one step in the manufacturing process, a laser exposes a photographic plate so that a musical signal is transferred into a digital signal coded with 0s and 1s. This process was statistically analyzed to identify the effects of different variables, such as the length of exposure and the thickness of the photographic emulsion. Methods of multiple regression showed that among all of the variables considered, four were most significant. The photographic process was adjusted for optimal results based on the four critical variables. As a result, the percentage of defective discs dropped and the tone quality was maintained. The use of multiple regression methods led to lower production costs and better control of the manufacturing process.

Using these guidelines in an attempt to find the best equation for predicting weights of bears, we find that for the data of Table 9-3, the best regression equation uses the single independent variable of chest size (CHEST). The best regression equation appears to be

$$\text{WEIGHT} = -195 + 11.4\ \text{CHEST}$$

or

$$\hat{y} = -195 + 11.4x_7$$

For cases involving a large number of independent variables, many of the large and powerful statistical software packages include a program for performing **stepwise regression,** whereby different combinations of independent variables are tried until the best model is obtained. Although stepwise regression can be easily accomplished with a computer, there are some serious problems associated with it, including these: It will not necessarily yield the best model if some predictor variables are highly correlated; it yields inflated values of R^2; it uses too much paper; and it allows us to not *think* about the problem. As always, we should be careful to use computer results as a tool that helps us make intelligent decisions; we should not let the computer become the decision-maker. Instead of relying solely on the result of a computer stepwise regression program, consider the preceding five factors when trying to identify the best multiple regression equation.

If we eliminate the variable AGE (as in guideline 1) and then run Minitab's stepwise regression program, we will get a display suggesting that the best regression equation is the one in which CHEST is the only independent variable. (If we include all six independent variables, Minitab selects a regression equation with the independent variables AGE, NECK, LENGTH, and CHEST, with an adjusted R^2 value of 0.997 and overall significance of 0.000.) It appears that we can estimate the weight of a bear based on its chest size, and the regression equation leads to this rule: The weight of a bear (in pounds) is estimated to be 11.4 times the chest size (in inches) minus 195.

When we discussed regression in Section 9-3, we listed four common errors that should be avoided when using regression equations to make predictions. These same errors should be avoided when using multiple regression equations. Be especially careful about concluding that a cause-effect relationship exists.

Using Technology

STATDISK: Select **Analysis,** then **Multiple Regression.** Either enter the data in the different columns, or use Copy/Paste to get the desired columns of data. Click on **Evaluate** and you will get a dialog box. Identify the columns that you want included, and identify the column representing the dependent (y) variable. STATDISK will provide the multiple regression equation along with other items, including the multiple coefficient of determination R^2 and the adjusted R^2.

Minitab: First enter the values in different columns. To avoid confusion among the different variables, we strongly recommend that you enter names for the different variables. Enter the names in the box atop each column of data. Select the main menu item **Statistics,** then select **Regression,** then **Regression** once again. In the dialog box, enter the variable to be used for the response (y) variable, and enter the variables you want included as x-variables. Click **OK.** The display will include the multiple regression equation, along with other items, including the multiple coefficient of determination R^2 and the adjusted R^2.

Excel: First enter the sample data in columns. Select **Tools** from the main menu, then select **Data Analysis** and **Regression.** In the dialog box, enter the range of values for the dependent Y-variable, then enter the range of values for the independent X-variables, which must be in adjacent columns. (Use Copy/Paste to move columns as desired.) The display will include the multiple coefficient of determination R^2, the adjusted R^2, and a list of the intercept and coefficient values used for the multiple regression equation.

9-5 Basic Skills and Concepts

Interpreting a Computer Display. *In Exercises 1–4, refer to the Minitab display given here and answer the given questions.*

1. Bear Measurements Identify the multiple regression equation that expresses weight in terms of age, head width, and neck size.

MINITAB

```
The regression equation is

WEIGHT = - 285 - 1.38 AGE - 11.2 HEADWDTH + 28.6 NECK

Predictor        Coef      Stdev    t-ratio       p
Constant      -285.21      78.45      -3.64   0.022
AGE           -1.3838     0.9022      -1.53   0.200
HEADWDTH       -11.24      20.88      -0.54   0.619
NECK           28.594      5.870       4.87   0.008

s = 32.49   R - sq = 96.9%   R - sq(adj) = 94.6%

Analysis of Variance

SOURCE        DF        SS        MS        F        p
Regression     3    132425     44142    41.81    0.002
Error          4      4223      1056
Total          7    136648

SOURCE        DF    SEQ SS
AGE            1     90527
HEADWDTH       1     16844
NECK           1     25054
```

2. Bear Measurements Identify the following.

a. The P-value corresponding to the overall significance of the multiple regression equation

b. The value of the multiple coefficient of determination R^2

c. The adjusted value of R^2

3. Bear Measurements Is the multiple regression equation usable for predicting a bear's weight based on its age, head width, and neck size? Why or why not?

4. Bear Measurements A 32-month-old bear is found to have a head width of 5.0 in. and a neck size of 21.5 in.

a. Find the predicted weight of the bear.

b. The bear in question actually weighed 180 lb. How accurate is the predicted weight from part (a)?

Bear Data: Finding and Using a Multiple Regression Equation. *In Exercises 5–8, refer to the bear data in Data Set 7 in Appendix B. Let the dependent variable be* `WEIGHT`*, and let the independent variables be those given in the exercise. (The data sets are already stored on the Disk included with this book.) Use software such as STATDISK or Minitab or Excel to answer these questions:*

a. Find the multiple regression equation that expresses the dependent variable `WEIGHT` *in terms of the given independent variables.*

b. *Identify the values of the multiple coefficient of determination R^2, the adjusted R^2, and (if Minitab or Excel is used) the P-value corresponding to the overall significance.*

c. *Does the multiple regression equation seem suitable for predicting the weight of a bear based on the given independent variables?*

5. LENGTH and CHEST

6. NECK, LENGTH, and CHEST

7. AGE, NECK, LENGTH, and CHEST

8. HEADLEN, LENGTH, and NECK

Movie Data: Finding and Using a Multiple Regression Equation. *In Exercises 9–12, refer to the movie data in Data Set 15 in Appendix B. We will try to find a multiple regression equation that the movie industry can use to make movies that people like (with higher viewer ratings), so let the dependent variable be Viewer Rating and let the independent variables be those given in the exercise. (The data sets are already stored on the Disk included with this book.) Use software such as STATDISK or Minitab or Excel to answer these questions:*

a. *Find the multiple regression equation that expresses the dependent variable Viewer Rating in terms of the given independent variables.*

b. *Identify the values of the multiple coefficient of determination R^2, the adjusted R^2, and (if Minitab or Excel is used) the P-value corresponding to the overall significance.*

c. *Does the multiple regression equation seem suitable for making predictions of viewer ratings based on the given independent variables?*

9. Budget: the amount of money budgeted for the movie

10. Length: the length (in minutes) of the movie

11. Budget and gross

12. Budget and length and gross

Diamond Data: Finding and Using a Multiple Regression Equation. *In Exercises 13–16, refer to the diamond data in Data Set 3 in Appendix B. We will try to find a multiple regression equation that consumers can use to buy diamonds, so let the dependent variable be Price and let the independent variables be those given in the exercise. (The data sets are already stored on the Disk included with this book.) Use software such as STATDISK or Minitab or Excel to answer these questions:*

a. *Find the multiple regression equation that expresses the dependent variable Price in terms of the given independent variables.*

b. *Identify the values of the multiple coefficient of determination R^2, the adjusted R^2, and (if Minitab or Excel is used) the P-value corresponding to the overall significance.*

c. *Does the multiple regression equation seem suitable for making predictions of price based on the given independent variables?*

13. Carat: weight in carats

14. Color

15. Carat and color

16. Carat and color and clarity

9-5 Beyond the Basics

17. Cigarette Nicotine: Finding Best Multiple Regression Equation Refer to Data Set 8 in Appendix B and find the best multiple regression equation with nicotine as the dependent variable. Is this "best" equation good for predicting the amount of nicotine in a cigarette based on the amount of tar and carbon monoxide?

18. Car Pollution: Finding Best Multiple Regression Equation Refer to Data Set 18 in Appendix B and find the best multiple regression equation that could be used to predict the amount of the greenhouse gases (GHG) emitted. For the independent variables, consider the highway fuel consumption rate (in mi/gal), the weight of the car, and the engine displacement.

19. Home Selling Price: Finding Best Multiple Regression Equation Refer to Data Set 16 in Appendix B and find the best multiple regression equation with selling price as the dependent variable. Is this "best" equation good for predicting the selling price of a home?

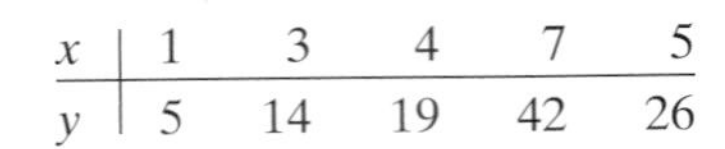

x	1	3	4	7	5
y	5	14	19	42	26

20. Using Multiple Regression for Equation of Parabola In some cases, the best-fitting multiple regression equation is of the form $\hat{y} = b_0 + b_1x + b_2x^2$. The graph of such an equation is a parabola. Using the data set listed in the margin, let $x_1 = x$, let $x_2 = x^2$, and find the multiple regression equation for the parabola that best fits the given data. Based on the value of the multiple coefficient of determination, how well does this equation fit the data?

9-6 Modeling

No, not that kind of modeling. This section introduces some basic concepts of developing a **mathematical model,** which is a mathematical function that "fits" or describes real-world data. For example, we might want a mathematical model consisting of an equation relating a variable for population size to another variable representing time. This is much like the methods of regression discussion in Section 9-3, except that we are no longer restricted to a model that must be linear. Also, instead of using randomly selected sample data, we will consider data collected periodically over time or some other basic unit of measurement. There are some powerful statistical methods that we could discuss (such as *time series*), but the main objective of this section is to describe briefly how technology can be used to find a good mathematical model.

The following are some generic models as listed in a menu from the TI-83 Plus calculator (press **STAT,** then select **CALC**):

Linear: $y = a + bx$

Quadratic: $y = ax^2 + bx + c$

Logarithmic: $y = a + b \ln x$

Exponential: $y = ab^x$

Power: $y = ax^b$

Logistic: $y = \dfrac{c}{1 + ae^{-bx}}$

The particular model that you select depends on the nature of the sample data, and a scatterplot can be very helpful in making that determination. The illustrations that follow are graphs of some common models displayed on a TI-83 Plus calculator.

TI-83 Plus

Linear: $y = 1 + 2x$

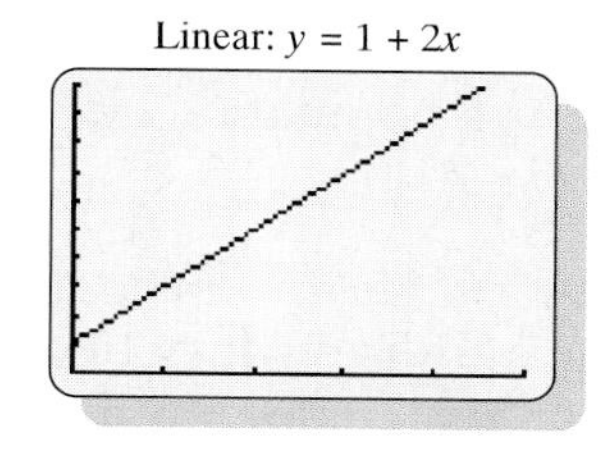

Quadratic: $y = 2x^2 - 8x + 9$

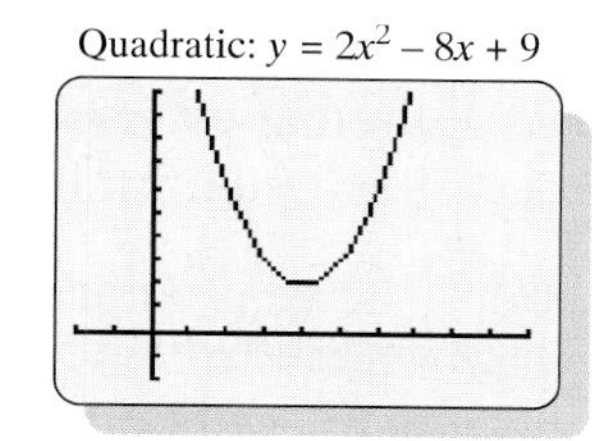

Logarithmic: $y = 1 + 2\ln x$

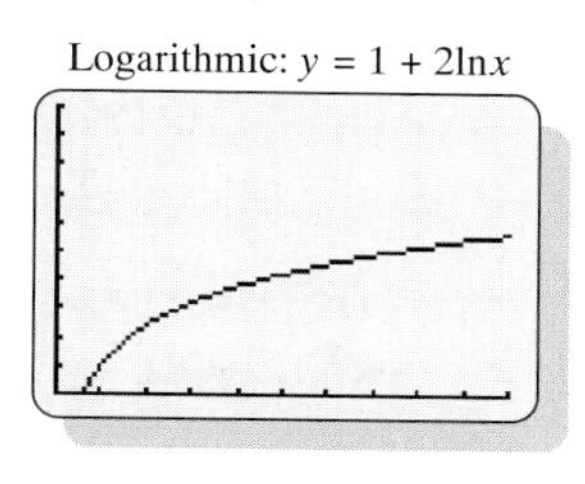

Exponential: $y = 2^x$

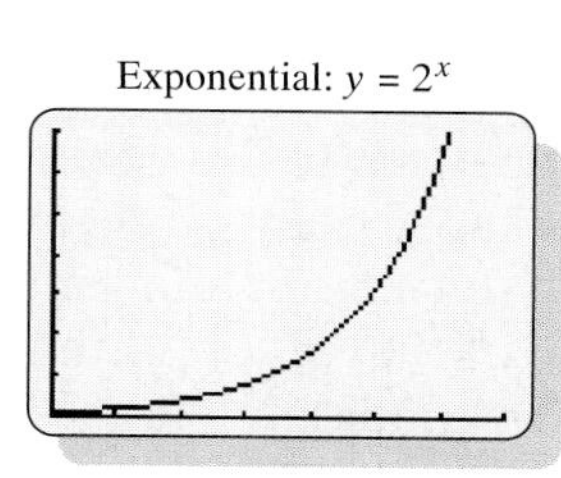

Power: $y = x^2$

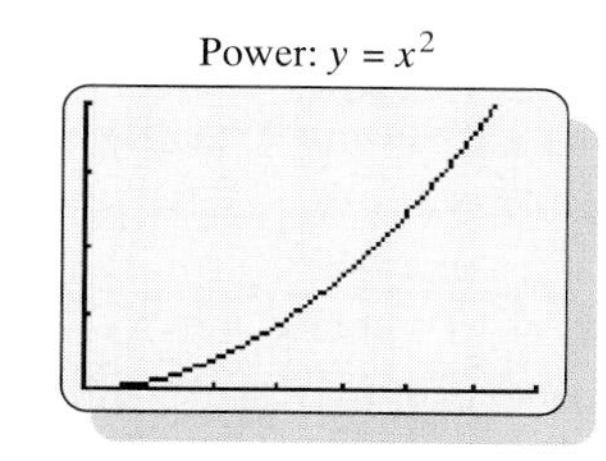

Logistic: $y = \frac{2}{1 + 50e^{-x}}$

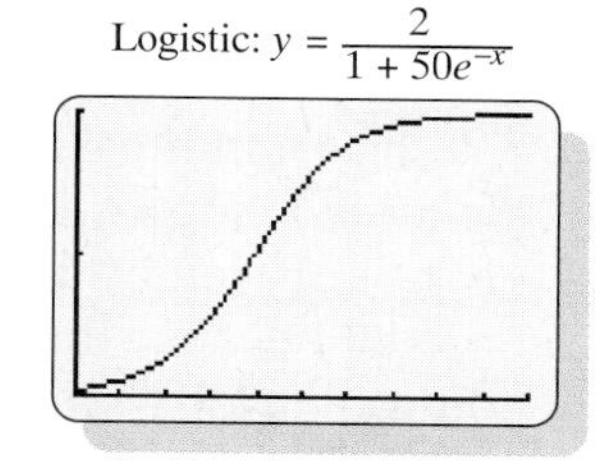

There are three basic rules for developing a good mathematical model:

1. *Look for a pattern in the graph.* Examine the graph of the plotted points and compare the basic pattern to the known generic graphs of a linear function, quadratic function, exponential function, power function, and so on. When trying to select a model, consider only those functions that visually appear to fit the observed points reasonably well.

2. *Find and compare values of R^2.* For each model being considered, use computer software or a TI-83 Plus calculator to find the value of the coefficient of determination R^2. Values of R^2 can be interpreted here the same way that they were interpreted in Section 9-5. When narrowing your possible models, select functions that result in larger values of R^2, because such larger values correspond to functions that better fit the observed points. However, don't place much importance on small differences, such as the difference between $R^2 = 0.984$ and $R^2 = 0.989$. (Another measurement used to assess the quality of a model is the sum of squares of the residuals. See Exercise 10.)

3. *Think.* Use common sense. Don't use a model that leads to predicted values known to be totally unrealistic. Use the model to calculate future values, past values, and values for missing years, then determine whether the results are realistic.

TABLE 9-5 Population (in millions) of the United States

Year	1800	1820	1840	1860	1880	1900	1920	1940	1960	1980
Coded year	1	2	3	4	5	6	7	8	9	10
Population	5	10	17	31	50	76	106	132	179	227

TI-83 Plus

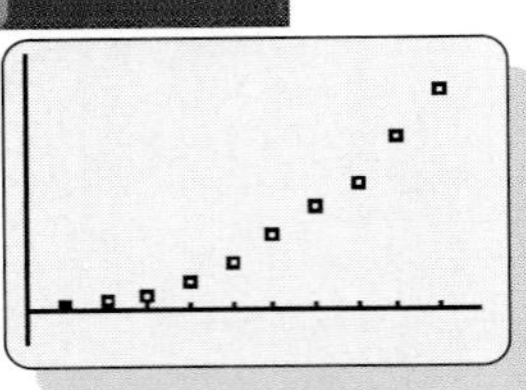

EXAMPLE Table 9-5 lists the population of the United States for different years. Find a good mathematical model for the population size, then predict the size of the U.S. population in the year 2020.

SOLUTION First, we "code" the year values by using 1, 2, 3 . . . instead of 1800, 1820, 1830. . . . The reason for this coding is to use values of x that are much smaller and much less likely to cause the computational difficulties that are likely to occur with really large x-values.

Look for a pattern in the graph: Examine the pattern of the data values as shown in the TI-83 Plus display and compare that pattern to the generic models shown earlier in this section. The pattern of those points is clearly not a straight line, so we rule out a linear model. We rule out a logistic model because the points don't show the "S" pattern of that graph, with a flattening of the graph occurring at the right. Good candidates for the model appear to be the quadratic, exponential, and power functions.

Find and compare values of R^2: The following displays show the TI-83 Plus results based on the quadratic, exponential, and power models. (If you are using a TI-83 Plus calculator, first press **2nd CATALOG,** then scroll down to **DiagnosticON** and press the **ENTER** key twice. With diagnostics on, press **STAT,** select **CALC,** and then select the desired model.) Comparing the values of the coefficient R^2, it appears that the quadratic model is best because it has the highest value of 0.9988, but the other displayed values are also quite high. If we select the quadratic function as the best model, we conclude that the equation $y = 2.67x^2 - 5.13x + 8.57$ best describes the relationship between the year x (coded with $x = 1$ representing 1800, $x = 2$ representing 1820, and so on) and the population y (in millions).

$y = 2.67x^2 - 5.13x + 8.57$

```
QuadReg
 y=ax²+bx+c
 a=2.674242424
 b=-5.131818182
 c=8.566666667
 R²=.998830762
```

$y = 4.77(1.52^x)$

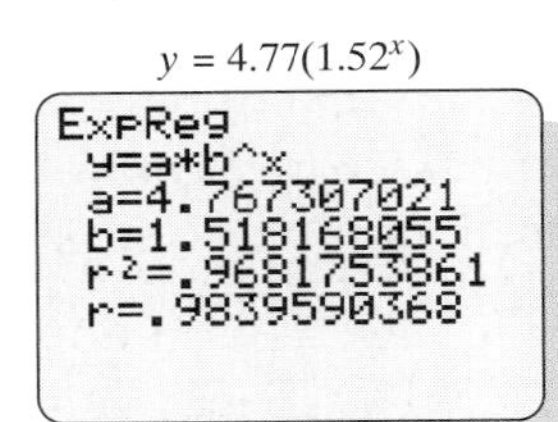

$y = 3.47x^{1.73}$

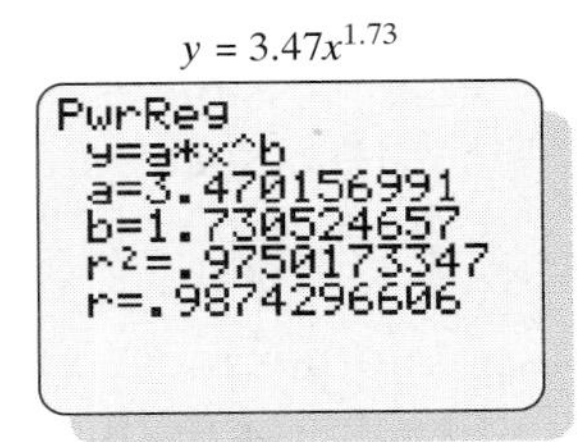

To predict the U.S. population for the year 2020, first note that the year 2020 is coded as $x = 12$ (see Table 9-5). Substituting $x = 12$ into the quadratic model of $y = 2.67x^2 - 5.13x + 8.57$, we get $y = 331$, which indicates that the U.S. population is estimated to be 331 million in the year 2020.

Think: The forecast result of 331 million in 2020 seems reasonable. (A U.S. Bureau of the Census projection suggests that the population in 2020 will be around 323 million.) However, there is considerable danger in making estimates for times way beyond the scope of the available data. For example, the quadratic model suggests that in 1492, the population was 636 million—an absurd result. For future estimates, only the logistic model shows this behavior typical of growing populations: The population begins to stabilize when it approaches the *carrying capacity of the environment*—the maximum population that can be supported by the limited resources. The quadratic model appears to be good for the available data (1800–1980), but other models might be better if it is absolutely necessary to make population estimates beyond that time frame.

In "Modeling the U.S. Population" (*AMATYC Review,* Vol. 20, No. 2), Sheldon Gordon uses more data than Table 9-5, and he uses much more advanced techniques to find better population models. In that article, he makes this important point:

> **"The best choice (of a model) depends on the set of data being analyzed and requires an exercise in judgment, not just computation."**

Using Technology

Any system capable of handling multiple regression can be used to generate some of the models described in this section. For example, STATDISK is not designed to work directly with the quadratic model, but its multiple regression feature can be used with the data in Table 9-5 to generate the quadratic model as follows: Select **Analysis,** then **Multiple Regression,** then proceed to enter the population values in column 1. Enter 1, 2, 3, . . . , 10 in column 2 and enter 1, 4, 9, . . . , 100 in column 3. After clicking on **Evaluate,** STATDISK generates the equation $y=8.5667 - 5.1318x + 2.6742x^2$ along with $R^2 = 0.99883$, which are the same results obtained from the TI-83 Plus calculator.

Minitab: First enter the matched data in columns C1 and C2, then select **Stat, Regression,** and **Fitted Line Plot.** You can choose a linear model, quadratic model, or cubic model. Displayed results include the equation, the value of R^2, and the sum of squares of the residuals.

TI-83 Plus: First turn on the diagnostics feature as follows: Press **2nd CATALOG,** then scroll down to **DiagnosticON** and press the **ENTER** key twice. Enter the matched data in lists L1 and L2. Press **STAT,** select **CALC,** and then select the desired model from the available options. Press **ENTER,** then enter L1, L2 (with the comma), and press **ENTER** again. The display includes the format of the equation along with the coefficients used in the equation; also the value of R^2 is included for many of the models.

9-6 Basic Skills and Concepts

Finding the Best Model. *In Exercises 1–8, construct a scatterplot and identify the mathematical model that best fits the given data. Assume that the model is to be used only for the scope of the given data, and consider only linear, quadratic, logarithmic, exponential, and power models.*

1.

x	1	2	3	4	5	6
y	8	2	0	2	8	18

2.

x	1	2	3	4	5	6
y	3	8	13	18	23	28

3.

x	1	2	3	4	5	6
y	3	9	27	80	245	725

4.

x	1	2	3	4	5	6
y	2.000	2.828	3.464	4.000	4.472	4.899

5. Shad Fish Harvest The accompanying table lists the weights (in pounds) of shad fish harvested in the Hudson River (based on data from the New York State Department of Environmental Conservation). What is the best predicted value for 1998?

Year	Shad (lb)
1980	1,313,100
1981	620,200
1982	378,900
1983	459,400
1984	701,400
1985	756,064
1986	798,768
1987	684,182
1988	782,932
1989	485,700
1990	463,529
1991	329,368
1992	265,598
1993	138,210
1994	157,672
1995	190,607
1996	135,629
1997	93,688

6. Swimming Records In "Beyond Modeling World Records with a Graphing Calculator: Assessing the Appropriateness of Models" (*Mathematics and Computer Education,* Vol. 32, No. 2), authors Martinez-Cruz and Ratliff include the given data for world record times in men's swimming. Is the best model much better than the others?

Year	1912	1924	1957	1968	1972	1976	1988	1994
Time (sec)	61.6	57.4	54.6	52.2	51.22	49.99	48.42	48.21

7. Return on Investment Kendra Korbin, owner and operator of the Cyber Video Game Store, records her business costs and revenue for different years, with the results listed below.

Amount invested (thousands of dollars)	1	2	5	11	20	31	41	46	48
Revenue (dollars)	2001	2639	3807	5219	6629	7899	8834	9250	9409

8. Distance/Time for Dropped Golf Ball In a physics experiment, a golf ball is dropped from a tall building and the distances (in feet) below the point of release are recorded for different times (in seconds) that the ball has fallen. The results are listed in the table. How well does the best model fit the data?

Time (sec)	0	0.5	1.0	1.5	2.0	2.5	3.0	3.5	4.0	4.5	5.0
Distance (ft)	0	4.0	15	35	63	100	143	194	253	320	396

9-6 Beyond the Basics

9. Using the Sum of Squares Criterion It was noted that in addition to the value of R^2, another measurement used to assess the quality of a model is the *sum of squares* of the residuals. A residual is the difference between an observed y-value and the value of y predicted from the model, which is denoted as $\hat{y}$. Better models have smaller sums of squares. Refer to the example in this section.

a. Find $\sum(y - \hat{y})^2$, the sum of squares of the residuals resulting from the linear model.

b. Find the sum of squares of residuals resulting from the quadratic model.

c. Verify that according to the sum of squares criterion, the quadratic model is better than the linear model.

10. Finding Sum of Squares and R^2 Using the data from Table 9-5, the logistic model is

$$y = \frac{383.852}{1 + 68.7950e^{-0.457556x}}$$

a. Find $\sum(y - \hat{y})^2$, the sum of squares of the residuals.

b. Find

$$R^2 = 1 - \frac{\sum(y - \hat{y})^2}{\sum(y - \bar{y})^2}$$

c. Verify that the quadratic model is better by comparing the R^2 values and the sums of the squares of the residuals.

Letter to the Editor

NEW YORK The *New York Times* published an article about sudden infant death syndrome, which is often abbreviated SIDS. In response to that article, Moorestown, New Jersey, resident Jean Mercer wrote a letter to the editor that was subsequently published by the *New York Times*. Her letter included this statement: "The article 'Clues on Sudden Infant Death Syndrome' stated that the practice of putting infants to sleep in the supine position has decreased deaths from SIDS. It would be more accurate to say that pediatricians advised the supine sleeping position during a time when the SIDS rate fell."

1. Explain what Jean Mercer meant when she claimed a lack of accuracy in the statement that the practice of putting infants to sleep in the supine position has decreased deaths from SIDS. In what sense is that statement inaccurate?

2. How does Jean Mercer's letter relate to correlation?

3. Find a specific passage in Section 9-2 that addresses the *common error* in the issue raised by Jean Mercer.

VOCABULARY LIST

bivariate data
correlation
bivariate normal distribution
scatterplot
scatter diagram
linear correlation coefficient
Pearson product moment correlation coefficient
lurking variable
centroid
regression equation
regression line
independent variable
predictor variable
dependent variable
response variable
marginal change
outlier
influential point
residual
least-squares property
total deviation
explained deviation
unexplained deviation
total variation
explained variation
unexplained variation
coefficient of determination
prediction interval
standard error of estimate
multiple regression equation
predictor variables
multiple coefficient of determination
adjusted coefficient of determination
stepwise regression
mathematical model

Review

This chapter presents basic methods for investigating relationships or correlations between two or more variables.

- Section 9-2 used scatter diagrams and the linear correlation coefficient to decide whether there is a linear correlation between two variables.
- Section 9-3 presented methods for finding the equation of the regression line that (by the least-squares criterion) best fits the paired data. When there is a significant linear correlation, the regression line can be used to predict the value of a variable, given some value of the other variable.
- Section 9-4 introduced the concept of total variation, with components of explained and unexplained variation. We defined the coefficient of determination r^2 to be the quotient obtained by dividing explained variation by total variation. We also developed methods for constructing prediction intervals, which are helpful in judging the accuracy of predicted values.
- In Section 9-5 we considered multiple regression, which allows us to investigate relationships among several variables. We discussed procedures for obtaining a multiple regression equation, as well as the value of the multiple coefficient of determination R^2, the adjusted R^2, and a P-value for the overall significance of the equation.
- In Section 9-6 we explored basic concepts of developing a mathematical model, which is a function that can be used to describe a relationship between two variables. Unlike the preceding sections of this chapter, Section 9-6 included nonlinear functions.

Review Exercises

Ice Cream Data: Understanding Correlation and Regression. *In Exercises 1–4, use the data in the accompanying table. The data come from a study of ice cream consumption that spanned the springs and summers of three years. The ice cream consumption is in pints per capita per week, price of the ice cream is in dollars, family income of consumers is in dollars per week, and temperature is in degrees Fahrenheit.*

Consumption	0.386	0.374	0.393	0.425	0.406	0.344	0.327	0.288	0.269	0.256
Price	1.35	1.41	1.39	1.40	1.36	1.31	1.38	1.34	1.33	1.39
Income	351	356	365	360	342	351	369	356	342	356
Temperature	41	56	63	68	69	65	61	47	32	24

Based on data from Kadiyala, *Econometrica,* Vol. 38.

1. a. Use a 0.05 significance level to test for a linear correlation between consumption and price.

b. What percentage of the variation in price can be explained by the variation in consumption?

c. Find the equation of the regression line that expresses consumption (y) in terms of price (x). $\hat{y} = -0.488 + 0.611x$

d. What is the best predicted consumption amount if the price is $1.38?

2. **a.** Use a 0.05 significance level to test for a linear correlation between consumption and income.
   b. What percentage of the variation in consumption can be explained by the variation in income?
   c. Find the equation of the regression line that expresses consumption (y) in terms of income (x).
   d. What is the best predicted consumption amount if the income is $365?

3. **a.** Use a 0.05 significance level to test for a linear correlation between consumption and temperature.
   b. What percentage of the variation in consumption can be explained by the variation in temperature?
   c. Find the equation of the regression line that expresses consumption (y) in terms of temperature (x).
   d. What is the best predicted consumption amount if the temperature is 32°F?

4. Use software such as STATDISK or Minitab or Excel to find the multiple regression equation of the form $\hat{y} = b_0 + b_1x_1 + b_2x_2 + b_3x_3$, where the dependent variable y represents consumption, x_1 represents price, x_2 represents income, and x_3 represents temperature. Also identify the value of the multiple coefficient of determination R^2, the adjusted R^2, and the P-value representing the overall significance of the multiple regression equation. Can the regression equation be used to predict ice cream consumption? Are any of the equations from Exercises 1–3 better?

Iowa Corn Data: Understanding Correlation and Regression. *In Exercises 5–8, use the sample data in the accompanying table. The data were collected from Iowa during a recent 10-year period. The amounts of precipitation are the annual totals (in inches). The average temperatures are annual averages (in degrees Fahrenheit). The corn values are the amounts of corn produced (in millions of bushels). The values of acres harvested are in thousands of acres.*

Year	1	2	3	4	5	6	7	8	9	10
Precipitation	32.4	41.8	36.5	37.5	31.6	40.3	33.3	21.6	24.7	39.4
Average temperature	49.80	46.93	48.47	48.28	46.67	49.31	51.87	49.34	47.06	49.88
Corn	1731	1578	744	1445	1707	1627	1320	899	1446	1562
Acres harvested	13,850	13,150	8,550	12,900	13,550	12,050	10,150	10,700	12,250	12,400

5. **a.** Use a 0.05 significance level to test for a significant linear correlation between the annual precipitation amounts and corn production amounts.
   b. Using the precipitation amounts and the corn production amounts, find the equation of the regression line. Let the precipitation amount be the independent variable.
   c. What is the best predicted amount of corn production for a year in which the precipitation amount is 29.3 in.?

6. **a.** Use a 0.05 significance level to test for a significant linear correlation between average annual temperature and corn production.
   b. Using the average annual temperatures and the corn production amounts, find the equation of the regression line. Let the average annual temperature be the independent variable.

(continued)

c. What is the best predicted amount of corn production for a year in which the average annual temperature amount is 48.86°F?

7. a. Use a 0.05 significance level to test for a significant linear correlation between the number of acres harvested and corn production amounts.
   b. Using the numbers of acres harvested and the corn production amounts, find the equation of the regression line. Let the acres harvested be the independent variable.
   c. What is the best predicted amount of corn production for a year in which 13,300,000 acres are harvested? (Remember, the table values of acres harvested are in thousands of acres.)

8. Iowa Weather/Corn Let y = corn production, x_1 = annual precipitation, x_2 = average annual temperature, and x_3 = acres harvested (in thousands of acres). Use software such as STATDISK or Minitab or Excel to find the multiple regression equation of the form $\hat{y} = b_0 + b_1x_1 + b_2x_2 + b_3x_3$. Also identify the value of the multiple coefficient of determination R^2, the adjusted R^2, and the P-value representing the overall significance of the multiple regression equation. Based on the results, should the multiple regression equation be used for making predictions? Why or why not?

Cumulative Review Exercises

1. Old Faithful Geyser Data: Hypothesis Test and Confidence Interval In 1970, the mean time between eruptions of the Old Faithful geyser was 66 min. Refer to the intervals (in minutes) between eruptions for the recent data listed in Data Set 11 in Appendix B.
   a. Test the claim of Yellowstone National Park geologist Rick Hutchinson that eruptions now occur at intervals that are longer than in 1970.
   b. Construct a 95% confidence interval for the mean time between eruptions.

2. Effects of Heredity and Environment on IQ In studying the effects of heredity and environment on intelligence, it has been helpful to analyze the IQs of identical twins who were separated soon after birth. Identical twins share identical genes inherited from the same fertilized egg. By studying identical twins raised apart, we can eliminate the variable of heredity and better isolate the effects of the environment. The accompanying table shows the IQs of pairs of identical twins (older twins are x) raised apart.
   a. Find the mean and standard deviation of the sample of older twins.
   b. Find the mean and standard deviation of the sample of younger twins.
   c. Based on the results from parts (a) and (b), does there appear to be a difference between the means of the two populations? In exploring the relationship between IQs of twins, is such a comparison of the two sample means the best approach? Why or why not?
   d. Is there a relationship between IQs of twins who were separated soon after birth? What method did you use? Write a summary statement about the effect of heredity and environment on intelligence, and note that your conclusions will be based on this relatively small sample of 12 pairs of identical twins.

x	107	96	103	90	96	113	86	99	109	105	96	89
y	111	97	116	107	99	111	85	108	102	105	100	93

Based on data from "IQs of Identical Twins Reared Apart," by Arthur Jensen, *Behavioral Genetics.*

3. Measuring Lung Volumes In a study of techniques used to measure lung volumes, physiological data were collected for 10 subjects. The values given in the accompanying table are in liters, representing the measured forced vital capacities of the 10 subjects in a sitting position and in a supine (lying) position. The issue we want to investigate is whether the position (sitting or supine) has an effect on the measured values.

a. If we test for a correlation between the sitting values and the supine values, will the result allow us to determine whether the position (sitting or supine) has an effect on the measured values? Why or why not?

b. Use an appropriate test for the claim that the position has no effect, so the mean difference is zero.

Subject	A	B	C	D	E	F	G	H	I	J
Sitting	4.66	5.70	5.37	3.34	3.77	7.43	4.15	6.21	5.90	5.77
Supine	4.63	6.34	5.72	3.23	3.60	6.96	3.66	5.81	5.61	5.33

Based on data from "Validation of Esophageal Balloon Technique at Different Lung Volumes and Postures," by Baydur et al., *Journal of Applied Physiology,* Vol. 62, No. 1.

COOPERATIVE GROUP ACTIVITIES

1. *Out-of-class activity:* Divide into groups of three or four people. Investigate the relationship between two variables by collecting your own paired sample data and using the methods of this chapter to determine whether there is a significant linear correlation. Also identify the regression equation and describe a procedure for predicting values of one of the variables when given values of the other variable. Suggested topics:

- Is there a relationship between taste and cost of different brands of chocolate chip cookies (or colas)? Taste can be measured on some number scale, such as 1 to 10.
- Is there a relationship between salaries of professional baseball (or basketball, or football) players and their season achievements?
- Rates versus weights: Is there a relationship between car fuel consumption rates and car weights? If so, what is it?
- Is there a relationship between the lengths of men's (or women's) feet and their heights?
- Is there a relationship between student grade-point averages and the amount of television watched? If so, what is it?

2. *In-class activity:* Divide into groups of 8 to 12 people. For each group member, *measure* height and arm span. For the arm span, the subject should stand with arms extended, like the wings on an airplane. It's easy to mark the height and arm span on a chalkboard, then measure the distances there. Using the paired sample data, is there a correlation between height and arm span? If so, find the regression equation with height expressed in terms of arm span. Can arm span be used as a reasonably good predictor of height?

3. *In-class activity:* Divide into groups of 8 to 12 people. For each group member, record the pulse rate by counting the number of heart beats in one minute. Also record height. Is there a relationship between pulse rate and height? If so, what is it?

4. *In-class activity:* Divide into groups of 8 to 12 people. For each group member, use a string and ruler to measure head circumference and forearm length. Is there a relationship between these two variables? If so, what is it?

5. *Out-of-class activity:* Divide into groups of three or four people for the purpose of studying the *regression effect,* first studied by Sir Francis Galton. He studied heredity and showed that tall fathers tend to have sons who are taller than average, but those sons are typically not as tall as their fathers. The heights of those sons tend to *regress,* or revert to a more typical mean height for people of the same gender. Similar results were found for sons of short fathers. Collect sample paired data on the heights of fathers and sons, and determine whether this regression effect continues to apply.

Technology Project

In Exercise 2 of the Cumulative Review Exercises, we noted that when studying the effects of heredity and environment on intelligence, it has been helpful to analyze the IQs of identical twins who were separated soon after birth. In this project, we will simulate 100 sets of twin births, but we will generate their IQ scores in a way that has no common genetic or environmental influences. Generate a list of 100 simulated IQ scores randomly selected from a normally distributed population having a mean of 100 and a standard deviation of 15. (Use the same procedure described in the Technology Project at the end of Chapter 5.) Now generate a second list of 100 simulated IQ scores that are also randomly selected from a normally distributed population with a mean of 100 and a standard deviation of 15. Even though the two lists were independently generated, treat them as paired data, so that the first score from each list represents the first set of twins, the second score from each list represents the second set of twins, and so on. Before doing any calculations, first estimate a value of the linear correlation coefficient that you would expect. Now test for a significant linear correlation and state your results. Given the way that the sample data were generated, what proportion of such trials should lead to the conclusion that there is a significant linear correlation? Is there a way to verify that the proportion is at least approximately correct? Describe a procedure for doing such a verification. If you're really ambitious, conduct the verification and write a brief report summarizing your results and conclusions.

from DATA to DECISION

Critical Thinking: Which home should you buy?

Refer to Data Set 16 in Appendix B. Construct a scatterplot and test for a correlation between selling price and list price. Describe a way to visually examine the graph of the scatterplot and identify homes that represent "good buys" in the sense that they sell for disproportionately less than the amount predicted by the regression equation.

Analyze the Results

What is the predicted selling price of a home that lists for $200,000? Is that predicted selling price likely to be reasonably accurate or a very rough approximation? Explain. Test for a correlation between selling price and tax bill. Assuming that the tax bill is supposed to be a fair assessment of a home's value, identify a home that appears to be assessed too low and identify another home that appears to be assessed too high.

Internet Project

Linear Regression

The linear correlation coefficient is a tool that can be used to measure the strength of the linear relationship between two variables. We can evaluate the linear correlation coefficient for any two sets of paired data, and then we can proceed to analyze the resulting values. Does it make sense for the two variables to be linearly correlated? Could a high correlation be caused by a third variable that is correlated with each of the two original variables? Go to the Web site for this book:

http://www.awlonline.com/triola

There you will find the Internet Project for this chapter. There you will be told how to find several paired data sets in the fields of sports, medicine, and economics. You will then apply the correlation and regression methods of this chapter to answer some of the questions that arise in this type of analysis.

Statistics at work

"Prospective employees should have a fundamental grasp of statistics and its implications in the business world."

Angela Gillespie

Traffic Analyst, Lycos.com

As a Traffic Analyst for Lycos, Inc., Angela reports on major and minor traffic metrics. She monitors changes in trends and behavior patterns for use, enhancing the site to increase reach and stickiness (the amount of time people spend online at any particular web site).

What is your job at Lycos?

I produce traffic reports of our site's activities each week. These are reviewed by our product group teams and senior management. They see what is increasing, what is decreasing, and make decisions about where our resources are spent.

My reports basically analyze trends on the sites and give projections for where we will be in a year or in any given time frame.

What concepts of statistics do you use?

Regression analysis and *R*-squared values.

How do you use statistics on the job?

To determine what is working and what is not working for our users. To determine the effectiveness of advertising campaigns, and to project future growth.

Please describe one specific example illustrating how the use of statistics was successful in improving a product or service.

At the end of our last fiscal year our CEO, Bob Davis, presented the company with an average daily pageview goal to be reached by the end of the next fiscal year. Using two years worth of pageview data, I put together a projection showing where we would be at the end of the next fiscal year if things remained static. Using an *R*-squared value gave these charts the oomph I needed to be effective. I updated the charts each week and presented them to the Product Management team. The data helped them understand what adjustments to make to their products and each week they got closer and closer to their goals. When Bob first presented the pageview goal, we all thought he had gone mad, but I am happy to say that at the end of the next fiscal year, we will have either reached our goal or be within 98% of it. Without the representation I supplied, product management would not have known where to focus their energy and resources. Because they were an efficient team, we have reached our unreachable goal.

Is your use of probability and statistics increasing, decreasing, or remaining stable?

As Lycos gets more sophisticated, they (management) expect more and more sophisticated reporting. It is increasing.

Do you feel job applicants are viewed more favorably if they have studied some statistics?

Absolutely, and not just within Lycos Reporting, but also in Product Marketing and Finance.

CHAPTER 10

Multinomial Experiments and Contingency Tables

CHAPTER PROBLEM

Titanic revisited: Women and children first?

The Chapter Problem for Chapter 3 included the table reproduced here as Table 10-1. This table summarizes the fate of the passengers and crew when the *Titanic* sank on Monday, April 15, 1912. If we examine the data, we see that 19.6% of the men (332 out of 1692) survived, 75.4% of the women (318 out of 422) survived, 45.3% of the boys (29 out of 64) survived, and 60% of the girls (27 out of 45) survived. These percentages are graphed in the accompanying Excel-generated displays. The first graph is a bar chart showing the percentage of survivors for each group. The second graph is a stacked bar chart; each bar represents the number who died stacked above the number who survived. The Excel graphs make it appear that there are differences, but are the differences really *significant?*

We will proceed to analyze the data in Table 10-1, and we will treat the 2223 people aboard the *Titanic* as a *sample.* We could take the position that the *Titanic* data in Table 10-1 constitute a *population* and therefore should not be treated as a sample, so that methods of inferential statistics do not apply. Let's stipulate that the Table 10-1 data are sample data randomly selected from the population of all theoretical people who would find themselves in the same conditions. Realistically, no other people will actually find themselves in the same conditions, but we will make that assumption for the purposes of this discussion and analysis. We can then determine whether the observed differences have statistical significance. (See also Paul Velleman's *ActivStats* software for the example involving the *Titanic.*)

TABLE 10-1 *Titanic* Mortality

	Men	Women	Boys	Girls	**Total**
Survived	332	318	29	27	**706**
Died	1360	104	35	18	**1517**
Total	**1692**	**422**	**64**	**45**	**2223**

EXCEL

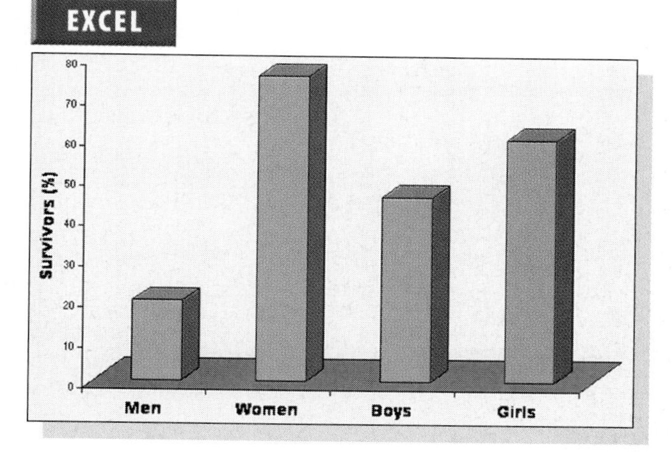

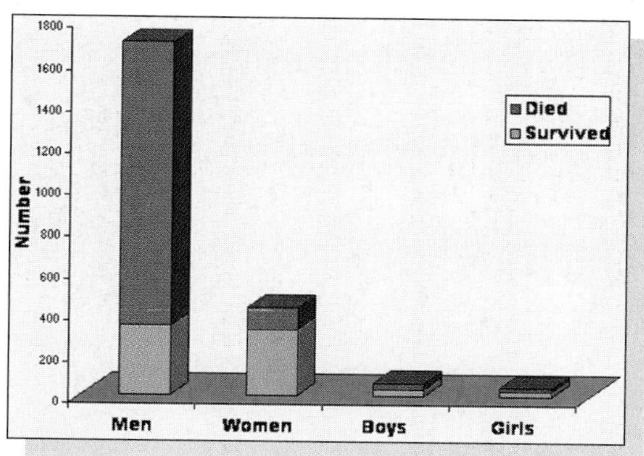

10-1 Overview

In this chapter we continue the pattern of applying inferential methods to different configurations of data. Recall from Chapter 1 that categorical (or qualitative, or attribute) data are those data that can be separated into different categories (often called **cells**) that are distinguished by some nonnumeric characteristic. For example, we might separate a sample of M&Ms into the color categories of red, orange, yellow, brown, blue, and green. After finding the frequency count for each category, we might proceed to test the claim that the frequencies fit (or agree with) the color distribution claimed by the manufacturer (Mars, Inc.). The main objective of this chapter is to test claims about categorical data consisting of frequency counts for different categories. In Section 10-2 we will consider multinomial experiments, which consist of observed frequency counts arranged in a single row or column (called a one-way frequency table), and we will test the claim that the observed frequency counts agree with some claimed distribution. In Section 10-3 we will consider contingency tables (or two-way frequency tables), which consist of frequency counts arranged in a table such as Table 10-1. We will use contingency tables for two types of very similar tests: (1) tests of independence, which test the claim that the row and column variables are independent; and (2) tests of homogeneity, which test the claim that different populations have the same proportion of some specified characteristic.

We will see that the methods of this chapter use the same χ^2 (chi-square) distribution that was first introduced in Section 6-6. Recall the following important properties of the chi-square distribution:

1. Unlike the normal and Student t distributions, the chi-square distribution is not symmetric. (See Figure 10-1.)
2. The values of the chi-square distribution can be 0 or positive, but they cannot be negative. (See Figure 10-1.)
3. The chi-square distribution is different for each number of degrees of freedom. (See Figure 10-2.)

Critical values of the chi-square distribution are found in Table A-4.

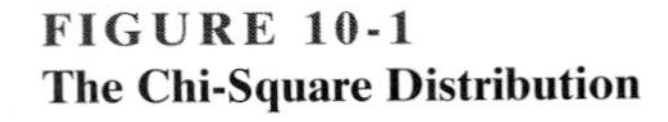

FIGURE 10-1
The Chi-Square Distribution

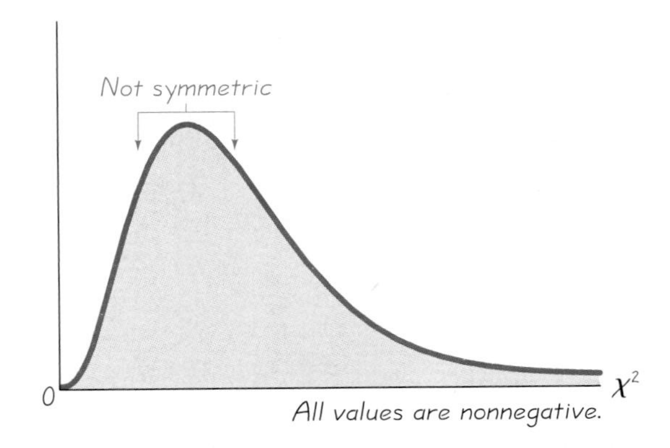

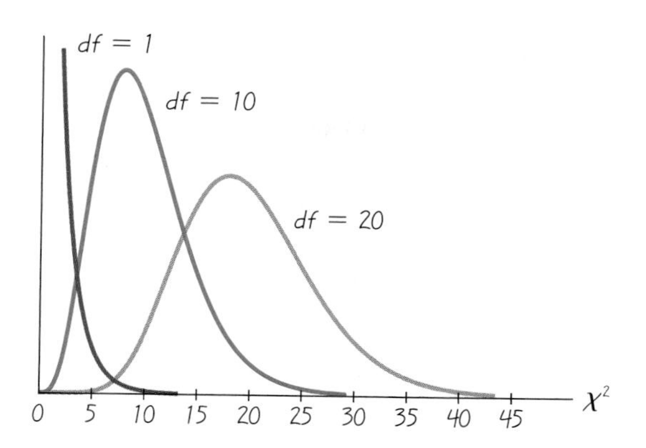

FIGURE 10-2
Chi-Square Distributions for 1, 10, and 20 Degrees of Freedom

10-2 Multinomial Experiments: Goodness-of-Fit

Each data set in this section consists of qualitative data that have been separated into different categories. The main objective is to determine whether the distribution agrees with or "fits" some claimed distribution.

The following assumptions apply when we test a hypothesis that the population proportion for each of the categories is as claimed.

Assumptions

1. The data have been randomly selected.
2. The sample data consist of frequency counts for each of the different categories.
3. For each category, the *expected* frequency is at least 5. (The expected frequency for a category is the frequency that would occur if the assumed distribution works exactly as it should. There is no requirement that the *observed* frequency for each category must be at least 5.)

We define a *multinomial experiment* the same way we defined a binomial experiment (Section 4-3), except that a multinomial experiment has more than two categories (unlike a binomial experiment, which has exactly two categories).

DEFINITION

A **multinomial experiment** is an experiment that meets the following conditions.

1. The number of trials is fixed.
2. The trials are independent.
3. All outcomes of each trial must be classified into exactly one of several different categories.
4. The probabilities for the different categories remain constant for each trial.

TABLE 10-2

Last-Digits of Mark McGwire's Home-Run Distances

Last Digit	Frequency
0	55
1	2
2	1
3	1
4	0
5	3
6	0
7	2
8	4
9	2

EXAMPLE **Last-Digit Analysis of Home-Run Distances** In 1998, Mark McGwire hit 70 home runs and broke a record that had been intact for about 30 years. *USA Today* published the distances of those home runs, and Table 10-2 summarizes the *last digits* of those distances. If such distances are actually measured, we usually expect that the last digits will occur with relative frequencies (or probabilities) that are roughly the same. In contrast, estimated values tend to have 0 or 5 occurring much more often as last digits. In Table 10-2, it appears that there is a disproportionate occurrence of 0s. Later, we will analyze the data, but at this point simply verify that the four conditions of a multinomial experiment are satisfied.

SOLUTION Here is the verification that the four conditions of a multinomial experiment are all satisfied:

1. The number of trials (last digits) is the fixed number 70.
2. The trials are independent, because the last digit of the length of a home run does not affect the last digit of the length of any other home run.
3. Each outcome (last digit) is classified into exactly 1 of 10 different categories. The categories are identified as 0, 1, 2, . . . , 9.
4. Finally, if we assume that the home-run distances are measured, the last digits should be equally likely, so that each possible digit has a probability of 1/10.

In this section we are presenting a method for testing a claim that in a multinomial experiment, the frequencies observed in the different categories fit a particular distribution. Because we test for how well an observed frequency distribution fits some specified theoretical distribution, this method is often called a *goodness-of-fit test.*

DEFINITION

A **goodness-of-fit test** is used to test the hypothesis that an observed frequency distribution fits (or conforms to) some claimed distribution.

For example, using the data in Table 10-2, we can test the hypothesis that the data fit a uniform distribution, with all of the digits being equally likely. Our goodness-of-fit tests will incorporate the following notation.

Notation

O represents the *observed frequency* of an outcome.

E represents the *expected frequency* of an outcome.

k represents the *number of different categories* or outcomes.

n represents the total *number of trials.*

Finding Expected Frequencies

In Table 10-2 we see that the observed frequencies are denoted by $O = 55$, $O = 2$, $O = 1$, and so on. The sum of the observed frequencies is 70, so $n = 70$. If we assume that the 70 digits were obtained from a population in which all digits are equally likely, then we *expect* that each digit should occur in 1/10 of the 70 trials, so each of the 10 expected frequencies is given by $E = 7$. If we generalize this result, we get an easy procedure for finding expected frequencies whenever we are assuming that all of the expected frequencies are equal: Simply divide the total number of observations by the number of different categories ($E = n/k$). In other cases where the expected frequencies are not all equal, we can often find the expected frequency for each category by multiplying the sum of all observed frequencies and the probability p for the category, so $E = np$. We summarize these two procedures here.

- **If all expected frequencies are equal, then each expected frequency is the sum of all observed frequencies divided by the number of categories, so that $E = n/k$.**
- **If the expected frequencies are not all equal, then each expected frequency is found by multiplying the sum of all observed frequencies by the probability for the category, so $E = np$ for each category.**

As good as these two formulas for E might be, it would be better to use an informal approach based on an understanding of the circumstances. Just ask yourself, "How can the observed frequencies be split up among the different categories so that there is perfect agreement with the claimed distribution?" See the two examples given later in this section.

We know that sample frequencies typically deviate somewhat from the values we theoretically expect, so we now present the key question: Are the differences between the actual *observed* values O and the theoretically *expected* values E statistically significant? We need a measure of the discrepancy between the O and E values, so we use the following test statistic. (Later, we will explain how this test statistic was developed, but you can see that it has differences of $O - E$ as a key component.)

Test Statistic for Goodness-of-Fit Tests in Multinomial Experiments

$$\chi^2 = \sum \frac{(O - E)^2}{E}$$

Critical Values

1. Critical values are found in Table A-4 by using $k - 1$ degrees of freedom, where k = number of categories.
2. Goodness-of-fit hypothesis tests are always *right-tailed.*

The form of the χ^2 test statistic is such that *close agreement* between observed and expected values will lead to a *small* value of χ^2 and a *large* P-value. A large discrepancy between observed and expected values will lead to a *large* value of χ^2 and a *small* P-value. The hypothesis tests of this section are therefore always right-tailed, because the critical value and critical region are located at the extreme right of the distribution. These relationships are summarized and illustrated in Figure 10-3.

Once we know how to find the value of the test statistic and the critical value, we can test hypotheses by using the procedure introduced in Chapter 7 and summarized in Figure 7-5.

EXAMPLE Last Digit Analysis of Home Runs: Equal Expected Frequencies Let's again refer to Table 10-2 for the last digits of Mark McGwire's home-run distances. The value of 0 seems to occur considerably more often, but is that really significant? Test the claim that the digits do not occur with the same frequency.

SOLUTION The claim that the digits do not occur with the same frequency is equivalent to the claim that the relative frequencies or probabilities of the 10 cells ($p_0, p_1, \ldots, p_9$) are not all equal. We will apply our standard procedure for testing hypotheses.

FIGURE 10-3
Relationships Among Components in Goodness-of-Fit Hypothesis Test

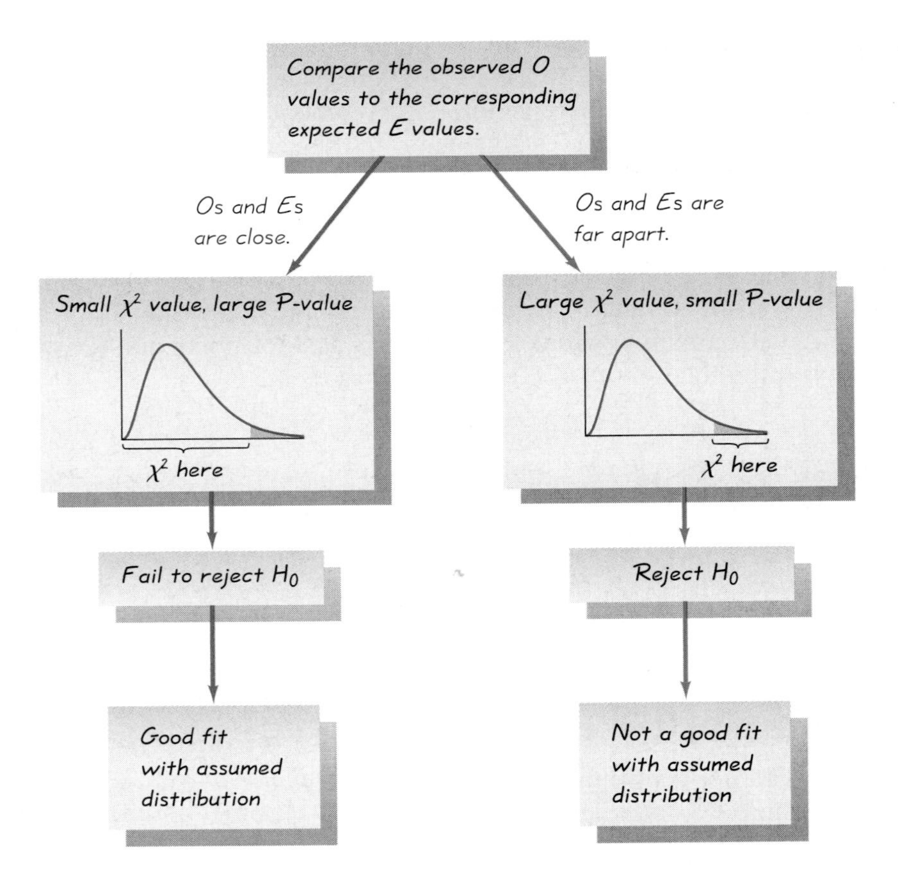

Step 1: The original claim is that the digits do not occur with the same frequency. That is, at least one of the probabilities $p_0, p_1, \ldots, p_9$ is different from the others.

Step 2: If the original claim is false, then all of the probabilities are the same. That is, $p_0 = p_1 = \cdots = p_9$.

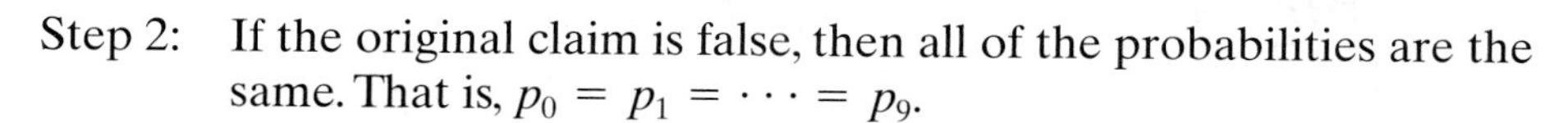

Step 3: The null hypothesis must contain the condition of equality, so we have

H_0: $p_0 = p_1 = p_2 = p_3 = p_4 = p_5 = p_6 = p_7 = p_8 = p_9$
H_1: At least one of the probabilities is different from the others.

Step 4: No significance level was specified, so we select $\alpha = 0.05$, a very common choice.

Step 5: Because we are testing a claim about the distribution of the last digits being a uniform distribution, we use the goodness-of-fit test described in this section. The χ^2 distribution is used with the test statistic given earlier.

Step 6: The observed frequencies O are listed in Table 10-2, and each corresponding expected frequency E is equal to 7 (because the 70 digits would be uniformly distributed through the 10 categories). Table 10-3 on the next page shows the computation of the χ^2 test statistic. The test statistic is $\chi^2 = 367.714$ (rounded). The critical value is $\chi^2 = 16.919$ (found in Table A-4 with $\alpha = 0.05$ in the right tail and degrees of freedom equal to $k - 1 = 9$). The test statistic and critical value are shown in Figure 10-4 on the next page.

Step 7: Because the test statistic falls within the critical region, there is sufficient evidence to reject the null hypothesis.

Step 8: There is sufficient evidence to support the claim that the last digits do not occur with the same relative frequency. We now have very strong evidence indicating that the home-run distances were not actually measured. It is reasonable to speculate that the distances are estimates instead of actual measurements.

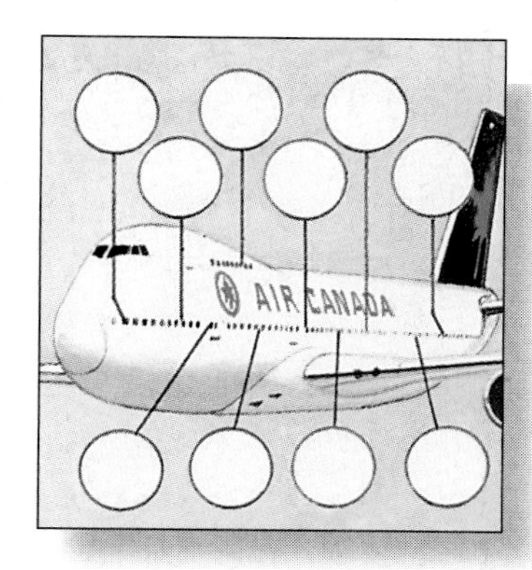

Safest Airplane Seats

Many of us believe that the rear seats are safest in an airplane crash. Safety experts do not agree that any particular part of an airplane is safer than others. Some planes crash nose first when they come down, but others crash tail first on takeoff. Matt McCormick, a survival expert for the National Transportation Safety Board, told *Travel* magazine that "There is no one safe place to sit." Goodness-of-fit tests can be used with a null hypothesis that all sections of an airplane are equally safe. Crashed airplanes could be divided into the front, middle, and rear sections. The observed frequencies of fatalities could then be compared to the frequencies that would be expected with a uniform distribution of fatalities. The χ^2 test statistic reflects the size of the discrepancies between observed and expected frequencies, and it would reveal whether some sections are safer than others.

The techniques in this section can be used to test whether an observed frequency distribution is a good fit with some theoretical frequency distribution. The preceding example tested for goodness-of-fit with a uniform distribution. Because many statistical analyses require a normally distributed population, we can use the chi-square test in this section to help determine whether given samples are drawn from normally distributed populations (see Exercise 22).

TABLE 10-3 Calculating the χ^2 Test Statistic for the Last Digits of Home-run Distances

Last Digit	Observed Frequency O	Expected Frequency E	$O - E$	$(O - E)^2$	$\frac{(O - E)^2}{E}$
0	55	7	48	2304	329.1429
1	2	7	−5	25	3.5714
2	1	7	−6	36	5.1429
3	1	7	−6	36	5.1429
4	0	7	−7	49	7.0000
5	3	7	−4	16	2.2857
6	0	7	−7	49	7.0000
7	2	7	−5	25	3.5714
8	4	7	−3	9	1.2857
9	2	7	−5	25	3.5714
	70	70			↑

(These two totals must agree.)

$$\chi^2 = \sum \frac{(O - E)^2}{E} = 367.7143$$

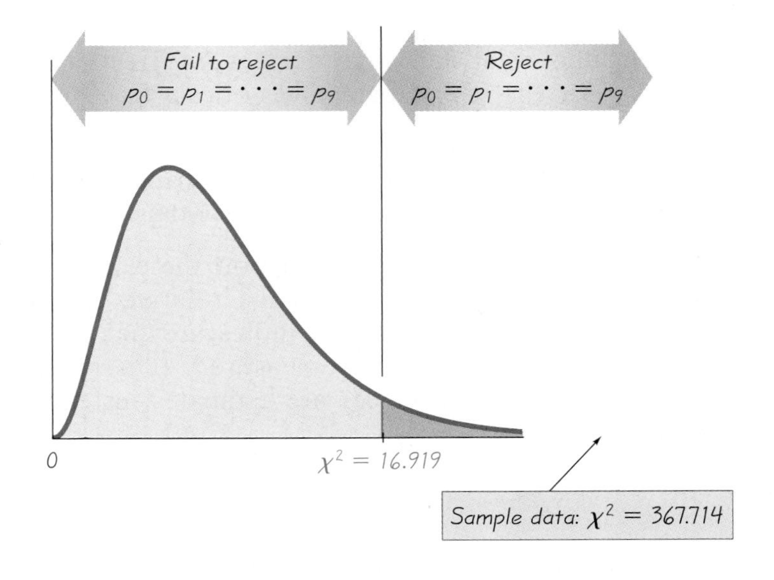

FIGURE 10-4
Goodness-of-Fit Test of
$p_0 = p_1 = p_2 = p_3 = p_4 = p_5 = p_6 = p_7 = p_8 = p_9$

The preceding example dealt with the null hypothesis that the probabilities for the different categories are all equal. The methods of this section can also be used when the hypothesized probabilities (or frequencies) are different, as shown in the next example.

EXAMPLE **M&Ms: Unequal Expected Frequencies** Mars, Inc., claims that its M&M plain candies are distributed with the following color percentages: 30% brown, 20% yellow, 20% red, 10% orange, 10% green, and 10% blue. The colors of the M&Ms listed in Data Set 10 in Appendix B are summarized in Table 10-4. Using the sample data and a 0.05 significance level, test the claim that the color distribution is as claimed by Mars, Inc.

SOLUTION We extended Table 10-4 to include the expected frequencies, which are calculated as follows. For n, we use the total number of trials (100), which is the total number of M&Ms observed in the sample. For the probabilities, we use the decimal equivalents of the claimed percentages (30%, 20%, . . . , 10%).

Brown: $E = np = (100)(0.30) = 30$

Yellow: $E = np = (100)(0.20) = 20$

⋮

Blue: $E = np = (100)(0.10) = 10$

In testing the given claim, Steps 1, 2, and 3 result in the following hypotheses:

H_0: $p_{\text{br}} = 0.3$ and $p_{\text{y}} = 0.2$ and $p_{\text{r}} = 0.2$ and $p_{\text{o}} = 0.1$ and $p_{\text{g}} = 0.1$ and $p_{\text{bl}} = 0.1$

H_1: At least one of the above proportions is different from the claimed value.

Steps 4, 5, and 6 lead us to use the goodness-of-fit test with a 0.05 significance level and a test statistic calculated from Table 10-5.

The test statistic is $\chi^2 = 5.950$. The critical value of χ^2 is 11.071, and it is found in Table A-4 (using $\alpha = 0.05$ in the right tail with $k - 1 = 5$ degrees of freedom). The test statistic and critical value are shown in Figure 10-5 on the next page. Because the test statistic does not fall within the critical region, there is not sufficient evidence to warrant rejection of the null hypothesis. There is not sufficient evidence to warrant rejection of the claim that the colors are distributed with the percentages given by Mars, Inc.

In Figure 10-6 on page 583 we graph the claimed proportions of 0.30, 0.20, 0.20, 0.10, 0.10 , 0.10, along with the observed proportions of 0.33,

TABLE 10-4 Frequencies of M&M Plain Candies

	Brown	Yellow	Red	Orange	Green	Blue
Observed frequency	33	26	21	8	7	5
Expected frequency	30	20	20	10	10	10

0.26, 0.21, 0.08, 0.07, and 0.05, so that we can visualize the discrepancy between the distribution that was claimed and the frequencies that were observed. The points along the purple line represent the claimed proportions, and the points along the green line represent the observed proportions. The corresponding pairs of points are all fairly close, showing that all of the expected frequencies are reasonably close to the corresponding observed frequencies. In general, graphs such as Figure 10-6 are helpful in visually comparing expected frequencies and observed frequencies, as well as suggesting which categories result in the major discrepancies.

TABLE 10-5 Calculating the χ^2 Test Statistic for M&M Data

Color Category	Observed Frequency O	Expected Frequency $E = np$	$O - E$	$(O - E)^2$	$\frac{(O - E)^2}{E}$
Brown	33	30	3	9	0.3000
Yellow	26	20	6	36	1.8000
Red	21	20	1	1	0.0500
Orange	8	10	−2	4	0.4000
Green	7	10	−3	9	0.9000
Blue	5	10	−5	25	2.5000
	100	100		$\chi^2 = \sum \frac{(O - E)^2}{E}$	$= 5.9500$

FIGURE 10-5
Goodness-of-Fit Test of $p_{br} = 0.3$ and $p_y = 0.2$ and $p_r = 0.2$ and $p_o = 0.1$ and $p_g = 0.1$ and $p_{bl} = 0.1$

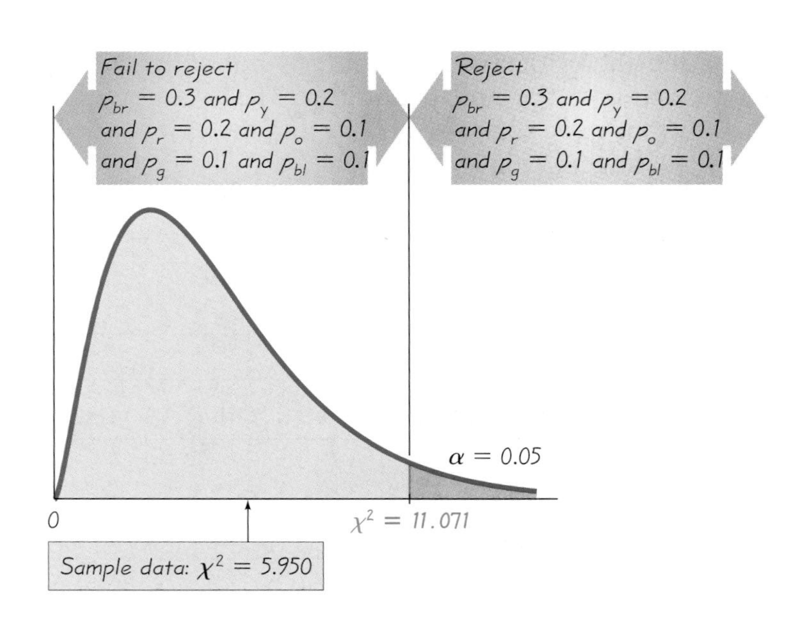

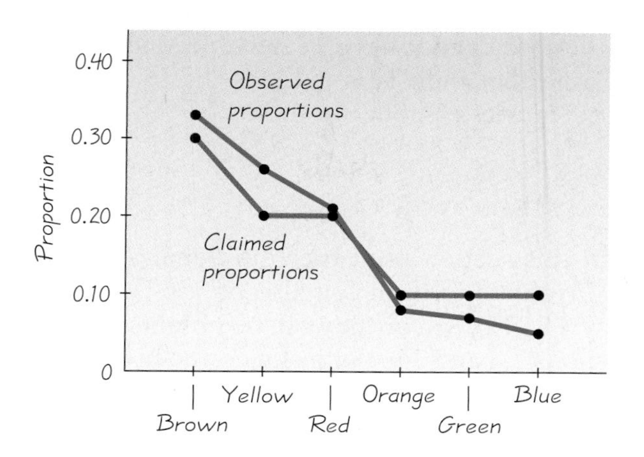

FIGURE 10-6
Comparison of Claimed and Observed Proportions

Rationale for the Test Statistic: The preceding examples should be helpful in developing a sense for the role of the χ^2 test statistic. It should be clear that we want to measure the amount of disagreement between observed and expected frequencies. Simply summing the differences between observed and expected values does not result in an effective measure because that sum is always 0, as shown below.

$$\Sigma(O - E) = \Sigma O - \Sigma E = n - n = 0$$

Squaring the $O - E$ values provides a better statistic, which reflects the differences between observed and expected frequencies. (The reasons for squaring the $O - E$ values are essentially the same as the reasons for squaring the $x - \overline{x}$ values in the formula for standard deviation.) The value of $\Sigma(O - E)^2$ measures only the magnitude of the differences, but we need to find the magnitude of the differences relative to what was expected. This relative magnitude is found through division by the expected frequencies, as in the test statistic.

The theoretical distribution of $\Sigma(O - E)^2/E$ is a discrete distribution because the number of possible values is limited. The distribution can be approximated by a chi-square distribution, which is continuous. This approximation is generally considered acceptable, provided that all expected values E are at least 5. We included this requirement with the assumptions that apply to this section. In Section 5-6 we saw that the continuous normal probability distribution can reasonably approximate the discrete binomial probability distribution, provided that np and nq are both at least 5. We now see that the continuous chi-square distribution can reasonably approximate the discrete distribution of $\Sigma(O - E)^2/E$, provided that all values of E are at least 5. (There are ways of circumventing the problem of an expected frequency that is less than 5, such as combining categories so that all expected frequencies are at least 5.)

The number of degrees of freedom reflects the fact that we can freely assign frequencies to $k - 1$ categories before the frequency for every category is determined. Yet, although we say that we can "freely" assign frequencies to $k - 1$ categories, we cannot have negative frequencies nor can we have frequencies so large that their sum exceeds the total of the observed frequencies for all categories combined.

P-Values

The examples in this section used the traditional approach to hypothesis testing, but the *P*-value approach can also be used. *P*-values are automatically provided by STATDISK, or they can be obtained by using the methods described in Sections 7-3 and 7-6. For instance, the preceding example resulted in a test statistic of $\chi^2 = 5.950$. That example had $k = 6$ categories, so there were $k - 1 = 5$ degrees of freedom. Referring to Table A-4, we see that for the row with 5 degrees of freedom, the test statistic of 5.950 is less than the lowest right-tailed critical value of 9.236, so the *P*-value is greater than 0.10. If the calculations for the preceding example are run on STATDISK, the display will include a *P*-value of 0.31111. A TI-83 Plus calculator can be used to determine that for $\chi^2 = 5.950$ and 5 degrees of freedom, the *P*-value is 0.311. The high *P*-value suggests that the null hypothesis should not be rejected. Remember, we reject the null hypothesis only when the *P*-value is equal to or less than the significance level.

Using Technology

STATDISK: Select **Analysis** from the main menu bar, then select the option **Multinomial Experiments.** Choose between "equal expected frequencies" and "unequal expected frequencies" and enter the data in the dialog box. If you choose "unequal expected frequencies," enter the expected values in the second column either as "counts" (with the actual expected frequencies) or as "proportions" (with the *probabilities* entered).

Minitab, Excel, TI-83 Plus: The methods of this section are not available as built-in procedures.

10-2 Basic Skills and Concepts

1. Testing Fairness of Roulette Wheel The author observed 500 spins of a roulette wheel at the Mirage Resort and Casino. (To the IRS: Isn't that Las Vegas trip now a tax deduction?) For each spin, the ball can land in any one of 38 different slots that are supposed to be equally likely. When STATDISK was used to test the claim that the slots are in fact equally likely, the test statistic $\chi^2 = 38.232$ was obtained.

a. Find the critical value assuming that the significance level is 0.10.

b. STATDISK displayed a *P*-value of 0.41331, but what do you know about the *P*-value if you must use only Table A-4 along with the given test statistic of 38.232, which results from the 38 spins? b. $0.10 < P\text{-value} < 0.90$

c. Write a conclusion about the claim that the 38 results are equally likely.

2. Testing Excel's Random Number Generator Excel can generate random numbers by using the Tools/Data Analysis/Random Number Generation options. The author used Excel to generate 1000 random digits (0, 1, . . . 9) by specifying that the results should be equally likely. Testing the claim of equally likely outcomes resulted in a test statistic of $\chi^2 = 10.400$ and a P-value of 0.31908.

a. Find the critical value assuming that the significance level is 0.025.

b. What can you conclude about the P-value if you know only the value of the test statistic and the number of categories (10)?

c. State a conclusion about the claim that Excel generates random digits in such a way that they are equally likely.

3. Flat Tire and Missed Class A classic tale involves four car-pooling students who missed a test and gave as an excuse a flat tire. On the makeup test, the instructor asked the students to identify the particular tire that went flat. If they really didn't have a flat tire, would they be able to identify the same tire? The author asked 41 other students to identify the tire they would select. The results are listed in the following table (except for one student who selected the spare). Use a 0.05 significance level to test the author's claim that the results fit a uniform distribution. What does the result suggest about the ability of the four students to select the same tire when they really didn't have a flat?

Tire	Left front	Right front	Left rear	Right rear
Number selected	11	15	8	6

4. Do Car Crashes Occur on Different Days with the Same Frequency? It is a common belief that more fatal car crashes occur on certain days of the week, such as Friday or Saturday. A sample of motor vehicle deaths for a recent year in Montana is randomly selected. The numbers of fatalities for the different days of the week are listed in the accompanying table. At the 0.05 significance level, test the claim that accidents occur with equal frequency on the different days.

Day	Sun	Mon	Tues	Wed	Thurs	Fri	Sat
Number of fatalities	31	20	20	22	22	29	36

Based on data from the Insurance Institute for Highway Safety.

5. Are DWI Fatalities the Result of Weekend Drinking? Many people believe that fatal DWI crashes occur because of casual drinkers who tend to binge on Friday and Saturday nights, whereas others believe that fatal DWI crashes are caused by people who drink every day of the week. In a study of fatal car crashes, 216 cases are randomly selected from the pool in which the driver was found to have a blood alcohol content over 0.10. These cases are broken down according to the day of the week, with the results listed in the accompanying table. At the 0.05 significance level, test the claim that such fatal crashes occur on the different days of the week with equal frequency. Does the evidence support the theory that fatal DWI car crashes are due to casual drinkers or that they are caused by those who drink daily?

Day	Sun	Mon	Tues	Wed	Thurs	Fri	Sat
Number	40	24	25	28	29	32	38

Based on data from the Dutchess County STOP-DWI Program.

6. Testing for Uniformly Distributed Industrial Accidents A study was made of 147 industrial accidents that required medical attention. Among those accidents, 31 occurred on Monday, 42 on Tuesday, 18 on Wednesday, 25 on Thursday, and 31 on Friday (based on results from "Counted Data CUSUM's," by Lucas, *Technometrics*, Vol. 27, No. 2). Test the claim that accidents occur with equal proportions on the five workdays. If the proportions are not the same, what factors might explain the differences?

7. Do Industrial Accidents Fit the Claimed Distribution? Use a 0.05 significance level and the industrial accident data from Exercise 6 to test the claim of a safety expert that accidents are distributed on workdays as follows: 30% on Monday, 15% on Tuesday, 15% on Wednesday, 20% on Thursday, and 20% on Friday. Does rejection of that claim provide any help in correcting the industrial accident problem?

8. Do Trucks Have the Same Color Distribution as Cars? A study of sport and compact car colors shows that 15.2% are green, 14.4% are white, 19.8% are red, 11.2% are black, and 39.4% are other colors (based on data from DuPont Automotive as reported in *USA Today*). When 100 trucks and vans are randomly selected, it is found that 16 are green, 24 are white, 16 are red, none are black, and 44 are other colors. Is there sufficient evidence to support the claim that the color distribution for trucks and vans is different from the color distribution for sport and compact cars?

9. Distribution of Digits in the Irrational Number Pi The number π is an irrational number with the property that when we try to express it in decimal form, it requires an infinite number of decimal places and there is no pattern of repetition. In the decimal representation of π, the first 100 digits occur with the frequencies described in the accompanying table. At the 0.05 significance level, test the claim that the digits are uniformly distributed.

Digit	0	1	2	3	4	5	6	7	8	9
Frequency	8	8	12	11	10	8	9	8	12	14

10. Distribution of Digits in the Rational Number 22/7 The number 22/7 is similar to π in the sense that they both require an infinite number of decimal places. However, 22/7 is a rational number because it can be expressed as the ratio of two integers, whereas π cannot. When rational numbers such as 22/7 are expressed in decimal form, there is a pattern of repetition. In the decimal representation of 22/7, the first 100 digits occur with the frequencies described in the accompanying table. At the 0.05 significance level, test the claim that the digits are uniformly distributed. How does the result differ from that found in Exercise 9?

Digit	0	1	2	3	4	5	6	7	8	9
Frequency	0	17	17	1	17	16	0	16	16	0

11. Car Crashes and Age Brackets Among drivers who have had a car crash in the last year, 88 are randomly selected and categorized by age, with the results listed in the accompanying table. If all ages have the same crash rate, we would expect (because of the age distribution of licensed drivers) the given categories to have 16%, 44%, 27%, and 13% of the subjects, respectively. At the 0.05 significance level, test the claim that the distribution of crashes conforms to the distribution of ages. Does any age group appear to have a disproportionate number of crashes?

Age	Under 25	25–44	45–64	Over 64
Drivers	36	21	12	19

Based on data from the Insurance Information Institute.

12. Do World War II Bomb Hits Fit a Poisson Distribution? In analyzing hits by V-1 buzz bombs in World War II, South London was subdivided into regions, each with an area of 0.25 km^2. In Section 4-5 we presented an example and included a table of actual frequencies of hits and the frequencies expected with the Poisson distribution. Use the values listed here and test the claim that the actual frequencies fit a Poisson distribution. Use a 0.05 significance level.

Number of bomb hits	0	1	2	3	4 or more
Actual number of regions	229	211	93	35	8
Expected number of regions (from Poisson distribution)	227.5	211.4	97.9	30.5	8.7

13. Post Position and Winning Horse Races Many people believe that when a horse races, it has a better chance of winning if its starting line-up position is closer to the rail on the inside of the track. The starting position of 1 is closest to the inside rail, followed by position 2, and so on. The accompanying table lists the numbers of wins for horses in the different starting positions. Test the claim that the probabilities of winning in the different post positions are not all the same.

Starting Position	1	2	3	4	5	6	7	8
Number of wins	29	19	18	25	17	10	15	11

Based on data from the *New York Post.*

14. Grade and Seating Location Do "A" students tend to sit in a particular part of the classroom? The author recorded the locations of the students who received grades of A, with these results: 17 sat in the front, 9 sat in the middle, and 5 sat in the back of the classroom. Is there sufficient evidence to support the claim that the "A" students are not evenly distributed throughout the classroom? If so, does that mean you can increase your likelihood of getting an A by sitting in the front?

15. Analyzing Rainfall Refer to Data Set 17 in Appendix B and use the last digits of the Boston rainfall amounts for Monday only. Is there sufficient evidence to support the claim that the digits do not occur with the same frequency? In the first example of this section, we saw that in some cases, the analysis of last digits can reveal whether the sample data have been obtained through actual measurements. Does that analysis apply to the last digits of the Monday rainfall amounts? Why or why not?

16. Analyzing Last Digits of Sammy Sosa Home-Run Distances Refer to Data Set 19 in Appendix B and use the last digits of the distances of home runs hit by Sammy Sosa. Is there sufficient evidence to conclude that the last digits do not occur with the same frequency? What does that conclusion suggest about how the home-run distances were obtained?

10-2 Beyond the Basics

17. Testing Effects of Outliers In doing a test for the goodness-of-fit as described in this section, does an outlier have much of an effect on the value of the χ^2 test statistic? Test for the effect of an outlier by repeating Exercise 3 after changing the frequency for the right rear tire from 6 to 60. Describe the general effect of an outlier.

18. Detecting Altered Experimental Data When Gregor Mendel conducted his famous hybridization experiments with peas, it appears that his gardening assistant knew the results that Mendel expected, and he altered the results to fit Mendel's expectations. Subsequent analysis of the results led to the conclusion that there is a probability of only 0.00004 that the expected results and reported results would agree so closely. How could the methods of this section be used to detect such results that are just too perfect to be realistic?

19. Equivalent Test In this exercise we will show that a hypothesis test involving a multinomial experiment with only two categories is equivalent to a hypothesis test for a proportion (Section 7-5). Assume that a particular multinomial experiment has only two possible outcomes, A and B, with observed frequencies of f_1 and f_2, respectively.

a. Find an expression for the χ^2 test statistic, and find the critical value for a 0.05 significance level. Assume that we are testing the claim that both categories have the same frequency, $(f_1 + f_2)/2$.

b. The test statistic

$$z = \frac{\hat{p} - p}{\sqrt{\frac{pq}{n}}}$$

is used to test the claim that a population proportion is equal to some value p. With the claim that $p = 0.5$, $\alpha = 0.05$, and

$$\hat{p} = \frac{f_1}{f_1 + f_2}$$

show that z^2 is equivalent to χ^2 [from part (a)]. Also show that the square of the critical z score is equal to the critical χ^2 value from part (a).

20. Testing Goodness-of-Fit with a Binomial Distribution An observed frequency distribution is as follows:

Number of successes	0	1	2	3
Frequency	89	133	52	26

a. Assuming a binomial distribution with $n = 3$ and $p = 1/3$, use the binomial probability formula to find the probability corresponding to each category of the table.

b. Using the probabilities found in part (a), find the expected frequency for each category.

c. Use a 0.05 significance level to test the claim that the observed frequencies fit a binomial distribution for which $n = 3$ and $p = 1/3$.

21. Testing Goodness-of-Fit with a Poisson Distribution In a recent year, there were 116 homicide deaths in Richmond, Virginia (based on "A Classroom Note On the Poisson Distribution: A Model for Homicidal Deaths In Richmond, VA for 1991," *Mathematics and Computer Education,* by Winston A. Richards). If the frequencies of deaths on different days conform to a Poisson distribution, they will be as shown in the accompanying table. Use a 0.05 significance level to test the claim that the actual frequencies fit a Poisson distribution. (*Caution:* Not all of the expected frequencies are at least 5.)

	Number of Homicides				
	0	1	2	3	4
Actual number of days	268	79	17	1	0
Expected number of days	265.6	84.4	13.4	1.4	0.1

22. Testing Goodness-of-Fit with a Normal Distribution An observed frequency distribution of sample IQ scores is as follows:

IQ score	Less than 80	80–95	96–110	111–120	More than 120
Frequency	20	20	80	40	40

a. Assuming a normal distribution with $\mu = 100$ and $\sigma = 15$, use the methods given in Chapter 5 to find the probability of a randomly selected subject belonging to each class. (Use class boundaries of 79.5, 95.5, 110.5, and 120.5.)

b. Using the probabilities found in part (a), find the expected frequency for each category.

c. Use a 0.01 significance level to test the claim that the IQ scores were randomly selected from a normally distributed population with $\mu = 100$ and $\sigma = 15$.

10-3 Contingency Tables: Independence and Homogeneity

The main objective of this section is to test claims about tables, such as Table 10-1 in the Chapter Problem. Tables similar to Table 10-1 are generally called *contingency tables,* or *two-way frequency tables.*

DEFINITIONS

A **contingency table** (or **two-way frequency table**) is a table in which frequencies correspond to two variables. (One variable is used to categorize rows, and a second variable is used to categorize columns.)

Table 10-1 has two variables: a row variable, which indicates whether the person survived or died; and a column variable, which lists the demographic categories—men, women, boys, girls. Unlike Section 10-2, which discussed frequencies listed in a single row (or column) according to category, this section discussed frequencies listed according to category, but here we consider cases with at least two rows and at least two columns.

Contingency tables are especially important because they are often used to analyze survey results. For example, we might ask subjects one question in

Delaying Death

University of California sociologist David Phillips has studied the ability of people to postpone their death until after some important event. Analyzing death rates of Jewish men who died near Passover, he found that the death rate dropped dramatically in the week before Passover, but rose the week after. He found a similar phenomenon occurring among Chinese-American women; their death rate dropped the week before their important Harvest Moon Festival, then rose the week after.

which they identify their gender (male/female), and we might ask another question in which they describe the frequency of their use of TV remote controls (often/sometimes/never). The methods of this section can then be used to determine whether the use of TV remote controls is independent of gender. (We probably already know the answer to that one.) Applications of this type are very numerous, so the methods presented in this section are among those most often used.

This section presents two types of hypothesis testing based on contingency tables. We first consider tests of independence, used to determine whether a contingency table's row variable is independent of its column variable. We then consider tests of homogeneity, used to determine whether different populations have the same proportions of some characteristic. Good news: Both types of hypothesis testing use the *same* basic methods. We begin with tests of independence.

Test of Independence

One of the two tests included in this section is a *test of independence* between the row variable and column variable.

DEFINITION

A **test of independence** tests the null hypothesis that the row variable and the column variable in a contingency table are not related. (The null hypothesis is the statement that the row and column variables are independent.)

It is very important to recognize that in this context, the word contingency refers to dependence, but this is only a statistical dependence, and it cannot be used to establish a direct cause-and-effect link between the two variables in question. For example, after analyzing the data in Table 10-1, we might conclude that whether a person survived the sinking of the Titanic is dependent on whether that person was a man, woman, boy, or girl, but that doesn't mean that the gender/age category has some direct causative effect on surviving.

When testing the null hypothesis of independence between the row and column variables in a contingency table, the following assumptions apply. (Note that these assumptions do not require that the parent population has a normal distribution or any other particular distribution.)

Assumptions

1. The sample data are randomly selected.
2. The null hypothesis H_0 is the statement that the row and column variables are *independent;* the alternative hypothesis H_1 is the statement that the row and column variables are dependent.
3. For every cell in the contingency table, the *expected* frequency E is at least 5. (There is no requirement that every *observed* frequency must be at least 5.)

Our test of independence between the row and column variables uses the following test statistic and critical values.

Test Statistic for a Test of Independence

$$\chi^2 = \sum \frac{(O - E)^2}{E}$$

Critical Values

1. The critical values are found in Table A-4 by using

$$\textbf{degrees of freedom} = (r - 1)(c - 1)$$

where r is the number of rows and c is the number of columns.

2. In a test of independence with a contingency table, the critical region is located in the *right tail only*.

The test statistic allows us to measure the degree of disagreement between the frequencies actually observed and those that we would theoretically expect when the two variables are independent. Small values of the χ^2 test statistic result from close agreement between frequencies observed and frequencies expected with independent row and column variables. Large values of the χ^2 test statistic are in the rightmost region of the chi-square distribution, and they reflect significant differences between observed and expected frequencies. In repeated large samplings, the distribution of the test statistic χ^2 can be approximated by the chi-square distribution, provided that all expected frequencies are at least 5. The number of degrees of freedom $(r - 1)(c - 1)$ reflects the fact that because we know the total of all frequencies in a contingency table, we can freely assign frequencies to only $r - 1$ rows and $c - 1$ columns before the frequency for every cell is determined. [However, we cannot have negative frequencies or frequencies so large that any row (or column) sum exceeds the total of the observed frequencies for that row (or column).]

In the preceding section we knew the corresponding probabilities and could easily determine the expected values, but the typical contingency table does not come with the relevant probabilities. For each cell in the frequency table, the expected frequency E can be calculated by applying the multiplication rule of probability for independent events. Assuming that the row and column variables are independent (which is the null hypothesis), the probability of a value being in a particular cell is the probability of being in the row containing the cell (namely, the row total divided by the sum of all frequencies) multiplied by the probability of being in the column containing the cell (namely, the column total divided by the sum of all frequencies) multiplied by the sum of all frequencies. Sound too complicated? The expected frequency for a cell can be simplified to the following equation.

Expected Frequency for a Contingency Table

$$\text{expected frequency} = \frac{(\text{row total})(\text{column total})}{(\text{grand total})}$$

Here *grand total* refers to the total of all observed frequencies in the table. For example, the expected frequency for the upper left cell of Table 10-6 (a duplicate of Table 10-1 with expected frequencies inserted in parentheses) is 537.360, which is found by noting that the total of all frequencies for the first row is 706, the total of the column frequencies is 1692, and the sum of all frequencies in the table is 2223, so we get an expected frequency of

$$E = \frac{(\text{row total})(\text{column total})}{(\text{grand total})} = \frac{(706)(1692)}{2223} = 537.360$$

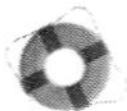

EXAMPLE **Finding Expected Frequency** The expected frequency for the upper left cell of Table 10-6 is 537.360. Find the expected frequency for the lower left cell, assuming independence between the row variable (whether the person survived) and the column variable (whether the person is a man, woman, boy, or girl).

SOLUTION The lower left cell lies in the second row (with total 1517) and the first column (with total 1692). The expected frequency is

$$E = \frac{(\text{row total})(\text{column total})}{(\text{grand total})} = \frac{(1517)(1692)}{2223} = 1154.640$$

INTERPRETATION To interpret this result for the lower left cell, we can say that although 1360 men actually died, we would have expected 1154.640 men to die if survivability is independent of whether the person is a man, woman, boy, or girl. There is a discrepancy between $O = 1360$ and $E = 1154.640$, and such discrepancies are key components of the test statistic.

TABLE 10-6 Observed Frequencies (and Expected Frequencies)

	Gender/Age Category				
	Men	Women	Boys	Girls	**Row totals**
Survived	332 (537.360)	318 (134.022)	29 (20.326)	27 (14.291)	**706**
Died	1360 (1154.640)	104 (287.978)	35 (43.674)	18 (30.709)	**1517**
Column totals:	**1692**	**422**	**64**	**45**	**Grand total: 2223**

To better understand the rationale for finding expected frequencies with this procedure, let's pretend that we know only the row and column totals and that we must fill in the cell expected frequencies by assuming independence (or no relationship) between the two variables involved—that is, we pretend that we know only the row and column totals shown in Table 10-6. Let's begin with the cell in the upper left corner. Because 706 of the 2223 persons survived, we have $P(\text{survived}) = 706/2223$. Similarly, 1692 of the people were men, so $P(\text{man}) = 1692/2223$. Because we are assuming independence between survivability and the column gender/age category, we can use the multiplication rule of probability to get

$$P(\text{survived and man}) = P(\text{survived}) \cdot P(\text{man}) = \frac{706}{2223} \cdot \frac{1692}{2223}$$

This equation is an application of the multiplication rule for independent events, which is expressed in general as follows: $P(A \text{ and } B) = P(A) \cdot P(B)$. Knowing the probability of being in the upper left cell, we can now find the *expected value* for that cell, which we get by multiplying the probability for that cell by the total number of people, as shown in the following equation.

$$E = n \cdot p = 2223\left(\frac{706}{2223} \cdot \frac{1692}{2223}\right) = 537.360$$

The form of this product suggests a general way to obtain the expected frequency of a cell:

$$\text{expected frequency } E = (\text{grand total}) \cdot \frac{(\text{row total})}{(\text{grand total})} \cdot \frac{(\text{column total})}{(\text{grand total})}$$

This expression can be simplified to

$$E = \frac{(\text{row total})(\text{column total})}{(\text{grand total})}$$

We can now proceed to use contingency table data for testing hypotheses, as in the following example, which uses the data given in the Chapter Problem.

Home Field Advantage

In the *Chance* magazine article "Predicting Professional Sports Game Outcomes from Intermediate Game Scores," authors Harris Cooper, Kristina DeNeve, and Frederick Mosteller used statistics to analyze two common beliefs: Teams have an advantage when they play at home, and only the last quarter of professional basketball games really counts. Using a random sample of hundreds of games, they found that for the four top sports, the home team wins about 58.6% of games. Also, basketball teams ahead after 3 quarters go on to win about 4 out of 5 times, but baseball teams ahead after 7 innings go on to win about 19 out of 20 times. The statistical methods of analysis included the chi-square distribution applied to a contingency table.

EXAMPLE ***Titanic* Sinking** At the 0.05 significance level, use the data in Table 10-1 to test the claim that when the *Titanic* sank, whether someone survived or died is independent of whether the person is a man, woman, boy, or girl. AS

SOLUTION The null hypothesis and alternative hypothesis are as follows:

H_0: Whether a person survived is independent of whether the person is a man, woman, boy, or girl.

H_1: Surviving the *Titanic* sinking and being a man, woman, boy, or girl are dependent.

The significance level is $\alpha = 0.05$.

Because the data are in the form of a contingency table, we use the χ^2 distribution with this test statistic:

$$\chi^2 = \sum \frac{(O - E)^2}{E}$$

$$= \frac{(332 - 537.360)^2}{537.360} + \frac{(318 - 134.022)^2}{134.022} + \frac{(29 - 20.326)^2}{20.326}$$

$$+ \frac{(27 - 14.291)^2}{14.291} + \frac{(1360 - 1154.640)^2}{1154.640} + \frac{(104 - 287.978)^2}{287.978}$$

$$+ \frac{(35 - 43.674)^2}{43.674} + \frac{(18 - 30.709)^2}{30.709}$$

$$= 78.481 + 252.555 + 3.702 + 11.302 + 36.525 + 117.536$$
$$+ 1.723 + 5.260$$
$$= 507.084$$

(The more accurate test statistic of 507.080 is obtained by carrying more decimal places in the intermediate calculations. STATDISK, Minitab, and the TI-83 Plus calculator all agree that 507.080 is a better result.) The critical value is $\chi^2 = 7.815$, and it is found from Table A-4 by noting that $\alpha = 0.05$ in the right tail and the number of degrees of freedom is given by $(r - 1)(c - 1) = (2 - 1)(4 - 1) = 3$. The test statistic and critical value are shown in Figure 10-7. Because the test statistic falls within the critical region, we reject the null hypothesis that whether a person survived is independent of whether the person is a man, woman, boy, or girl. It appears that whether a person survived the *Titanic* and whether that person is a man, woman, boy, or girl are dependent variables.

FIGURE 10-7
Test of Independence for Table 10-1 Data

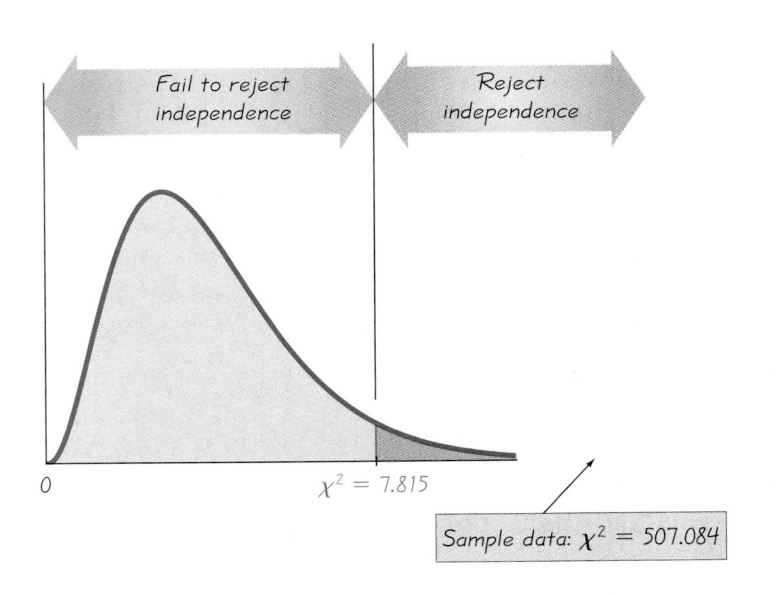

P-Values

The preceding example used the traditional approach to hypothesis testing, but we can easily use the *P*-value approach. STATDISK, Minitab, Excel, and the TI-83 Plus calculator all provide *P*-values for tests of independence in contingency tables. If you don't have a suitable calculator or statistical software package, you can estimate *P*-values from Table A-4 in Appendix A. Locate the appropriate number of degrees of freedom to isolate a particular row in that table. Find where the test statistic falls in that row, and you can identify a range of possible *P*-values by referring to the areas given at the top of each column. In the preceding example, there are 3 degrees of freedom, so go to the third row of Table A-4. Now use the test statistic of $\chi^2 = 507.084$ to see that this test statistic is greater than (and farther to the right of) every critical value of χ^2 that is in the third row, so the *P*-value is less than 0.005. On the basis of this small *P*-value, we again reject the null hypothesis and conclude that there is sufficient sample evidence to warrant rejection of the null hypothesis of independence.

As in Section 10-2, if observed and expected frequencies are close, the χ^2 test statistic will be small and the *P*-value will be large. If observed and expected frequencies are far apart, the χ^2 test statistic will be large and the *P*-value will be small. These relationships are summarized and illustrated in Figure 10-8.

Test of Homogeneity

In the preceding example, we illustrated a test of independence by using a sample of 2223 people who were aboard the *Titanic*. We were treating those 2223 people as a random sample drawn from *one* hypothetical population of

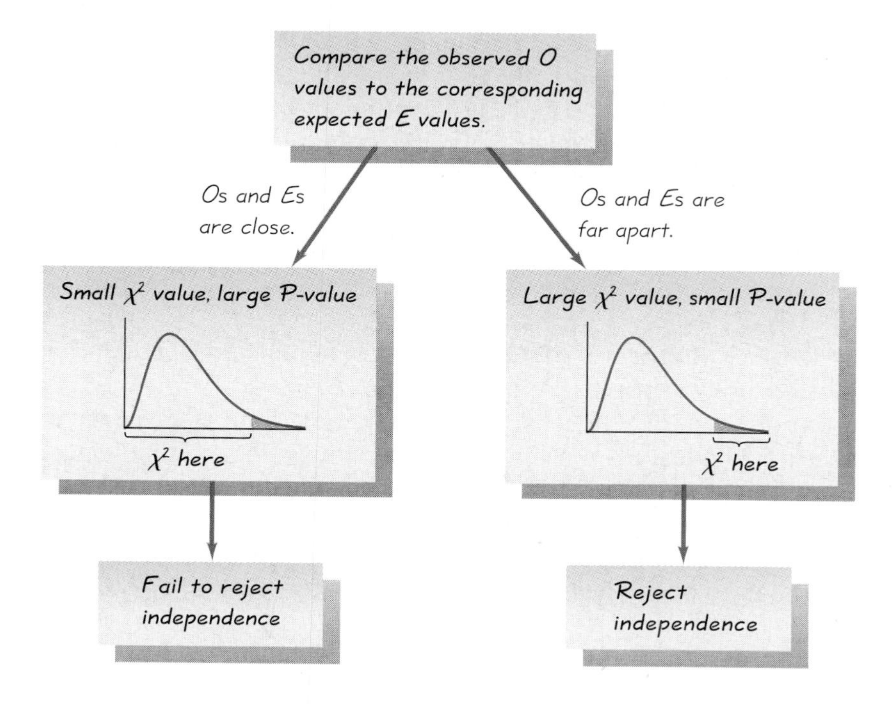

FIGURE 10-8
Relationships Among Components in χ^2 Test of Independence

all people who would find themselves in similar circumstances. However, some other samples are drawn from *different* populations, and we want to determine whether those populations have the same proportions of the characteristics being considered. The *test of homogeneity* can be used in such cases. (The word *homogeneous* means "having the same quality," and in this context, we are testing to determine whether the proportions are the same.)

DEFINITION

In a **test of homogeneity,** we test the claim that *different populations* have the same proportions of some characteristics.

Because a test of homogeneity uses data sampled from different populations, we have predetermined totals for either the rows or the columns in the contingency table. Consequently, a test of homogeneity involves random selections made in such a way that either the row totals are predetermined or the column totals are predetermined. In trying to distinguish between a test for homogeneity and a test for independence, we can therefore pose the following question:

Were *predetermined* sample sizes used for different populations (test of homogeneity), or was one big sample drawn so both row and column totals were determined randomly (test of independence)?

As an example of a test of homogeneity, suppose we want to test the claim that the proportion of college students who take a statistics course is the same in public four-year colleges, public two-year colleges, private four-year colleges, and private two-year colleges. Suppose we decide to sample 200 people from public four-year colleges, 300 people from public two-year colleges, 400 people from private four-year colleges, and 500 people from private two-year colleges. Then the contingency table summarizing the results will have either the row totals or the column totals (whichever represent the different colleges) predetermined as 200, 300, 400, and 500.

In conducting a test of homogeneity, we can use the same procedures already presented in this section, as illustrated in the following example.

EXAMPLE **Influence of Gender** Does a pollster's gender have an effect on poll responses by men? A *U.S. News & World Report* article about polls stated: "On sensitive issues, people tend to give 'acceptable' rather than honest responses; their answers may depend on the gender or race of the interviewer." To support that claim, data were provided for an Eagleton Institute poll in which surveyed men were asked if they agreed with this statement: "Abortion is a private matter that should be left to the woman to decide without government intervention." We will analyze the effect of gender on male survey subjects only. Table 10-7 is based on the responses of surveyed men. Assume that the survey was designed so

TABLE 10-7 Gender and Survey Responses

	Gender of Interviewer	
	Man	Woman
Men who agree	560	308
Men who disagree	240	92

that male interviewers were instructed to obtain 800 responses from male subjects, and female interviewers were instructed to obtain 400 responses from male subjects. Using a 0.05 significance level, test the claim that the proportions of agree/disagree responses are the same for the subjects interviewed by men and the subjects interviewed by women.

SOLUTION Because we have predetermined column totals of 800 subjects interviewed by men and 400 subjects interviewed by women, we test for homogeneity with these hypotheses:

H_0: The proportions of agree/disagree responses are the same for the subjects interviewed by men and the subjects interviewed by women.

H_1: The proportions are different.

The significance level is $\alpha = 0.05$. We use the same χ^2 test statistic described earlier, and it is calculated by using the same procedure. Instead of listing the details of that calculation, we provide the Minitab display that results from the data in Table 10-7.

The Minitab display shows the expected frequencies of 578.67, 289.33, 221.33, and 110.67. The display also includes the test statistic of $\chi^2 = 6.529$ and the P-value of 0.011. Using the P-value approach to hypothesis testing, we reject the null hypothesis of equal (homogeneous)

MINITAB

```
Expected counts are printed below observed counts
             C1          C2      Total
1           560         308        868
         578.67      289.33

2           240          92        332
         221.33      110.67

Total       800         400       1200
ChiSq = 0.602 + 1.204 +
        1.574 + 3.149 = 6.529
df = 1, p = 0.011
```

proportions. There is sufficient evidence to warrant rejection of the claim that the proportions are the same. It appears that response and the gender of the interviewer are dependent. It seems that men are influenced by the gender of the interviewer, although this conclusion is a statement of causality that is not justified by the statistical analysis.

Using Technology

STATDISK: Select **Analysis** from the main menu bar, then select **Contingency Tables,** and proceed to enter the frequencies as they appear in the contingency table. Click on **Evaluate.** The STATDISK results include the test statistic, critical value, *P*-value, and conclusion.

Minitab: First enter the observed frequencies in columns, then select **Stat** from the main menu bar. Next select the option **Tables,** then select **Chi Square Test** and proceed to enter the names of the columns containing the observed frequencies, such as C1 C2 C3 C4. Minitab provides the test statistic and *P*-value.

TI-83 Plus: First enter the contingency table as a *matrix* by pressing **2nd** x^{-1} to get the **MATRIX** menu (or the **MATRIX** key on the TI-83). Select **EDIT,** and press **ENTER.** Enter the dimensions of the matrix (rows by columns) and proceed to enter the individual frequencies. When finished, press **STAT,** select **TESTS,** and then select the option χ^2**-Test.** Be sure that the observed matrix is the one you entered, such as matrix A. The expected frequencies will be automatically calculated and stored in the separate matrix identified as "Expected." Scroll down to **Calculate** and press **ENTER** to get the test statistic, *P*-value, and number of degrees of freedom.

Excel: You must enter the observed frequencies, and you must also determine and enter the expected frequencies. When finished, click on the **fx** icon in the menu bar, select the function category **Statistical,** and then select the function name **CHITEST.** You must enter the range of values for the observed frequencies and the range of values for the expected frequencies. Only the *P*-value is provided.

10-3 Basic Skills and Concepts

1. Drug Testing: Is Treatment Independent of Reaction? Nicorette is a chewing gum designed to help people stop smoking cigarettes. Tests for adverse reactions yielded the results given in the accompanying table. Using this table, the

Minitab display includes the information shown here. At the 0.05 significance level, test the claim that the treatment (drug or placebo) is independent of the reaction (whether or not mouth or throat soreness was experienced). If you are thinking about using Nicorette as an aid to stop smoking, should you be concerned about mouth or throat soreness?

	Drug	Placebo
Mouth or throat soreness	43	35
No mouth or throat soreness	109	118

Based on data from Merrell Dow Pharmaceuticals, Inc.

MINITAB

```
Chi-Sq = 0.438 + 0.435 +
         0.151 + 0.150 = 1.174
DF = 1, P-Value = 0.279
```

2. Testing Effectiveness of Bicycle Helmets A study was conducted of 531 persons injured in bicycle crashes, and randomly selected sample results are summarized in the accompanying table. The TI-83 Plus results also are shown. At the 0.05 significance level, test the claim that wearing a helmet has no effect on whether facial injuries are received. Based on these results, does a helmet seem to be effective in helping to prevent facial injuries in a crash?

TI-83 Plus

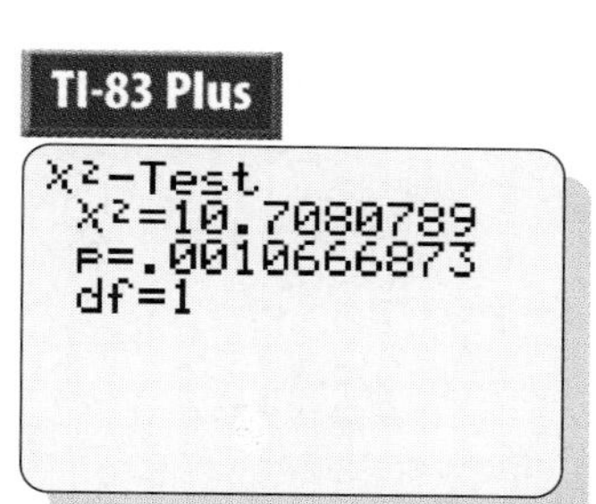

	Helmet Worn	No Helmet
Facial injuries received	30	182
All injuries nonfacial	83	236

Based on data from "A Case-Control Study of the Effectiveness of Bicycle Safety Helmets in Preventing Facial Injury," by Thompson, Thompson, Rivara, and Wolf, *American Journal of Public Health,* Vol. 80, No. 12.

3. Is Gender Independent of Opinion about Car Seizure? Considerable controversy arose when New York City introduced a program of keeping the cars belonging to people charged with drunk driving. The Associated Press conducted a poll, and the accompanying table is based on the results. Use a 0.01 significance level to test the claim that the opinion about car seizure is independent of the gender of the person being surveyed. Identify at least one key factor that might help explain the result.

	Should Car Be Seized?	
	Yes	No
Men	391	425
Women	480	256

4. Testing for Independence Between Early Discharge and Rehospitalization of Newborn Is it safe to discharge newborns from the hospital early after their births? The accompanying table shows results from a study of this issue. Use a 0.05 significance level to test the claim that whether the newborn was discharged early or late is independent of whether the newborn was rehospitalized

within a week of discharge. Does the conclusion change if the significance level is changed to 0.01?

	Rehospitalized within Week of Discharge?	
	Yes	No
Early discharge (less than 30 hours)	622	3997
Late discharge (30–78 hours)	631	4660

Based on data from "The Safety of Newborn Early Discharge," by Liu and others, *Journal of the American Medical Association,* Vol. 278, No. 4.

5. Testing Influence of Gender Table 10-7 summarizes data for male survey subjects, but the accompanying table summarizes data for a sample of women. Using a 0.01 significance level, and assuming that the sample sizes of 800 men and 400 women are predetermined, test the claim that the proportions of agree/disagree responses are the same for the subjects interviewed by men and the subjects interviewed by women.

	Gender of Interviewer	
	Man	Woman
Women who agree	512	336
Women who disagree	288	64

Based on data from the Eagleton Institute.

6. Testing for Discrimination In the judicial case *United States* v. *City of Chicago,* fair employment practices were challenged. A minority group (group A) and a majority group (group B) took the Fire Captain Examination. Assume that the study began with predetermined sample sizes of 24 minority candidates (Group A) and 562 majority candidates (Group B), with the results as shown in the table. At the 0.05 significance level, test the claim that the proportion of minority candidates who pass is the same as the proportion of majority candidates who pass. Based on the results, does the test appear to discriminate?

	Pass	Fail
Group A	10	14
Group B	417	145

7. Is Gender Independent of Confidence in Police? A survey was conducted to determine whether there is a gender gap in the confidence people have in police. The sample results are listed in the accompanying table. Use a 0.05 significance level to test the claim that there is such a gender gap.

	Confidence in Police		
	Great Deal	Some	Very Little or None
Men	115	56	29
Women	175	94	31

Based on data from the U.S. Department of Justice and the Gallup Organization.

8. Is Scanner Accuracy the Same for Specials? In a study of store checkout scanning systems, samples of purchases were used to compare the scanned prices to the posted prices. The accompanying table summarizes results for a sample of 819 items. When stores use scanners to check out items, are the error rates the same for regular-priced items as they are for advertised-special items? How might the behavior of consumers change if they believe that disproportionately more overcharges occur with advertised-special items?

	Regular-Priced Items	Advertised-Special Items
Undercharge	20	7
Overcharge	15	29
Correct price	384	364

Based on data from "UPC Scanner Pricing Systems: Are They Accurate?" by Ronald Goodstein, *Journal of Marketing*, Vol. 58.

9. Boeing 777s and Airline Companies The Boeing Commercial Airplane Group conducted a survey of 133 planes being flown or on order from three airline companies based in different countries. The sample data are summarized in the accompanying table. Is there sufficient evidence to support the claim that the distribution of planes flying and planes ordered is independent of the airline company? Use a 0.01 significance level.

	United	British Air	Singapore
Boeing 777s flying	36	22	14
Boeing 777s ordered	16	23	22

10. Firearm Training and Safety Does firearm training result in safer practices by gun owners? In one study, randomly selected subjects were surveyed with the results given in the accompanying table. Use a 0.05 significance level to test the claim that formal firearm training is independent of how firearms are stored. Does the formal training appear to have a positive effect?

	Guns Stored Loaded and Unlocked?	
	Yes	No
Had formal firearm training	122	329
Had no formal firearm training	49	299

Based on data from "Firearm Training and Storage," by Hemenway, Solnick, Azrael, *Journal of the American Medical Association*, Vol. 273, No. 1.

11. Crime and Strangers The accompanying table lists survey results obtained from a random sample of different crime victims. At the 0.05 significance level, test the claim that the type of crime is independent of whether the criminal is a stranger. How might the results affect the strategy police officers use when they investigate crimes?

	Homicide	Robbery	Assault
Criminal was a stranger	12	379	727
Criminal was acquaintance or relative	39	106	642

Based on data from the U.S. Department of Justice.

12. Is Seat-Belt Use Independent of Cigarette Smoking? A study of seat-belt users and nonusers yielded the randomly selected sample data summarized in the table on the next page. Test the claim that the amount of smoking is independent of seat-belt use. A plausible theory is that people who smoke more are less concerned about their health and safety and are therefore less inclined to wear seat belts. Is this theory supported by the sample data?

	Number of Cigarettes Smoked per Day			
	0	1–14	15–34	35 and over
Wear seat belts	175	20	42	6
Don't wear seat belts	149	17	41	9

Based on data from "What Kinds of People Do Not Use Seat Belts?" by Helsing and Comstock, *American Journal of Public Health,* Vol. 67, No. 11.

13. Is Sentence Independent of Plea? Many people believe that criminals who plead guilty tend to get lighter sentences than those who are convicted in trials. The accompanying table summarizes randomly selected sample data for San Francisco defendants in burglary cases. All of the subjects had prior prison sentences. At the 0.05 significance level, test the claim that the sentence (sent to prison or not sent to prison) is independent of the plea. If you were an attorney defending a guilty defendant, would these results suggest that you should encourage a guilty plea?

	Guilty Plea	Not-Guilty Plea
Sent to prison	392	58
Not sent to prison	564	14

Based on data from "Does It Pay to Plead Guilty? Differential Sentencing and the Functioning of the Criminal Courts," by Brereton and Casper, *Law and Society Review,* Vol. 16, No. 1.

14. Is the Home Field Advantage Independent of the Sport? Winning team data were collected for teams in different sports, with the results given in the accompanying table. Use a 0.10 significance level to test the claim that home/visitor wins are independent of the sport. Given that among the four sports included here, baseball is the only sport in which the home team can modify field dimensions to favor its own players, does it appear that baseball teams are effective in using this advantage?

	Basketball	Baseball	Hockey	Football
Home-team wins	127	53	50	57
Visiting-team wins	71	47	43	42

Based on data from "Predicting Professional Sports Game Outcomes from Intermediate Game Scores," by Copper, DeNeve, and Mosteller, *Chance,* Vol. 5, No. 3–4.

15. Is Drinking Independent of Type of Crime? The accompanying table lists sample data that statistician Karl Pearson used in 1909. Does the type of crime appear to be related to whether the criminal drinks or abstains? Are there any crimes that appear to be associated with drinking?

	Arson	Rape	Violence	Stealing	Coining (Counterfeiting)	Fraud
Drinker	50	88	155	379	18	63
Abstainer	43	62	110	300	14	144

16. Survey Refusals and Age Bracket A study of people who refused to answer survey questions provided the randomly selected sample data shown in the table on the next page. At the 0.01 significance level, test the claim that the cooperation of the subject (response or refusal) is independent of the age category. Does any particular age group appear to be particularly uncooperative?

	Age					
	18–21	22–29	30–39	40–49	50–59	60 and over
Responded	73	255	245	136	138	202
Refused	11	20	33	16	27	49

Based on data from "I Hear You Knocking But You Can't Come In," by Fitzgerald and Fuller, *Sociological Methods and Research,* Vol. 11, No. 1.

17. Is Smoking Independent of Gender? Refer to Data Set 14 in Appendix B and test the claim that the gender of statistics students is independent of whether they smoke.

18. Is Exercise Independent of Gender? Refer to Data Set 14 in Appendix B and test the claim that whether statistics students exercise is independent of gender.

10-3 Beyond the Basics

19. Using Yates' Correction for Continuity The chi-square distribution is continuous, whereas the test statistic used in this section is discrete. Some statisticians use *Yates' correction for continuity* in cells with an expected frequency of less than 10 or in all cells of a contingency table with two rows and two columns. With Yates' correction, we replace

$$\sum \frac{(O - E)^2}{E} \quad \text{with} \quad \sum \frac{(|O - E| - 0.5)^2}{E}$$

Given the contingency table in Exercise 1, find the value of the χ^2 test statistic with and without Yates' correction. In general, what effect does Yates' correction have on the value of the test statistic?

20. Equivalent Tests Assume that a contingency table has two rows and two columns with frequencies of a and b in the first row and frequencies of c and d in the second row.

a. Verify that the test statistic can be expressed as

$$\chi^2 = \frac{(a + b + c + d)(ad - bc)^2}{(a + b)(c + d)(b + d)(a + c)}$$

b. Let $\hat{p}_1 = a/(a + c)$ and let $\hat{p}_2 = b/(b + d)$. Show that the test statistic

$$z = \frac{(\hat{p}_1 - \hat{p}_2) - 0}{\sqrt{\dfrac{\bar{p}\,\bar{q}}{n_1} + \dfrac{\bar{p}\,\bar{q}}{n_2}}}$$

where $$\bar{p} = \frac{a + b}{a + b + c + d}$$

and $$\bar{q} = 1 - \bar{p}$$

is such that $z^2 = \chi^2$ [the same result as in part (a)]. This result shows that the chi-square test involving a 2×2 table is equivalent to the test for the difference between two proportions, as described in Section 8-4.

Race and the Future

NEW YORK Among the many uses of polls, some are designed to help companies improve marketing strategy, some are designed to help political candidates identify areas of strength and weakness, and some are designed to measure changing perceptions about culture, society, and the future. A *Newsweek*-sponsored poll included a question in which subjects were asked to look ahead 10 years into the future and identify how they think they will fare (better, worse, same) with respect to different factors. The table includes sample results for those who responded with "better." *Newsweek* noted that "Princeton Research Associates conducted telephone interviews with 751 adults including 264 African-Americans on April 16–19."

Responses of "Better"

	Blacks	Whites
Family income	71%	59%
Quality of education in local schools	64%	55%
Job opportunities for your family	57%	48%

1. Use the reported sample sizes to convert the given percentages to frequency counts.

2. The *Newsweek* article included this statement: "Though blacks see some serious problems yet to be resolved, they are even more optimistic about the future than white Americans." Test that claim. Is there sufficient sample evidence to support the claim?

VOCABULARY LIST

cells
multinomial experiment
goodness-of-fit test
contingency table
two-way frequency table
test of independence
test of homogeneity

REVIEW

In this chapter we worked with data in the form of frequency counts for different categories. In Section 10-2 we described methods for testing goodness-of-fit in a multinomial experiment, which is similar to a binomial experiment except that there are more than two categories of outcomes. Multinomial experiments result in frequency counts arranged in a single row or column, and we tested to determine whether the observed sample frequencies agree with some claimed distribution.

In Section 10-3 we described methods for testing claims involving contingency tables (or two-way frequency tables), which have at least two rows and two columns. Contingency tables incorporate two variables: One variable is used for determining the row that describes a sample value, and the second variable is used for determining the column that describes a sample value. Section 10-3 included two types of hypothesis test: (1) a test of independence between the row and column variables; (2) a test of homogeneity to decide whether different populations have the same proportions of some characteristics. The following are some key components of the methods discussed in this chapter.

- *Section 10-2 (Test for goodness-of-fit):*

  Test statistic is $\chi^2 = \sum \frac{(O - E)^2}{E}$

  Test is right-tailed with $k - 1$ degrees of freedom. All expected frequencies must be at least 5.

- *Section 10-3 (Contingency table test of independence or homogeneity):*

  Test statistic is $\chi^2 = \sum \frac{(O - E)^2}{E}$

  Test is right-tailed with $(r - 1)(c - 1)$ degrees of freedom. All expected frequencies must be at least 5.

Review Exercises

1. NRA Membership and Gun Safety The accompanying table describes sample data obtained from a survey of randomly selected gun owners. Use a 0.01 significance level to test the claim that whether gun owners store their guns loaded and unlocked is independent of whether they are a member of the National Rifle Association (NRA). If it appears that these two variables are dependent, identify at least one relevant factor that might explain the result.

	Guns Stored Loaded and Unlocked?	
	Yes	No
NRA member	40	80
Not a member of NRA	129	548

Based on data from "Firearm Training and Storage," by Hemenway, Solnick, Azrael, *Journal of the American Medical Association,* Vol. 273, No. 1

2. Is the Experiment Fair? The self-proclaimed amazing mentalist Bob claims that he has the power to read minds and he boldly challenges anyone to prove otherwise. If Bob wins the challenge, the challenger must pay him $100,000, and Bob agrees to disavow any mental powers if he loses the challenge. An experiment is devised whereby a neutral third party rolls a die and concentrates on the outcome for one minute; then Bob must identify the outcome while in another room. This is repeated for a total of 30 different trials, and the die results are listed in the accompanying table. Bob lost the challenge because he correctly identified only four outcomes, whereas someone making random guesses typically gets five of them correct. However, Bob charges that the experiment is unfair because the outcomes of the die are not uniform. Use a 0.05 significance level to test Bob's claim that there is sufficient evidence to conclude that the die is not fair, because all outcomes are not equally likely.

Die outcome	1	2	3	4	5	6
Frequency	2	8	4	3	7	6

3. Do Gunfire Deaths Occur More Often on Weekends? When *Time* magazine tracked U.S. deaths by gunfire during a one-week period, the results shown in the accompanying table were obtained. At the 0.05 significance level, test the claim that gunfire death rates are the same for the different days of the week. Is there any support for the theory that more gunfire deaths occur on weekends when more people are at home?

Weekday	Mon	Tues	Wed	Thurs	Fri	Sat	Sun
Number of deaths by gunfire	74	60	66	71	51	66	76

4. Are the Proportions of On-Time Flights the Same? The accompanying table summarizes results for 100 flights randomly selected from each of three different airline companies. Use a 0.05 significance level to test the claim that USAir, American, and Delta have the same proportions of on-time flights.

	USAir	American	Delta
Arrived on time	80	77	76
Arrived late	20	23	24

Based on data from the Department of Transportation.

5. Testing Adverse Reactions to Seldane Clinical tests of the allergy drug Seldane yielded results summarized in the accompanying table. At the 0.05 significance level, test the claim that the occurrence of headaches is independent of the group (Seldane, placebo, control). Based on these results, should Seldane users be concerned about getting headaches?

	Seldane Users	Placebo Users	Control
Headache	49	49	24
No headache	732	616	602

Based on data from Merrell Dow Pharmaceuticals, Inc.

CUMULATIVE REVIEW EXERCISES

TABLE 10-8

	A	B	C	D
x	66	80	82	75
y	77	89	94	84

1. Finding Statistics Assume that in Table 10-8, the row and column titles have no meaning so that the table contains test scores for eight randomly selected prisoners. Find the mean, median, range, variance, standard deviation, and 5-number summary.

2. Finding Probability Assume that in Table 10-8, the letters A, B, C, and D represent the choices on the first question of a multiple-choice quiz. Also assume that x represents men and y represents women and that the table entries are frequency counts, so 66 men chose answer A, 77 women chose answer A, 80 men chose answer B, and so on.
   - **a.** If one response is randomly selected, find the probability that it is response C.
   - **b.** If one response is randomly selected, find the probability that it was made by a man.
   - **c.** If one response is randomly selected, find the probability that it is C or was made by a man.
   - **d.** If two different responses are randomly selected, find the probability that they were both made by a woman.

3. Testing for Equal Proportions Using the same assumptions as in Exercise 2, test the claim that men and women choose the different answers in the same proportions.

4. Testing for a Relationship Assume that Table 10-8 lists test scores for four people, where the x-score is from a test of memory and the y-score is from a test of reasoning. Test the claim that there is a relationship between the x- and y-scores.

5. Testing for Effectiveness of Training Assume that Table 10-8 lists test scores for four people, where the x-score is from a pretest taken before a training session on memory improvement and the y-score is from a posttest taken after the training. Test the claim that the training session is effective in raising scores.

6. Testing for Equality of Means Assume that in Table 10-8, the letters A, B, C, and D represent different versions of the same test of reasoning. The x-scores were obtained by four randomly selected men and the y-scores were obtained by four randomly selected women. Test the claim that men and women have the same mean score.

Cooperative Group Activities

1. *Out-of-class activity:* Divide into groups of four or five students. Each group member should survey at least 15 male students and 15 female students at the same college by asking two questions: (1) Which political party does the subject favor most? (2) If the subject were to make up an absence excuse of a flat tire, which tire would he or she say went flat if the instructor asked? (See Exercise 3 in Section 10-2.) Ask the subject to write the two responses on an index card, and also record the gender of the subject and whether the subject wrote with the right or left hand. Use the methods of this chapter to analyze the data collected. Include these tests:

   - Political party choice is independent of the gender of the subject.
   - The tire identified as being flat is independent of the gender of the subject.
   - Political party choice is independent of whether the subject is right- or left-handed.
   - The tire identified as being flat is independent of whether the subject is right- or left-handed.
   - Gender is independent of whether the subject is right- or left-handed.
   - Political party choice is independent of the tire identified as being flat.

2. *Out-of-class activity:* Divide into groups of four or five students. Each group member should select about 15 other students and first ask them to "randomly" select four digits each. After the four digits have been recorded, ask each subject to write the last four digits of his or her social security number. Take the "random" sample results and mix them into one big sample, then mix the social security digits into a second big sample. Using the "random" sample set, test the claim that students select digits randomly. Then use the social security digits to test the claim that they come from a population of random digits. Compare the results. Does it appear that students can randomly select digits? Are they likely to select any digits more often than others? Are they likely to select any digits less often than others? Do the last digits of social security numbers appear to be randomly selected?

3. *In-class activity:* Divide into groups of three or four students. Each group should be given a die along with the instruction that it should be tested for "fairness." Is the die fair or is it biased? Describe the analysis and results.

4. *Out-of-class activity:* Divide into groups of two or three students. Some examples and exercises of this chapter were based on the analysis of last digits of values. (See the examples in Section 10-2 and Exercises 15 and 16 in Section 10-2.) It was noted that the analysis of last digits can sometimes reveal whether values are the results of actual measurements or whether they are reported estimates. Refer to an almanac and find the lengths of rivers in the world, then analyze the last digits to determine whether those lengths appear to be actual measurements or whether they appear to be reported estimates. (Instead of lengths of rivers, you could use heights of mountains, heights of the tallest buildings, lengths of bridges, and so on.)

TECHNOLOGY PROJECT

Use STATDISK, or Minitab, or Excel, or a TI-83 Plus calculator, or any other software package or calculator capable of generating equally likely random digits between 0 and 9 inclusive. Generate 500 digits and record the results in the accompanying table. Use a 0.05 significance level to test the claim that the sample digits come from a population with a uniform distribution (so that all digits are equally likely).

Digit	0	1	2	3	4	5	6	7	8	9
Frequency										

from DATA *to* DECISION

Critical Thinking: Are noncombat mortality rates the same for military personnel in a combat zone as for military personnel not so deployed?

After the military operations of Desert Shield and Desert Storm in the Persian Gulf, there were claims of dramatic increases in death rates of U.S. troops who served in that region. One study compared noncombat mortality rates for U.S. military personnel who were deployed in combat situations to those not deployed. Table 10-9 summarizes sample data for deaths in three different categories. The table entries show that among those who died from unintentional injuries (from cars, aircraft, explosions, and so on), 19% were deployed; among those who died from illness, 10% were deployed; among those who died from homicide or suicide, 3% were deployed.

Analyze the Results

Are those differences significant? Analyze the sample data and form conclusions. If the cause of death and deployment appear to be dependent, describe the effect of deployment and try to identify at least one factor that might explain the dependency. Make a recommendation that might improve the situation for deployed personnel.

TABLE 10-9 Noncombat Causes of Death for Deployed and Nondeployed Military Personnel

	Cause of Death		
	Unintentional Injury	Illness	Homicide or Suicide
Deployed	183	30	11
Not deployed	784	264	308

Table values are based on data from "Comparative Mortality Among U.S. Military Personnel in the Persian Gulf Region and Worldwide During Operations Desert Shield and Desert Storm," by Writer, DeFraites, and Brundage, *Journal of the American Medical Association,* Vol. 275, No. 2.

Internet Project

Contingency Tables

In all of the examples in Section 10-3, data were sorted into categories formed by the rows and columns of a contingency table. We used contingency tables along with the associated chi-square test to form conclusions regarding qualitative population data. Characteristics such as gender, race, and geographic location become fair game for formal hypothesis testing procedures. Go to the Internet Project for this chapter, located at

http://www.awlonline.com/triola

After collecting demographic data for a number of populations, you will apply the techniques of this chapter to form conclusions about the independence of interesting pairings of characteristics.

Statistics at work

"Even if you're not a numbers cruncher, [statistical] knowledge can be helpful in any situation that requires prediction, decision making, or evaluation."

Nabil Lebbos

Graphics Illustrator, Published Image

As analyst for Standard & Poor's *Published Image,* Nabil's studies on investment performance are published in newspapers read by over one million investors.

Please describe your occupation.

I work for *Published Image* where I use statistics to generate the charts and data that we use in our financial publications—using loads of statistics and applications. We write newsletters for banks and mutual funds.

What concepts of statistics do you use?

I use standard deviation to measure risk, regression to measure an investment's relationship to its benchmark, and correlation to determine an investment's movement in relation to other investments.

How do you use statistics on the job?

I start with a given set of raw data. These are usually monthly, daily, or annual returns on an investment. I then use Excel to chart the data so I can get a picture of what I'm dealing with. From there I proceed to perform an analysis. Sometimes, the results do not back up a point that the accompanying article is trying to make strongly enough. In such situations, I look at other possibilities.

Please describe one specific example illustrating how the use of statistics was successful in improving a product or service.

One of our clients wanted to make the point that although their mutual fund did not outperform all others, it did succeed in consistently avoiding large negative returns. I ran some tests on skewness and downside risk and showed that, in fact, the fund's returns were positively skewed. We created histograms comparing this fund with an average of all funds, and that clearly made the point.

In terms of statistics, what would you recommend for prospective employees?

It's a logical tool that, when used informatively, can convince you and your audience of the point you're trying to make much more effectively than words. Even if you're not a numbers cruncher, [statistical] knowledge can be helpful in any situation that requires prediction, decision making, or evaluation.

Do you feel job applicants are viewed more favorably if they have studied some statistics?

Yes.

While a college student, did you expect to be using statistics on the job?

No. I studied architecture as an undergrad and business as a grad student.

Do you recommend statistics for today's college students? Why?

Yes. If not for work, then for life. One of my favorite quotes is "You can never run into a contradiction. If you do, then check your premises, and you will find that one of them is wrong." In doing so, statistics can be extremely useful, if not through literal statistical analysis, then at least using the concepts of statistics. Statistics is extremely useful in making everyday decisions and analyzing situations.

Which other skills are important for today's college students?

It largely depends on what line of work they intend to follow, but communications is a skill that is important to all fields.

CHAPTER 11

Analysis of Variance

CHAPTER PROBLEM

Crash test dummies: Does the weight category of a car make a difference?

Table 11-1 on the following page lists recent data from car crash experiments conducted by the National Transportation Safety Administration. New cars were purchased and crashed into a fixed barrier at 35 mi/h, and measurements were recorded for the dummy in the driver's seat. (We've all seen a dummy in a driver's seat, but these dummies were not human.) Does the weight class (subcompact, compact, midsize, full size) make a difference? Let's begin by focusing on head injuries. In the spirit of exploring the data, we first list the statistics for the head injury measurements.

Head Injuries

	Subcompact	Compact	Midsize	Full Size
n	5	5	5	5
$\bar{x}$	668.8	555.8	486.8	537.8
s	242.0	91.0	167.7	154.6

Examination of the head injury values shows that there are no outliers. With only five values in each sample, there isn't a real need for 5-number summaries, but the STATDISK display of the boxplots is shown. Again, those boxplots aren't too meaningful because they are based on very small samples. The Excel display shows the four sample means. We should usually explore the distributions of the sample data, but we can't get a good sense of their distributions with only five values in each sample.

There appear to be considerable differences, such as the difference between the means of 668.8 and 486.8. But is there sufficient sample evidence to support the conclusion that the weight category makes a difference? We will consider this problem in Section 11-2.

STATDISK

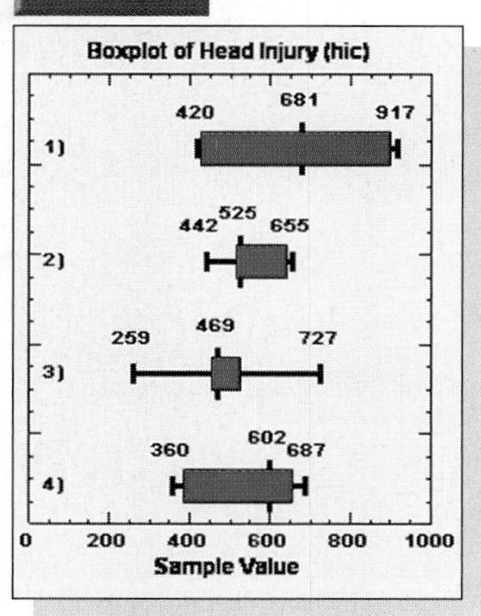

EXCEL

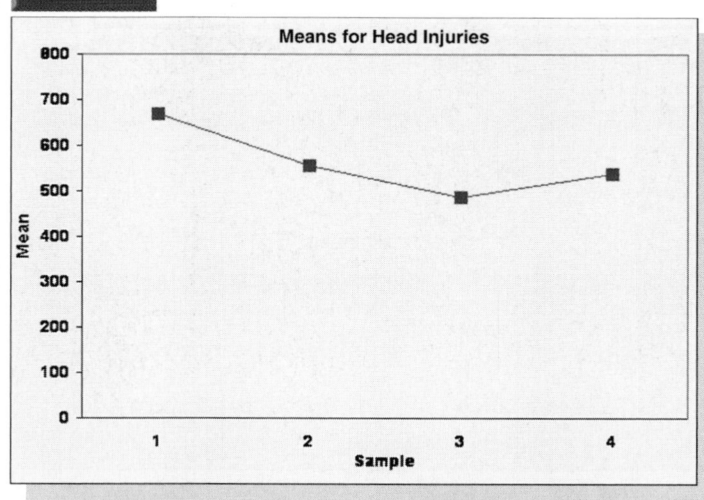

TABLE 11-1 Injuries to Car Crash Test Dummies

	Head Injury (hic)	Chest Deceleration (g)	Left Femur Load (lb)
Subcompact Cars			
Ford Escort	681	55	595
Honda Civic	428	47	1063
Hyundai Accent	917	59	885
Nissan Sentra	898	49	519
Saturn SL4	420	42	422
Compact Cars			
Chevrolet Cavalier	643	57	1051
Dodge Neon	655	57	1193
Mazda 626 DX	442	46	946
Pontiac Sunfire	514	54	984
Subaru Legacy	525	51	584
Midsize Cars			
Chevrolet Camaro	469	45	629
Dodge Intrepid	727	53	1686
Ford Mustang	525	49	880
Honda Accord	454	51	181
Volvo S70	259	46	645
Full-Size Cars			
Audi A8	384	44	1085
Cadillac Deville	656	45	971
Ford Crown Victoria	602	39	996
Oldsmobile Aurora	687	58	804
Pontiac Bonneville	360	44	1376

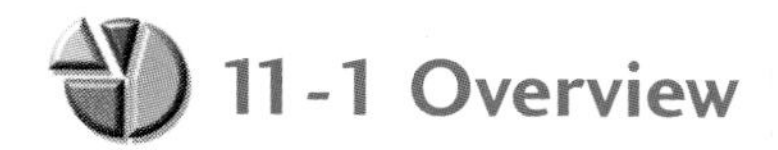

11-1 Overview

In Chapter 8 we developed procedures for testing the hypothesis that two population means are equal. In Section 11-2 we introduce a procedure for testing the hypothesis that *three or more* population means are equal. (The methods used in this chapter can also be used to test for equality between two population means, but the methods used in Chapter 8 are more efficient.) A typical null hypothesis in Section 11-2 will be H_0: $\mu_1 = \mu_2 = \mu_3 = \mu_4$; the alternative hypothesis is H_1: At least one mean is different. The method we use is based on an analysis of sample variances.

DEFINITION

Analysis of variance (ANOVA) is a method of testing the equality of three or more population means by analyzing sample variances.

ANOVA is used in applications such as the following:

- When three different groups of people (such as smokers, nonsmokers exposed to environmental tobacco smoke, and nonsmokers not so exposed) are measured for cotinine (a marker of nicotine), we can test to determine whether they have different mean levels of cotinine.
- If we treat one group with two aspirin tablets each day and a second group with one aspirin tablet each day, while a third group is given a placebo each day, we can test to determine if there is sufficient evidence to support the claim that the three groups have different mean blood pressure levels.

Why Can't We Just Test Two Samples at a Time? Why do we need a new procedure when we can test for equality of two means by using the methods presented in Chapter 8? For example, if we want to use the sample data from Table 11-1 to test the claim that the four populations have the same mean, why not simply pair them off and do two at a time by testing H_0: $\mu_1 = \mu_2$, then H_0: $\mu_2 = \mu_3$, and so on? This approach (doing two at a time) requires six different hypothesis tests, so the degree of confidence could be as low as 0.95^6 (or 0.735). In general, as we increase the number of individual tests of significance, we increase the likelihood of finding a difference by chance alone (instead of a real difference in the means). The risk of a type I error—finding a difference in one of the pairs when no such difference actually exists—is far too high. The method of analysis of variance lets us avoid that particular pitfall (rejecting a true null hypothesis) by using one test for equality of several means.

F Distribution

The ANOVA methods of this chapter require the *F* distribution, which was first introduced in Section 8-5. In Section 8-5 we noted that the *F* distribution has the following important properties (see Figure 11-1):

1. The *F* distribution is not symmetric; it is skewed to the right.
2. The values of *F* can be 0 or positive, but they cannot be negative.
3. There is a different *F* distribution for each pair of degrees of freedom for the numerator and denominator.

Critical values of *F* are given in Table A-5.

Analysis of variance (ANOVA) is based on a comparison of two different estimates of the variance common to the different populations. Those estimates (the *variance between samples* and the *variance within samples*) will be described in Section 11-2. The term *one-way* is used because the sample data are separated into groups according to one characteristic, or factor. For example, the head injury measurements listed in Table 11-1 are separated into four different groups according to the one characteristic (or factor) of weight class (subcompact, compact, midsize, full size). In Section 11-3 we will introduce two-way analysis of variance, which allows us to compare populations separated into categories using two characteristics (or factors). For example, we might separate heights of people using the following two factors: (1) gender (male or female) and (2) right- or left-handedness.

Suggested Study Strategy: Because the procedures used in this chapter require complicated calculations, we will emphasize the use and interpretation of computer software, such as STATDISK, Minitab, and Excel, along with the TI-83 Plus calculator. We suggest that you begin Section 11-2 by focusing on this key concept: We are using a procedure to test a claim that three or more means are equal. Although the procedure is complicated, our conclusion will be based on a *P*-value. If the *P*-value is small,

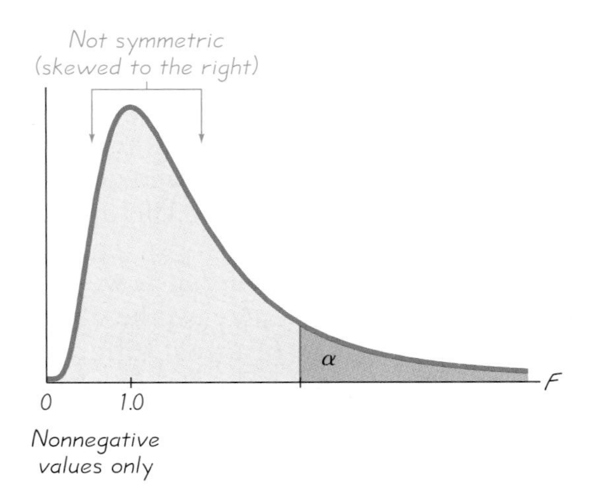

FIGURE 11-1
The *F* Distribution

There is a different *F* distribution for each different pair of degrees of freedom for numerator and denominator.

such as 0.05 or lower, reject equality of equal means. Otherwise, fail to reject equality of equal means. After understanding that basic and simple procedure, proceed to understand the underlying rationale.

11-2 One-Way ANOVA

In this section we consider tests of hypotheses that three or more population means are all equal, as in H_0: $\mu_1 = \mu_2 = \mu_3$. The calculations are very complicated, so we recommend the following approach:

1. Develop an understanding of how to interpret computer or calculator results.
2. Develop an understanding of the underlying rationale by studying the example in this section.
3. Become acquainted with the nature of the SS (sum of square) and MS (mean square) values and their role in determining the F test statistic, but use statistical software packages or a calculator for finding those values.

The following assumptions apply when testing the null hypothesis that three or more samples come from populations with the same mean:

Assumptions

1. The populations have distributions that are approximately normal.
2. The populations have the same variance σ^2 (or standard deviation σ).
3. The samples are simple random samples. (That is, samples of the same size have the same probability of being selected.)
4. The samples are independent of each other. (The samples are not matched or paired in any way.)
5. The different samples are from populations that are categorized in only one way. (This is the basis for the name of the method: *one-way* analysis of variance.)

The requirements of normality and equal variances are somewhat relaxed, because the methods in this section work reasonably well unless a population has a distribution that is very nonnormal or the population variances differ by large amounts. University of Wisconsin statistician George E. P. Box showed that as long as the sample sizes are equal (or nearly equal), the variances can differ by amounts that make the largest up to nine times the smallest and the results of ANOVA will continue to be essentially reliable. (If the samples appear to come from populations that are very nonnormal, we cannot use the methods of this section, but we can use the Kruskal-Wallis test presented in Section 13-5.)

Poll Resistance

Surveys based on relatively small samples can be quite accurate, provided the sample is random or representative of the population. However, increasing survey refusal rates are now making it more difficult to obtain random samples. The Council of American Survey Research Organizations reported that in a recent year, 38% of consumers refused to respond to surveys. The head of one market research company said, "Everyone is fearful of self-selection and worried that generalizations you make are based on cooperators only." Results from the multibillion-dollar market research industry affect the products we buy, the television shows we watch, and many other facets of our lives.

The method we use is called **one-way analysis of variance** (or **single-factor analysis of variance**) because we use a single property, or characteristic, for categorizing the populations. This characteristic is sometimes referred to as a *treatment*, or *factor*.

DEFINITION

A **treatment** (or **factor**) is a property, or characteristic, that allows us to distinguish the different populations from one another.

For example, the head injury measurements listed in Table 11-1 are sample data drawn from four different populations that are distinguished according to the treatment (or factor) of weight class (subcompact, compact, midsize, full size). The term *treatment* is used because early applications of analysis of variance involved agricultural experiments in which different plots of farmland were treated with different fertilizers, seed types, insecticides, and so on.

Interpreting Computer and Calculator Results

Because the calculations required for analysis of variance are extremely involved, almost everyone uses computer software or a calculator (except for one statistician living in the Nevada desert). STATDISK, Minitab, Excel, and the TI-83 Plus calculator all provide *P*-values, so we will use the *P*-value approach for testing hypotheses.

Procedure for Testing H_0: $\mu_1 = \mu_2 = \mu_3 = \cdots$

1. **Use STATDISK, Minitab, Excel, or a TI-83 Plus calculator to obtain results.**
2. **Identify the *P*-value from the display.**
3. **Form a conclusion based on these criteria:**
   - **If *P*-value $\leq \alpha$, reject the null hypothesis of equal means.**
   - **If *P*-value $> \alpha$, fail to reject the null hypothesis of equal means.**

We will describe the nature of the *F* test statistic after the following example, but Figure 11-2 shows the relationships among the key components of the ANOVA test.

EXAMPLE **Head Injuries in Car Crashes** Given the sample data in Table 11-1 and a significance level of $\alpha = 0.05$, use STATDISK, Minitab, Excel, or a TI-83 PLUS calculator to test the claim that the four samples come from populations with the same mean ($\mu_1 = \mu_2 = \mu_3 = \mu_4$).

SOLUTION At the end of this section we will describe specific procedures for obtaining computer or calculator displays, but we will now consider the displayed results. A key element is the *P*-value, which all

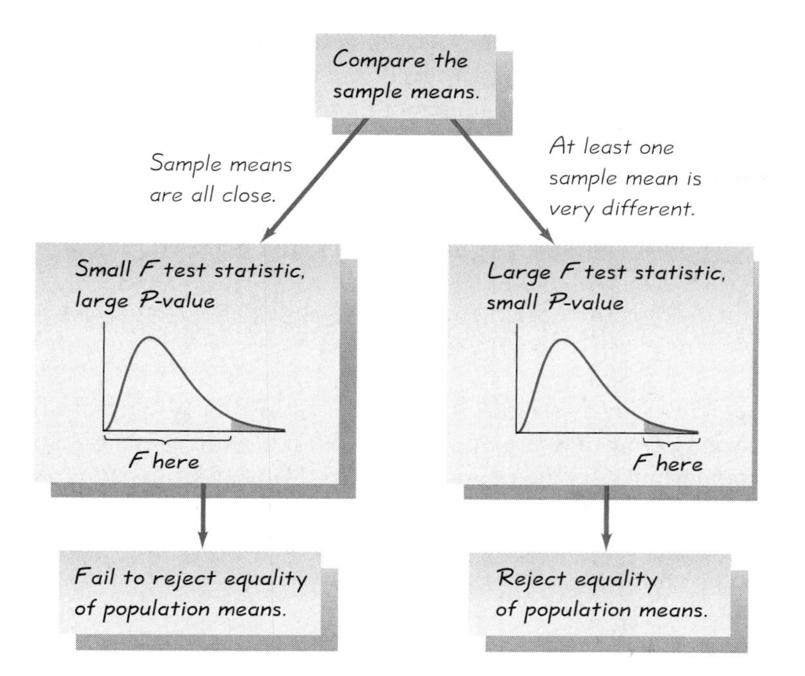

FIGURE 11-2
Relationships Among Components of ANOVA

displays show as 0.422 when rounded. Because the P-value is greater than the significance level of $\alpha = 0.05$, we fail to reject the null hypothesis of equal means.

STATDISK

Equal Length Samples	
Total Num Values	20
Upper Deg Free	3
Lower Deg Free	16
SS(treatment)	88425
SS(error)	475323
SS(total)	563748
MS(treatment)	29475
MS(error)	29708
MS(total)	29671
Test Statistic, F	0.99217
Critical F	3.2389
P-Value	0.42157

Fail to Reject the Null Hypothesis

Data does not provide enough evidence to indicate that the sample means are unequal

EXCEL

	A	B	C	D	E	F	G
1	Anova: Single Factor						
2							
3	SUMMARY						
4	Groups	Count	Sum	Average	Variance		
5	Column 1	5	3344	668.8	58542.7		
6	Column 2	5	2779	555.8	8272.7		
7	Column 3	5	2434	486.8	28110.2		
8	Column 4	5	2689	537.8	23905.2		
9							
10							
11	ANOVA						
12	Source of Variation	SS	df	MS	F	P-value	F crit
13	Between Groups	88425	3	29475	0.992167	0.42157	3.238867
14	Within Groups	475323.2	16	29707.7			
15							
16	Total	563748.2	19				

TI-83 Plus

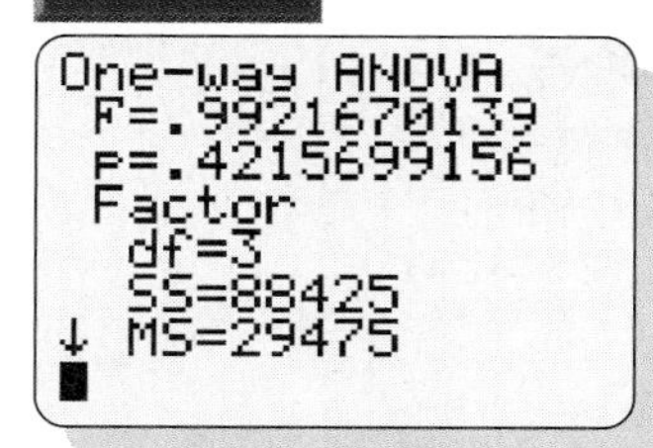


```
One-way ANOVA
↑ MS=29475
 Error
  df=16
  SS=475323.2
  MS=29707.7
 Sxp=172.359218
```

MINITAB

```
Analysis of Variance
Source    DF       SS       MS       F       P
Factor     3    88425    29475    0.99   0.422
Error     16   475323    29708
Total     19   563748
```

INTERPRETATION There is not sufficient evidence to warrant rejection of the claim that the four samples come from populations with the same mean. We might feel safer in larger cars, and that feeling might be justified by the reality that occupants of larger cars have less severe head injuries in crashes, but the data in Table 11-1 do not provide enough evidence to support that claim. We would need to crash test more cars, a time-consuming and obviously expensive process.

Rationale

The method of analysis of variance is based on this fundamental concept: With the assumption that the populations all have the same variance σ^2, we estimate the common value of σ^2 using two different approaches. The F test statistic is the ratio of those estimates, so that a significantly *large* F test statistic (located far to the right in the F distribution graph) is evidence against equal population means. (See Figure 11-2.) The two approaches for estimating the common value of σ^2 are as follows.

1. The **variance between samples** (also called **variation due to treatment**) is an estimate of the common population variance σ^2 that is based on the variability among the sample *means*.

2. The **variance within samples** (also called **variation due to error**) is an estimate of the common population variance σ^2 based on the sample *variances*.

Test Statistic for One-Way ANOVA

$$F = \frac{\text{variance between samples}}{\text{variance within samples}}$$

The numerator of the test statistic F measures variation between sample means. The estimate of variance in the denominator depends only on the sample variances and is not affected by differences among the sample means. Consequently, sample means that are close in value result in an F test statistic that is close to 1, and we conclude that there are no significant differences among the sample means. But if the value of F is excessively *large*, then we reject the claim of equal means. (The vague terms "close to 1" and "excessively large" are made objective by the corresponding P-value, which tells us whether the F test statistic is or is not in the critical region.) Because excessively large values of F reflect unequal means, the test is right-tailed.

Calculations with Equal Sample Sizes

If the data sets all have the same sample size (as in Table 11-1), the required calculations aren't overwhelmingly difficult. First, find the variance between samples by evaluating $ns_{\bar{x}}^2$, where $s_{\bar{x}}^2$ is the variance of the sample means.

That is, consider the sample means to be an ordinary set of values and calculate the variance. (From the central limit theorem, $\sigma_{\bar{x}} = \sigma/\sqrt{n}$ can be solved for σ to get $\sigma = \sqrt{n} \cdot \sigma_{\bar{x}}$, so that we can estimate σ^2 with $ns_{\bar{x}}^2$. For example, the sample means in Table 11-1 are 668.8, 555.8, 486.8, and 537.8. Those four values have a standard deviation of $s_{\bar{x}} = 76.7789$, so that

$$\text{variance between samples} = ns_{\bar{x}}^2 = 5(76.7789)^2 = 29{,}475$$

Refer to the computer/calculator displays to find the value of 29,475.

Next, estimate the variance within samples by calculating s_p^2, which is the pooled variance obtained by finding the mean of the sample variances. The sample standard deviations in Table 11-1 are 242.0, 91.0, 167.7, and 154.6, so that

$$\begin{aligned}\text{variance within samples} &= s_p^2 \\ &= \frac{242.0^2 + 91.0^2 + 167.7^2 + 154.6^2}{4} \\ &= 29{,}717 \text{ (or 29,708 using more precise values)}\end{aligned}$$

Finally, evaluate the F test statistic as follows.

$$F = \frac{\text{variance between samples}}{\text{variance within samples}} = \frac{29{,}475.0}{29{,}717} = 0.9919$$

Carrying more decimal places would result in the more accurate test statistic of $F = 0.9922$. See the computer/calculator displays in the previous example and locate this value of the F test statistic.

The critical value of F is found by assuming a right-tailed test, because large values of F correspond to significant differences among means. With k samples each having n values, the numbers of degrees of freedom are computed as follows:

Degrees of Freedom with k Samples of the Same Size n

$$\text{numerator degrees of freedom} = k - 1$$
$$\text{denominator degrees of freedom} = k(n - 1)$$

For the head injury sample data in Table 11-1, $k = 4$ and $n = 5$, so the degrees of freedom are 3 for the numerator and $4(4) = 16$ for the denominator. With $\alpha = 0.05$, 3 degrees of freedom for the numerator, and 16 degrees of freedom for the denominator, the critical value from Table A-5 is $F = 3.2389$. If we were to use the traditional method of hypothesis testing, we would see that this right-tailed test has a test statistic of $F = 0.9919$ and a critical value of $F = 3.2389$, so the test statistic is not in the critical region and we therefore fail to reject the null hypothesis of equal means.

Calculations with Unequal Sample Sizes

While the calculations required for cases with equal sample sizes are reasonable, they become really complicated when the sample sizes are not all the same. The same basic reasoning applies because we calculate an F test statistic that is the ratio of two different estimates of the common population variance σ^2, but those estimates involve *weighted* measures that take the sample sizes into account, as shown below.

$$F = \frac{\text{variance between samples}}{\text{variance within samples}} = \frac{\left[\dfrac{\Sigma n_i(\bar{x}_i - \bar{\bar{x}})^2}{k - 1}\right]}{\left[\dfrac{\Sigma(n_i - 1)s_i^2}{\Sigma(n_i - 1)}\right]}$$

where $\bar{\bar{x}}$ = mean of all sample values combined
k = number of population means being compared
n_i = number of values in the ith sample
$\bar{x}_i$ = mean of values in the ith sample
s_i^2 = variance of values in the ith sample

Note that the numerator in this test statistic is really a form of the formula for variance that was given in Chapter 2:

$$s^2 = \frac{\Sigma(x - \bar{x})^2}{n - 1}$$

The factor of n_i is included so that larger samples carry more weight. The denominator of the test statistic is simply the mean of the sample variances, but it is a weighted mean with the weights based on the sample sizes.

Because calculating this test statistic can lead to large rounding errors, the various software packages typically use a different (but equivalent) expression that involves SS (for sum of squares) and MS (for mean square) notation. Although the following notation and components are complicated and involved, the basic idea is the same: The test statistic F is a ratio with a numerator reflecting variation *between* the means of the samples and a denominator reflecting variation *within* the samples. If the populations have equal means, the F ratio tends to be close to 1, but if the population means are not equal, the F ratio tends to be significantly larger than 1. Key components in our ANOVA method are identified in the following boxes.

SS(total), or total sum of squares, is a measure of the total variation (around $\bar{\bar{x}}$) in all of the sample data combined.

Formula 11-1 $$\text{SS(total)} = \Sigma(x - \bar{\bar{x}})^2$$

SS(total) can be broken down into two components, SS(treatment) and SS(error), described as follows:

SS(treatment) is a measure of the variation between the sample means. In one-way ANOVA, SS(treatment) is sometimes referred to as SS(factor). Because it is a measure of variability *between* the sample means, it is also referred to as SS(between groups) or SS(between samples).]

Formula 11-2

$$\begin{aligned} SS(\text{treatment}) &= n_1(\bar{x}_1 - \bar{\bar{x}})^2 + n_2(\bar{x}_2 - \bar{\bar{x}})^2 + \cdots + n_k(\bar{x}_k - \bar{\bar{x}})^2 \\ &= \Sigma n_i(\bar{x} - \bar{\bar{x}})^2 \end{aligned}$$

If the population means $(\mu_1, \mu_2, \ldots, \mu_k)$ are equal, then the sample means $\bar{x}_1, \bar{x}_2, \ldots, \bar{x}_k$ will all tend to be close together and also close to $\bar{\bar{x}}$. The result will be a relatively small value of SS(treatment). If the population means are not all equal, however, then at least one of $\bar{x}_1, \bar{x}_2, \ldots, \bar{x}_k$ will tend to be far apart from the others and also far apart from $\bar{\bar{x}}$. The result will be a relatively large value of SS(treatment).

SS(error) is a sum of squares representing the variability that is assumed to be common to all the populations being considered.

Formula 11-3

$$\begin{aligned} SS(\text{error}) &= (n_1 - 1)s_1^2 + (n_2 - 1)s_2^2 + \cdots + (n_k - 1)s_k^2 \\ &= \Sigma(n_i - 1)s_i^2 \end{aligned}$$

Because SS(error) is a measure of the variance within groups, it is sometimes denoted as SS(within groups) or SS(within samples). Given the preceding expressions for SS(total), SS(treatment), and SS(error), the following relationship will always hold.

Formula 11-4 $\quad SS(\text{total}) = SS(\text{treatment}) + SS(\text{error})$

SS(treatment) and SS(error) are both sums of squares, and if we divide each by its corresponding number of degrees of freedom, we get *mean* squares. Some of the following expressions for mean squares include the notation N:

$$N = \text{total number of values in all samples combined}$$

MS(treatment) is a mean square for treatment, obtained as follows:

Formula 11-5 $\quad MS(\text{treatment}) = \dfrac{SS(\text{treatment})}{k - 1}$

MS(error) is a mean square for error, obtained as follows:

Formula 11-6 $$\text{MS(error)} = \frac{\text{SS(error)}}{N - k}$$

MS(total) is a mean square for the total variation, obtained as follows:

Formula 11-7 $$\text{MS(total)} = \frac{\text{SS(total)}}{N - 1}$$

The SS and MS values are used for determining the F test statistic that applies when the samples do not all have the same size.

Test Statistic for ANOVA with Unequal Sample Sizes

In testing the null hypothesis H_0: $\mu_1 = \mu_2 = \cdots = \mu_k$ against the alternative hypothesis that these means are not all equal, the test statistic

Formula 11-8 $$F = \frac{\text{MS(treatment)}}{\text{MS(error)}}$$

has an F distribution (when the null hypothesis H_0 is true) with degrees of freedom given by

$$\text{numerator degrees of freedom} = k - 1$$
$$\text{denominator degrees of freedom} = N - k$$

This test statistic is essentially the same as the one given earlier, and its interpretation is also the same as described earlier. The denominator depends only on the sample variances that measure variation within the treatments and is not affected by the differences among the sample means. In contrast, the numerator does depend on differences among the sample means. If the differences among the sample means are extreme, they will cause the numerator to be excessively large, so F will also be excessively large. Consequently, very large values of F suggest unequal means, and the ANOVA test is therefore right-tailed.

Tables are a convenient format for summarizing key results in ANOVA calculations, and Table 11-2 has a format often used in computer displays. (See the preceding Minitab and Excel displays.) The entries in Table 11-2 result from the head injury data in Table 11-1.

If we use ANOVA methods and conclude that there is sufficient evidence to reject the claim of equal population means, we cannot conclude that any particular mean is different from the others. There are several other

TABLE 11-2 ANOVA Table for Head Injury

Source of Variation	Sum of Squares (SS)	Degrees of Freedom	Mean Square (MS)	*F* Test Statistic
Treatments	88,425	3	29,475	0.9922
Error	475,323	16	29,708	
Total	563,748	19		

tests that can be used to identify the specific means that are different, and those procedures are called **multiple comparison procedures.** Comparison of confidence intervals, the Scheffé test, the extended Tukey test, and the Bonferroni test are four common multiple comparison procedures that are usually included in more advanced texts.

Designing the Experiment: When we use one-way (or single-factor) analysis of variance and conclude that the differences among the means are significant, we can't be absolutely sure that the given factor is responsible for the differences. It is possible that the variation of some other unknown factor is responsible. One way to reduce the effect of the extraneous factors is to design the experiment so that it has a **completely randomized design,** in which each element is given the same chance of belonging to the different categories, or treatments. For example, you might assign subjects to a treatment group, placebo group, and control group through a process of random selection equivalent to picking slips from a bowl. Another way to reduce the effect of extraneous factors is to use a **rigorously controlled design,** in which elements are carefully chosen so that all other factors have no variability. For example, you might treat a healthy seven-year-old girl from Texas, while another healthy seven-year-old girl from Texas is given a placebo, while a third healthy seven-year-old girl from Texas is put in a control group that is given nothing. But in addition to health, age, gender, and state of residence, you might have to identify other relevant factors that should be considered. In general, good results require that the experiment be carefully designed and executed.

Using Technology

STATDISK: Select **Analysis** from the main menu bar, then select **One-Way Analysis of Variance,** and proceed to enter the sample data. Click **Evaluate** when done.

Minitab: First enter the sample data in columns C1, C2, C3, ... Next, select **Stat, ANOVA, ONEWAY (UNSTACKED),** and enter C1 C2 C3 . . . in the box identified as "Responses" (in separate columns).

Excel: First enter the data in columns A, B, C, ... Next select **Tools** from the main menu bar, then select **Data Analysis,** followed by **Anova: Single Factor.** In the dialog box, enter the range containing the sample data. (For example, enter A1:D5 if the first value is in row 1 of column A and the last entry is in row 5 of column D.)

TI-83 Plus: First enter the data as lists in L1, L2, L3, . . . , then press **STAT,** select **TESTS,** and choose the option **ANOVA.** Enter the column labels. For example, if the data are in columns L1, L2, L3, and L4, enter those columns to get **ANOVA (L1, L2, L3, L4),** and press the **ENTER** key.

11-2 Basic Skills and Concepts

1. Interpreting Computer Display The accompanying table shows the analysis of variance results from a Minitab display. The sample values are measured consumer reactions to television commercials for different products. Assume that we want to use a 0.05 significance level in testing the null hypothesis that the different products have commercials resulting in the same mean reaction score.

a. How many different samples are included in this study?
b. Identify the value of the test statistic.
c. Find the critical value from Table A-5.
d. Identify the P-value.
e. Based on the preceding results, what do you conclude about equality of the population means?

Source	DF	SS	MS	F	P
Factor	5	50.00	10.00	3.33	0.032
Error	15	5.00	3.00		
Total	20	55.00			

2. Interpreting Computer Display Repeat Exercise 1 assuming that the Minitab results are as shown here.

Source	DF	SS	MS	F	P
Factor	4	120.00	30.00	1.50	0.289
Error	8	160.00	20.00		
Total	12	280.00			

3. Mean Body Temperatures for Different Age Groups Do different age groups have different mean body temperatures? The accompanying table lists the body temperatures of five randomly selected subjects from each of three different age groups.

a. Find the variance *between* samples by evaluating $ns_{\bar{x}}^2$ (Recall that $s_{\bar{x}}^2$ is the variance of the three sample means.)
b. Find the variance *within* samples by evaluating the pooled variance s_p^2, which is the mean of the sample variances.
c. Use the results from parts (a) and (b) to find the F test statistic.
d. Find the F critical value. [Recall that with equal sample sizes, the degrees of freedom are $k - 1$ for the numerator and $k(n - 1)$ for the denominator.]

e. Based on the preceding results, what do you conclude about the claim that the three age-group populations have the same mean body temperature?

Body Temperatures (°F) Categorized by Age

18–20	21–29	30 and older
98.0	99.6	98.6
98.4	98.2	98.6
97.7	99.0	97.0
98.5	98.2	97.5
97.1	97.9	97.3
$n_1 = 5$	$n_2 = 5$	$n_3 = 5$
$\bar{x}_1 = 97.940$	$\bar{x}_2 = 98.580$	$\bar{x}_3 = 97.800$
$s_1 = 0.568$	$s_2 = 0.701$	$s_3 = 0.752$

Based on data from Dr. Philip Mackowiak, Dr. Steven Wasserman, and Dr. Myron Levine of the University of Maryland.

4. Waste Disposal of Metal, Paper, Plastic, and Glass The City Resource Recovery Company (CRRC) collects the waste discarded by households in a region. Discarded waste must be separated into categories of metal, paper, plastic, and glass. In planning for the equipment needed to collect and process the garbage, CRRC refers to the data we have summarized in Data Set 5 in Appendix B (provided by the Garbage Project at the University of Arizona). The results (weights in pounds) are summarized in the table that follows. At the 0.05 significance level, test the claim that the four specific populations of garbage have the same mean. Based on the results, does it appear that these four categories require the same collection and processing resources? Does any single category seem to be a particularly large part of the waste-management problem?

Metal	Paper	Plastic	Glass
$n = 62$	$n = 62$	$n = 62$	$n = 62$
$\bar{x} = 2.218$	$\bar{x} = 9.428$	$\bar{x} = 1.911$	$\bar{x} = 3.752$
$s = 1.091$	$s = 4.168$	$s = 1.065$	$s = 3.108$

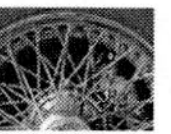

5. Crash Test Dummies: Mean Chest Deceleration Value The example of this section used the head injury data in Table 11-1 in a test of the null hypothesis that the means are the same for the different weight categories (subcompact, compact, midsize, full size). Use a 0.05 significance level to test the null hypothesis that the different weight categories have the same mean chest deceleration value.

6. Crash Test Dummies: Mean Left Femur Load The example of this section used the head injury data in Table 11-1 in a test of the null hypothesis that the means are the same for the different weight categories (subcompact, compact, midsize, full size). Use a 0.05 significance level to test the null hypothesis that the different weight categories have the same mean left femur load.

7. Archeology: Skull Breadths from Different Epochs The values in the table on the next page are measured maximum breadths of male Egyptian skulls from different epochs (based on data from *Ancient Races of the Thebaid,* by Thomson and Randall-Maciver). Changes in head shape over time suggest that interbreeding occurred with immigrant populations. Use a 0.05 significance level to test the claim that the mean is the same for the different epochs.

4000 B.C.	1850 B.C.	150 A.D.
131	129	128
138	134	138
125	136	136
129	137	139
132	137	141
135	129	142
132	136	137
134	138	145
138	134	137

8. Solar Energy in Different Weather A student of the author lives in a home with a solar electric system. At the same time each day, she collected voltage readings from a meter connected to the system and the results are listed in the accompanying table. Use a 0.05 significance level to test the claim that the mean voltage reading is the same under the three different types of day. Is there sufficient evidence to support a claim of different population means? We might expect that a solar system would provide more electrical energy on sunny days than on cloudy or rainy days. Can we conclude that sunny days result in greater amounts of electrical energy?

Sunny Days	Cloudy Days	Rainy Days
13.5	12.7	12.1
13.0	12.5	12.2
13.2	12.6	12.3
13.9	12.7	11.9
13.8	13.0	11.6
14.0	13.0	12.2

Laboratory

1	2	3	4	5
2.9	2.7	3.3	3.3	4.1
3.1	3.4	3.3	3.2	4.1
3.1	3.6	3.5	3.4	3.7
3.7	3.2	3.5	2.7	4.2
3.1	4.0	2.8	2.7	3.1
4.2	4.1	2.8	3.3	3.5

9. Fabric Flammability Tests in Different Laboratories Flammability tests were conducted on children's sleepwear. The Vertical Semirestrained Test was used, in which pieces of fabric were burned under controlled conditions. After the burning stopped, the length of the charred portion was measured and recorded. Results are given in the margin for the same fabric tested at different laboratories. Because the same fabric was used, the different laboratories should have obtained the same results. Is there sufficient evidence to support the claim that the means for the different laboratories are the same? (The data were provided by Minitab, Inc.)

10. Mean Weights of Different Colas Refer to Data Set 1 in Appendix B. Is there sufficient evidence to support the claim that cans of regular Coke, diet Coke, regular Pepsi, and diet Pepsi have different mean weights?

11. Mean Weights of M&Ms Refer to Data Set 10 in Appendix B. At the 0.05 significance level, test the claim that the mean weight of M&Ms is the same for each of the six different color populations. If it is the intent of Mars, Inc., to make the candies so that the different color populations have the same mean weight, do these results suggest that the company has a problem requiring corrective action?

12. Second-Hand Smoke in Different Groups Refer to Data Set 9 in Appendix B. Use a 0.01 significance level to test the claim that the mean cotinine level is different for these three groups: nonsmokers who are not exposed to environmental tobacco smoke, nonsmokers who are exposed to tobacco smoke, and people who smoke. What do the results suggest about second-hand smoke?

11-2 Beyond the Basics

13. Effects of Data Transformations How are the analysis of variance results from Exercise 3 affected in each of the following cases?

a. 2°F is added to each temperature listed for the 18–20 age group.

b. All of the original temperatures are converted from the Fahrenheit scale to the Celsius scale. In general, how are ANOVA results affected by the scale used?

c. The same constant is added to every one of the original sample values.

d. Each of the original sample values is multiplied by the same constant.

e. The order of the samples is changed.

14. Using t Test Five independent samples of 50 values each are randomly drawn from populations that are normally distributed with equal variances. We wish to test the claim that $\mu_1 = \mu_2 = \mu_3 = \mu_4 = \mu_5$.

a. If we used only the methods given in Chapter 8, we would test the individual claims $\mu_1 = \mu_2$, $\mu_1 = \mu_3$, . . . , $\mu_4 = \mu_5$. How many ways can we pair off five means?

b. Assume that for each test of equality between two means, there is a 0.95 probability of not making a type I error. If all possible pairs of means are tested for equality, what is the probability of making no type I errors? (Although the tests are not actually independent, assume that they are.)

c. If we use analysis of variance to test the claim that $\mu_1 = \mu_2 = \mu_3 = \mu_4 = \mu_5$ at the 0.05 significance level, what is the probability of not making a type I error?

d. Compare the results of parts (b) and (c). Which approach is better in the sense of giving us a greater chance of not making a type I error?

15. Equivalent Tests In this exercise you will verify that when you have two sets of sample data, the t test for independent samples and the ANOVA method of this section are equivalent. Refer to the head injury measurements in Table 11-1, but use only the data for subcompact and compact cars.

a. Use a 0.05 significance level and the method of Section 8-6 to test the claim that the two samples come from populations with the same mean. (Assume that both populations have the same variance.)

b. Use a 0.05 significance level and the ANOVA method of this section to test the claim made in part (a).

c. Verify that the squares of the t test statistic and critical value from part (a) are equal to the F test statistic and critical value from part (b).

16. Components in ANOVA Table Complete the ANOVA table assuming that there are three samples with sizes of 5, 7, and 7, respectively.

Source of Variation	Sum of Squares (SS)	Degrees of Freedom	Mean Square (MS)	Test Statistic
Treatments	?	?	?	$F = ?$
Error	100.00	?	?	
Total	123.45	?		

11-3 Two-Way ANOVA

In Section 11-2 we used analysis of variance to decide whether three or more populations have the same mean. That section used procedures referred to as *one*-way analysis of variance (or single-factor analysis of variance) because the data are categorized into groups according to a *single* factor (or treatment). Recall that a factor, or treatment, is a property that is the basis for categorizing the different groups of data. For example, in Table 11-3, 60 SAT scores are separated into three categories according to the color of the M&M candy used as a treatment. The three populations are the people who eat red M&Ms, the people who eat green M&Ms, and the people who eat blue M&Ms. The SAT scores are based on data from the College Board, and the M&M element is a product of author whimsy. If we use one-way analysis of variance to test that the three M&M color populations have equal means, we get the Minitab display shown here.

MINITAB

```
Analysis of Variance for SAT
Source     DF        SS       MS      F       P
M&M         2     80310    40155   0.86   0.427
Error      57   2646987    46438
Total      59   2727297
```

TABLE 11-3 SAT Scores with Single Factor: Treatment with M&M Candies (Red, Green, Blue)

Red	1130	621	813	996	1030	1257	898	743	921	1179
	1092	855	896	858	1095	1133	896	1190	908	699
Green	996	630	583	828	1121	993	1025	907	1111	1147
	780	916	793	1188	499	1180	1229	1450	1071	1153
Blue	706	1068	1013	892	1370	1611	939	1004	821	915
	866	848	1408	793	1097	1244	996	1131	1039	1159

This Minitab display includes a *P*-value of 0.427, so we fail to reject the null hypothesis that the three populations have the same mean. This is pretty much what we expect: The colors of the M&Ms consumed don't appear to have an effect on SAT scores.

Two-way analysis of variance involves *two* factors, such as the Table 11-4 factors of (1) color of M&M candies used as treatment, and (2) gender. Using the two factors of M&M color and gender, we separate the data into six categories, as shown in Table 11-4. Such subcategories are often called *cells,* so Table 11-4 has six cells containing 10 scores each.

The following assumptions apply when testing with two-way analysis of variance.

Assumptions

1. For each cell, the sample values come from a population with a distribution that is approximately normal.
2. The populations have the same variance σ^2 (or standard deviation σ).
3. The samples are simple random samples. (That is, samples of the same size have the same probability of being selected.)
4. The samples are independent of each other. (The samples are not matched or paired in any way.)
5. The sample values are categorized two ways. (This is the basis for the name of the method: *two-way* analysis of variance.)
6. All of the cells have the same number of sample values. (This is called a *balanced* design.)

In analyzing the sample data in Table 11-4, we have already discussed the one-way analysis of variance for the single factor of M&M color, so it might seem reasonable to simply proceed with another one-way ANOVA for the factor of gender. Unfortunately, that approach wastes information and totally ignores a very important element: the effect from an interaction between the two factors.

TABLE 11-4 SAT Scores with Two Factors: Gender and Treatment with M&M Candies (Red, Green, Blue)

	Female					Male				
Red	1130	621	813	996	1030	1257	898	743	921	1179
	1092	855	896	858	1095	1133	896	1190	908	699
Green	996	630	583	828	1121	993	1025	907	1111	1147
	780	916	793	1188	499	1180	1229	1450	1071	1153
Blue	706	1068	1013	892	1370	1611	939	1004	821	915
	866	848	1408	793	1097	1244	996	1131	1039	1159

DEFINITION

There is an **interaction** between two factors if the effect of one of the factors changes for different categories of the other factor.

As an example of an *interaction* between two factors, consider the pairings of food and wine at a quality restaurant. It is known that certain foods and wines interact well while others interact poorly. There is a good interaction between Chablis wine and oysters; the limestone in the soil where Chablis is made leaves a residue in the wine that interacts well with oysters. In contrast, red wine and turkey interact in a way that results in a bad taste. Willing to make any sacrifices to improve the quality of this book, the author experimented with different food/wine pairings and recorded taste test results in Table 11-5. The scores in the body of the table reflect these effects: (1) Red wine interacts in a positive way with red meat (see the high scores of 10, 9, and 10 for the pairing of red wine and red meat); (2) white wine goes better with chicken or fish (see the high scores of 6, 5, 7 and 8, 6, 7). There appears to be an effect due to an interaction between the type of wine and the type of food.

TABLE 11-5 Taste Test Scores for Food/Wine Pairings

	Red Meat	Chicken	Fish
Red wine	10, 9, 10	4, 4, 2	3, 2, 4
White wine	3, 3, 2	6, 5, 7	8, 6, 7

Now that we understand the effects of an interaction, let's return to the SAT scores in Table 11-4. In using two-way ANOVA for the data of Table 11-4, we consider the effects of an interaction between M&M color and gender, as well as the effects of M&M color and the effects of gender on SAT scores. The calculations are quite involved, so *we will assume that a software package is being used.* (Minitab and Excel can do two-way ANOVA, but STATDISK and the TI-83 Plus calculator cannot. Procedures for using Minitab and Excel are described at the end of this section.)

The Minitab display includes SS (sum of squares) components similar to those described in Section 11-2. Because the circumstances of Section 11-2 involved only a single factor, we used SS(treatment) as a measure of the variation due to the different treatment categories, and we used SS(error) as a measure of the variation due to sampling error. We now use SS(Gender) as a measure of variation among the gender means. We use SS(M&M) as a measure of variation among the M&M color means. We continue to use SS(error) as a measure of variation due to sampling error. Similarly, we use MS(Gender) and MS(M&M) for the two different mean squares and continue to use MS(error) as before. Also, we use df(Gender) and df(M&M) for the two different degrees of freedom.

MINITAB

```
Analysis of Variance for SAT
Source        DF        SS        MS       F         P
M&M            2     80310     40155    1.00     0.376
Gender         1    289537    289537    7.19     0.010
Interaction    2    181727     90863    2.26     0.115
Error         54   2175723     40291
Total         59   2727297
```

Table 11-6 compares the analyses of the data in Tables 11-3 and 11-4. Because Tables 11-3 and 11-4 have the same data in the M&M color categories, the calculated values of SS(M&M), MS(M&M), and df(M&M) are also the same in both cases. But note that as we go from the one-factor case (Table 11-3) to the two-factor case (Table 11-4) by further partitioning the data according to the second factor of gender, the value of SS(error) is partitioned into SS(Gender), SS(interaction), and SS(error). Also, df(error) is partitioned into df(Gender), df(interaction), and df(error), which can be easily verified by noting that $57 = 1 + 2 + 54$. Similar partitioning does not apply to MS(error); the value of MS(error) is 46,438, which does not equal $289{,}537 + 90{,}863 + 40{,}291$.

TABLE 11-6 Comparison of Components Used in One-Way ANOVA and Two-Way ANOVA

One-Way ANOVA	Two-Way ANOVA
Sample data: Table 11-3 One factor: M&M Color	Sample data: Table 11-4 Two factors: M&M Color and Gender
SS(M&M) = 80,310	Same
MS(M&M) = 40,155	Same
df(M&M) = 2	Same
SS(total) = 2,727,297	Same
df(total) = 59	Same
SS(error) = 2,646,987 →	SS(Gender) = 289,537 SS(interaction) = 181,727 SS(error) = 2,175,723
df(error) = 57 →	df(Gender) = 1 df(interaction) = 2 df(error) = 54
MS(error) = 46,438	MS(Gender) = 289,537 MS(interaction) = 90,863 MS(error) = 40,291

When conducting a two-way analysis of variance, we consider three effects:

1. The effect due to the *interaction* between the two factors of M&M color and gender
2. The effect due to the *row* factor (M&M color)
3. The effect due to the *column* factor (gender)

We now summarize the basic procedure for two-way analysis of variance. This procedure is very similar to the procedure presented in Section 11-2. We form conclusions about equal means by analyzing two estimates of variance, and the test statistic F is the ratio of those two estimates. A significantly large value for F indicates that there is a statistically significant difference in means. The following procedure for two-way ANOVA is summarized in Figure 11-3.

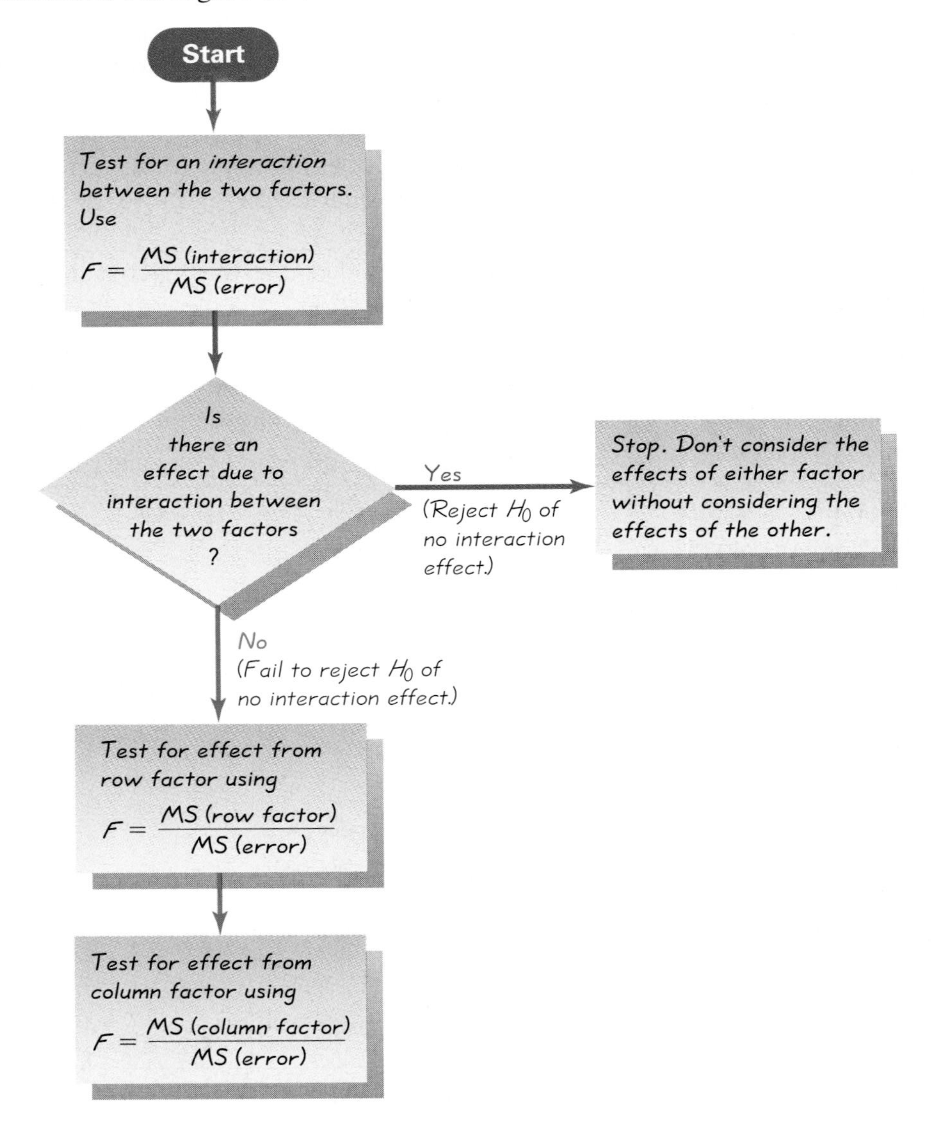

FIGURE 11-3
Procedure for Two-Way ANOVA

Procedure for Two-Way ANOVA

Step 1: *Interaction Effect:* In two-way analysis of variance, begin by testing the null hypothesis that there is no interaction between the two factors. Using Minitab for the data in Table 11-4, we get the following test statistic:

$$F = \frac{\text{MS(interaction)}}{\text{MS(error)}} = \frac{90{,}863}{40{,}291} = 2.26$$

Interpretation: The corresponding P-value is shown in the Minitab display as 0.115, so we fail to reject the null hypothesis of no interaction between the two factors. As we might expect, it does not appear that SAT scores are affected by an interaction between the color of consumed M&Ms and the gender of the subject.

Step 2: *Row/Column Effects:* If we do reject the null hypothesis of no interaction between factors, then we should stop now; we should not proceed with the two additional tests. (If there is an interaction between factors, we shouldn't consider the effects of either factor without considering those of the other.)

If we fail to reject the null hypothesis of no interaction between factors, then we should proceed to test the following two hypotheses:

H_0: There are no effects from the row factor (that is, the row means are equal).

H_0: There are no effects from the column factor (that is, the column means are equal).

In Step 1, we failed to reject the null hypothesis of no interaction between factors, so we proceed with the next two hypothesis tests identified in Step 2.

For the row factor,

$$F = \frac{\text{MS(M\&M)}}{\text{MS(error)}} = \frac{40{,}155}{40{,}291} = 1.00$$

Interpretation: This value is not significant because the corresponding P-value is shown in the Minitab display as 0.376. We fail to reject the null hypothesis of no effects from M&M color. The color of consumed M&Ms does not appear to have an effect on SAT scores. No big surprise there.

For the column factor,

$$F = \frac{\text{MS(Gender)}}{\text{MS(error)}} = \frac{289{,}537}{40{,}291} = 7.19$$

Interpretation: This value is significant because the corresponding P-value is shown as 0.010. We therefore reject the null hypothesis of no effects from gender. The gender of the subject does appear to have an effect on the SAT score. Based on the sample data in Table 11-4, we conclude that SAT scores do appear to have unequal means for men and women, but treating test subjects with

M&Ms of different colors does not appear to have an effect. The red, green, and blue treatment groups appear to have equal mean SAT scores.

Special Case: One Observation per Cell and No Interaction Table 11-4 contains 10 observations per cell. If our sample data consist of only one observation per cell, we lose MS(interaction), SS(interaction), and df(interaction) because those values are based on sample variances computed for each individual cell. If there is only one observation per cell, there is no variation within individual cells and those sample variances cannot be calculated. Here's how we proceed when there is one observation per cell: *If it seems reasonable to assume (based on knowledge about the circumstances) that there is no interaction between the two factors, make that assumption and then proceed as before to test the following two hypotheses separately:*

H_0: There are no effects from the row factor.

H_0: There are no effects from the column factor.

As an example, suppose that we have only the first score in each cell of Table 11-4. Using only those first scores, the three row means are 1193.5, 994.5, and 1158.5. Are those differences significant, suggesting that there is an effect due to M&M color? Again using only the first score in each cell, the two column means are 944.0 and 1287.0. Is that difference significant, suggesting that there is an effect due to gender? It is reasonable to believe that SAT scores are not affected by some interaction between the color of consumed M&Ms and the gender of the subject. (If we believe there is an interaction, the method described here does not apply.) Following is the Minitab display for the data in Table 11-4, with only the first score from each cell.

MINITAB

Analysis of Variance for SAT

Source	DF	SS	MS	F	P
M&M	2	45148	22574	0.19	0.842
Gender	1	176474	176474	1.46	0.350
Error	2	241108	120554		
Total	5	462729			

We first use the results from the Minitab display to test the null hypothesis of no effects from the row factor of M&M color.

$$F = \frac{\text{MS(M\&M)}}{\text{MS(error)}} = \frac{22{,}574}{120{,}554} = 0.19$$

This test statistic is not significant, because the corresponding P-value in the Minitab display is 0.842. We fail to reject the null hypothesis; it appears that the color of the M&Ms consumed does not affect SAT scores.

We now use the Minitab display to test the null hypothesis of no effect from the column factor of gender. The test statistic is

$$F = \frac{\text{MS(Gender)}}{\text{MS(error)}} = \frac{176{,}474}{120{,}554} = 1.46$$

This test statistic is not significant because the corresponding P-value is given in the Minitab display as 0.350. We fail to reject the null hypothesis, so it appears that SAT score is not affected by gender. Using only the first score from each cell, we are concluding that SAT scores do not appear to be affected by either M&M colors or genders, but when we used 10 scores from each cell, we concluded that SAT scores appeared to be affected by gender. Such is the power of larger samples.

In this section we have briefly discussed an important branch of statistics. We have emphasized the interpretation of computer displays while omitting the manual calculations and formulas, which are quite formidable.

Using Technology

STATDISK and the TI-83 Plus calculator do not provide results for two-way analysis of variance.

Minitab: First enter all of the sample values in column C1. Enter the corresponding row numbers in column C2. Enter the corresponding column numbers in column C3. From the main menu bar, select **Stat,** then select **ANOVA,** then **Two-Way.** In the dialog box, enter C1 for Response, enter C2 for Row factor, and enter C3 for Column factor. Click **OK.** *Hint:* Avoid confusion by labeling the columns C1, C2, and C3 with meaningful names.

Excel: For two-way tables with more than one entry per cell: Entries from the same cell must be listed down a column, not across a row. Enter the labels corresponding to the data set in column A and row 1, as in this example, which corresponds to Table 11-4:

	A	B	C
1		Female	Male
2	Red	1130	1257
3	Red	621	898
⋮	⋮	⋮	⋮

After entering the sample data and labels, select **Tools** from the main menu bar, then **Data Analysis,** then **Anova: Two-Factor With Replication.** In the dialog box, enter the input range. For the data in Table 11-4, enter A1:C31. For "rows per sample," enter the number of values in each cell; enter 10 for the data in Table 11-4. Click **OK.**

For two-way tables with exactly one entry per cell: The labels are not required. Enter the sample data as they appear in the table. Select **Tools**, then **Data Analysis,** then **Anova: Two-factor Without Replication.** In the dialog box, enter the input range of the sample values only; do not include labels in the input range. Click **OK.**

11-3 Basic Skills and Concepts

Interpreting a Computer Display. *In Exercises 1–3, use the Minitab display, which results from the scores listed in the accompanying table. The sample data are SAT scores on the verbal and math portions of SAT-I and are based on reported statistics from the College Board.*

Verbal

Female	646	539	348	623	478	429	298	782	626	533
Male	562	525	512	576	570	480	571	555	519	596

Math

Female	484	489	436	396	545	504	574	352	365	350
Male	547	678	464	651	645	673	624	624	328	548

MINITAB

Analysis of Variance for SAT

Source	DF	SS	MS	F	P
Gender	1	52635	52635	5.03	0.031
Ver/Math	1	6027	6027	0.58	0.453
Interaction	1	31528	31528	3.01	0.091
Error	36	376748	10465		
Total	39	466938			

1. Interaction Effect Test the null hypothesis that SAT scores are not affected by an interaction between gender and test (verbal/math). What do you conclude?
2. Effect of Gender Assume that SAT scores are not affected by an interaction between gender and the type of test (verbal/math). Is there sufficient evidence to support the claim that gender has an effect on SAT scores?
3. Effect of Type of SAT Test Assume that SAT scores are not affected by an interaction between gender and the type of test (verbal/math). Is there sufficient evidence to support the claim that the type of test (verbal/math) has an effect on SAT scores?

Interpreting a Computer Display. *In Exercises 4–6, use the Minitab display, which results from the values listed in the accompanying table. The sample data are student estimates (in feet) of the length of their classroom. The actual length of the classroom is 24 ft 7.5 in.*

	Major								
	Math			Business			Liberal Arts		
Female	28	25	30	35	25	20	40	21	30
Male	25	30	20	30	24	25	25	20	32

MINITAB

Analysis of Variance for Length

Source	DF	SS	MS	F	P
Gender	1	29.4	29.4	0.78	0.395
Major	2	10.1	5.1	0.13	0.876
Interaction	2	14.1	7.1	0.19	0.832
Error	12	453.3	37.8		
Total	17	506.9			

4. Interaction Effect Test the null hypothesis that the estimated lengths are not affected by an interaction between gender and major.

5. Effect of Gender Assume that estimated lengths are not affected by an interaction between gender and major. Is there sufficient evidence to support the claim that estimated length is affected by gender?

6. Effect of Major Assume that estimated lengths are not affected by an interaction between gender and major. Is there sufficient evidence to support the claim that estimated length is affected by major?

Interpreting a Computer Display. *In Exercises 7 and 8, refer to the given Minitab display. This display results from a study in which 24 subjects were given hearing tests using four different lists of words. The 24 subjects had normal hearing and the tests were conducted with no background noise. The main objective was to determine whether the four lists are equally difficult to understand. In the original table of hearing test scores, each cell has one entry. The original data are from* A Study of the Interlist Equivalency of the CID W-22 Word List Presented in Quiet and in Noise, *by Faith Loven, University of Iowa. The original data are available on the Internet through DASL (Data and Story Library).*

MINITAB

Analysis of Variance for Hearing

Source	DF	SS	MS	F	P
Subject	23	3231.6	140.5	3.87	0.000
List	3	920.5	306.8	8.45	0.000
Error	69	2506.5	36.3		
Total	95	6658.6			

7. Hearing Tests: Effect of Subject Assuming that there is no effect on hearing test scores from an interaction between subject and list, is there sufficient evidence to support the claim that the choice of subject has an effect on the hearing test score? Interpret the result by explaining why it makes practical sense.

8. Hearing Tests: Effect of Word List Assuming that there is no effect on hearing test scores from an interaction between subject and list, is there sufficient evidence to support the claim that the choice of word list has an effect on the hearing test score?

Interpreting a Computer Display. *Exercises 9 and 10 refer to the sample data in the given table and the corresponding Minitab display. The table entries are the numbers of support beams manufactured by four different operators using each of three different machines. Assume that there is no interaction effect from operator and machine.*

9. Effect of Machine Operator Using a 0.05 significance level, test the hypothesis that the four operators have the same mean production output. Interpret the result.

10. Effect of Machine Using a 0.05 significance level, test the claim that the choice of machine has no effect on the production output. Interpret the result.

		Machine		
		1	2	3
Operator	1	66	74	67
	2	58	67	68
	3	65	71	65
	4	60	64	66

MINITAB

Analysis of Variance for Beams

Source	DF	SS	MS	F	P
Operator	3	59.58	19.86	2.47	0.159
Machine	2	93.17	46.58	5.80	0.040
Error	6	48.17	8.03		
Total	11	200.92			

11. Interaction Between Gender and Smoking Refer to Data Set 14 in Appendix B and construct a table with pulse rates categorized according to the two factors of gender and whether the individual smokes. Select nine values for each cell and test the null hypothesis of no interaction between gender and smoking.

12. Effect of Gender on Pulse Rates Use the same data collected for Exercise 11, assume that pulse rates are not affected by an interaction between gender and smoking, and test the null hypothesis that gender has no effect on pulse rates.

13. Effect of Smoking on Pulse Rates Use the same data collected for Exercise 11, assume that pulse rates are not affected by an interaction between gender and smoking, and test the null hypothesis that smoking has no effect on pulse rates.

11-3 Beyond the Basics

14. Transformations of Data Assume that two-way ANOVA is used to analyze sample data consisting of more than one entry per cell. How are the ANOVA results affected in each of the following cases?
    a. The same constant is added to each sample value.
    b. Each sample value is multiplied by the same nonzero constant.

c. The format of the table is transposed, so that the row and column factors are interchanged.
d. The first sample value in the first cell is changed so that it becomes an outlier.

15. Changing Interaction Effect In analyzing Table 11-4, we concluded that SAT scores are not affected by an interaction between gender and the color of M&Ms consumed, and SAT scores are not affected by color of M&Ms consumed, but they are affected by gender.
a. Change the table entries so that there is an effect from the interaction between gender and color of consumed M&Ms.
b. Change the table entries so that there is no effect from the interaction between gender and color of consumed M&Ms, but there is an effect from the color of consumed M&Ms.
c. Change the table entries so that there is no effect from the interaction between gender and color of consumed M&Ms, there is no effect from color of consumed M&Ms, and there is no effect from gender.

Birth Order Affects Height

NEW YORK *New York Times* reporter C. Claiborne Ray wrote that "many studies have found a relationship between birth order and birth weight, with the later-born children in a family tending to be larger, and birth weight has some correlation with eventual height." In referring to a German study, he noted that "the researchers found that the first-born siblings had a statistically significant deficit in average height as compared with the average of all siblings, but the gap amounted to only a fraction of an inch. That study also found that men tended to increase in height with increasing birth order, while women did not, beyond the second born."

1. Describe in detail the design of an observational study that could be used to support or refute the key point that older siblings tend to be taller. Include a sampling plan and the method of analysis. How do you control for the inherent differences in heights between men and women? How might analysis of variance be used? How do you handle children who have no brothers or sisters?

2. One approach to designing an observational study might be to collect heights only from families in which there are exactly three children, all males. Among families with exactly three children, what proportion of them are all males?

3. Consider families with exactly three male children. Claiborne noted that younger siblings tend to be shorter, "but the gap amounted to only a fraction of an inch." If the differences in mean heights of the first, second, and third males are very small, how does the smallness of those differences affect the size of the sample needed to support the statistical significance of the differences?

4. Suppose someone argues that the stated results cannot be true because he knows a family in which the three children are all males, but their heights decrease with birth order. How do you respond?

Vocabulary List

analysis of variance (ANOVA)
one-way analysis of variance
single-factor analysis of variance
treatment
factor
variance between samples
variation due to treatment
variance within samples
variation due to error
multiple comparison procedures
completely randomized design
rigorously controlled design
two-way analysis of variance
interaction

Review

In Chapter 8 we used t distributions and normal distributions to test for equality between *two* population means, but in Section 11-2 we used analysis of variance (or ANOVA) to test for equality of three or more population means. This method requires (1) normally distributed populations, (2) populations with the same standard deviation (or variance), and (3) simple random samples that are independent of each other. The methods of one-way analysis of variance are used when we have three or more samples taken from populations that are characterized according to a single factor. The following are key features of one-way analysis of variance:

- The F test statistic is based on the ratio of two different estimates of the common population variance σ^2, as shown below.

$$F = \frac{\text{variance between samples}}{\text{variance within samples}} = \frac{\text{MS(treatment)}}{\text{MS(error)}}$$

- Critical values of F can be found in Table A-5, but we focused on the interpretation of P-values that are included as part of a computer display.

In Section 11-3 we considered two-way analysis of variance, characterized by data categorized according to two different factors. The method for two-way analysis of variance is summarized in Figure 11-3. We also considered two-way analysis of variance with one observation per cell.

Because of the nature of the calculations required throughout this chapter, we emphasized the interpretation of computer displays.

Review Exercises

1. Car Fuel Consumption: Testing for Interaction Twelve different 4-cylinder cars were tested for fuel consumption (in mi/gal) after being driven under identical highway conditions; the results are listed in the table and accompanying Minitab

MINITAB

Analysis of Variance for MPG

Source	DF	SS	MS	F	P
Transmis	1	40.3	40.3	3.56	0.108
Engine	2	43.2	21.6	1.90	0.229
Interaction	2	1.2	0.6	0.05	0.950
Error	6	68.0	11.3		
Total	11	152.7			

display. At the 0.05 significance level, test the claim that fuel consumption is not affected by an *interaction* between engine size and transmission type.

Highway Fuel Consumption (mi/gal)
of Different 4-Cylinder Compact Cars

	Engine Size (liters)		
	1.5	2.2	2.5
Automatic transmission	31, 32	28, 26	31, 23
Manual transmission	33, 36	33, 30	27, 34

2. Effect of Engine Size Refer to the data used in Exercise 1 and assume that fuel consumption is not affected by an interaction between engine size and type of transmission. Use a 0.05 significance level to test the claim that fuel consumption is not affected by engine size.

3. Effect of Transmission Refer to the data used in Exercise 1 and assume that fuel consumption is not affected by an interaction between engine size and type of transmission. Use a 0.05 significance level to test the claim that fuel consumption is not affected by type of transmission.

4. Location, Location, Location The accompanying list shows selling prices (in thousands of dollars) for homes located on Long Beach Island in New Jersey. Different mean selling prices are expected for the different locations. Do these sample data support the claim of different mean selling prices? Use a 0.05 significance level.

Oceanside:	235	395	547	469	369	279
Oceanfront:	538	446	435	639	499	399
Bayside:	199	219	239	309	399	190
Bayfront:	695	389	489	489	599	549

5. Auto Pollution The accompanying table lists the amounts of greenhouse gases emitted by different cars in one year. (See Data Set 18 in Appendix B.) The Minitab display on the next page results from this table.

a. Assuming that there is no interaction effect, is there sufficient evidence to support the claim that amounts of emitted greenhouse gases are affected by the type of transmission (automatic/manual)?

b. Assuming that there is no interaction effect, is there sufficient evidence to support the claim that amounts of emitted greenhouse gases are affected by the number of cylinders?

c. Based on the results from parts (a) and (b), can we conclude that greenhouse gas emissions are not affected by the type of transmission or the number of cylinders? Why or why not?

Emission of Greenhouse Gases (tons/year)

	4 Cylinders	6 Cylinders	8 Cylinders
Automatic	10	12	14
Manual	10	12	12

MINITAB

Analysis of Variance for Gases

Source	DF	SS	MS	F	P
Transmis	1	0.667	0.667	1.00	0.423
Cylinder	2	9.333	4.667	7.00	0.125
Error	2	1.333	0.667		
Total	5	11.333			

6. Drinking and Driving The Associated Insurance Institute sponsors studies of the effects of drinking on driving. In one such study, three groups of adult men were randomly selected for an experiment designed to measure their blood alcohol levels after consuming five drinks. Members of group A were tested after one hour, members of group B were tested after two hours, and members of group C were tested after four hours. The results are given in the accompanying table; the Minitab display for these data is also shown. At the 0.05 significance level, test the claim that the three groups have the same mean level.

MINITAB

Analysis of Variance

Source	DF	SS	MS	F	P
Factor	2	0.0076571	0.0038286	46.90	0.000
Error	14	0.0011429	0.0000816		
Total	16	0.0088000			

A	B	C
0.11	0.08	0.04
0.10	0.09	0.04
0.09	0.07	0.05
0.09	0.07	0.05
0.10	0.06	0.06
		0.04
		0.05

CUMULATIVE REVIEW EXERCISES

1. Boston Rainfall Statistics Refer to the Boston rainfall amounts for Monday, as listed in Data Set 17 in Appendix B.
   a. Find the mean.
   b. Find the standard deviation.
   c. Find the 5-number summary.

d. Identify any outliers.
e. Construct a histogram.
f. Assume that you want to test the null hypothesis that the mean amount of rainfall is the same for the seven days of the week. Can you use one-way ANOVA? Why or why not?
g. Based on the sample data, estimate the probability that precipitation will fall on a randomly selected Monday in Boston.

2. Advertising: Comparing Mean Reactions The Rocky Mountain Brewing Company plans to launch a major media campaign. Three advertising companies prepared trial commercials in an attempt to win a $2 million contract. The commercials were tested on randomly selected consumers, whose reactions were measured; the results are summarized in the accompanying table. (Higher scores indicate more positive reactions to the commercial.)
   a. Construct a boxplot for each of the three samples. Use the same scale so that the boxplots can be compared. Do the boxplots reveal any notable differences?
   b. Find the mean and standard deviation for each of the three sets of sample data.
   c. Use the methods of Section 8-5 to test the claim that the population of Barnum scores has a mean that is equal to the population mean for Solomon & Ford scores. Use a 0.05 significance level.
   d. For each of the three samples, construct a 95% confidence interval estimate of the population mean μ. Do the results suggest any notable differences?
   e. At the 0.05 significance level, test the claim that the three populations have the same mean reaction score. If you were responsible for advertising at this brewery, which company would you select on the basis of these results? Why?

Barnum Advertising Co.	Solomon & Ford Advertising	Diaz and Florio Advertising
52 68 75 40	69 73 82 59	42 73 69 53
77 63 55 72	66 84 75 70	57 61 73 74

3. Weights of Babies: Finding Probabilities In the United States, weights of newborn babies are normally distributed with a mean of 7.54 lb and a standard deviation of 1.09 lb (based on data from "Birth Weight and Prenatal Mortality," by Wilcox, Skjaerven, Buekens, and Kiely, *Journal of the American Medical Association,* Vol. 273, No. 9).
   a. If a newborn baby is randomly selected, what is the probability that he or she weighs more than 8.00 lb?
   b. If 16 newborn babies are randomly selected, what is the probability that their mean weight is more than 8.00 lb?
   c. What is the probability that each of the next three babies will have a birth weight greater than 7.54 lb?

Cooperative Group Activities

1. *In-class activity:* Begin by asking each student in the class to estimate the length of the classroom. Specify that the length is the distance between the chalkboard and the opposite wall. On the same sheet of paper, each student should also write his or her gender (male/female) and major. (See Exercises 4–6 in Section

11-3.) Then divide into groups of three or four, and use the data from the entire class to address these questions:

- Is there a significant difference between the mean estimate for males and the mean estimate for females?
- Is there sufficient evidence to reject equality of the mean estimates for different majors? Describe how the majors were categorized.
- Does an interaction between gender and major have an effect on the estimated length?
- Does gender appear to have an effect on estimated length?
- Does major appear to have an effect on estimated length?

2. *Out-of-class activity:* Divide into groups of three or four students. Each group should survey other students at the same college by asking them to identify their major and gender. You might include other factors, such as employment (none, part-time, full-time) and age (under 21, 21–30, over 30). For each surveyed subject, determine the accuracy of the time on his or her wristwatch. First set your own watch to the correct time using an accurate and reliable source ("At the tone, the time is . . ."). For watches that are ahead of the correct time, record positive times. For watches that are behind the correct time, record negative times. Use the sample data to address questions such as these:

- Does gender appear to have an effect on the accuracy of the wristwatch?
- Does major have an effect on wristwatch accuracy?
- Does an interaction between gender and major have an effect on wristwatch accuracy?

3. *In-class activity:* Divide into groups of five or six students. This activity is a contest of reaction times to determine which group is fastest, if there is a "fastest" group. Use the reaction timer included with the Cooperative Group Activities in Chapter 5. Test and record the reaction time using the dominant hand of each group member. (Only one try per person.) Each group should calculate the values of n, $\bar{x}$, and s, and record those summary statistics on the chalkboard along with the original list of reaction times. After all groups have reported their results, identify the group with the fastest mean reaction time. But is that group actually fastest? Use ANOVA to determine whether the means are significantly different. If they are not, there is no real "winner."

4. *Out-of-class activity:* The *World Almanac and Book of Facts* includes a section called "Noted Personalities," with subsections comprised of architects, artists, business leaders, cartoonists, social scientists, military leaders, philosophers, political leaders, scientists, writers, composers, entertainers, and others. Design and conduct an observational study that begins with the selection of samples from select groups, to be followed by a comparison of life spans of people from the different categories. Do any particular groups appear to have life spans that are different from the other groups? Can you explain such differences?

Technology Project

For U.S. presidents and the popes and British monarchs since 1690, the accompanying table lists the numbers of years that they lived after their inauguration, election, or coronation. Use boxplots and analysis of variance to determine whether the survival times for the different groups differ. Conduct the analysis of variance by using STATDISK, Minitab, Excel, a TI-83 Plus calculator, or some other statistical software package. Obtain printed copies of the computer displays and write your observations and conclusions.

Presidents		Popes		Kings and Queens	
Washington	10	Alex VIII	2	James II	17
J. Adams	29	Innoc XII	9	Mary II	6
Jefferson	26	Clem XI	21	William III	13
Madison	28	Innoc XIII	3	Anne	12
Monroe	15	Ben XIII	6	George I	13
J. Q. Adams	23	Clem XII	10	George II	33
Jackson	17	Ben XIV	18	George III	59
Van Buren	25	Clem XIII	11	George IV	10
Harrison	0	Clem XIV	6	William IV	7
Tyler	20	Pius VI	25	Victoria	63
Polk	4	Pius VII	23	Edward VII	9
Taylor	1	Leo XII	6	George V	25
Fillmore	24	Pius VIII	2	Edward VIII	36
Pierce	16	Greg XVI	15	George VI	15
Buchanan	12	Pius IX	32		
Lincoln	4	Leo XIII	25		
A. Johnson	10	Pius X	11		
Grant	17	Ben XV	8		
Hayes	16	Pius XII	17		
Garfield	0	Pius XIII	19		
Arthur	7	John XXIII	5		
Cleveland	24	Paul VI	15		
Harrison	12	John Paul I	0		
McKinley	4				
T. Roosevelt	18				
Taft	21				
Wilson	11				
Harding	2				
Coolidge	9				
Hoover	36				
F. Roosevelt	12				
Truman	28				
Kennedy	3				
Eisenhower	16				
L. Johnson	9				
Nixon	25				

Based on data from *Computer-Interactive Data Analysis,* by Lunn and McNeil, John Wiley & Sons.

Critical Thinking: Is Old Faithful becoming less faithful?

One of the most popular natural attractions in the United States is the Old Faithful geyser in Yellowstone National Park. This geyser was named for its predictable time intervals between eruptions, but is Old Faithful changing? The table lists time intervals (in minutes) between eruptions for four different years. The four samples have means of 63.3 min, 74.3 min, 81.7 min, and 83.8 min, respectively, so it appears that the mean time interval between eruptions is not the same for the four years.

Analyze the Results

Are the time intervals for the different years really different? Analyze the data using measures of center, variation, boxplots, and histograms. Identify any outliers, and use methods of analysis of variance.

Time Intervals (in min) Between Eruptions of the Old Faithful Geyser

1951	1985	1995	1996
74	89	86	88
60	90	86	86
74	60	62	85
42	65	104	89
74	82	62	83
52	84	95	85
65	54	79	91
68	85	62	68
62	58	94	91
66	79	79	56
62	57	86	89
60	88	85	94
$n_1 = 12$	$n_2 = 12$	$n_3 = 12$	$n_4 = 12$
$\bar{x}_1 = 63.3$	$\bar{x}_2 = 74.3$	$\bar{x}_3 = 81.7$	$\bar{x}_4 = 83.8$
$s_1 = 9.4$	$s_2 = 14.2$	$s_3 = 13.7$	$s_4 = 10.9$

Based on data from geologist Rick Hutchinson and the National Park Service.

Analysis of Variance

Go to the Web site for this book at

http://www.awlonline.com/triola

and link to the Internet Project for this chapter. There you will find instructions for finding data sets. Each data set can be divided naturally into samples from different populations so that ANOVA methods may be applied. In applying ANOVA methods, you will be conducting analyses in areas as varied as the functioning of the human body, consumer product labeling, and performance in sports.

Statistics *at work*

"Basic knowledge of statistics is critical in finance."

Joseph Marvan
Portfolio Manager

Principal and the Unit Head of Portfolio Management and Trading for the bond group at State Street Global Advisors (SSGA)

As a portfolio manager, Joseph specializes in the trading of fixed income derivatives. In his work, he uses statistical methods in assessing relative values among various financial instruments. SSGA is one of the county's largest institutional money managers with over $580 billion in assets under management.

What is your job?

I am Portfolio Manager and Unit Head of Portfolio Management for State Street Global Advisor (SSGA). SSGA is one of the nation's largest investment management companies and is a subsidiary of State Street Corporation. With over $580 billion in assets under management, SSGA's primary focus is managing the assets of public and private pension and retirement accounts. Portfolio Managers at SSGA look to add value through the use of quantitative modeling and objective relative value assessment. The Bond group (my group) manages portfolios with bonds that have a maturity greater than one year. Our objective is to trade bonds to provide a return on our client's investments that is higher than the return that our competitors can provide.

What concepts of statistics do you use?

We actively use probability analysis and hypothesis testing. We also use both linear and non-linear regression analysis. For portfolio construction we use mean variance optimization. We use these statistics in assessing a bond's value. Without a good understanding of the basics and foundations of statistics, I would not be able to effectively perform my responsibilities. Basic knowledge of statistics is critical in finance.

Please describe one specific example illustrating how the use of statistics was successful in improving a product or service.

In determining which bond to buy, we have used simple hypothesis testing. The yield difference between a corporate bond's yield to maturity and the risk free rate, as represented by a similar maturity U. S. Treasury bond, represents a bond's risk premium, or "spread." A bond's spread is what we as managers use to compare the value of one bond to another. Bond spreads tend to be "mean reverting" over time, which results in a somewhat normal distribution. Assuming mean reversion and a normal distribution, we can use simple hypothesis testing to look for statistical significance. In other words, if a bond's spread is statistically significantly wide versus another bond's spread, with all else being equal, we might consider the bond to be "cheap." A wider spread means a higher yield. When spreads tighten, the price or value of a bond goes up.

Managers in the bond market increase their returns by buying bonds that have spreads that are expected to tighten. We want to buy bonds that have spreads that are 1.0 to 2.0 standard deviations wider than the average and sell bonds that have spreads that are 1.0 to 2.0 standard deviations away (tighter) from the average. Simple hypothesis testing!

*I*s your use of probability and statistics increasing, decreasing, or remaining stable?

Our use and desire to learn probability and statistical analysis is always increasing. The markets are very efficient, which reduces the amount of predictable information available from historical data. To maintain our competitive advantage, we need to constantly update our methods and statistical tools.

*I*n terms of statistics, what would you recommend for prospective employees?

Probability and distribution analysis. I would also recommend analyzing non-normal distributions. In addition, an understanding of linear and some non-linear regression analysis is essential.

*D*o you recommend statistics for today's college students? Why?

I absolutely recommend it to ALL college students. Statistics provides students with an excellent foundation for making better decisions. In positions dealing with economics and finance, individuals with statistics backgrounds are looked at more favorably.

CHAPTER 12

Statistical Process Control

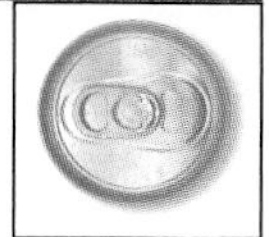

Is the process of cola can production out of control?

An *axial load* of an aluminum can is the maximum weight supported by its sides. It is important to have an axial load high enough so that the can isn't crushed when the top lid is pressed into place. The sample data in Table 12-1 come from a population of cans with a 0.0109-in. thickness. During each day of production, seven cans were randomly selected and tested. (The data are from a real manufacturing process, and they were provided by a student who used an earlier edition of this book.) In this chapter we will analyze the manufacturing process by focusing on the pattern of axial loads over time.

TABLE 12-1 Axial Loads (in pounds) of Aluminum Cans

Day	Axial Load (pounds)							$\bar{x}$	Median	Range	s
1	270	273	258	204	254	228	282	252.7	258	78	27.6
2	278	201	264	265	223	274	230	247.9	264	77	29.7
3	250	275	281	271	263	277	275	270.3	275	31	10.6
4	278	260	262	273	274	286	236	267.0	273	50	16.3
5	290	286	278	283	262	277	295	281.6	283	33	10.7
6	274	272	265	275	263	251	289	269.9	272	38	11.8
7	242	284	241	276	200	278	283	257.7	276	84	31.4
8	269	282	267	282	272	277	261	272.9	272	21	7.9
9	257	278	295	270	268	286	262	273.7	270	38	13.5
10	272	268	283	256	206	277	252	259.1	268	77	25.9
11	265	263	281	268	280	289	283	275.6	280	26	10.1
12	263	273	209	259	287	269	277	262.4	269	78	25.3
13	234	282	276	272	257	267	204	256.0	267	78	27.8
14	270	285	273	269	284	276	286	277.6	276	17	7.3
15	273	289	263	270	279	206	270	264.3	270	83	27.0
16	270	268	218	251	252	284	278	260.1	268	66	22.3
17	277	208	271	208	280	269	270	254.7	270	72	32.2
18	294	292	289	290	215	284	283	278.1	289	79	28.1
19	279	275	223	220	281	268	272	259.7	272	61	26.5
20	268	279	217	259	291	291	281	269.4	279	74	25.9
21	230	276	225	282	276	289	288	266.6	276	64	27.2
22	268	242	283	277	285	293	248	270.9	277	51	19.3
23	278	285	292	282	287	277	266	281.0	282	26	8.4
24	268	273	270	256	297	280	256	271.4	270	41	14.3
25	262	268	262	293	290	274	292	277.3	274	31	14.1

12-1 Overview

In Chapter 2 we noted that when describing, exploring, or comparing data sets, the following characteristics are usually extremely important.

1. *Center:* Measure of center, which is a representative or average value that gives us an indication of where the middle of the data set is located
2. *Variation:* A measure of the amount that the values vary among themselves
3. *Distribution:* The nature or shape of the distribution of the data, such as bell-shaped, uniform, or skewed
4. *Outliers:* Sample values that lie very far away from the vast majority of the other sample values
5. *Time:* Changing characteristics of the data over time

The main objective of this chapter is to address the fifth item: changing characteristics of data over time. By monitoring this characteristic, we are better able to control the production of goods and services, thereby ensuring better quality.

There is currently a strong trend toward trying to improve the quality of American goods and services, and the methods presented in this chapter are being used by growing numbers of businesses. Evidence of the increasing importance of quality is found in its greater role in advertising and the growing number of books and articles that focus on the issue of quality. In many cases, job applicants (you?) have a definite advantage when they can tell employers that they have studied statistics and methods of quality control. This chapter will present some of the basic tools commonly used to monitor quality.

Minitab, Excel, and other software packages include programs for automatically generating charts of the type discussed in this chapter, and we will include several examples of such displays.

12-2 Control Charts for Variation and Mean

The main objective of this section is to monitor important features of data over time. Such data are often referred to as *process data.*

DEFINITION

Process data are data arranged according to some time sequence. They are measurements of a characteristic of goods or services that results from some combination of equipment, people, materials, methods, and conditions.

For example, Table 12-1 includes process data consisting of the measured axial loads of aluminum cans over 25 consecutive days of production. Each day, seven cans were randomly selected and tested. Because the data in

Table 12-1 are arranged according to the time at which they were selected, they are process data. It is very important to recognize this point:

Important characteristics of process data can change over time.

In making aluminum cans, a manufacturer might use a good machine that is correctly calibrated, but if the machine wears with use, the cans might begin to become defective. Companies have gone bankrupt because they unknowingly allowed manufacturing processes to deteriorate without constant monitoring.

Run Charts

There are various methods that can be used to monitor a process to ensure that the important desired characteristics don't change—analysis of a *run chart* is one such method.

DEFINITION

A **run chart** is a sequential plot of *individual* data values over time. One axis (usually the vertical axis) is used for the data values, and the other axis (usually the horizontal axis) is used for the time sequence.

EXAMPLE **Manufacturing Cola Cans** Treating the 175 axial loads in Table 12-1 as a string of consecutive measurements, construct a run chart by using a vertical axis for the axial loads and a horizontal axis to identify the order of the sample data.

SOLUTION Figure 12-1 is the Minitab-generated run chart for the data in Table 12-1. The vertical scale is designed to be suitable for axial loads ranging from 200 lb to 297 lb, which are the minimum and maximum values in Table 12-1. The horizontal scale is designed to include the 175 values

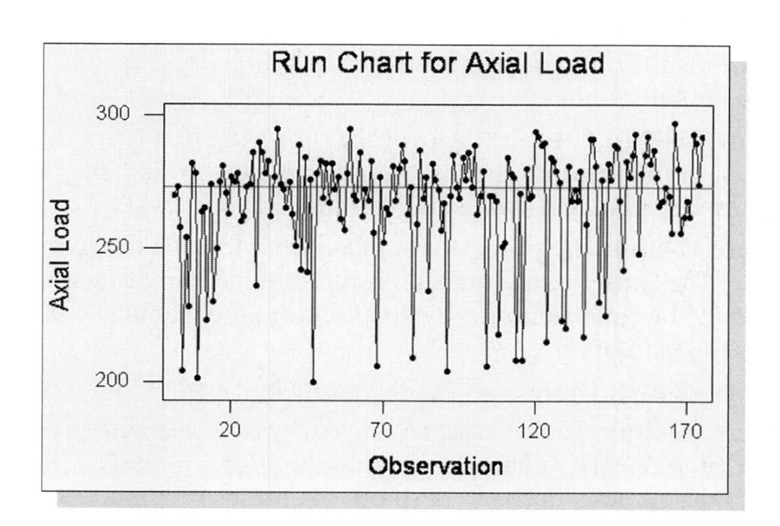

FIGURE 12-1
Run Chart of Axial Loads in Table 12-1

arranged in sequence. The first point represents the first value of 270 lb, the second point represents the second value of 273 lb, and so on.

In Figure 12-1, the horizontal scale identifies the sample number, so the number 20 indicates the 20th sample item. The vertical scale represents the measured axial load, so 250 indicates a load of 250 lb. Now examine Figure 12-1 and observe that there aren't any dramatic patterns that jump out begging for attention. Figure 12-1 does not appear to reveal any problems that need correction.

The Flynn Effect: Upward Trend in IQ Scores

A run chart or control chart of IQ scores would reveal that they exhibit an upward trend, because IQ scores have been steadily increasing since they began to be used about 70 years ago. The trend is worldwide, and it is the same for different types of IQ tests, even those that rely heavily on abstract and nonverbal reasoning with minimal cultural influence. This upward trend has been named the Flynn effect, because political scientist James R. Flynn discovered the trend in his studies of U. S. military recruits. The amount of the increase is quite substantial: Based on a current mean IQ score of 100, it is estimated that the mean IQ in 1920 would be about 77. The typical student of today is therefore brilliant when compared to his or her great-grandparents. So far, there is no generally accepted explanation for the Flynn effect.

Interpreting Run Charts Only when a process is *statistically stable* can its data be treated as if they came from a population with a constant mean, standard deviation, distribution, and other characteristics.

DEFINITION

A process is **statistically stable** (or **within statistical control**) if it has only natural variation, with no patterns, cycles, or unusual points.

Figure 12-2 illustrates typical patterns showing ways in which the process of filling 16-oz soup cans may not be statistically stable.

- **Figure 12-2(a):** There is an obvious *upward trend* that corresponds to values that are increasing over time. If the filling process were to follow this type of pattern, the cans would be filled with more and more soup until they began to overflow, eventually leaving the employees swimming in soup.
- **Figure 12-2(b):** There is an obvious *downward trend* that corresponds to steadily decreasing values. The cans would be filled with less and less soup until they were extremely underfilled. Such a process would require a complete reworking of the cans in order to get them full enough for distribution to consumers.
- **Figure 12-2(c):** There is an *upward shift.* A run chart such as this one might result from an adjustment to the filling process, making all subsequent values higher.
- **Figure 12-2(d):** There is a *downward shift*—the first few values are relatively stable, and then something happened so that the last several values are relatively stable, but at a much lower level.
- **Figure 12-2(e):** The process is stable except for one *exceptionally high value.* The cause of that unusual value should be investigated. Perhaps the cans became temporarily stuck and one particular can was filled twice instead of once.
- **Figure 12-2(f):** There is an *exceptionally low value.*
- **Figure 12-2(g):** There is a *cyclical pattern* (or repeating cycle). This pattern is clearly nonrandom and therefore reveals a statistically unstable process. Perhaps periodic overadjustments are being made

to the machinery, with the effect that some desired value is continually being chased but never quite captured.

- **Figure 12-2(h):** The *variation is increasing over time.* This is a common problem in quality control. The net effect is that products vary more and more until almost all of them are worthless. For example, some soup cans will be overflowing with wasted soup, and some will be underfilled and unsuitable for distribution to consumers.

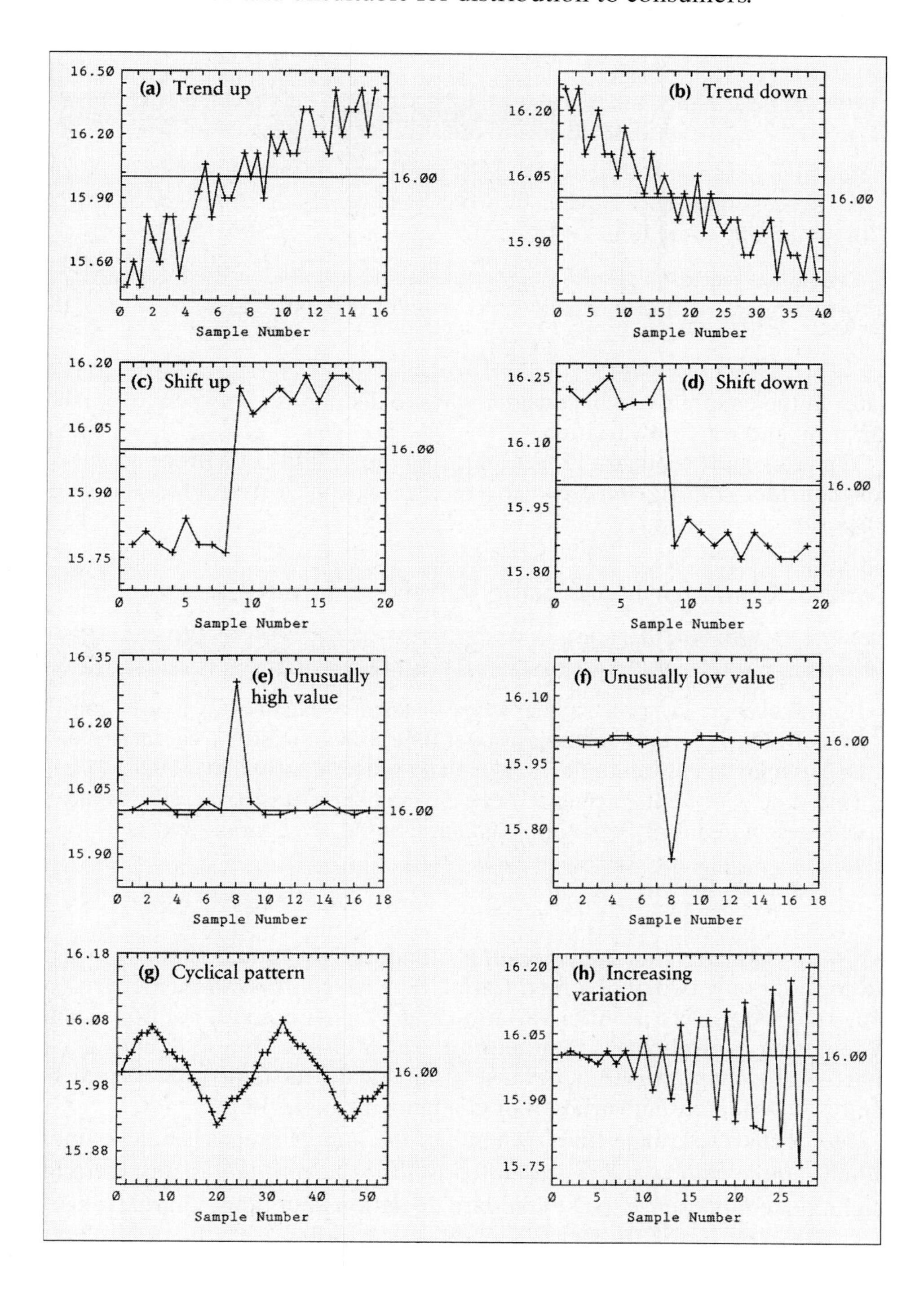

FIGURE 12-2
Processes with Patterns That Aren't Statistically Stable

A common goal of many different methods of quality control is this: *reduce variation* in the product or service. For example, Ford became concerned with variation when it found that its transmissions required significantly more warranty repairs than the same type of transmissions made by Mazda in Japan. A study showed that the Mazda transmissions had substantially less variation in the gearboxes; that is, crucial gearbox measurements varied much less in the Mazda transmissions. Although the Ford transmissions were built within the allowable limits, the Mazda transmissions were more reliable because of their lower variation. Variation in a process can result from two types of causes.

DEFINITIONS

Random variation is due to chance; it is the type of variation inherent in any process that is not capable of producing every good or service exactly the same way every time.

Assignable variation results from causes that can be identified (such factors as defective machinery, untrained employees, and so on).

Later in the chapter we will consider ways to distinguish between assignable variation and random variation.

The run chart is one tool for monitoring the stability of a process. We will now consider *control charts,* which are also extremely useful for that same purpose.

Control Chart for Monitoring Variation: The *R* Chart

DEFINITIONS

A **control chart** of a process characteristic (such as mean or variation) consists of values plotted sequentially over time, and it includes a **centerline** as well as a **lower control limit** (LCL) and an **upper control limit** (UCL). The centerline represents a central value of the characteristic measurements, whereas the control limits are boundaries used to separate and identify any points considered to be *unusual.*

We will assume that the population standard deviation σ is not known as we consider only two of several different types of *control charts: R* charts (or range charts) used to monitor variation and $\overline{x}$ charts used to monitor means. When using control charts to monitor a process, it is common to consider R charts and $\overline{x}$ charts together, because a statistically unstable process may be the result of increasing variation or changing means or both.

An ***R* chart** (or **range chart**) is a plot of the sample ranges instead of individual sample values, and it is used to monitor the *variation* in a process. (It might make more sense to use standard deviations, but range charts are used more often in practice. This is a carryover from times when calculators and computers were not available.) In addition to plotting the range values, we include a centerline located at $\overline{R}$, which denotes the mean of all sample ranges, as well as another line for the lower control limit and a third line for the upper control limit. Following is a summary of notation for the components of the R chart.

Don't Tamper!

Nashua Corp. had trouble with its paper-coating machine and considered spending a million dollars to replace it. The machine was working well with a stable process, but samples were taken every so often and, based on the results, adjustments were made. These overadjustments, called tampering, caused shifts away from the distribution that had been good. The effect was an increase in defects. When statistician and quality expert W. Edwards Deming studied the process, he recommended that no adjustments be made unless warranted by a signal that the process had shifted or had become unstable. The company was better off with no adjustments than with the tampering that took place.

Notation

Given: Process data consisting of a sequence of samples all of the same size, and the distribution of the process data is essentially normal.

n = size of each sample, or *subgroup*

$\overline{\overline{x}}$ = mean of the sample means, which is equivalent to the mean of all sample values combined

$\overline{R}$ = mean of the sample ranges (that is, the sum of the sample ranges divided by the number of samples)

Monitoring Process Variation: Control Chart for R

Points plotted: Sample ranges

Centerline: $\overline{R}$

Upper control limit (UCL): $D_4\overline{R}$

Lower control limit (LCL): $D_3\overline{R}$

where the values of D_4 and D_3 are found in Table 12-2. (See the next page.)

The values of D_4 and D_3 were computed by quality-control experts, and they are intended to simplify calculations. The upper and lower control limits of $D_4\overline{R}$ and $D_3\overline{R}$ are values that are roughly equivalent to 99.7% confidence interval limits. It is therefore highly unlikely that values would fall beyond those limits. If a value does fall beyond the control limits, it's very likely that something is wrong with the process.

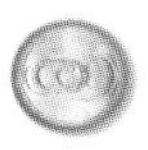

EXAMPLE Manufacturing Cola Cans Refer to the axial loads in Table 12-1. Using the samples of size $n = 7$ collected each day of manufacturing, construct a control chart for R.

SOLUTION We begin by finding the value of $\overline{R}$, the mean of the sample ranges.

$$\overline{R} = \frac{78 + 77 + \cdots + 31}{25} = 54.96$$

TABLE 12-2 Control Chart Constants

Observations in Subgroup, n	$\bar{x}$ A_2	A_3	s B_3	B_4	R D_3	D_4
2	1.880	2.659	0.000	3.267	0.000	3.267
3	1.023	1.954	0.000	2.568	0.000	2.574
4	0.729	1.628	0.000	2.266	0.000	2.282
5	0.577	1.427	0.000	2.089	0.000	2.114
6	0.483	1.287	0.030	1.970	0.000	2.004
7	0.419	1.182	0.118	1.882	0.076	1.924
8	0.373	1.099	0.185	1.815	0.136	1.864
9	0.337	1.032	0.239	1.761	0.184	1.816
10	0.308	0.975	0.284	1.716	0.223	1.777
11	0.285	0.927	0.321	1.679	0.256	1.744
12	0.266	0.886	0.354	1.646	0.283	1.717
13	0.249	0.850	0.382	1.618	0.307	1.693
14	0.235	0.817	0.406	1.594	0.328	1.672
15	0.223	0.789	0.428	1.572	0.347	1.653
16	0.212	0.763	0.448	1.552	0.363	1.637
17	0.203	0.739	0.466	1.534	0.378	1.622
18	0.194	0.718	0.482	1.518	0.391	1.608
19	0.187	0.698	0.497	1.503	0.403	1.597
20	0.180	0.680	0.510	1.490	0.415	1.585
21	0.173	0.663	0.523	1.477	0.425	1.575
22	0.167	0.647	0.534	1.466	0.434	1.566
23	0.162	0.633	0.545	1.455	0.443	1.557
24	0.157	0.619	0.555	1.445	0.451	1.548
25	0.153	0.606	0.565	1.435	0.459	1.541

Source: Adapted from *ASTM Manual on the Presentation of Data and Control Chart Analysis,* © 1976 ASTM, pp. 134–136. Reprinted with permission of American Society for Testing and Materials.

The centerline for our R chart is therefore located at $\overline{R} = 54.96$. To find the upper and lower control limits, we must first find the values of D_3 and D_4. Referring to Table 12-2 for $n = 7$, we get $D_3 = 0.076$ and $D_4 = 1.924$, so the control limits are as follows:

$$\text{Upper control limit: } D_4\overline{R} = (1.924)(54.96) = 105.74$$
$$\text{Lower control limit: } D_3\overline{R} = (0.076)(54.96) = 4.18$$

Using a centerline value of $\overline{R} = 54.96$ and control limits of 105.74 and 4.18, we now proceed to plot the sample ranges. The result is shown in the Minitab display that follows. (There is a very small discrepancy between the control limits calculated here and those shown in the Minitab display.)

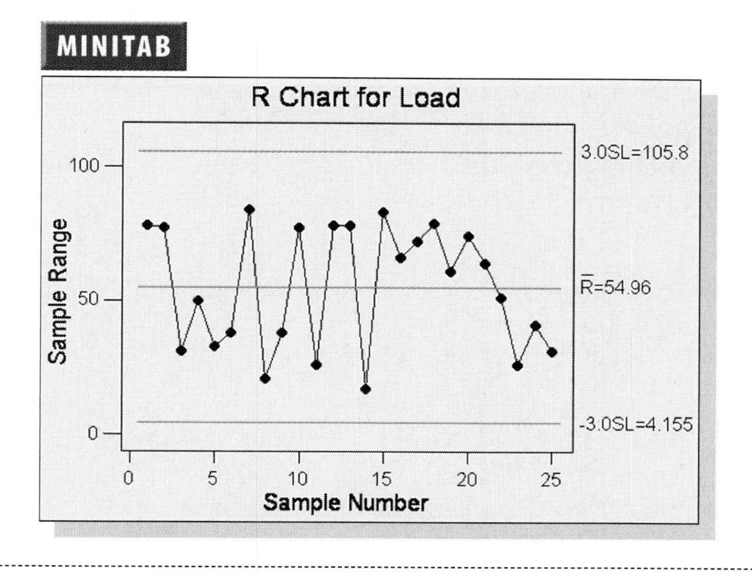

Interpreting Control Charts

Here is an extremely important point:

> **Upper and lower control limits of a control chart are based on the *actual* behavior of the process, not the *desired* behavior.**

For example, control charts might suggest that 16-oz soup cans are being filled with a statistically stable process, but if they are being filled to a level of only 8 oz, then there is a major problem because the statistically stable *actual* behavior is very different from the *desired* behavior. Also, we should clearly understand the specific criteria for determining whether a process is in statistical control (that is, whether it is statistically stable). So far, we have noted that a process is not statistically stable if its pattern resembles any of the patterns shown in Figure 12-2. This criterion is included with some others in the following list.

Criteria for Determining When a Process Is Not Statistically Stable (Out of Statistical Control)

1. There is a pattern, trend, or cycle that is obviously not random (such as those depicted in Figure 12-2).
2. There is a point lying beyond the upper or lower control limits.
3. *Run of 8 Rule:* There are eight consecutive points all above or all below the centerline. (With a statistically stable process, there is a 0.5 probability that a point will be above or below the centerline, so it is very unlikely that eight consecutive points will all be above the centerline or all below it.)

Bribery Detected with Control Charts

Control charts were used to help convict a person who bribed Florida jai alai players to lose. (See "Using Control Charts to Corroborate Bribery in Jai Alai," by Charnes and Gitlow, The American Statistician, Vol. 49, No. 4.) An auditor for one jai alai facility noticed that abnormally large sums of money were wagered for certain types of bets, and some contestants didn't win as much as expected when those bets were made. R charts and $\bar{x}$ charts were used in court as evidence of highly unusual patterns of betting. Examination of the control charts clearly shows points well beyond the upper control limit, indicating that the process of betting was out of statistical control. The statistician was able to identify a date at which assignable variation appeared to stop, and prosecutors knew that it was the date of the suspect's arrest.

We will use only the three out-of-control criteria listed above, but some businesses use additional criteria such as these:

- There are six consecutive points all increasing or all decreasing.
- There are 14 consecutive points all alternating between up and down (such as up, down, up, down, and so on).
- Two out of three consecutive points are beyond control limits that are 1 standard deviation away from the centerline.
- Four out of five consecutive points are beyond control limits that are 2 standard deviations away from the centerline.

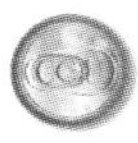

EXAMPLE Statistical Process Control Examine the R chart shown in the Minitab display for the preceding example and determine whether the process variation is within statistical control.

SOLUTION We can interpret control charts for R by applying the three out-of-control criteria just listed. Applying the three criteria to the Minitab display of the R chart, we conclude that variation in this process is within statistical control because of the following.

1. There is no pattern, trend, or cycle that is obviously not random.
2. No point lies beyond the upper or lower control limits.
3. There are not eight consecutive points all above or all below the centerline.

INTERPRETATION We therefore conclude that the variation (not necessarily the mean) of the process is within statistical control. No action is required to correct the *variation* among the axial loads of the cans.

Control Chart for Monitoring Means: The $\bar{x}$ Chart

An **$\bar{x}$ chart** is a plot of the sample means, and it is used to monitor the *center* in a process. In addition to plotting the sample means, we include a centerline located at $\bar{\bar{x}}$, which denotes the mean of all sample means (equal to the mean of all sample values combined), as well as another line for the lower control limit and a third line for the upper control limit. Using the approach common in business and industry, the centerline and control limits are based on ranges (instead of standard deviations).

Monitoring Process Mean: Control Chart for $\bar{x}$

Points plotted: Sample means

Centerline: $\bar{\bar{x}}$

Upper control limit (UCL): $\bar{\bar{x}} + A_2\bar{R}$

Lower control limit (LCL): $\bar{\bar{x}} - A_2\bar{R}$

where the values of A_2 are found in Table 12-2.

EXAMPLE Manufacturing Cola Cans Refer to the axial loads in Table 12-1. Using samples of size $n = 7$ collected each working day, construct a control chart for $\bar{x}$. Based on the control chart for $\bar{x}$ only, determine whether the process mean is within statistical control.

SOLUTION Before plotting the 25 points corresponding to the 25 values of $\bar{x}$, we must first find the value for the centerline and the values for the control limits. We get

$$\bar{\bar{x}} = \frac{252.7 + 247.9 + \cdots + 277.3}{25} = 267.12$$

$$\bar{R} = \frac{78 + 77 + \cdots + 31}{25} = 54.96$$

Referring to Table 12-2, we find that for $n = 7$, $A_2 = 0.419$. Knowing the values of $\bar{\bar{x}}$, A_2, and $\bar{R}$, we are now able to find the control limits.

Upper control limit: $\bar{\bar{x}} + A_2\bar{R} = 267.12 + (0.419)(54.96) = 290.15$

Lower control limit: $\bar{\bar{x}} - A_2\bar{R} = 267.12 - (0.419)(54.96) = 244.09$

INTERPRETATION The resulting control chart for $\bar{x}$ will be as shown in the accompanying Excel display. Examination of the control chart shows that the process mean is within statistical control because none of the three out-of-control criteria are satisfied. Again, no corrective action is required.

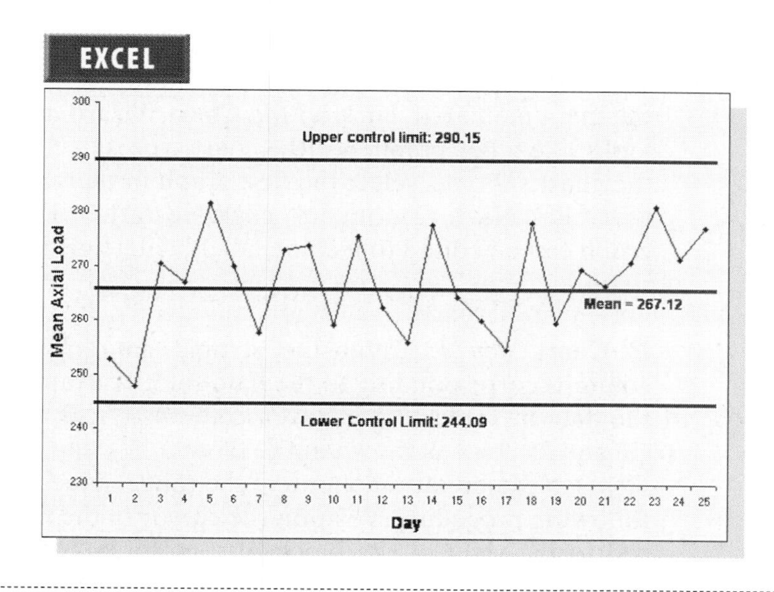

Using Technology

Minitab: **Run Chart** To construct a run chart, such as the one shown in Figure 12-1, begin by entering all of the sample data in column C1. Select the option **Stat,** then **Quality Tools,** then **Run Chart.** In the indicated boxes, enter C1 for the single column variable, enter 1 for the subgroup size, and then click on **OK.**

***R* Chart** First enter the individual sample values sequentially in column C1. Next, select the options **Stat, Control Charts,** and **R.** Enter C1 in the "single column" box, enter the sample size in the box for the subgroup size, and click on **estimate.** Select **Rbar.** (Selection of the *R* bar estimate causes the variation of the population distribution to be estimated with the sample ranges instead of the sample standard deviations, which is the default.) Click **OK** twice.

$\bar{x}$ Chart First enter the individual sample values sequentially in column C1. Next, select the options **Stat, Control Charts,** and **Xbar.** Enter C1 in the "single column" box, enter the size of each of the samples in the "subgroup size box," and click on **estimate;** then select **Rbar.** Click **OK** twice.

Excel: To use the Data Desk XL add-in, click on **DDXL** and select **Process Control.** Proceed to select the type of chart you want. (You must first enter the data in column A with sample identifying codes entered in column B. For the data of Table 12-1, for example, enter a 1 in column B adjacent to each value from day 1, enter a 2 for each value from day 2, and so on.)

To use Excel's built-in graphics features instead of Data Desk XL, see the following.

Run chart Enter all of the sample data in column A. On the main menu bar, click on the **Chart Wizard** icon, which looks like a bar graph. For the chart type, select **Line.** For the chart subtype, select the first graph in the second row, then click **Next.** Continue to click **Next,** then **Finish.** The graph can be edited to include labels, delete grid lines, and so on.

***R* Chart** *Step 1:* Enter the sample data in rows and columns corresponding to the data set. For example, enter the data in Table 12-1 in seven columns (A, B, C, D, E, F, G) and 25 rows as shown in the table.

Step 2: Next, create a column of the range values using the following procedure. Position the cursor in the first empty cell to the right of the block of sample data, then enter this expression in the formula box: =MAX(A1:G1)-MIN(A1:G1), where the range A1:G1 should be modified to describe the first row of your data set. After pressing the **Enter** key, the range for the first row should appear. Use the mouse to click and drag the lower right corner of this cell, so that the whole column fills up with the ranges for the different rows.

Step 3: Next, produce a graph by following the same procedure described for the run charts, but be sure to refer to the column of *ranges* when entering the input range. You can insert the required centerline and upper and lower control limits by editing the graph. Click on the line on the bottom of the screen, then click and drag to position the line correctly.

$\bar{x}$ Chart *Step 1:*Enter the sample data in rows and columns corresponding to the data set. For example, enter the data in Table 12-1 in seven columns (A, B, C, D, E, F, G) and 25 rows as shown in the table.
Step 2: Next, create a column of the sample means using the following procedure. Position the cursor in the first empty cell to the right of the block of sample data, then enter this expression in the formula box: = AVERAGE(A1:G1), where the range A1:G1 should be modified to describe the first row of your data set. After pressing the **Enter** key, the mean for the first row should appear. Use the mouse to click and drag the lower right corner of this cell, so that the whole column fills up with the means for the different rows.
Step 3: Next, produce a graph by following the same procedure described for the run chart, but be sure to refer to the column of *means* when entering the input range. You can insert the required centerline and upper and lower control limits by editing the graph. Click on the line on the bottom of the screen, then click and drag to position the line correctly.

12-2 Basic Skills and Concepts

Constructing Control Charts for Eruptions of Old Faithful. *In Exercises 1–3, use the following information:*

> *The Old Faithful Geyser was monitored for 25 recent and consecutive years. In each year, six intervals (in minutes) between eruptions were recorded, with the results given in the accompanying table on the next page.*

1. Old Faithful Geyser: Constructing a Run Chart Construct a run chart for the 150 values. Does there appear to be a pattern suggesting that the process is not within statistical control? What are the implications for tourists wishing to see Old Faithful erupt?

2. Old Faithful Geyser: Constructing an R Chart Construct an R chart and determine whether the process variation is within statistical control. If it is not, identify which of the three out-of-control criteria lead to rejection of statistically stable variation. How would tourists be affected by out-of-control variation?

3. Old Faithful Geyser: Constructing an $\bar{x}$ Chart Construct an $\bar{x}$ chart and determine whether the process mean is within statistical control. If it is not, identify which of the three out-of-control criteria lead to rejection of a statistically stable mean. How would tourists be affected by an out-of-control mean?

Intervals Between Eruptions

Year	Interval (min)						Mean	Range
1	65	72	60	69	65	67	66.3	12
2	74	65	60	69	68	59	65.8	15
3	68	66	69	64	70	73	68.3	9
4	73	65	71	77	63	77	71.0	14
5	79	67	64	61	81	77	71.5	20
6	74	76	65	69	76	64	70.7	12
7	70	73	74	77	65	73	72.0	12
8	71	68	70	79	75	82	74.2	14
9	62	63	61	48	59	77	61.7	29
10	60	74	77	57	52	78	66.3	26
11	67	73	47	81	92	57	69.5	45
12	79	84	79	72	61	80	75.8	23
13	83	78	83	74	61	68	74.5	22
14	57	68	72	75	56	79	67.8	23
15	59	76	78	86	64	72	72.5	27
16	63	63	71	77	81	65	70.0	18
17	67	84	72	75	70	70	73.0	17
18	93	83	85	79	90	74	84.0	19
19	81	74	80	65	70	84	75.7	19
20	83	67	71	67	97	88	78.8	30
21	62	61	57	86	70	77	68.8	29
22	67	75	67	89	93	81	78.7	26
23	86	65	70	74	83	74	75.3	21
24	74	67	99	75	41	83	73.2	58
25	97	93	73	81	85	90	86.5	24

Constructing Control Charts for Minted Quarters. *In Exercises 4–6, use the following information:*

The U.S. Mint has a goal of making quarters with a weight of 5.670 g, but any weight between 5.443 g and 5.897 g is considered acceptable. A new minting machine is placed into service and the weights are recorded for a quarter randomly selected every 12 min for 20 consecutive hours. The results are listed in the accompanying table on the next page.

4. Minting Quarters: Constructing a Run Chart Construct a run chart for the 100 values. Does there appear to be a pattern suggesting that the process is not within statistical control? What are the practical implications of the run chart?

5. Minting Quarters: Constructing an R Chart Construct an R chart and determine whether the process variation is within statistical control. If it is not, identify which of the three out-of-control criteria lead to rejection of statistically stable variation.

6. Minting Quarters: Constructing an $\overline{x}$ Chart Construct an $\overline{x}$ chart and determine whether the process mean is within statistical control. If it is not, identify which of the three out-of-control criteria lead to rejection of a statistically stable mean. Does this process need corrective action?

Weights (in grams) of Minted Quarters

Hour	Weight (g)					$\bar{x}$	s	Range
1	5.639	5.636	5.679	5.637	5.691	5.6564	0.0265	0.055
2	5.655	5.641	5.626	5.668	5.679	5.6538	0.0211	0.053
3	5.682	5.704	5.725	5.661	5.721	5.6986	0.0270	0.064
4	5.675	5.648	5.622	5.669	5.585	5.6398	0.0370	0.090
5	5.690	5.636	5.715	5.694	5.709	5.6888	0.0313	0.079
6	5.641	5.571	5.600	5.665	5.676	5.6306	0.0443	0.105
7	5.503	5.601	5.706	5.624	5.620	5.6108	0.0725	0.203
8	5.669	5.589	5.606	5.685	5.556	5.6210	0.0545	0.129
9	5.668	5.749	5.762	5.778	5.672	5.7258	0.0520	0.110
10	5.693	5.690	5.666	5.563	5.668	5.6560	0.0534	0.130
11	5.449	5.464	5.732	5.619	5.673	5.5874	0.1261	0.283
12	5.763	5.704	5.656	5.778	5.703	5.7208	0.0496	0.122
13	5.679	5.810	5.608	5.635	5.577	5.6618	0.0909	0.233
14	5.389	5.916	5.985	5.580	5.935	5.7610	0.2625	0.596
15	5.747	6.188	5.615	5.622	5.510	5.7364	0.2661	0.678
16	5.768	5.153	5.528	5.700	6.131	5.6560	0.3569	0.978
17	5.688	5.481	6.058	5.940	5.059	5.6452	0.3968	0.999
18	6.065	6.282	6.097	5.948	5.624	6.0032	0.2435	0.658
19	5.463	5.876	5.905	5.801	5.847	5.7784	0.1804	0.442
20	5.682	5.475	6.144	6.260	6.760	6.0642	0.5055	1.285

Constructing Control Charts for Home Energy Consumption. *In Exercises 7–9, use the following information:*

The author recorded his electrical energy consumption (in kilowatt-hours) for his home in upstate New York for two-month intervals over four years, and the results are listed in the table.

	Jan.–Feb.	Mar.–Apr.	May–June	July–Aug.	Sept.–Oct.	Nov.–Dec.
Year 1	4762	3875	2657	4358	2201	3187
Year 2	4504	3237	2198	2511	3020	2857
Year 3	3952	2785	2118	2658	2139	3071
Year 4	3863	3013	2023	2953	3456	2647

7. Energy Consumption: Constructing a Run Chart Construct a run chart for the 24 values. Does there appear to be a pattern suggesting that the process is not within statistical control? Is there any pattern or variation that can be explained?

8. Energy Consumption: Constructing an R Chart Use samples of size 3 by combining the first three values for each year and combining the last three values for each year. Construct an R chart and determine whether the process variation is within statistical control. If it is not, identify which of the three out-of-control criteria lead to rejection of statistically stable variation.

9. Energy Consumption: Constructing an $\bar{x}$ Chart Use samples of size 3 by combining the first three values for each year and combining the last three values for each year. Construct an $\overline{x}$ chart and determine whether the process mean is within statistical control. If it is not, identify which of the three out-of-control criteria lead to rejection of a statistically stable mean. What is a practical effect of not having this process in statistical control? Give an example of a cause that would make the process go out of statistical control.

Constructing Control Charts for Boston Rainfall. *In Exercises 10–12, refer to the daily amounts of rainfall in Boston for one year, as listed in Data Set 17 in Appendix B. Omit the last entry for Wednesday so that each day of the week has exactly 52 values.*

10. Boston Rainfall: Constructing a Run Chart Using only the 52 rainfall amounts for Monday, construct a run chart. Does the process appear to be within statistical control?

11. Boston Rainfall: Constructing an R Chart Using the 52 samples of seven values each, construct an R chart and determine whether the process variation is within statistical control. If it is not, identify which of the three out-of-control criteria lead to rejection of statistically stable variation.

12. Boston Rainfall: Constructing an $\overline{x}$ Chart Using the 52 samples of seven values each, construct an $\overline{x}$ chart and determine whether the process mean is within statistical control. If it is not, identify which of the three out-of-control criteria lead to rejection of a statistically stable mean. If not, what can be done to bring the process within statistical control?

12-2 Beyond the Basics

13. Constructing an s Chart In this section we described control charts for R and $\overline{x}$ based on ranges. Control charts for monitoring variation and center (mean) can also be based on standard deviations. An *s chart* for monitoring variation is made by plotting sample standard deviations with a centerline at $\overline{s}$ (the mean of the sample standard deviations) and control limits at $B_4\overline{s}$ and $B_3\overline{s}$, where B_4 and B_3 are found in Table 12-2. Construct an s chart for the data of Table 12-1. Compare the result to the R chart given in this section.

14. Constructing an $\overline{x}$ Chart Based on Standard Deviations An $\overline{x}$ chart based on standard deviations (instead of ranges) is made by plotting sample means with a centerline at $\overline{\overline{x}}$ and control limits at $\overline{\overline{x}} + A_3\overline{s}$ and $\overline{\overline{x}} - A_3\overline{s}$, where A_3 is found in Table 12-2 and $\overline{s}$ is the mean of the sample standard deviations. Use the data in Table 12-1 to construct an $\overline{x}$ chart based on standard deviations. Compare the result to the $\overline{x}$ chart based on sample ranges (shown in this section).

12-3 Control Charts for Attributes

The main objective of this section is to develop the ability to monitor an attribute by constructing and interpreting an appropriate control chart. In Section 12-2 we monitored *quantitative* data, but we now consider *qualitative* data, investigating questions such as whether an item is defective, whether an item weighs less than a prescribed amount, or whether an item is nonconforming. (A good or a service is nonconforming if it doesn't meet specifications or requirements; nonconforming goods are sometimes discarded, repaired, or called "seconds" and sold at reduced prices.) As in Section 12-2, we select samples of size n at regular time intervals and plot points in a sequential graph with a centerline and control limits. (There are ways to deal with samples of different sizes, but we don't consider them here.) The **control chart for p** (or **p chart**) is a control chart used to monitor the proportion p for some attribute. The notation and control chart values are as follows (where the attribute of "defective" can be replaced by any other relevant attribute).

Quality Control at Perstorp

Perstorp Components, Inc., uses a computer that automatically generates control charts to monitor the thicknesses of the floor insulation the company makes for Ford Rangers and Jeep Grand Cherokees. The $20,000 cost of the computer was offset by a first-year savings of $40,000 in labor, which had been used to manually generate control charts to ensure that insulation thicknesses were between the specifications of 2.912 mm and 2.988 mm. Through the use of control charts and other quality-control methods, Perstorp reduced its waste by more than two-thirds.

Notation

$\bar{p}$ = pooled estimate of the proportion of defective items in the process

$$= \frac{\text{total number of defects found among all items sampled}}{\text{total number of items sampled}}$$

$\bar{q}$ = pooled estimate of the proportion of process items that are *not* defective

$= 1 - \bar{p}$

n = size of each sample (not the number of samples)

Control Chart for p

Centerline: $\bar{p}$

Upper control limit: $\bar{p} + 3\sqrt{\frac{\bar{p}\,\bar{q}}{n}}$

Lower control limit: $\bar{p} - 3\sqrt{\frac{\bar{p}\,\bar{q}}{n}}$

(If the calculation for the lower control limit results in a negative value, use 0 instead. If the calculation for the upper control limit exceeds 1, use 1 instead.)

We use $\bar{p}$ for the centerline because it is the best estimate of the proportion of defects from the process. The expressions for the control limits correspond to 99.7% confidence interval limits as described in Section 6-5.

EXAMPLE **Deaths from Infectious Diseases** Physicians report that infectious diseases should be carefully monitored over time because they are much more likely to have sudden changes in trends than are other diseases, such as cancer. In each of 13 consecutive and recent years, 100,000 subjects were randomly selected and the number who died from respiratory tract infections is recorded, with the results given here (based on data from "Trends in Infectious Diseases Mortality in the United States," by Pinner et al., *Journal of the American Medical Association,* Vol. 275, No. 3). Construct a control chart for p and determine whether the process is within statistical control. If not, identify which of the three out-of-control criteria apply.

Number of deaths: 25 24 22 25 27 30 31 30 33 32 33 32 31

Six Sigma in Industry

Sigma (σ) denotes the standard deviation of a population; Six Sigma is the term used in industry to describe a process that results in a rate of no more than 3.4 defects out of a million. The reference to Six Sigma suggests six standard deviations away from the center of a normal distribution, but the assumption of a perfectly stable process is replaced with the assumption of a process that drifts slightly, so the defect rate is no more than 3 or 4 defects per million. This defect rate is so low that defects are considered to be almost eliminated.

Adopting the Six Sigma goal of quality, Motorola saved more than $940 million in three years. Allied Signal reported a savings of $1.5 billion. GE, Polaroid, and Texas Instruments are other major companies that have adopted the Six Sigma goal.

SOLUTION The centerline for our control chart is located by the value of $\bar{p}$:

$$\bar{p} = \frac{\text{total number of deaths from all samples combined}}{\text{total number of subjects sampled}}$$

$$= \frac{25 + 24 + 22 + \cdots + 31}{13 \cdot 100{,}000} = \frac{375}{1{,}300{,}000} = 0.000288$$

Because $\bar{p} = 0.000288$, it follows that $\bar{q} = 1 - \bar{p} = 0.999712$. Using $\bar{p} = 0.000288$, $\bar{q} = 0.999712$, and $n = 100{,}000$, we find the control limits as follows:

Upper control limit:

$$\bar{p} + 3\sqrt{\frac{\bar{p}\,\bar{q}}{n}} = 0.000288 + 3\sqrt{\frac{(0.000288)(0.999712)}{100{,}000}} = 0.000449$$

Lower control limit:

$$\bar{p} - 3\sqrt{\frac{\bar{p}\,\bar{q}}{n}} = 0.000288 - 3\sqrt{\frac{(0.000288)(0.999712)}{100{,}000}} = 0.000127$$

Having found the values for the centerline and control limits, we can proceed to plot the yearly proportion of deaths from respiratory tract infections. The Excel control chart for p is shown in the accompanying display.

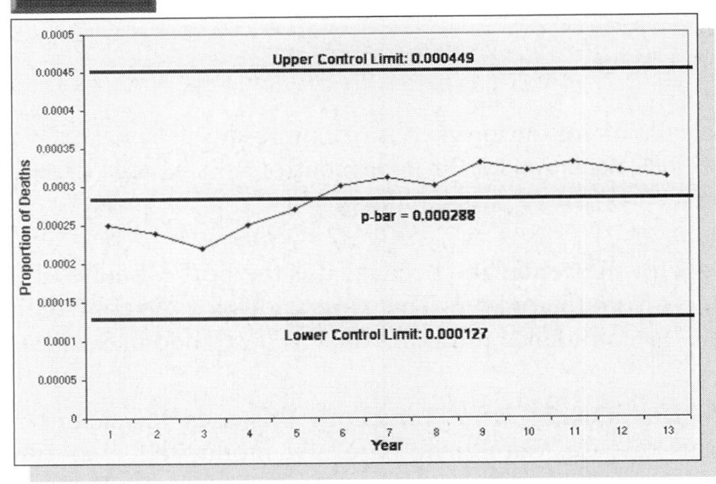

INTERPRETATION We can interpret the control chart for p by considering the three out-of-control criteria listed in Section 12-2. Using those criteria, we conclude that this process is out of statistical control for these reasons: There appears to be an upward trend, and there are eight consecutive points all lying above the centerline (Run of 8 Rule). Based on these data, public health policies affecting respiratory tract infections should be modified to cause a decrease in the death rate.

Using Technology

Minitab: Enter the numbers of defects (or items with any particular attribute) in column C1. Select the option **Stat,** then **Control Charts,** then **P.** Enter C1 in the box identified as `variable`, and enter the size of the samples in the box identified as `subgroup size`, then click **OK.**

Excel:

Using DDXL To use the DDXL add-in, begin by entering the numbers of defects or successes in column A, and enter the sample sizes in column B. For the example of this section, the first three items would be entered in the Excel spreadsheet as shown below.

	A	B
1	25	100000
2	24	100000
3	22	100000

Click on **DDXL**, select **Process Control**, then select **Summ Prop Control Chart** (for summary proportions control chart). A dialog box should appear. Click on the pencil icon for "Success Variable" and enter the range of values for column A, such as A1:A13. Click on the pencil icon for "Totals Variable" and enter the range of values for column B, such as B1:B13. Click **OK**. Next click on the **Open Control Chart** bar and the control chart will be displayed.

Using Excel's Chart Wizard: Enter the sample proportions in column A. (You could enter the actual numbers of defects in column A, then use Excel to create a column B consisting of the proportions. In the formula box, enter $=\mathrm{A}1/n$, where n is replaced by the size of each sample. After pressing Enter, cell B1 should contain the first sample proportion. Click and drag the lower right corner of cell B1 so that the entire B column has sample proportions corresponding to the actual numbers of defects in column A.) Having the data entered, proceed to generate the graph by first clicking on the **Chart Wizard** icon, which looks like a bar graph. For the chart type, select **Line.** For the chart subtype, select the first graph in the second row, then click **Next.** Continue to click **Next,** then **Finish.** The graph can be edited to include labels, delete grid lines, and so on. You can insert the required centerline and upper and lower control limits by editing the graph. Click on the line on the bottom of the screen, then click and drag to position the line correctly.

12-3 Basic Skills and Concepts

Determining Whether a Process Is in Control. *In Exercises 1–4, examine the given control chart for p and determine whether the process is within statistical control. If it is not, identify which of the three out-of-control criteria apply.*

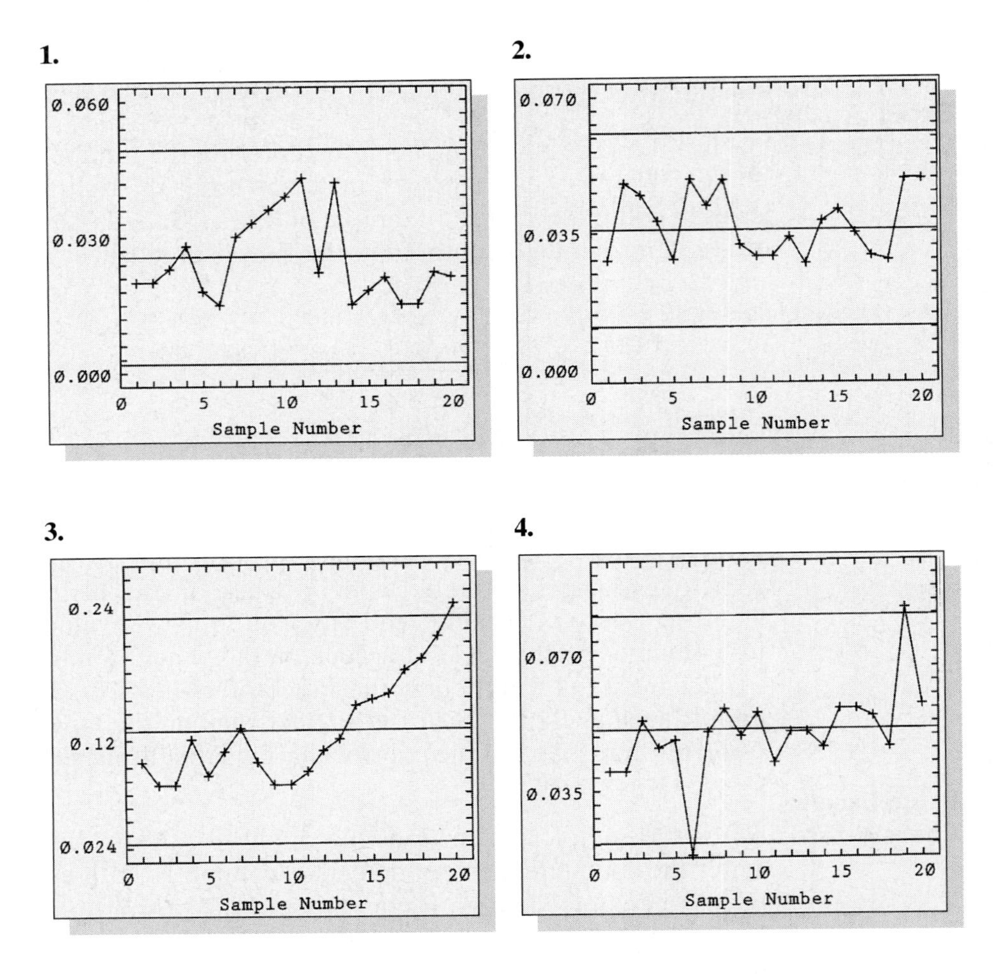

Constructing Control Charts for p. *In Exercises 5–8, use the given process data to construct a control chart for p. In each case, use the three out-of-control criteria listed in Section 12-2 and determine whether the process is within statistical control. If it is not, identify which of the three out-of-control criteria apply.*

5. *p* Chart for Deaths from Infectious Diseases In each of 13 consecutive and recent years, 100,000 children aged 0–4 years were randomly selected and the number who died from infectious diseases is recorded, with the results given below (based on data from "Trends in Infectious Diseases Mortality in the United States," by Pinner et al., *Journal of the American Medical Association,* Vol. 275, No. 3). Do the results suggest a problem that should be corrected?

Number who died: 30 29 29 27 23 25 25 23 24 25 25 24 23

6. *p* Chart for Victims of Crime In each of 20 consecutive and recent years, 1000 adults were randomly selected and surveyed. Each value below is the number who were victims of violent crime (based on data from the U.S. Department of Justice, Bureau of Justice Statistics). Do the data suggest a problem that should be corrected?

29	33	24	29	27	33	36	22	25	24
31	31	27	23	30	35	26	31	32	24

7. *p* Chart for Boston Rainfall Refer to the Boston rainfall amounts in Data Set 17 in Appendix B. For each of the 52 weeks, let the sample proportion be the proportion of days that it rained. (Delete the 53rd value for Wednesday). In the first week, for example, the sample proportion is $3/7 = 0.429$. Do the data represent a statistically stable process?

8. *p* Chart for Marriage Rates Use *p* charts to compare the statistical stability of the marriage rates of Japan and the United States. In each year, 10,000 people in each country were randomly selected, and the numbers of marriages are given for eight consecutive and recent years (based on United Nations data).

Japan:	58	60	61	64	63	63	64	63
United States:	98	94	92	90	91	89	88	87

12-3 Beyond the Basics

9. Constructing an *np* Chart A variation of the control chart for *p* is the **np chart** in which the *actual numbers* of defects are plotted instead of the *proportions* of defects. The *np* chart will have a centerline value of $n\bar{p}$, and the control limits will have values of $n\bar{p} + 3\sqrt{n\bar{p}\,\bar{q}}$ and $n\bar{p} - 3\sqrt{n\bar{p}\,\bar{q}}$. The *p* chart and the *np* chart differ only in the scale of values used for the vertical axis. Construct the *np* chart for the example given in this section. Compare the result with the control chart for *p* given in this section.

10. Identifying Effect of Sample Size on *p* Chart

a. Identify the locations of the centerline and control limits for a *p* chart representing a process that has been having a 5% rate of nonconforming items, based on samples of size 100.

b. Repeat part (a) after changing the sample size to 300.

c. Compare the two sets of results. Name an advantage and a disadvantage of using the larger sample size. Which chart would be better in detecting a shift from 5% to 10%?

Global Warming

NEW YORK Global warming is the rise in the earth's temperatures, supposedly caused in large part by actions taken by us humans. We are changing the chemical makeup of our atmosphere and, in the process, causing a buildup of the "greenhouse gases": carbon monoxide, methane, and nitrous oxide. These gases trap some of the energy from the sun, and this trapped heat is responsible for our rising temperatures. Consequences of rising world temperatures include rising sea levels, a host of new health problems, and climactic changes such as an increase in world precipitation and a greater frequency of intense rainstorms.

We can think of temperature measurements as process data, because they are arranged according to a time sequence. According to many respected scientists, global warming is a real phenomenon that must be corrected before the bad consequences become too serious. But there are skeptics. Harvard University meteorologist Dr. Richard Lindzen says, "We don't have any evidence that this (global warming) is a serious problem." See the chart below for changes in global temperature (in degrees Fahrenheit) every 10 years starting with 1861.

1. Construct an appropriate run or control chart for the given data, which are mean changes in the earth's average temperature. Does the process appear to be within statistical control? If not, identify which criteria lead to rejection of statistical stability.

2. Stricter auto emission controls constitute one step taken to help control global warming. Identify another such step.

3. Which type of variation is more responsible for global warming: random variation or assignable variation? Explain your choice.

Year	1861	1871	1881	1891	1901	1911	1921	1931	1941	1951	1961	1971	1981	1991
Change	−0.8	−0.5	−0.4	−0.6	−0.4	−0.8	−0.5	−0.3	0.1	−0.2	0	−0.1	0.2	0.5

VOCABULARY LIST

process data
run chart
statistically stable
within statistical control
random variation
assignable variation
control chart
centerline
lower control limit
upper control limit
R chart
range chart
$\overline{x}$ chart
control chart for *p*
p chart
np chart

Review

Whereas earlier chapters of this book focused on the important data characteristics of center, variation, distribution, and outliers, this chapter focused on pattern over time. Process data were defined to be data arranged according to some time sequence, and such data can be analyzed with run charts and control charts. Control charts have a centerline, an upper control limit, and a lower control limit. A process is statistically stable (or within statistical control) if it has only natural variation with no patterns, cycles, or unusual points. Decisions about statistical stability are based on how a process is actually behaving, not how we might like it to behave because of such factors as manufacturer specifications. The following graphs were described:

- *Run chart:* a sequential plot of *individual* data values over time
- *R chart:* a control chart that uses ranges in an attempt to monitor the *variation* in a process
- $\bar{x}$ *chart:* a control chart used to determine whether the process *mean* is within statistical control
- *p chart:* a control chart used to monitor the proportion of some process *attribute,* such as whether items are defective

Review Exercises

Constructing Control Charts for Acid Rain. *In Exercises 1–3, use the following information.*

As part of a study monitoring acid rain, measurements of sulfate deposits (kg/hectare) are recorded for different locations on the East Coast (based on data from the U.S. Department of Agriculture). The results are listed in the following table.

Acid Rain: Sulfate Deposits (kg/hectare)

Year	Location 1	Location 2	Location 3	Location 4	Location 5
1980	11.94	13.09	7.96	17.29	12.12
1981	11.28	10.88	12.84	13.87	11.21
1982	10.38	12.19	7.38	13.64	9.95
1983	8.00	10.75	7.26	12.37	8.77
1984	12.12	17.21	10.12	15.73	11.68
1985	10.27	10.26	8.89	13.21	9.71
1986	14.80	15.49	11.60	17.94	15.59
1987	13.52	11.61	9.02	11.22	13.05
1988	10.55	10.53	7.78	10.57	11.77
1989	9.81	12.50	8.70	13.29	9.37
1990	11.27	9.94	10.50	11.28	10.54

1. Sulfate Deposits: Constructing a Run Chart Construct a run chart for the 55 values. Does there appear to be a pattern suggesting that the process is not within statistical control?

2. Sulfate Deposits: Constructing an R Chart Construct an R chart and determine whether the process variation is within statistical control. If it is not, identify which of the three out-of-control criteria lead to rejection of statistically stable variation.

3. Sulfate Deposits: Constructing an $\bar{x}$ Chart Construct an $\bar{x}$ chart and determine whether the process mean is within statistical control. Does the process appear to be statistically stable? How should this process behave if we implement effective programs to reduce the amount of acid rain?

4. Constructing a Control Chart for Infectious Diseases In each of 13 consecutive and recent years, 100,000 adults 65 years of age or older were randomly selected and the number who died from infectious diseases is recorded, with the results given below (based on data from "Trends in Infectious Diseases Mortality in the United States," by Pinner et al., *Journal of the American Medical Association,* Vol. 275, No. 3). Construct an appropriate control chart and determine whether the process is within statistical control. If not, identify which criteria lead to rejection of statistical stability.

   Number who died: 270 264 250 278 302 334 348 347 377 357 362 351 343

5. Constructing a Control Chart for Voter Turnout In a continuing study of voter turnout, 1000 people of voting age are randomly selected in each year when there is a national election, and the numbers who actually voted are listed below (based on data from the *Time Almanac*). Construct an appropriate control chart and determine whether the process is within statistical control. If not, identify which criteria lead to rejection of statistical stability.

   Number who voted: 608 466 552 382 536 372 526 398 531 364 501 365 551 388 491

CUMULATIVE REVIEW EXERCISES

1. Analyzing Fuse Production Process The Telektronic Company produces 20-amp fuses used to protect car radios from too much electrical power. Each day 400 fuses are randomly selected and tested; the results (numbers of defects per 400 fuses tested) for 20 consecutive days are as follows:

   10 8 7 6 6 9 12 5 4 7 9 6 11 4 6 5 10 5 9 11

   a. Use a control chart for p to verify that the process is within statistical control, so the data can be treated as coming from a population with fixed variation and mean.
   b. Using all of the data combined, construct a 95% confidence interval for the proportion of defects.
   c. Using a 0.05 significance level, test the claim that the rate of defects is 1% or less.

2. Using Probability in Control Charts When interpreting control charts, one of the three out-of-control criteria is that there are eight consecutive points all above or all below the centerline. For a statistically stable process, there is a 0.5 probability that a point will be above the centerline and there is a 0.5 probability that a point will be below the centerline. In each of the following, assume that sample values are independent and the process is statistically stable.
   a. Find the probability that when eight consecutive points are randomly selected, they are all above the centerline.

b. Find the probability that when eight consecutive points are randomly selected, they are all below the centerline.
c. Find the probability that when eight consecutive points are randomly selected, they are all above or all below the centerline.

3. Using Control Charts for Temperatures In Exercises 7–9 in Section 12-2, the amounts of electrical energy consumption were listed for the author's home during a period of four recent years. The accompanying table lists the average temperature (in degrees Fahrenheit) for the same time periods. Use appropriate control or run charts to determine whether the data appear to be part of a statistically stable process.

	Jan.–Feb.	Mar.–Apr.	May–June	July–Aug.	Sept.–Oct.	Nov.–Dec.
Year 1	32	35	59	76	66	42
Year 2	22	33	56	70	63	42
Year 3	30	38	55	71	61	38
Year 4	32	40	57	72	65	45

4. Relationship Between Energy Consumption and Temperature Refer to the data in Exercise 3 and the data used for Exercises 7–9 in Section 12-2. Match the data in pairs according to the corresponding time periods.
a. Is there a significant linear correlation between the amounts of electrical energy consumption and the temperatures? Explain.
b. Identify the linear regression equation that relates electrical energy consumption (y) and the temperature (x).
c. What is the best predicted amount of electrical energy consumption for a two-month period with an average temperature of 60 °F?

Cooperative Group Activities

1. *Out-of-class activity:* Collect your own set of process data and analyze them using the methods of this section. It would be ideal to collect data from a real manufacturing process, but that is usually difficult to accomplish. Instead, consider using a simulation or referring to published data, such as those found in an almanac. Here are some suggestions:

   - Shoot five basketball foul shots (or shoot five crumpled sheets of paper into a wastebasket) and record the number of shots made; then repeat this procedure 20 times, and use a p chart to test for statistical stability in the proportion of shots made.
   - Go through newspapers for the past 12 weeks and record the closing of the Dow Jones Industrial Average for each business day. Use run and control charts to explore the statistical stability of the Dow Jones Industrial Average. Identify at least one practical consequence of having this process statistically stable, and identify at least one practical consequence of having this process out of statistical control.
   - Find the divorce rate in terms of divorces per 1000 population for several years. (See the *Information Please Almanac* or the *Statistical Abstract of the United States*.) Assume that in each year 1000 people were randomly selected

and surveyed to determine whether they became divorced. Use a p chart to test for statistical stability of the divorce rate. (Other possible rates: marriage, birth, death, accident fatality.)

Obtain a printed copy of computer results, and write a report summarizing your conclusions.

2. *In-class activity:* If the instructor can distribute the numbers of absences for each class meeting, groups of three or four students can analyze them for statistical stability and make recommendations based on the conclusions.

TECHNOLOGY PROJECT

a. Simulate the following process for 20 days: Each day, 200 heart pacemakers are manufactured with a 1% rate of defective units, and the proportion of defects is recorded for each of the 20 days. The pacemakers for one day are simulated by randomly generating 200 numbers, where each number is between 1 and 100. Consider an outcome of 1 to be a defect, with 2 through 99 being acceptable. This corresponds to a 1% rate of defects.

STATDISK: Select **Data** from the main menu bar, then select **Uniform Generator** and proceed to generate 200 values with a minimum of 1 and a maximum of 100. Display the sorted data by using the **Format** menu in **Sample Editor.** Repeat this procedure until results for 20 days have been simulated.

Minitab: From the main menu bar, select **Calc,** then **Random Data,** then **Integer.** Enter 200 in the box for the number of rows of data, enter C1 as the column to be used for storing the data, enter 1 for the minimum value, and enter 100 for the maximum value. Repeat this procedure until results for 20 days have been simulated.

Excel: Click on the **fx** icon on the main menu bar, then select the function category **Math & Trig,** followed by **RANDBETWEEN.** In the dialog box, enter 1 for bottom and 100 for top. A random value should appear in the first row of column A. Use the mouse to click and drag the lower right corner of that cell, then pull down the cell to cover the first 100 rows of column A. When you release the mouse button, column A should contain 100 random numbers. You can also click/drag the lower right corner of the bottom cell by moving the mouse to the right so that you get 20 columns of 100 numbers each. The different columns represent the different days of manufacturing.

TI-83 Plus: Press the **MATH** key. Select **PRB,** then select the 5th menu item, **randInt(**, and proceed to enter 1, 100, 200; then press the **ENTER** key. Press **STO** and **L1** to store the data in list L1. After recording the number of defects, repeat this procedure until results for 20 days have been simulated.

b. Construct a p chart for the proportion of defective pacemakers, and determine whether the process is within statistical control. Since we know the process is actually stable with $p = 0.01$, the conclusion that it is not stable would be a type I error; that is, we would have a false positive signal, causing us to believe that the process needed to be adjusted when in fact it should be left alone.

c. The result from part (a) is a simulation of 20 days. Now simulate another 10 days of manufacturing pacemakers, but modify these last 10 days so that the nonconforming rate is 3% instead of 1%.

d. Combine the data generated from parts (a) and (c) to represent a total of 30 days of sample results. Construct a p chart for this combined data set. Is the process out of control? If we concluded that the process was not out of control, we would be making a type II error; that is, we would believe that the process was okay when in fact it should be repaired or adjusted to correct the shift to the 3% nonconforming rate.

Critical Thinking: Are the axial loads within statistical control?

Is the process of manufacturing cans proceeding as it should?

The Chapter Problem listed process data from a New York company that manufactures 0.0109-in. thick aluminum cans for a major beverage supplier, and those data were analyzed in Section 12-2. Refer to Data Set 12 in Appendix B and conduct an analysis of the process data for the cans that are 0.0111 in. thick. The values in the data set are the measured axial loads of cans, and the top lids are pressed into place with pressures that vary between 158 lb and 165 lb.

Analyze the Results

Should you take any corrective action? Write a report summarizing your conclusions. Address not only the issue of statistical stability, but also the ability of the cans to withstand the pressures applied when the top lids are pressed into place. Also compare the behavior of the 0.0111-in. cans to the behavior of the 0.0109-in. cans and recommend which thickness should be used.

Internet Project

Control Charts

This chapter introduced different charting techniques for summarizing and analyzing process data. The run chart is based on individual values, but the others are based on summary statistics. The R chart is constructed from sample ranges, the $\bar{x}$ chart is based on sample means, and the p chart is based on sample proportions. Except for the run chart, we do not need the original sample data; we need only the satatistics that summzrize each sample. Thus, data sets described with summary statistics can be used for determining whether the process is within statistical control.

Go to the Internet Site for this book:

http://www.awlonline.com/triola

Find the Internet Project for this chapter. There you will be directed to several data sets to be used for constructing appropriate control charts. From your charts you will be asked to interpret and discuss trends in the underlying process.

Statistics at work

"There is a certain amount of respect that is given to someone who knows statistics and can explain it to someone who doesn't know it."

Dan O'Toole

Account Executive, A.C. Nielsen

In his work in the Advanced Analytics Group at A.C. Nielsen, Dan develops statistical solutions to help clients like Polaroid, Ocean Spray, and Gillete understand which of their marketing vehicles drives sales most profitably. Dan has a Masters Degree in Business Economics from Bentley College.

What is your job and what do you do?

I am an Account Executive and a member of A.C. Nielsen's Analytic's department. We help manufacturers and retailers with their marketing objectives for their products—how they should price them, where they should position them, how they should be advertised, etc.

What concepts of statistics do you use?

In my one year here at A.C. Nielsen, I have worked with analyses as simple as correlation and general significance tests, all the way to multiple regression, factor analysis, correspondence analysis, and cluster analysis.

How do you use statistics on the job?

My job is to discover or uncover client issues, and then find out if we can apply one of our statistical techniques to their specific issue. You really have to know the statistics behind an analysis so that you have a strong understanding of a technique's limitations. If a technique won't help a client, then you need to know that. An example of how I use stats: A client may say, I sell product "X", whether it is juice, bread, or a camera. Right now, they may control 20% of the market. They may come to us to see if they can increase market share by lowering their price. My job would be to design a study to analyze this question. To do this, I have to design a study that will take into account everything that affects the sales of a product. Using techniques like regression, if I am able to create a model with good significance, I will be able to isolate specific influences on the sales and offer recommendations. Things like seasonality, distribution, as well as any marketing efforts that may have taken place, must be included. In addition, I will have to take into account complementary products' price (butter is complementary to bread; while film is for a camera) and also competitive products. For instance bread may compete with English muffins (I know it does for me).

Could you describe one specific example illustrating how the use of statistics was successful in improving a product or service?

We were modeling a juice product. The client felt that store brands, or private label brands, affected their sales, and the client felt they needed to drop price to save their market share. When we finished modeling, it seemed that the two products did not compete with each other based on price. So, if store brands lowered their price, it wouldn't affect their sales. This seemed not to make sense. How could this be true? What we discovered was that, when Private Label entered a market, they would steal all of the value customers—the people who will buy the lowest product no matter what. However, everyone else would remain. So, although the client lost some

of their customers, if they were to lower the price, it wouldn't gain them any more sales or buyers.

Is your use of probability and statistics increasing, decreasing, or remaining stable?

It definitely is increasing. In this business (consulting), you are constantly challenged to learn a new technique or look at an old technique in order to improve it. In addition, since we are constantly coming out with new products, our understanding of statistics has to increase to use these techniques effectively.

How beneficial do you find your knowledge of statistics for performing your responsibilities?

It is not a question of beneficial, but rather it is a necessity. In fact, we find that we have to know it so well, so that we can explain it in "layman's" terms to our clients.

In terms of statistics, what would you recommend for prospective employees?

Many people have taken statistics, but the level of retention of most concepts is low. If you had to focus on some core concepts. I would say you would need to understand correlation and linear regression—understanding these concepts, will help you interpret other concepts you bump into, like multiple regression. I would also go with factor analysis—just a general understanding would even put you ahead of the curve in most respects.

Do you feel job applicants are viewed more favorably if they have studied some statistics?

By far. There is a certain amount of respect that is given to someone who knows statistics and can explain it to someone who doesn't know it (because it means you really know it and aren't reciting from a textbook). If you know some stats, you will be perceived quickly as intelligent. Almost every job uses statistics (particularly correlations and regressions). People will, say things like, "Oh, check if they're correlated". "What's that relationship?", etc. Your bosses won't know how to do it themselves, but they know enough to know that it can be done. That's why you'll need it.

CHAPTER 13

Nonparametric Statistics

CHAPTER PROBLEM

It rains more on weekends!?

In an article for the Knight Ridder News Service, Usha Lee McFarling writes that "Your worst weather fears are true. It does rain more on weekends. Scientists who pored through years of rainfall data have discovered a clear and dismaying pattern. Fridays, Saturdays, and Sundays are the rainiest days of the week all along America's eastern coast, from Maine to Florida." The article refers to a study conducted by Arizona State University scientists Randall S. Cerveny and Robert C. Balling. But are their conclusions correct? Have they been reported and interpreted correctly?

Data Set 17 in Appendix B includes rainfall amounts for a recent year in Boston. In selecting a city for verifying the weekend rain phenomenon, Boston should be a good choice because it is located on the eastern coast. If we use STATDISK to enter the rainfall amount for each day of the week, we get the boxplots shown on the next page. Going from top to bottom, the boxplots represent Monday, Tuesday, . . . , Sunday. The Monday boxplot at the top is unusual because it appears to be missing the box, but the Monday rainfall amounts have so many 0s that the minimum, first quartile, and median are all 0s, with the result that the box gets crunched to the left. Greater amounts of rainfall on Friday, Saturday, and Sunday would be visible with the distributions positioned farther to the right. Is that really the case? Are the differences really significant?

We might consider using methods presented in earlier chapters to investigate this issue. Analysis of variance (Section 11-2) might come to mind, but that method requires that the samples come from populations having a normal distribution. The histogram for the Monday rainfall amounts shows clearly that these values do not come from a normally distributed population. A major advantage of the methods discussed in this chapter is that they do not require a particular distribution.

Can we use the Boston data to support the claim of more rainfall on weekends? Are the rainfall amounts for the different days significantly different? We will address these questions later in the chapter.

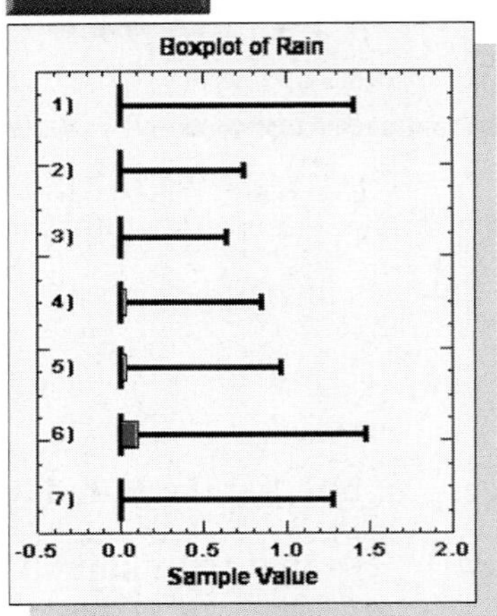

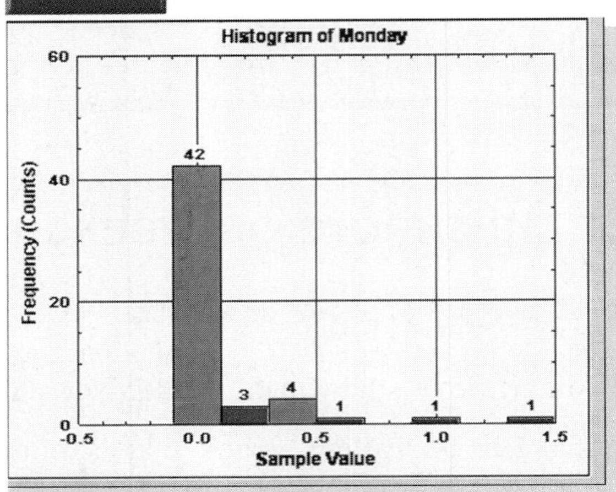

13-1 Overview

The methods of inferential statistics presented in Chapters 6, 7, 8, 9, and 11 are called *parametric methods* because they are based on sampling from a population with specific parameters, such as the mean μ, standard deviation σ, or proportion p. Those parametric methods usually must conform to some fairly strict conditions, such as a requirement that the sample data come from a normally distributed population. This chapter introduces nonparametric methods, which do not have such strict requirements.

DEFINITIONS

Parametric tests require assumptions about the nature or shape of the populations involved; **nonparametric tests** do not require such assumptions. Consequently, nonparametric tests of hypotheses are often called **distribution-free tests.**

Although the term *nonparametric* suggests that the test is not based on a parameter, there are some nonparametric tests that do depend on a parameter such as the median. The nonparametric tests do not, however, require a particular distribution, so they are sometimes referred to as *distribution-free* tests. Although *distribution-free* is a more accurate description, the term *nonparametric* is more commonly used. The following are major advantages and disadvantages of nonparametric methods.

Advantages of Nonparametric Methods

1. Nonparametric methods can be applied to a wide variety of situations because they do not have the more rigid requirements of the correspon-

ding parametric methods. In particular, nonparametric methods do not require normally distributed populations.

2. Unlike parametric methods, nonparametric methods can often be applied to nonnumerical data, such as the genders of survey respondents.
3. Nonparametric methods usually involve simpler computations than the corresponding parametric methods and are therefore easier to understand and apply.

Disadvantages of Nonparametric Methods

1. Nonparametric methods tend to waste information because exact numerical data are often reduced to a qualitative form. For example, in the nonparametric sign test (described in Section 13-2), weight losses by dieters are recorded simply as negative signs; the actual magnitudes of the weight losses are ignored.
2. Nonparametric tests are not as efficient as parametric tests, so with a nonparametric test we generally need stronger evidence (such as a larger sample or greater differences) before we reject a null hypothesis.

When the requirements of population distributions are satisfied, nonparametric tests are generally less efficient than their parametric counterparts, but the reduced efficiency can be compensated for by an increased sample size. For example, Section 13-6 will present a concept called *rank correlation,* which has an efficiency rating of 0.91 when compared to the linear correlation presented in Chapter 9. This means that with all other things being equal, nonparametric rank correlation requires 100 sample observations to achieve the same results as 91 sample observations analyzed through parametric linear correlation, assuming the stricter requirements for using the parametric method are met. Table 13-1 lists the nonparametric methods covered in this chapter, along with the corresponding parametric approach and **efficiency** rating. Table 13-1 shows that several nonparametric tests have

TABLE 13-1 Efficiency: Comparison of Parametric and Nonparametric Tests

Application	Parametric Test	Nonparametric Test	Efficiency Rating of Nonparametric Test with Normal Population
Matched pairs of sample data	t test or z test	Sign test	0.63
		Wilcoxon signed-ranks test	0.95
Two independent samples	t test or z test	Wilcoxon rank-sum test	0.95
Several independent samples	Analysis of variance (F test)	Kruskal-Wallis test	0.95
Correlation	Linear correlation	Rank correlation test	0.91
Randomness	No parametric test	Runs test	No basis for comparison

efficiency ratings above 0.90, so the lower efficiency might not be a critical factor in choosing between parametric and nonparametric methods. However, because parametric tests do have higher efficiency ratings than their nonparametric counterparts, it's generally better to use the parametric tests when their required assumptions are satisfied.

Ranks

Sections 13-3 through 13-6 use methods based on ranks. Let's now briefly describe them.

DEFINITION

Data are *sorted* when they are arranged according to some criterion, such as smallest to largest or best to worst. A **rank** is a number assigned to an individual sample item according to its order in the sorted list. The first item is assigned a rank of 1, the second item is assigned a rank of 2, and so on.

EXAMPLE The numbers 5, 3, 40, 10, and 12 can be sorted (arranged from lowest to highest) as 3, 5, 10, 12, and 40, and these numbers have ranks of 1, 2, 3, 4, and 5, respectively:

5	3	40	10	12	Original values
3	5	10	12	40	Values arranged in order
↑	↑	↑	↑	↑	
1	2	3	4	5	Ranks

Handling ties in ranks: If a tie in ranks occurs, the usual procedure is to find the mean of the ranks involved and then assign this mean rank to each of the tied items, as in the following example.

EXAMPLE The numbers 3, 5, 5, 10, and 12 are given ranks of 1, 2.5, 2.5, 4, and 5, respectively. In this case, ranks 2 and 3 were tied, so we found the mean of 2 and 3 (which is 2.5) and assigned it to the values that created the tie:

3	5	5	10	12	Original values
↑	↑	↑	↑	↑	
1	2.5	2.5	4	5	Ranks
	↑	↑			
	2 and 3 are tied				

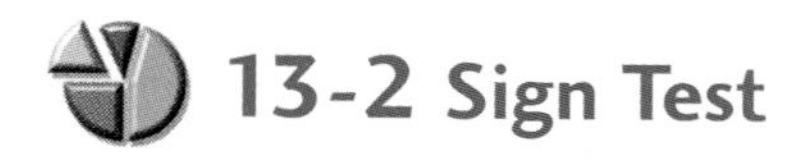

13-2 Sign Test

The main objective of this section is to understand the *sign test* procedure, which is among the easiest for nonparametric tests.

DEFINITION

The **sign test** is a nonparametric (distribution-free) test that uses plus and minus signs to test different claims, including:

1. Claims involving matched pairs of sample data
2. Claims involving nominal data
3. Claims about the median of a single population

Assumptions

1. The sample data have been randomly selected.
2. There is *no* requirement that the sample data come from a population with a particular distribution, such as a normal distribution.

Basic Concept of the Sign Test The basic idea underlying the sign test is to analyze the frequencies of the plus and minus signs to determine whether they are significantly different. For example, suppose that we test a treatment designed to lower blood pressure. If 100 subjects are treated and 51 of them experience lower blood pressure while the other 49 have increased blood pressure, common sense suggests that there is not sufficient evidence to say that the drug is effective, because 51 decreases out of 100 is not significant. But what about 52 decreases and 48 increases? Or 90 decreases and 10 increases? The sign test allows us to determine when such results are significant. Figure 13-1 on the next page summarizes the sign test procedure, which will be illustrated with examples that follow.

For consistency and ease, we will stipulate the following notation.

Notation for the Sign Test

x = the number of times the *less frequent* sign occurs

n = the total number of positive and negative signs combined

FIGURE 13-1
Sign Test Procedure

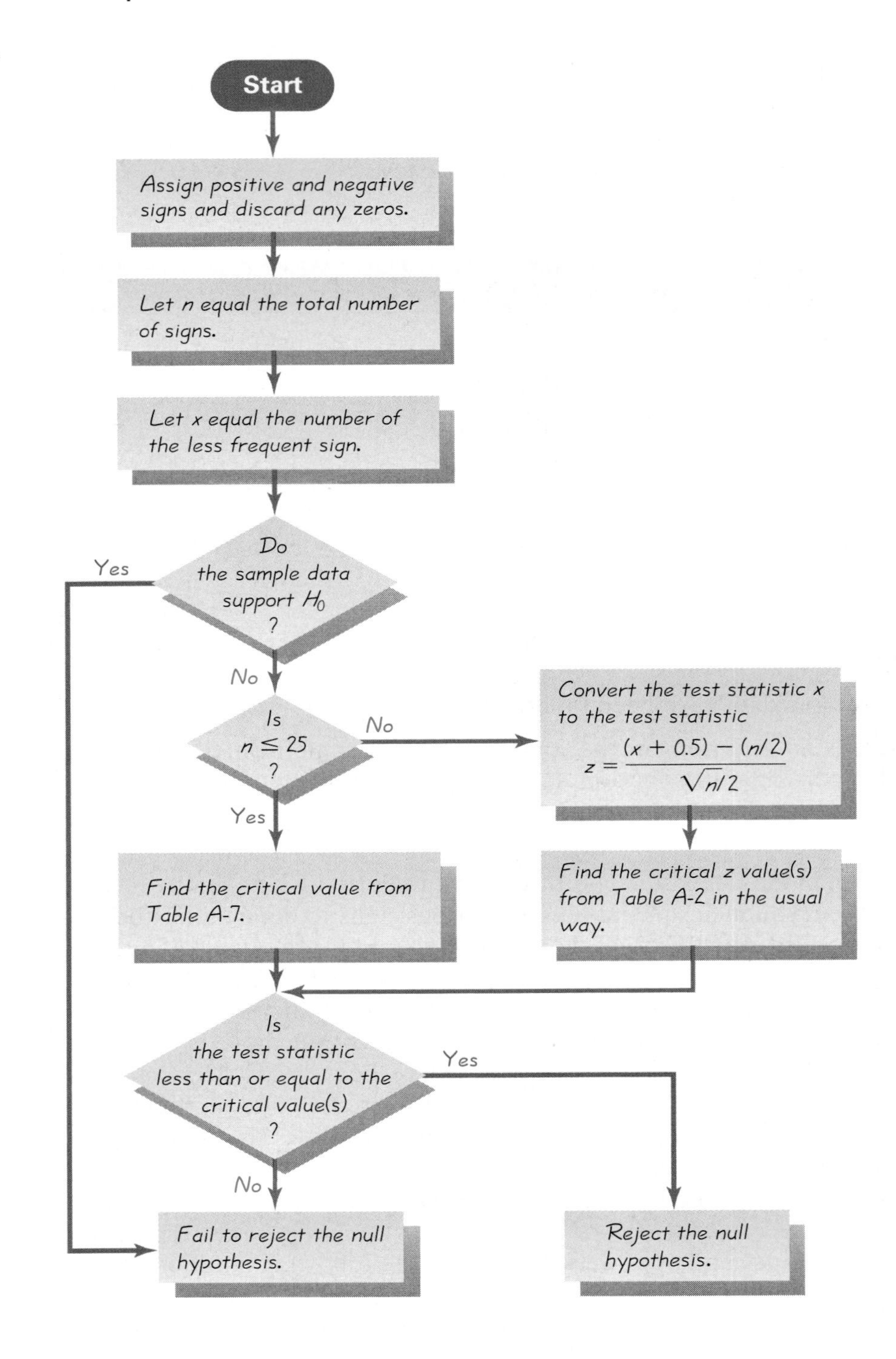

Test Statistic for the Sign Test

For $n \leq 25$: x (the number of times the less frequent sign occurs)

For $n > 25$: $$z = \frac{(x + 0.5) - \left(\frac{n}{2}\right)}{\frac{\sqrt{n}}{2}}$$

Critical values:

1. For $n \leq 25$, critical x values are in Table A-7.
2. For $n > 25$, critical z values are in Table A-2.

Caution: When applying the sign test in a one-tailed test, we need to be very careful to avoid making the wrong conclusion when one sign occurs significantly more often than the other, but the sample data are *consistent with* the null hypothesis. If the sense of the data is consistent with (instead of conflicting with) the null hypothesis, then we fail to reject the null hypothesis and don't proceed with the sign test. Figure 13-1 summarizes the procedure for the sign test and includes this check: Do the sample data "support" (in the sense of being consistent with) H_0? If the answer is yes, fail to reject the null hypothesis. *It is always important to think about the data and to avoid relying on blind calculations or computer results.*

Claims Involving Matched Pairs

When using the sign test with data that are matched by pairs, we convert the raw data to plus and minus signs as follows:

1. We subtract each value of the second variable from the corresponding value of the first variable.
2. We record only the *sign* of the difference found in Step 1. We *exclude ties:* that is, we exclude any matched pairs in which both values are equal. (For other ways to handle ties, see Exercise 13.)

The key concept underlying this use of the sign test is this:

If the two sets of data have equal medians, the number of positive signs should be approximately equal to the number of negative signs.

EXAMPLE **Reported and Measured Heights** The data in Table 13-2 on the next page are matched pairs of heights (in inches) obtained from a random sample of 12 male statistics students. Each student reported his height, then his height was measured. Use a 0.05 significance level to test the claim that there is no difference between reported height and measured height. (This same data set was used for an example in Section 8-3.)

TABLE 13-2 Reported and Measured Heights of Male Statistics Students

Reported height	68	74	82.25	66.5	69	68	71	70	70	67	68	70
Measured height	66.8	73.9	74.3	66.1	67.2	67.9	69.4	69.9	68.6	67.9	67.6	68.8
Difference	1.2	0.1	7.95	0.4	1.8	0.1	1.6	0.1	1.4	−0.9	0.4	1.2
Sign of difference	+	+	+	+	+	+	+	+	+	−	+	+

SOLUTION Here's the basic idea: If male statistics students accurately report their heights, their reported heights and measured heights should be about the same, so the numbers of positive and negative signs should be approximately equal—but in Table 13-2 we have 11 positive signs and 1 negative sign. Are the numbers of positive and negative signs approximately equal, or are they significantly different? We follow the same basic steps for testing hypotheses as outlined in Figure 7-5.

Steps 1, 2, 3: The null hypothesis is the claim of no difference between reported heights and measured heights, and the alternative hypothesis is the claim that there is a difference.

H_0: There is no difference. (The median of the differences is equal to 0.)

H_1: There is a difference. (The median of the differences is not equal to 0.)

Step 4: The significance level is $\alpha = 0.05$.

Step 5: We are using the nonparametric sign test.

Step 6: The test statistic x is the number of times the less frequent sign occurs. Table 13-2 includes differences with 11 positive signs and 1 negative sign, so we let x equal the smaller of 11 and 1 or $x = 1$. Also, $n = 12$ (the total number of positive and negative signs combined). Our test is two-tailed with $\alpha = 0.05$, and reference to Table A-7 shows that the critical value is 2. (See Figure 13-1.)

Step 7: With a test statistic of $x = 1$ and a critical value of 2, we reject the null hypothesis of no difference. [See Note 2 included with Table A-7: "The null hypothesis is rejected if the number of the less frequent sign (x) is less than or equal to the value in the table." Because $x = 1$ is less than or equal to 2, we reject the null hypothesis.]

Step 8: There is sufficient evidence to warrant rejection of the claim that the median of the differences is equal to 0; that is, there is sufficient evidence to warrant rejection of the claim that there is no difference between the reported and measured heights.

This is the same conclusion reached in Section 8-3, where we used a parametric t test, but sign test results do not always agree with parametric test results.

Class Attendance and Grades

In a study of 424 undergraduates at the University of Michigan, it was found that students with the worst attendance records tended to get the lowest grades. (Is anybody surprised?) Those who were absent less than 10% of the time tended to receive grades of B or above. The study also showed that students who sit in the front of the class tend to get significantly better grades.

Claims Involving Nominal Data

Recall that nominal data consist of names, labels, or categories only. Although such a nominal data set limits the calculations that are possible, we can identify the *proportion* of the sample data that belong to a particular category, and we can test claims about the corresponding population proportion p. The following example uses nominal data consisting of genders (male/female). The sign test is used by representing men with positive $(+)$ signs and women with negative $(-)$ signs. (Those signs are chosen arbitrarily, honest.) Also note the procedure for handling cases in which $n > 25$.

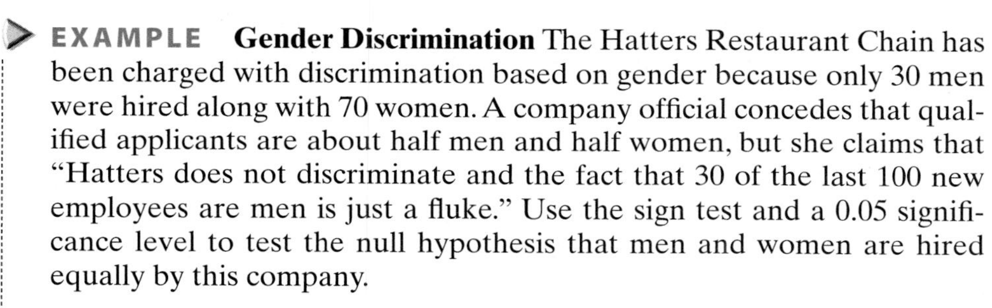

EXAMPLE **Gender Discrimination** The Hatters Restaurant Chain has been charged with discrimination based on gender because only 30 men were hired along with 70 women. A company official concedes that qualified applicants are about half men and half women, but she claims that "Hatters does not discriminate and the fact that 30 of the last 100 new employees are men is just a fluke." Use the sign test and a 0.05 significance level to test the null hypothesis that men and women are hired equally by this company.

SOLUTION Let p denote the population proportion of hired men. The claim of no discrimination implies that the proportions of hired men and women are both equal to 0.5, so that $p = 0.5$. The null and alternative hypotheses can therefore be stated as follows.

H_0: $p = 0.5$ (the proportion of hired men is equal to 0.5)

H_1: $p \neq 0.5$

Denoting hired men by $+$ and hired women by $-$, we have 30 positive signs and 70 negative signs. Refer now to the sign test procedure summarized in Figure 13-1. The test statistic x is the smaller of 30 and 70, so $x = 30$. This test involves two tails because a disproportionately low number of either gender will cause us to reject the claim of equality. The sample data do not support the null hypothesis because 30 and 70 are not precisely equal. (That is, the sample data are not consistent with the null hypothesis.) Continuing with the procedure in Figure 13-1, we note that the value of $n = 100$ is above 25, so the test statistic x is converted (using a correction for continuity) to the test statistic z as follows:

$$z = \frac{(x + 0.5) - \left(\frac{n}{2}\right)}{\frac{\sqrt{n}}{2}}$$

FIGURE 13-2
Testing the Claim That Hiring Practices Are Fair

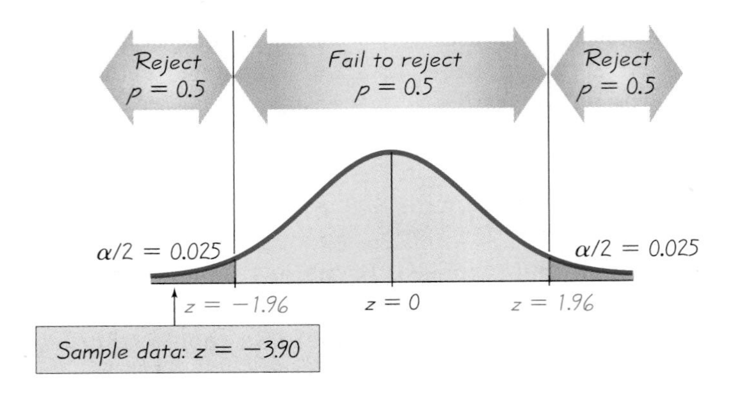

$$= \frac{(30 + 0.5) - \left(\frac{100}{2}\right)}{\frac{\sqrt{100}}{2}} = -3.90$$

With $\alpha = 0.05$ in a two-tailed test, the critical values are $z = \pm 1.96$. The test statistic $z = -3.90$ is less than -1.96 (see Figure 13-2), so we reject the null hypothesis that the proportion of hired men is equal to 0.5. There is sufficient sample evidence to warrant rejection of the claim that the hiring practices are fair, with the proportions of hired men and women both equal to 0.5. This company appears to discriminate by not hiring equal proportions of men and women.

Claims About the Median of a Single Population

The next example illustrates the procedure for using the sign test in testing a claim about the median of a single population. See how the negative and positive signs are based on the claimed value of the median.

EXAMPLE **Body Temperatures** In Chapter 7 we used sample data to test the claim that the mean body temperature of healthy adults is not 98.6°F. With the same data, use the sign test to test the claim that the median is less than 98.6°F. The data set used in Chapter 7 has 106 subjects—68 subjects with temperatures below 98.6°F, 23 subjects with temperatures above 98.6°F, and 15 subjects with temperatures equal to 98.6°F.

SOLUTION The claim that the median is less than 98.6°F is the alternative hypothesis, while the null hypothesis is the claim that the median is at least 98.6°F.

H_0: Median is at least 98.6°F. (median $\geq$ 98.6°F)

H_1: Median is less than 98.6°F. (median $<$ 98.6°F)

Following the procedure outlined in Figure 13-1, we discard the 15 zeros, we use the negative sign (−) to denote each temperature that is below 98.6°F, and we use the positive sign (+) to denote each temperature that is above 98.6°F. We therefore have 68 negative signs and 23 positive signs, so $n = 91$ and $x = 23$ (the number of the less frequent sign). The sample data thus conflict with the null hypothesis, because fewer than half of the 91 temperatures are at least 98.6°F. We must now proceed to determine whether this conflict is significant. (If the sample data did not conflict with the null hypothesis, we could immediately terminate the test by concluding that we fail to reject the null hypothesis.) The value of n exceeds 25, so we convert the test statistic x to the test statistic z.

$$z = \frac{(x + 0.5) - \left(\frac{n}{2}\right)}{\frac{\sqrt{n}}{2}}$$

$$= \frac{(23 + 0.5) - \left(\frac{91}{2}\right)}{\frac{\sqrt{91}}{2}} = -4.61$$

In this one-tailed test with $\alpha = 0.05$, we use Table A-2 to get the critical z value of -1.645. From Figure 13-3 we can see that the test statistic of $z = -4.61$ does fall within the critical region. We therefore reject the null hypothesis. On the basis of the available sample evidence, we support the claim that the median body temperature of healthy adults is less than 98.6°F.

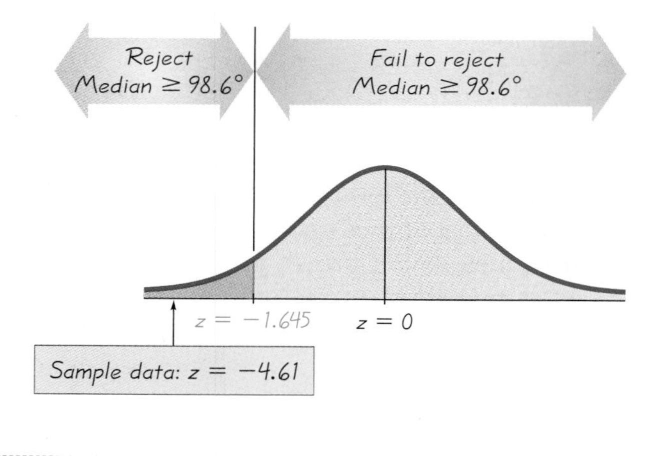

FIGURE 13-3
Testing the Claim That the Median Is Less Than 98.6°F

In this sign test of the claim that the median is below 98.6°F, we get a test statistic of $z = -4.61$, but a parametric test of the claim that $\mu < 98.6$°F results in a test statistic of $z = -6.64$. Because -4.61 isn't as extreme as -6.64, we see that the sign test isn't as sensitive as the parametric test. Both tests lead to rejection of the null hypothesis, but the sign test doesn't consider the sample data to be as extreme, partly because the sign test uses only information about the *direction* of the data, ignoring the *magnitudes* of the data values. The next section introduces the Wilcoxon signed-ranks test, which largely overcomes that disadvantage.

Rationale for the test statistic used when $n > 25$: When finding critical values for the sign test, we use Table A-7 only for n up to 25. When $n > 25$, the test statistic z is based on a normal approximation to the binomial probability distribution with $p = q = 1/2$. Recall that in Section 5-6 we saw that the normal approximation to the binomial distribution is acceptable when both $np \geq 5$ and $nq \geq 5$. Recall also that in Section 4-4 we saw that $\mu = np$ and $\sigma = \sqrt{npq}$ for binomial probability distributions. Because this sign test assumes that $p = q = 1/2$, we meet the $np \geq 5$ and $nq \geq 5$ prerequisites whenever $n \geq 10$. Also, with the assumption that $p = q = 1/2$, we get $\mu = np = n/2$ and $\sigma = \sqrt{npq} = \sqrt{n/4} = \sqrt{n}/2$, so

$$z = \frac{x - \mu}{\sigma}$$

becomes

$$z = \frac{x - \left(\frac{n}{2}\right)}{\frac{\sqrt{n}}{2}}$$

Finally, we replace x by $x + 0.5$ as a correction for continuity. That is, the values of x are discrete, but because we are using a continuous probability distribution, a discrete value such as 10 is actually represented by the interval from 9.5 to 10.5. Because x represents the less frequent sign, we act conservatively by concerning ourselves only with $x + 0.5$; we thus get the test statistic z, as given in the equation and in Figure 13-1.

Using Technology

STATDISK: Select **Analysis** from the main menu bar, then select **Sign Test.** Select the option **Given Number of Signs** if you know the number of plus and minus signs, or select **Given Pairs of Values** if you prefer to enter matched pairs of data. After making the required entries in the dialog box, the displayed results will include the test statistic, critical value, and conclusion.

Minitab: You must first create a column of values representing the differences between matched pairs of data or the number of plus and minus signs. (See the *Minitab Student Laboratory Manual and Workbook* for details.) Select **Stat,** then **Nonparametrics,** then **1-Sample Sign.** Click on the button for **Test Median.** Enter the median value and select the type of test, then click **OK.** Minitab will provide the P-value, so reject the null hypothesis if the P-value is less than or equal to the significance level. Otherwise, fail to reject the null hypothesis.

Excel: Excel does not have a built-in function dedicated to the sign test, but you can use Excel's BINOMDIST function to find the P-value for a sign test. Click **fx** on the main menu bar, then select the function category **Statistical** and then **BINOMDIST.** In the dialog box, first enter x, then the number of trials n, and then a probability of 0.5. Enter **TRUE** in the box for "cumulative." The resulting value is the probability of getting x or fewer successes among n trials. *Double this value for two-tailed tests.* The final result is the P-value, so reject the null hypothesis if the P-value is less than or equal to the significance level. Otherwise, fail to reject the null hypothesis.

TI-83 Plus: The TI-83 Plus calculator does not have a built-in function dedicated to the sign test, but you can use the binomcdf function to find the P-value for a sign test. Press **2nd, VARS** (to get the **DISTR** menu); then scroll down to select **binomcdf.** Complete the entry of **binomcdf(n, p, x)** with n for the total number of plus and minus signs, 0.5 for p, and the number of the less frequent sign for x. Now press **ENTER,** and the result will be the probability of getting x or fewer successes among n trials. *Double this value for two-tailed tests.* The final result is the P-value, so reject the null hypothesis if the P-value is less than or equal to the significance level. Otherwise, fail to reject the null hypothesis.

13-2 Basic Skills and Concepts

In Exercises 1–12, use the sign test.

1. Testing for a Difference Between Reported and Measured Female Heights This section used reported heights and measured heights for a sample of *male* statistics students. Listed here are heights (in inches) collected from a random sample of *female* statistics students. Is there sufficient evidence to support the claim that there is a difference between reported and measured heights of female statistics students? Use a 0.05 significance level.

Reported height	64	63	64	65	64	64	63	59	66	64
Measured height	63.5	63.1	63.8	63.4	62.1	64.4	62.7	59.3	65.4	62.2

2. Testing for a Difference Between the Qwerty and Dvorak Keyboards The following sentence, which contains every letter of the alphabet, was used for testing typewriters: "The quick brown fox jumped over the lazy red dog." If we use the word rating procedure described in the Chapter Problem for Chapter 2, we find that the 10 words in that sentence have Qwerty ratings of 2, 5, 7, 3, 5, 5, 2, 3, 2, 1. The corresponding Dvorak ratings are 0, 5, 5, 3, 5, 3, 0, 4, 1, 1. Is there sufficient evidence to support the claim that the Dvorak keyboard has lower ratings?

3. Nominal Data: Survey of Voters In a survey of 1002 people, 701 said that they voted in the recent presidential election (based on data from ICR Research Group). Is there sufficient evidence to support the claim that the majority of people say that they voted in the election?

4. Nominal Data: Smoking and Nicotine Patches In one study of 71 smokers who tried to quit smoking with nicotine patch therapy, 41 were smoking one year after the treatment (based on data from "High-Dose Nicotine Patch Therapy," by Dale et al., *Journal of the American Medical Association,* Vol. 274, No. 17). Use a 0.05 significance level to test the claim that among smokers who try to quit with nicotine patch therapy, the majority are smoking a year after the treatment.

5. Testing for Median Volume of Coke Cans Refer to Data Set 1 in Appendix B and use the volumes of regular Coke. Test the claim that cans of regular Coke have volumes with a median greater than 12 oz. Does it appear that the Coke cans are being filled correctly?

6. Testing for Median Amount of Domino Sugar in Packets Refer to Data Set 4 in Appendix B and use the sample data to test the claim that the median amount of sugar in the packets is equal to 3.5 oz. Does it appear that the sugar packets are being filled correctly?

7. Testing for Median Weight of M&Ms Using the weights of only the *brown* M&Ms listed in Data Set 10 in Appendix B, test the claim that the median is greater than 0.9085 g. (For the 1498 M&Ms to produce a total package weight of 1361 g, the mean weight must be at least 0.9085 g.) Use a 0.05 significance level. Based on the result, does it appear that the package is labeled correctly?

8. Testing for Median Time Interval for Old Faithful Geyser Refer to Data Set 11 in Appendix B. Test the claim that the intervals between eruptions of the Old Faithful geyser have a median greater than 77 min, which was the median about 20 years ago.

9. Testing for Difference Between Right/Left Reaction Times The accompanying table lists sample data obtained when 14 subjects were tested for reaction times with their left and right hands. (Only right-handed subjects were used.) Use a 0.05 significance level to test the claim of no difference between the right- and left-hand reaction times.

Right	191	97	116	165	116	129	171	155	112	102	188	158	121	133
Left	224	171	191	207	196	165	177	165	140	188	155	219	177	174
Sign of difference (right − left)	−	−	−	−	−	−	−	−	−	−	+	−	−	−

10. Testing Effectiveness of SAT Prep Courses Does it pay to take preparatory courses for standardized tests such as the SAT? Using a 0.05 significance level, test the claim that the Allan Preparation Course has no effect on SAT scores. Use the sample data in the accompanying table.

Subject	A	B	C	D	E	F	G	H	I	J
SAT score before course	700	840	830	860	840	690	830	1180	930	1070
SAT score after course	720	840	820	900	870	700	800	1200	950	1080

Based on data from the College Board and "An Analysis of the Impact of Commercial Test Preparation Courses on SAT Scores," by Sesnowitz, Bernhardt, and Knain, *American Educational Research Journal,* Vol. 19, No. 3.

11. Testing for Effect of Drinking on Reaction Time The Life Trust Insurance Company funded a university study of drinking and driving. After 30 randomly selected drivers were tested for reaction times, they were given two drinks and tested again, with the result that 22 had slower reaction times, 6 had faster reaction times, and 2 had the same reaction times as before the drinks. At the 0.01 significance level, test the claim that the drinks had no effect on the reaction times. Based on these very limited results, does it appear that the insurance company is justified in charging higher rates for those who drink and drive?

12. Nominal Data: Testing for Gender Discrimination The Diaz Tool and Die Company claims that hiring is done without any gender bias. Among the last 54 new employees hired, 1/3 are women. Job applicants are about half men and half women, who are all qualified. Is there sufficient evidence to charge gender bias? Use a 0.01 significance level, because we don't want to make such a serious charge unless there is very strong evidence.

13-2 Beyond the Basics

13. Procedures for Handling Ties In the sign test procedure described in this section, we excluded ties (represented by 0 instead of a sign of + or −). A second approach is to treat half of the 0s as positive signs and half as negative signs. (If the number of 0s is odd, exclude one so that they can be divided equally.) With a third approach, in two-tailed tests make half of the 0s positive and half negative; in one-tailed tests make all 0s either positive or negative, whichever supports the null hypothesis. Assume that in using the sign test on a claim that the median value is at least 100, we get 60 values below 100, 40 values above 100, and 21 values equal to 100. Identify the test statistic and conclusion for the three different ways of handling differences of 0. Assume a 0.05 significance level in all three cases.

14. IQ Scores: Agreement Between Sign Test and t Test A sample of 10 randomly selected IQ scores is obtained from the population of professional athletes. We want to test the claim that the population is centered at the IQ score of 100.

a. Is it possible to have sample data such that the sign test leads to rejection of the null hypothesis that the median is equal to 100 while the t test (Section 7-4) conclusion is failure to reject the null hypothesis of $\mu = 100$? If so, construct such a data set.

b. Is it possible to have sample data such that the sign test conclusion is failure to reject the null hypothesis that the median is equal to 100 while the t test (Section 7-4) conclusion is rejection of the null hypothesis that $\mu = 100$? If so, construct such a data set.

15. Finding Critical Values Table A-7 lists critical values for limited choices of α. Use Table A-1 to add a new column in Table A-7 (down to $n = 15$) that represents a significance level of 0.03 in one tail or 0.06 in two tails. For any particular n, use $p = 0.5$, because the sign test requires the assumption that

$$P(\text{positive sign}) = P(\text{negative sign}) = 0.5$$

The probability of x or fewer like signs is the sum of the probabilities for values up to and including x.

13-3 Wilcoxon Signed-Ranks Test for Matched Pairs

In Section 13-2 we used the sign test to analyze data, including sample data consisting of matched pairs. The sign test used only the signs of the differences and did not use their actual magnitudes (how large the numbers are). This section introduces the *Wilcoxon signed-ranks test,* which is also used with sample paired data. By using ranks, this test takes the magnitudes of the differences into account. (See Section 13-1 for a description of ranks.) Because the Wilcoxon signed-ranks test incorporates and uses more information than the sign test, it tends to yield conclusions that better reflect the true nature of the data.

DEFINITION

The **Wilcoxon signed-ranks test** is a nonparametric test that uses ranks of sample data consisting of matched pairs. It is used to test for differences in the population distributions, so the null and alternative hypotheses are as follows:

H_0: The two samples come from populations with the same distribution.

H_1: The two samples come from populations with different distributions.

(The Wilcoxon signed-ranks test can also be used to test the claim that a sample comes from a population with a specified median. See Exercise 10 for this application.)

Assumptions

1. The sample data have been randomly selected
2. The population of differences (found from the pairs of data) has a distribution that is approximately *symmetric,* meaning that the left half of its histogram is roughly a mirror image of its right half. (There is *no* requirement that the data have a normal distribution.)

Procedure for Finding the Value of the Test Statistic

Step 1: For each pair of data, find the difference d by subtracting the second value from the first value. Keep the signs, but discard any pairs for which $d = 0$.

Step 2: *Ignore the signs of the differences,* then sort the differences from lowest to highest and replace the differences by the corresponding rank value (as described in Section 13-1). When differences have the same numerical value, assign to them the mean of the ranks involved in the tie.

Step 3: Attach to each rank the sign of the difference from which it came. That is, insert those signs that were ignored in Step 2.

Step 4: Find the sum of the absolute values of the negative ranks. Also find the sum of the positive ranks.

Step 5: Let T be the *smaller* of the two sums found in Step 4. Either sum could be used, but for a simplified procedure we arbitrarily select the smaller of the two sums. (See the notation for T in the Notation box.)

Step 6: Let n be the number of pairs of data for which the difference d is not 0.

Step 7: Determine the test statistic and critical values based on the sample size, as shown in the Test Statistic box.

Step 8: When forming the conclusion, reject the null hypothesis if the sample data lead to a test statistic that is in the critical region—that is, the test statistic is less than or equal to the critical value(s). Otherwise, fail to reject the null hypothesis.

Notation

T = the smaller of the following two sums:

1. The sum of the absolute values of the negative ranks
2. The sum of the positive ranks

Test Statistic for the Wilcoxon Signed-Ranks Test for Matched Pairs

For $n \leq 30$: T

For $n > 30$: $$z = \frac{T - \frac{n(n+1)}{4}}{\sqrt{\frac{n(n+1)(2n+1)}{24}}}$$

Critical values:

1. If $n \leq 30$, the critical T value is found in Table A-8.
2. If $n > 30$, the critical z values are found in Table A-2.

EXAMPLE **Reported and Measured Heights** The data in Table 13-3 on the next page are matched pairs of heights (in inches) obtained from a random sample of 12 male statistics students. Each student reported his height, then his height was measured. Use the Wilcoxon signed-ranks test and a 0.05 significance level to test the claim that there is no difference between reported heights and measured heights.

SOLUTION The null and alternative hypotheses are as follows:

H_0: There is no difference between reported and measured heights.

H_1: There is a difference between reported and measured heights.

The significance level is $\alpha = 0.05$. We are using the Wilcoxon signed-ranks test procedure, so the test statistic is calculated by using the eight-step procedure presented earlier in this section.

TABLE 13-3 Reported and Measured Heights of Male Statistics Students

Reported height	68	74	82.25	66.5	69	68	71	70	70	67	68	70
Measured height	66.8	73.9	74.3	66.1	67.2	67.9	69.4	69.9	68.6	67.9	67.6	68.8
Differences d	1.2	0.1	7.95	0.4	1.8	0.1	1.6	0.1	1.4	−0.9	0.4	1.2
Ranks of differences	7.5	2	12	4.5	11	2	10	2	9	6	4.5	7.5
Signed ranks	+7.5	+2	+12	+4.5	+11	+2	+10	+2	+9	−6	+4.5	+7.5

Step 1: In Table 13-3, the row of differences is obtained by computing

$$d = \text{reported height} - \text{measured height}$$

for each pair of data.

Step 2: Ignoring their signs, we rank the absolute differences from lowest to highest. Note that ties in ranks are handled by assigning the mean of the involved ranks to each of the tied values. For example, the first three values in the sorted list are each 0.1, so we find the mean of ranks 1, 2, 3 and assign that mean of 2 to each of the 0.1 values.

Step 3: The bottom row of Table 13-3 is created by attaching to each rank the sign of the corresponding difference. If there really is no difference between reported and measured heights (as in the null hypothesis), we expect the number of positive ranks to be approximately equal to the number of negative ranks.

Step 4: We now find the sum of the absolute values of the negative ranks, and we also find the sum of the positive ranks.

Sum of absolute values of negative ranks: 6

Sum of positive ranks: 72

(which is $7.5 + 2 + 12 + 4.5 + 11 + 2 + 10 + 2 + 9 + 4.5 + 7.5$)

Step 5: Letting T be the smaller of the two sums found in Step 4, we find that $T = 6$.

Step 6: Letting n be the number of pairs of data for which the difference d is not 0, we have $n = 12$.

Step 7: Because $n = 12$, we have $n \leq 30$, so we use a test statistic of $T = 6$ (and we do not calculate a z test statistic). Also, because $n \leq 30$, we use Table A-8 to find the critical value of 14.

Step 8: The test statistic $T = 6$ is less than or equal to the critical value of 14, so we reject the null hypothesis. It appears that there is a difference between reported and measured heights.

If we use the sign test with the preceding example, we will arrive at the same conclusion. Although the sign test and the Wilcoxon signed-ranks test agree in this particular case, there are other cases in which they do not agree.

Rationale: In this example the unsigned ranks of 1 through 12 have a total of 78, so if there are no significant differences, each of the two signed-rank totals should be around $78 \div 2$, or 39. That is, the negative ranks and positive ranks should split up as 39-39 or something close, such as 38-40. The table of critical values shows that at the 0.05 significance level with 12 pairs of data, a 14-64 split represents a significant departure from the null hypothesis, and any split that is farther apart (such as 13-65 or 12-66 or 6-72) will also represent a significant departure from the null hypothesis. Conversely, splits like 38-40 or 15-63 do not represent significant departures from a 39-39 split, and they would not be a basis for rejecting the null hypothesis. The Wilcoxon signed-ranks test is based on the lower rank total, so instead of analyzing both numbers constituting the split, we consider only the lower number.

The sum $1 + 2 + 3 + \cdots + n$ of all the ranks is equal to $n(n + 1)/2$; if this is a rank sum to be divided equally between two categories (positive and negative), each of the two totals should be near $n(n + 1)/4$, which is half of $n(n + 1)/2$. Recognition of this principle helps us understand the test statistic used when $n > 30$. The denominator in that expression represents a standard deviation of T and is based on the principle that

$$1^2 + 2^2 + 3^3 + \cdots + n^2 = \frac{n(n + 1)(2n + 1)}{6}$$

The Wilcoxon signed-ranks test can only be used for matched pairs of data. The next section will describe a rank-sum test that can be applied to two sets of independent data that are not matched in pairs.

Using Technology

STATDISK: Select **Analysis** from the main menu bar, then select **Wilcoxon Tests.** Now select **Signed-Ranks Test,** and proceed to enter the matched sample data. Click on **Evaluate.** The STATDISK display will include the test statistic, critical value, and conclusion.

Minitab: Enter the paired data in columns C1 and C2. Click on **Editor,** then **Enable Command Editor,** and enter the command **LET C3 = C1 − C2.** Press the **Enter** key. Select the options **Stat, Nonparametrics,** and **1-Sample Wilcoxon.** Enter C3 for the variable and click on the button for **Test Median.** The Minitab display will include the P-value. Reject the null hypothesis of equal distributions if the P-value is less than or equal to the significance level. Fail to reject the null hypothesis if the P-value is greater than the significance level.

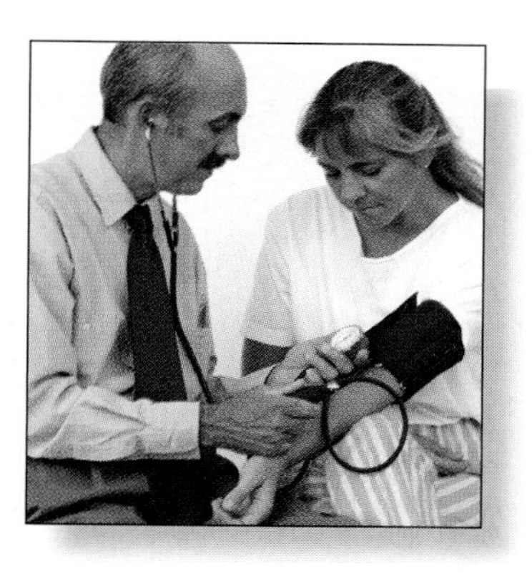

Gender Gap in Drug Testing

A study of the relationship between heart attacks and doses of aspirin involved 22,000 male physicians. This study, like many others, excluded women. The General Accounting Office recently criticized the National Institutes of Health for not including both sexes in many studies because results of medical tests on males do not necessarily apply to females. For example, women's hearts are different from men's in many important ways. When forming conclusions based on sample results, we should be wary of an inference that extends to a population larger than the one from which the sample was drawn.

Excel: Excel is not programmed for the Wilcoxon signed-ranks test.

TI-83 Plus: The TI-83 Plus calculator is not programmed for the Wilcoxon signed-ranks test.

13-3 Basic Skills and Concepts

Using the Wilcoxon Signed-Ranks Test. *In Exercises 1–4, refer to the sample data for the given exercises in Section 13-2. Instead of the sign test, use the Wilcoxon signed-ranks test to test the claim that both samples come from populations having the same distribution. Use the significance level α that is given.*

1. Exercise 1; $\alpha = 0.01$ **2.** Exercise 2; $\alpha = 0.05$

3. Exercise 9; $\alpha = 0.05$ **4.** Exercise 10; $\alpha = 0.01$

In Exercises 5–8, use the Wilcoxon signed-ranks test.

5. Intelligence in Children: Testing for a Difference Between Trials Mental measurements of young children are often made by giving them blocks and telling them to build a tower as tall as possible. One experiment of block building was repeated a month later, with the times (in seconds) listed in the accompanying table. Is there sufficient evidence to support the claim that there is a difference between the two times? Use a 0.01 significance level.

Child	A	B	C	D	E	F	G	H	I	J	K	L	M	N	O
First trial	30	19	19	23	29	178	42	20	12	39	14	81	17	31	52
Second trial	30	6	14	8	14	52	14	22	17	8	11	30	14	17	15

Based on data from "Tower Building," by Johnson and Courtney, *Child Development,* Vol. 3.

6. Testing for Difference Between Sitting Measurements and Lying Measurements In a study of techniques used to measure lung volumes, physiological data were collected for 10 subjects. The values given in the table are in liters and represent the measured functional residual capacities of the 10 subjects both in a sitting position and in a supine (lying) position. At the 0.05 significance level, test the claim that there is no significant difference between the measurements taken in the two positions.

Sitting	2.96	4.65	3.27	2.50	2.59	5.97	1.74	3.51	4.37	4.02
Supine	1.97	3.05	2.29	1.68	1.58	4.43	1.53	2.81	2.70	2.70

Based on "Validation of Esophageal Balloon Technique at Different Lung Volumes and Postures," by Baydur, Cha, and Sassoon, *Journal of Applied Physiology,* Vol. 62, No. 1.

7. Testing for Drug Effectiveness Captopril is a drug designed to lower systolic blood pressure. When subjects were tested with this drug, their systolic blood pressure readings (in mm of mercury) were measured before and after the drug was taken, with the results given in the accompanying table. Is there sufficient evidence to support the claim that the drug has an effect? Does Captopril appear to lower systolic blood pressure?

Subject	A	B	C	D	E	F	G	H	I	J	K	L
Before	200	174	198	170	179	182	193	209	185	155	169	210
After	191	170	177	167	159	151	176	183	159	145	146	177

Based on data from "Essential Hypertension: Effect of an Oral Inhibitor of Angiotensin-Converting Enzyme," by MacGregor et al., *British Medical Journal,* Vol. 2.

8. Testing for Difference Between Morning and Night Body Temperatures Refer to Data Set 6 in Appendix B. Use the paired data consisting of body temperatures of women at 8:00 A.M. and at 12:00 A.M. on Day 2. Is there sufficient evidence to support the claim that morning and night body temperatures are different?

13-3 Beyond the Basics

9. Finding Critical Values Assume that the Wilcoxon signed-ranks test is being used for a two-tailed hypothesis test with a 0.05 significance level.

a. With $n = 10$ pairs of data, find the lowest and highest possible values of T.

b. With $n = 50$ pairs of data, find the lowest and highest possible values of T.

c. If there are $n = 100$ pairs of data with no differences of 0 and no tied ranks, find the critical value of T.

10. Using the Wilcoxon Signed-Ranks Test for Claims About a Median The Wilcoxon signed-ranks test can be used to test the claim that a sample comes from a population with a specified median. The procedure used is the same as the one described in this section, except that the differences (Step 1) are obtained by subtracting the value of the hypothesized median from each value. Use the sample data consisting of the 106 body temperatures listed in Table 7-1. (See the Chapter Problem at the beginning of Chapter 7.) At the 0.05 significance level, test the claim that healthy adults have a median body temperature that is equal to 98.6°F.

13-4 Wilcoxon Rank-Sum Test for Two Independent Samples

This section introduces the *Wilcoxon rank-sum test,* which is a nonparametric test that can be applied to two independent sets of sample data. Two samples are independent if the sample values selected from one population are not related or somehow matched or paired with the sample values from the other population.

DEFINITION

The **Wilcoxon rank-sum test** is a nonparametric test that uses ranks of sample data from two independent populations. It is used to test the null hypothesis that the two independent samples come from populations with the same distribution. (That is, the two populations are identical.) The alternative hypothesis is the claim that the two population distributions are different in some way.

H_0: The two samples come from populations with the same distribution. (That is, the two populations are identical.)

H_1: The two samples come from populations with different distributions. (That is, the two populations are different in some way.)

The Wilcoxon rank-sum test is equivalent to the **Mann-Whitney U test** (see Exercise 11), which is included in some other textbooks and software packages (such as Minitab). The key idea underlying the Wilcoxon rank-sum test is this: If two samples are drawn from identical populations and the individual values are all ranked as one combined collection of values, then the high and low ranks should fall evenly between the two samples. If the low ranks are found predominantly in one sample and the high ranks are found predominantly in the other sample, we suspect that the two populations are not identical.

Assumptions

1. There are two independent samples that were randomly selected.
2. Each of the two samples has more than 10 values. (For samples with 10 or fewer values, special tables are available in reference books, such as *CRC Standard Probability and Statistics Tables and Formulae,* published by CRC Press.)
3. There is *no* requirement that the two populations have a normal distribution or any other particular distribution.

Note that unlike the corresponding hypothesis tests in Section 8-6, the Wilcoxon rank-sum test does *not* require normally distributed populations. Also, the Wilcoxon rank-sum test can be used with data at the ordinal level of measurement, such as data consisting of ranks. In contrast, the parametric methods of Sections 8-2 and 8-6 cannot be used with data at the ordinal level of measurement. In Table 13-1 we noted that the Wilcoxon rank-sum test has a 0.95 efficiency rating when compared with the parametric t test or z test. Because this test has such a high efficiency rating and involves easier calculations, it is often preferred over the parametric tests presented in Sections 8-2 and 8-6, even when the requirement of normality is satisfied.

Procedure for Finding the Value of the Test Statistic

1. Temporarily combine the two samples into one big sample, then replace each sample value with its rank. (The lowest value gets a rank of 1, the next lowest value gets a rank of 2, and so on. See Section 13-1 for a description of ranks.)
2. Find the sum of the ranks for either one of the two samples.
3. Calculate the value of the z test statistic as shown in the Test Statistic box, where either sample can be used as "sample 1."

Notation for the Wilcoxon Rank-Sum Test

n_1 = size of sample 1

n_2 = size of sample 2

R_1 = sum of ranks for sample 1

R_2 = sum of ranks for sample 2

R = same as R_1 (sum of ranks for sample 1)

(continued)

μ_R = mean of the sample R values that is expected when the two populations are identical

σ_R = standard deviation of the sample R values that is expected when the two populations are identical

If testing the null hypothesis of identical populations and if both sample sizes are greater than 10, then the sampling distribution of R is approximately normal with mean μ_R and standard deviation σ_R, and the test statistic is as follows.

Test Statistic for the Wilcoxon Rank-Sum Test for Two Independent Samples

$$z = \frac{R - \mu_R}{\sigma_R}$$

where

$$\mu_R = \frac{n_1(n_1 + n_2 + 1)}{2}$$

$$\sigma_R = \sqrt{\frac{n_1 n_2 (n_1 + n_2 + 1)}{12}}$$

n_1 = size of the sample from which the rank sum R is found

n_2 = size of the other sample

R = sum of ranks of the sample with size n_1

Critical values: Critical values can be found in Table A-2 (because the test statistic is based on the normal distribution).

The expression for μ_R is based on the following result of mathematical induction: The sum of the first n positive integers is given by $1 + 2 + 3 + \cdots + n = n(n + 1)/2$. The expression for σ_R is based on a result stating that the integers $1, 2, 3, \ldots, n$ have standard deviation $\sqrt{(n^2 - 1)/12}$.

EXAMPLE **M&Ms** Samples of M&M plain candies are randomly selected, and the red and yellow M&Ms are weighed, with the results listed in Table 13-4 on the next page (from Data Set 10 in Appendix B). Use a 0.05 significance level to test the claim that weights of red M&Ms and yellow M&Ms have the same distribution.

SOLUTION The null and alternative hypotheses are as follows:

H_0: Red and yellow M&M plain candies have weights with the same distribution.

H_1: The two populations have weight distributions that are different in some way.

TABLE 13-4

Weights (in grams) of M&Ms

Red		Yellow	
0.870	(2)	0.906	(19)
0.933	(35)	0.978	(45)
0.952	(42)	0.926	(34)
0.908	(21)	0.868	(1)
0.911	(24.5)	0.876	(5)
0.908	(21)	0.968	(44)
0.913	(27)	0.921	(30)
0.983	(46)	0.893	(15)
0.920	(29)	0.939	(38)
0.936	(37)	0.886	(10.5)
0.891	(13)	0.924	(32)
0.924	(32)	0.910	(23)
0.874	(4)	0.877	(6)
0.908	(21)	0.879	(7.5)
0.924	(32)	0.941	(40)
0.897	(16)	0.879	(7.5)
0.912	(26)	0.940	(39)
0.888	(12)	0.960	(43)
0.872	(3)	0.989	(47)
0.898	(17)	0.900	(18)
0.882	(9)	0.917	(28)
		0.911	(24.5)
		0.892	(14)
		0.886	(10.5)
		0.949	(41)
		0.934	(36)
$n_1 = 21$		$n_2 = 26$	
$R_1 = 469.5$		$R_2 = 658.5$	

Rank all 47 weights combined, beginning with a rank of 1 (assigned to the lowest weight of 0.868 g). Ties in ranks are handled as described in Section 13-1: Find the mean of the ranks involved and assign this mean rank to each of the tied values. The ranks corresponding to the individual sample values are shown in parentheses in Table 13-4. R denotes the sum of the ranks for the sample we choose as sample 1. If we choose the red M&Ms, we get

$$R = 2 + 35 + 42 + \cdots + 9 = 469.5$$

Because there are 21 red M&Ms, we have $n_1 = 21$. Also, $n_2 = 26$, because there are 26 yellow M&Ms. We can now determine the values of μ_R, σ_R, and the test statistic z.

$$\mu_R = \frac{n_1(n_1 + n_2 + 1)}{2} = \frac{21(21 + 26 + 1)}{2} = 504$$

$$\sigma_R = \sqrt{\frac{n_1 n_2(n_1 + n_2 + 1)}{12}} = \sqrt{\frac{(21)(26)(21 + 26 + 1)}{12}} = 46.73$$

$$z = \frac{R - \mu_R}{\sigma_R} = \frac{469.5 - 504}{46.73} = -0.74$$

The test is two-tailed because a large positive value of z would indicate that the higher ranks are found disproportionately in the first sample, and a large negative value of z would indicate that the first sample had a disproportionate share of lower ranks. In either case, we would have strong evidence against the claim that the two samples come from populations with the same distribution.

The significance of the test statistic z can be treated in the same manner as in previous chapters. We are now testing (with $\alpha = 0.05$) the hypothesis that the two populations have the same distribution, so we have a two-tailed test with critical z values of 1.96 and -1.96. The test statistic of $z = -0.74$ does not fall within the critical region, so we fail to reject the null hypothesis that red and yellow M&Ms have the same weights. It appears that red and yellow M&Ms do come from populations with the same distribution.

We can verify that if we interchange the two sets of weights and consider the sample of yellow M&Ms to be first, $R = 658.5$, $\mu_R = 624$, $\sigma_R = 46.73$, and $z = 0.74$, so the conclusion is the same.

EXAMPLE **Wednesday and Saturday Rainfall** The Chapter Problem referred to the Boston rainfall amounts listed in Data Set 17 in Appendix B. The Chapter Problem included boxplots of the rainfall amounts for the seven days of the week, starting with Monday at the top. Comparisons of those boxplots show that Wednesday and Saturday appear to be the two days that differ most. But are those differences significant? Use the Wilcoxon rank-sum test to test the claim that the rainfall amounts for Wednesdays and Saturdays come from the same distribution.

SOLUTION The null and alternative hypotheses are as follows:

H_0: The Wednesday and Saturday rainfall amounts come from populations with the same distribution.

H_1: The two distributions are different in some way.

Instead of manually calculating the rank sums, we refer to the Minitab display shown here. In that Minitab display, "ETA1" and "ETA2" denote the median of the first sample and the median of the second sample, respectively. The display suggests that we are testing the null hypothesis of equal medians, but the Wilcoxon rank-sum test is based on the entire distributions, not just the medians. Here are the key components of the Minitab display: The rank sum for Wednesday is W = 2639.0, the *P*-value is 0.2773 (or 0.1992 after an adjustment for ties), and the conclusion is that we cannot reject (the null hypothesis) with a significance level of 0.05. Bottom line: The differences between Wednesday and Saturday are not significant. This seems to contradict the media reports that it rains more on weekends, but we will consider this issue more in the following section.

MINITAB

```
Wed     N = 53          Median =        0.0000
Sat     N = 52          Median =        0.0000

Point estimate for ETA1-ETA2 is         0.0000
95.1 Percent CI for ETA1-ETA2 is (0.0000, 0.0000)

W = 2639.0

Test of ETA1 = ETA2 vs ETA1 not = ETA2 is significant at 0.2773
The test is significant at 0.1992 (adjusted for ties)

Cannot reject at alpha = 0.05
```

Using Technology

STATDISK: Select **Analysis** from the main menu bar, then select **Wilcoxon Tests,** followed by the option **Rank-Sum Test.** Enter the sample data in the dialog box, then click on **Evaluate** to get a display that includes the rank sums, sample size, test statistic, critical value, and conclusion.

Minitab: First enter the two sets of sample data in columns C1 and C2. Then select the options **Stat, Nonparametrics,** and **Mann-Whitney,** and proceed to enter C1 for the first sample and C2 for the second sample. The confidence level of 95.0 corresponds to a significance level of $\alpha = 0.05$, and the "alternate: not equal" box refers to the alternative hypothesis, where "not equal" corresponds to a two-tailed hypothesis test. Minitab provides the P-value and conclusion. See the sample Minitab display included with the preceding example.

Excel: Excel is not programmed for the Wilcoxon rank-sum test.

TI-83 Plus: The TI-83 calculator is not programmed for the Wilcoxon rank-sum test.

13-4 Basic Skills and Concepts

Using the Wilcoxon Rank-Sum Test. *In Exercises 1–8, use the Wilcoxon rank-sum test.*

1. Testing Red and Brown M&Ms for Identical Populations The example in this section tested the null hypothesis that red and yellow M&M plain candies have weights with the same distribution. Refer to Data Set 10 in Appendix B and test the claim that red and brown M&M plain candies have weights with the same distribution. That is, test the claim that the populations of red and brown M&M plain candies are identical. Use a 0.05 significance level.

2. Testing the Anchoring Effect Randomly selected statistics students were given five seconds to estimate the value of a product of numbers with the results given in the accompanying table. (See the Cooperative Group Activities at the end of Chapter 2.) Is there sufficient evidence to support the claim that the two samples come from populations with different distributions?

Estimates from Students Given $1 \times 2 \times 3 \times 4 \times 5 \times 6 \times 7 \times 8$

1560	169	5635	25	842	40,320	5000	500	1110	10,000
200	1252	4000	2040	175	856	42,200	49,654	560	800

Estimates from Students Given $8 \times 7 \times 6 \times 5 \times 4 \times 3 \times 2 \times 1$

100,000	2000	42,000	1500	52,836	2050	428	372	300	225	64,582
23,410	500	1200	400	49,000	4000	1876	3600	354	750	640

3. Are Severe Psychiatric Disorders Related to Biological Factors? One study used X-ray computed tomography (CT) to collect data on brain volumes for a group of patients with obsessive-compulsive disorders and a control group of healthy persons. The accompanying list shows sample results (in milliliters) for volumes of the right cordate (based on data from "Neuroanatomical Abnormalities in Obsessive-Compulsive Disorder Detected with Quantitative X-Ray Computed Tomography," by Luxenberg et al., *American Journal of Psychiatry,* Vol. 145, No. 9).

Obsessive-compulsive patients				Control group			
0.308	0.210	0.304	0.344	0.519	0.476	0.413	0.429
0.407	0.455	0.287	0.288	0.501	0.402	0.349	0.594
0.463	0.334	0.340	0.305	0.334	0.483	0.460	0.445

At the 0.01 significance level, test the claim that obsessive-compulsive patients and healthy persons have the same brain volumes. Based on this result, can we conclude that obsessive-compulsive disorders have a biological basis?

4. Does the Arrangement of Test Items Affect the Score? The arrangement of test items was studied for its effect on anxiety. Sample results are as follows:

Easy to difficult				Difficult to easy			
24.64	39.29	16.32	32.83	33.62	34.02	26.63	30.26
28.02	33.31	20.60	21.13	35.91	26.68	29.49	35.32
26.69	28.90	26.43	24.23	27.24	32.34	29.34	33.53
7.10	32.86	21.06	28.89	27.62	42.91	30.20	32.54
28.71	31.73	30.02	21.96				
25.49	38.81	27.85	30.29				
30.72							

At the 0.05 significance level, test the claim that the two samples come from populations with the same scores. (The data are based on "Item Arrangement, Cognitive Entry Characteristics, Sex and Test Anxiety as Predictors of Achievement Examination Performance," by Klimko, *Journal of Experimental Education,* Vol. 52, No. 4.)

5. Are Weights of Regular Coke and Diet Coke Different? Refer to Data Set 1 in Appendix B. Is there sufficient evidence to support the claim that weights of regular Coke and diet Coke are different? How might such a difference be explained?

6. Are Weights of Regular Coke and Regular Pepsi Different? Refer to Data Set 1 in Appendix B. Is there sufficient evidence to support the claim that weights of regular Coke and regular Pepsi are different?

7. Are Textbook Prices Different at the Two Colleges? Refer to Data Set 2 in Appendix B. Is there sufficient evidence to support the claim that new textbook prices are different at the two colleges?

8. Testing Effect of Second-Hand Smoke Refer to Data Set 9 in Appendix B. Use a 0.01 significance level to test the claim that the population of cotinine levels is the same for these two groups: nonsmokers who are not exposed to environmental tobacco smoke (labeled as NOETS), and nonsmokers who are exposed to tobacco smoke (labeled as ETS). What do the results suggest about second-hand smoke?

13-4 Beyond the Basics

9. How Do Units of Measurement Affect Results? When using the Wilcoxon rank-sum test, describe the effect of changing the unit of measurement for each sample value. Using the M&M data from the example in this section, how are the results affected if each weight is changed from grams to ounces?

10. How Do Outliers Affect Results? When using the Wilcoxon rank-sum test, is the effect of an outlier small or large? Explain.

11. Using the Mann-Whitney U Test The Mann-Whitney U test is equivalent to the Wilcoxon rank-sum test for independent samples in the sense that they both apply to the same situations and always lead to the same conclusions. In the Mann-Whitney U test we calculate

$$z = \frac{U - \frac{n_1 n_2}{2}}{\sqrt{\frac{n_1 n_2 (n_1 + n_2 + 1)}{12}}}$$

where

$$U = n_1 n_2 + \frac{n_1(n_1 + 1)}{2} - R$$

Using the sample data from the M&M example in this section, find the z test statistic for the Mann-Whitney U test and compare it to the z test statistic of -0.74 that was found using the Wilcoxon rank-sum test.

12. Finding Critical Values Assume that we have two treatments (A and B) that produce measurable results, and we have only two observations for treatment A and two observations for treatment B. We cannot use the test statistic given in this section because both sample sizes do not exceed 10.

Rank				Rank sum for
1	2	3	4	treatment A
A	A	B	B	3

a. Complete the accompanying table by listing the five rows corresponding to the other five cases, and enter the corresponding rank sums for treatment A.
b. List the possible values of R, along with their corresponding probabilities. [Assume that the rows of the table from part (a) are equally likely.]
c. Is it possible, at the 0.10 significance level, to reject the null hypothesis that there is no difference between treatments A and B? Explain.

13-5 Kruskal-Wallis Test

This section introduces the *Kruskal-Wallis test,* which is used to test the claim that several different independent samples come from identical populations. In Section 11-2 we used one-way analysis of variance (ANOVA) to test hypotheses that differences in means among several samples are significant, but that method requires that all of the involved populations have normal distributions. The Kruskal-Wallis test does not require normal distributions.

DEFINITION

The **Kruskal-Wallis Test** (also called the ***H* test**) is a nonparametric test that uses ranks of sample data from three or more independent populations. It is used to test the null hypothesis that the independent samples come from populations with the same distribution; the alternative hypothesis is the claim that the population distributions are different in some way.

H_0: The samples come from populations with the same distribution.

H_1: The samples come from populations with different distributions.

Assumptions

1. We have at least three independent samples, all of which are randomly selected.
2. Each sample has at least five observations. (If samples have fewer than five observations, refer to special tables of critical values, such as *CRC Standard Probability and Statistics Tables and Formulae,* published by CRC Press.)
3. There is *no* requirement that the populations have a normal distribution or any other particular distribution.

In applying the Kruskal-Wallis test, we compute the *test statistic H, which has a distribution that can be approximated by the chi-square distribution as long as each sample has at least five observations.* When we use the chi-square distribution in this context, the number of degrees of freedom is $k - 1$, where k is the number of samples. (For a quick review of the key features of the chi-square distribution, see Section 6-6.)

Procedure for Finding the Value of the Test Statistic

1. Temporarily combine all samples into one big sample and assign a rank to each sample value. (Sort the values from lowest to highest, and in cases of ties, assign to each observation the mean of the ranks involved.)
2. For each sample, find the sum of the ranks and find the sample size.
3. Calculate H by using the results of Step 2 and the following notation and test statistic.

Notation for the Kruskal-Wallis Test

N = total number of observations in all samples combined

k = number of samples

R_1 = sum of ranks for sample 1

n_1 = number of observations in sample 1

For sample 2, the sum of ranks is R_2 and the number of observations is n_2, and similar notation is used for the other samples.

Test Statistic for the Kruskal-Wallis Test

$$H = \frac{12}{N(N+1)}\left(\frac{R_1^2}{n_1} + \frac{R_2^2}{n_2} + \cdots + \frac{R_k^2}{n_k}\right) - 3(N+1)$$

where degrees of freedom $= k - 1$

Critical values:

1. The test is *right-tailed.*
2. Use Table A-4 (because the test statistic H can be approximated by a chi-square distribution).
3. Degrees of freedom $= k - 1$

The test statistic H is basically a measure of the variance of the rank sums R_1, $R_2, \ldots, R_k$. If the ranks are distributed evenly among the sample groups, then H should be a relatively small number. If the samples are very different, then the ranks will be excessively low in some groups and high in others, with the net effect that H will be large. Consequently, only large values of H lead to rejection of the null hypothesis that the samples come from identical populations. *The Kruskal-Wallis test is therefore a right-tailed test.*

EXAMPLE **Crash Test Dummies** We will use some of the same data that were included with the Chapter Problem for Chapter 11. See Table 13-5 for car crash results obtained from the National Transportation Safety Administration. Using the *head injury* measurements, is there sufficient evidence to conclude that head injuries for the four weight categories are not all the same?

SOLUTION We will follow the hypothesis testing procedure summarized in Figure 7-5.

Steps 1, 2, and 3: The null and alternative hypotheses are as follows.

H_0: The four weight categories have head injury measurements with identical populations.

H_1: The four populations are not identical.

Step 4: No significance level was specified. In the absence of any overriding circumstances, we use $\alpha = 0.05$.

Step 5: In Chapter 11 we used an F test for analysis of variance, but here we illustrate the method of the Kruskal-Wallis test.

Step 6: In determining the value of the test statistic H, we must first rank all of the data. We begin with the lowest value of 259, which is assigned a rank of 1. Ranks are shown in parentheses with the original head injury data in Table 13-5. Next we find the sample size, n, and sum of ranks, R, for each sample. If we

TABLE 13-5 Injuries to Car Crash Test Dummies

	Head Injury (hic)		Chest Deceleration (g)	Left Femur Load (lb)
Subcompact Cars				
Ford Escort	681	(16)	55	595
Honda Civic	428	(5)	47	1063
Hyundai Accent	917	(20)	59	885
Nissan Sentra	898	(19)	49	519
Saturn SL4	420	(4)	42	422
Compact Cars				
Chevrolet Cavalier	643	(13)	57	1051
Dodge Neon	655	(14)	57	1193
Mazda 626 DX	442	(6)	46	946
Pontiac Sunfire	514	(9)	54	984
Subaru Legacy	525	(10.5)	51	584
Midsize Cars				
Chevrolet Camaro	469	(8)	45	629
Dodge Intrepid	727	(18)	53	1686
Ford Mustang	525	(10.5)	49	880
Honda Accord	454	(7)	51	181
Volvo S70	259	(1)	46	645
Full Size Cars				
Audi A8	384	(3)	44	1085
Cadillac Deville	656	(15)	45	971
Ford Crown Victoria	602	(12)	39	996
Oldsmobile Aurora	687	(17)	58	804
Pontiac Bonneville	360	(2)	44	1376

add the five ranks for the five subcompact cars only, we get $16 + 5 + 20 + 19 + 4 = 64$. We list this result with the others.

Subcompact	**Compact**	**Midsize**	**Full Size**
$n_1 = 5$	$n_2 = 5$	$n_3 = 5$	$n_4 = 5$
$R_1 = 64$	$R_2 = 52.5$	$R_3 = 44.5$	$R_4 = 49$

Because the total number of observations is 20, we have $N = 20$. We can now evaluate the test statistic as follows:

$$H = \frac{12}{N(N+1)}\left(\frac{R_1^2}{n_1} + \frac{R_2^2}{n_2} + \cdots + \frac{R_k^2}{n_k}\right) - 3(N+1)$$

$$= \frac{12}{20(20+1)}\left(\frac{64^2}{5} + \frac{52.5^2}{5} + \frac{44.5^2}{5} + \frac{49^2}{5}\right) - 3(20+1)$$

$$= 1.1914$$

Because each sample has at least five observations, the distribution of H is approximately a chi-square distribution with $k - 1$ degrees of freedom. The number of samples is $k = 4$, so we have $4 - 1 = 3$ degrees of freedom. Refer to Table A-4 to find the critical value of 7.815, which corresponds to 3 degrees of freedom and a 0.05 significance level.

Step 7: The test statistic $H = 1.1914$ is not in the critical region bounded by 7.815, so we fail to reject the null hypothesis of identical populations. (In Section 11-2, we failed to reject the null hypothesis of equal means.)

Step 8: There is not sufficient evidence to support the claim that head injuries for the four weight categories are not all the same. Because the double negative in that conclusion can be confusing, we might simply say that on the basis of the available evidence, the hypothesis that the four weight categories appear to have the same distribution of head injury measurements is reasonable.

EXAMPLE **Rains More on Weekends?** In the Chapter Problem we noted that media reported that it rains more on weekends all along America's eastern coast, from Maine to Florida. Data Set 17 in Appendix B includes rainfall amounts for a recent year in Boston. Using that data set, test the claim that the seven week days have distributions that are not all the same.

SOLUTION Data Set 17 seems like it could be analyzed using analysis of variance methods introduced in Section 11-2, but those methods require that the sample values come from populations having distributions that are approximately normal. The Chapter Problem includes a histogram for the Monday rainfall amounts, and it is obvious that this is not a normal distribution. The histograms for the other days of the week have the same basic shape as Monday. Because the data do not come from normal distributions, analysis of variance cannot be used and the Kruskal-Wallis test is an ideal alternative. Data Set 17 includes data for each of 365 days, so we are dealing with large data sets and manual calculations would be too cumbersome. We will use software instead. Shown below are the bottom two lines of the Minitab display. (See Exercise 11 for the correction to be used when there are many tied sample values.)

MINITAB

```
H = 2.78    DF = 6    P = 0.836
H = 3.85    DF = 6    P = 0.697 (adjusted for ties)
```

We fail to reject the null hypothesis of identical distributions because the Minitab P-value is greater than a reasonable significance level of 0.05. STATDISK yields a test statistic of $H = 2.7806$, a critical value of 12.592, and includes the conclusion of "fail to reject the null hypothesis." There isn't enough evidence to support a claim that rainfall amounts on the seven week days have distributions that are not all the same. The rainfall amounts appear to be the same on the different days of the week.

INTERPRETATION Based on the Boston rainfall amounts, there does not appear to be evidence to support the claim that it rains more on weekends. So how did the newspaper, magazine, and television reports lead us to believe that it does rain more on weekends? The original study was conducted by Arizona State University scientists Randall S. Cerveny, and Robert C. Balling. Can they be blamed for providing false information? No. The author contacted Randall Cerveny who stated that the original paper concerned rainfall *off the coast* of the Atlantic seaboard—not the rainfall associated with any particular city. Cerveny and Balling used satellite precipitation estimates and found that *areas in the ocean* and near the coast did get more rainfall on weekends, and they explain this phenomenon by its relationship with pollution coming from coastal regions. Their findings are interesting and significant. The media misinterpreted the Cerveny/Balling conclusions with reports implying that it rains more on weekends for those of us who live along the Atlantic coast. It's an interesting case of media distortion.

Rationale: The test statistic H, as presented earlier, is the rank version of the test statistic F used in the analysis of variance discussed in Chapter 11. When we deal with ranks R instead of original values x, many components are predetermined. For example, the sum of all ranks can be expressed as $N(N + 1)/2$, where N is the total number of values in all samples combined. The expression

$$H = \frac{12}{N(N+1)} \Sigma n_i (\overline{R}_i - \overline{\overline{R}})^2$$

where

$$\overline{R}_i = \frac{R_i}{n_i} \qquad \overline{\overline{R}} = \frac{\Sigma R_i}{\Sigma n_i}$$

combines weighted variances of ranks to produce the test statistic H given here. This expression for H is algebraically equivalent to the expression for H given earlier as the test statistic. The earlier form of H (not the one given here) is easier to work with. In comparing the procedures of the parametric F test for analysis of variance and the nonparametric Kruskal-Wallis test, we see that in the absence of computer software, the Kruskal-Wallis test is much simpler to apply. We need not compute the sample variances and sample means. We do not require normal population distributions. Life becomes so much easier. However, the Kruskal-Wallis test is not as efficient as the F test, so it might require more dramatic differences for the null hypothesis to be rejected.

Using Technology

STATDISK: Select **Analysis** from the main menu bar, then select **Kruskal-Wallis Test** and proceed to enter or copy the sample data into the dialog box. STATDISK will display the sum of the ranks for each sample, the H test statistic, the critical value, and the conclusion.

Minitab: Refer to the *Minitab Student Laboratory Manual and Workbook* for the procedure required to use the options **Stat, Nonparametrics,** and **Kruskal-Wallis.** The basic idea is to list all of the sample data in one big column, with another column identifying the sample for the corresponding values. For the head injury data of Table 13-5, enter the 20 measurements in Minitab's column C1; enter the five subcompact values, followed by the five compact values, and so on. In column C2, enter 1, 1, 1, 1, 1, 2, 2, 2, 2, 2, 3, 3, 3, 3, 3, 4, 4, 4, 4, 4. Now select **Stat, Nonparametrics,** and **Kruskal-Wallis.** In the dialog box, enter C1 for response, C2 for factor, then click OK. The Minitab display includes the H test statistic and the P-value.

Excel: Excel is not programmed for the Kruskal-Wallis test.

TI-83 Plus: The TI-83 Plus calculator is not programmed for the Kruskal-Wallis test.

13-5 Basic Skills and Concepts

Using the Kruskal-Wallis Test. *In Exercises 1–8, use the Kruskal-Wallis test.*

1. Is Old Faithful Changing Over Time? The accompanying table below lists the time intervals (in minutes) between eruptions of the Old Faithful geyser for four different years. The table also includes the ranks in parentheses and the sample sizes and rank sums at the bottom. Use the Kruskal-Wallis test to test the null hypothesis that the different years have time intervals with identical populations.

1951		1985		1995		1996	
74	(21)	89	(40)	86	(34.5)	88	(37.5)
60	(8)	90	(42)	86	(34.5)	86	(34.5)
74	(21)	60	(8)	62	(12)	85	(30.5)
42	(1)	65	(15.5)	104	(48)	89	(40)
74	(21)	82	(26)	62	(12)	83	(27)
52	(2)	84	(28)	95	(47)	85	(30.5)
65	(15.5)	54	(3)	79	(24)	91	(43.5)
68	(18.5)	85	(30.5)	62	(12)	68	(18.5)
62	(12)	58	(6)	94	(45.5)	91	(43.5)
66	(17)	79	(24)	79	(24)	56	(4)
62	(12)	57	(5)	86	(34.5)	89	(40)
60	(8)	88	(37.5)	85	(30.5)	94	(45.5)
$n_1 = 12$		$n_2 = 12$		$n_3 = 12$		$n_4 = 12$	
$R_1 = 157$		$R_2 = 265.5$		$R_3 = 358.5$		$R_4 = 395$	

Data from geologist Rick Hutchinson and the National Park Service.

2. Does Weight of a Car Affect Chest Injuries in a Crash? In an example in this section, we used the head injury measurements from Table 13-5 to test for differences in the four car weight categories. Use the sample data in Table 13-5 to test for differences in chest deceleration measurements among the four weight categories. Is there sufficient evidence to conclude that chest deceleration measurements for the four car weight categories are not all the same? If not, does this show that the weight of a car appears to have no effect on chest deceleration in a crash?

3. Does Weight of a Car Affect Leg Injuries in a Crash? In an example in this section, we used the head injury measurements from Table 13-5 to test for differences in the four car weight categories. Use the sample data in Table 13-5 to test for differences in left femur load measurements among the four weight categories. Is there sufficient evidence to conclude that left femur loads for the four car weight categories are not all the same? If not, does this show that the weight of a car appears to have no effect on left femur load in a crash?

4. Is Solar Energy the Same Every Day? A student of the author lives in a home with a solar electric system. At the same time each day, she collected voltage readings from a meter connected to the system and the results are listed in the accompanying table. Use a 0.05 significance level to test the claim that voltage readings are the same for the three different types of day. Is there sufficient evidence to support a claim of different population distributions? We might expect that a solar system would provide more electrical energy on sunny days than on cloudy or rainy days. Can we conclude that sunny days result in greater amounts of electrical energy?

Sunny Days	Cloudy Days	Rainy Days
13.5	12.7	12.1
13.0	12.5	12.2
13.2	12.6	12.3
13.9	12.7	11.9
13.8	13.0	11.6
14.0	13.0	12.2

5. Testing for Skull-Breadth Differences in Different Times The accompanying values are measured maximum breadths of male Egyptian skulls from different epochs. Changes in head shape over time suggest that interbreeding occurred with immigrant populations. Use a 0.05 significance level to test the claim that the three samples come from identical populations.

4000 B.C.	1850 B.C.	150 A.D.
131	129	128
138	134	138
125	136	136
129	137	139
132	137	141
135	129	142
132	136	137
134	138	145
138	134	137

Based on data from *Ancient Races of the Thebaid*, by Thomson and Randall-Maciver.

Laboratory				
1	2	3	4	5
2.9	2.7	3.3	3.3	4.1
3.1	3.4	3.3	3.2	4.1
3.1	3.6	3.5	3.4	3.7
3.7	3.2	3.5	2.7	4.2
3.1	4.0	2.8	2.7	3.1
4.2	4.1	2.8	3.3	3.5
3.7	3.8	3.2	2.9	2.8
3.9	3.8	2.8	3.2	
3.1	4.3	3.8	2.9	
3.0	3.4	3.5		
2.9	3.3			

Data provided by Minitab, Inc.

6. Laboratory Testing of Flammability of Children's Sleepwear Flammability tests were conducted on children's sleepwear. The Vertical Semirestrained Test was used, in which pieces of fabric were burned under controlled conditions. After the burning stopped, the length of the charred portion was measured and recorded. Results are given in the margin for the same fabric tested at different laboratories. Because the same fabric was used, the different laboratories should have obtained the same results. Did they?

7. Do All Colors of M&Ms Weigh the Same? Refer to Data Set 10 in Appendix B. At the 0.05 significance level, test the claim that the weights of M&Ms are the same for each of the six different color populations. If it is the intent of Mars, Inc., to make the candies so that the different color populations are the same, do your results suggest that the company has a problem that requires corrective action?

8. Testing the Effect of Second-Hand Smoke Refer to Data Set 9 in Appendix B. Use a 0.01 significance level to test the claim that the populations of cotinine levels are identical for these three groups: nonsmokers who are not exposed to environmental tobacco smoke, nonsmokers who are exposed to tobacco smoke, and people who smoke. What do the results suggest about second-hand smoke?

13-5 Beyond the Basics

9. Testing the Effect of Transforming the Sample Data

a. In general, how is the value of the test statistic H affected if a constant is added to (or subtracted from) each sample value?

b. In general, how is the value of the test statistic H affected if each sample value is multiplied (or divided) by a positive constant?

c. In general, how is the value of the test statistic H affected if a single sample value is changed to a value that causes it to become an outlier?

10. Finding Values of the Test Statistic For three samples, each of size 5, find the largest and smallest possible values of the test statistic H.

11. Correcting the Test Statistic for Ties In using the Kruskal-Wallis test, there is a correction factor that should be applied whenever there are many ties: Divide H by

$$1 - \frac{\Sigma T}{N^3 - N}$$

For each group of tied observations, calculate $T = t^3 - t$, where t is the number of observations that are tied within the individual group. Find t for each group of tied values, then compute the value of T for each group, then add the T values to get ΣT. The total number of observations in all samples combined is N. For Exercise 1 presented in this section, use this procedure to find the corrected value of H. Does the corrected value of H differ substantially from the value found in Exercise 1?

12. Equivalent Tests Show that for the case of two samples, the Kruskal-Wallis test is equivalent to the Wilcoxon rank-sum test. This can be done by showing that for the case of two samples, the test statistic H equals the square of the test statistic z used in the Wilcoxon rank-sum test. Also note that with 1 degree of freedom, the critical values of χ^2 correspond to the square of the critical z score.

13-6 Rank Correlation

In this section we describe how the nonparametric method of rank correlation is used with paired data to test for an association between two variables. In Chapter 9 we used paired sample data to compute values for the linear correlation coefficient r, but in this section we use *ranks* as the basis for measuring the strength of the correlation between two variables.

DEFINITION

The **rank correlation test** (or **Spearman's rank correlation test**) is a nonparametric test that uses ranks of sample data consisting of matched pairs. It is used to test for an association between two variables, so the null and alternative hypotheses are as follows (where ρ_s denotes the rank correlation coefficient for the entire population):

H_0: $\rho_s = 0$ (There is *no* correlation between the two variables.)

H_1: $\rho_s \neq 0$ (There is a correlation between the two variables.)

Assumptions

1. The sample data have been randomly selected.
2. Unlike the parametric methods of Section 9-2, there is *no* requirement that the sample pairs of data have a bivariate normal distribution (as described in Section 9-2). There is *no* requirement of a normal distribution for any population.

Advantages: Rank correlation has some distinct advantages over the parametric methods discussed in Chapter 9:

1. The nonparametric method of rank correlation can be used in a wider variety of circumstances than the parametric method of linear correlation. With rank correlation, we can analyze paired data that are ranks or can be converted to ranks. For example, if two judges rank 30 different gymnasts, we can use rank correlation, but not linear correlation.
2. Rank correlation can be used to detect some (not all) relationships that are not linear. (An example will be given later in this section.)
3. The computations for rank correlation are much simpler than the computations for linear correlation, as can be readily seen by comparing

Direct Link Between Smoking and Cancer

When we find a statistical correlation between two variables, we must be extremely careful to avoid the mistake of concluding that there is a cause-effect link. The tobacco industry has consistently emphasized that correlation does not imply causality. However, Dr. David Sidransky of Johns Hopkins University now says that "we have such strong molecular proof that we can take an individual cancer and potentially, based on the patterns of genetic change, determine whether cigarette smoking was the cause of that cancer." Based on his findings, he also said that "The smoker had a much higher incidence of the mutation, but the second thing that nailed it was the very distinct pattern of mutations . . . so we had the smoking gun." Although statistical methods cannot prove that smoking *causes* cancer, such proof can be established with physical evidence of the type described by Dr. Sidransky.

the formulas used to compute these statistics. If calculators or computers are not available, the rank correlation coefficient is easier to compute.

Disadvantage: A disadvantage of rank correlation is its efficiency rating of 0.91, as described in Section 13-1. This efficiency rating shows that with all other circumstances being equal, the nonparametric approach of rank correlation requires 100 pairs of sample data to achieve the same results as only 91 pairs of sample observations analyzed through the parametric approach, assuming the stricter requirements of the parametric approach are met.

We will use the following notation, which closely parallels the notation used in Chapter 9 for linear correlation. (Recall from Chapter 9 that r denotes the linear correlation coefficient for sample paired data, ρ denotes the linear correlation coefficient for all paired data in the entire population, and n denotes the number of pairs of data.)

Notation

r_s = rank correlation coefficient for sample paired data (r_s is a sample statistic)

ρ_s = rank correlation coefficient for all the population data (ρ_s is a population parameter)

n = number of pairs of sample data

d = difference between ranks for the two values within a pair

We use the notation r_s for the rank correlation coefficient so that we don't confuse it with the linear correlation coefficient r. The subscript s has nothing to do with standard deviation; it is used in honor of Charles Spearman (1863–1945), who originated the rank correlation approach. In fact, r_s is often called **Spearman's rank correlation coefficient.**

The procedure for using the rank correlation procedure is summarized in Figure 13-4. That figure includes the following test statistic.

Test Statistic for the Rank Correlation Coefficient

$$r_s = 1 - \frac{6\Sigma d^2}{n(n^2 - 1)}$$

where each value of d is a difference between the ranks for a pair of sample data.

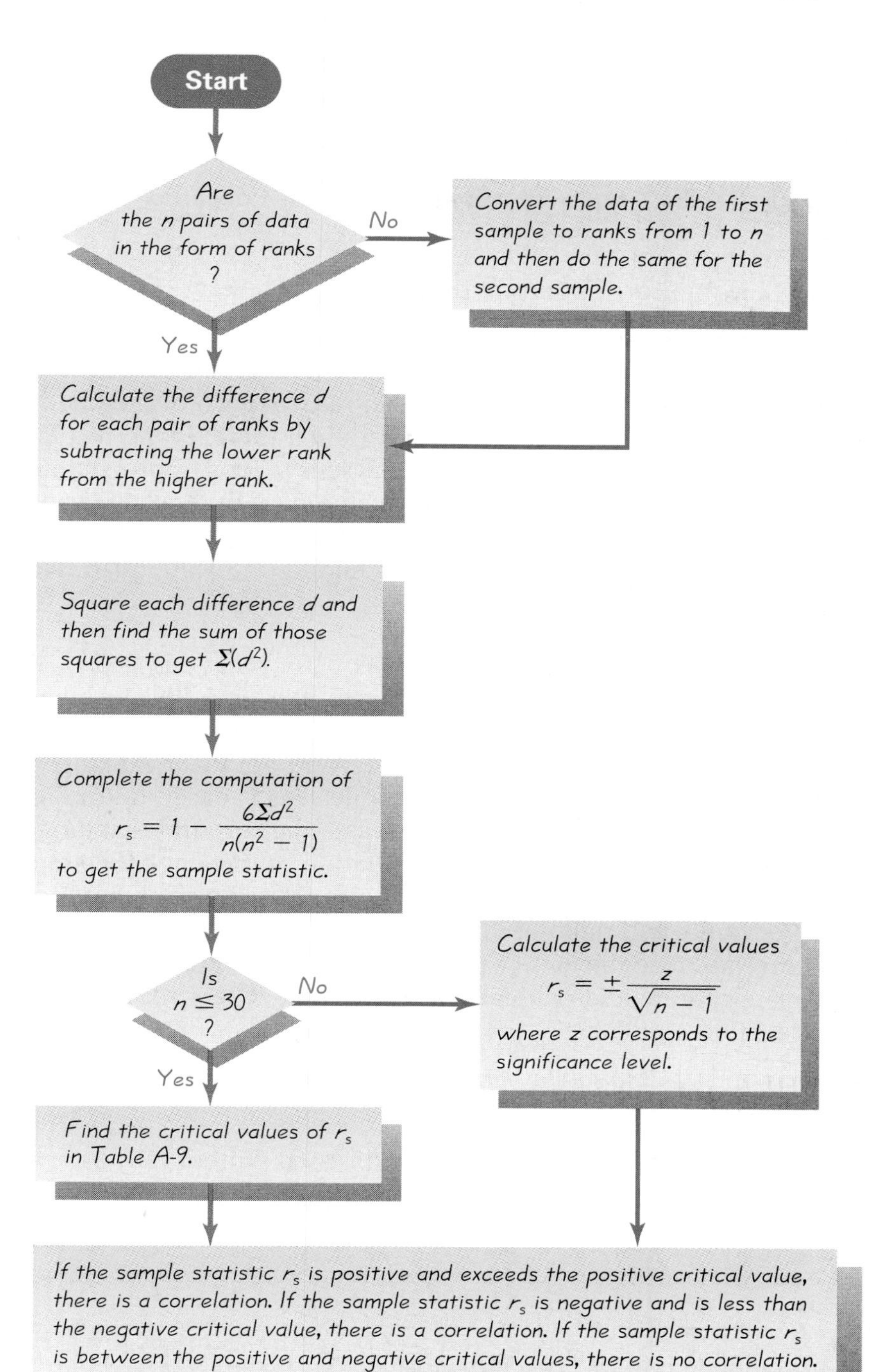

FIGURE 13-4
Rank Correlation for Testing H_0: $\rho_s = 0$

Critical values:

1. If $n \leq 30$, critical values are found in Table A-9.
2. If $n > 30$, critical values of r_s are found by using Formula 13-1.

Formula 13-1 $$r_s = \frac{\pm z}{\sqrt{n-1}} \quad \text{(critical values when } n > 30\text{)}$$

where the value of z corresponds to the significance level.

Handling Ties: Ties in ranks of the original sample values can be handled as in the preceding sections of this chapter: Find the mean of the ranks involved in the tie, and then assign the mean rank to each of the tied items. This test statistic yields the exact value of r_s only if there are no ties. With a relatively small number of ties, the test statistic is a good approximation of r_s. (When ties occur, we can get an exact value of r_s by first replacing the original sample values by their corresponding ranks; then we use Formula 9-1 for the linear correlation coefficient. After finding the value of r_s, we can proceed with the methods of this section.)

EXAMPLE **Business School Rankings** *Business Week* magazine ranked business schools two different ways. Corporate rankings were based on surveys of corporate recruiters, and graduate rankings were based on surveys of MBA graduates. Table 13-6 is based on the results for 10 schools. Is there a correlation between the corporate rankings and the graduate rankings? The linear correlation coefficient r (Section 9-2) should not be used because it requires normal distributions, but the data consist of ranks, and ranks are not normally distributed. Instead, use the rank correlation coefficient to test the claim that there is a relationship between corporate and graduate rankings (that is, $\rho_s \neq 0$). Use a significance level of $\alpha = 0.05$.

SOLUTION We follow the same basic steps for testing hypotheses as outlined in Figure 7-5.

Step 1: The claim of a correlation is expressed symbolically as $\rho_s \neq 0$.

Step 2: The negation of the claim in Step 1 is $\rho_s = 0$.

TABLE 13-6 Rankings of Business Schools

School	PA	NW	Chi	Sfd	Hvd	MI	IN	Clb	UCLA	MIT
Corporate ranking	**1**	**2**	**4**	**5**	**3**	**6**	**8**	**7**	**10**	**9**
Graduate ranking	**3**	**5**	**4**	**1**	**10**	**7**	**6**	**8**	**2**	**9**
d	2	3	0	4	7	1	2	1	8	0
d^2	4	9	0	16	49	1	4	1	64	0 → Total = 148

Step 3: Because the null hypothesis must contain the condition of equality, we have

H_0: $\rho_s = 0$

H_1: $\rho_s \neq 0$ (Original claim)

Step 4: The significance level is $\alpha = 0.05$.

Step 5: As we noted, we cannot use the linear correlation approach of Section 9-2 because ranks do not satisfy the requirement of a normal distribution. We will use the rank correlation approach instead.

Step 6: We now find the value of the test statistic. Table 13-6 shows the calculation of the differences d and their squares d^2, which results in a value of $\Sigma d^2 = 148$. With $n = 10$ (for 10 pairs of data) and $\Sigma d^2 = 148$, we can find the value of the test statistic r_s as follows:

$$r_s = 1 - \frac{6\Sigma d^2}{n(n^2 - 1)} = 1 - \frac{6(148)}{10(10^2 - 1)}$$

$$= 1 - \frac{888}{990} = 0.103$$

Now we refer to Table A-9 to determine that the critical values are ± 0.648 (based on $\alpha = 0.05$ and $n = 10$). Because the test statistic $r_s = 0.103$ does not exceed the critical value of 0.648, we fail to reject the null hypothesis. There is not sufficient evidence to support the claim of a correlation between corporate and graduate rankings of business schools. It appears that corporate recruiters and business school graduates have different perceptions of the qualities of the schools.

Manatees Saved

Manatees are large mammals that like to float just below the water's surface, where they are in danger from powerboat propellers. A Florida study of the number of powerboat registrations and the numbers of accidental manatee deaths confirmed that there was a significant positive correlation. As a result, Florida created coastal sanctuaries where powerboats are prohibited so that manatees could thrive. (See Program 1 from the series *Against All Odds: Inside Statistics* for a discussion of this case.) This is one of many examples of the beneficial use of statistics.

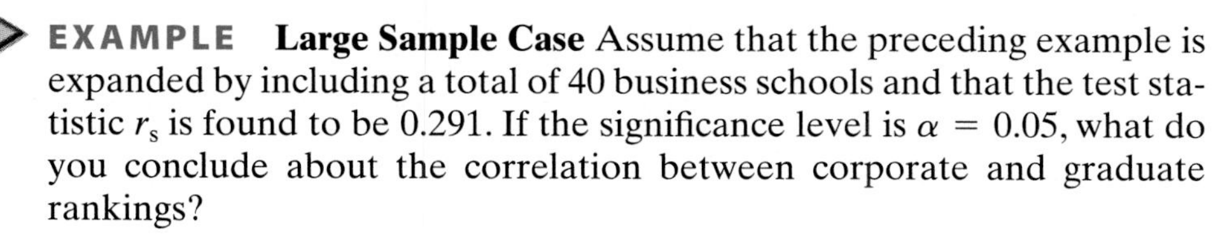

EXAMPLE **Large Sample Case** Assume that the preceding example is expanded by including a total of 40 business schools and that the test statistic r_s is found to be 0.291. If the significance level is $\alpha = 0.05$, what do you conclude about the correlation between corporate and graduate rankings?

SOLUTION Because there are 40 pairs of data, we have $n = 40$. Because n exceeds 30, we find the critical values from Formula 13-1 instead of Table A-9. With $\alpha = 0.05$ in two tails, we let $z = 1.96$ to get

$$r_s = \frac{\pm 1.96}{\sqrt{40 - 1}} = \pm 0.314$$

The test statistic of $r_s = 0.291$ does not exceed the critical value of 0.314, so we fail to reject the null hypothesis. There is not sufficient evidence to support the claim of a correlation between the corporate rankings and graduate rankings.

The next example is intended to illustrate the principle that rank correlation can sometimes be used to detect relationships that are not linear.

EXAMPLE **Learning Tasks** A *Raiders of the Lost Ark* pinball machine (model L-7) is used to measure learning that results from repeating manual functions. Subjects were selected so that they are similar in important characteristics of age, gender, intelligence, education, and so on. Table 13-7 lists the numbers of games played and the last scores (in millions) for subjects randomly selected from the group with similar characteristics. We expect that there should be an association between the number of games played and the pinball score. Is there sufficient evidence to support the claim that there is such an association?

SOLUTION

We will test the null hypothesis of no rank correlation ($\rho_s = 0$).

H_0: $\rho_s = 0$ (no correlation)
H_1: $\rho_s \neq 0$ (correlation)

Refer to Figure 13-4, which we follow in this solution. The original scores are not ranks, so we converted them to ranks and entered the results in parentheses in Table 13-7. (Section 13-1 describes the procedure for converting scores into ranks.)

After expressing all data as ranks, we calculate the differences, d, and then square them. The sum of the d^2 values is 4. We now calculate

$$r_s = 1 - \frac{6\Sigma d^2}{n(n^2 - 1)} = 1 - \frac{6(4)}{10(10^2 - 1)}$$
$$= 1 - \frac{24}{990} = 0.976$$

Proceeding with Figure 13-4, we have $n = 10$, so we answer yes when asked if $n \leq 30$. We use Table A-9 to get the critical values of ± 0.648. Finally, the sample statistic of 0.976 exceeds 0.648, so we conclude that

TABLE 13-7 Pinball Scores (Ranks in parentheses)

Number of games played	9 (4)	13 (6)	21 (7)	5 (1)	6 (2)	25 (8)	7 (3)	33 (9)	11 (5)	104 (10)
Score	22 (3)	62 (6)	70 (7)	2 (1)	10 (2)	82 (9)	26 (4)	78 (8)	58 (5)	86 (10)
d	1	0	0	0	0	1	1	1	0	0
d^2	1	0	0	0	0	1	1	1	0	0

there is significant correlation. Higher numbers of games played appear to be associated with higher scores. Subjects appeared to better learn the game by playing more.

In the preceding example, if we compute the linear correlation coefficient r (using Formula 9-1) for the original data, we get $r = 0.629$, which leads to the conclusion that there is not enough evidence to support the claim of a significant linear correlation at the 0.05 significance level. If we examine the Excel scatter diagram, we can see that the pattern of points is not a straight-line pattern. This last example illustrates these two advantages of the nonparametric approach over the parametric approach: (1) With rank correlation, we can sometimes detect relationships that are not linear; and (2) Spearman's rank correlation coefficient r_s is less sensitive to an outlier, such as the subject who played 104 games.

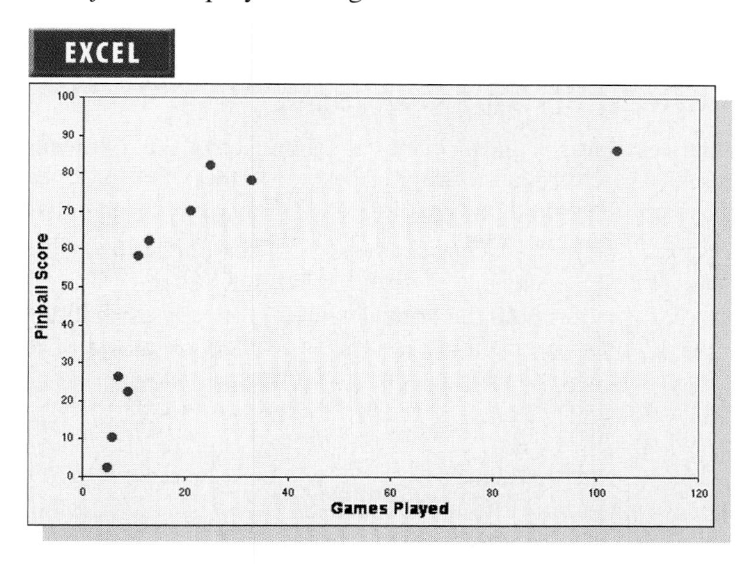

Using Technology

STATDISK: Select **Analysis** from the main menu bar, then select **Rank Correlation.** Enter the paired sample data in the dialog box, then click **Evaluate.** The STATDISK results include the test statistic, critical value, and conclusion.

Minitab: Enter the paired data in columns C1 and C2. If the data are not already ranks, use Minitab's **Manip** and **Rank** options to convert the data to ranks, then select **Stat,** followed by **Basic Statistics,** followed by **Correlation.** Minitab will display a one-line statement of the value for r_s, such as the following:
`Correlation of C1 and C2 = 0.976.`

Excel: Excel does not have a function that calculates the rank correlation coefficient from original sample values, but that value can be found as follows. First replace each of the original sample values by its corresponding rank. Enter those ranks in columns A and B. Click on the **fx** function key located on the main menu bar. Select the function category **Statistical** and the function name **CORREL,** then click **OK.** In the dialog box, enter the cell range of values for x, such as A1:A6. Also enter the cell range of values for y, such as B1:B6. Excel will display the value of the rank correlation coefficient.

TI-83 Plus: If using a TI-83 Plus calculator or any other calculator with 2-variable statistics, you can find the value of r_s as follows: (1) Replace each sample value by its corresponding rank, then (2) calculate the value of the linear correlation coefficient r with the same procedures used in Section 9-2. Enter the paired ranks in lists L1 and L2, then press **STAT** and select **TESTS.** Using the option **LinRegTTest** will result in several displayed values, including the value of the correlation coefficient.

13-6 Basic Skills and Concepts

1. Finding the Test Statistic and Critical Value For each of the following samples of paired ranks, sketch a scatter diagram, estimate the value of r_s, calculate the value of r_s, and state whether there appears to be a correlation between x and y.

a.

x	1	3	5	4	2
y	1	3	5	4	2

b.

x	1	2	3	4	5
y	5	4	3	2	1

c.

x	1	2	3	4	5
y	2	5	3	1	4

2. Finding Critical Values Find the critical value(s) for r_s by using either Table A-9 or Formula 13-1, as appropriate. Assume two-tailed cases, where α represents the significance level and n represents the number of pairs of data.

a. $n = 20,\ \alpha = 0.05$ **b.** $n = 50,\ \alpha = 0.05$
c. $n = 40,\ \alpha = 0.02$ **d.** $n = 15,\ \alpha = 0.01$
e. $n = 82,\ \alpha = 0.04$

Testing for Rank Correlation. *In Exercises 3–14, use the rank correlation coefficient to test for a correlation between the two variables. Use a significance level of $\alpha = 0.05$.*

3. Correlation Between Salary and Stress The accompanying table lists salary rankings and stress rankings for randomly selected jobs. Does it appear that salary increases as stress increases?

Job	Salary Rank	Stress Rank
Stockbroker	2	2
Zoologist	6	7
Electrical engineer	3	6
School principal	5	4
Hotel manager	7	5
Bank officer	10	8
Occ. safety inspector	9	9
Home economist	8	10
Psychologist	4	3
Airline pilot	1	1

Based on data from *The Jobs Rated Almanac.*

4. Correlation Between Salary and Physical Demand Exercise 3 includes paired salary and stress level ranks for 10 randomly selected jobs. The physical demands of the jobs were also ranked; the salary and physical demand ranks are given below. Does there appear to be a relationship between the salary of a job and its physical demands?

Salary	2	6	3	5	7	10	9	8	4	1
Physical demand	5	2	3	8	10	9	1	7	6	4

Based on data from *The Jobs Rated Almanac.*

5. Correlation Between Stress and Physical Demand Ten jobs were randomly selected and ranked according to stress level and physical demand, with the results given below. Does there appear to be a relationship between the stress levels of jobs and their physical demands?

Stress level	2	7	6	4	5	8	9	10	3	1
Physical demand	5	2	3	8	10	9	1	7	6	4

Based on data from *The Jobs Rated Almanac.*

6. Correlation Between Weapons and Murder Rate The accompanying table lists rankings for the number of registered automatic weapons (in thousands), along with the murder rate (in murders per 100,000), for randomly selected states. Automatic weapons are guns that continue to fire repeatedly while the trigger is held back. Are firearm murders often committed with automatic weapons? Does a significant correlation imply that increased numbers of automatic weapons result in more murders?

Automatic weapons	8	7	5	1	6	3	2	4
Murder rate	8	6	5	2	7	4	1	3

Data provided by the FBI and the Bureau of Alcohol, Tobacco, and Firearms.

7. Correlation Between Restaurant Bills and Tips In Section 9-2 we included an example in which we tested for a correlation between the amount of the restaurant bill and the amount of the tip that was left. The sample data are reproduced here (based on data from students of the author). Use rank correlation to determine whether there is a correlation between the amount of the bill and the amount of the tip.

Bill (dollars)	33.46	50.68	87.92	98.84	63.60	107.34
Tip (dollars)	5.50	5.00	8.08	17.00	12.00	16.00

8. Correlation Between Heights and Weights of Supermodels Listed below are heights (in inches) and weights (in pounds) for supermodels Niki Taylor, Nadia Avermann, Claudia Schiffer, Elle MacPherson, Christy Turlington, Bridget Hall, Kate Moss, Valerie Mazza, and Kristy Hume.

Height	71	70.5	71	72	70	70	66.5	70	71
Weight	125	119	128	128	119	127	105	123	115

9. Correlation Between Age and BAC of DWI Prisoners A study was conducted to investigate a relationship between age (in years) and BAC (blood alcohol concentration) measured when convicted DWI jail inmates were first arrested. Sample data are given on the next page for randomly selected subjects. Based on the result, does the BAC level seem to be related to the age of the person tested?

Age	17.2	43.5	30.7	53.1	37.2	21.0	27.6	46.3
BAC	0.19	0.20	0.26	0.16	0.24	0.20	0.18	0.23

Based on data from the Dutchess County STOP-DWI Program.

10. Correlation Between Car Weight and Fuel Consumption The accompanying table lists weights (in hundreds of pounds) and highway fuel consumption amounts (in mi/gal) for a sample of domestic new cars. Based on the result, can you expect to pay more for gas if you buy a heavier car? How do the results change if the weights are entered as 2900, 3500, . . . , 2400?

x Weight	29	35	28	44	25	34	30	33	28	24
y Fuel	31	27	29	25	31	29	28	28	28	33

Based on data from the EPA.

11. Old Faithful Geyser Eruption Data Refer to Data Set 11 in Appendix B.

a. Use the paired data for durations and intervals after eruptions of the geyser. Is there a significant correlation, suggesting that the interval after an eruption is related to the duration of the eruption?

b. Use the paired data for intervals after eruptions and heights of eruptions of the Old Faithful geyser. Is there a significant correlation, suggesting that the interval after an eruption is related to the height of the eruption?

c. Assume that you want to develop a method for predicting the time interval to the next eruption. Based on the results from parts (a) and (b), which factor would be more relevant: eruption duration or eruption height? Why?

12. Cigarette Nicotine, Tar, Carbon Monoxide Refer to Data Set 8 in Appendix B.

a. Use the paired data consisting of tar and nicotine. Based on the result, does there appear to be a significant correlation between cigarette tar and nicotine? If so, can researchers reduce their laboratory expenses by measuring only one of these two variables?

b. Use the paired data consisting of carbon monoxide and nicotine. Based on the result, does there appear to be a significant correlation between cigarette nicotine and carbon monoxide? If so, can researchers reduce their laboratory expenses by measuring only one of these two variables?

c. Assume that researchers want to develop a method for predicting the amount of nicotine, and they want to measure only one other item. In choosing between tar and carbon monoxide, which is the better choice? Why?

13. Diamond Prices, Weights, Color Refer to Data Set 3 in Appendix B.

a. Use the paired data consisting of price and carat (weight). Is there a significant correlation between the price of a diamond and its weight in carats?

b. Use the paired price/color data. Is there a significant correlation between the price of a diamond and its color?

c. Assume that you are planning to buy a diamond engagement ring. In considering the value of a diamond, which characteristic should you consider to be more important: the carat weight or the color? Why?

14. Movie Gross, Budget, and Viewer Rating Refer to Data Set 15 in Appendix B.

a. Use the paired data consisting of the amount of money grossed and the amount of money budgeted. Is there a significant correlation between those two variables?

b. Use the paired data consisting of the amount of money grossed and the viewer rating. Is there a significant correlation between those two variables?

(continued)

c. Based on the preceding results, if you plan to invest in a company that makes movies, which is the better choice: a company that makes big budget movies or a company that makes movies that viewers rate highly? Why?

13-6 Beyond the Basics

15. Finding Critical Values One alternative to using Table A-9 to find critical values is to compute them using this approximation:

$$r_s = \pm\sqrt{\frac{t^2}{t^2 + n - 2}}$$

Here t is the t score from Table A-3 corresponding to the significance level and $n - 2$ degrees of freedom. Apply this approximation to find critical values of r_s for the following cases.

a. $n = 8,\ \alpha = 0.05$
b. $n = 15,\ \alpha = 0.05$
c. $n = 30,\ \alpha = 0.05$
d. $n = 30,\ \alpha = 0.01$
e. $n = 8,\ \alpha = 0.01$

16. Effects of Scales and Rankings

a. How is r_s affected if the scale for one of the variables is changed from feet to inches?

b. How is r_s affected if one variable is ranked from low to high while the other variable is ranked from high to low?

c. How is r_s affected if the two variables are interchanged?

d. One researcher ranks both variables from low to high, while another researcher ranks both variables from high to low. How will their values of r_s compare?

e. How is r_s affected if each value of x is replaced by $\log x$?

13-7 Runs Test for Randomness

The main objective of this section is to understand how the runs test for randomness can be used to determine whether the sample data in a sequence are in a random order. The importance of randomness has been stressed throughout this book, and we now address one method for determining whether that characteristic is present.

DEFINITIONS

A **run** is a sequence of data having the same characteristic; the sequence is preceded and followed by data with a different characteristic or by no data at all.

The **runs test** uses the number of runs in a sequence of sample data to test for randomness in the order of the data.

Assumptions

1. The sample data are arranged according to some ordering scheme, such as the order in which the sample values were obtained.

Sports Hot Streaks

It is a common belief that athletes often have "hot streaks"—that is, brief periods of extraordinary success. Stanford University psychologist Amos Tversky and other researchers used statistics to analyze the thousands of shots taken by the Philadelphia 76ers for one full season and half of another. They found that the number of "hot streaks" was no different than you would expect from random trials with the outcome of each trial independent of any preceding results. That is, the probability of a hit doesn't depend on the preceding hit or miss.

2. Each data value can be categorized into one of *two* separate categories.
3. The runs test for randomness is based on the *order* in which the data occur; it is *not* based on the *frequency* of the data. (For example, a sequence of 3 men and 20 women might appear to be random, but the issue of whether 3 men and 20 women constitute a *biased* sample is not addressed by the runs test.)

Notation

n_1 = number of elements in the sequence that have one particular characteristic (The characteristic chosen for n_1 is arbitrary.)

n_2 = number of elements in the sequence that have the other characteristic

G = number of runs

EXAMPLE **Market Research** Manufacturers of cola conduct market research to determine consumer preferences. When 10 consumers are asked whether they prefer diet cola or regular cola, the responses of the first 10 subjects are listed in the order in which they were obtained. We use D to denote a consumer who prefers *diet* cola and R to denote a consumer who prefers *regular* cola. Find the number of runs G, and also identify the values of n_1 and n_2.

$$\underbrace{\text{D D D D}}_{\text{1st run}}\quad\underbrace{\text{R R}}_{\text{2nd run}}\quad\underbrace{\text{D D D}}_{\text{3rd run}}\quad\underbrace{\text{R}}_{\text{4th run}}$$

SOLUTION There are exactly four runs, as shown, so $G = 4$. Letting n_1 represent the number of diet (D) colas, we have $n_1 = 7$ because there are seven diet colas present. It follows that the number of regular (R) colas is described by $n_2 = 3$.

Fundamental Principle of the Runs Test

The fundamental principle of the runs test can be briefly stated as follows:

Reject randomness if the number of runs is very low or very high.

- Example: DDDDDRRRRR is not random because it has 2 runs, so the number of runs is very *low*.
- Example: DRDRDRDRDR is not random because there are 10 runs, which is very *high*.

We reject randomness if the number of runs G is too small or too large, but how do we determine exactly which values of G are too small or too large?

We use the following criterion:

5% Cutoff Criterion

Reject randomness if the number of runs G is so small or so large that in repeated samplings, a value at least as extreme as G will occur 5% of the time or less.

Although this criterion wins no prizes for simplicity, it's quite easy to apply if we use Table A-10. Table A-10 identifies those values of G that are so small or so large that they belong to the category of exceptional sequences that happen 5% of the time or less. Using Table A-10, the 5% criterion can be simply restated as follows:

Simplified 5% Cutoff Criterion

Reject randomness if the number of runs G is

- **less than or equal to the smaller entry in Table A-10**
- **or greater than or equal to the larger entry in Table A-10.**

The sequence

D D D D R R R R R

results in these values: $n_1 = 4$, $n_2 = 5$, and $G = 2$ runs. With $n_1 = 4$ and $n_2 = 5$, Table A-10 indicates that we should reject randomness if the number of runs is 2 or less or 9 or greater. With $G = 2$ runs, we therefore reject randomness according to the simplified 5% cutoff criterion based on Table A-10.

Rationale: How do we find the cutoff values separating random sequences from those that are not? That is, how do we find the critical values in Table A-10? Let's consider the sequence DDDDRRRRR, with $G = 2$, $n_1 = 4$, and $n_2 = 5$. If we refer to Table A-10, we find that $n_1 = 4$ and $n_2 = 5$ correspond to cutoff values of 2 and 9. If we had an abundance of time and patience, we could list all possible sequences of 4 diet colas and 5 regular colas. Examination of that list would reveal these facts:

- There are 126 different possible sequences of 4 diet and 5 regular colas.
- Among the 126 different possible cases, 2 cases have 2 runs, 7 cases have 3 runs, and so on, as summarized in Table 13-8.

Based on the 5% cutoff criterion, and based on the 126 cases summarized in Table 13-8, we should reject randomness if the number of runs G is 2 or 9 because, with randomly selected data, we will get 2 runs or 9 runs only 2.38% of the time.

TABLE 13-8 Frequency Table of the Number of Runs

Number of runs	2	3	4	5	6	7	8	9
Frequency in 126 cases	2	7	24	30	36	18	8	1

(With only 2 cases having 2 runs and with only 1 case having 9 runs, the number of runs G is excessively low or high in 3 cases, which is 2.38% of the total number of cases. We arrived at 2.38% by converting $3 \div 126$ to 0.0238, which is equivalent to 2.38%.) It is easy to get 3, 4, 5, 6, 7, or 8 runs, because these values occur more than 95% of the time; but it is unusual to get 2 or 9 runs, because these values occur only 2.38% of the time. By analyzing the 126 cases summarized in Table 13-8, we can see how to find the cutoff values of 2 and 9 that are included in Table A-10. Although this detailed procedure for finding cutoff values is perfectly valid, it's too awkward to apply in practice, so we use Table A-10 instead.

EXAMPLE **Basketball Foul Shots** In the course of a game, WNBA player Cynthia Cooper shoots 12 free throws. Denoting shots made by H (for "hit") and denoting missed shots by M, her results are as follows: H, H, H, M, H, H, H, H, M, M, M, H. Use the (simplified) 5% cutoff criterion to test for randomness in the sequence of hits and misses.

SOLUTION

We must first find the values of n_1, n_2, and the number of runs G.

$$n_1 = \text{number of shots made (H)} = 8$$
$$n_2 = \text{number of shots missed (M)} = 4$$
$$G = \text{number of runs} = 5$$

With $G = 5$, we refer to Table A-10 to find the cutoff values of 3 and 10. Because $G = 5$ is not less than or equal to 3, nor is it greater than or equal to 10, *we do not reject randomness.* There is not sufficient evidence to warrant rejection of the claim that the hits and misses occur randomly.

Large Sample Cases

Table A-10 applies when the following three conditions are all met:

1. We are using 5% as the cutoff for sequences that have too few or too many runs,
2. $n_1 \leq 20$, and
3. $n_2 \leq 20$.

If we wish to use the runs test for randomness but $n_1 > 20$ or $n_2 > 20$ or $\alpha \neq 0.05$, we use the property that the number of runs G has a distribution that is approximately normal with mean μ_G and standard deviation σ_G described as follows:

Formula 13-2
$$\mu_G = \frac{2n_1n_2}{n_1 + n_2} + 1$$

Formula 13-3
$$\sigma_G = \sqrt{\frac{(2n_1n_2)(2n_1n_2 - n_1 - n_2)}{(n_1 + n_2)^2(n_1 + n_2 - 1)}}$$

After finding the values of μ_G and σ_G, the test statistic can be computed as $z = (G - \mu_G)/\sigma_G$. The normal approximation (with test statistic z) is quite good. If the entire table of critical values (Table A-10) had been computed using this normal approximation, no critical value would be off by more than one unit.

Test Statistic for the Runs Test for Randomness

If $\alpha = 0.05$ and $n_1 \leq 20$ and $n_2 \leq 20$, the test statistic is G.
If $\alpha \neq 0.05$ or $n_1 > 20$ or $n_2 > 20$, the test statistic is

$$z = \frac{G - \mu_G}{\sigma_G}$$

where μ_G and σ_G are from Formulas 13-2 and 13-3.

Critical values:

1. If the test statistic is G, critical values are found in Table A-10.
2. If the test statistic is z, critical values are found in Table A-2 by using the procedures introduced in Chapter 6.

Figure 13-5 on the next page summarizes the procedures for the runs test for randomness and includes cases in which the test statistic is G as well as cases in which the test statistic is z.

EXAMPLE **Boston Rainfall On Mondays** Refer to the rainfall amounts for Boston as listed in Data Set 17 in Appendix B. Is there sufficient evidence to support the claim that rain on Mondays is not random?

SOLUTION Let D (for dry) represent Mondays with no rain (indicated by values of 0.00), and let R represent Mondays with some rain (any value greater than 0.00). The 52 consecutive Mondays are represented by this sequence:

DDDDRDRDDRDDRDDDRDDRRRDDDDRDRDRRRDRDDDR
DDDRDRDDRDDDR

The null and alternative hypotheses are as follows:

H_0: The sequence is random.
H_1: The sequence is not random.

The significance level is $\alpha = 0.05$; we are using the runs test for randomness. The test statistic is obtained by first finding the number of Ds, the number of Rs, and the number of runs. It's easy to examine the sequence to find that

n_1 = number of Ds = 33
n_2 = number of Rs = 19
G = number of runs = 30

Minitab: Minitab will do a runs test with a sequence of numerical data only, but see the *Minitab Student Laboratory Manual and Workbook* for ways to circumvent that constraint. Enter numerical data in column C1, then select **Stat, Nonparametrics,** and **Runs Test.** In the dialog box, enter C1 for the variable, then either choose to test for randomness above and below the mean, or enter a value to be used. Click **OK.** The Minitab results include the number of runs and the *P*-value ("test is significant at . . . ").

Excel: Excel is not programmed for the runs test for randomness.

TI-83 PLUS: The TI-83 Plus calculator is not programmed for the runs test for randomness.

13-7 Basic Skills and Concepts

Identifying Runs and Finding Critical Values. *In Exercises 1–4, use the given sequence to determine the values of n_1, n_2, the number of runs G, and the 5% cutoff values from Table A-10, and use those results to determine whether the sequence appears to be random.*

1. D D D D D R R R R R D D D D D

2. T T T T T T T T T T T T T T T F F F F F F F F F F F F F F F

3. A A B B B A B B B B A A A A A B B B B B

4. Y Y N N Y Y N N Y Y N N Y Y N N Y Y N N Y Y N N

Using the Runs Test for Randomness. *In Exercises 5–12, use the runs test of this section to determine whether the given sequence is random. Use a significance level of $\alpha = 0.05$. (All data are listed in order by row.)*

5. Randomness of Roulette Wheel Outcomes In conducting research for this book, the author recorded the outcomes of a roulette wheel in the Stardust Casino. (Yes, it was hard work, but somebody had to do it.) Test for randomness of odd (O) and even (E) numbers for the results given in the following sequence. What would a lack of randomness mean to the author? To the casino?

O O E E E E E O O E O E O O O O O O O E O E

6. Testing for Randomness of Survey Respondents A pollster for the Newton Institute for Opinion Research was instructed to randomly select people walking through New York City's Grand Central Station and ask them several questions about brand recognition for toothpaste. She was also instructed to record the gender of each survey subject, and her results are listed below. Does it appear that the pollster is following her instructions to make random selections?

M M M M M F F F F F F M M M M M F F F F F

7. Testing for Randomness in Dating Prospects Fred has had difficulty getting dates with women, so he is abandoning his strategy of careful selection and replacing it with a desperate strategy of random selection. In pursuing dates with randomly selected women, Fred finds that some of them are unavailable because they are married. Fred, who has an abundance of time for such activities, records and analyzes his observations. Given the results listed below (where M denotes married and S denotes single), what should Fred conclude about the randomness of the women he selects?

M S S S S S S M M S M S M M S S M M S S M M M M M M M S

8. Testing for Randomness of Baseball World Series Victories Test the claim that the sequence of World Series wins by American League and National League teams is random. Given below are recent results with American and National league teams represented by A and N, respectively. What does the result suggest about the abilities of the two leagues?

A N A N A A A N N A A N N N N A A A N A N A N A A A N A N A

9. Testing for Randomness of Presidential Election Winners For a recent sequence of presidential elections, the political party of the winner is indicated by D for Democrat and R for Republican. Does it appear that we elect Democrat and Republican candidates in a sequence that is random?

R R D R D R R R R D D R R R D D D D D R R D D R R D R R R D D

10. Stock Market: Testing for Randomness Above and Below the Median Trends in business and economics applications are often analyzed with the runs test. The accompanying list shows (in order by row) the annual high points of the Dow Jones Industrial Average for a recent sequence of years. First find the median of the values, then replace each value by A if it is above the median and B if it is below the median. Then apply the runs test to the resulting sequence of As and Bs. What does the result suggest about the stock market as an investment consideration?

969	842	951	1036	1052	892	882	1015	1000	908
898	1000	1024	1071	1287	1287	1553	1956	2722	2184
2791	3000	3169	3413	3794	3978	5216	6561	8259	9338

11. Testing for Randomness of Selected Military Draft Dates Men were once drafted into the U.S. Army by using a process that was supposed to randomly select birthdays. Suppose the first few selections are as listed below. Test the sequence for randomness before and after the middle of the year.

Nov. 27	July 7	Aug. 3	Oct. 19	Dec. 19	Sept. 21	May 3
Mar. 5	June 10	May 15	June 27	Jan. 5		

12. Testing Lottery Results for Randomness Listed below are numbers selected in consecutive drawings from the Maryland Pick Three lottery. Test for randomness of odd (O) and even (E) numbers. What would lack of randomness mean for those who play the lottery?

0 0 0 7 1 3 3 6 4 6 8 6 2 4 7 7 6 9 6 2 5 6 7 7 6 1 1 3 3 3 8 2 2

13. Testing for Randomness of Odd and Even Digits in Pi A *New York Times* article about the calculation of decimal places of π noted that "mathematicians are pretty sure that the digits of π are indistinguishable from any random sequence." Given below are the first 100 decimal places of π. Test for randomness of odd (O) and even (E) digits.

1 4 1 5 9 2 6 5 3 5 8 9 7 9 3 2 3 8 4 6 2 6 4 3 3 8 3 2 7 9 5 0 2 8 8 4 1 9 7 1

6 9 3 9 9 3 7 5 1 0 5 8 2 0 9 7 4 9 4 4 5 9 2 3 0 7 8 1 6 4 0 6 2 8 6 2 0 8 9 9

8 6 2 8 0 3 4 8 2 5 3 4 2 1 1 7 0 6 7 9

14. Testing for Randomness Above and Below the Median of 4.5 in the Digits of Pi If digits 0 through 9 are generated in a way that is random, the mean and median should both be 4.5. Using the first 100 decimal places of π from Exercise 13, test for randomness above (A) and below (B) the value of 4.5.

15. Testing for Randomness Above and Below the Median in the Home-Run Distances of Mark McGwire Refer to the Mark McGwire home run-distances listed in Data Set 19 in Appendix B. Test for randomness above and below the median distance, which is 420 ft. In Chapter 2 we used a frequency table (Table 2-7) of the last digits to show that these distances were probably estimated instead of being measured. What does this runs test for randomness suggest?

16. Large Sample: Testing for Randomness of Baseball World Series Victories Test the claim that the sequence of World Series wins by American League and National League teams is random. Given below are recent results, with American and National League teams represented by A and N, respectively.

A N A N N N A A A A N A A A A N A N

N A A N N A A A A N A N N A A A A A

N A N A N A N A A A A A A A N N A N

A N N A A N N N A N A N A N A A A N

N A A N N N N A A A N A N A N A A A

N A N A

13-7 Beyond the Basics

17. Finding Critical Numbers of Runs Using the elements A, A, B, B, what is the minimum number of possible runs that can be arranged? What is the maximum number of runs? Now refer to Table A-10 to find the 5% cutoff G values for $n_1 = n_2 = 2$. What do you conclude about this case?

18. Finding Critical Values

a. Using all of the elements A, A, A, B, B, B, B, B, B, list the 84 different possible sequences.

b. Find the number of runs for each of the 84 sequences.

c. Use the results from parts (a) and (b) to find your own 5% cutoff values for G.

d. Compare your results to those given in Table A-10.

Search for Extraterrestrial Life Continues

NEW SOUTH WALES Project Phoenix is now operational with a radio telescope located in New South Wales, Australia. This is the largest radio telescope in the Southern Hemisphere. The objective is to detect radio signals being transmitted by extraterrestrial civilizations. Radio telescopes around the world monitor signals on millions of different channels. Because of the great number of the signals received, computers are used to search for signals that seem to come from an intelligent source instead of being random noise. The ability to distinguish between an actual message and random noise is essential to such projects.

Past attempts to identify extraterrestrial intelligent life have involved efforts to send radio messages carrying information about us earthlings. Dr. Frank Drake of Cornell University developed such a radio message that could be transmitted as a series of 1271 pulses and gaps. The pulses and gaps can be thought of as 1s and 0s. If we factor 1271 into the prime numbers of 41 and 31 and then make a 41×31 grid and put a dot at those positions corresponding to a pulse or 1, we get the pattern shown in the accompanying figure. This pattern contains information including the location of Earth in the solar system; the symbols for hydrogen, carbon, and oxygen; and drawings of a man, woman, child, fish, and water.

1. Examine the accompanying figure and identify the man, woman, child, fish, and ocean waves.

2. Suppose we send this message and it is intercepted by extraterrestrial intelligent life. Will they think that the pattern is random? Use the methods of this chapter to determine whether the sequence appears to be random so that it would be mistaken for "random noise."

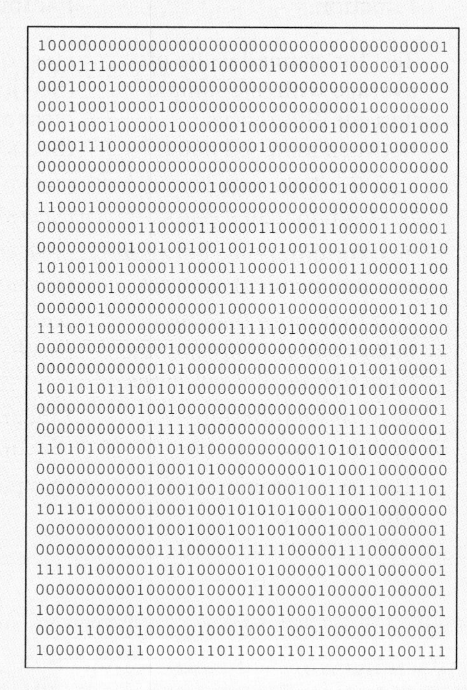

```
10000000000000000000000000000000000000001
000011100000000000100000100000010000010000
00010001000000000000000000000000000000000
00010001000010000000000000000000100000000
00010001000001000000100000000010001000
00001110000000000000001000000000001000000
00000000000000000000000000000000000000000
00000000000000000100000100000010000010000
11000100000000000000000000000000000000000
00000000011000011000011000011000011000001
00000000100100100100100100100100100100010
10100100100001100001100001100001100001100
00000001000000000000111110100000000000000000
00000010000000000001000001000000000000010110
11100100000000000000111110100000000000000000
00000000000001000000000000000000001000100111
00000000000010100000000000000000010100100001
10010101110010100000000000000000010100100001
00000000010010000000000000000000010010000001
00000000000111110000000000000000000111110000001
11010100000010101000000000000001010100000001
00000000001000101000000000001010001000000000
00000000001000100100010001001101100111101
10110100000100010001010101000100010000000
00000000001000100010010010001000100000001
00000000000111000000111110000001110000000001
11110100000101010000010100000100010000001
00000000010000010000111000010000001000001
10000000001000001000100010001000001000001
00001100001000001000100010001000001000001
10000000011000001101100011011000001100111
```

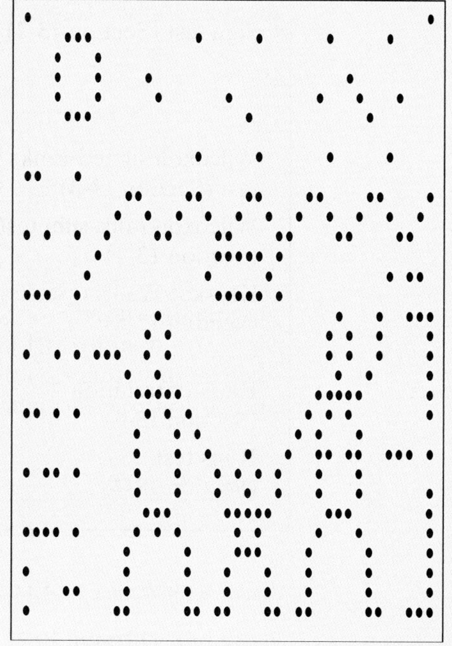

populations with the same distribution. Is there sufficient evidence to support the claim that height is an advantage to candidates?

Winner	76	66	70	70	74	71.5	73	74
Runner-up	64	71	72	72	68	71	69.5	74

Cumulative Review Exercises

1. Analyzing Poll Results A market researcher for American Airlines was instructed to randomly select passengers waiting to board. The passengers were asked several questions about the airline service. Responses were recorded along with their genders, which are listed in the order in which they were selected.

a. At the 0.05 significance level, test the claim that the sequence is random.

b. At the 0.05 significance level, test the claim that the proportion of women is different from 0.5.

c. Use the sample data to construct a 95% confidence interval for the proportion of women.

d. What do the preceding results suggest? Is the sample biased against either gender? Was the sample obtained in a random sequence? If you are the manager, do you have any problems with these results?

M M M M M F F M M M F M M F F

F M F M M F M M F M M F M M M

2. Constructing a Histogram of Sunday Rainfall Amounts In Section 13-5 we used the Kruskal-Wallis test to test the null hypothesis that the rainfall amounts for Boston are the same for each of the seven days of the week. We noted that analysis-of-variance methods described in Section 11-2 cannot be used because the rainfall amounts do not appear to be normally distributed, and we illustrated this with a histogram of the Monday rainfall amounts. (See the Chapter Opening problem.) Refer to Data Set 17 in Appendix B and construct a histogram of the 52 rainfall amounts for Sunday. Compare the result to the histogram for the Monday rainfall amounts. Do the Sunday rainfall amounts appear to come from a population having a normal distribution?

3. Finding Descriptive Statistics for Sunday Rainfall Amounts Refer to the Sunday rainfall amounts in Data Set 17 in Appendix B and find the following.

a. Mean

b. Median

c. Mode

d. Range

e. Standard deviation

f. Any outliers

4. Proportions of Rainy Days Refer to Data Set 17 in Appendix B. Is there sufficient evidence to support the claim that the proportions of rainy days are not the same for all seven days of the week? (There are 53 Wednesdays, but delete the last value for Wednesday and then assume that all seven days of the week occur 52 times.)

5. Testing Effectiveness of SAT Coaching The Block Preparation Program is designed to help students achieve better scores on the SAT exam. The accompanying table lists results for randomly selected students. Use the indicated test with a 0.05 significance level to test the hypothesis that the course has no effect on SAT scores.

a. Sign test
b. Wilcoxon signed-ranks test
c. t test for a claim about matched pairs of sample data (Section 8-3)
d. How do the preceding results support the statement that nonparametric tests lack the sensitivity of parametric tests (so stronger evidence is required before a null hypothesis is rejected)?

Subject	A	B	C	D	E	F	G	H	I	J
SAT score before course	700	840	830	860	840	690	830	1180	930	1070
SAT score after course	800	840	820	980	980	800	800	1270	1080	1220

COOPERATIVE GROUP ACTIVITIES

1. *In-class activity:* Use the existing seating arrangement in your class and apply the runs test to determine whether the students are arranged randomly according to gender. After recording the seating arrangement, analysis can be done in subgroups of three or four students.

2. *In-class activity:* Divide into groups of 8 to 12 people. For each group member, *measure* his or her height and *measure* his or her arm span. For the arm span, the subject should stand with arms extended, like the wings on an airplane. It's easy to mark the height and arm span on a chalkboard, then measure the distances there. Divide the following tasks among subgroups of three or four people.

a. Use rank correlation with the paired sample data to determine whether there is a correlation between height and arm span.
b. Use the sign test to test for a difference between the two variables.
c. Use the Wilcoxon signed-ranks test to test for a difference between the two variables.

3. *In-class activity:* Do activity 2 using pulse rate instead of arm span. Measure pulse rates by counting the number of heartbeats in 1 minute.

4. *In-class activity:* Divide into groups of about 10 or 12 students and use the reaction timer included with the Chapter 5 Cooperative Group Activities. Each group member should be tested for right-hand reaction time and left-hand reaction time. Analyze the results using methods from this chapter. State the methods used and the conclusions reached.

5. *In-class activity:* Divide into groups of five or six students. This activity is a contest of reaction times to determine which group is fastest, if there is a "fastest" group. Use the reaction timer included with the Cooperative Group Activities in Chapter 5. Test and record the reaction time using the dominant hand of each group member. (Only one try per person.) After all groups have reported their results, identify the group with the fastest reaction time, and identify the criterion used to determine the fastest group. Use the Kruskal-Wallis test to determine if differences among the groups are significantly different.

6. *Out-of-class activity:* Divide into groups of three or four students. Investigate the relationship between two variables by collecting your own paired sample data and using the methods of Section 13-6 to determine whether there is a significant correlation. Suggested topics:

- Is there a relationship between taste and cost of different brands of chocolate chip cookies (or colas)? (Taste can be measured on some number scale, such as 1 to 10.)
- Is there a relationship between salaries of professional baseball (or basketball, or football) players and their season achievements?
- Rates versus weights: Is there a relationship between car fuel consumption rates and car weights?
- Is there a relationship between the lengths of men's (or women's) feet and their heights?
- Is there a relationship between student grade-point averages and the amount of television watched?
- Is there a relationship between heights of fathers (or mothers) and heights of their first sons (or daughters)?

7. *Out-of-class activity:* See the "From Data to Decision" project, which involves analysis of the 1970 lottery used for drafting men into the U.S. Army. Because the 1970 results raised concerns about the randomness of selecting draft priority numbers, design a new procedure for generating the 366 priority numbers. Use your procedure to generate the 366 numbers and test your results by using the techniques suggested in parts (a), (b), and (c) of the "From Data to Decision" project. How do your results compare to those obtained in 1970? Does your random selection process appear to be better than the one used in 1970? Write a report that clearly describes the process you designed. Also include your analyses and conclusions.

8. *Out-of-class activity:* Divide into groups of three or four. Survey students by asking them to identify their major and gender. For each surveyed subject, determine the accuracy of the time on his or her wristwatch. First set your own watch to the correct time using an accurate and reliable source ("At the tone, the time is . . . "). For watches that are ahead of the correct time, record positive times. For watches that are behind the correct time, record negative times. Use the sample data to address these questions:

- Do the errors appear to be the same for both genders?
- Do the errors appear to be the same for the different majors?

TECHNOLOGY PROJECT

Randomly generated digits are commonly used for a variety of different applications. Pollsters often select telephone numbers by using random digits, so that unlisted numbers can be included. Simulations, such as those discussed in Section 3-6, often require the random generation of digits. First use the appropriate nonparametric method from this chapter to determine whether the randomly generated digits listed here come from identical populations. Then generate your own data by using the technology available to you, and proceed to test for randomness.

STATDISK:	1 7 8 3 0 0 4 9 7 5 6 5 9 6 1 1 5 5 8 5
Minitab:	1 2 5 1 1 3 1 8 8 5 3 5 5 3 5 3 5 2 3 3
Excel:	3 1 5 9 8 8 4 8 1 2 0 0 1 2 0 2 3 5 3 9
TI-83 Plus:	5 8 4 9 5 4 0 1 4 2 4 3 1 4 9 7 5 3 7 5

Critical Thinking: Was the draft lottery random?

In 1970, a lottery was used to determine who would be drafted into the U.S. Army. The 366 dates in the year were placed in individual capsules. First, the 31 January capsules were placed in a box; then the 29 February capsules were added and the two months were mixed. Then the 31 March capsules were added and the three months were mixed. This process continued until all months were included. The first capsule selected was September 14, so men born on that date were drafted first. The accompanying list shows the 366 dates in the order of selection.

Analyze the Results

a. Use the runs test to test the sequence for randomness above and below the median of 183.5.

b. Use the Kruskal-Wallis test to test the claim that the 12 months had priority numbers drawn from the same population.

c. Calculate the 12 monthly means. Then plot those 12 means on a graph. (The horizontal scale lists the 12 months, and the vertical scale ranges from 100 to 260.) Note any pattern suggesting that the original priority numbers were not randomly selected.

d. Based on the results from parts (a), (b), and (c), decide whether this particular draft lottery was fair. Write a statement either supporting your position that it was fair or explaining why you believe that it was not fair. If you decided that this lottery was unfair, describe a process for selecting lottery numbers that would have been fair.

Jan:	305	159	251	215	101	224	306	199	194	325	329	221	318	238	017	121
	235	140	058	280	186	337	118	059	052	092	355	077	349	164	211	
Feb:	086	144	297	210	214	347	091	181	338	216	150	068	152	004	089	212
	189	292	025	302	363	290	057	236	179	365	205	299	285			
Mar:	108	029	267	275	293	139	122	213	317	323	136	300	259	354	169	166
	033	332	200	239	334	265	256	258	343	170	268	223	362	217	030	
Apr:	032	271	083	081	269	253	147	312	219	218	014	346	124	231	273	148
	260	090	336	345	062	316	252	002	351	340	074	262	191	208		
May:	330	298	040	276	364	155	035	321	197	065	037	133	295	178	130	055
	112	278	075	183	250	326	319	031	361	357	296	308	226	103	313	
Jun:	249	228	301	020	028	110	085	366	335	206	134	272	069	356	180	274
	073	341	104	360	060	247	109	358	137	022	064	222	353	209		
Jul:	093	350	115	279	188	327	050	013	277	284	248	015	042	331	322	120
	098	190	227	187	027	153	172	023	067	303	289	088	270	287	193	
Aug:	111	045	261	145	054	114	168	048	106	021	324	142	307	198	102	044
	154	141	311	344	291	339	116	036	286	245	352	167	061	333	011	
Sep:	225	161	049	232	082	006	008	184	263	071	158	242	175	001	113	207
	255	246	177	063	204	160	119	195	149	018	233	257	151	315		
Oct:	359	125	244	202	024	087	234	283	342	220	237	072	138	294	171	254
	288	005	241	192	243	117	201	196	176	007	264	094	229	038	079	
Nov:	019	034	348	266	310	076	051	097	080	282	046	066	126	127	131	107
	143	146	203	185	156	009	182	230	132	309	047	281	099	174		
Dec:	129	328	157	165	056	010	012	105	043	041	039	314	163	026	320	096
	304	128	240	135	070	053	162	095	084	173	078	123	016	003	100	

Internet Project

Nonparametric Tests

This chapter introduced several different nonparametric or distribution-free tests. Such tests allow you to test hypotheses without making assumptions about the distributions underlying the associated data sets. The Internet Project for this chapter, found at

http://www.awlonline.com/triola

will ask you to revisit some of the hyotheses tested in earlier Internet Projects, but this time you will apply an appropriate nonparametric test. In the second part of the project, you will be asked to apply the runs test to determine randomness of a sequence

Statistics at work

"You can't effectively communicate here unless you use a common language, and that common language happens to be statistics."

Anthony DiUglio

Nuclear Analyst, Probabilistic Risk Assessment, Consolidated Edison Company of New York, Inc.

Anthony DiUglio works in the Probabilistic Risk Assessment (PRA) Group at Consolidated Edison's Indian Point Unit #2 nuclear generating facility in Buchanan, New York, In his work as a Nuclear Analyst, Tony develops probabilities that are used in quantifying various aspects of the plant-specific risk assessment. He is a former student of the author.

What is your job?

In PRA we are concerned with three basic questions about risk: What can happen, how likely is it to happen, and what are the consequences of its happening? We apply these questions about risk to the safe, reliable, and continuous operation of our power plant. When we quantify risk, we obtain numbers that are probabilities. If someone suggests a modification to a plant safety system, we analyze it from a risk perspective. Is the modification better for the system? Does it affect the operation of the plant or put the public health and safety at risk?

How do you use probability and/or statistics?

They're our primary tools. Our PRA requires that we quantify plant-specific rates for all safety-related components in our plant. In developing component-repair rates for pumps and valves, we look at industrywide data (generic) and our plant-specific data. We combine this information together, under uncertainty, and end up with component-specific repair probabilities.

How do you use probability and/or statistics in other departments at Indian Point?

Our Performance Department measures various plant parameters, such as heat rate, megawatt generation, cost per kilowatt of generation, etc. These parameters are all analyzed by use of statistics. The statistical tools used are data trending, normal statistical curves, standard deviations, histograms, etc. Financial Planning makes extensive use of statistics in projecting budget needs and determining its constraints. Our corporate forecasters use probability theory to predict power demands at different times during the year (e.g., winter and summer, one, three, and five years down the road). We have so many people using statistics in their everyday work that statistics has now become a tool for engineers, planners, forecasters, and those of us in Risk Assessment.

In terms of statistics, what would you recommend for prospective employees?

They should have a good understanding of probability, statistics, and their applications. Because PRA is still a relatively new area, we often deal with problems that haven't been addressed before, so many of the problems we address require creative problem solving. Once you have the basic tools, your time is efficiently spent. You can't effectively communicate here unless you use a common language, and that common language happens to be statistics.

Has your work been helpful in convincing the public that your plant is safe?

Safety is always our first concern. In the early 1980s there was a series of public hearings conducted by the Nuclear

Regulatory Commission (NRC) to discuss whether or not the plant should continue operation. Con Edison maintained that the plant was safe, and we were able to help justify the continued operation of our plant through the use of our PRA. At the conclusion of those hearings the NRC agreed with our position, and we continued to operate.

Who was your best math teacher?

Professor Mario Triola.

Is your use of probability and statistics increasing, decreasing, or remaining stable?

It's increasing all the time. We are very much involved with plant performance indicators as parameters for efficient plant operation. With PRA we now have a tool that allows us to focus our attention on the more important plant components and functions. In the case where three components all need maintenance, PRA allows us to identify which component should be returned to service first. In engineering, if we have several components that should be improved, PRA allows us to identify which component should be improved first. We can quantify the effects and thereby target our resources better, thus making the plant safer.

CHAPTER 14

Projects, Procedures, Perspectives

 14-1 A Statistics Group Project

The main objective of this section is to provide some suggestions for a study that can be used as a capstone project for the introductory statistics course. One terrific advantage of this course is that it deals with skills and concepts that can be applied immediately to the real world. After only one fun semester, students are able to conduct their own studies. Some of the suggested topics can be addressed by actually conducting experiments, whereas others might be observational studies that require research of results already available. For example, testing the effectiveness of air bags by actually crashing cars is strongly discouraged, but destructive taste tests of chocolate chip cookies can be an easy and somewhat enjoyable experiment. Here is a suggested format, followed by a list of suggested topics.

Group/Individual Topics can be assigned to individuals, but group projects are particularly effective because they help develop the interpersonal skills that are so necessary in today's working environment.

Oral Report A 10- to 15-minute-long class presentation should involve all group members in a coordinated effort to clearly describe the important components of the study.

Written Report The main objective of the project is not to produce a written document equivalent to a term paper, but a written report should be submitted, and it should include the following components.

1. List of data collected
2. Description of the method of analysis
3. Relevant graphs and/or statistics, including STATDISK, Minitab, Excel, or TI-83 displays

51. Analysis of times that McDonald's' patrons wait in line
52. Analysis of times cars require for refueling
53. Is the state lottery a wise investment?
54. Comparison of casino games: craps versus roulette
55. Starting with $1, is it easier to win a million dollars by playing casino craps or by playing a state lottery?
56. Bold versus cautious strategies of gambling: When gambling with $100, does it make any difference if you bet $1 at a time or if you bet the whole $100 at once?
57. Designing and analyzing results from a test for extrasensory perception
58. Analyzing paired data consisting of heights of fathers (or mothers) and heights of their first sons (or daughters)
59. Gender differences in preferences of dinner partners among the options of Brad Pitt, the President, Nicole Kidman, Cameron Diaz, Sandra Bullock, Queen Margaret, and the Pope
60. Gender differences in preferences of activities among the options of dinner, movie, watching television, reading a book, golf, tennis, swimming, attending a baseball game, attending a football game

14-2 Which Procedure Applies?

Data Collection You can collect your own data through experiments or observational studies. It is absolutely essential to analyze the method used to collect the data, because data carelessly collected may be so completely useless that no amount of statistical torturing can salvage them. Look carefully for bias in the way data are collected, as well as bias on the part of the person or group collecting the data. Many of the procedures in this book are based on the assumption that we are working with a simple random sample, meaning that every possible sample of the same size has the same chance of being selected. If a sample is self-selected (voluntary response), it is worthless for making inferences about a population.

Exploring, Comparing, Describing After collecting data, first consider exploring, describing, and comparing data sets using the basic tools included in Chapter 2. Be sure to address the following:

1. *Center:* Find the mean and median, which are measures of center that are representative or average values giving us an indication of where the middle of the data set is located.
2. *Variation:* Find the range and standard deviation, which are measures of the amount that the sample values vary among themselves.
3. *Distribution:* Construct a histogram to see the nature or shape of the distribution of the data, and determine if the distribution is bell-shaped, uniform, or skewed.

4. *Outliers:* Identify any sample values that lie very far away from the vast majority of the other sample values.
5. *Time:* Determine if the population is stable or if its characteristics are changing over time.

Inferences: Estimating Parameters and Hypothesis Testing When trying to use sample data for making inferences about a population, it is often difficult to choose the particular procedure that should be applied. This text includes a wide variety of procedures that apply to many different circumstances. Here are some key questions that should be answered.

- What is the level of measurement (nominal, ordinal, interval, ratio) of the data?
- Does the study involve one, two, or more populations?
- Is there a claim to be tested or a parameter to be estimated?
- What is the relevant parameter (mean, standard deviation, proportion)?
- Is the sample large ($n > 30$) or small?
- Is there reason to believe that the population is normally distributed?
- What is the basic question or issue that you want to address?

In Figure 14-1 on the next page we list the major methods included in this book, along with a scheme for determining which of those methods should be used. To use Figure 14-1, start at the extreme left side of the figure and begin by identifying the level of measurement of the data. Proceed to follow the path suggested by the level of measurement, the number of populations, and the claim or parameter being considered.

Note: This figure applies to a fixed population. If the data are from a process that may change over time, construct a control chart (see Chapter 12) to determine whether the process is statistically stable. This figure applies to process data only if the process is statistically stable.

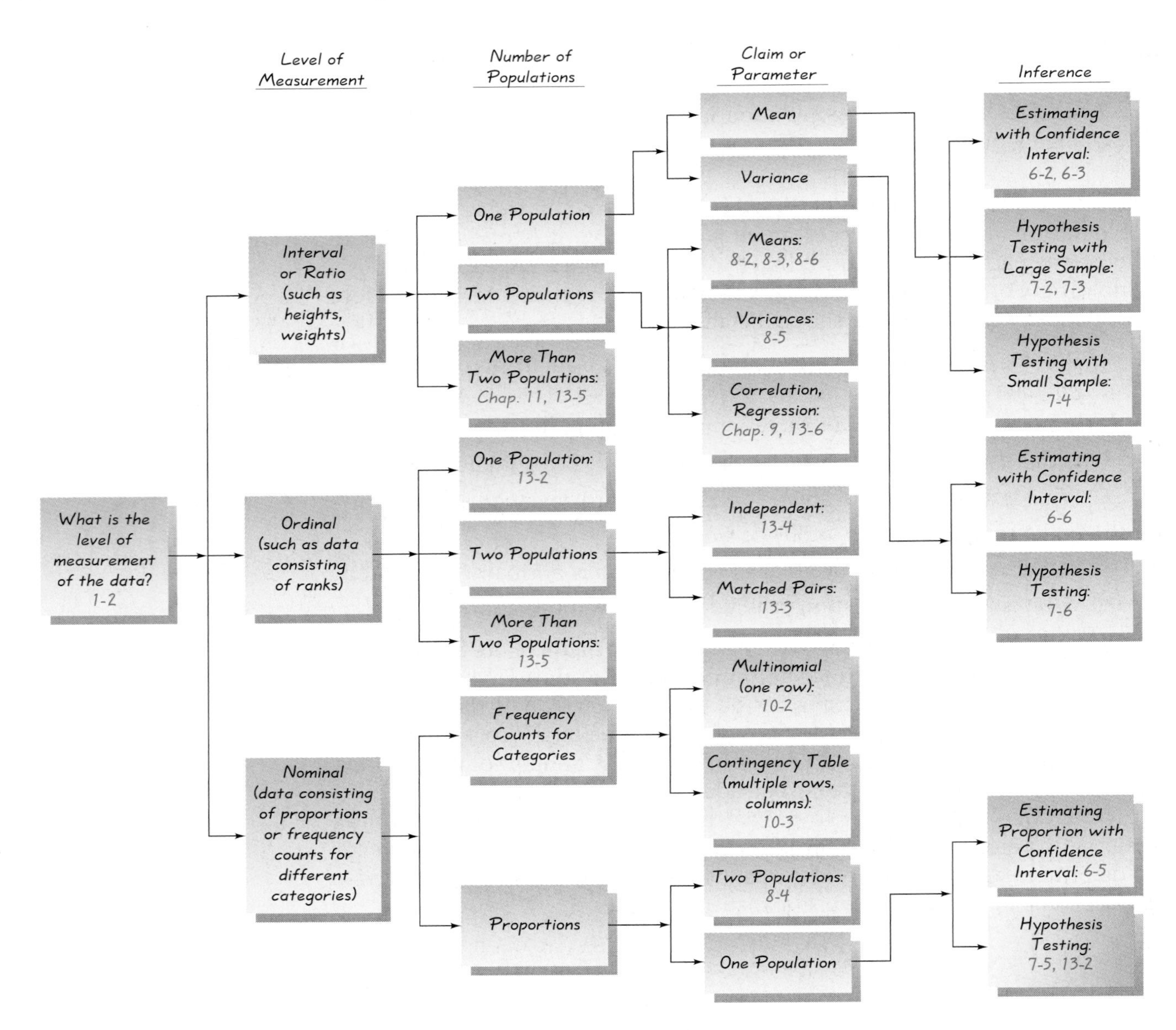

FIGURE 14-1
Inferential Statistics: Applicable Text Sections

14-3 A Perspective

No one expects a single introductory statistics course to transform anyone into an expert statistician. After studying several of the chapters in this book, it is natural for students to feel that they have not yet mastered the material to the extent necessary for confidently using statistics in real applications. Many important topics (such as factor analysis and discriminant analysis) are not included in this text because they are too advanced for this introductory level. Some easier topics (such as time series) have been excluded for other reasons. It is important to know that professional help is available from expert statisticians, and this introductory statistics course will help you in discussions with one of these experts.

Although this course is not designed to make you an expert statistician, it is designed to make you a better educated person with improved job marketability. You should know and understand the basic concepts of probability and chance. You should know that in attempting to gain insight into a set of data, it is important to investigate measures of center (such as mean and median), measures of variation (such as range and standard deviation), the nature of the distribution (via a frequency table or graph), the presence of outliers, and whether the population is stable or is changing over time. You should know and understand the importance of estimating population parameters (such as a mean, standard deviation, and proportion), as well as testing claims made about population parameters. You should realize that the nature and configuration of the data have a dramatic effect on the particular statistical procedures that are used.

Throughout this text we have emphasized the importance of good sampling. You should recognize that a bad sample may be beyond repair by even the most expert statisticians, using the most sophisticated techniques. There are many mail, magazine, and telephone call-in surveys that allow respondents to be "self-selected." The results of such surveys are generally worthless when judged according to the criteria of sound statistical methodology. Keep this in mind when you are exposed to self-selected surveys, so that you don't let them affect your beliefs and decisions. You should also recognize, however, that many surveys and polls obtain very good results, even though the sample sizes might seem to be relatively small. Although many people refuse to believe it, a nationwide survey of only 1700 voters can provide good results if the sampling is carefully planned and executed.

At one time a person was considered educated if he or she could read, but we are in a new millennium that is much more demanding. Today, an educated person must be able to read, write, understand the significance of the Renaissance, operate a computer, and do algebra. The larger picture combines disciplines with common goals, including the quest for truth. The study of statistics helps us see the truth that is sometimes distorted by others or concealed by data that are disorganized or perhaps not yet collected. Understanding statistics is now essential for employees, employers, and citizens. H. G. Wells once said that "statistical thinking will one day be as necessary for efficient citizenship as the ability to read and write." That day is now.

Appendices

TABLE A-1 Binomial Probabilities

n	x	p .01	.05	.10	.20	.30	.40	.50	.60	.70	.80	.90	.95	.99	x
2	0	.980	.902	.810	.640	.490	.360	.250	.160	.090	.040	.010	.002	0+	0
	1	.020	.095	.180	.320	.420	.480	.500	.480	.420	.320	.180	.095	.020	1
	2	0+	.002	.010	.040	.090	.160	.250	.360	.490	.640	.810	.902	.980	2
3	0	.970	.857	.729	.512	.343	.216	.125	.064	.027	.008	.001	0+	0+	0
	1	.029	.135	.243	.384	.441	.432	.375	.288	.189	.096	.027	.007	0+	1
	2	0+	.007	.027	.096	.189	.288	.375	.432	.441	.384	.243	.135	.029	2
	3	0+	0+	.001	.008	.027	.064	.125	.216	.343	.512	.729	.857	.970	3
4	0	.961	.815	.656	.410	.240	.130	.062	.026	.008	.002	0+	0+	0+	0
	1	.039	.171	.292	.410	.412	.346	.250	.154	.076	.026	.004	0+	0+	1
	2	.001	.014	.049	.154	.265	.346	.375	.346	.265	.154	.049	.014	.001	2
	3	0+	0+	.004	.026	.076	.154	.250	.346	.412	.410	.292	.171	.039	3
	4	0+	0+	0+	.002	.008	.026	.062	.130	.240	.410	.656	.815	.961	4
5	0	.951	.774	.590	.328	.168	.078	.031	.010	.002	0+	0+	0+	0+	0
	1	.048	.204	.328	.410	.360	.259	.156	.077	.028	.006	0+	0+	0+	1
	2	.001	.021	.073	.205	.309	.346	.312	.230	.132	.051	.008	.001	0+	2
	3	0+	.001	.008	.051	.132	.230	.312	.346	.309	.205	.073	.021	.001	3
	4	0+	0+	0+	.006	.028	.077	.156	.259	.360	.410	.328	.204	.048	4
	5	0+	0+	0+	0+	.002	.010	.031	.078	.168	.328	.590	.774	.951	5
6	0	.941	.735	.531	.262	.118	.047	.016	.004	.001	0+	0+	0+	0+	0
	1	.057	.232	.354	.393	.303	.187	.094	.037	.010	.002	0+	0+	0+	1
	2	.001	.031	.098	.246	.324	.311	.234	.138	.060	.015	.001	0+	0+	2
	3	0+	.002	.015	.082	.185	.276	.312	.276	.185	.082	.015	.002	0+	3
	4	0+	0+	.001	.015	.060	.138	.234	.311	.324	.246	.098	.031	.001	4
	5	0+	0+	0+	.002	.010	.037	.094	.187	.303	.393	.354	.232	.057	5
	6	0+	0+	0+	0+	.001	.004	.016	.047	.118	.262	.531	.735	.941	6
7	0	.932	.698	.478	.210	.082	.028	.008	.002	0+	0+	0+	0+	0+	0
	1	.066	.257	.372	.367	.247	.131	.055	.017	.004	0+	0+	0+	0+	1
	2	.002	.041	.124	.275	.318	.261	.164	.077	.025	.004	0+	0+	0+	2
	3	0+	.004	.023	.115	.227	.290	.273	.194	.097	.029	.003	0+	0+	3
	4	0+	0+	.003	.029	.097	.194	.273	.290	.227	.115	.023	.004	0+	4
	5	0+	0+	0+	.004	.025	.077	.164	.261	.318	.275	.124	.041	.002	5
	6	0+	0+	0+	0+	.004	.017	.055	.131	.247	.367	.372	.257	.066	6
	7	0+	0+	0+	0+	0+	.002	.008	.028	.082	.210	.478	.698	.932	7
8	0	.923	.663	.430	.168	.058	.017	.004	.001	0+	0+	0+	0+	0+	0
	1	.075	.279	.383	.336	.198	.090	.031	.008	.001	0+	0+	0+	0+	1
	2	.003	.051	.149	.294	.296	.209	.109	.041	.010	.001	0+	0+	0+	2
	3	0+	.005	.033	.147	.254	.279	.219	.124	.047	.009	0+	0+	0+	3
	4	0+	0+	.005	.046	.136	.232	.273	.232	.136	.046	.005	0+	0+	4
	5	0+	0+	0+	.009	.047	.124	.219	.279	.254	.147	.033	.005	0+	5
	6	0+	0+	0+	.001	.010	.041	.109	.209	.296	.294	.149	.051	.003	6
	7	0+	0+	0+	0+	.001	.008	.031	.090	.198	.336	.383	.279	.075	7
	8	0+	0+	0+	0+	0+	.001	.004	.017	.058	.168	.430	.663	.923	8

NOTE: 0+ represents a positive probability less than 0.0005.

(continued)

TABLE A-1 Binomial Probabilities *(continued)*

n	x	p = .01	.05	.10	.20	.30	.40	.50	.60	.70	.80	.90	.95	.99	x
9	0	.914	.630	.387	.134	.040	.010	.002	0+	0+	0+	0+	0+	0+	0
	1	.083	.299	.387	.302	.156	.060	.018	.004	0+	0+	0+	0+	0+	1
	2	.003	.063	.172	.302	.267	.161	.070	.021	.004	0+	0+	0+	0+	2
	3	0+	.008	.045	.176	.267	.251	.164	.074	.021	.003	0+	0+	0+	3
	4	0+	.001	.007	.066	.172	.251	.246	.167	.074	.017	.001	0+	0+	4
	5	0+	0+	.001	.017	.074	.167	.246	.251	.172	.066	.007	.001	0+	5
	6	0+	0+	0+	.003	.021	.074	.164	.251	.267	.176	.045	.008	0+	6
	7	0+	0+	0+	0+	.004	.021	.070	.161	.267	.302	.172	.063	.003	7
	8	0+	0+	0+	0+	0+	.004	.018	.060	.156	.302	.387	.299	.083	8
	9	0+	0+	0+	0+	0+	0+	.002	.010	.040	.134	.387	.630	.914	9
10	0	.904	.599	.349	.107	.028	.006	.001	0+	0+	0+	0+	0+	0+	0
	1	.091	.315	.387	.268	.121	.040	.010	.002	0+	0+	0+	0+	0+	1
	2	.004	.075	.194	.302	.233	.121	.044	.011	.001	0+	0+	0+	0+	2
	3	0+	.010	.057	.201	.267	.215	.117	.042	.009	.001	0+	0+	0+	3
	4	0+	.001	.011	.088	.200	.251	.205	.111	.037	.006	0+	0+	0+	4
	5	0+	0+	.001	.026	.103	.201	.246	.201	.103	.026	.001	0+	0+	5
	6	0+	0+	0+	.006	.037	.111	.205	.251	.200	.088	.011	.001	0+	6
	7	0+	0+	0+	.001	.009	.042	.117	.215	.267	.201	.057	.010	0+	7
	8	0+	0+	0+	0+	.001	.011	.044	.121	.233	.302	.194	.075	.004	8
	9	0+	0+	0+	0+	0+	.002	.010	.040	.121	.268	.387	.315	.091	9
	10	0+	0+	0+	0+	0+	0+	.001	.006	.028	.107	.349	.599	.904	10
11	0	.895	.569	.314	.086	.020	.004	0+	0+	0+	0+	0+	0+	0+	0
	1	.099	.329	.384	.236	.093	.027	.005	.001	0+	0+	0+	0+	0+	1
	2	.005	.087	.213	.295	.200	.089	.027	.005	.001	0+	0+	0+	0+	2
	3	0+	.014	.071	.221	.257	.177	.081	.023	.004	0+	0+	0+	0+	3
	4	0+	.001	.016	.111	.220	.236	.161	.070	.017	.002	0+	0+	0+	4
	5	0+	0+	.002	.039	.132	.221	.226	.147	.057	.010	0+	0+	0+	5
	6	0+	0+	0+	.010	.057	.147	.226	.221	.132	.039	.002	0+	0+	6
	7	0+	0+	0+	.002	.017	.070	.161	.236	.220	.111	.016	.001	0+	7
	8	0+	0+	0+	0+	.004	.023	.081	.177	.257	.221	.071	.014	0+	8
	9	0+	0+	0+	0+	.001	.005	.027	.089	.200	.295	.213	.087	.005	9
	10	0+	0+	0+	0+	0+	.001	.005	.027	.093	.236	.384	.329	.099	10
	11	0+	0+	0+	0+	0+	0+	0+	.004	.020	.086	.314	.569	.895	11
12	0	.886	.540	.282	.069	.014	.002	0+	0+	0+	0+	0+	0+	0+	0
	1	.107	.341	.377	.206	.071	.017	.003	0+	0+	0+	0+	0+	0+	1
	2	.006	.099	.230	.283	.168	.064	.016	.002	0+	0+	0+	0+	0+	2
	3	0+	.017	.085	.236	.240	.142	.054	.012	.001	0+	0+	0+	0+	3
	4	0+	.002	.021	.133	.231	.213	.121	.042	.008	.001	0+	0+	0+	4
	5	0+	0+	.004	.053	.158	.227	.193	.101	.029	.003	0+	0+	0+	5
	6	0+	0+	0+	.016	.079	.177	.226	.177	.079	.016	0+	0+	0+	6
	7	0+	0+	0+	.003	.029	.101	.193	.227	.158	.053	.004	0+	0+	7
	8	0+	0+	0+	.001	.008	.042	.121	.213	.231	.133	.021	.002	0+	8
	9	0+	0+	0+	0+	.001	.012	.054	.142	.240	.236	.085	.017	0+	9
	10	0+	0+	0+	0+	0+	.002	.016	.064	.168	.283	.230	.099	.006	10
	11	0+	0+	0+	0+	0+	0+	.003	.017	.071	.206	.377	.341	.107	11
	12	0+	0+	0+	0+	0+	0+	0+	.002	.014	.069	.282	.540	.886	12

NOTE: 0+ represents a positive probability less than 0.0005.

(continued)

TABLE A-1 Binomial Probabilities *(continued)*

n	x	.01	.05	.10	.20	.30	.40	.50	.60	.70	.80	.90	.95	.99	x
13	0	.878	.513	.254	.055	.010	.001	0+	0+	0+	0+	0+	0+	0+	0
	1	.115	.351	.367	.179	.054	.011	.002	0+	0+	0+	0+	0+	0+	1
	2	.007	.111	.245	.268	.139	.045	.010	.001	0+	0+	0+	0+	0+	2
	3	0+	.021	.100	.246	.218	.111	.035	.006	.001	0+	0+	0+	0+	3
	4	0+	.003	.028	.154	.234	.184	.087	.024	.003	0+	0+	0+	0+	4
	5	0+	0+	.006	.069	.180	.221	.157	.066	.014	.001	0+	0+	0+	5
	6	0+	0+	.001	.023	.103	.197	.209	.131	.044	.006	0+	0+	0+	6
	7	0+	0+	0+	.006	.044	.131	.209	.197	.103	.023	.001	0+	0+	7
	8	0+	0+	0+	.001	.014	.066	.157	.221	.180	.069	.006	0+	0+	8
	9	0+	0+	0+	0+	.003	.024	.087	.184	.234	.154	.028	.003	0+	9
	10	0+	0+	0+	0+	.001	.006	.035	.111	.218	.246	.100	.021	0+	10
	11	0+	0+	0+	0+	0+	.001	.010	.045	.139	.268	.245	.111	.007	11
	12	0+	0+	0+	0+	0+	0+	.002	.011	.054	.179	.367	.351	.115	12
	13	0+	0+	0+	0+	0+	0+	0+	.001	.010	.055	.254	.513	.878	13
14	0	.869	.488	.229	.044	.007	.001	0+	0+	0+	0+	0+	0+	0+	0
	1	.123	.359	.356	.154	.041	.007	.001	0+	0+	0+	0+	0+	0+	1
	2	.008	.123	.257	.250	.113	.032	.006	.001	0+	0+	0+	0+	0+	2
	3	0+	.026	.114	.250	.194	.085	.022	.003	0+	0+	0+	0+	0+	3
	4	0+	.004	.035	.172	.229	.155	.061	.014	.001	0+	0+	0+	0+	4
	5	0+	0+	.008	.086	.196	.207	.122	.041	.007	0+	0+	0+	0+	5
	6	0+	0+	.001	.032	.126	.207	.183	.092	.023	.002	0+	0+	0+	6
	7	0+	0+	0+	.009	.062	.157	.209	.157	.062	.009	0+	0+	0+	7
	8	0+	0+	0+	.002	.023	.092	.183	.207	.126	.032	.001	0+	0+	8
	9	0+	0+	0+	0+	.007	.041	.122	.207	.196	.086	.008	0+	0+	9
	10	0+	0+	0+	0+	.001	.014	.061	.155	.229	.172	.035	.004	0+	10
	11	0+	0+	0+	0+	0+	.003	.022	.085	.194	.250	.114	.026	0+	11
	12	0+	0+	0+	0+	0+	.001	.006	.032	.113	.250	.257	.123	.008	12
	13	0+	0+	0+	0+	0+	0+	.001	.007	.041	.154	.356	.359	.123	13
	14	0+	0+	0+	0+	0+	0+	0+	.001	.007	.044	.229	.488	.869	14
15	0	.860	.463	.206	.035	.005	0+	0+	0+	0+	0+	0+	0+	0+	0
	1	.130	.366	.343	.132	.031	.005	0+	0+	0+	0+	0+	0+	0+	1
	2	.009	.135	.267	.231	.092	.022	.003	0+	0+	0+	0+	0+	0+	2
	3	0+	.031	.129	.250	.170	.063	.014	.002	0+	0+	0+	0+	0+	3
	4	0+	.005	.043	.188	.219	.127	.042	.007	.001	0+	0+	0+	0+	4
	5	0+	.001	.010	.103	.206	.186	.092	.024	.003	0+	0+	0+	0+	5
	6	0+	0+	.002	.043	.147	.207	.153	.061	.012	.001	0+	0+	0+	6
	7	0+	0+	0+	.014	.081	.177	.196	.118	.035	.003	0+	0+	0+	7
	8	0+	0+	0+	.003	.035	.118	.196	.177	.081	.014	0+	0+	0+	8
	9	0+	0+	0+	.001	.012	.061	.153	.207	.147	.043	.002	0+	0+	9
	10	0+	0+	0+	0+	.003	.024	.092	.186	.206	.103	.010	.001	0+	10
	11	0+	0+	0+	0+	.001	.007	.042	.127	.219	.188	.043	.005	0+	11
	12	0+	0+	0+	0+	0+	.002	.014	.063	.170	.250	.129	.031	0+	12
	13	0+	0+	0+	0+	0+	0+	.003	.022	.092	.231	.267	.135	.009	13
	14	0+	0+	0+	0+	0+	0+	0+	.005	.031	.132	.343	.366	.130	14
	15	0+	0+	0+	0+	0+	0+	0+	0+	.005	.035	.206	.463	.860	15

NOTE: 0+ represents a positive probability less than 0.0005.

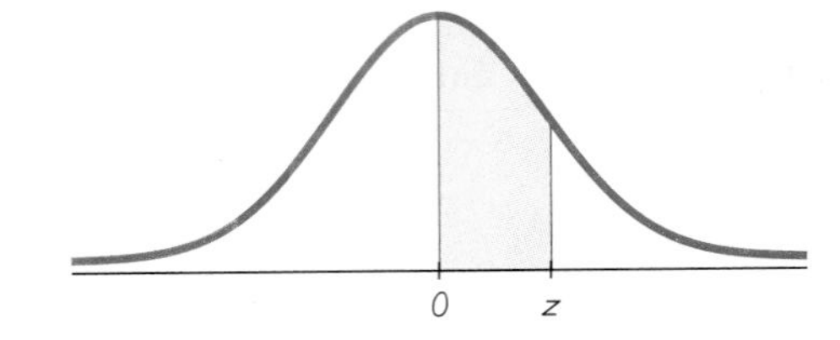

TABLE A-2 Standard Normal (z) Distribution

z	.00	.01	.02	.03	.04	.05	.06	.07	.08	.09
0.0	.0000	.0040	.0080	.0120	.0160	.0199	.0239	.0279	.0319	.0359
0.1	.0398	.0438	.0478	.0517	.0557	.0596	.0636	.0675	.0714	.0753
0.2	.0793	.0832	.0871	.0910	.0948	.0987	.1026	.1064	.1103	.1141
0.3	.1179	.1217	.1255	.1293	.1331	.1368	.1406	.1443	.1480	.1517
0.4	.1554	.1591	.1628	.1664	.1700	.1736	.1772	.1808	.1844	.1879
0.5	.1915	.1950	.1985	.2019	.2054	.2088	.2123	.2157	.2190	.2224
0.6	.2257	.2291	.2324	.2357	.2389	.2422	.2454	.2486	.2517	.2549
0.7	.2580	.2611	.2642	.2673	.2704	.2734	.2764	.2794	.2823	.2852
0.8	.2881	.2910	.2939	.2967	.2995	.3023	.3051	.3078	.3106	.3133
0.9	.3159	.3186	.3212	.3238	.3264	.3289	.3315	.3340	.3365	.3389
1.0	.3413	.3438	.3461	.3485	.3508	.3531	.3554	.3577	.3599	.3621
1.1	.3643	.3665	.3686	.3708	.3729	.3749	.3770	.3790	.3810	.3830
1.2	.3849	.3869	.3888	.3907	.3925	.3944	.3962	.3980	.3997	.4015
1.3	.4032	.4049	.4066	.4082	.4099	.4115	.4131	.4147	.4162	.4177
1.4	.4192	.4207	.4222	.4236	.4251	.4265	.4279	.4292	.4306	.4319
1.5	.4332	.4345	.4357	.4370	.4382	.4394	.4406	.4418	.4429	.4441
1.6	.4452	.4463	.4474	.4484	.4495 *	.4505	.4515	.4525	.4535	.4545
1.7	.4554	.4564	.4573	.4582	.4591	.4599	.4608	.4616	.4625	.4633
1.8	.4641	.4649	.4656	.4664	.4671	.4678	.4686	.4693	.4699	.4706
1.9	.4713	.4719	.4726	.4732	.4738	.4744	.4750	.4756	.4761	.4767
2.0	.4772	.4778	.4783	.4788	.4793	.4798	.4803	.4808	.4812	.4817
2.1	.4821	.4826	.4830	.4834	.4838	.4842	.4846	.4850	.4854	.4857
2.2	.4861	.4864	.4868	.4871	.4875	.4878	.4881	.4884	.4887	.4890
2.3	.4893	.4896	.4898	.4901	.4904	.4906	.4909	.4911	.4913	.4916
2.4	.4918	.4920	.4922	.4925	.4927	.4929	.4931	.4932	.4934	.4936
2.5	.4938	.4940	.4941	.4943	.4945	.4946	.4948	.4949 *	.4951	.4952
2.6	.4953	.4955	.4956	.4957	.4959	.4960	.4961	.4962	.4963	.4964
2.7	.4965	.4966	.4967	.4968	.4969	.4970	.4971	.4972	.4973	.4974
2.8	.4974	.4975	.4976	.4977	.4977	.4978	.4979	.4979	.4980	.4981
2.9	.4981	.4982	.4982	.4983	.4984	.4984	.4985	.4985	.4986	.4986
3.0	.4987	.4987	.4987	.4988	.4988	.4989	.4989	.4989	.4990	.4990
3.10 and higher	.4999									

NOTE: For values of z above 3.09, use 0.4999 for the area.
*Use these common values that result from interpolation:

z score	Area
1.645	0.4500
2.575	0.4950

From Frederick C. Mosteller and Robert E. K. Rourke, *Sturdy Statistics,* 1973, Addison-Wesley Publishing Co., Reading, MA. Reprinted with permission of Frederick Mosteller.

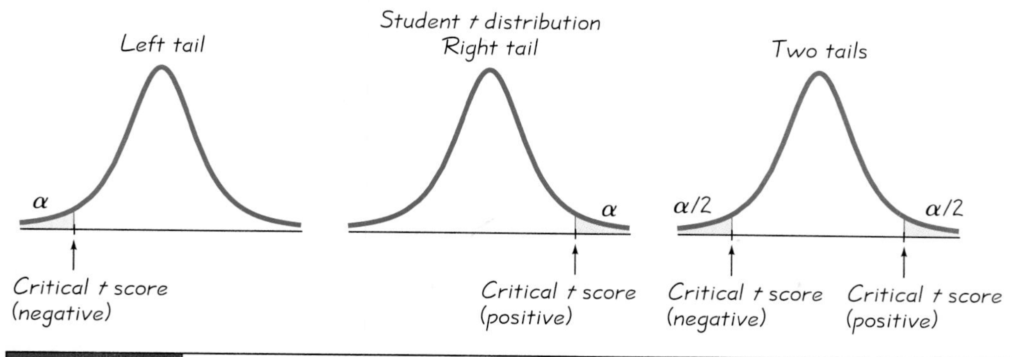

TABLE A-3 *t* Distribution

Degrees of Freedom	α: .005 (one tail) .01 (two tails)	.01 (one tail) .02 (two tails)	.025 (one tail) .05 (two tails)	.05 (one tail) .10 (two tails)	.10 (one tail) .20 (two tails)	.25 (one tail) .50 (two tails)
1	63.657	31.821	12.706	6.314	3.078	1.000
2	9.925	6.965	4.303	2.920	1.886	.816
3	5.841	4.541	3.182	2.353	1.638	.765
4	4.604	3.747	2.776	2.132	1.533	.741
5	4.032	3.365	2.571	2.015	1.476	.727
6	3.707	3.143	2.447	1.943	1.440	.718
7	3.500	2.998	2.365	1.895	1.415	.711
8	3.355	2.896	2.306	1.860	1.397	.706
9	3.250	2.821	2.262	1.833	1.383	.703
10	3.169	2.764	2.228	1.812	1.372	.700
11	3.106	2.718	2.201	1.796	1.363	.697
12	3.054	2.681	2.179	1.782	1.356	.696
13	3.012	2.650	2.160	1.771	1.350	.694
14	2.977	2.625	2.145	1.761	1.345	.692
15	2.947	2.602	2.132	1.753	1.341	.691
16	2.921	2.584	2.120	1.746	1.337	.690
17	2.898	2.567	2.110	1.740	1.333	.689
18	2.878	2.552	2.101	1.734	1.330	.688
19	2.861	2.540	2.093	1.729	1.328	.688
20	2.845	2.528	2.086	1.725	1.325	.687
21	2.831	2.518	2.080	1.721	1.323	.686
22	2.819	2.508	2.074	1.717	1.321	.686
23	2.807	2.500	2.069	1.714	1.320	.685
24	2.797	2.492	2.064	1.711	1.318	.685
25	2.787	2.485	2.060	1.708	1.316	.684
26	2.779	2.479	2.056	1.706	1.315	.684
27	2.771	2.473	2.052	1.703	1.314	.684
28	2.763	2.467	2.048	1.701	1.313	.683
29	2.756	2.462	2.045	1.699	1.311	.683
Large (*z*)	2.575	2.326	1.960	1.645	1.282	.675

TABLE A-4 Chi-Square (χ^2) Distribution

Degrees of Freedom	Area to the Right of the Critical Value									
	0.995	0.99	0.975	0.95	0.90	0.10	0.05	0.025	0.01	0.005
1	—	—	0.001	0.004	0.016	2.706	3.841	5.024	6.635	7.879
2	0.010	0.020	0.051	0.103	0.211	4.605	5.991	7.378	9.210	10.597
3	0.072	0.115	0.216	0.352	0.584	6.251	7.815	9.348	11.345	12.838
4	0.207	0.297	0.484	0.711	1.064	7.779	9.488	11.143	13.277	14.860
5	0.412	0.554	0.831	1.145	1.610	9.236	11.071	12.833	15.086	16.750
6	0.676	0.872	1.237	1.635	2.204	10.645	12.592	14.449	16.812	18.548
7	0.989	1.239	1.690	2.167	2.833	12.017	14.067	16.013	18.475	20.278
8	1.344	1.646	2.180	2.733	3.490	13.362	15.507	17.535	20.090	21.955
9	1.735	2.088	2.700	3.325	4.168	14.684	16.919	19.023	21.666	23.589
10	2.156	2.558	3.247	3.940	4.865	15.987	18.307	20.483	23.209	25.188
11	2.603	3.053	3.816	4.575	5.578	17.275	19.675	21.920	24.725	26.757
12	3.074	3.571	4.404	5.226	6.304	18.549	21.026	23.337	26.217	28.299
13	3.565	4.107	5.009	5.892	7.042	19.812	22.362	24.736	27.688	29.819
14	4.075	4.660	5.629	6.571	7.790	21.064	23.685	26.119	29.141	31.319
15	4.601	5.229	6.262	7.261	8.547	22.307	24.996	27.488	30.578	32.801
16	5.142	5.812	6.908	7.962	9.312	23.542	26.296	28.845	32.000	34.267
17	5.697	6.408	7.564	8.672	10.085	24.769	27.587	30.191	33.409	35.718
18	6.265	7.015	8.231	9.390	10.865	25.989	28.869	31.526	34.805	37.156
19	6.844	7.633	8.907	10.117	11.651	27.204	30.144	32.852	36.191	38.582
20	7.434	8.260	9.591	10.851	12.443	28.412	31.410	34.170	37.566	39.997
21	8.034	8.897	10.283	11.591	13.240	29.615	32.671	35.479	38.932	41.401
22	8.643	9.542	10.982	12.338	14.042	30.813	33.924	36.781	40.289	42.796
23	9.260	10.196	11.689	13.091	14.848	32.007	35.172	38.076	41.638	44.181
24	9.886	10.856	12.401	13.848	15.659	33.196	36.415	39.364	42.980	45.559
25	10.520	11.524	13.120	14.611	16.473	34.382	37.652	40.646	44.314	46.928
26	11.160	12.198	13.844	15.379	17.292	35.563	38.885	41.923	45.642	48.290
27	11.808	12.879	14.573	16.151	18.114	36.741	40.113	43.194	46.963	49.645
28	12.461	13.565	15.308	16.928	18.939	37.916	41.337	44.461	48.278	50.993
29	13.121	14.257	16.047	17.708	19.768	39.087	42.557	45.722	49.588	52.336
30	13.787	14.954	16.791	18.493	20.599	40.256	43.773	46.979	50.892	53.672
40	20.707	22.164	24.433	26.509	29.051	51.805	55.758	59.342	63.691	66.766
50	27.991	29.707	32.357	34.764	37.689	63.167	67.505	71.420	76.154	79.490
60	35.534	37.485	40.482	43.188	46.459	74.397	79.082	83.298	88.379	91.952
70	43.275	45.442	48.758	51.739	55.329	85.527	90.531	95.023	100.425	104.215
80	51.172	53.540	57.153	60.391	64.278	96.578	101.879	106.629	112.329	116.321
90	59.196	61.754	65.647	69.126	73.291	107.565	113.145	118.136	124.116	128.299
100	67.328	70.065	74.222	77.929	82.358	118.498	124.342	129.561	135.807	140.169

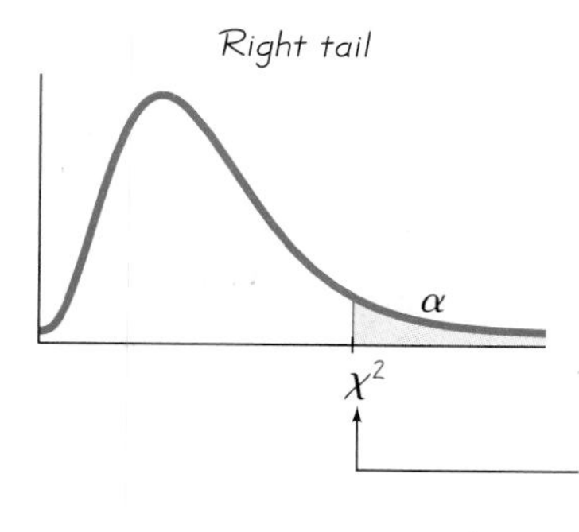

To find this value, use the column with the area α given at the top of the table.

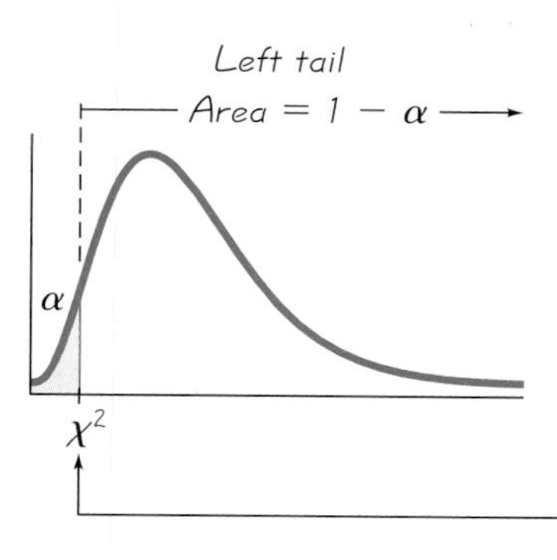

To find this value, determine the area of the region to the right of this boundary (the unshaded area) and use the column with this value at the top. If the left tail has area α, use the column with the value of $1 - \alpha$ at the top of the table.

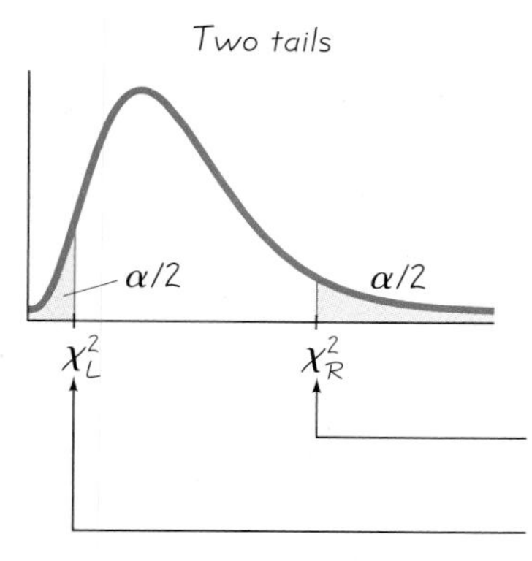

To find this value, use the column with area $\alpha/2$ at the top of the table.

To find this value, use the column with area $1 - \alpha/2$ at the top of the table.

TABLE A-5 *F* Distribution ($\alpha = 0.01$ in the right tail)

Denominator degrees of freedom (df_2)	Numerator degrees of freedom (df_1): 1	2	3	4	5	6	7	8	9
1	4052.2	4999.5	5403.4	5624.6	5763.6	5859.0	5928.4	5981.1	6022.5
2	98.503	99.000	99.166	99.249	99.299	99.333	99.356	99.374	99.388
3	34.116	30.817	29.457	28.710	28.237	27.911	27.672	27.489	27.345
4	21.198	18.000	16.694	15.977	15.522	15.207	14.976	14.799	14.659
5	16.258	13.274	12.060	11.392	10.967	10.672	10.456	10.289	10.158
6	13.745	10.925	9.7795	9.1483	8.7459	8.4661	8.2600	8.1017	7.9761
7	12.246	9.5466	8.4513	7.8466	7.4604	7.1914	6.9928	6.8400	6.7188
8	11.259	8.6491	7.5910	7.0061	6.6318	6.3707	6.1776	6.0289	5.9106
9	10.561	8.0215	6.9919	6.4221	6.0569	5.8018	5.6129	5.4671	5.3511
10	10.044	7.5594	6.5523	5.9943	5.6363	5.3858	5.2001	5.0567	4.9424
11	9.6460	7.2057	6.2167	5.6683	5.3160	5.0692	4.8861	4.7445	4.6315
12	9.3302	6.9266	5.9525	5.4120	5.0643	4.8206	4.6395	4.4994	4.3875
13	9.0738	6.7010	5.7394	5.2053	4.8616	4.6204	4.4410	4.3021	4.1911
14	8.8616	6.5149	5.5639	5.0354	4.6950	4.4558	4.2779	4.1399	4.0297
15	8.6831	6.3589	5.4170	4.8932	4.5556	4.3183	4.1415	4.0045	3.8948
16	8.5310	6.2262	5.2922	4.7726	4.4374	4.2016	4.0259	3.8896	3.7804
17	8.3997	6.1121	5.1850	4.6690	4.3359	4.1015	3.9267	3.7910	3.6822
18	8.2854	6.0129	5.0919	4.5790	4.2479	4.0146	3.8406	3.7054	3.5971
19	8.1849	5.9259	5.0103	4.5003	4.1708	3.9386	3.7653	3.6305	3.5225
20	8.0960	5.8489	4.9382	4.4307	4.1027	3.8714	3.6987	3.5644	3.4567
21	8.0166	5.7804	4.8740	4.3688	4.0421	3.8117	3.6396	3.5056	3.3981
22	7.9454	5.7190	4.8166	4.3134	3.9880	3.7583	3.5867	3.4530	3.3458
23	7.8811	5.6637	4.7649	4.2636	3.9392	3.7102	3.5390	3.4057	3.2986
24	7.8229	5.6136	4.7181	4.2184	3.8951	3.6667	3.4959	3.3629	3.2560
25	7.7698	5.5680	4.6755	4.1774	3.8550	3.6272	3.4568	3.3239	3.2172
26	7.7213	5.5263	4.6366	4.1400	3.8183	3.5911	3.4210	3.2884	3.1818
27	7.6767	5.4881	4.6009	4.1056	3.7848	3.5580	3.3882	3.2558	3.1494
28	7.6356	5.4529	4.5681	4.0740	3.7539	3.5276	3.3581	3.2259	3.1195
29	7.5977	5.4204	4.5378	4.0449	3.7254	3.4995	3.3303	3.1982	3.0920
30	7.5625	5.3903	4.5097	4.0179	3.6990	3.4735	3.3045	3.1726	3.0665
40	7.3141	5.1785	4.3126	3.8283	3.5138	3.2910	3.1238	2.9930	2.8876
60	7.0771	4.9774	4.1259	3.6490	3.3389	3.1187	2.9530	2.8233	2.7185
120	6.8509	4.7865	3.9491	3.4795	3.1735	2.9559	2.7918	2.6629	2.5586
∞	6.6349	4.6052	3.7816	3.3192	3.0173	2.8020	2.6393	2.5113	2.4073

(*continued*)

TABLE A-5 *F* Distribution ($\alpha = 0.01$ in the right tail) *(continued)*

Denominator degrees of freedom (df_2)	Numerator degrees of freedom (df_1)									
	10	12	15	20	24	30	40	60	120	∞
1	6055.8	6106.3	6157.3	6208.7	6234.6	6260.6	6286.8	6313.0	6339.4	6365.9
2	99.399	99.416	99.433	99.449	99.458	99.466	99.474	99.482	99.491	99.499
3	27.229	27.052	26.872	26.690	26.598	26.505	26.411	26.316	26.221	26.125
4	14.546	14.374	14.198	14.020	13.929	13.838	13.745	13.652	13.558	13.463
5	10.051	9.8883	9.7222	9.5526	9.4665	9.3793	9.2912	9.2020	9.1118	9.0204
6	7.8741	7.7183	7.5590	7.3958	7.3127	7.2285	7.1432	7.0567	6.9690	6.8800
7	6.6201	6.4691	6.3143	6.1554	6.0743	5.9920	5.9084	5.8236	5.7373	5.6495
8	5.8143	5.6667	5.5151	5.3591	5.2793	5.1981	5.1156	5.0316	4.9461	4.8588
9	5.2565	5.1114	4.9621	4.8080	4.7290	4.6486	4.5666	4.4831	4.3978	4.3105
10	4.8491	4.7059	4.5581	4.4054	4.3269	4.2469	4.1653	4.0819	3.9965	3.9090
11	4.5393	4.3974	4.2509	4.0990	4.0209	3.9411	3.8596	3.7761	3.6904	3.6024
12	4.2961	4.1553	4.0096	3.8584	3.7805	3.7008	3.6192	3.5355	3.4494	3.3608
13	4.1003	4.9603	3.8154	3.6646	3.5868	3.5070	3.4253	3.3413	3.2548	3.1654
14	3.9394	3.8001	3.6557	3.5052	3.4274	3.3476	3.2656	3.1813	3.0942	3.0040
15	3.8049	3.6662	3.5222	3.3719	3.2940	3.2141	3.1319	3.0471	2.9595	2.8684
16	3.6909	3.5527	3.4089	3.2587	3.1808	3.1007	3.0182	2.9330	2.8447	2.7528
17	3.5931	3.4552	3.3117	3.1615	3.0835	3.0032	2.9205	2.8348	2.7459	2.6530
18	3.5082	3.3706	3.2273	3.0771	2.9990	2.9185	2.8354	2.7493	2.6597	2.5660
19	3.4338	3.2965	3.1533	3.0031	2.9249	2.8442	2.7608	2.6742	2.5839	2.4893
20	3.3682	3.2311	3.0880	2.9377	2.8594	2.7785	2.6947	2.6077	2.5168	2.4212
21	3.3098	3.1730	3.0300	2.8796	2.8010	2.7200	2.6359	2.5484	2.4568	2.3603
22	3.2576	3.1209	2.9779	2.8274	2.7488	2.6675	2.5831	2.4951	2.4029	2.3055
23	3.2106	3.0740	2.9311	2.7805	2.7017	2.6202	2.5355	2.4471	2.3542	2.2558
24	3.1681	3.0316	2.8887	2.7380	2.6591	2.5773	2.4923	2.4035	2.3100	2.2107
25	3.1294	2.9931	2.8502	2.6993	2.6203	2.5383	2.4530	2.3637	2.2696	2.1694
26	3.0941	2.9578	2.8150	2.6640	2.5848	2.5026	2.4170	2.3273	2.2325	2.1315
27	3.0618	2.9256	2.7827	2.6316	2.5522	2.4699	2.3840	2.2938	2.1985	2.0965
28	3.0320	2.8959	2.7530	2.6017	2.5223	2.4397	2.3535	2.2629	2.1670	2.0642
29	3.0045	2.8685	2.7256	2.5742	2.4946	2.4118	2.3253	2.2344	2.1379	2.0342
30	2.9791	2.8431	2.7002	2.5487	2.4689	2.3860	2.2992	2.2079	2.1108	2.0062
40	2.8005	2.6648	2.5216	2.3689	2.2880	2.2034	2.1142	2.0194	1.9172	1.8047
60	2.6318	2.4961	2.3523	2.1978	2.1154	2.0285	1.9360	1.8363	1.7263	1.6006
120	2.4721	2.3363	2.1915	2.0346	1.9500	1.8600	1.7628	1.6557	1.5330	1.3805
∞	2.3209	2.1847	2.0385	1.8783	1.7908	1.6964	1.5923	1.4730	1.3246	1.0000

From Maxine Merrington and Catherine M. Thompson, "Tables of Percentage Points of the Inverted Beta (F) Distribution," *Biometrika 33* (1943): 80–84. Reproduced with permission of the Biometrika Trustees.

(continued)

TABLE A-5 *F* Distribution ($\alpha = 0.025$ in the right tail)

Numerator degrees of freedom (df_1); rows: Denominator degrees of freedom (df_2)

df_2	1	2	3	4	5	6	7	8	9
1	647.79	799.50	864.16	899.58	921.85	937.11	948.22	956.66	963.28
2	38.506	39.000	39.165	39.248	39.298	39.331	39.335	39.373	39.387
3	17.443	16.044	15.439	15.101	14.885	14.735	14.624	14.540	14.473
4	12.218	10.649	9.9792	9.6045	9.3645	9.1973	9.0741	8.9796	8.9047
5	10.007	8.4336	7.7636	7.3879	7.1464	6.9777	6.8531	6.7572	6.6811
6	8.8131	7.2599	6.5988	6.2272	5.9876	5.8198	5.6955	5.5996	5.5234
7	8.0727	6.5415	5.8898	5.5226	5.2852	5.1186	4.9949	4.8993	4.8232
8	7.5709	6.0595	5.4160	5.0526	4.8173	4.6517	4.5286	4.4333	4.3572
9	7.2093	5.7147	5.0781	4.7181	4.4844	4.3197	4.1970	4.1020	4.0260
10	6.9367	5.4564	4.8256	4.4683	4.2361	4.0721	3.9498	3.8549	3.7790
11	6.7241	5.2559	4.6300	4.2751	4.0440	3.8807	3.7586	3.6638	3.5879
12	6.5538	5.0959	4.4742	4.1212	3.8911	3.7283	3.6065	3.5118	3.4358
13	6.4143	4.9653	4.3472	3.9959	3.7667	3.6043	3.4827	3.3880	3.3120
14	6.2979	4.8567	4.2417	3.8919	3.6634	3.5014	3.3799	3.2853	3.2093
15	6.1995	4.7650	4.1528	3.8043	3.5764	3.4147	3.2934	3.1987	3.1227
16	6.1151	4.6867	4.0768	3.7294	3.5021	3.3406	3.2194	3.1248	3.0488
17	6.0420	4.6189	4.0112	3.6648	3.4379	3.2767	3.1556	3.0610	2.9849
18	5.9781	4.5597	3.9539	3.6083	3.3820	3.2209	3.0999	3.0053	2.9291
19	5.9216	4.5075	3.9034	3.5587	3.3327	3.1718	3.0509	2.9563	2.8801
20	5.8715	4.4613	3.8587	3.5147	3.2891	3.1283	3.0074	2.9128	2.8365
21	5.8266	4.4199	3.8188	3.4754	3.2501	3.0895	2.9686	2.8740	2.7977
22	5.7863	4.3828	3.7829	3.4401	3.2151	3.0546	2.9338	2.8392	2.7628
23	5.7498	4.3492	3.7505	3.4083	3.1835	3.0232	2.9023	2.8077	2.7313
24	5.7166	4.3187	3.7211	3.3794	3.1548	2.9946	2.8738	2.7791	2.7027
25	5.6864	4.2909	3.6943	3.3530	3.1287	2.9685	2.8478	2.7531	2.6766
26	5.6586	4.2655	3.6697	3.3289	3.1048	2.9447	2.8240	2.7293	2.6528
27	5.6331	4.2421	3.6472	3.3067	3.0828	2.9228	2.8021	2.7074	2.6309
28	5.6096	4.2205	3.6264	3.2863	3.0626	2.9027	2.7820	2.6872	2.6106
29	5.5878	4.2006	3.6072	3.2674	3.0438	2.8840	2.7633	2.6686	2.5919
30	5.5675	4.1821	3.5894	3.2499	3.0265	2.8667	2.7460	2.6513	2.5746
40	5.4239	4.0510	3.4633	3.1261	2.9037	2.7444	2.6238	2.5289	2.4519
60	5.2856	3.9253	3.3425	3.0077	2.7863	2.6274	2.5068	2.4117	2.3344
120	5.1523	3.8046	3.2269	2.8943	2.6740	2.5154	2.3948	2.2994	2.2217
∞	5.0239	3.6889	3.1161	2.7858	2.5665	2.4082	2.2875	2.1918	2.1136

(continued)

TABLE A-5 *F* Distribution ($\alpha = 0.025$ in the right tail) *(continued)*

Numerator degrees of freedom (df_1); rows: Denominator degrees of freedom (df_2)

df_2	10	12	15	20	24	30	40	60	120	∞
1	968.63	976.71	984.87	993.10	997.25	1001.4	1005.6	1009.8	1014.0	1018.3
2	39.398	39.415	39.431	39.448	39.456	39.465	39.473	39.481	39.490	39.498
3	14.419	14.337	14.253	14.167	14.124	14.081	14.037	13.992	13.947	13.902
4	8.8439	8.7512	8.6565	8.5599	8.5109	8.4613	8.4111	8.3604	8.3092	8.2573
5	6.6192	6.5245	6.4277	6.3286	6.2780	6.2269	6.1750	6.1225	6.0693	6.0153
6	5.4613	5.3662	5.2687	5.1684	5.1172	5.0652	5.0125	4.9589	4.9044	4.8491
7	4.7611	4.6658	4.5678	4.4667	4.4150	4.3624	4.3089	4.2544	4.1989	4.1423
8	4.2951	4.1997	4.1012	3.9995	3.9472	3.8940	3.8398	3.7844	3.7279	3.6702
9	3.9639	3.8682	3.7694	3.6669	3.6142	3.5604	3.5055	3.4493	3.3918	3.3329
10	3.7168	3.6209	3.5217	3.4185	3.3654	3.3110	3.2554	3.1984	3.1399	3.0798
11	3.5257	3.4296	3.3299	3.2261	3.1725	3.1176	3.0613	3.0035	2.9441	2.8828
12	3.3736	3.2773	3.1772	3.0728	3.0187	2.9633	2.9063	2.8478	2.7874	2.7249
13	3.2497	3.1532	3.0527	2.9477	2.8932	2.8372	2.7797	2.7204	2.6590	2.5955
14	3.1469	3.0502	2.9493	2.8437	2.7888	2.7324	2.6742	2.6142	2.5519	2.4872
15	3.0602	2.9633	2.8621	2.7559	2.7006	2.6437	2.5850	2.5242	2.4611	2.3953
16	2.9862	2.8890	2.7875	2.6808	2.6252	2.5678	2.5085	2.4471	2.3831	2.3163
17	2.9222	2.8249	2.7230	2.6158	2.5598	2.5020	2.4422	2.3801	2.3153	2.2474
18	2.8664	2.7689	2.6667	2.5590	2.5027	2.4445	2.3842	2.3214	2.2558	2.1869
19	2.8172	2.7196	2.6171	2.5089	2.4523	2.3937	2.3329	2.2696	2.2032	2.1333
20	2.7737	2.6758	2.5731	2.4645	2.4076	2.3486	2.2873	2.2234	2.1562	2.0853
21	2.7348	2.6368	2.5338	2.4247	2.3675	2.3082	2.2465	2.1819	2.1141	2.0422
22	2.6998	2.6017	2.4984	2.3890	2.3315	2.2718	2.2097	2.1446	2.0760	2.0032
23	2.6682	2.5699	2.4665	2.3567	2.2989	2.2389	2.1763	2.1107	2.0415	1.9677
24	2.6396	2.5411	2.4374	2.3273	2.2693	2.2090	2.1460	2.0799	2.0099	1.9353
25	2.6135	2.5149	2.4110	2.3005	2.2422	2.1816	2.1183	2.0516	1.9811	1.9055
26	2.5896	2.4908	2.3867	2.2759	2.2174	2.1565	2.0928	2.0257	1.9545	1.8781
27	2.5676	2.4688	2.3644	2.2533	2.1946	2.1334	2.0693	2.0018	1.9299	1.8527
28	2.5473	2.4484	2.3438	2.2324	2.1735	2.1121	2.0477	1.9797	1.9072	1.8291
29	2.5286	2.4295	2.3248	2.2131	2.1540	2.0923	2.0276	1.9591	1.8861	1.8072
30	2.5112	2.4120	2.3072	2.1952	2.1359	2.0739	2.0089	1.9400	1.8664	1.7867
40	2.3882	2.2882	2.1819	2.0677	2.0069	1.9429	1.8752	1.8028	1.7242	1.6371
60	2.2702	2.1692	2.0613	1.9445	1.8817	1.8152	1.7440	1.6668	1.5810	1.4821
120	2.1570	2.0548	1.9450	1.8249	1.7597	1.6899	1.6141	1.5299	1.4327	1.3104
∞	2.0483	1.9447	1.8326	1.7085	1.6402	1.5660	1.4835	1.3883	1.2684	1.0000

From Maxine Merrington and Catherine M. Thompson, "Tables of Percentage Points of the Inverted Beta (*F*) Distribution," *Biometrika 33* (1943): 80–84. Reproduced with permission of the Biometrika Trustees.

(continued)

TABLE A-5 *F* Distribution ($\alpha = 0.05$ in the right tail)

Denominator degrees of freedom (df_2)	Numerator degrees of freedom (df_1)								
	1	2	3	4	5	6	7	8	9
1	161.45	199.50	215.71	224.58	230.16	233.99	236.77	238.88	240.54
2	18.513	19.000	19.164	19.247	19.296	19.330	19.353	19.371	19.385
3	10.128	9.5521	9.2766	9.1172	9.0135	8.9406	8.8867	8.8452	8.8123
4	7.7086	6.9443	6.5914	6.3882	6.2561	6.1631	6.0942	6.0410	6.9988
5	6.6079	5.7861	5.4095	5.1922	5.0503	4.9503	4.8759	4.8183	4.7725
6	5.9874	5.1433	4.7571	4.5337	4.3874	4.2839	4.2067	4.1468	4.0990
7	5.5914	4.7374	4.3468	4.1203	3.9715	3.8660	3.7870	3.7257	3.6767
8	5.3177	4.4590	4.0662	3.8379	3.6875	3.5806	3.5005	3.4381	3.3881
9	5.1174	4.2565	3.8625	3.6331	3.4817	3.3738	3.2927	3.2296	3.1789
10	4.9646	4.1028	3.7083	3.4780	3.3258	3.2172	3.1355	3.0717	3.0204
11	4.8443	3.9823	3.5874	3.3567	3.2039	3.0946	3.0123	2.9480	2.8962
12	4.7472	3.8853	3.4903	3.2592	3.1059	2.9961	2.9134	2.8486	2.7964
13	4.6672	3.8056	3.4105	3.1791	3.0254	2.9153	2.8321	2.7669	2.7144
14	4.6001	3.7389	3.3439	3.1122	2.9582	2.8477	2.7642	2.6987	2.6458
15	4.5431	3.6823	3.2874	3.0556	2.9013	2.7905	2.7066	2.6408	2.5876
16	4.4940	3.6337	3.2389	3.0069	2.8524	2.7413	2.6572	2.5911	2.5377
17	4.4513	3.5915	3.1968	2.9647	2.8100	2.6987	2.6143	2.5480	2.4943
18	4.4139	3.5546	3.1599	2.9277	2.7729	2.6613	2.5767	2.5102	2.4563
19	4.3807	3.5219	3.1274	2.8951	2.7401	2.6283	2.5435	2.4768	2.4227
20	4.3512	3.4928	3.0984	2.8661	2.7109	2.5990	2.5140	2.4471	2.3928
21	4.3248	3.4668	3.0725	2.8401	2.6848	2.5727	2.4876	2.4205	2.3660
22	4.3009	3.4434	3.0491	2.8167	2.6613	2.5491	2.4638	2.3965	2.3419
23	4.2793	3.4221	3.0280	2.7955	2.6400	2.5277	2.4422	2.3748	2.3201
24	4.2597	3.4028	3.0088	2.7763	2.6207	2.5082	2.4226	2.3551	2.3002
25	4.2417	3.3852	2.9912	2.7587	2.6030	2.4904	2.4047	2.3371	2.2821
26	4.2252	3.3690	2.9752	2.7426	2.5868	2.4741	2.3883	2.3205	2.2655
27	4.2100	3.3541	2.9604	2.7278	2.5719	2.4591	2.3732	2.3053	2.2501
28	4.1960	3.3404	2.9467	2.7141	2.5581	2.4453	2.3593	2.2913	2.2360
29	4.1830	3.3277	2.9340	2.7014	2.5454	2.4324	2.3463	2.2783	2.2229
30	4.1709	3.3158	2.9223	2.6896	2.5336	2.4205	2.3343	2.2662	2.2107
40	4.0847	3.2317	2.8387	2.6060	2.4495	2.3359	2.2490	2.1802	2.1240
60	4.0012	3.1504	2.7581	2.5252	2.3683	2.2541	2.1665	2.0970	2.0401
120	3.9201	3.0718	2.6802	2.4472	2.2899	2.1750	2.0868	2.0164	1.9588
∞	3.8415	2.9957	2.6049	2.3719	2.2141	2.0986	2.0096	1.9384	1.8799

0.05
F

(continued)

Table A-5 *F* Distribution ($\alpha = 0.05$ in the right tail) *(continued)*

Numerator degrees of freedom (df_1)

Denominator degrees of freedom (df_2)	10	12	15	20	24	30	40	60	120	∞
1	241.88	243.91	245.95	248.01	249.05	250.10	251.14	252.20	253.25	254.31
2	19.396	19.413	19.429	19.446	19.454	19.462	19.471	19.479	19.487	19.496
3	8.7855	8.7446	8.7029	8.6602	8.6385	8.6166	8.5944	8.5720	8.5494	8.5264
4	5.9644	5.9117	5.8578	5.8025	5.7744	5.7459	5.7170	5.6877	5.6581	5.6281
5	4.7351	4.6777	4.6188	4.5581	4.5272	4.4957	4.4638	4.4314	4.3985	4.3650
6	4.0600	3.9999	3.9381	3.8742	3.8415	3.8082	3.7743	3.7398	3.7047	3.6689
7	3.6365	3.5747	3.5107	3.4445	3.4105	3.3758	3.3404	3.3043	3.2674	3.2298
8	3.3472	3.2839	3.2184	3.1503	3.1152	3.0794	3.0428	3.0053	2.9669	2.9276
9	3.1373	3.0729	3.0061	2.9365	2.9005	2.8637	2.8259	2.7872	2.7475	2.7067
10	2.9782	2.9130	2.8450	2.7740	2.7372	2.6996	2.6609	2.6211	2.5801	2.5379
11	2.8536	2.7876	2.7186	2.6464	2.6090	2.5705	2.5309	2.4901	2.4480	2.4045
12	2.7534	2.6866	2.6169	2.5436	2.5055	2.4663	2.4259	2.3842	2.3410	2.2962
13	2.6710	2.6037	2.5331	2.4589	2.4202	2.3803	2.3392	2.2966	2.2524	2.2064
14	2.6022	2.5342	2.4630	2.3879	2.3487	2.3082	2.2664	2.2229	2.1778	2.1307
15	2.5437	2.4753	2.4034	2.3275	2.2878	2.2468	2.2043	2.1601	2.1141	2.0658
16	2.4935	2.4247	2.3522	2.2756	2.2354	2.1938	2.1507	2.1058	2.0589	2.0096
17	2.4499	2.3807	2.3077	2.2304	2.1898	2.1477	2.1040	2.0584	2.0107	1.9604
18	2.4117	2.3421	2.2686	2.1906	2.1497	2.1071	2.0629	2.0166	1.9681	1.9168
19	2.3779	2.3080	2.2341	2.1555	2.1141	2.0712	2.0264	1.9795	1.9302	1.8780
20	2.3479	2.2776	2.2033	2.1242	2.0825	2.0391	1.9938	1.9464	1.8963	1.8432
21	2.3210	2.2504	2.1757	2.0960	2.0540	2.0102	1.9645	1.9165	1.8657	1.8117
22	2.2967	2.2258	2.1508	2.0707	2.0283	1.9842	1.9380	1.8894	1.8380	1.7831
23	2.2747	2.2036	2.1282	2.0476	2.0050	1.9605	1.9139	1.8648	1.8128	1.7570
24	2.2547	2.1834	2.1077	2.0267	1.9838	1.9390	1.8920	1.8424	1.7896	1.7330
25	2.2365	2.1649	2.0889	2.0075	1.9643	1.9192	1.8718	1.8217	1.7684	1.7110
26	2.2197	2.1479	2.0716	1.9898	1.9464	1.9010	1.8533	1.8027	1.7488	1.6906
27	2.2043	2.1323	2.0558	1.9736	1.9299	1.8842	1.8361	1.7851	1.7306	1.6717
28	2.1900	2.1179	2.0411	1.9586	1.9147	1.8687	1.8203	1.7689	1.7138	1.6541
29	2.1768	2.1045	2.0275	1.9446	1.9005	1.8543	1.8055	1.7537	1.6981	1.6376
30	2.1646	2.0921	2.0148	1.9317	1.8874	1.8409	1.7918	1.7396	1.6835	1.6223
40	2.0772	2.0035	1.9245	1.8389	1.7929	1.7444	1.6928	1.6373	1.5766	1.5089
60	1.9926	1.9174	1.8364	1.7480	1.7001	1.6491	1.5943	1.5343	1.4673	1.3893
120	1.9105	1.8337	1.7505	1.6587	1.6084	1.5543	1.4952	1.4290	1.3519	1.2539
∞	1.8307	1.7522	1.6664	1.5705	1.5173	1.4591	1.3940	1.3180	1.2214	1.0000

From Maxine Merrington and Catherine M. Thompson, "Tables of Percentage Points of the Inverted Beta (*F*) Distribution," *Biometrika 33* (1943): 80–84. Reproduced with permission of the Biometrika Trustees.

TABLE A-6 Critical Values of the Pearson Correlation Coefficient r

n	$\alpha = .05$	$\alpha = .01$
4	.950	.999
5	.878	.959
6	.811	.917
7	.754	.875
8	.707	.834
9	.666	.798
10	.632	.765
11	.602	.735
12	.576	.708
13	.553	.684
14	.532	.661
15	.514	.641
16	.497	.623
17	.482	.606
18	.468	.590
19	.456	.575
20	.444	.561
25	.396	.505
30	.361	.463
35	.335	.430
40	.312	.402
45	.294	.378
50	.279	.361
60	.254	.330
70	.236	.305
80	.220	.286
90	.207	.269
100	.196	.256

NOTE: To test H_0: $\rho = 0$ against H_1: $\rho \neq 0$, reject H_0 if the absolute value of r is greater than the critical value in the table.

TABLE A-7 Critical Values for the Sign Test

	α			
n	.005 (one tail) .01 (two tails)	.01 (one tail) .02 (two tails)	.025 (one tail) .05 (two tails)	.05 (one tail) .10 (two tails)
1	*	*	*	*
2	*	*	*	*
3	*	*	*	*
4	*	*	*	*
5	*	*	*	0
6	*	*	0	0
7	*	0	0	0
8	0	0	0	1
9	0	0	1	1
10	0	0	1	1
11	0	1	1	2
12	1	1	2	2
13	1	1	2	3
14	1	2	2	3
15	2	2	3	3
16	2	2	3	4
17	2	3	4	4
18	3	3	4	5
19	3	4	4	5
20	3	4	5	5
21	4	4	5	6
22	4	5	5	6
23	4	5	6	7
24	5	5	6	7
25	5	6	7	7

NOTES:

1. * indicates that it is not possible to get a value in the critical region.
2. Reject the null hypothesis if the number of the less frequent sign (x) is less than or equal to the value in the table.
3. For values of n greater than 25, a normal approximation is used with

$$z = \frac{(x + 0.5) - \left(\frac{n}{2}\right)}{\frac{\sqrt{n}}{2}}$$

TABLE A-8 Critical Values of T for the Wilcoxon Signed-Ranks Test

	α			
n	.005 (one tail) .01 (two tails)	.01 (one tail) .02 (two tails)	.025 (one tail) .05 (two tails)	.05 (one tail) .10 (two tails)
5	*	*	*	1
6	*	*	1	2
7	*	0	2	4
8	0	2	4	6
9	2	3	6	8
10	3	5	8	11
11	5	7	11	14
12	7	10	14	17
13	10	13	17	21
14	13	16	21	26
15	16	20	25	30
16	19	24	30	36
17	23	28	35	41
18	28	33	40	47
19	32	38	46	54
20	37	43	52	60
21	43	49	59	68
22	49	56	66	75
23	55	62	73	83
24	61	69	81	92
25	68	77	90	101
26	76	85	98	110
27	84	93	107	120
28	92	102	117	130
29	100	111	127	141
30	109	120	137	152

NOTES:

1. * indicates that it is not possible to get a value in the critical region.
2. Reject the null hypothesis if the test statistic T is less than or equal to the critical value found in this table. Fail to reject the null hypothesis if the test statistic T is greater than the critical value found in the table.

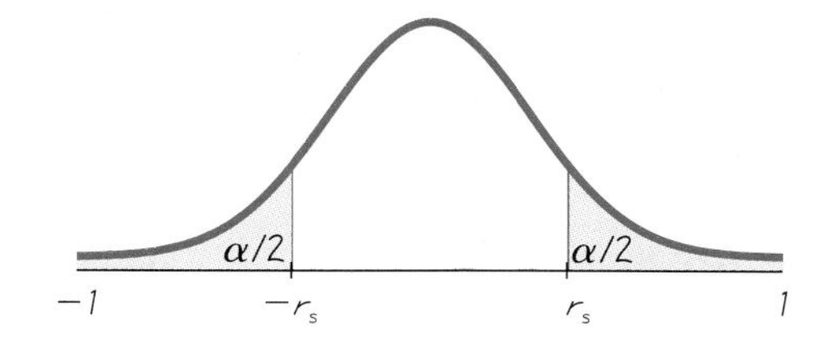

TABLE A-9 Critical Values of Spearman's Rank Correlation Coefficient r_s

n	$\alpha = 0.10$	$\alpha = 0.05$	$\alpha = 0.02$	$\alpha = 0.01$
5	.900	—	—	—
6	.829	.886	.943	—
7	.714	.786	.893	—
8	.643	.738	.833	.881
9	.600	.683	.783	.833
10	.564	.648	.745	.794
11	.523	.623	.736	.818
12	.497	.591	.703	.780
13	.475	.566	.673	.745
14	.457	.545	.646	.716
15	.441	.525	.623	.689
16	.425	.507	.601	.666
17	.412	.490	.582	.645
18	.399	.476	.564	.625
19	.388	.462	.549	.608
20	.377	.450	.534	.591
21	.368	.438	.521	.576
22	.359	.428	.508	.562
23	.351	.418	.496	.549
24	.343	.409	.485	.537
25	.336	.400	.475	.526
26	.329	.392	.465	.515
27	.323	.385	.456	.505
28	.317	.377	.448	.496
29	.311	.370	.440	.487
30	.305	.364	.432	.478

NOTE: For $n > 30$, use $r_s = \pm z/\sqrt{n-1}$, where z corresponds to the level of significance. For example, if $\alpha = 0.05$, then $z = 1.96$.

To test H_0: $\rho_s = 0$

against H_1: $\rho_s \neq 0$

From "Distribution of sums of squares of rank differences to small numbers of individuals," *The Annals of Mathematical Statistics,* Vol. 9, No. 2. Reprinted with permission of the Institute of Mathematical Statistics.

TABLE A-10 Critical Values for Number of Runs G

Value of n_1 \ Value of n_2	2	3	4	5	6	7	8	9	10	11	12	13	14	15	16	17	18	19	20
2	1 6	1 6	1 6	1 6	1 6	1 6	1 6	1 6	1 6	1 6	2 6	2 6	2 6	2 6	2 6	2 6	2 6	2 6	2 6
3	1 6	1 8	1 8	1 8	2 8	2 8	2 8	2 8	2 8	2 8	2 8	2 8	2 8	3 8	3 8	3 8	3 8	3 8	3 8
4	1 6	1 8	1 9	2 9	2 9	2 10	3 10	3 10	3 10	3 10	3 10	3 10	3 10	3 10	4 10	4 10	4 10	4 10	4 10
5	1 6	1 8	2 9	2 10	3 10	3 11	3 11	3 12	3 12	4 12	4 12	4 12	4 12	4 12	4 12	4 12	5 12	5 12	5 12
6	1 6	2 8	2 9	3 10	3 11	3 12	3 12	4 13	4 13	4 13	4 13	5 14	5 14	5 14	5 14	5 14	5 14	6 14	6 14
7	1 6	2 8	2 10	3 11	3 12	3 13	4 13	4 14	5 14	5 14	5 14	5 15	5 15	6 15	6 16	6 16	6 16	6 16	6 16
8	1 6	2 8	3 10	3 11	3 12	4 13	4 14	5 14	5 15	5 15	6 16	6 16	6 16	6 16	6 17	7 17	7 17	7 17	7 17
9	1 6	2 8	3 10	3 12	4 13	4 14	5 14	5 15	5 16	6 16	6 16	6 17	7 17	7 18	7 18	7 18	8 18	8 18	8 18
10	1 6	2 8	3 10	3 12	4 13	5 14	5 15	5 16	6 16	6 17	7 17	7 18	7 18	7 18	8 19	8 19	8 19	8 20	9 20
11	1 6	2 8	3 10	4 12	4 13	5 14	5 15	6 16	6 17	7 17	7 18	7 19	8 19	8 19	8 20	9 20	9 20	9 21	9 21
12	2 6	2 8	3 10	4 12	4 13	5 14	6 16	6 16	7 17	7 18	7 19	8 19	8 20	8 20	9 21	9 21	9 21	10 22	10 22
13	2 6	2 8	3 10	4 12	5 14	5 15	6 16	6 17	7 18	7 19	8 19	8 20	9 20	9 21	9 21	10 22	10 22	10 23	10 23
14	2 6	2 8	3 10	4 12	5 14	5 15	6 16	7 17	7 18	8 19	8 20	9 20	9 21	9 22	10 22	10 23	10 23	11 23	11 24
15	2 6	3 8	3 10	4 12	5 14	6 15	6 16	7 18	7 18	8 19	8 20	9 21	9 22	10 22	10 23	11 23	11 24	11 24	12 25
16	2 6	3 8	4 10	4 12	5 14	6 16	6 17	7 18	8 19	8 20	9 21	9 21	10 22	10 23	11 23	11 24	11 25	12 25	12 25
17	2 6	3 8	4 10	4 12	5 14	6 16	7 17	7 18	8 19	9 20	9 21	10 22	10 23	11 23	11 24	11 25	12 25	12 26	13 26
18	2 6	3 8	4 10	5 12	5 14	6 16	7 17	8 18	8 19	9 20	9 21	10 22	10 23	11 24	11 25	12 25	12 26	13 26	13 27
19	2 6	3 8	4 10	5 12	6 14	6 16	7 17	8 18	8 20	9 21	10 22	10 23	11 23	11 24	12 25	12 26	13 26	13 27	13 27
20	2 6	3 8	4 10	5 12	6 14	6 16	7 17	8 18	9 20	9 21	10 22	10 23	11 24	12 25	12 25	13 26	13 27	13 27	14 28

NOTE:

1. The entries in this table are the critical G values, assuming a two-tailed test with a significance level of $\alpha = 0.05$.
2. The null hypothesis of randomness is rejected if the total number of runs G is less than or equal to the smaller entry or greater than or equal to the larger entry.

From "Tables for testing randomness of groupings in a sequence of alternatives," *The Annals of Mathematical Statistics,* Vol. 14, No. 1. Reprinted with permission of the Institute of Mathematical Statistics.

Appendix B: Data Sets

The Appendix B Data Sets (except Data Set 6) are available on the CD-ROM included with this book.

Data Set 1: Weights (in pounds) and Volumes (in ounces) of Cola

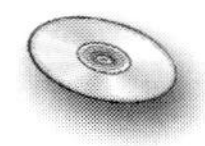

STATDISK: Variable names are CKREGWT, CKREGVOL, CKDIETWT, CKDTVOL, PPREGWT, PPREGVOL, PPDIETWT, and PPDTVOL.
Minitab: Worksheet name is COLA.MTW.
Excel: Workbook name is COLA.XLS.
TI-83 Plus and text files: CRGWT, CRGVL, CDTWT, CDTVL, PRGWT, PRGVL, PDTWT, and PDTVL.

WEIGHT REGULAR COKE	VOLUME REGULAR COKE	WEIGHT DIET COKE	VOLUME DIET COKE	WEIGHT REGULAR PEPSI	VOLUME REGULAR PEPSI	WEIGHT DIET PEPSI	VOLUME DIET PEPSI
0.8192	12.3	0.7773	12.1	0.8258	12.4	0.7925	12.3
0.8150	12.1	0.7758	12.1	0.8156	12.2	0.7868	12.2
0.8163	12.2	0.7896	12.3	0.8211	12.2	0.7846	12.2
0.8211	12.3	0.7868	12.3	0.8170	12.2	0.7938	12.3
0.8181	12.2	0.7844	12.2	0.8216	12.2	0.7861	12.2
0.8247	12.3	0.7861	12.3	0.8302	12.4	0.7844	12.2
0.8062	12.0	0.7806	12.2	0.8192	12.2	0.7795	12.2
0.8128	12.1	0.7830	12.2	0.8192	12.2	0.7883	12.3
0.8172	12.2	0.7852	12.2	0.8271	12.3	0.7879	12.2
0.8110	12.1	0.7879	12.3	0.8251	12.3	0.7850	12.3
0.8251	12.3	0.7881	12.3	0.8227	12.2	0.7899	12.3
0.8264	12.3	0.7826	12.3	0.8256	12.3	0.7877	12.2
0.7901	11.8	0.7923	12.3	0.8139	12.2	0.7852	12.2
0.8244	12.3	0.7852	12.3	0.8260	12.3	0.7756	12.1
0.8073	12.1	0.7872	12.3	0.8227	12.2	0.7837	12.2
0.8079	12.1	0.7813	12.2	0.8388	12.5	0.7879	12.2
0.8044	12.0	0.7885	12.3	0.8260	12.3	0.7839	12.2
0.8170	12.2	0.7760	12.1	0.8317	12.4	0.7817	12.2
0.8161	12.2	0.7822	12.2	0.8247	12.3	0.7822	12.2
0.8194	12.2	0.7874	12.3	0.8200	12.2	0.7742	12.1
0.8189	12.2	0.7822	12.2	0.8172	12.2	0.7833	12.2
0.8194	12.2	0.7839	12.2	0.8227	12.3	0.7835	12.2
0.8176	12.2	0.7802	12.1	0.8244	12.3	0.7855	12.2
0.8284	12.4	0.7892	12.3	0.8244	12.2	0.7859	12.2
0.8165	12.2	0.7874	12.2	0.8319	12.4	0.7775	12.1
0.8143	12.2	0.7907	12.3	0.8247	12.3	0.7833	12.2
0.8229	12.3	0.7771	12.1	0.8214	12.2	0.7835	12.2
0.8150	12.2	0.7870	12.2	0.8291	12.4	0.7826	12.2
0.8152	12.2	0.7833	12.3	0.8227	12.3	0.7815	12.2
0.8244	12.3	0.7822	12.2	0.8211	12.3	0.7791	12.1
0.8207	12.2	0.7837	12.3	0.8401	12.5	0.7866	12.3
0.8152	12.2	0.7910	12.4	0.8233	12.3	0.7855	12.2
0.8126	12.1	0.7879	12.3	0.8291	12.4	0.7848	12.2
0.8295	12.4	0.7923	12.4	0.8172	12.2	0.7806	12.2
0.8161	12.2	0.7859	12.3	0.8233	12.4	0.7773	12.1
0.8192	12.2	0.7811	12.2	0.8211	12.3	0.7775	12.1

Data Set 2: Textbook Prices

*Denotes Paperback.

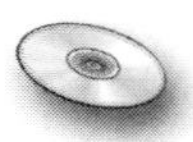

STATDISK: Variable names are NEWUMASS, USDUMASS, NEWDCC, and USEDDCC.
Minitab: Worksheet name is TEXTBOOK.MTW.
Excel: Workbook name is TEXTBOOK.XLS.
TI-83 Plus and text files: NWUMA, USUMA, NWDCC, and USDCC.

UNIVERSITY OF MASSACHUSETTS

	NEW	USED
Accounting	84.30	66.55
Advertising	83.35	62.55
Algebra	80.55	60.45
Anthropology*	41.95	31.50
Art. Intell.	67.00	50.25
Astronomy	52.00	39.00
Biology	94.00	70.50
Bus. Management	67.70	50.80
Bus. Statistics	75.35	56.55
Calculus (Hvd)*	72.70	54.55
Calculus (Stewart)	102.95	77.25
Chinese*	27.95	21.00
Cooking	54.70	41.05
Differential Equations	86.70	65.05
Discrete Math	78.00	58.50
Drawing*	57.35	43.05
Economics	86.00	64.50
Education	73.00	54.75
Elementary Statistics	76.00	57.00
Engineering	88.00	66.00
Engineering Design	45.60	36.00
Finance	77.00	57.75
History*	26.70	22.00
History*	24.00	18.00
Inorganic Chemistry	92.00	69.00
Literature*	50.70	38.05
Marketing	82.00	61.50
Micro Biology	89.35	67.05
Music*	38.70	29.05
Nutrition*	60.00	45.00
Organic Chemistry	79.20	59.40
Organic Chemistry*	48.70	36.55
Photography*	60.00	45.00
Physics	95.00	71.25
Prob. Statistics	92.70	69.55
Psychology	70.00	52.50
Public Speaking*	41.60	31.20
Religion*	18.95	14.25
Russian*	28.00	21.00
Sociology*	34.95	26.25

DUTCHESS COMMUNITY COLLEGE

	NEW	USED
Accounting I	54.35	40.75
Accounting II	80.00	60.00
Architecture	65.65	49.25
Art History	69.35	50.25
Graphics Design*	50.65	36.00
Banking	60.00	45.00
Biology	75.95	56.95
Business	75.00	56.25
Heredity*	62.00	46.50
Marketing	41.95	31.45
Chemistry*	61.00	45.75
Systems Analysis	75.65	53.00
Music	33.35	25.00
Nutrition	58.65	44.00
Microeconomics*	51.35	38.50
Literature	44.65	33.50
Poetry*	28.35	21.25
Programming*	44.00	33.00
Engineering	80.00	60.00
French	53.35	40.00
Government	57.55	43.15
Sexuality	61.00	45.75
History	63.60	47.70
Italian	59.35	44.50
Precalculus	64.55	48.40
Calculus*	60.00	45.00
Parasitology	38.45	28.85
Philosophy	53.35	40.00
Physics	88.00	66.00
Psychology	77.80	58.35
Gerontology	61.00	45.75
Speech*	39.15	27.85
Social Problems	50.15	37.60
Typographic Design*	42.65	32.00
Writing*	35.00	26.25

Data Set 3: Diamonds

Price is in dollars. Depth is 100 times the ratio of height to diameter. Table is size of the upper flat surface. (Depth and table determine "cut.") Color indices are on a standard scale, with 1 = colorless and increasing numbers indicating more yellow. On the clarity scale, 1 = flawless and 6 indicates inclusions that can be seen by eye.

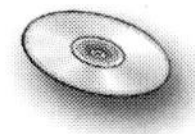

STATDISK: Variable names are DMDPRICE, DMDCARAT, DMDDEPTH, DMDTABLE, DMDCOLOR, and DMDCLRTY.
Minitab: Worksheet name is DIAMONDS.MTW.
Excel: Workbook name is DIAMONDS.XLS.
TI-83 Plus and text files: PRICE, CARAT, DEPTH, TABLE, COLOR, and CLRTY.

PRICE	CARAT	DEPTH	TABLE	COLOR	CLARITY
6958	1.00	60.5	65	3	4
5885	1.00	59.2	65	5	4
6333	1.01	62.3	55	4	4
4299	1.01	64.4	62	5	5
9589	1.02	63.9	58	2	3
6921	1.04	60.0	61	4	4
4426	1.04	62.0	62	5	5
6885	1.07	63.6	61	4	3
5826	1.07	61.6	62	5	5
3670	1.11	60.4	60	9	4
7176	1.12	60.2	65	2	3
7497	1.16	59.5	60	5	3
5170	1.20	62.6	61	6	4
5547	1.23	59.2	65	7	4
18596	1.25	61.2	61	1	2
7521	1.29	59.6	59	6	2
7260	1.50	61.1	65	6	4
8139	1.51	63.0	60	6	4
12196	1.67	58.7	64	3	5
14998	1.72	58.5	61	4	3
9736	1.76	57.9	62	8	2
9859	1.80	59.6	63	5	5
12398	1.88	62.9	62	6	2
25322	2.03	60.1	62	2	3
11008	2.03	62.0	63	8	3
38794	2.06	58.2	63	2	2
66780	3.00	63.3	62	1	3
46769	4.01	57.1	51	3	4
28800	4.01	63.0	63	6	5
28868	4.05	59.3	60	7	4

Data Set 4: Weights (in grams) of Domino Sugar Packets

STATDISK: Variable name is SUGAR.
Minitab: Worksheet name is SUGAR.MTW.
Excel: Workbook name is SUGAR.XLS.
TI-83 Plus and text file: SUGAR

3.647	3.638	3.635	3.645	3.521	3.617	3.666
3.588	3.545	3.590	3.621	3.532	3.511	3.516
3.531	3.678	3.643	3.583	3.723	3.673	3.588
3.600	3.611	3.580	3.667	3.506	3.632	3.450
3.660	3.569	3.573	3.526	3.494	3.601	3.604
3.407	3.522	3.598	3.585	3.577	3.522	3.464
3.604	3.508	3.718	3.635	3.643	3.507	3.687
3.582	3.622	3.654	3.482	3.494	3.475	3.492
3.542	3.625	3.688	3.468	3.639	3.582	3.491
3.535	3.548	3.671	3.665	3.726	3.576	3.725

Data Set 5: Weights (in pounds) of Discarded Household Garbage for One Week

HHSIZE = household size
METAL = weight of discarded metals
PAPER = weight of discarded paper goods
PLAS = weight of discarded plastic goods
GLASS = weight of discarded glass products
FOOD = weight of discarded food items
YARD = weight of discarded yard waste
TEXT = weight of discarded textile goods
OTHER = weight of discarded goods not included in the above categories
TOTAL = total weight of discarded materials

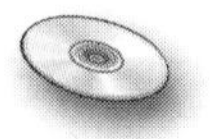

STATDISK: Variable names are HHSIZE, METAL, PAPER, PLAS, GLASS, FOOD, YARD, TEXT, OTHER, and TOTAL.
Minitab: Worksheet name is GARBAGE.MTW.
Excel: Workbook name is GARBAGE.XLS.
TI-83 Plus and text files: HHSIZ, METAL, PAPER, PLAS, GLASS, FOOD, YARD, TEXT, OTHER, and TOTAL.

HOUSEHOLD	HHSIZE	METAL	PAPER	PLAS	GLASS	FOOD	YARD	TEXT	OTHER	TOTAL
1	2	1.09	2.41	0.27	0.86	1.04	0.38	0.05	4.66	10.76
2	3	1.04	7.57	1.41	3.46	3.68	0.00	0.46	2.34	19.96
3	3	2.57	9.55	2.19	4.52	4.43	0.24	0.50	3.60	27.60
4	6	3.02	8.82	2.83	4.92	2.98	0.63	2.26	12.65	38.11
5	4	1.50	8.72	2.19	6.31	6.30	0.15	0.55	2.18	27.90
6	2	2.10	6.96	1.81	2.49	1.46	4.58	0.36	2.14	21.90
7	1	1.93	6.83	0.85	0.51	8.82	0.07	0.60	2.22	21.83
8	5	3.57	11.42	3.05	5.81	9.62	4.76	0.21	10.83	49.27
9	6	2.32	16.08	3.42	1.96	4.41	0.13	0.81	4.14	33.27
10	4	1.89	6.38	2.10	17.67	2.73	3.86	0.66	0.25	35.54
11	4	3.26	13.05	2.93	3.21	9.31	0.70	0.37	11.61	44.44
12	7	3.99	11.36	2.44	4.94	3.59	13.45	4.25	1.15	45.17
13	3	2.04	15.09	2.17	3.10	5.36	0.74	0.42	4.15	33.07
14	5	0.99	2.80	1.41	1.39	1.47	0.82	0.44	1.03	10.35
15	6	2.96	6.44	2.00	5.21	7.06	6.14	0.20	14.43	44.44
16	2	1.50	5.86	0.93	2.03	2.52	1.37	0.27	9.65	24.13
17	4	2.43	11.08	2.97	1.74	1.75	14.70	0.39	2.54	37.60
18	4	2.97	12.43	2.04	3.99	5.64	0.22	2.47	9.20	38.96
19	3	1.42	6.05	0.65	6.26	1.93	0.00	0.86	0.00	17.17
20	3	3.60	13.61	2.13	3.52	6.46	0.00	0.96	1.32	31.60
21	2	4.48	6.98	0.63	2.01	6.72	2.00	0.11	0.18	23.11
22	2	1.36	14.33	1.53	2.21	5.76	0.58	0.17	1.62	27.56
23	4	2.11	13.31	4.69	0.25	9.72	0.02	0.46	0.40	30.96
24	1	0.41	3.27	0.15	0.09	0.16	0.00	0.00	0.00	4.08

(continued)

Data Set 5 (*continued*)

HOUSEHOLD	HHSIZE	METAL	PAPER	PLAS	GLASS	FOOD	YARD	TEXT	OTHER	TOTAL
25	4	2.02	6.67	1.45	6.85	5.52	0.00	0.68	0.03	23.22
26	6	3.27	17.65	2.68	2.33	11.92	0.83	0.28	4.03	42.99
27	11	4.95	12.73	3.53	5.45	4.68	0.00	0.67	19.89	51.90
28	3	1.00	9.83	1.49	2.04	4.76	0.42	0.54	0.12	20.20
29	4	1.55	16.39	2.31	4.98	7.85	2.04	0.20	1.48	36.80
30	3	1.41	6.33	0.92	3.54	2.90	3.85	0.03	0.04	19.02
31	2	1.05	9.19	0.89	1.06	2.87	0.33	0.01	0.03	15.43
32	2	1.31	9.41	0.80	2.70	5.09	0.64	0.05	0.71	20.71
33	2	2.50	9.45	0.72	1.14	3.17	0.00	0.02	0.01	17.01
34	4	2.35	12.32	2.66	12.24	2.40	7.87	4.73	0.78	45.35
35	6	3.69	20.12	4.37	5.67	13.20	0.00	1.15	1.17	49.37
36	2	3.61	7.72	0.92	2.43	2.07	0.68	0.63	0.00	18.06
37	2	1.49	6.16	1.40	4.02	4.00	0.30	0.04	0.00	17.41
38	2	1.36	7.98	1.45	6.45	4.27	0.02	0.12	2.02	23.67
39	2	1.73	9.64	1.68	1.89	1.87	0.01	1.73	0.58	19.13
40	2	0.94	8.08	1.53	1.78	8.13	0.36	0.12	0.05	20.99
41	3	1.33	10.99	1.44	2.93	3.51	0.00	0.39	0.59	21.18
42	3	2.62	13.11	1.44	1.82	4.21	4.73	0.64	0.49	29.06
43	2	1.25	3.26	1.36	2.89	3.34	2.69	0.00	0.16	14.95
44	2	0.26	1.65	0.38	0.99	0.77	0.34	0.04	0.00	4.43
45	3	4.41	10.00	1.74	1.93	1.14	0.92	0.08	4.60	24.82
46	6	3.22	8.96	2.35	3.61	1.45	0.00	0.09	1.12	20.80
47	4	1.86	9.46	2.30	2.53	6.54	0.00	0.65	2.45	25.79
48	4	1.76	5.88	1.14	3.76	0.92	1.12	0.00	0.04	14.62
49	3	2.83	8.26	2.88	1.32	5.14	5.60	0.35	2.03	28.41
50	3	2.74	12.45	2.13	2.64	4.59	1.07	0.41	1.14	27.17
51	10	4.63	10.58	5.28	12.33	2.94	0.12	2.94	15.65	54.47
52	3	1.70	5.87	1.48	1.79	1.42	0.00	0.27	0.59	13.12
53	6	3.29	8.78	3.36	3.99	10.44	0.90	1.71	13.30	45.77
54	5	1.22	11.03	2.83	4.44	3.00	4.30	1.95	6.02	34.79
55	4	3.20	12.29	2.87	9.25	5.91	1.32	1.87	0.55	37.26
56	7	3.09	20.58	2.96	4.02	16.81	0.47	1.52	2.13	51.58
57	5	2.58	12.56	1.61	1.38	5.01	0.00	0.21	1.46	24.81
58	4	1.67	9.92	1.58	1.59	9.96	0.13	0.20	1.13	26.18
59	2	0.85	3.45	1.15	0.85	3.89	0.00	0.02	1.04	11.25
60	4	1.52	9.09	1.28	8.87	4.83	0.00	0.95	1.61	28.15
61	2	1.37	3.69	0.58	3.64	1.78	0.08	0.00	0.00	11.14
62	2	1.32	2.61	0.74	3.03	3.37	0.17	0.00	0.46	11.70

Data provided by Masakuza Tani, the Garbage Project, University of Arizona.

Data Set 6: Body Temperatures (in degrees Fahrenheit) of Healthy Adults

This is the only Appendix B data set not available on disk.

SUBJECT	AGE	SEX	SMOKE	Temperature Day 1 8 AM	Temperature Day 1 12 AM	Temperature Day 2 8 AM	Temperature Day 2 12 AM
1	22	M	Y	98.0	98.0	98.0	98.6
2	23	M	Y	97.0	97.6	97.4	—
3	22	M	Y	98.6	98.8	97.8	98.6
4	19	M	N	97.4	98.0	97.0	98.0
5	18	M	N	98.2	98.8	97.0	98.0
6	20	M	Y	98.2	98.8	96.6	99.0
7	27	M	Y	98.2	97.6	97.0	98.4
8	19	M	Y	96.6	98.6	96.8	98.4
9	19	M	Y	97.4	98.6	96.6	98.4
10	24	M	N	97.4	98.8	96.6	98.4
11	35	M	Y	98.2	98.0	96.2	98.6
12	25	M	Y	97.4	98.2	97.6	98.6
13	25	M	N	97.8	98.0	98.6	98.8
14	35	M	Y	98.4	98.0	97.0	98.6
15	21	M	N	97.6	97.0	97.4	97.0
16	33	M	N	96.2	97.2	98.0	97.0
17	19	M	Y	98.0	98.2	97.6	98.8
18	24	M	Y	—	—	97.2	97.6
19	18	F	N	—	—	97.0	97.7
20	22	F	Y	—	—	98.0	98.8
21	20	M	Y	—	—	97.0	98.0
22	30	F	Y	—	—	96.4	98.0
23	29	M	N	—	—	96.1	98.3
24	18	M	Y	—	—	98.0	98.5
25	31	M	Y	—	98.1	96.8	97.3
26	28	F	Y	—	98.2	98.2	98.7
27	27	M	Y	—	98.5	97.8	97.4
28	21	M	Y	—	98.5	98.2	98.9
29	30	M	Y	—	99.0	97.8	98.6
30	27	M	N	—	98.0	99.0	99.5
31	32	M	Y	—	97.0	97.4	97.5
32	33	M	Y	—	97.3	97.4	97.3
33	23	M	Y	—	97.3	97.5	97.6
34	29	M	Y	—	98.1	97.8	98.2
35	25	M	Y	—	—	97.9	99.6
36	31	M	N	—	97.8	97.8	98.7
37	25	M	Y	—	99.0	98.3	99.4
38	28	M	N	—	97.6	98.0	98.2
39	30	M	Y	—	97.4	—	98.0
40	33	M	Y	—	98.0	—	98.6
41	28	M	Y	98.0	97.4	—	98.6
42	22	M	Y	98.8	98.0	—	97.2
43	21	F	Y	99.0	—	—	98.4
44	30	M	N	—	98.6	—	98.6

(continued)

Data Set 6 (*continued*)

SUBJECT	AGE	SEX	SMOKE	Temperature Day 1 8 AM	Temperature Day 1 12 AM	Temperature Day 2 8 AM	Temperature Day 2 12 AM
45	22	M	Y	—	98.6	—	98.2
46	22	F	N	98.0	98.4	—	98.0
47	20	M	Y	—	97.0	—	97.8
48	19	M	Y	—	—	—	98.0
49	33	M	N	—	98.4	—	98.4
50	31	M	Y	99.0	99.0	—	98.6
51	26	M	N	—	98.0	—	98.6
52	18	M	N	—	—	—	97.8
53	23	M	N	—	99.4	—	99.0
54	28	M	Y	—	—	—	96.5
55	19	M	Y	—	97.8	—	97.6
56	21	M	N	—	—	—	98.0
57	27	M	Y	—	98.2	—	96.9
58	29	M	Y	—	99.2	—	97.6
59	38	M	N	—	99.0	—	97.1
60	29	F	Y	—	97.7	—	97.9
61	22	M	Y	—	98.2	—	98.4
62	22	M	Y	—	98.2	—	97.3
63	26	M	Y	—	98.8	—	98.0
64	32	M	N	—	98.1	—	97.5
65	25	M	Y	—	98.5	—	97.6
66	21	F	N	—	97.2	—	98.2
67	25	M	Y	—	98.5	—	98.5
68	24	M	Y	—	99.2	97.0	98.8
69	25	M	Y	—	98.3	97.6	98.7
70	35	M	Y	—	98.7	97.5	97.8
71	23	F	Y	—	98.8	98.8	98.0
72	31	M	Y	—	98.6	98.4	97.1
73	28	M	Y	—	98.0	98.2	97.4
74	29	M	Y	—	99.1	97.7	99.4
75	26	M	Y	—	97.2	97.3	98.4
76	32	M	N	—	97.6	97.5	98.6
77	32	M	Y	—	97.9	97.1	98.4
78	21	F	Y	—	98.8	98.6	98.5
79	20	M	Y	—	98.6	98.6	98.6
80	24	F	Y	—	98.6	97.8	98.3
81	21	F	Y	—	99.3	98.7	98.7
82	28	M	Y	—	97.8	97.9	98.8
83	27	F	N	98.8	98.7	97.8	99.1
84	28	M	N	99.4	99.3	97.8	98.6
85	29	M	Y	98.8	97.8	97.6	97.9
86	19	M	N	97.7	98.4	96.8	98.8
87	24	M	Y	99.0	97.7	96.0	98.0
88	29	M	N	98.1	98.3	98.0	98.7
89	25	M	Y	98.7	97.7	97.0	98.5
90	27	M	N	97.5	97.1	97.4	98.9
91	25	M	Y	98.9	98.4	97.6	98.4
92	21	M	Y	98.4	98.6	97.6	98.6
93	19	M	Y	97.2	97.4	96.2	97.1

(continued)

Data Set 6 (*continued*)

SUBJECT	AGE	SEX	SMOKE	Temperature Day 1 8 AM	Temperature Day 1 12 AM	Temperature Day 2 8 AM	Temperature Day 2 12 AM
94	27	M	Y	—	—	96.2	97.9
95	32	M	N	98.8	96.7	98.1	98.8
96	24	M	Y	97.3	96.9	97.1	98.7
97	32	M	Y	98.7	98.4	98.2	97.6
98	19	F	Y	98.9	98.2	96.4	98.2
99	18	F	Y	99.2	98.6	96.9	99.2
100	27	M	N	—	97.0	—	97.8
101	34	M	Y	—	97.4	—	98.0
102	25	M	N	—	98.4	—	98.4
103	18	M	N	—	97.4	—	97.8
104	32	M	Y	—	96.8	—	98.4
105	31	M	Y	—	98.2	—	97.4
106	26	M	N	—	97.4	—	98.0
107	23	M	N	—	98.0	—	97.0

Data provided by Dr. Steven Wasserman, Dr. Philip Mackowiak, and Dr. Myron Levine of the University of Maryland.

Data Set 7: Bears (wild bears anesthetized)

Age: Months
Month: Month of measurement (1 = Jan., 2 = Feb., etc.)
Sex: 1 = male, 2 = female
HEADLEN: Length (inches) of head
HEADWTH: Width (inches) of head
NECK: Distance (inches) around neck
LENGTH: Length (inches) of body
CHEST: Distance (inches) around the chest
WEIGHT: Measured weight (pounds)

STATDISK: Variable names are BEARAGE, MONTH, BEARSEX, HEADLEN, HEADWTH, NECK, BEARLEN, CHEST, BEARWT

Minitab: Worksheet name is BEARS.MTW.

Excel: Workbook name is BEARS.XLS.

TI-83 Plus and text files: BAGE, BMNTH, BSEX, BHDLN, BHDWD, BNECK, BLEN, BCHST, and BWGHT.

AGE	MONTH	SEX	HEADLEN	HEADWTH	NECK	LENGTH	CHEST	WEIGHT
19	7	1	11.0	5.5	16.0	53.0	26.0	80
55	7	1	16.5	9.0	28.0	67.5	45.0	344
81	9	1	15.5	8.0	31.0	72.0	54.0	416
115	7	1	17.0	10.0	31.5	72.0	49.0	348
104	8	2	15.5	6.5	22.0	62.0	35.0	166
100	4	2	13.0	7.0	21.0	70.0	41.0	220
56	7	1	15.0	7.5	26.5	73.5	41.0	262
51	4	1	13.5	8.0	27.0	68.5	49.0	360
57	9	2	13.5	7.0	20.0	64.0	38.0	204
53	5	2	12.5	6.0	18.0	58.0	31.0	144
68	8	1	16.0	9.0	29.0	73.0	44.0	332
8	8	1	9.0	4.5	13.0	37.0	19.0	34
44	8	2	12.5	4.5	10.5	63.0	32.0	140
32	8	1	14.0	5.0	21.5	67.0	37.0	180
20	8	2	11.5	5.0	17.5	52.0	29.0	105
32	8	1	13.0	8.0	21.5	59.0	33.0	166
45	9	1	13.5	7.0	24.0	64.0	39.0	204
9	9	2	9.0	4.5	12.0	36.0	19.0	26
21	9	1	13.0	6.0	19.0	59.0	30.0	120
177	9	1	16.0	9.5	30.0	72.0	48.0	436
57	9	2	12.5	5.0	19.0	57.5	32.0	125
81	9	2	13.0	5.0	20.0	61.0	33.0	132
21	9	1	13.0	5.0	17.0	54.0	28.0	90
9	9	1	10.0	4.0	13.0	40.0	23.0	40
45	9	1	16.0	6.0	24.0	63.0	42.0	220
9	9	1	10.0	4.0	13.5	43.0	23.0	46
33	9	1	13.5	6.0	22.0	66.5	34.0	154
57	9	2	13.0	5.5	17.5	60.5	31.0	116
45	9	2	13.0	6.5	21.0	60.0	34.5	182
21	9	1	14.5	5.5	20.0	61.0	34.0	150
10	10	1	9.5	4.5	16.0	40.0	26.0	65
82	10	2	13.5	6.5	28.0	64.0	48.0	356
70	10	2	14.5	6.5	26.0	65.0	48.0	316
10	10	1	11.0	5.0	17.0	49.0	29.0	94
10	10	1	11.5	5.0	17.0	47.0	29.5	86
34	10	1	13.0	7.0	21.0	59.0	35.0	150
34	10	1	16.5	6.5	27.0	72.0	44.5	270
34	10	1	14.0	5.5	24.0	65.0	39.0	202
58	10	2	13.5	6.5	21.5	63.0	40.0	202
58	10	1	15.5	7.0	28.0	70.5	50.0	365
11	11	1	11.5	6.0	16.5	48.0	31.0	79
23	11	1	12.0	6.5	19.0	50.0	38.0	148
70	10	1	15.5	7.0	28.0	76.5	55.0	446
11	11	2	9.0	5.0	15.0	46.0	27.0	62
83	11	2	14.5	7.0	23.0	61.5	44.0	236
35	11	1	13.5	8.5	23.0	63.5	44.0	212
16	4	1	10.0	4.0	15.5	48.0	26.0	60
16	4	1	10.0	5.0	15.0	41.0	26.0	64
17	5	1	11.5	5.0	17.0	53.0	30.5	114
17	5	2	11.5	5.0	15.0	52.5	28.0	76
17	5	2	11.0	4.5	13.0	46.0	23.0	48
8	8	2	10.0	4.5	10.0	43.5	24.0	29
83	11	1	15.5	8.0	30.5	75.0	54.0	514
18	6	1	12.5	8.5	18.0	57.3	32.8	140

Data from Gary Alt and Minitab, Inc.

Data Set 8: Cigarette Tar, Nicotine, and Carbon Monoxide

All measurements are in milligrams per cigarette, and all cigarettes are 100 mm long, filtered, and not menthol or light types.

STATDISK: Variable names are TAR, NICOTINE, CO.
Minitab: Worksheet name is CIGARET.MTW.
Excel: Workbook name is CIGARET.XLS.
TI-83 Plus and text files: TAR, NICOT, and CO.

BRAND	TAR	NICOTINE	CO
American Filter	16	1.2	15
Benson & Hedges	16	1.2	15
Camel	16	1.0	17
Capri	9	0.8	6
Carlton	1	0.1	1
Cartier Vendome	8	0.8	8
Chelsea	10	0.8	10
GPC Approved	16	1.0	17
Hi-Lite	14	1.0	13
Kent	13	1.0	13
Lucky Strike	13	1.1	13
Malibu	15	1.2	15
Marlboro	16	1.2	15
Merit	9	0.7	11
Newport Stripe	11	0.9	15
Now	2	0.2	3
Old Gold	18	1.4	18
Pall Mall	15	1.2	15
Players	13	1.1	12
Raleigh	15	1.0	16
Richland	17	1.3	16
Rite	9	0.8	10
Silva Thins	12	1.0	10
Tareyton	14	1.0	17
Triumph	5	0.5	7
True	6	0.6	7
Vantage	8	0.7	11
Viceroy	18	1.4	15
Winston	16	1.1	18

Based on data from the Federal Trade Commission.

Data Set 9: Passive and Active Smoke

All values are measured levels of serum cotinine (in ng/ml), a metabolite of nicotine. (When nicotine is absorbed by the body, cotinine is produced.)

NOETS: Subjects are nonsmokers who have no environmental tobacco smoke (ETS) exposure at home or work.

ETS: Subjects are nonsmokers who are exposed to environmental tobacco smoke at home or work.

SMOKERS: Subjects report tobacco use.

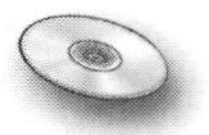

STATDISK: Variable names are NOETS, ETS, and SMOKERS.
Minitab: Worksheet name is COTININE.MTW.
Excel: Workbook name is COTININE.XLS.
TI-83 Plus and text files: NOETS, ETS, and SMKRS.

NOETS	ETS	SMOKERS
0.03	0.03	0.08
0.05	0.07	0.14
0.05	0.08	0.27
0.06	0.08	0.44
0.06	0.09	0.51
0.06	0.09	1.78
0.07	0.10	2.55
0.08	0.11	3.03
0.08	0.12	3.44
0.08	0.12	4.98
0.08	0.14	6.87
0.08	0.17	11.12
0.08	0.20	12.58
0.08	0.23	13.73
0.08	0.27	14.42
0.08	0.28	18.22
0.09	0.30	19.28
0.09	0.33	20.16
0.10	0.37	23.67
0.10	0.38	25.00
0.10	0.44	25.39
0.10	0.49	29.41
0.12	0.51	30.71
0.13	0.51	32.54
0.13	0.68	32.56
0.15	0.82	34.21
0.15	0.97	36.73
0.16	1.12	37.73
0.16	1.23	39.48
0.18	1.37	48.58
0.19	1.40	51.21
0.20	1.67	56.74
0.20	1.98	58.69
0.20	2.33	72.37
0.22	2.42	104.54
0.24	2.66	114.49
0.25	2.87	145.43
0.28	3.13	187.34
0.30	3.54	226.82
0.32	3.76	267.83
0.32	4.58	328.46
0.37	5.31	388.74
0.41	6.20	405.28
0.46	7.14	415.38
0.55	7.25	417.82
0.69	10.23	539.62
0.79	10.83	592.79
1.26	17.11	688.36
1.58	37.44	692.51
8.56	61.33	983.41

Data are based in measurements from the National Health and Nutrition Examination Survey (National Institutes of Health).

Data Set 10: Weights (in grams) of a Sample of M&M Plain Candies

STATDISK: Variable names are RED, ORANGE, YELLOW, BROWN, BLUE, GREEN.
Minitab: Worksheet name is M&M.MTW.
Excel: Workbook name is M&M.XLS.
TI-83 Plus and text files: RED, ORNG, YLLW, BROWN, BLUE, and GREEN.

RED	ORANGE	YELLOW	BROWN	BLUE	GREEN
0.870	0.903	0.906	0.932	0.838	0.911
0.933	0.920	0.978	0.860	0.875	1.002
0.952	0.861	0.926	0.919	0.870	0.902
0.908	1.009	0.868	0.914	0.956	0.930
0.911	0.971	0.876	0.914	0.968	0.949
0.908	0.898	0.968	0.904		0.890
0.913	0.942	0.921	0.930		0.902
0.983	0.897	0.893	0.871		
0.920		0.939	1.033		
0.936		0.886	0.955		
0.891		0.924	0.876		
0.924		0.910	0.856		
0.874		0.877	0.866		
0.908		0.879	0.858		
0.924		0.941	0.988		
0.897		0.879	0.936		
0.912		0.940	0.930		
0.888		0.960	0.923		
0.872		0.989	0.867		
0.898		0.900	0.965		
0.882		0.917	0.902		
		0.911	0.928		
		0.892	0.900		
		0.886	0.889		
		0.949	0.875		
		0.934	0.909		
			0.976		
			0.921		
			0.898		
			0.897		
			0.902		
			0.920		
			0.909		

Data Set 11: Old Faithful Geyser

Durations (in seconds), time intervals (in minutes) to the next eruption, and heights (in feet) of eruptions of the Old Faithful geyser in Yellowstone National Park.

STATDISK: Variable names are DURATION, INTERVAL, GEYSERHT.
Minitab: Worksheet name is OLDFAITH.MTW.
Excel: Workbook name is OLDFAITH.XLS.
TI-83 Plus and text files: OFDTN, OFINT, and OFHT.

DURATION	INTERVAL	HEIGHT
240	86	140
237	86	154
122	62	140
267	104	140
113	62	160
258	95	140
232	79	150
105	62	150
276	94	160
248	79	155
243	86	125
241	85	136
214	86	140
114	58	155
272	89	130
227	79	125
237	83	125
238	82	139
203	84	125
270	82	140
218	78	140
226	91	135
250	89	141
245	79	140
120	57	139
267	100	110
103	62	140
270	87	135
241	70	140
239	88	135
233	82	140
238	83	139
102	56	100
271	81	105
127	74	130
275	102	135
140	61	131
264	83	135
134	73	153
268	97	155
124	67	140
270	90	150
249	84	153
237	82	120
235	81	138
228	78	135
265	89	145
120	69	130
275	98	136
241	79	150

Data courtesy of the National Park Service and research geologist Rick Hutchinson.

Data Set 12: Aluminum Cans

Axial loads (in pounds) of aluminum cans with 0.0109 in. thickness and 0.0111 in. thickness

STATDISK: Variable names are CANS109 and CANS111.
Minitab: Worksheet name is CANS.MTW.
Excel: Workbook name is CANS.XLS.
TI-83 Plus and text files: CN109 and CN111.

Sample	Aluminum cans 0.0109 in. Load (pounds)						
1	270	273	258	204	254	228	282
2	278	201	264	265	223	274	230
3	250	275	281	271	263	277	275
4	278	260	262	273	274	286	236
5	290	286	278	283	262	277	295
6	274	272	265	275	263	251	289
7	242	284	241	276	200	278	283
8	269	282	267	282	272	277	261
9	257	278	295	270	268	286	262
10	272	268	283	256	206	277	252
11	265	263	281	268	280	289	283
12	263	273	209	259	287	269	277
13	234	282	276	272	257	267	204
14	270	285	273	269	284	276	286
15	273	289	263	270	279	206	270
16	270	268	218	251	252	284	278
17	277	208	271	208	280	269	270
18	294	292	289	290	215	284	283
19	279	275	223	220	281	268	272
20	268	279	217	259	291	291	281
21	230	276	225	282	276	289	288
22	268	242	283	277	285	293	248
23	278	285	292	282	287	277	266
24	268	273	270	256	297	280	256
25	262	268	262	293	290	274	292

Sample	Aluminum cans 0.0111 in. Load (pounds)						
1	287	216	260	291	210	272	260
2	294	253	292	280	262	295	230
3	283	255	295	271	268	225	246
4	297	302	282	310	305	306	262
5	222	276	270	280	288	296	281
6	300	290	284	304	291	277	317
7	292	215	287	280	311	283	293
8	285	276	301	285	277	270	275
9	290	288	287	282	275	279	300
10	293	290	313	299	300	265	285
11	294	262	297	272	284	291	306
12	263	304	288	256	290	284	307
13	273	283	250	244	231	266	504
14	284	227	269	282	292	286	281
15	296	287	285	281	298	289	283
16	247	279	276	288	284	301	309
17	284	284	286	303	308	288	303
18	306	285	289	292	295	283	315
19	290	247	268	283	305	279	287
20	285	298	279	274	205	302	296
21	282	300	284	281	279	255	210
22	279	286	293	285	288	289	281
23	297	314	295	257	298	211	275
24	247	279	303	286	287	287	275
25	243	274	299	291	281	303	269

Data Set 13: Weights (in grams) of Quarters

STATDISK: Variable name is QUARTERS.
Minitab: Worksheet name is QUARTERS.MTW.
Excel: Workbook name is QUARTERS.XLS.
TI-83 Plus and text files: QRTRS.

5.60	5.63	5.58	5.56	5.66	5.58	5.57	5.59	5.67	5.61	5.84
5.73	5.53	5.58	5.52	5.65	5.57	5.71	5.59	5.53	5.63	5.68
5.62	5.60	5.53	5.58	5.60	5.58	5.59	5.66	5.73	5.59	5.63
5.66	5.67	5.60	5.74	5.57	5.62	5.73	5.60	5.60	5.57	5.71
5.62	5.72	5.57	5.70	5.60	5.49					

Data Set 14: Survey of 100 Statistics Students

Sex: 1 = male, 2 = female

Age: Years

Height: Inches

Coins: Value (in cents) of coins in possession of respondent

Keys: Number of keys in possession of respondent

Credit: Number of credit cards in possession of respondent

Pulse: Number of heartbeats in one minute

Exercise: 1 = yes, 2 = no (for vigorous exercise consisting of at least 20 minutes at least twice a week)

Smoke: 1 = yes, 2 = no

Color: 1 = yes, 2 = no ("Are you color blind?")

Hand: 1 = left-handed, 2 = right-handed, 3 = ambidextrous

STATDISK: Variable names are STATSEX, STATAGE, STATHT, COINS, KEYS, CREDIT, PULSE, EXERCISE, SMOKE, COLOR, HAND.
Minitab: Worksheet name is STATSURV.MTW.
Excel: Workbook name is STATSURV.XLS.
TI-83 Plus and text files: SVSEX, SVAGE, SVHT, SVCNS, SVKEY, SVCRD, SVPLS, SVEXR, SVSMK, SVCLR, and SVHND.

SEX	AGE	HEIGHT	COINS	KEYS	CREDIT	PULSE	EXERCISE	SMOKE	COLOR	HAND
2	19	64	0	3	0	97	2	2	2	2
2	28	67.5	100	5	0	88	1	2	2	2
1	19	68	0	0	1	69	1	2	1	2
1	20	70.5	23	4	2	67	1	2	2	3
2	18	65	35	5	5	83	1	2	2	2
2	17	63	185	6	0	77	1	2	2	2
1	18	75	0	3	0	66	2	2	2	2
2	48	64	0	3	0	60	2	2	2	3
2	19	68.75	43	3	0	78	2	2	2	3
2	17	57	35	3	0	73	1	1	2	1
2	35	63	250	10	2	8	1	2	2	2
2	18	64	178	5	10	67	1	1	2	2
1	19	72	10	2	1	55	1	2	2	2
2	28	67	90	5	0	72	1	1	2	2
2	24	62.5	0	8	14	82	1	1	2	1
2	30	63	200	1	10	70	1	2	2	2
1	21	69	0	5	0	47	1	2	2	1
1	19	68	40	4	2	63	1	1	2	2
1	19	68	73	2	0	52	2	1	2	2
1	24	68	20	2	1	55	1	2	2	2
2	22	5	500	4	1	67	1	2	2	2
1	21	69	0	2	1	75	2	2	2	1
1	19	69	0	3	3	76	1	2	2	2
2	19	60	35	10	0	60	2	2	2	2
1	20	69	130	3	0	84	1	2	2	2

(continued)

Data Set 14 (*continued*)

SEX	AGE	HEIGHT	COINS	KEYS	CREDIT	PULSE	EXERCISE	SMOKE	COLOR	HAND
1	30	73	62	10	1	40	1	2	2	2
1	33	74	5	7	8	64	1	2	2	2
1	19	67	0	3	0	72	2	2	2	2
1	18	70	0	5	1	72	1	2	2	2
1	20	70	0	3	5	75	1	2	1	2
2	18	76	0	3	0	80	2	2	2	2
1	20	68	32	2	4	63	1	2	2	2
1	50	72	74	8	4	72	2	2	2	2
2	20	65	14	4	4	90	1	1	2	2
1	18	68	25	2	0	70	2	2	2	2
2	18	64	0	2	0	100	2	2	2	2
2	18	64	25	1	0	69	1	2	2	2
1	22	68	0	5	5	64	1	2	2	2
2	21	64	27	2	2	80	2	2	2	2
1	41	72	76	2	0	60	2	1	2	2
1	18	68	160	3	0	66	2	1	2	2
1	21	68	34	26	0	78	1	2	2	2
2	17	60	75	3	0	60	2	1	2	2
2	40	64	20	10	5	68	1	2	2	2
1	19	74	0	1	3	72	1	2	2	2
1	19	69	0	5	0	60	1	2	2	2
2	28	68	453	5	7	88	1	1	2	2
2	28	64	0	3	0	58	1	2	2	2
2	19	63	79	6	0	88	1	2	2	1
2	41	63	100	6	3	80	2	2	2	2
1	18	71	25	2	1	61	1	2	2	2
1	18	73	181	6	0	67	2	2	2	2
1	21	71	72	3	4	60	1	1	2	2
1	18	73	0	5	0	80	1	2	2	2
1	22	69	75	12	5	60	1	1	2	2
2	22	69	0	3	5	80	1	2	2	2
1	21	72	0	4	0	68	1	1	2	2
2	26	60	25	15	0	78	1	2	2	2
1	21	72	97	2	0	54	1	2	2	2
1	20	54	0	5	5	81	2	2	2	2
2	19	65	0	8	0	67	2	2	2	2
1	22	66	30	8	4	70	1	2	2	2
1	20	76	0	6	1	63	1	2	2	2
1	19	71	0	7	0	90	1	2	1	2
1	36	73	18	11	4	70	2	2	2	2
1	20	71	0	5	1	69	2	2	1	2
1	19	71	50	7	0	69	2	1	2	2
2	18	67.5	0	4	0	75	2	1	2	2
2	19	64	0	3	1	80	1	1	2	2
1	52	71.7	51	4	5	92	2	2	2	2
2	41	68	800	7	4	72	2	2	2	2
1	20	69	0	3	3	63	1	1	2	2
1	20	72	85	3	1	60	1	2	2	2
2	30	63	111	5	0	78	2	2	2	2
1	21	73	77	5	0	77	1	2	2	2
1	21	70	35	5	4	71	1	1	2	2

(continued)

Data Set 14 (*continued*)

SEX	AGE	HEIGHT	COINS	KEYS	CREDIT	PULSE	EXERCISE	SMOKE	COLOR	HAND
2	34	65.5	300	4	2	15	2	2	2	2
1	19	69	36	5	0	83	2	1	2	2
1	20	69	0	2	0	80	1	2	2	2
1	20	69	0	1	1	71	1	2	2	2
2	20	66	45	4	0	86	2	2	2	2
2	36	64.5	116	3	8	65	1	2	2	2
1	19	68	52	4	2	70	1	2	2	2
2	20	67	358	7	7	76	1	2	2	2
1	19	71	0	4	0	78	1	1	2	2
1	19	72	15	10	1	63	1	2	2	3
1	19	71	1	0	0	52	1	1	2	2
1	25	71	25	4	1	78	2	2	2	1
2	29	64	26	6	3	92	2	2	2	2
1	19	81	0	4	3	48	1	2	2	2
2	17	67.5	0	3	0	68	1	2	2	2
2	19	68	0	8	1	85	2	2	2	2
1	24	73	0	14	0	64	1	2	2	1
2	19	63	0	3	4	65	2	2	2	2
2	18	64	25	3	0		1	2	2	2
2	23	69	0	3	26		2	1	2	1
1	19	60	50	4	0		1	2	2	2
1	19	72	83	6	2		1	1	2	2
2	21	67	50	3	8		1	2	2	2
1	20	74	0	6	0		2	1	2	1

Data Set 15: Movies

STATDISK: Variable names are MOVBUDG, MOVGROSS, MOVLEN, and MOVRATNG.
Minitab: Worksheet name is MOVIES.MTW.
Excel: Workbook name is MOVIES.XLS.
TI-83 Plus and text files: MVBUD, MVGRS, MVLEN, and MVRAT.

Title	Year	Rating	Budget ($) in Millions	Gross ($) in Millions	Length in Minutes	Viewer Rating
Aliens	1986	R	18.5	81.843	137	8.2
Armageddon	1998	PG-13	140	194.125	144	6.7
As Good As It Gets	1997	PG-13	50	147.54	138	8.1
Braveheart	1995	R	72	75.6	177	8.3
Chasing Amy	1997	R	0.25	12.006	105	7.9
Contact	1997	PG	90	100.853	153	8.3
Dante's Peak	1997	PG-13	104	67.155	112	6.7
Deep Impact	1998	PG-13	75	140.424	120	6.4
Executive Decision	1996	R	55	68.75	129	7.3
Forrest Gump	1994	PG-13	55	329.691	142	7.7
Ghost	1990	PG-13	22	217.631	128	7.1
Gone with the Wind	1939	G	3.9	198.571	222	8.0
Good Will Hunting	1997	R	10	138.339	126	8.5
Grease	1978	PG	6	181.28	110	7.3
Halloween	1978	R	0.325	47	93	7.7
Hard Rain	1998	R	70	19.819	95	5.2
I Know What You Did Last Summer	1997	R	17	72.219	100	6.5
Independence Day	1996	PG-13	75	306.124	142	6.6
Indiana Jones and the Last Crusade	1989	PG-13	39	197.171	127	7.8
Jaws	1975	PG	12	260	124	7.8
Men in Black	1997	PG-13	90	250.147	98	7.4
Multiplicity	1996	PG-13	45	20.1	117	6.8
Pulp Fiction	1994	R	8	107.93	154	8.3
Raiders of the Lost Ark	1981	PG	20	242.374	115	8.3
Saving Private Ryan	1998	R	70	178.091	170	9.1
Schindler's List	1993	R	25	96.067	197	8.6
Scream	1996	R	15	103.001	111	7.7
Speed 2: Cruise Control	1997	PG-13	110	48.068	121	4.3
Terminator	1984	R	6.4	36.9	108	7.7
The American President	1995	PG-13	62	65	114	7.6
The Fifth Element	1997	PG-13	90	63.54	126	7.8
The Game	1997	R	50	48.265	128	7.6
The Man in the Iron Mask	1998	PG-13	35	56.876	132	6.5
Titanic	1997	PG-13	200	600.743	195	8.4
True Lies	1994	R	100	146.261	144	7.2
Volcano	1997	PG-13	90	47.474	102	5.8

Data Set 16: Homes Sold in Dutchess County

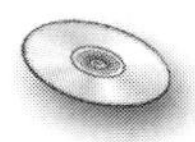

STATDISK: Variable names are HOMESELL, HOMELIST, HOMELA, HOMERMS, HOMEBRS, HOMEBTHS, HOMEAGE, HOMEACRE, and HOMETAX.
Minitab: Worksheet name is HOMES.MTW.
Excel: Workbook name is HOMES.XLS.
TI-83 Plus and text files: HMSP, HMLST, HMLA, HMRMS, HMBRS, HMBTH, HMAGE, HMACR, and HMTAX.

Selling Price (thousands)	List Price (thousands)	Living Area (hundreds of sq. ft.)	Rooms	Bedrooms	Bathrooms	Age (years)	Acres	Taxes (dollars)
142.0	160	28	10	5	3	60	0.28	3167
175.0	180	18	8	4	1	12	0.43	4033
129.0	132	13	6	3	1	41	0.33	1471
138.0	140	17	7	3	1	22	0.46	3204
232.0	240	25	8	4	3	5	2.05	3613
135.0	140	18	7	4	3	9	0.57	3028
150.0	160	20	8	4	3	18	4.00	3131
207.0	225	22	8	4	2	16	2.22	5158
271.0	285	30	10	5	2	30	0.53	5702
89.0	90	10	5	3	1	43	0.30	2054
153.0	157	22	8	3	3	18	0.38	4127
86.5	90	16	7	3	1	50	0.65	1445
234.0	238	25	8	4	2	2	1.61	2087
105.5	116	20	8	4	1	13	0.22	2818
175.0	180	22	8	4	2	15	2.06	3917
165.0	170	17	8	4	2	33	0.46	2220
166.0	170	23	9	4	2	37	0.27	3498
136.0	140	19	7	3	1	22	0.63	3607
148.0	160	17	7	3	2	13	0.36	3648
151.0	153	19	8	4	2	24	0.34	3561
180.0	190	24	9	4	2	10	1.55	4681
293.0	305	26	8	4	3	6	0.46	7088
167.0	170	20	9	4	2	46	0.46	3482
190.0	193	22	9	5	2	37	0.48	3920
184.0	190	21	9	5	2	27	1.30	4162
157.0	165	20	8	4	2	7	0.30	3785
110.0	115	16	8	4	1	26	0.29	3103
135.0	145	18	7	4	1	35	0.43	3363
567.0	625	64	11	4	4	4	0.85	12192
180.0	185	20	8	4	2	11	1.00	3831
183.0	188	17	7	3	2	16	3.00	3564
185.0	193	20	9	3	2	56	6.49	3765
152.0	155	17	8	4	1	33	0.70	3361
148.0	153	13	6	3	2	22	0.39	3950
152.0	159	15	7	3	1	25	0.59	3055
146.0	150	16	7	3	1	31	0.36	2950
170.0	190	24	10	3	2	33	0.57	3346

(continued)

Data Set 16 (continued)

Selling Price (thousands)	List Price (thousands)	Living Area (hundreds of sq. ft.)	Rooms	Bedrooms	Bathrooms	Age (years)	Acres	Taxes (dollars)
127.0	130	20	8	4	1	65	0.40	3334
265.0	270	36	10	6	3	33	1.20	5853
157.0	163	18	8	4	2	12	1.13	3982
128.0	135	17	9	4	1	25	0.52	3374
110.0	120	15	8	4	2	11	0.59	3119
123.0	130	18	8	4	2	43	0.39	3268
212.0	230	39	12	5	3	202	4.29	3648
145.0	145	18	8	4	2	44	0.22	2783
129.0	135	10	6	3	1	15	1.00	2438
143.0	145	21	7	4	2	10	1.20	3529
247.0	252	29	9	4	2	4	1.25	4626
111.0	120	15	8	3	1	97	1.11	3205
133.0	145	26	7	3	1	42	0.36	3059

Data Set 17: Rainfall (in inches) in Boston for One Year

STATDISK: Variable names are RAINMON, RAINTUES, RAINWED, RAINTHUR, RAINFRI, RAINSAT, and RAINSUN.
Minitab: Worksheet name is BOSTRAIN.MTW.
Excel: Workbook name is BOSTRAIN.XLS.
TI-83 Plus and text files: RNMON, RNTUE, RNWED, RNTHU, RNFRI, RNSAT, and RNSUN.

MON	TUES	WED	THURS	FRI	SAT	SUN
0	0	0	0.04	0.04	0	0.05
0	0	0	0.06	0.03	0.1	0
0	0	0	0.71	0	0	0
0	0.44	0.14	0.04	0.04	0.64	0
0.05	0	0	0	0.01	0.05	0
0	0	0.64	0	0	0	0
0.01	0	0	0	0.3	0.05	0
0	0	0.01	0	0	0	0
0	0.01	0.01	0.16	0	0	0.09
0.12	0.06	0.18	0.39	0	0.1	0
0	0	0	0	0.78	0.49	0
0	0.02	0	0	0.01	0.17	0
1.41	0.65	0.31	0	0	0.54	0
0	0	0	0	0	0	0
0	0	0	0	0	0.4	0.28
0	0	0	0.3	0.87	0.49	0
0.47	0	0	0	0	0	0
0	0.09	0	0.24	0	0.05	0
0	0.14	0	0	0.04	0.07	0
0.92	0.36	0.02	0.09	0.27	0	0
0.01	0	0.06	0	0	0	0.27
0.01	0	0	0	0	0	0.01
0	0	0	0	0	0	0
0	0	0	0	0.71	0	0
0	0	0.27	0.08	0	0	0.33
0	0	0	0	0	0	0
0.03	0	0.08	0.14	0	0	0
0	0.11	0.06	0.02	0	0	0
0.01	0.05	0	0.01	0	0	0
0	0	0	0	0.12	0	0
0.11	0.03	0	0	0	0	0.44
0.01	0.01	0	0	0.11	0.18	0
0.49	0	0.64	0.01	0	0	0.01
0	0	0.08	0.85	0.01	0	0
0.01	0.02	0	0	0.03	0	0
0	0	0.12	0	0	0	0
0	0	0.01	0.04	0.26	0.04	0
0	0	0	0	0	0.4	0
0.12	0	0	0	0	0	0
0	0	0	0	0.24	0	0.23
0	0	0	0.02	0	0	0

(continued)

Data Set 17 (*continued*)

MON	TUES	WED	THURS	FRI	SAT	SUN
0	0	0	0.02	0	0	0
0.59	0	0	0	0	0.68	0
0	0.01	0	0	0	1.48	0.21
0.01	0	0	0	0.05	0.69	1.28
0	0	0	0	0.96	0	0.01
0	0	0	0	0	0.79	0.02
0.41	0	0.06	0.01	0	0	0.28
0	0	0	0.08	0.04	0	0
0	0	0	0	0	0	0
0	0.74	0	0	0	0	0
0.43	0.3	0	0.26	0	0.02	0.01
		0				

Data Set 18: Cars

CITY: City fuel consumption in mi/gal
HWY: Highway fuel consumption in mi/gal
WEIGHT: Weight of car in pounds
CYLINDER: Number of cylinders
DISPLACEMENT: Engine displacement in liters
MAN/AUT: M = manual transmission;
A = automatic transmission
GHG: Greenhouse gases emitted (in tons/yr)
NOX: Tailpipe emissions of NO_x (in lb/yr)

STATDISK: Variable names are CARCITY, CARHWY, CARWT, CARCYL, CARDISP, CARGHG, and CARNOX.
Minitab: Worksheet name is CARS.MTW.
Excel: Worksheet name is CARS.XLS.
TI-83 Plus and text files: CRCTY, CRHWY, CRWT, CRCYL, CRDSP, CRGHG, and CRNOX.

CAR	CITY	HWY	WEIGHT	CYLINDER	DISPLACEMENT	MAN/AUT	GHG	NOX
Chev. Camaro	19	30	3545	6	3.8	M	12	34.4
Chev. Cavalier	23	31	2795	4	2.2	A	10	25.1
Dodge Neon	23	32	2600	4	2	A	10	25.1
Ford Taurus	19	27	3515	6	3	A	12	25.1
Honda Accord	23	30	3245	4	2.3	A	11	25.1
Lincoln Cont.	17	24	3930	8	4.6	A	14	25.1
Mercury Mystique	20	29	3115	6	2.5	A	12	34.4
Mitsubishi Eclipse	22	33	3235	4	2	M	10	25.1
Olds. Aurora	17	26	3995	8	4	A	13	34.4
Pontiac Grand Am	22	30	3115	4	2.4	A	11	25.1
Toyota Camry	23	32	3240	4	2.2	M	10	25.1
Cadillac DeVille	17	26	4020	8	4.6	A	13	34.4
Chev. Corvette	18	28	3220	8	5.7	M	12	34.4
Chrysler Sebring	19	27	3175	6	2.5	A	12	25.1
Ford Mustang	20	29	3450	6	3.8	M	12	34.4
BMW 3-Series	19	27	3225	6	2.8	A	12	34.4
Ford Crown Victoria	17	24	3985	8	4.6	A	14	25.1
Honda Civic	32	37	2440	4	1.6	M	8	25.1
Mazda Protege	29	34	2500	4	1.6	A	9	25.1
Hyundai Accent	28	37	2290	4	1.5	A	9	34.4

Data Set 19: Mark McGwire and Sammy Sosa Home-run Distances (in feet)

STATDISK: Variable names are MCGWIRE and SOSA.
Minitab: Worksheet name is HOMERUNS.MTW.
Excel: Workbook name is HOMERUNS.XLS.
TI-83 Plus and text files: MCGWR and SOSA.

MCGWIRE		SOSA	
360	380	371	420
370	470	350	360
370	398	430	368
430	409	420	430
420	385	430	433
340	369	434	388
460	460	370	440
410	390	420	414
440	510	440	482
410	500	410	364
380	450	420	370
360	470	460	400
350	430	400	405
527	458	430	433
380	380	410	390
550	430	370	480
478	341	370	480
420	385	410	434
390	410	380	344
420	420	340	410
425	380	350	420
370	400	420	
480	440	410	
390	377	415	
430	370	430	
388		380	
423		380	
410		366	
360		500	
410		380	
450		390	
350		400	
450		364	
430		430	
461		450	
430		440	
470		365	
440		420	
400		350	
390		420	
510		400	
430		380	
450		380	
452		400	
420		370	

Data Set 20: Wristwatch Errors (in sec)

STATDISK: Variable name is WATCH.
Minitab: Worksheet name is WATCH.MTW.
Excel: Workbook name is WATCH.XLS.
TI-83 Plus and text files: WATCH.

140	−125	105	−241
−85	41	186	−151
325	80	27	20
20	30	−65	36
305	211	168	115
205	−30	555	265
20	−12	324	361
−93	323	143	−56
282	190	188	−100
15	368	570	33

Appendix C: TI-83 Plus

CLEAR — To CLEAR data in list L_1:

```
                                    L1
STAT    [4:ClrList]    2nd    1     ENTER
```

ENTER — To ENTER data in list L_1:

```
STAT    [1:Edit]    ENTER value, press ENTER,...
                                                  Quit
When all data have been ENTERed, press  2nd      Mode
```

STATS — To get STATISTICS for data in list L_1:

```
                                               L1
STAT    [CALC]    [1:1-Var Stats]       2nd    1    ENTER
```

Notes: Sx is the sample standard deviation s.
Q_1 and Q_3 may be different from textbook.

GRAPH — To get HISTOGRAM or BOXPLOT for data in L_1:

```
            STAT PLOT
1. 2nd      Y=        ENTER    ENTER
2. Select "Type" (for boxplot, middle of second row).
3. ZOOM     [9: ZoomStat]
```

FREQ. TABLE — To get STATISTICS FROM A FREQUENCY TABLE:

```
                                            L1         L2
1. Clear L1 and L2: STAT [4:ClrList] 2nd    1  ,  2nd  2  ENTER
2. ENTER the data in L1 and L2: ENTER CLASS MIDPOINTS IN L1.
                                ENTER FREQUENCIES IN L2.
3. To get the statistics:
                                            L1         L2
   STAT    [CALC]    [1:1-VarStats]    2nd  1  ,  2nd  2  ENTER
```

BINOM. — To find BINOMIAL PROBABILITIES:

```
                                            L1         L2
1. Clear L1 and L2: STAT [4:ClrList] 2nd    1  ,  2nd  2  ENTER
        DISTR     number of trials  prob.
2. 2nd  Vars   [0:binompdf(]      n   ,   p     ENTER
                L2
   STO→   2nd   2   ENTER
3. Now use STAT Edit to view the probabilities in list L2 and to
   ENTER the x-values (such as 0, 1, 2, ...) in L1.
4. You can get the mean μ and the standard deviation σ with
                                            L1         L2
   STAT  [CALC]  [1:1-Var Stats]   2nd      1   ,  2nd  2   ENTER
```

NORMAL

NORMAL DISTRIBUTION:

2nd | DISTR Vars | to get Area → 2: normalcdf (lower score, upper score, μ, σ)

to get Score → 3: invNorm (LEFT area, μ, σ)

↑ Total area to *LEFT* of score.

CONF INT

To construct **CONFIDENCE INTERVALS:**

If σ isn't known, use s for σ.

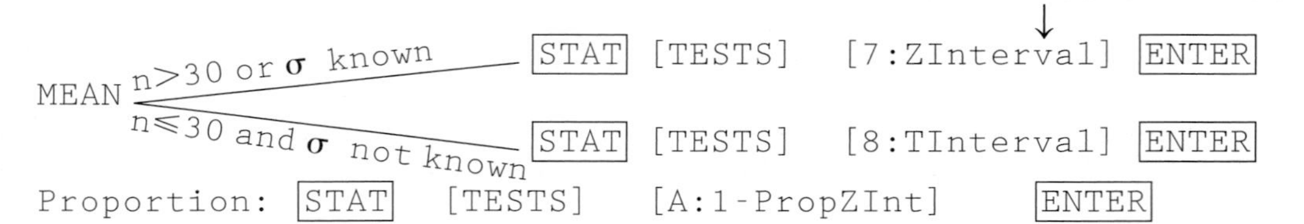

Proportion: STAT [TESTS] [A:1-PropZInt] ENTER

HYP TEST

HYPOTHESES TESTING

If σ isn't known, use s for σ.

MEAN

- n>30 or σ known: STAT [TESTS] [1:Z-Test] ENTER
- n≤30 and σ not known: STAT [TESTS] [2:T-Test] ENTER

PROPORTION: STAT [TESTS] [5:1-PropZInt} ENTER

St. dev. or variance: You're on your own: $\chi^2 = \frac{(n-1)s^2}{\sigma^2}$

and Table A-4

CORR REG

CORRELATION and REGRESSION

1. Enter **PAIRED** data in lists L_1 and L_2.
2. STAT [TESTS] [E:LinRegTTest] Choose Freq. 1 and ≠ 0
3. Interpret: Correlation: If P-value < α, there IS a significant linear correlation.
   Regression: Get equation y = a + bx
   (fill in values for a and b)

CONTIN. TABLE

CONTINGENCY TABLE

1. Enter Table as a matrix: 2nd MATRIX x^{-1} [EDIT] ENTER
   then press 2nd QUIT MODE when done.
2. STAT [TESTS] [C:χ^2-Test]

Appendix D: Glossary

Absolute deviation The measure of variation equal to the sum of the deviations of each value from the mean, divided by the number of values

Acceptance sampling Sampling items without replacement and rejecting the whole batch based on the number of defects obtained

Actual odds against The ratio $P(\overline{A})/P(A)$, usually expressed in the form of *a*:*b* (or "*a* to *b*")

Actual odds in favor The reciprocal of the actual odds against an event

Addition rule Rule for determining the probability that, on a single trial, either event *A* occurs, or event *B* occurs, or they both occur

Adjusted coefficient of determination Multiple coefficient of determination R^2 modified to account for the number of variables and sample size

Alpha (α) Symbol used to represent the probability of a type I error. *See also* significance level.

Alternative hypothesis Statement that is equivalent to the negation of the null hypothesis; denoted by H_1

Analysis of variance Method of analyzing population variances in order to test hypotheses about means of populations

ANOVA *See* analysis of variance.

Arithmetic mean Sum of a set of values divided by the number of values; usually referred to as the mean

Assignable variation Type of variation in a process that results from causes that can be identified

Attribute data Data that can be separated into different categories distinguished by some nonnumeric characteristic

Average Any one of several measures designed to reveal the center of a collection of data

Beta (β) Symbol used to represent the probability of a type II error

Bimodal Having two modes

Binomial experiment Experiment with a fixed number of independent trials, where each outcome falls into exactly one of two categories

Binomial probability formula Expression used to calculate probabilities in a binomial experiment (see Formula 4-5 in Section 4-3)

Bivariate data Data arranged in pairs

Bivariate normal distribution Distribution of paired data in which, for any fixed value of one variable, the values of the other variable are normally distributed

Blinding Procedure used in experiments whereby the subject doesn't know whether he or she is receiving a treatment or a placebo

Block In analysis of variance, a group of similar individuals

Box-and-whisker diagram *See* boxplot.

Boxplot Graphical representation of the spread of a set of data

Categorical data Data that can be separated into different categories that are distinguished by some nonnumeric characteristic

Cell Category used to separate qualitative (or attribute) data

Census Collection of data from every element in a population

Center line Line used in a control chart to represent a central value of the characteristic measurements

Central limit theorem Theorem stating that sample means tend to be normally distributed with mean μ and standard deviation $\sigma/\sqrt{n}$

Centroid The point $(\overline{x}, \overline{y})$ determined from a collection of bivariate data

Chebyshev's theorem Theorem that uses the standard deviation to provide information about the distribution of data

Chi-square distribution A continuous probability distribution (first introduced in Section 6-6)

Class boundaries Values obtained from a frequency table by increasing the upper class limits and decreasing the lower class limits by the same amount so that there are no gaps between consecutive classes

Classical approach to probability Approach in which the probability of an event is determined by dividing the number of ways the event can occur by the total number of possible outcomes

Classical method of testing hypotheses Method of testing hypotheses based on a comparison of the test statistic and critical values

Class midpoint In a class of a frequency table, the value midway between the lower class limit and the upper class limit

Class width The difference between two consecutive lower class limits in a frequency table

Cluster sampling Dividing the population area into sections (or clusters), then randomly selecting a few of those sections, and then choosing *all* the members from those selected sections

Coefficient of determination Amount of the variation in *y* that is explained by the regression line

Combinations rule Rule for determining the number of different combinations of selected items

Complement of an event All outcomes in which the original event does not occur

Completely randomized design Procedure in an experiment whereby each element is given the same chance of belonging to the different categories or treatments

Compound event Combination of simple events

Conditional probability The probability of an event, given that some other event has already occurred

Confidence coefficient Probability that a population parameter is contained within a particular confidence interval; also called level of confidence or degree of confidence

Confidence interval Range of values used to estimate some population parameter with a specific level of confidence; also called an interval estimate

Confidence interval limits Two numbers that are used as the high and low boundaries of a confidence interval

Confounding A situation that occurs when the effects from two or more variables cannot be distinguished from each other

Contingency table Table of observed frequencies where the rows correspond to one variable of classification and the columns correspond to another variable of classification; also called a two-way table

Continuity correction Adjustment made when a discrete random variable is being approximated by a continuous random variable (Section 5-6)

Continuous data Data resulting from infinitely many possible values that correspond to some continuous scale that covers a range of values without gaps, interruptions, or jumps.

Continuous random variable A random variable with infinite values that can be associated with points on a continuous line interval

Control chart Any one of several types of charts (Chapter 12) depicting some characteristic of a process in order to determine whether there is statistical stability

Control group A group of subjects in an experiment who are not given a particular treatment

Control limit Boundary used in a control chart for identifying unusual points

Convenience sampling Sampling in which data are selected because they are readily available

Correlation Statistical association between two variables

Correlation coefficient Measurement of the strength of the relationship between two variables

Critical region The set of all values of the test statistic that would cause rejection of the null hypothesis

Critical value Value separating the critical region from the values of the test statistic that would not lead to rejection of the null hypothesis

Cumulative frequency Sum of the frequencies for a class and all preceding classes

Cumulative frequency table Frequency table in which each class and frequency represents cumulative data up to and including that class

Data Numbers or information describing some characteristic

Decile The nine values that divide sorted data into ten groups with approximately 10% of the values in each group

Degree of confidence Probability that a population parameter is contained within a particular confidence interval; also called level of confidence

Degrees of freedom Number of values that are free to vary after certain restrictions have been imposed on all values

Denominator degrees of freedom Degrees of freedom corresponding to the denominator of the F test statistic

Density curve Graph of a continuous probability distribution

Dependent events Events for which the occurrence of any one event affects the probabilities of the occurrences of the other events

Dependent sample Sample whose values are related to the values in another sample

Dependent variable y variable in a regression or multiple regression equation

Descriptive statistics Methods used to summarize the key characteristics of known data

Deviation Amount of difference between a value and the mean; expressed as $x - \bar{x}$

Discrete data Data with the property that the number of possible values is either a finite number or a "countable" number, which results in 0 possibilities, or 1 possibility, or 2 possibilities, and so on

Discrete random variable Random variable with either a finite number of values or a countable number of values

Distribution-free tests Tests not requiring a particular distribution, such as the normal distribution. *See also* nonparametric tests.

Dotplot Graph in which each data value is plotted as a point (or dot) along a scale of values

Double-blind Procedure used in an experiment whereby the subject doesn't know whether he or she is receiving a treatment or placebo, and the person administering the treatment also does not know

Efficiency Measure of the sensitivity of a nonparametric test in comparison to a corresponding parametric test

Empirical rule Rule that uses standard deviation to provide information about data with a bell-shaped distribution (Section 2-5)

Estimate Specific value or range of values used to approximate some population parameter

Estimator Sample statistic (such as the sample mean $\bar{x}$) used to approximate a population parameter

Event Result or outcome of an experiment

Expected frequency Theoretical frequency for a cell of a contingency table or multinomial table

Expected value For a discrete random variable, the mean value of the outcomes

Experiment Application of some treatment followed by observation of its effects on the subjects

Experimental units Subjects in an experiment

Explained deviation For one pair of values in a collection of bivariate data, the difference between the predicted y value and the mean of the y values

Explained variation Sum of the squares of the explained deviations for all pairs of bivariate data in a sample

Exploratory data analysis (EDA) Branch of statistics emphasizing the investigation of data

Factor In analysis of variance, a property or characteristic that allows us to distinguish the different populations from one another

Factorial rule Rule stating that n different items can be arranged $n!$ different ways

***F* distribution** Continuous probability distribution first introduced in Section 8-5

Finite population correction factor Factor for correcting the standard error of the mean when a sample size exceeds 5% of the size of a finite population

Five-number summary Minimum value, maximum value, median, and the first and third quartiles of a set of data

Fractiles Numbers that partition data into parts that are approximately equal in size

Frequency polygon Graphical representation of the distribution of data using connected straight-line segments

Frequency table List of categories of values along with their corresponding frequencies

Fundamental counting rule Rule stating that, for a sequence of two events in which the first event can occur m ways and the second can occur n ways, the events together can occur a total of $m \cdot n$ ways

Goodness-of-fit test Test for how well some observed frequency distribution fits some theoretical distribution

Histogram Graph of vertical bars representing the frequency distribution of a set of data

***H* test** The nonparametric Kruskal-Wallis test

Hypothesis Statement or claim about some property of a population

Hypothesis test Method for testing claims made about populations; also called test of significance

Independent events Events for which the occurrence of any one of the events does not affect the probabilities of the occurrences of the other events

Independent sample Sample whose values are not related to the values in another sample

Independent variable The x variable in a regression equation, or one of the x variables in a multiple regression equation

Inferential statistics Methods involving the use of sample data to make generalizations or inferences about a population

Influential point Point that strongly affects the graph of a regression line

Interaction In two-way analysis of variance, the effect when one of the factors changes for different categories of the other factor

Interquartile range The difference between the first and third quartiles

Interval Level of measurement of data; characterizes data that can be arranged in order and for which differences between data values are meaningful

Interval estimate Range of values used to estimate some population parameter with a specific level of confidence; also called a confidence interval

Kruskal-Wallis test Nonparametric hypothesis test used to compare three or more independent samples; also called an *H* test

Least-squares property Property stating that, for a regression line, the sum of the squares of the vertical deviations of the sample points from the regression line is the smallest sum possible

Left-tailed test Hypothesis test in which the critical region is located in the extreme left area of the probability distribution

Level of confidence Probability that a population parameter is contained within a particular confidence interval; also called degree of confidence

Linear correlation coefficient Measure of the strength of the relationship between two variables

Lower class limits Smallest numbers that can actually belong to the different classes in a frequency table

Lower control limit Boundary used in a control chart to separate points that are unusually low

Lurking variable Variable that affects the variables being studied, but is not itself included in the study

Mann-Whitney *U* test Hypothesis test equivalent to the Wilcoxon rank-sum test for two independent samples

Marginal change For variables related by a regression equation, the amount of change in the dependent variable when one of the independent variables changes by one unit and the other independent variables remain constant

Margin of error Maximum likely (with probability $1 - \alpha$) difference between the observed sample statistic and the true value of the population parameter

Matched pairs Data from dependent samples; also called paired samples

Mathematical model Mathematical function that "fits" or describes real-world data

Maximum error of estimate *See* margin of error.

Mean The sum of a set of values divided by the number of values

Mean deviation Measure of variation equal to the sum of the deviations of each value from the mean, divided by the number of values

Measure of center Value intended to indicate the center of the values in a collection of data

Measure of variation Any of several measures designed to reflect the amount of variation or spread for a set of values

Median Middle value of a set of values arranged in order of magnitude

Midquartile One-half of the sum of the first and third quartiles

Midrange One-half the sum of the highest and lowest values

Mode Value that occurs most frequently

MS(error) Mean square for error; used in analysis of variance

MS(total) Mean square for total variation; used in analysis of variance

MS(treatment) Mean square for treatments; used in analysis of variance

Multimodal Having more than two modes

Multinomial experiment Experiment with a fixed number of independent trials, where each outcome falls into exactly one of several categories

Multiple coefficient of determination Measure of how well a multiple regression equation fits the sample data

Multiple comparison procedures Procedures for identifying which particular means are different, after concluding that three or more means are not all equal

Multiple regression Study of linear relationships among three or more variables

Multiple regression equation Equation that expresses a linear relationship between a dependent variable y and two or more independent variables $(x_1, x_2, \ldots, x_k)$

Multiplication rule Rule for determining the probability that event A will occur on one trial and event B will occur on a second trial

Mutually exclusive events Events that cannot occur simultaneously

Negatively skewed Skewed to the left

Nominal Level of measurement of data; characterizes data that consist of names, labels, or categories only

Nonparametric tests Statistical procedures for testing hypotheses or estimating parameters, where there are no required assumptions about the nature or shape of population distributions; also called distribution-free tests

Nonsampling errors Errors from external factors not related to sampling

Normal distribution Bell-shaped probability distribution described algebraically by Formula 5-1 in Section 5-1

Normal quantile plot Graph of points (x, y), where each x value is from the original set of sample data, and each y value is a z score corresponding to a quantile value of the standard normal distribution

***np* chart** Control chart in which numbers of defects are plotted so that a process can be monitored

Null hypothesis Claim made about some population characteristic, usually involving the case of no difference; denoted by H_0

Numerator degrees of freedom Degrees of freedom corresponding to the numerator of the F test statistic

Numerical data Data consisting of numbers representing counts or measurements

Observational study Study in which we observe and measure specific characteristics, but don't attempt to manipulate or modify the subjects being studied

Observed frequency Actual frequency count recorded in one cell of a contingency table or multinomial table

Odds against Ratio of the probability of an event not occurring to the event occurring, usually expressed in the form of $a{:}b$ where a and b are integers having no common factors

Odds in favor Ratio of the probability of an event occurring to the event not occurring, usually expressed as the ratio of two integers with no common factors

Ogive Graphical representation of a cumulative frequency table

One-way analysis of variance Analysis of variance involving data classified into groups according to a single criterion only

Ordinal Level of measurement of data; characterizes data that may be arranged in order, but differences between data values either cannot be determined or are meaningless

Outliers Values that are very unusual in the sense that they are very far away from most of the data

Paired samples Two samples that are dependent in the sense that the data values are matched by pairs

Parameter Measured characteristic of a population

Parametric tests Statistical procedures, based on population parameters, for testing hypotheses or estimating parameters

Pareto chart Bar graph for qualitative data, with the bars arranged in order according to frequencies

Payoff odds Ratio of net profit (if you win) to the amount bet

***p* chart** Control chart used to monitor the proportion p for some attribute in a process

Pearson's product moment correlation coefficient *See* linear correlation coefficient.

Percentile The 99 values that divide ranked data into 100 groups with approximately 1% of the values in each group

Permutations rule Rule for determining the number of different arrangements of selected items

Pie chart Graphical representation of data in the form of a circle containing wedges

Placebo effect Effect that occurs when an untreated subject incorrectly believes that he or she is receiving a real treatment and reports an improvement in symptoms

Point estimate Single value that serves as an estimate of a population parameter

Poisson distribution Discrete probability distribution that applies to occurrences of some event over a specified interval of time, distance, area, volume, or some similar unit

Pooled estimate of p_1 and p_2 Probability obtained by combining the data from two sample proportions and dividing the total number of successes by the total number of observations

Pooled estimate of σ^2 Estimate of the variance σ^2 that is common to two populations, found by computing a weighted average of the two sample variances

Population Complete and entire collection of elements to be studied

Positively skewed Skewed to the right

Power of a test Probability $(1 - \beta)$ of rejecting a false null hypothesis

Predicted values Values of a dependent variable found by using values of independent variables in a regression equation

Prediction interval Confidence interval estimate of a predicted value of y

Predictor variables Independent variables in a regression equation

Probability Measure of the likelihood that a given event will occur; expressed as a number between 0 and 1

Probability distribution Collection of values of a random variable along with their corresponding probabilities

Probability histogram Histogram with outcomes listed along the horizontal axis and probabilities listed along the vertical axis

Probability value *See* *P*-value.

Process data Data, arranged according to some time sequence, that measure a characteristic of goods or services resulting from some combination of equipment, people, materials, methods, and conditions

***P*-value** Probability that a test statistic in a hypothesis test is at least as extreme as the one actually obtained

Qualitative data Data that can be separated into different categories distinguished by some nonnumeric characteristic

Quantitative data Data consisting of numbers representing counts or measurements

Quartiles The three values that divide ranked data into four groups with approximately 25% of the values in each group

Randomized block design Design in which a measurement is obtained for each treatment on each of several individuals matched according to similar characteristics

Random sample Sample selected in a way that allows every member of the population to have the same chance of being chosen

Random selection Selection of sample elements in such a way that all elements available for selection have the same chance of being selected

Random variable Variable (typically represented by x) that has a single numerical value (determined by chance) for each outcome of an experiment

Random variation Type of variation in a process that is due to chance; the type of variation inherent in any process not capable of producing every good or service exactly the same way every time

Range The measure of variation that is the difference between the highest and lowest values

Range chart Control chart based on sample ranges; used to monitor variation in a process

Range rule of thumb Rule based on the principle that for typical data sets, the difference between the lowest typical value and the highest typical value is approximately 4 standard deviations ($4s$)

Rank Numerical position of an item in a sample set arranged in order

Rank correlation coefficient Measure of the strength of the relationship between two variables; based on the ranks of the values

Rare event rule If, under a given assumption, the probability of a particular observed result is extremely small, we conclude that the assumption is probably not correct.

Ratio Level of measurement of data; characterizes data that can be arranged in order, for which differences between data values are meaningful, and there is an inherent zero starting point

***R* chart** Control chart based on sample ranges; used to monitor variation in a process

Regression equation Algebraic equation describing the relationship among variables

Regression line Straight line that best fits a collection of points representing paired sample data

Relative frequency Frequency for a class, divided by the total of all frequencies

Relative frequency approximation of probability Estimated value of probability based on actual observations

Relative frequency histogram Variation of the basic histogram in which frequencies are replaced by relative frequencies

Relative frequency table Variation of the basic frequency table in which the frequency for each class is divided by the total of all frequencies

Replication Repetition of an experiment

Residual Difference between an observed sample y value and the value of y that is predicted from a regression equation

Right-tailed test Hypothesis test in which the critical region is located in the extreme right area of the probability distribution

Rigorously controlled design Design of experiment in which all factors are forced to be constant so that effects of extraneous factors are eliminated

Run Sequence of data exhibiting the same characteristic; used in runs test for randomness

Run chart Sequential plot of individual data values over time, where one axis (usually the vertical axis) is used for the data values and the other axis (usually the horizontal axis) is used for the time sequence

Runs test Nonparametric method used to test for randomness

Sample Subset of a population

Sample size Number of items in a sample

Sample space Set of all possible outcomes or events in an experiment that cannot be further broken down

Sampling distribution of sample means Distribution of the sample means that is obtained when we repeatedly draw samples of the same size from the same population

Sampling error Difference between a sample result and the true population result; results from chance sample fluctuations

Scatter diagram Graphical display of paired (x, y) data

***s* chart** Control chart, based on sample standard deviations, that is used to monitor variation in a process

Self-selected sample Sample in which the respondents themselves decide whether to be included; also called voluntary response sample

Semi-interquartile range One-half of the difference between the first and third quartiles

Significance level Probability of making a type I error when conducting a hypothesis test

Sign test Nonparametric hypothesis test used to compare samples from two populations

Simple event Experimental outcome that cannot be further broken down

Simple random sample Sample of a particular size selected so that every possible sample of the same size has the same chance of being chosen

Simulation Process that behaves in a way that is similar to some experiment so that similar results are produced

Single factor analysis of variance *See* one-way analysis of variance.

Skewed Not symmetric and extending more to one side than the other

Slope Measure of steepness of a straight line

Spearman's rank correlation coefficient *See* rank correlation coefficient.

SS(error) Sum of squares representing the variability that is assumed to be common to all the populations being considered; used in analysis of variance

SS(total) Measure of the total variation (around $\bar{\bar{x}}$) in all of the sample data combined; used in analysis of variance

SS(treatment) Measure of the variation between the sample means; used in analysis of variance

Standard deviation Measure of variation equal to the square root of the variance

Standard error of estimate Measure of spread of sample points about the regression line

Standard error of the mean Standard deviation of all possible sample means $\bar{x}$

Standard normal distribution Normal distribution with a mean of 0 and a standard deviation equal to 1

Standard score Number of standard deviations that a given value is above or below the mean; also called z score

Statistic Measured characteristic of a sample

Statistically stable process Process with only natural variation and no patterns, cycles, or unusual points

Statistical process control (SPC) Use of statistical techniques such as control charts to analyze a process or its outputs so as to take appropriate actions to achieve and maintain a state of statistical control and to improve the process capability

Statistics Collection of methods for planning experiments, obtaining data, organizing, summarizing, presenting, analyzing, interpreting, and drawing conclusions based on data

Stem-and-leaf plot Method of sorting and arranging data to reveal the distribution

Stepwise regression Process of using different combinations of variables until the best model is obtained; used in multiple regression

Stratified sampling Sampling in which samples are drawn from each stratum (class)

Student t distribution *See* t distribution.

Subjective probability Guess or estimate of a probability based on knowledge of relevant circumstances

Symmetric Property of data for which the distribution can be divided into two halves that are approximately mirror images by drawing a vertical line through the middle

Systematic sampling Sampling in which every kth element is selected

t distribution Bell-shaped distribution usually associated with small sample experiments; also called the Student t distribution

10–90 percentile range Difference between the 10th and 90th percentiles

Test of homogeneity Test of the claim that different populations have the same proportion of some characteristic

Test of independence Test of the null hypothesis that for a contingency table, the row variable and column variable are not related

Test of significance *See* hypothesis test.

Test statistic Sample statistic based on the sample data; used in making the decision about rejection of the null hypothesis

Total deviation Sum of the explained deviation and unexplained deviation for a given pair of values in a collection of bivariate data

Total variation Sum of the squares of the total deviation for all pairs of bivariate data in a sample

Traditional method of testing hypotheses Method of testing hypotheses based on a comparison of the test statistic and critical values

Treatment Property or characteristic that allows us to distinguish the different populations from one another; used in analysis of variance

Treatment group Group of subjects given some treatment in an experiment

Tree diagram Graphical depiction of the different possible outcomes in a compound event

Two-tailed test Hypothesis test in which the critical region is divided between the left and right extreme areas of the probability distribution

Two-way analysis of variance Analysis of variance involving data classified according to two different factors

Two-way table *See* contingency table.

Type I error Mistake of rejecting the null hypothesis when it is true

Type II error Mistake of failing to reject the null hypothesis when it is false

Unbiased estimator Sample statistic that tends to target the population parameter that it is used to estimate

Unexplained deviation For one pair of values in a collection of bivariate data, the difference between the y coordinate and the predicted value

Unexplained variation Sum of the squares of the unexplained deviations for all pairs of bivariate data in a sample

Uniform distribution Probability distribution in which every value of the random variable is equally likely

Upper class limits Largest numbers that can belong to the different classes in a frequency table

Upper control limit Boundary used in a control chart to separate points that are unusually high

Variance Measure of variation equal to the square of the standard deviation

Variance between samples In analysis of variance, the variation among the different samples

Variation due to error *See* variation within samples.

Variation due to treatment *See* variance between samples.

Variation within samples In analysis of variance, the variation that is due to chance

Weighted mean Mean of a collection of values that have been assigned different degrees of importance

Wilcoxon rank-sum test Nonparametric hypothesis test used to compare two independent samples

Wilcoxon signed-ranks test Nonparametric hypothesis test used to compare two dependent samples

Within statistical control *See* statistically stable process.

$\bar{x}$ chart Control chart used to monitor the mean of a process

y-intercept Point at which a straight line crosses the y-axis

z score Number of standard deviations that a given value is above or below the mean

Appendix E: Bibliography

***An asterisk denotes a book recommended for reading. Other books are recommended as reference texts.**

Andrews D., and A. Herzberg. 1985. *Data: A Collection of Problems from Many Fields for the Student and Research Worker.* New York: Springer.

Beyer, W. 1991. *CRC Standard Probability and Statistics Tables and Formulae.* Boca Raton, Fla.: CRC Press.

*Campbell, S. 1974. *Flaws and Fallacies in Statistical Thinking.* Englewood Cliffs, N.J.: Prentice-Hall.

*Crossen, C. 1994. *Tainted Truth: The Manipulation of Fact in America.* New York: Simon & Schuster.

Devore, J., and R. Peck. 1997. *Statistics: The Exploration and Analysis of Data.* 3rd ed. St. Paul, Minn.: West Publishing.

*Fairley, W., and F. Mosteller. 1977. *Statistics and Public Policy.* Reading, Mass.: Addison-Wesley.

Fisher, R. 1966. *The Design of Experiments.* 8th ed. New York: Hafner.

*Freedman, D., R. Pisani, R. Purves, and A. Adhikari. 1991. *Statistics.* 2nd ed. New York: Norton.

*Gonick, L., and W. Smith. 1993. *The Cartoon Guide to Statistics.* New York: HarperCollins.

Halsey, J., and E. Reda. 2001. *Excel Student Laboratory Manual and Workbook.* Reading, Mass.: Addison-Wesley.

Hoaglin, D., F. Mosteller, and J. Tukey, eds. 1983. *Understanding Robust and Exploratory Data Analysis.* New York: Wiley.

*Hollander, M., and F. Proschan. 1984. *The Statistical Exorcist: Dispelling Statistics Anxiety.* New York: Marcel Dekker.

*Holmes, C. 1990. *The Honest Truth About Lying with Statistics.* Springfield, Ill.: Charles C Thomas.

*Hooke, R. 1983. *How to Tell the Liars from the Statisticians.* New York: Marcel Dekker.

*Huff, D. 1993. *How to Lie with Statistics.* New York: Norton.

*Jaffe, A., and H. Spirer. 1987. *Misused Statistics.* New York: Marcel Dekker.

*Kimble, G. 1978. *How to Use (and Misuse) Statistics.* Englewood Cliffs, N.J.: Prentice-Hall.

Kotz, S., and D. Stroup. 1983. *Educated Guessing—How to Cope in an Uncertain World.* New York: Marcel Dekker.

*Loyer, M. 2001. *Student Solutions Manual to Accompany Elementary Statistics.* 8th ed. Reading, Mass.: Addison-Wesley.

*Moore, D. 1997. *Statistics: Concepts and Controversies.* 4th ed. San Francisco: Freeman.

*Morgan, L. 2001. *The TI-83 Companion to Accompany Elementary Statistics.* 8th ed. Reading, Mass.: Addison-Wesley.

Mosteller, F., R. Rourke, and G. Thomas, Jr. 1970. *Probability with Statistical Applications.* 2nd ed. Reading, Mass.: Addison-Wesley.

Ott, L., and W. Mendenhall. 1994. *Understanding Statistics.* 6th ed. Boston: Duxbury Press.

Owen, D. 1962. *Handbook of Statistical Tables.* Reading, Mass.: Addison-Wesley.

*Paulos, J. 1988. *Innumeracy: Mathematical Illiteracy and Its Consequences.* New York: Hill and Wang.

Peck, R. 2001. *SPSS Student Laboratory Manual and Workbook.* Reading, Mass.: Addison-Wesley.

*Reichard, R. 1974. *The Figure Finaglers.* New York: McGraw-Hill.

*Reichmann, W. 1962. *Use and Abuse of Statistics.* New York: Oxford University Press.

*Rossman, A. 1996. *Workshop Statistics: Discovery with Data.* New York: Springer.

Ryan, T., B. Joiner, and B. Ryan. 1995. *MINITAB Handbook.* 3rd ed. Boston: Duxbury.

Schaeffer, R., M. Gnanadesikan, A. Watkins, and J. Witmer. 1996. *Activity-Based Statistics: Student Guide.* New York: Springer.

Schmid, C. 1983. *Statistical Graphics.* New York: Wiley.

Sheskin, D. 1997. *Handbook of Parametric and Nonparametric Statistical Procedures*. Boca Raton, Fla.: CRCPress.

Simon, J. 1992. *Resampling: The New Statistics.* Belmont, Calif.: Duxbury Press.

Smith, G. 1995. *Statistical Process Control and Quality Improvement.* 2nd ed. Columbus: Merrill.

*Stigler, S. 1986. *The History of Statistics.* Cambridge, Mass.: Harvard University Press.

*Tanur, J., ed. 1989. *Statistics: A Guide to the Unknown.* 3rd ed. Belmont, Calif.: Wadsworth.

Triola, M. 2001. *Minitab Student Laboratory Manual and Workbook.* 8th ed. Reading, Mass.: Addison-Wesley.

Triola, M. 2001. *STATDISK 8.0 Student Laboratory Manual and Workbook.* 8th ed. Reading, Mass.: Addison-Wesley.

Triola, M., and L. Franklin. 1994. *Business Statistics.* Reading, Mass.: Addison-Wesley.

*Tufte, E. 1983. *The Visual Display of Quantitative Information.* Cheshire, Conn.: Graphics Press.

Tukey, J. 1977. *Exploratory Data Analysis.* Reading, Mass.: Addison-Wesley.

Utts, J. 1996. *Seeing Through Statistics.* Belmont, Calif.: Wadsworth.

Appendix F: Answers to Odd-Numbered Exercises (and ALL Review Exercises and Cumulative Review Exercises)

Chapter 1 Answers

Section 1-2

1. Statistic
3. Parameter
5. Discrete
7. Discrete
9. Ratio
11. Interval
13. Interval
15. Ordinal
17. Ratio
19. With no natural starting point, the temperatures are at the interval level of measurement; ratios such as "twice" are meaningless.

Section 1-3

1. Because the viewers themselves decided whether to be included in the survey, we have a self-selected survey and the results cannot be used to conclude anything about the general population. In fact, Ted Koppel reported that a "scientific" poll of 500 people showed that 72% of us want the United Nations to stay in the United States. In this poll of 500 people, respondents were randomly selected by the pollster so that the results are much more likely to reflect the true opinion of the general population.
3. a. $525
   b. $498.75; no
5. Mothers who eat lobsters tend to be wealthier and can afford better health care.
7. Motorcyclists who were killed
9. First, it requires a calculation, which will result in some errors. Second, by asking for heights instead of measuring them, we tend to get desired values instead of actual values.
11. No, two consecutive 5% price cuts is equivalent to a 9.75% price cut, so a 10% price cut results in a lower price.
13. Because the groups consist of 20 mice each, all percentages of success should be multiples of 5. The given percentages cannot be correct.
15. a. Nothing is left.
    b. According to the *New York Times*, "It would have to remove all the plaque, remove it again and then remove it for a third time plus some more still."

Section 1-4

1. Observational study
3. Experiment
5. Random
7. Convenience
9. Stratified
11. Cluster
13. Stratified
15. Systematic
17. Answer varies.
19. Answer varies.
21. a. An advantage of open questions is that they provide the subject and the interviewer with a much wider variety of responses; a disadvantage is that open questions can be very difficult to analyze.
    b. An advantage of closed questions is that they reduce the chance of misinterpreting the topic; a disadvantage is that closed questions prevent the inclusion of valid responses the pollster might not have considered.
    c. Closed questions are easier to analyze with formal statistical procedures.
23. Answers vary.
25. a. Not always; no
    b. Not always; no

Chapter 1 Review Exercises

1. a. Discrete
   b. Ratio
   c. Stratified
   d. Statistic
   e. The largest values because they represent stockholders that could potentially gain control of the company.
   f. The self-selected sample is likely to be biased.
2. a. Systematic
   b. Convenience
   c. Cluster
   d. Random
   e. Stratified
3. Answers vary.
4. a. Design the experiment so that the subjects don't know whether they are using Sleepeze or a placebo, and also design it so that those who observe and evaluate the subjects do not know which subjects are using Sleepeze and which are using a placebo.
   b. Blinding will help to distinguish between the effectiveness of Sleepeze and the placebo effect, whereby subjects and evaluators tend to believe that improvements are occurring just because some treatment is given.
   c. Subjects are put into different groups through a process of *random selection*.

d. Subjects are very *carefully chosen* for the different groups so that those groups are made to be similar in the ways that are important.
e. Replication is used when the experiment is repeated. The sample size should be large enough so that we can see the true nature of any effects. It is important so that we are not misled by erratic behavior of samples that are too small.

5. It includes only students who are present on campus, and it excludes those dropouts who are not on campus.
6. a. Ratio
   b. Ordinal
   c. Nominal
   d. Ordinal
   e. Ratio

Chapter 1 Cumulative Review Exercises

1. 4.56
2. 4.3588989
3. 1475.1744
4. −6.6423420
5. 2,042,975
6. 0.47667832
7. 0.89735239
8. 18.647867
9. 0.00045555497
10. 152,587,890,625
11. 19,770,609,664
12. 0.0009765625

Chapter 2 Answers

Section 2-2

1. Class width: 5. Class midpoints: 57, 62, 67, 72, 77. Class boundaries: 54.5, 59.5, 64.5, 69.5, 74.5, 79.5.
3. Class width: 0.5. Class midpoints: 0.245, 0.745, 1.245, 1.745, 2.245, 2.745, 3.245, 3.745. Class boundaries: −0.005, 0.495, 0.995, 1.495, 1.995, 2.495, 2.995, 3.495, 3.995.
5.

Height of Men (in.)	Relative Frequency
55–59	1%
60–64	3%
65–69	49%
70–74	46%
75–79	1%

7.

GPA	Relative Frequency
0.00–0.49	6.3%
0.50–0.99	2.0%
1.00–1.49	4.1%
1.50–1.99	11.8%
2.00–2.49	25.3%
2.50–2.99	24.2%
3.00–3.49	17.7%
3.50–3.99	8.5%

9.

Height of Men (in.)	Cumulative Frequency
Less than 60	1
Less than 65	4
Less than 70	53
Less than 75	99
Less than 80	100

11.

GPA	Cumulative Frequency
Less than 0.50	72
Less than 1.00	95
Less than 1.50	142
Less than 2.00	277
Less than 2.50	565
Less than 3.00	841
Less than 3.50	1043
Less than 4.00	1140

13.

Weight (lb)	Frequency
0.7900–0.7949	1
0.7950–0.7999	0
0.8000–0.8049	1
0.8050–0.8099	3
0.8100–0.8149	4
0.8150–0.8199	17
0.8200–0.8249	6
0.8250–0.8299	4

15. Roughly similar to Exercise 13; no notable difference.

Weight (lb)	Frequency
0.8100–0.8149	1
0.8150–0.8199	6
0.8200–0.8249	16
0.8250–0.8299	8
0.8300–0.8349	3
0.8350–0.8399	1
0.8400–0.8449	1

17.

Weight (lb)	Frequency
0–49	6
50–99	10

23.

Height	(a) Frequency	(b) Frequency
66	1	3

31.

Actors	Stem	Actresses
	2	1466678
998776532221	3	00113344445557789
88765543322100	4	111249
66531	5	0
2100	6	011
6	7	4
	8	0

Section 2-4

1. $\bar{x} = 16.0$; median = 14.5; mode = 14; midrange = 17.5
3. $\bar{x} = 81.0$ sec; median = 85.5 sec; mode: none; midrange = 77.0 sec
5. Jefferson Valley: $\bar{x} = 7.15$; median = 7.20; mode = 7.7; midrange = 7.10
   Providence: same results as Jefferson Valley
7. Coke: $\bar{x} = 0.81907$; median = 0.81865; mode: none; midrange = 0.81985
   Pepsi: $\bar{x} = 0.82188$; median = 0.82135; mode: none; midrange = 0.82290
   There does not appear to be a significant difference.
9. 74.4 min
11. 46.8 mi/h
13. Author: $\bar{x} = \$57.624$; median = \$59.350
    UMASS: $\bar{x} = \$65.117$; median = \$71.350
    Textbooks at the author's college appear to cost less, possibly because they apply to only the first two years and do not include junior and senior courses that are more expensive because they are used by smaller audiences.
15. Thursday: $\bar{x} = 0.069$ in.; median = 0.000 in.
    Sunday: $\bar{x} = 0.068$ in.; median = 0.000 in.
    There does not appear to be a substantial difference.
17. 48.0 mi/h
19. 62.9 volts
21. a. They all change by the same amount k.
    b. They are all multiplied by k.
23. a. 182.9 lb
    b. 171.0 lb
    c. 159.2 lb
    The results differ by substantial amounts, suggesting that the mean of the original set of weights is strongly affected by extreme scores.

Section 2-5

1. range = 19.0; $s^2 = 28.9$; $s = 5.4$
3. range = 84.0 sec; $s^2 = 699.3$ sec^2; $s = 26.4$ sec
5. Jefferson Valley: range = 1.20; $s^2 = 0.23$; $s = 0.48$
   Providence: range = 5.80; $s^2 = 3.32$; $s = 1.82$
7. Coke: range = 0.00970; $s^2 = 0.00001$; $s = 0.00349$
   Pepsi: range = 0.01460; $s^2 = 0.00003$; $s = 0.00545$
9. 14.7 min
11. 4.1 mi/h
13. Author: \$14.683; UMASS: \$23.076
15. Thursday: 0.167 in.; Sunday: 0.200 in.
17. Approximately 0.75 year (assuming a minimum of 16 and maximum of 19)
19. 51, 99; yes
21. a. 68%
    b. 99.7%
23. Percentage is at least 75%.
25. All values are the same.
27. Everlast batteries are better in the sense that they are more consistent and predictable.
29. Section 1: range = 19.0; $s = 5.7$
    Section 2: range = 17.0; $s = 6.7$
    The ranges suggest that Section 2 has less variation, but the standard deviations suggest that Section 1 has less variation.
31. a. 65%
    b. 1.5
33. 4.4
35. a. 2/3
    b. 2/3
    c. 1/3
    d. Part b because the mean is 2/3, the value of σ^2. Use division by $n - 1$, as in part b.
    e. No

Section 2-6

1. a. 30
   b. 2.00
   c. 2.00
   d. Same
3. a. 5.71
   b. −1.79
   c. 0.26
5. 2.56; unusual
7. 4.52; yes; patient is ill.
9. Psychology test, because $z = -0.38$ is greater than $z = -0.42$.
11. −3.56; yes
13. 92
15. 61
17. 0.8229
19. 0.8189
21. 0.8209
23. 0.8073
25. 44
27. 80
29. 344

31. 86
33. 360
35. 150
37. The z score remains the same.
39. a. Uniform
    b. Bell-shaped
    c. The shape of the distribution remains the same.
41. a. 0.00625
    b. 0.81778
    c. 0.0178
    d. Yes; yes
    e. No; no

Section 2-7

1. These employees are much older than typical employees.

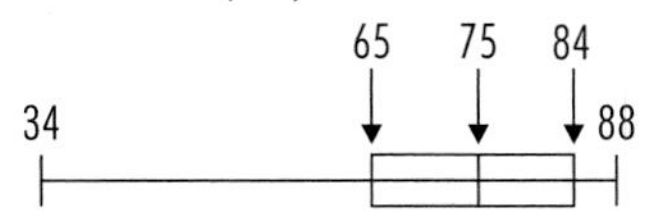

3. Outliers of 8 and 15 must be wrong values.

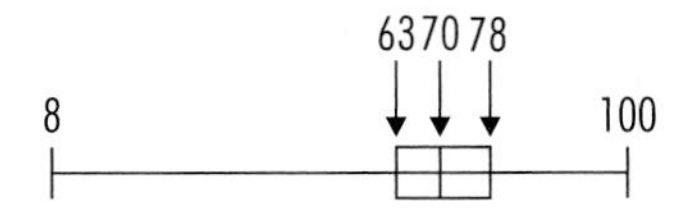

5. The value of 504 is an outlier.

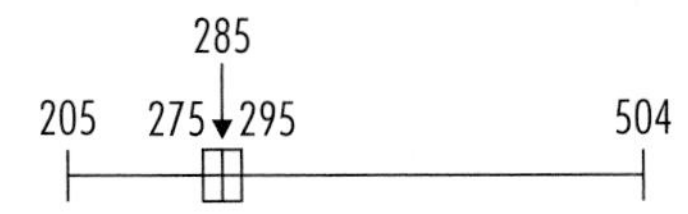

7. Actresses tend to be considerably younger.

Actors:

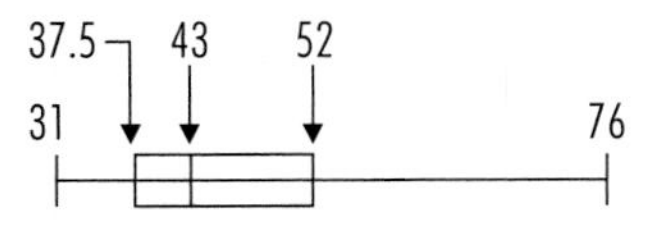

Actresses:

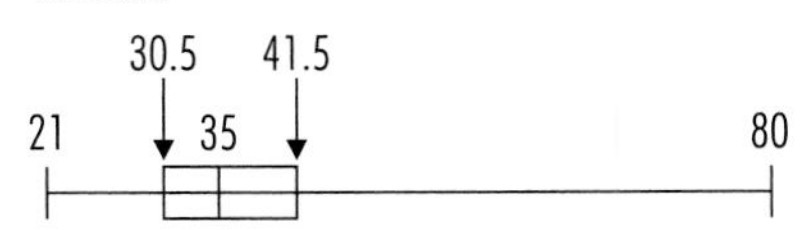

9. UMASS prices tend to be higher.

Author:

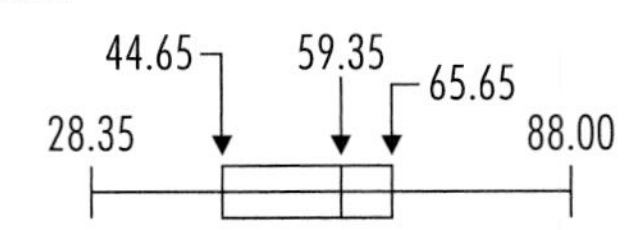

UMASS:

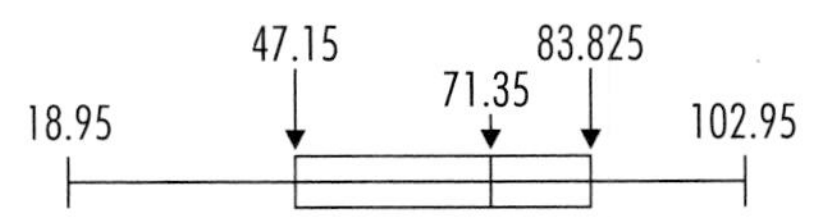

11. $D = 20$. Mild outliers: 215, 216, 222, 225, 227, 230, 231, 243, 244; extreme outliers: 205, 210, 210, 211, 504

Chapter 2 Review Exercises

1. a. 54.9
   b. 55.0
   c. 51
   d. 55.5
   e. 27.0
   f. 6.3
   g. 39.6
   h. 51
   i. 51
   j. 57
2. a. -2.05 (using unrounded mean and st. dev.: -2.04)
   b. Yes, because the z score is less than -2, indicating that his age is more than two standard deviations below the mean.
   c. 68, 69 (because usual values are between 42.3 and 67.5)
3.

Age	Frequency
40–44	2
45–49	6
50–54	12
55–59	12
60–64	7
65–69	3

4. Bell-shaped

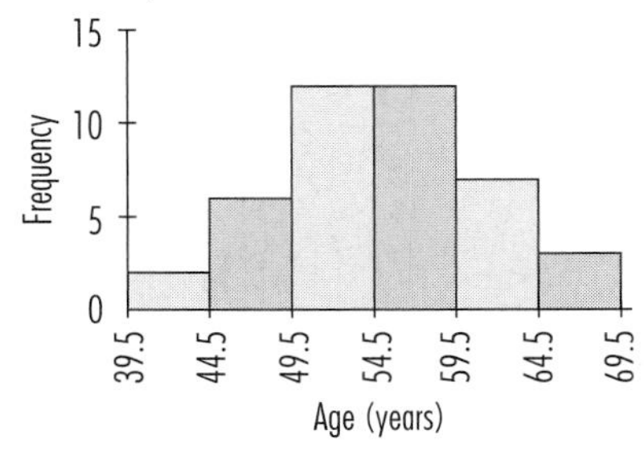

5. 42 51 55 58 69

6. a. The percentage is 68%.
   b. The percentage is 95%.
7. The score of 19 is better, because $z = -0.20$ is greater than $z = -0.67$.
8. a. Answer varies, but roughly 2 years.
   b. Answer varies, but roughly (5 – 0) / 4 = 1.25.
9. a. 17.3 min
   b. 4.0 min
   c. 16.0 square minutes

10.

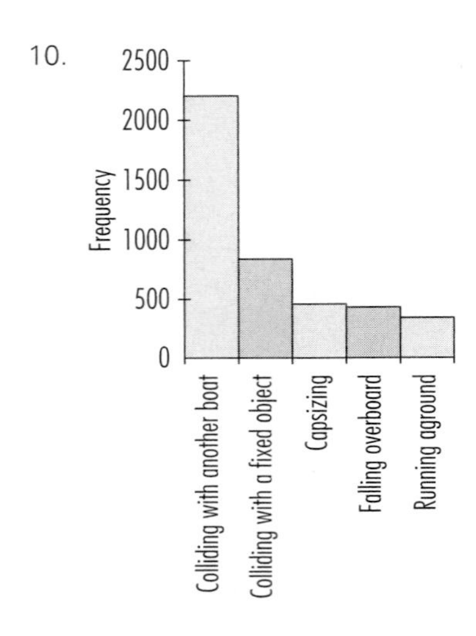

Chapter 2 Cumulative Review Exercises

1. a. $\bar{x} = 12.6$; median = 12.0; mode = 7, 12; midrange = 15.0
   b. $s = 8.6$; $s^2 = 73.3$; range = 30.0
   c. The original exact ages are continuous, but the given data appear to have been rounded to discrete values.
   d. Ratio
2. a. Mode, because the other measures of central tendency require calculations that cannot (or should not) be done with data at the nominal level of measurement.
   b. Convenience
   c. Cluster
3. No, the 50 values should be weighted, with the state populations used as weights.

Chapter 3 Answers

Section 3-2

1. −1, 2, 5/3, $\sqrt{2}$
3. 1/2
5. a. 0.0961
   b. 0.08
   c. They agree approximately.
7. 0.138
9. a. 1/36
   b. Yes
11. a. 1/10
    b. No
13. a. 0.14
    b. No
15. a. 0.0000000774
    b. Yes
17. 0.0997; yes
19. 0.45
21. a. 1/365
    b. 31/365
23. 0.130
25. 0.34; yes
27. 0.159
29. a. bb, bg, gb, gg
    b. 1/4
    c. 1/2
31. a. 14:3
    b. \$16
33. Because the probability of showing improvement with an ineffective drug is so small (0.04), it appears that the drug is effective.
35. 5/8
37. 0.250
39. 1

Section 3-3

1. a. No
   b. No
   c. Yes
3. a. 0.95
   b. 0.487
5. 4/5
7. a. 0.210
   b. 0.364
9. 364/365
11. 0.707
13. 0.600
15. 0.490
17. 0.140
19. 0.870
21. 0.290
23. 0.580
25. a. They are mutually exclusive.
    b. They are not mutually exclusive.
27. $P(A \text{ or } B) = P(A) + P(B) - 2P(A \text{ and } B)$

Section 3-4

1. a. independent
   b. dependent
   c. independent
3. a. 1/9
   b. 1/11
5. 0.350
7. 0.512
9. a. 1/133,225 or 0.00000751
   b. 1/365
11. 0.694
13. 1/1024; yes, because the probability of getting 10 girls by chance is so small.

15. a. 0.001
    b. Because the probability of getting 20 left-handed students is so small, either the students are lying or there was some factor causing the class to attract left-handed students.
17. 1/64
19. 0.360; no
21. 0.00277
23. 0.00000885
25. a. 0.992
    b. 0.973
    c. 0.431
27. 0.0192

Section 3-5

1. All three are boys.
3. At least one student does not get an A.
5. 7/8
7. 0.598
9. 1/2
11. 1
13. 0.999999
15. 0.271
17. 0.897
19. 0.0793
21. a. They are all tripled.
    b. Same result: 0.679
    c. No
23. a. 0.431
    b. 0.569

Section 3-6

1. Using the first four digits of each row: 6/20 = 0.3.
3. In each row, start at the left and include the first three digits that are 1, 2, 3, 4, 5, or 6. Omit the four rows that do not have three such digits. $P(10) = 2/16 = 1/8$.
5. Approximately 0.25.
7. Approximately 0.125.
9. Switch: $P(\text{win}) = 2/3$; stick: $P(\text{win}) = 1/3$.
11. No; no

Section 3-7

1. 87,178,291,200
3. 117,600
5. 1/13,983,816
7. 1/7,059,052
9. 1/12,966,811,200
11. a. 11,880
    b. 495
13. 10
15. a. 24
    b. 1/24
17. 1/20,160
19. 14,348,907
21. a. 1,048,576
    b. 184,756
    c. 0.176
    d. With a probability of 0.176, the result is common, but it should not happen consistently.
23. 1/479,001,600; yes
25. a. 0.0455
    b. 0.159
    c. 0.318
27. 0.00198
29. 2,095,681,645,538 (about 2 trillion)
31. a. Calculator: 3.0414093×10^{64}; approximation: 3.0363452×10^{64}
    b. 615

Chapter 3 Review Exercises

1. 0.498
2. 0.744
3. 0.613
4. 0.859
5. 0.0648
6. 0.251
7. 0.229
8. 0.643
9. a. 0.73
   b. 0.0729
   c. 0.611
10. 0.0777
11. 31/32
12. a. 1/120
    b. 720
13. a. 9/19
    b. 10:9
    c. $5
14. 0.000000531; no
15. 0.979
16. a. 0.698
    b. 0.272
    c. 0.0292

Chapter 3 Cumulative Review Exercises

1. a. 11.8 years
   b. 11.5 years
   c. 7.0 years
   d. 49.2 years squared
   e. No, a new plane with age 0 is within two standard deviations of the mean, so it is not unusual.

f. ratio
g. 1/4
h. 2/3
i. 1/11
j. 1/9
2. a. 63.6 in.
b. 1/4
c. 3/4
d. 1/16
e. 5/16

Chapter 4 Answers

Section 4-2

1. a. continuous
b. discrete
c. continuous
d. continuous
e. discrete
3. $\mu = 1.5$, $\sigma = 0.9$
5. $\mu = 2.0$, $\sigma = 1.0$
7. $\mu = 0.7$, $\sigma = 0.9$
9. $\mu = 0.8$, $\sigma = 0.7$
11. $\mu = 1.1$, $\sigma = 0.9$
13. a. −26¢
b. −26¢
c. Don't bet, because the expected value with no bet is 0, which is better than −26¢.
15. a. Lives: −$250 (a loss); dies: $99,750 (a gain)
b. −$100
c. $150
17. Yes (more than 2 standard deviations below the mean)
19. a. 0.161
b. Yes, because the probability of at least three 7s is 0.035, which is low (less than 0.05).
21. 0.090; no, because the probability of at least 10 girls is 0.090, which is high (greater than 0.05).
23. Same as table in Exercise 5; $\mu = 2.0$, $\sigma = 1.0$
25. a. Yes
b. Yes
c. No, $\Sigma P(x) > 1$.
d. Yes
27. a. $\mu = 4.5$, $\sigma = 2.9$
b. $\mu = 0$, $\sigma = 1$
c. Yes
29. Put 1, 2, 3, 4, 5, 6 on one die and put 0, 0, 0, 6, 6, 6 on the other die.

Section 4-3

1. Binomial
3. Not binomial; more than two outcomes
5. Not binomial; more than two outcomes
7. Not binomial; more than two outcomes
9. a. 0.128
b. WWC, WCW, CWW; 0.128 for each
c. 0.384
11. 0.066
13. 0.124
15. 0+
17. 0.0879
19. 0.228
21. 0.2415
23. 0.8585
25. a. 0+ (or 0.00000980)
b. 0+ (or 0.00000985)
c. They are probably being targeted.
27. a. 0.347
b. 0.544
c. No, there is a good chance (0.544) of getting at most one wrong number if the rate is actually 15%.
29. 0.0833
31. a. 0.882
b. Yes, because the probability of 7 (or fewer) graduates is small (less than 0.05).
33. 0.007 (or 0.00637); study
35. 0.751
37. 0.0524
39. 0.000535

Section 4-4

1. $\mu = 25.0$, $\sigma = 4.3$, minimum = 16.4, maximum = 33.6
3. $\mu = 158.0$, $\sigma = 7.3$, minimum = 143.4, maximum = 172.6
5. a. $\mu = 10.0$, $\sigma = 2.2$
b. No, because 12 is within two standard deviations of the mean.
7. a. $\mu = 2.6$, $\sigma = 1.6$
b. No, because 0 wins is within two standard deviations of the mean.
9. a. $\mu = 1025.0$, $\sigma = 13.6$
b. Yes, because 1107 is more than two standard deviations above the mean. The region attracts people with higher levels of education.
11. a. $\mu = 15.0$, $\sigma = 3.7$
b. If the program has no effect, the result of 12 could easily occur by chance because it is within two standard deviations of the mean.
13. a. $\mu = 27.2$, $\sigma = 5.1$
b. Yes, it appears that the training program had an effect.
15. a. $\mu = 3.0$, $\sigma = 1.7$
b. Yes, 16 is unusual because it is more than two standard deviations above the mean. It appears that more

than 6% of commercials are now more than one minute long.

17. a. Yes; see histogram.
    b. Probability is 0.95.
    c. Probablity is 0.997
    d. At least 75% of such groups of 100 will have between 40 and 60 girls.
19. 10

Section 4-5

1. 0.168
3. 0.0552
5. a. 0.100
   b. 0.231
   c. 0.265
   d. 0.203
   e. 0.0206
7. 0.997; there will be about one day each year when the single team is not sufficient.
9. a. 0.728
   b. 0.231
   c. 0.0368
   d. 0.00389
   e. 0.000309
   Using the computed probabilities, the expected frequencies are 266, 84, 13, 1.4, and 0.1, and they agree quite well with the actual frequencies.
11. 4.82×10^{-64} is so small that, for all practical purposes, we can consider it to be zero.

Chapter 4 Review Exercises

1. a. A random variable is a variable that has a single numerical value (determined by chance) for each outcome of some procedure.
   b. A probability distribution gives the probability for each value of the random variable.
   c. Yes, because each probability value is between 0 and 1 and the sum of the probabilities is 1.
   d. 3.4
   e. 0.7
   f. Yes, because the probability of 0.0004 is very low.
2. a. 6.8
   b. 6.8
   c. 2.1
   d. 0.138
   e. Yes, because 12 is more than two standard deviations above the mean.
3. a. 0.026
   b. 0.992 (or 0.994)
   c. $\mu = 8.0$, $\sigma = 1.3$
   d. No, because 6 is within two standard deviations of the mean.
4. a. 0.00361
   b. This company appears to be very different because the event of at least four firings is so unlikely with a probability of only 0.00361.
5. a. 0.581
   b. 0.0944
   c. 0.0183; no

Chapter 4 Cumulative Review Exercises

1. a. $\bar{x} = 1.2$, $s = 2.7$
   b.

x	Relative Frequency
0	78.6%
1	2.9%
2	1.4%
3	1.4%
4	0%
5	4.3%
6	0%
7	2.9%
8	5.7%
9	2.9%

   c. $\mu = 4.5$, $\sigma = 2.9$
   d. The excessively large number of zeros suggests that the distances were estimated, not measured. The digits do not appear to be randomly selected.
2. a. $\bar{x} = 9.7$, $s = 2.9$
   b. 0.4; very different from 0.0278
   c. 0.431
   d. Claim that sample is too small; claim that the results reflect variation due to chance.

Chapter 5 Answers

Section 5-2

1. 0.8
3. 0.2
5. 1/3
7. 1/2
9. 0.4332
11. 0.4750
13. 0.0367
15. 0.0202
17. 0.2417
19. 0.1359
21. 0.9474; no , change policy.
23. 0.6064
25. 0.4892
27. 0.5

29. 0.9500
31. 0.9950
33. 1.28
35. −0.67
37. −1.645
39. −1.88
41. a. 68.26%
 b. 95%
 c. 99.74%
 d. 81.85%
 e. 4.56%
43. a. 1.23
 b. 1.50
 c. 1.52
 d. −2.42
 e. −0.13

Section 5-3

1. 0.3413
3. 0.4052
5. 0.1844
7. 35.93%
9. 31.21%
11. 12.30%; yes by a small amount for individual packets, but not in general.
13. 0.0038; either a very rare event has occurred or the husband is not the father.
15. 4.18%; no
17. 96.32%; no
19. 0.31%
21. 0.0013
23. a. 0.0179
 b. 1343
25. The z scores are real numbers that have no units of measurement.
27. μ = 64.9 kg, σ = 13.2 kg, distribution is normal.

Section 5-4

1. 12.56
3. 8.32
5. 162 lb
7. 113 lb
9. 67.7 in.
11. 1058
13. a. 0.0018
 b. 5.64 years
15. a. 0.9082
 b. 19.2 lb
17. 242 days
19. a. 0.0222
 b. 254.6
21. a. 75, 5
 b. No, the conversion should also account for variation.
 c. 31.4, 27.6, 22.4, 18.6
 d. Part c, because variation is included.
23. a. 85 lb, 201 lb
 b. 95.44%
 c. 86.2 lb, 199.8 lb

Section 5-5

1. a. 0.1591
 b. 0.4934
3. a. 0.2776
 b. 0.0001
5. a. 0.1175
 b. If the original population has a normal distribution, the central limit theorem provides good results for any sample size.
7. 0.0070; yes
9. a. 0.0051
 b. Yes
11. a. 0.2676
 b. 0.0250
 c. If the original population has a normal distribution, the central limit theorem provides good results for any sample size.
13. a. 0.0001
 b. No, but consumers are not being cheated because the cans are being overfilled, not underfilled.
15. a. 0.2358
 b. 0.0019
 c. If the original population has a normal distribution, the central limit theorem provides good results for any sample size.
 d. Yes, because the probability of 0.0019 shows that it is highly unlikely that by chance, a randomly selected group would get a mean as high as 590.
17. 0.0069; level is acceptable.
19. a. 0.5675
 b. 0.9999
 c. Yes
21. 0.0202
23. They are all 0.975.

Section 5-6

1. The area to the right of 57.5
3. The area to the left of 56.5
5. The area to the left of 24.5
7. The area between 15.5 and 22.5
9. Table: 0.201; normal approximation is not suitable.
11. Table: 0.610; approximation: 0.6026
13. 0.3821

15. 0.1020
17. 0.0239; error rate is probably less than 15%.
19. Yes, because the probability is only 0.0032.
21. 0.0708; yes
23. 0.0080; yes
25. 0.0526; no
27. 0.6368
29. 6; 0.4602
31. 0.2946
33. a. 0.4129 − 0.3264 = 0.0865
    b. 0.3192 − 0.2643 = 0.0549
    c. 0.0256 − 0.0228 = 0.0028
    As n gets larger, the difference becomes smaller.

Section 5-7

1. Not normal
3. Not normal
5. Not normal
7. Not normal
9. Not normal
11. Not normal
13. a. 1/14, 3/14, 5/14, 7/14, 9/14, 11/14, 13/14
    b. −1.47, −0.79, −0.37, 0, 0.37, 0.79, 1.47
    c.

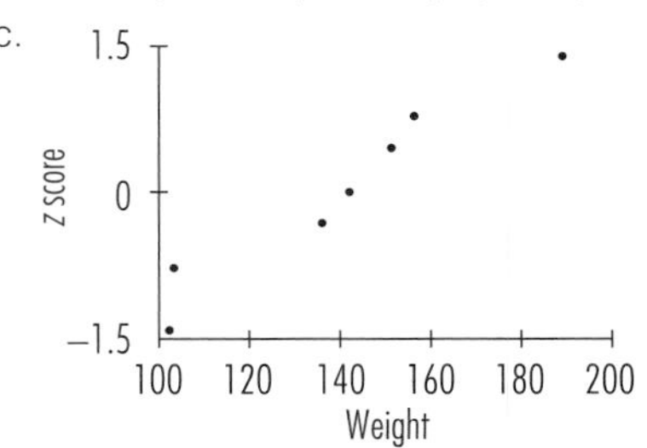

Chapter 5 Review Exercises

1. a. 98.74%
   b. 57.8 in., 69.4 in.
   c. 0.7177
2. a. 3.67%
   b. 0.9876
   c. 63.3 in., 74.7 in.
   d. 0.9633
3. a. 0.0853
   b. 128.1 mm
   c. 0.2670
4. 0.6940
5. a. 0.3413
   b. 0.0228
   c. 0.9332
   d. 0.0994
   e. −0.25
6. a. 0.5
   b. 0.5
   c. 0.25
7. a. 61.41%
   b. 25.65%
   c. 10.5
   d. 7.7
   e. 0.6103
8. 0.3050
9. a. 0.9049
   b. 0.8133
   c. 27,563 mi
10. 0.0018; yes, because it is so unlikely that he would get 13 correct by guessing. (But his powers of prediction might be helped with tricks.)

Chapter 5 Cumulative Review Exercises

1. a. 61.0 mm
   b. 61.0 mm
   c. 60 mm
   d. 3.6 mm
   e. Yes, the histogram is roughly bell-shaped.
   f. 1.4
   g. 72%
   h. 71%
   i. ratio
   j. continuous
2. a. 0.001
   b. 0.271
   c. The requirement that $np \geq 5$ is not satisfied, indicating that the normal approximation would result in errors that are too large.
   d. 5.0
   e. 2.1
   f. No, 8 is within two standard deviations of the mean and is within the range of values that could easily occur by chance.

Chapter 6 Answers

Section 6-2

1. $\bar{x} = 77.6$; $E = 5.0$
3. $\bar{x} = 209.5$; $E = 2.5$
5. 2.575
7. 2.17
9. a. \$3124
   b. $\$82{,}554 < \mu < \$88{,}802$
11. a. 0.57 sec
    b. $7.03 < \mu < 8.17$
13. We are 95% confident that the interval from 133.35 mm to 135.65 mm actually does contain the true population mean.

15. a. $194.5 < \mu < 216.5$
 b. $176.2 < \mu < 193.8$
 c. It appears that scores on the oiled lanes are significantly higher.
17. $\$95.045 < \mu < \$230.899$
19. $0.9499 \text{ mm} < \mu < 0.9513 \text{ mm}$
21. a. $0.81437 \text{ lbs} < \mu < 0.81927 \text{ lbs}$
 b. $0.78336 \text{ lbs} < \mu < 0.78622 \text{ lbs}$
 c. The cans of diet Coke weigh significantly less, probably because they contain less sugar.
23. Plastic: $1.674 \text{ lbs} < \mu < 2.148 \text{ lbs}$;
 Paper: $8.502 \text{ lbs} < \mu < 10.354 \text{ lbs}$
 By weight, paper is a significantly larger ecological problem.
25. a. $90.16 < \mu < 122.99$
 b. The effects of an outlier can be extreme. The confidence interval limits can be very sensitive to outliers.
 c. Outliers should be carefully examined, and they should definitely be discarded if they are found to be errors.
27. a. 450
 b. 102
 c. $424 < \mu < 476$
 d. 92%

Section 6-3

1. $t_{\alpha/2} = 2.093$
3. $t_{\alpha/2} = 2.977$
5. Neither applies.
7. $t_{\alpha/2} = 2.896$
9. $E = 59.8$; $436 < \mu < 556$
11. $99 < \mu < 109$; we have 95% confidence that the interval from 99 to 109 actually does contain the true population mean.
13. $111 < \mu < 137$; with manual shoveling, we obtained $164 < \mu < 186$, so the use of an electric snow thrower appears to result in significantly lower heart rates.
15. $\$16{,}142 < \mu < \$36{,}312$; we have 95% confidence that the interval from \$16,142 to \$36,312 actually does contain the true population mean.
17. $0.075 < \mu < 0.168$
19. $8.0 \text{ yr} < \mu < 8.4 \text{ yr}$; no
21. a. $0.9059 \text{ g} < \mu < 0.9286 \text{ g}$
 b. $0.9015 \text{ g} < \mu < 0.9241 \text{ g}$
 c. The results agree reasonably well.
23. a. $\$71.985 < \mu < \$85.811$
 b. $\$31.449 < \mu < \$52.851$
 c. The paperback books appear to cost significantly less than the hardcover books.
25. The confidence interval limits are closer than they should be.
27. 90%

Section 6-4

1. 543
3. 164
5. 601
7. 236
9. Answer varies, but using 0 and 8 for minimum and maximum yields 269.
11. a. 246
 b. 544
 c. Because the standard deviations differ by considerable amounts, the sample sizes differ by considerable amounts.
13. 147
15. 0.005

Section 6-5

1. $\hat{p} = 0.820$; $E = 0.020$
3. $\hat{p} = 0.4435$; $E = 0.0115$
5. 0.0351
7. 0.0296
9. 0.0661
11. $0.162 < p < 0.238$
13. $0.144 < p < 0.183$
15. 2653
17. 545
19. a. We have 99% confidence that the interval from 0.931 to 0.949 actually contains the true population proportion.
 b. Based on the result from part a, about 5% to 7% of the population does not have a telephone, so those people are missed.
21. a. 9.00%
 b. $7.40\% < p < 10.6\%$
23. a. $4.72\% < p < 5.67\%$
 b. Yes, explosion or fire is a serious risk factor.
25. a. $0.134\% < p < 6.20\%$ using $x = 7$, $n = 221$
 b. Ziac does not appear to cause dizziness as an adverse reaction.
27. a. 321
 b. 543
29. a. 473
 b. 982
31. a. $0.0355 < p < 0.139$
 b. 373
 c. Yes
33. $13.0\% < p < 29.0\%$; yes
35. 4506
37. $78.3\% < p < 81.7\%$; the results are not valid because the sample is self-selected.
39. $p > 0.818$; 81.8%
41. $0.894 < p < 1.006$; use an upper limit of 1.

Section 6-6

1. 16.047, 45.722
3. 27.991, 79.490 (approximately)
5. $79 < \sigma < 170$
7. $10 < \sigma < 16$
9. 767
11. 1401
13. $38.2 \text{ mL} < \sigma < 95.7 \text{ mL}$; no, the fluctuation appears to be too high.
15. a. 0.04425
    b. Use $s = 0.03951697$; $0.0313 \text{ g} < \sigma < 0.0578 \text{ g}$
    c. Yes
17. a. $0.33 \text{ min} < \sigma < 0.87 \text{ min}$
    b. $1.25 \text{ min} < \sigma < 3.33 \text{ min}$
    c. The variation appears to be significantly lower with a single line. The single line appears to be better.
19. a. $\$10.364 < \sigma < \$20.623$
    b. $\$10.943 < \sigma < \$27.366$
    c. There is not a significant difference.
21. a. 98%
    b. 27.0

Chapter 6 Review Exercises

1. a. $3.5 < \mu < 5.3$
   b. $1.2 < \mu < 2.3$
   c. Ratings on the Dvorak keyboard appear to be significantly lower.
   d. The populations appear to be skewed instead of being normally distributed.
2. 2944
3. a. 291
   b. The sample is likely to be biased with regional characteristics.
4. a. $5.47 \text{ years} < \mu < 8.55 \text{ years}$
   b. $2.92 \text{ years} < \sigma < 5.20 \text{ years}$
5. $55.2\% < p < 74.5\%$; we have 95% confidence that the interval from 55.2% to 74.5% actually does contain the true population percentage.
6. 221
7. $71.29 \text{ cm} < \mu < 74.04 \text{ cm}$
8. $514 < \mu < 602$
9. 404
10. $21.6\% < p < 26.5\%$

Chapter 6 Cumulative Review Exercises

1. a. 121.0 lbs
   b. 123.0 lbs
   c. 119 lbs, 128 lbs
   d. 116.5 lbs
   e. 23.0 lbs
   f. 56.8 lbs^2
   g. 7.5 lbs
   h. 119.0 lbs
   i. 123.0 lbs
   j. 127.0 lbs
   k. ratio
   l.

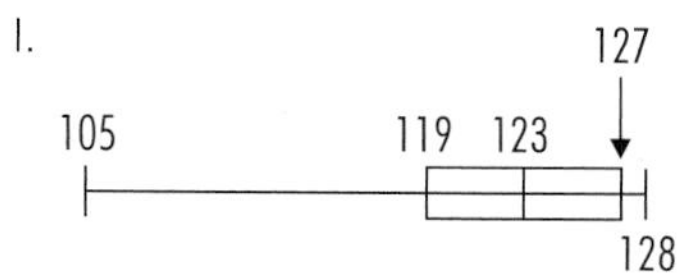

   m. $112.6 \text{ lbs} < \mu < 129.4 \text{ lbs}$
   n. $4.5 \text{ lbs} < \sigma < 18.4 \text{ lbs}$
   o. 95
   p. The individual supermodel weights do not appear to be considerably different from weights of randomly selected women, because they are all within 1.31 standard deviations of the mean of 143 lbs. However, when considered as a group, their mean is significantly less than the mean of 143 lbs (see part m).
2. a. 0.0089
   b. $0.260 < p < 0.390$
   c. Because the confidence interval limits do not contain 0.25, it is unlikely that the expert is correct.

Chapter 7 Answers

Section 7-2

1. Reject the claim that the coin is fair.
3. Reject the claim that women who eat blue M&M candies have a better chance of having a baby boy.
5. Claim: Lisa Kerr's car theft rate is exceptionally high. It appears that her rate is, in fact, exceptionally high.
7. Claim: The experimental vaccine is effective in preventing Lyme disease. There is not sufficient sample evidence to support that claim.
9. H_0: $\mu \leq 30$ years
   H_1: $\mu > 30$ years
11. H_0: $p \geq 0.5$
    H_1: $p < 0.5$
13. H_0: $\sigma \geq 2.8$ in.
    H_1: $\sigma < 2.8$ in.
15. H_0: $\mu \geq 12$ oz
    H_1: $\mu < 12$ oz
17. -1.96, 1.96
19. 2.05
21. -2.05, 2.05
23. 1.56
25. 14.61
27. -38.82

29. H_0: $\mu = 60.0$ in.
Conclusion: There is sufficient evidence to warrant rejection of the claim that women have a mean height equal to 60.0 in.
31. H_0: $\mu \geq 12$ oz
Conclusion: There is not sufficient evidence to support the claim that the mean is less than 12 oz.
33. Type I error: Reject the claim (null hypothesis) that the mean is equal to 12 oz when the mean actually does equal 12 oz.
Type II error: Fail to reject the claim that the mean is equal to 12 oz when the mean is actually different from 12 oz.
35. Type I error: Reject the claim (null hypothesis) that the mean IQ is 100 or greater when the mean is actually 100 or greater.
Type II error: Fail to reject the claim that the mean is 100 or greater when the mean is actually less than 100.
37. There must be a fixed value of μ so that a specific normal distribution can be used for calculation of the test statistic.
39. There are no finite critical values corresponding to $\alpha = 0$. With $\alpha = 0$, the null hypothesis will never be rejected.

Section 7-3

1. 0.3030
3. 0.0001
5. a. $z = 2.98$
b. $z = -1.96$, $z = 1.96$
c. 0.0028
d. There is sufficient evidence to warrant rejection of the claim that the mean score is equal to 75.
7. a. $z = 1.53$
b. $z = 2.05$
c. 0.0630
d. There is not sufficient evidence to support the claim that the mean grade point average is greater than 2.25.
9. Test statistic: $z = -6.64$. Critical value: $z = -2.33$.
Reject H_0: $\mu \geq 98.6°$ F. There is sufficient evidence to support the claim that the mean body temperature is less than 98.6° F.
11. Test statistic: $z = -0.60$. Critical values: $z = \pm 1.96$.
Fail to reject H_0: $\mu = 92.84$ in. There is not sufficient evidence to support the claim that the new balls have bounce heights with a mean different from 92.84 in. The new balls do not appear to be significantly different.
13. Test statistic: $z = -2.11$. Critical values: $z = \pm 2.575$.
Fail to reject H_0: $\mu = 600$ mg. There is not sufficient evidence to warrant rejection of the claim that the mean is equal to 600 mg.
15. Test statistic: $z = 1.37$. P-value: 0.0853.
Fail to reject H_0: $\mu \leq 0$. There is not sufficient evidence to support the claim that the population mean increase is greater than 0.
17. Test statistic: $z = 2.65$. P-value: 0.0080.
Reject H_0: $\mu = 69.5$ years. There is sufficient evidence to support the claim that the mean is different from 69.5 years.
19. Test statistic: $z = -4.99$. P-value: 0.0002.
Reject H_0: $\mu = 5.670$ g. There is sufficient evidence to warrant rejection of the claim that the mean weight of quarters in circulation is 5.670 g. The coins in circulation lose weight as they wear.
21. Test statistic: $z = 9.72$. Critical values depend on choice of significance level. P-value: 0.0002.
Reject H_0: $\mu = 3.5$ g. There is sufficient evidence to warrant rejection of the claim that the mean is equal to 3.5 g.
23. Test statistic: $z = 19.31$. Critical value depends on the significance level. P-value: 0.0001.
Reject H_0: $\mu \leq 12$ oz. There is sufficient evidence to support the claim that the mean is greater than 12 oz.
25. P-value: 0.73. Fail to reject H_0: $\mu \geq 0.9085$ g. There is not sufficient evidence to support the claim that the mean is less than 0.9085 g. Consumers are not being cheated.
27. P-value: 0.00011. Reject H_0: $\mu = 98.6$. There is sufficient evidence to warrant rejection of the claim that the mean is equal to 98.6. The sample data is questionable because of the large number of missing values.
29. 11.0
31. a. 0.6178
b. 0.0868

Section 7-4

1. Test statistic: $t = 0.533$. Critical values: $t = \pm 2.132$.
Fail to reject H_0: $\mu = 75$. There is not sufficient evidence to warrant rejection of the claim that the mean score is equal to 75.
3. Test statistic: $t = 2.449$. Critical value: $t = 2.500$.
Fail to reject H_0: $\mu \leq 2.00$. There is not sufficient evidence to support the claim that the mean grade point average is greater than 2.00.
5. P-value > 0.5
7. $0.01 < P$-value < 0.025
9. Test statistic: $t = -3.214$. Critical values: $t = \pm 2.064$.
Reject H_0: $\mu = 98.6$°F. There is sufficient evidence to warrant rejection of the claim that the mean body temperature is equal to 98.6°F.
11. Test statistic: $t = -1.827$. Critical value: $t = -2.132$.
Fail to reject H_0: $\mu \geq \$1000$. There is not sufficient evidence to support the claim that the mean is less than \$1000.
13. Test statistic: $t = -9.740$. Critical values: $t = \pm 2.080$.
Reject H_0: $\mu = 243.5$. There is sufficient evidence to support the claim that the mean for the VDT subjects differs from the mean of 243.5.

15. Test statistic: $t = -4.021$. Critical values: $t = \pm 2.977$. Reject H_0: $\mu = 41.9$. The third-grade sample mean does appear to differ significantly from 41.9.
17. Test statistic: $z = 0.63$. Critical value: $z = 1.645$. Fail to reject H_0: $\mu \leq 0.9085$ g. There is not sufficient evidence to support the claim that the mean is greater than 0.9085 g. We cannot conclude that the packages contain more than the claimed weight printed on the label.
19. Test statistic: $t = 3.164$. Critical value: $t = 2.896$. Reject H_0: $\mu \leq 62.2$ oz. There is sufficient evidence to support the claim that with the enriched feed, the mean weight is greater than 62.2 oz.
21. Test statistic: $z = 20.49$. The critical value depends on the significance level, but it will be less than the test statistic for any reasonable choices. Reject H_0: $\mu \leq 12$ oz. There is sufficient evidence to support the claim that the mean is greater than 12 oz.
23. Test statistic: $t = 1.734$. Critical values: $t = \pm 2.132$. Fail to reject H_0: $\mu = 3.39$ kg. There is not sufficient evidence to warrant rejection of the claim that the mean is equal to 3.39 kg. There is not sufficient evidence to state that the vitamin supplement has an effect on birth weight.
25. *P*-value: 0.08666. Assuming a 0.05 significance level, fail to reject H_0: $\mu \geq 1000$ mg. There is not sufficient evidence to support the claim that consumers are being cheated (by getting a mean less than 1000 mg).
27. *P*-value: 0.025. Fail to reject $\mu \leq 12.00$. There is not sufficient evidence to support the claim that the mean is greater than 12.00.
29. The *P*-value becomes 0.0068.
31. With $z = 1.645$, the table and approximation both result in $t = 1.833$.

Section 7-5

1. Test statistic: $z = -2.28$. Critical value: $z = -1.645$. *P*-value: 0.0113. Reject H_0: $p \geq 0.20$. There is sufficient evidence to support the claim that less than 20% of U.S. households use e-mail. The conclusion is not valid today because the population is changing.
3. Test statistic: $z = 8.16$. Critical value: $z = 2.05$. *P*-value: 0.0001. Reject H_0: $p \geq 0.5$. There is sufficient evidence to support the claim that most (more than 50%) drivers change tapes or CDs while driving.
5. Test statistic: $z = 3.70$. Critical values: $z = \pm 2.05$. *P*-value: 0.0002. Reject H_0: $p = 0.43$. There is sufficient evidence to warrant rejection of the claim that the percentage of voters who say that they voted for the winning candidate is equal to 43%.
7. Test statistic: $z = 2.19$. Critical values: $z = \pm 1.96$. *P*-value: 0.0286. Reject H_0: $p = 0.01$. There is sufficient evidence to warrant rejection of the claim that 1% of sales are overcharges.
9. Test statistic: $z = 1.33$. Critical value: $z = 1.645$. *P*-value: 0.0918. Fail to reject H_0: $p \leq 0.5$. There is not sufficient evidence to support the claim that most students don't know what "Holocaust" refers to.
11. Test statistic: $z = 0.83$. Critical value: $z = 1.28$. *P*-value: 0.2033. Fail to reject H_0: $p \leq 0.5$. There is not sufficient evidence to support the claim that the majority are smoking a year after treatment.
13. Test statistic: $z = 3.61$. Critical values depend on significance level, but the test statistic is in the critical region for any reasonable choice. *P*-value: 0.0002. Reject H_0: $p = 3/4$. There is sufficient evidence to warrant rejection of the claim that the proportion is equal to 3/4.
15. Test statistic: $z = -2.00$. Critical value: $z = -2.33$. *P*-value: 0.0228. Fail to reject H_0: $p \geq 0.10$. There is not sufficient evidence to support the claim that less than 10% of all U.S. college students carry a gun, knife, or other such weapon. The sample is from one college, so the results do not necessarily apply to all U.S. college students.
17. The test statistic is $z = 8.26$ and the *P*-value is 0.0000 when rounded to four decimal places. Reject H_0: $p \leq 0.75$. There is sufficient evidence to support the claim that among aircraft accidents involving spatial disorientation, more than three-fourths result in fatalities.
19. Test statistic: $z = -1.67$. Critical values assuming a 0.05 significance level: $z = \pm 1.96$. *P*-value: 0.0950. Fail to reject H_0: $p = 0.10$. There is not sufficient evidence to warrant rejection of the claim that 10% of the plain M&M candies are blue.
21. a. With $n = 80$ and $p = 0.0025$, the conditions $np \geq 5$ and $nq \geq 5$ are not both satisfied.
    b. 0.00000000165 is found by using the binomial probability distribution.
    c. If the probability of a man being colorblind is 0.0025, the probability of getting at least 7 colorblind men among 80 is extremely small, so it appears that the colorblind rate is actually greater than 0.0025 or 0.25%. There is sufficient evidence to support the claim that the rate for men is greater than 0.25%.
23. 1.977%
25. 47% is not a possible result because, with 20 mice, the only possible success rates are 0%, 5%, 10%, . . . , 100%.

Section 7-6

1. a. 1.735, 23.589
   b. 45.642
   c. 10.851

3. Test statistic: $\chi^2 = 12.960$. Critical values: $\chi^2 = 2.700$, 19.023. Fail to reject H_0: $\sigma = 15$. There is not sufficient evidence to warrant rejection of the claim that the population standard deviation equals 15.
5. Test statistic: $\chi^2 = 2342.438$. Critical value: $\chi^2 = 63.691$ (approximately). Reject H_0: $\sigma \leq 0.04$ g. There is sufficient evidence to support the claim that peanut M&M candies have weights that vary more.
7. Test statistic: $\chi^2 = 114.586$. Critical values: $\chi^2 = 57.153$, 106.629. Reject H_0: $\sigma = 43.7$ ft. There is sufficient evidence to support the claim that the standard deviation is different from 43.7 ft. Because there appears to be more variation, the new production method appears to be worse.
9. Test statistic: $\chi^2 = 11.311$. Critical value: $\chi^2 = 12.198$. Reject H_0: $\sigma \geq 14.1$. There is sufficient evidence to support the claim that the current class has less variation than past classes. A lower standard deviation does not suggest that the class is doing better.
11. Test statistic: $\chi^2 = 69.135$. Critical value: $\chi^2 = 67.505$ (approximately). Reject H_0: $\sigma \leq 19.7$. There is sufficient evidence to support the claim that women have more variation than men.
13. $s = 2.340$. Test statistic: $\chi^2 = 15.626$. Critical values: $\chi^2 = 12.401$, 39.364 (assuming a significance level of 0.05). Fail to reject H_0: $\sigma = 2.9$ in. There is not sufficient evidence to support the claim that men aged 45–54 have heights with a standard deviation different from 2.9 in.
15. $s = 7.533$ lb. Test statistic: $\chi^2 = 0.540$. Critical value: $\chi^2 = 1.646$. Reject H_0: $\sigma \geq 29$ lb. There is sufficient evidence to support the claim that weights of female supermodels have a standard deviation less than 29 lb.
17. a. P-value > 0.20
    b. P-value < 0.005
    c. P-value < 0.005
19. a. Estimated values: 74.216, 129.565; Table A-4 values: 74.222, 129.561
    b. 117.093, 184.690
21. a. The standard deviation will be lower.
    b. The requirement of a normal distribution is not satisfied.

Chapter 7 Review Exercises

1. a. 0.001
   b. $\mu < 114.8$
   c. $\mu \geq 90$
   d. There is not sufficient evidence to support the claim that the mean is greater than 12 oz.
   e. rejecting a true null hypothesis
2. a. $t = -1.833$
   b. $z = \pm 2.575$
   c. $\chi^2 = 8.907, 32.852$
   d. $z = \pm 1.645$
   e. $z = -2.33$
3. Test statistic: $z = 3.12$. Critical value: $z = 2.33$. P-value: 0.0001. Reject H_0: $p \leq 0.5$. There is sufficient evidence to support the claim that the majority of gun owners favor stricter gun laws.
4. Test statistic: $t = -4.741$. Critical value: $t = -1.714$ (assuming a 0.05 significance level). Reject H_0: $\mu \geq 12$. There is sufficient evidence to support the claim that the company is cheating consumers. Argument is not valid.
5. Test statistic: $\chi^2 = 24.576$. Critical value: $\chi^2 = 43.188$. Reject H_0: $\sigma \geq 0.75$ lb. There is sufficient evidence to support the claim that the weights are more consistent (with a lower standard deviation).
6. Test statistic: $z = -2.34$. Critical value: $z = -2.33$. P-value: 0.0096. Reject H_0: $p \geq 0.10$. There is sufficient evidence to support the claim that the true percentage is less than 10%. The phrase "almost 1 out of 10" is not justified.
7. Test statistic: $z = -10.02$. Critical value: $z = -2.33$. P-value: 0.0001. Reject H_0: $\mu \geq 7124$ mi. There is sufficient evidence to support the claim that the population mean for women in the 16–24 age bracket is less than 7124 mi.
8. Test statistic: $z = -14.02$. Critical value: $z = -2.33$. P-value: 0.0001. Reject H_0: $p \geq 0.5$. There is sufficient evidence to reject the claim that at least half of all adults eat their fruitcakes.
9. $\bar{x} = 23.29$, $s = 4.53$. Test statistic: $t = 3.240$. Critical values: $t = \pm 2.861$. Reject H_0: $\mu = 20.0$ mg. There is sufficient evidence to warrant rejection of the claim that the mean equals 20 mg. These pills are not acceptable.
10. Test statistic: $z = 0.10$. Critical value: $z = 2.33$. P-value: 0.4602. Fail to reject H_0: $\mu \leq \$30,000$. There is not sufficient evidence to support the claim that the mean is greater than \$30,000.

Chapter 7 Cumulative Review Exercises

1. a. 562.5
   b. 530.0
   c. 136.7
   d. 18,678.6
   e. 370.0
   f. $448.2 < \mu < 676.8$
   g. Test statistic: $t = 1.376$. Critical values: $t = \pm 2.365$. Fail to reject H_0: $\mu = 496$. There is not sufficient evidence to warrant rejection of the claim that the mean equals 496.
   h. The applicant's scores are not significantly different from those of the general population.

2. a. 0.4840
   b. 0.0266 (from 0.4840^5)
   c. 0.4681
   d. 634
3. a. $\bar{x} = 124.23$, $s = 22.52$
   b. Test statistic: $t = 1.675$. Critical values: $t = \pm 2.132$. Fail to reject H_0: $\mu = 114.8$. There is not sufficient evidence to warrant rejection of the claim that the population mean equals 114.8.
   c. $112.23 < \mu < 136.23$; yes
   d. Test statistic: $\chi^2 = 44.339$. Critical values: $\chi^2 = 6.262, 27.488$. Reject H_0: $\sigma = 13.1$. There is sufficient evidence to warrant rejection of the claim that the standard deviation is equal to 13.1.
   e. Variation appears to be affected.
4. a. 6.3
   b. 2.2
   c. 0.0034; using normal approx: 0.0019
   d. Based on the low probability value in part (c), reject H_0: $p = 0.25$. There is sufficient evidence to reject the claim that the subject made random guesses.
   e. 423

Chapter 8 Answers

Section 8-2

1. Test statistic: $z = 4.08$. Critical values: $z = \pm 1.96$. P-value: 0.0002. Reject H_0: $\mu_1 = \mu_2$. There is sufficient evidence to warrant rejection of the claim that the two populations have the same mean.
3. $0.52 < \mu_1 - \mu_2 < 1.48$; no; there is a significant difference.
5. a. Test statistic: $z = 22.10$. Critical values: $z = \pm 2.575$. P-value: 0.0002. Reject H_0: $\mu_1 = \mu_2$. There is sufficient evidence to warrant rejection of the claim that regular Coke and diet Coke have the same mean weight.
   b. $0.028298 < \mu_1 - \mu_2 < 0.035762$
7. Test statistic: $z = 2.88$. Critical value: $z = 2.33$. P-value: 0.0020. Reject H_0: $\mu_1 = \mu_2$. There is sufficient evidence to support the claim that the mean stress score is lower than the mean nonstress score.
9. a. Test statistic: $z = 8.56$. Critical value: $z = 2.33$. P-value: 0.0001. Reject H_0: $\mu_1 \leq \mu_2$. There is sufficient evidence to support the claim that the older men come from a population with a mean that is less than the mean for the younger men.
   b. $8 < \mu_1 - \mu_2 < 16$; no; significant difference
11. Test statistic: $z = 2.17$. Critical value: $z = 1.645$. P-value: 0.0150. Reject H_0: $\mu_1 \leq \mu_2$. There is sufficient evidence to support the claim that zinc supplementation results in a higher mean birth weight.
13. UMASS: $n_1 = 40$, $\bar{x}_1 = 65.1175$, $s_1 = 23.0757$; DCC: $n_2 = 35$, $\bar{x}_2 = 57.6243$, $s_2 = 14.6830$. Test statistic: $z = 1.70$. Critical values: $z = \pm 1.96$ (assuming a 0.05 significance level). P-value: 0.0892. Fail to reject H_0: $\mu_1 = \mu_2$. There is not sufficient evidence to warrant rejection of the claim that the two colleges have the same mean price.
15. NOETS: $n_1 = 50$, $\bar{x}_1 = 0.405000$, $s_1 = 1.21290$; ETS: $n_2 = 50$, $\bar{x}_2 = 4.0976$, $s_2 = 10.21176$. Test statistic: $z = -2.54$. Critical values: $z = \pm 1.96$ (assuming a 0.05 significance level). P-value: 0.0110. Reject H_0: $\mu_1 = \mu_2$. There is sufficient evidence to warrant rejection of the claim that the two populations have the same mean serum cotinine level.
17. Test statistic: $z = -0.7858$. Critical value: $z = -1.644853$. P-value: 0.215991. There is not sufficient evidence to support the claim that the treatment population has a mean less than the mean for the placebo population.
19. Test statistic: $z = 9.96$. P-value: 0.0000 (rounded). There is sufficient evidence to support the claim that the mean height of men is greater than the mean height of women.
21. Cans 0.0111 in.: $n_1 = 175$, $\bar{x}_1 = 281.8$, $s_1 = 27.8$; cans 0.0109 in.: $n_2 = 175$, $\bar{x}_2 = 267.1$, $s_2 = 22.1$. Test statistic: $z = 5.48$. Critical values: $z = -1.645$ (assuming a 0.05 significance level). P-value: 0.0001. Reject H_0: $\mu_1 \leq \mu_2$. There is sufficient evidence to support the claim that the 0.0111 in. cans have a mean axial load that is greater than the mean axial load of the 0.0109 in. cans. If the outlier is excluded, the test statistic changes to 5.66, so the conclusion will be the same. The 95% confidence interval limits change from (9.4, 20.0) to (8.8, 18.0), so those differences are not substantial. The five-numbers summary used for the boxplot of the 0.0111 in. cans changes from 205, 275, 285, 295, 504 to 205, 275, 285, 294, 317, so the only major change is shortening of the right whisker.
23. a. 50/3
    b. 2/3
    c. 52/3
    d. 52/3 = 50/3 + 2/3
    e. The range of the $x - y$ values equals the range of the x values plus the range of the y values.

Section 8-3

1. a. $\bar{d} = 2.8$
   b. $s_d = 3.6$
   c. $t = 1.757$
   d. $t = \pm 2.776$
3. $-1.6 < \mu_d < 7.2$
5. a. Test statistic: $t = 2.238$. Critical value: $t = 1.833$. Reject H_0: $\mu_d \leq 0$. There is sufficient evidence to support the claim that the reported heights are greater than the measured heights.
   b. $-0.007 < \mu_d < 1.2$

7. a. Test statistic: $t = -1.718$. Critical value: $t = -1.833$. Fail to reject H_0: $\mu_d \geq 0$. There is not sufficient evidence to conclude that the preparatory course is effective in raising scores.
   b. $-25.5 < \mu_d < 3.5$; we have 95% confidence that the interval from -25.5 to 3.5 actually contains the true population mean difference.
9. a. $0.69 < \mu_d < 5.56$
   b. Test statistic: $t = 3.036$. Critical value: $t = 1.895$. Reject H_0: $\mu_d \leq 0$. There is sufficient evidence to support the claim that the sensory measurements are lower after hypnosis.
   c. Yes
11. a. $-1.40 < \mu_d < -0.17$
   b. Test statistic: $t = -2.840$. Critical values: $t = \pm 2.228$. Reject H_0: $\mu_d = 0$. There is sufficient evidence to warrant rejection of the claim that the mean difference is 0. Morning and night body temperatures do not appear to be about the same.
13. a. Test statistic: $t = -0.41$. P-value: 0.69 Fail to reject H_0: $\mu_d = 0$. There is not sufficient evidence to support the claim that astemizole has an effect.
   b. 0.345; there is not sufficient evidence to support the claim that astemizole prevents motion sickness.
   c. $-48.8 < \mu_d < 33.8$; yes
15. In the hypothesis test, the test statistic changes from 2.701 to 1.975. The confidence interval limits change from (0.12, 1.23) to (-0.15, 2.70). The effect of an outlier can be dramatic.
17. 0.025

Section 8-4

1. 117
3. 85
5. a. 0.556
   b. -0.67
   c. ± 1.96
   d. 0.5028
7. Test statistic: $z = -12.39$. Critical value: $z = -1.645$ (assuming a 0.05 significance level). P-value: 0.0001. Reject H_0: $p_1 \geq p_2$. There is sufficient evidence to support the claim that the flu rate is lower for those with the vaccine.
9. Test statistic: $z = 1.07$. Critical value: $z = 1.645$. P-value: 0.1423. Fail to reject H_0: $p_1 \leq p_2$. There is not sufficient evidence to support the claim that the rate of severe injuries is lower for children wearing seat belts.
11. With a test statistic of $z = -6.74$ and a P-value of 0.0000 (rounded), there is sufficient evidence to support the claim that the polio rate is lower for those given the Salk vaccine.
13. $-0.0144 < p_1 - p_2 < 0.0086$; yes
15. a. Test statistic: $z = -3.06$. Critical values: $z = \pm 1.96$ (assuming a 0.05 significance level). P-value: 0.0022. Reject H_0: $p_1 = p_2$. There is sufficient evidence to support the claim that the two population percentages are different.
   b. $-0.0823 < p_1 - p_2 < -0.00713$
17. Test statistic: $z = 3.86$. Critical value: $z = 2.33$. P-value: 0.0001. Reject H_0: $p_1 \leq p_2$. There is sufficient evidence to support the claim that the proportion of drinkers among convicted arsonists is greater than the proportion of drinkers convicted of fraud.
19. Test statistic: $z = -2.82$. Critical value: $z = -1.645$. P-value: 0.0024. Reject H_0: $p_1 \geq p_2$. There is sufficient evidence to support the claim that the infection rate is lower for those who are warmed.
21. Test statistic: $z = 1.26$. Critical values: $z = \pm 1.96$. P-value: 0.2076. Fail to reject H_0: $p_1 - p_2 = 0.10$. There is not sufficient evidence to warrant rejection of the claim that the headache rate of Viagra users is 10 percentage points more than the percentage for those who use a placebo.
23. 2135

Section 8-5

1. Test statistic: $F = 4.0000$. Critical value: $F = 2.3248$. Reject H_0: $\sigma_1^2 = \sigma_2^2$. There is sufficient evidence to support the claim that the treatment and placebo populations have different variances.
3. Test statistic: $F = 2.9228$. Critical value: $F = 1.9752$ (approximately). Reject H_0: $\sigma_1 = \sigma_2$. There is sufficient evidence to support the claim that weights of regular Coke and diet Coke have different standard deviations.
5. Test statistic: $F = 1.2949$. Critical value: $F = 1.6928$ (approximately). Fail to reject H_0: $\sigma_1 = \sigma_2$. There is not sufficient evidence to support the claim that the two populations have different standard deviations.
7. Test statistic: $F = 1.0110$. Critical value of F is less than 1.3519 (assuming a 0.05 significance level). Although the conclusion is not clear from the test statistic and critical value, the values of the standard deviations (3.67, 3.65) suggest that the difference is not significant. Fail to reject H_0: $\sigma_1^2 \leq \sigma_2^2$. There is not sufficient evidence to support the claim that the faculty cars vary less than the student cars.
9. Test statistic: $F = 2.590782$. Critical value from Table A-5: $F = 2.7006$. Fail to reject H_0: $\sigma_1^2 = \sigma_2^2$. There is not sufficient evidence to warrant rejection of the claim that the two populations have the same variance.

11. Diet Coke: $n_1 = 36$, $s_1 = 0.00439090$; diet Pepsi: $n_2 = 36$, $s_2 = 0.00436161$. Test statistic: $F = 1.0135$. Critical F value is between 1.8752 and 2.0739. Fail to reject H_0: $\sigma_1 = \sigma_2$. There is not sufficient evidence to support the claim that diet Coke and diet Pepsi have weights with different standard deviations.
13. P-value < 0.02
15. $0.41 < \sigma_1^2 / \sigma_2^2 < 4.0$ (approximately)

Section 8-6

1. Test statistic: $t = -21.366$. Critical values: $t = \pm 2.048$. Reject H_0: $\mu_1 = \mu_2$. There is sufficient evidence to warrant rejection of the claim that the treatment and placebo groups have the same mean.
3. $-7.0 < \mu_1 - \mu_2 < -5.8$
5. a. Test statistic: $t = 3.288$. Critical values: $t = \pm 3.500$. Fail to reject H_0: $\mu_1 = \mu_2$. There is not sufficient evidence to warrant rejection of the claim that men and women have the same mean height.
   b. Test statistic: $t = 3.479$. Critical values: $t = \pm 5.841$. Fail to reject H_0: $\mu_1 = \mu_2$. There is not sufficient evidence to warrant rejection of the claim that men and women have the same mean height.
   c. Both procedures lead to the same conclusion.
   d. Although men and women actually do have different mean heights, the given sample data are not enough evidence to support that claim.
7. Test statistic: $t = -3.075$. Critical values: $t = \pm 2.878$. Reject H_0: $\mu_1 = \mu_2$. There is sufficient evidence to warrant rejection of the claim that the two populations have the same mean.
9. a. $-17.16 < \mu_1 - \mu_2 < 260.40$
   b. $-27.80 < \mu_1 - \mu_2 < 271.04$
   c. Part (a) is better because the sample standard deviations are not very different.
11. F-test: Fail to reject the claim of equal population variances, because the test statistic is $F = 2.9459$ and the critical value is $F = 3.0074$. Test for equal means: Test Statistic: $t = -1.099$. Critical values: $t = \pm 2.052$. Fail to reject H_0: $\mu_1 = \mu_2$. There is not sufficient evidence to warrant rejection of the claim that the two populations have the same mean.
13. Filtered: $n = 21$, $\bar{x} = 12.857$, $s = 3.071$.
    Nonfiltered: $n = 8$, $\bar{x} = 15.625$, $s = 1.188$.
    Assuming unequal variances: Test statistic: $t = -3.500$. Critical values: $t = \pm 2.365$. Reject H_0: $\mu_1 = \mu_2$. There is sufficient evidence to warrant rejection of the claim that the two populations have the same mean. The filtered cigarettes appear to have a lower mean amount of carbon monoxide.
15. Assuming unequal variances: Test statistic: $t = 22.698$. Critical values: $t = \pm 2.086$ (assuming a 0.05 significance level). Reject H_0: $\mu_1 = \mu_2$. There is sufficient evidence to warrant rejection of the claim that the two populations have the same mean.
17. Test statistic: $t = 15.322$. Critical values: $t = \pm 2.080$. Reject H_0: $\mu_1 = \mu_2$. There is sufficient evidence to warrant rejection of the claim that the two populations have the same mean.
19. a. Degrees of freedom changes from 7 to 26; test statistic is the same; critical value changes from -1.895 to -1.706; confidence interval limits change from $(-13.1, -8.3)$ to $(-12.8, -8.6)$.
    b. The test statistic changes to -12.30; the confidence interval limits change to $(-12.6, -8.8)$.

Chapter 8 Review Exercises

1. a. Test statistic: $z = 4.48$. Critical value: $z = 1.645$ (assuming a 0.05 significance level). Reject H_0: $p_1 \leq p_2$. There is sufficient evidence to support the claim that dyspepsia occurs at a higher rate among Viagra users.
   b. $0.0277 < p_1 - p_2 < 0.0699$; no
2. Experimental: $n = 20$, $\bar{x} = 66.3805$, $s = 20.9635$.
   Control: $n = 16$, $\bar{x} = 103.7675$, $s = 24.3371$.
   a. Test statistic: $F = 1.3478$. Critical value: $F = 2.6171$. Fail to reject H_0: $\sigma_1^2 = \sigma_2^2$. There is not sufficient evidence to warrant rejection of the claim that the two populations have the same variance.
   b. Test statistic: $t = -4.951$. Critical values: $t = \pm 1.96$ (assuming a 0.05 significance level). Reject H_0: $\mu_1 = \mu_2$. There is sufficient evidence to warrant rejection of the claim that both groups come from populations with the same mean.
   c. $-52.188 < \mu_1 - \mu_2 < -22.586$
3. Test statistic: $z = 2.41$. Critical value: $z = 1.645$. P-value: 0.0080. Reject H_0: $p_1 \leq p_2$. There is sufficient evidence to support the stated claim.
4. 23 min $< \mu_1 - \mu_2 <$ 65 min; no
5. Test statistic: $t = 2.301$. Critical values: $t = \pm 2.262$ (assuming a 0.05 significance level). Reject H_0: $\mu_d = 0$. There is sufficient evidence to conclude that there is a difference between the pretraining and posttraining weights.
6. $0.0 < \mu_d < 4.0$
7. a. Test statistic: $z = -12.44$. Critical value: $z = -2.33$. P-value: 0.0001. Reject H_0: $\mu_1 \geq \mu_2$. There is sufficient evidence to support the claim that women with a college degree have incomes with a higher mean.
   b. $\$14,415 < \mu_1 - \mu_2 < \$21,939$; no

8. a. Test statistic: $F = 3.8643$. Critical value: $F = 2.9249$ (approximately). Reject H_0: $\sigma_1 \leq \sigma_2$. There is sufficient evidence to support the claim that waiting times for the single line have a lower standard deviation.
   b. Test statistic: $t = -0.677$. Critical values: $t = \pm 2.861$. Fail to reject H_0: $\mu_1 = \mu_2$. There is not sufficient evidence to warrant rejection of the claim that the two populations have the same mean.
9. a. Test statistic: $t = -3.847$. Critical value: $t = -1.796$. Reject H_0: $\mu_d \geq 0$. There is sufficient evidence to conclude that the Dozenol tablets are more soluble after the storage period.
   b. $12.4 < \mu_d < 45.6$
10. F-test: Test statistic: $F = 1.3110$. Critical value: $F = 2.9685$ (assuming a 0.05 significance level). Fail to reject H_0: $\sigma_1^2 = \sigma_2^2$.
    Test for equality of means: Test statistic: $t = -0.627$. Critical values: $t = \pm 1.96$ (assuming a 0.05 significance level). Fail to reject H_0: $\mu_1 = \mu_2$. There is not sufficient evidence to warrant rejection of the claim that yellow M&M's and green M&M's have the same mean weight.

Chapter 8 Cumulative Review Exercises

1. a. 0.0707
   b. 0.369
   c. 0.104
   d. 0.0540
   e. Test statistic: $z = -2.52$. Critical value: $z = -1.645$. P-value: 0.0059. Reject H_0: $p_1 \geq p_2$. There is sufficient evidence to support the claim that the percentage of women ticketed for speeding is less than the percentage for men.
2. There must be an error, because the rates of 13.7% and 10.6% are not possible with sample sizes of 100.
3. a. Test statistic: $z = 4.92$. Critical value: $z = 1.645$. P-value: 0.0001. Reject H_0: $p \leq 0.5$. There is sufficient evidence to support the claim that the majority of the prepared students answered correctly.
   b. Test statistic: $z = 3.67$. Critical value: $z = 1.645$. P-value: 0.0001. Reject H_0: $p_1 \leq p_2$. There is sufficient evidence to support the stated claim.
4. a. Test statistic: $F = 1.3786$. Critical value: $F = 1.8363$ (approximately). Fail to reject H_0: $\sigma_1 = \sigma_2$. There is not sufficient evidence to warrant rejection of the claim that both groups come from populations with the same standard deviation.
   b. Test statistic: $z = 0.78$. Critical values: $z = \pm 2.33$. P-value: 0.4354. Fail to reject H_0: $\mu_1 = \mu_2$. There is not sufficient evidence to warrant rejection of the claim that the two populations have the same mean.
   c. $506.56 < \mu < 571.08$

Chapter 9 Answers

Section 9-2

1. a. Yes, the critical values of r are ± 0.312
   b. 0.161
3. $r = 0.238$; critical values: $r = \pm 0.878$ (assuming a 0.05 significance level). There appears to be a correlation that is not linear.
5. $r = 0.796$; critical values: $r = \pm 0.666$; significant linear correlation.
7. $r = 0.842$; critical values: $r = \pm 0.707$; significant linear correlation.
9. $r = -0.069$; critical values: $r = \pm 0.707$; no significant linear correlation.
11. a. $r = 0.870$; critical values: $r = \pm 0.279$; significant linear correlation.
    b. $r = -0.010$; critical values: $r = \pm 0.279$; no significant linear correlation.
    c. Duration
13. a. $r = 0.767$; critical values: $r = \pm 0.361$; significant linear correlation.
    b. $r = -0.441$; critical values: $r = \pm 0.361$; significant linear correlation.
    c. Part (a) because there is a higher correlation.
15. a. $r = 0.399$; critical values: $r = \pm 0.335$ (approximately); significant linear correlation.
    b. $r = 0.282$; critical values: $r = \pm 0.335$ (approximately); no significant linear correlation.
    c. Invest in big budget movies.
17. With a linear correlation coefficient very close to 0, there does not appear to be a correlation, but the conclusion suggests that there is a correlation.
19. Although there is no *linear* correlation, the variables may be related in some other *nonlinear* way.
21. $r = 0.819$ (approximately); critical values: $r = \pm 0.553$; significant linear correlation.
23. a. ± 0.272
    b. ± 0.189
    c. -0.378
    d. 0.549
    e. 0.658
25. No, correlation is used only to determine whether there is *some* relationship between two variables. It would not reveal whether the reported heights are significantly greater than the measured heights.

Section 9-3

1. $\hat{y} = 5 + x$
3. $\hat{y} = 11.8 + 0.891x$
5. $\hat{y} = -152 + 3.88x$; 116 lb
7. $\hat{y} = 0.549 + 1.48x$; 1.3 persons
9. $\hat{y} = 0.214 - 0.000182x$; 0.21
11. a. $\hat{y} = 41.9 + 0.179x$; 79 min
    b. $\hat{y} = 81.9 - 0.009x$; 81 min
    c. Part (a) because there is a significant linear correlation between durations and intervals.
13. a. $\hat{y} = -6158 + 12{,}201x$; $12,144
    b. $\hat{y} = 28{,}647 - 3044x$; $19,515
    c. Part (a) because the carat (weight) has a higher correlation with price.
15. a. $\hat{y} = 82.6 + 1.03x$; $98 million
    b. $\hat{y} = -107 + 33.1x$; $138 million (the mean gross amount)
    c. Yes, because there is a significant linear correlation between budget and gross.
17. a. 10.00
    b. 2.50
19. Outlier: yes; influential point: no
21. $\hat{y} = -182 + 0.000351x$; $\hat{y} = -182 + 0.351x$. The slope is multiplied by 1000 and the y-intercept doesn't change. If each y entry is divided by 1000, the slope and the y-intercept are both divided by 1000.
23. The equation $\hat{y} = -49.9 + 27.2x$ is better because it has $r = 0.997$, which is higher than $r = 0.963$ for $\hat{y} = -103.2 + 134.9 \ln x$.

Section 9-4

1. 0.09; 9%
3. 0.107; 10.7%
5. $r = 0.963$; critical values: $r = \pm 0.279$ (approximately); significant linear correlation.
7. 362.74 lb
9. a. 154.8
   b. 0
   c. 154.8
   d. 1
   e. 0
11. a. 13.836615
    b. 5.663385
    c. 19.5
    d. 0.70957
    e. 0.971544
13. a. 14
    b. $s_e = 0$; $E = 0$; no interval estimate.
15. a. 4.2 persons
    b. $1.6 < y < 6.9$
17. $-\$3.25 < y < \$17.42$
19. $-\$1.73 < y < \$24.81$
21. $-\$11.27 < \beta_0 < \$10.60$; $0.00901 < \beta_1 < 0.288$; predicted value can vary widely
23. a. $(n - 2)s_e^2$
    b. $\dfrac{r^2 \cdot (\text{unexplained variation})}{1 - r^2}$
    c. $r = -0.949$

Section 9-5

1. $\hat{y} = -285 - 1.38x_1 - 11.2x_2 + 28.6x_3$
3. Yes, with a P-value of 0.002, the equation has overall significance.
5. a. $\hat{y} = -274 + 0.426x_1 + 12.1x_2$
   b. 0.928; 0.925; 0.000
   c. Yes
7. a. $\hat{y} = -235 + 0.403x_1 + 5.11x_2 - 0.555x_3 + 9.19x_4$
   b. 0.942; 0.938; 0.000
   c. Yes
9. a. $\hat{y} = 7.74 - 0.00585x_1$
   b. 0.070; 0.043; 0.119
   c. No
11. a. $\hat{y} = 7.41 - 0.00992x_1 + 0.00394x_2$
    b. 0.249; 0.204; 0.009
    c. Yes
13. a. $\hat{y} = -6158 + 12{,}201x_1$
    b. 0.589; 0.574; 0.000
    c. Yes
15. a. $\hat{y} = 8221 + 12{,}297x_1 - 3116x_2$
    b. 0.793; 0.778; 0.000
    c. Yes
17. $\hat{y} = 0.154 + 0.0651x_1$ (where x_1 represents tar); yes.
19. $\hat{y} = 7.26 + 0.914x_1$ (where x_1 represents list price); by using more variables, adjusted R^2 can be raised from 0.995 to 0.996, but the small increase in adjusted R^2 does not justify the inclusion of additional variables.

Section 9-6

1. Quadratic: $y = 2x^2 - 12x + 18$
3. Exponential: $y = 3^x$
5. Exponential: $y = (1{,}203{,}014)(0.886814)^x$, where x is coded as 1 for 1980, 2 for 1981, and so on. Predicted value: 122,769 lb.
7. Power: $y = 2000x^{0.4}$. With $R^2 = 0.999999938$, the model fits almost perfectly.
9. a. 3837.41
   b. 61.3758
   c. The quadratic sum of squares is lower.

Chapter 9 Review Exercises

1. a. $r = 0.338$; critical values: $r = \pm 0.632$; no significant linear correlation.
   b. 11%
   c. $\hat{y} = -0.488 + 0.611x$
   d. 0.347
2. a. $r = 0.116$; critical values: $r = \pm 0.632$; no significant linear correlation.
   b. 1.4%
   c. $\hat{y} = 0.0657 + 0.000792x$
   d. 0.347
3. a. $r = 0.777$; critical values: $r = \pm 0.632$; significant linear correlation.
   b. 60%
   c. $\hat{y} = 0.193 + 0.00293x$
   d. 0.286
4. $\hat{y} = -0.0526 + 0.747x_1 - 0.00220x_2 + 0.00303x_3$; $R^2 = 0.726$; adjusted $R^2 = 0.589$; P-value $= 0.040$. Because the overall P-value of 0.040 is less than 0.05, the equation can be used to predict ice cream consumption. Using the consumption/temperature data, the adjusted R^2 is 0.554. Although the adjusted R^2 is slightly higher using all three variables, the slight increase in adjusted R^2 does not justify the inclusion of additional variables, so the best regression equation appears to result from using temperature as the only independent variable.
5. a. $r = 0.340$; critical values: $r = \pm 0.632$; no significant linear correlation.
   b. $\hat{y} = 826 + 17.1x$
   c. 1405.9 (mean)
6. a. $r = -0.157$; critical values: $r = \pm 0.632$; no significant linear correlation.
   b. $\hat{y} = 2990 - 32.5x$
   c. 1405.9 (mean)
7. a. $r = 0.892$; critical values: $r = \pm 0.632$; significant linear correlation.
   b. $\hat{y} = -724 + 0.178x$
   c. 1643 million bushels
8. $\hat{y} = -3145 + 10.1x_1 + 40.4x_2 + 0.187x_3$; 0.874; 0.811; 0.004; yes.

Chapter 9 Cumulative Review Exercises

1. a. $n = 50$, $\bar{x} = 80.660$, $s = 11.977$. Test statistic: $z = 8.66$. Critical value: $z = 1.645$ (assuming a 0.05 significance level). Reject H_0: $\mu \leq 66$. There is sufficient evidence to support the claim that intervals between eruptions now have a mean greater than 66 min.
   b. 77.3 min $< \mu <$ 84.0 min
2. a. $\bar{x} = 99.1$, $s = 8.5$
   b. $\bar{x} = 102.8$, $s = 8.7$
   c. No, but a better comparison would involve treating the data as matched pairs instead of two independent samples.
   d. Yes. $r = 0.702$ and the critical values are $r = \pm 0.576$ (assuming a 0.05 significance level). There is a significant linear correlation.
3. a. No
   b. Test statistic: $t = 1.185$. Critical values: $t = \pm 2.262$. Fail to reject H_0: $\mu_d = 0$. There is not sufficient evidence to warrant rejection of the claim that position has no effect.

Chapter 10 Answers

Section 10-2

1. a. df = 37, so $\chi^2 = 51.805$ (approximately).
   b. $0.10 < P\text{-value} < 0.90$
   c. There is not sufficient evidence to warrant rejection of the claim that the roulette slots are equally likely.
3. Test statistic: $\chi^2 = 4.600$. Critical value: $\chi^2 = 7.815$. There is sufficient evidence to warrant rejection of the claim that the results fit a uniform distribution. Students do not have the ability to select the same tire.
5. Test statistic: $\chi^2 = 7.417$. Critical value: $\chi^2 = 12.592$. There is not sufficient evidence to warrant rejection of the claim that fatal crashes occur on the different days of the week with equal frequency.
7. Test statistic: $\chi^2 = 23.431$. Critical value: $\chi^2 = 9.488$. There is sufficient evidence to warrant rejection of the claim that the accidents are distributed as claimed.
9. Test statistic: $\chi^2 = 4.200$. Critical value: $\chi^2 = 16.919$. There is not sufficient evidence to warrant rejection of the claim that the digits are uniformly distributed.
11. Test statistic: $\chi^2 = 53.051$. Critical value: $\chi^2 = 7.815$. There is sufficient evidence to warrant rejection of the claim that the distribution of crashes is the same as the distribution of ages.
13. Test statistic: $\chi^2 = 16.333$. Critical value: $\chi^2 = 14.067$ (assuming a 0.05 significance level). There is sufficient evidence to support the claim that the probabilities of winning in the different post positions are not all the same.
15. Test statistic: $\chi^2 = 180.308$. Critical value: $\chi^2 = 16.919$ (assuming a 0.05 significance level). There is sufficient evidence to support the claim that the last digits do not occur with the same frequency. Because of the large number of days with no rain, we expect many zeros, so we cannot conclude that the amounts were estimated; they may have been measured.

17. The test statistic changes from 4.600 to 76.638, so the outlier has a dramatic effect.
19. a. Critical value is $\chi^2 = 3.841$, and test statistic is

$$\chi^2 = \frac{\left(f_1 - \frac{f_1 + f_2}{2}\right)^2}{\frac{f_1 + f_2}{2}} + \frac{\left(f_2 - \frac{f_1 + f_2}{2}\right)^2}{\frac{f_1 + f_2}{2}} = \frac{(f_1 - f_2)^2}{f_1 + f_2}$$

   b. Critical values: The χ^2 critical value is 3.841, and it is approximately equal to the square of $z = 1.96$.
21. Combine the last three cells.
    Test statistic: $\chi^2 = 1.012$. Critical value: $\chi^2 = 5.991$. Fail to reject H_0: The frequencies fit a Poisson distribution. There is not sufficient evidence to warrant rejection of the claim that the frequencies fit a Poisson distribution.

Section 10-3

1. Test statistic: $\chi^2 = 1.174$. *P*-value: 0.279. There is not sufficient evidence to warrant rejection of the claim that treatment is independent of the reaction.
3. Test statistic: $\chi^2 = 47.035$. Critical value: $\chi^2 = 6.635$. There is sufficient evidence to warrant rejection of the claim that opinion and gender are independent.
5. Test statistic: $\chi^2 = 51.458$. Critical value: $\chi^2 = 6.635$. There is sufficient evidence to warrant rejection of the claim that the proportions of agree/disagree responses are the same for the subjects interviewed by men and the subjects interviewed by women.
7. Test statistic: $\chi^2 = 2.195$. Critical value: $\chi^2 = 5.991$. There is not sufficient evidence to support the claim that there is a gender gap.
9. Test statistic: $\chi^2 = 8.642$. Critical value: $\chi^2 = 9.210$. There is not sufficient evidence to warrant rejection of the claim that the distribution of planes flying and planes ordered is independent of the airline company.
11. Test statistic: $\chi^2 = 119.330$. Critical value: $\chi^2 = 5.991$. There is sufficient evidence to warrant rejection of the claim that the type of crime is independent of whether the criminal is a stranger.
13. Test statistic: $\chi^2 = 42.557$. Critical value: $\chi^2 = 3.841$. There is sufficient evidence to warrant rejection of the claim that the sentence is independent of the plea.
15. Test statistic: $\chi^2 = 49.731$. Critical value: $\chi^2 = 11.071$ (assuming a 0.05 significance level). There is sufficient evidence to support the claim that the type of crime is related to whether the criminal drinks or abstains.
17. Test statistic: $\chi^2 = 0.122$. Critical value: $\chi^2 = 3.841$ (assuming a 0.05 significance level). There is not sufficient evidence to warrant rejection of the claim that gender of statistics students is independent of whether they smoke.
19. Without Yates' correction: $\chi^2 = 1.174$. With Yates' correction: $\chi^2 = 0.907$. Yates' correction decreases the test statistic so that sample data must be more extreme to be considered significant.

Chapter 10 Review Exercises

1. Test statistic: $\chi^2 = 12.438$. Critical value: $\chi^2 = 6.635$. There is sufficient evidence to warrant rejection of the claim that whether gun owners store their guns loaded and unlocked is independent of whether they are a member of the NRA.
2. Test statistic: $\chi^2 = 5.600$. Critical value: $\chi^2 = 11.071$. There is not sufficient evidence to support the claim that the outcomes are not equally likely.
3. Test statistic: $\chi^2 = 6.780$. Critical value: $\chi^2 = 12.592$. There is not sufficient evidence to support the theory that more gunfire deaths occur on weekends.
4. Test statistic: $\chi^2 = 0.500$. Critical value: $\chi^2 = 5.991$. There is not sufficient evidence to warrant rejection of the claim that USAir, American, and Delta have the same proportions of on-time flights.
5. Test statistic: $\chi^2 = 7.607$. Critical value: $\chi^2 = 5.991$. There is sufficient evidence to warrant rejection of the claim that the occurrence of headaches is independent of the group (Seldane, placebo, control).

Chapter 10 Cumulative Review Exercises

1. $\bar{x} = 80.9$; median: 81.0; range: 28.0; $s^2 = 74.4$; $s = 8.6$; 5–number summary: 66, 76.0, 81.0, 86.5, 94.
2. a. 0.272
   b. 0.468
   c. 0.614
   d. 0.282
3. Contingency table; see Section 10-3. Test statistic: $\chi^2 = 0.055$. Critical value: $\chi^2 = 7.815$ (assuming a 0.05 significance level). There is not sufficient evidence to warrant rejection of the claim that men and women choose the different answers in the same proportions.
4. Use correlation; see Section 9-2. Test statistic: $r = 0.978$. Critical values: $r = \pm 0.950$ (assuming a 0.05 significance level). There is sufficient evidence to support the claim that there is relationship between the memory and reasoning scores.
5. Use the test for matched pairs; see Section 8-3. $\bar{d} = -10.25$; $s_d = 1.5$. Test statistic: $t = -13.667$. Critical value: $t = -2.353$ (assuming a 0.05 significance level). Reject H_0: $\mu_d \geq 0$. There is sufficient evidence to

support the claim that the training session is effective in raising scores.

6. Test for the difference between two independent samples; see Section 8-6. The sample standard deviations are 7.1 and 7.3, so assume that the two populations have equal variances. Test statistic: $t = -2.014$. Critical values: $t = \pm 2.447$ (assuming a 0.05 significance level). Fail to reject H_0: $\mu_1 = \mu_2$. There is not sufficient evidence to warrant rejection of the claim that men and women have the same mean score.

Chapter 11 Answers

Section 11-2

1. a. 6
   b. $F = 3.33$
   c. $F = 2.9013$
   d. 0.032
   e. There is sufficient evidence to warrant rejection of the claim that the different products have commercials resulting in the same mean reaction score.
3. a. 0.864667
   b. 0.46
   c. $F = 1.8797$
   d. $F = 3.8853$ (assuming a 0.05 significance level)
   e. There is not sufficient evidence to warrant rejection of the claim that the three age-group populations have the same mean body temperature.
5. Test statistic: $F = 1.3432$. Critical value: $F = 3.2389$. Fail to reject H_0: $\mu_1 = \mu_2 = \mu_3 = \mu_4$. There is not sufficient evidence to warrant rejection of the claim that the populations of cars in the different weight categories have the same mean chest deceleration value.
7. Test statistic: $F = 4.0497$. Critical value: $F = 3.4028$. Reject H_0: $\mu_1 = \mu_2 = \mu_3$. There is sufficient evidence to warrant rejection of the claim that the mean breadth is the same for the different epochs.
9. Test statistic: $F = 2.9493$. Critical value: $F = 2.6060$. Reject H_0: $\mu_1 = \mu_2 = \mu_3 = \mu_4 = \mu_5$. There is sufficient evidence to warrant rejection of the claim that the means for the different laboratories are the same.
11. Test statistic: $F = 0.5083$. Critical value: $F = 2.2899$ (approximately). Fail to reject H_0: $\mu_1 = \mu_2 = \mu_3 = \mu_4 = \mu_5 = \mu_6$. There is not sufficient evidence to warrant rejection of the claim that the populations of different colors of M&Ms have the same mean.
13. a. The variance between samples changes to 5.864667, the variance within samples remains the same, the test statistic changes to $F = 12.7493$, and the critical value remains the same.
    b. The F test statistic is not affected by the scale used.
    c. The F test statistic does not change.
    d. The F test statistic does not change.
    e. The F test statistic does not change.
15. a. Test statistic: $t = 0.978$. Critical values: $t = \pm 2.306$. There is not sufficient evidence to warrant rejection of the claim that the two populations have the same mean.
    b. Test statistic: $F = 0.9555$. Critical value: $F = 5.3177$. Fail to reject H_0: $\mu_1 = \mu_2$. There is not sufficient evidence to warrant rejection of the claim that the two populations have the same mean.
    c. Test statistics: $t^2 = F = 0.956$. Critical values: $t^2 = F = 5.32$.

Section 11-3

1. Test statistic: $F = 3.01$. P-value: 0.091. Fail to reject the null hypothesis of no interaction. There does not appear to be a significant effect from the interaction between gender and test (math/verbal).
3. Test statistic: $F = 0.58$. P-value: 0.453. Because the P-value is greater than 0.05, fail to reject the null hypothesis that the type of test (math/verbal) has no effect on SAT scores. There is not sufficient evidence to support the claim that the type of test (math/verbal) has an effect on SAT scores.
5. Test statistic: $F = 0.78$. P-value: 0.395. Because the P-value is greater than 0.05, fail to reject the null hypothesis that gender has no effect on estimated length. There is not sufficient evidence to support the claim that gender has an effect on estimated length.
7. Test statistic: $F = 3.87$. P-value: 0.000. Because the P-value is less than 0.05, reject the null hypothesis that the choice of subject has no effect on the hearing test score. There is sufficient evidence to support the claim that the choice of subject has an effect on the hearing test score.
9. Test statistic: $F = 2.47$. P-value: 0.159. Because the P-value is greater than 0.05, fail to reject the null hypothesis that the four machine operators have the same mean production output. The different operators appear to be working at about the same level.
11. (The given results use the first nine values for each cell, but the pulse rate of 8 was excluded as being not feasible.) Test statistic: $F = 0.4200$. Critical value: $F = 4.1709$ (approximately, assuming a 0.05 significance level). There does not appear to be an interaction between gender and smoking.
13. Test statistic: $F = 0.1814$. Critical value: $F = 4.1709$ (approximately, assuming a 0.05 significance level). Smoking does not appear to have an effect on pulse rates. (This is not too surprising, because college students have not been smoking very long.)
15. Answers vary.

Chapter 11 Review Exercises

1. Test statistic: $F = 0.05$. Critical value: $F = 5.1433$. Fuel consumption does not appear to be affected by an interaction between transmission type and engine size.
2. Test statistic: $F = 1.90$. Critical value: $F = 5.1433$. The size of the engine does not appear to have an effect on fuel consumption.
3. Test statistic: $F = 3.56$. Critical value: $F = 5.9874$. The type of transmission does not appear to have an effect on fuel consumption.
4. Test statistic: $F = 9.4827$. Critical value: $F = 3.0984$. Reject the null hypothesis of equal means. There is sufficient evidence to support the claim of different mean selling prices.
5. a. Test statistic: $F = 1.00$. *P*-value: 0.423. There is not sufficient evidence to support the claim that amounts of emitted greenhouse gases are affected by the type of transmission.
   b. Test statistic: $F = 7.00$. *P*-value: 0.125. There is not sufficient evidence to support the claim that amounts of greenhouse gases are affected by the number of cylinders.
   c. Perhaps greenhouse gases *are* affected by the type of transmission and/or the number of cylinders; however, the given sample data do not provide sufficient evidence to support such claims.
6. Test statistic: $F = 46.90$. *P*-value: 0.000. Reject the null hypothesis that the three populations have equal means. There is sufficient evidence to warrant rejection of the claim of equal population means.

Chapter 11 Cumulative Review Exercises

1. a. 0.100 in.
   b. 0.263 in.
   c. 0.00, 0.00, 0.00, 0.010, 1.41
   d. 0.92 in., 1.41 in.
   e. Answer varies, depending on the number of classes used, but the histogram should depict a distribution that is skewed to the right.
   f. No, because the data do not appear to come from a normally distributed population.
   g. 19/52 or 0.365
2. a.

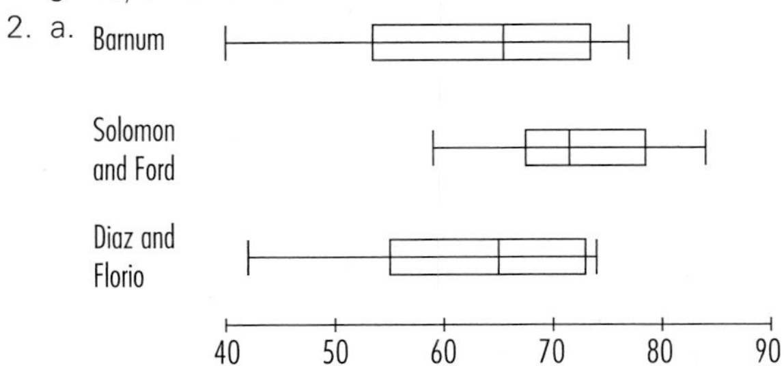

   b. Barnum: $\bar{x} = 62.8$, $s = 12.9$; Solomon & Ford: $\bar{x} = 72.3$, $s = 8.2$; Diaz & Florio: $\bar{x} = 62.8$, $s = 11.6$
   c. Based on the sample standard deviations, assume that the two populations have equal variances. Test statistic: $t = -1.760$. Critical values: $t = \pm 2.145$. Fail to reject the null hypothesis of equal means. There is not sufficient evidence to warrant rejection of the claim that the two populations have equal means.
   d. Barnum: $52.0 < \mu < 73.5$; Solomon & Ford: $65.4 < \mu < 79.1$; Diaz & Florio: $53.1 < \mu < 72.4$. Solomon & Ford seem slightly higher.
   e. Test statistic: $F = 1.9677$. Critical value: $F = 3.4668$. Fail to reject the null hypothesis of equal means. There is not sufficient evidence to warrant rejection of the claim that the three populations have the same mean.
3. a. 0.3372
   b. 0.0455
   c. 1/8

Chapter 12 Answers

Section 12-2

1. There is an upward trend indicating that the intervals between eruptions are increasing, with the result that tourists must wait longer. There is increasing variation, with the result that predicted times of eruptions are becoming less reliable.

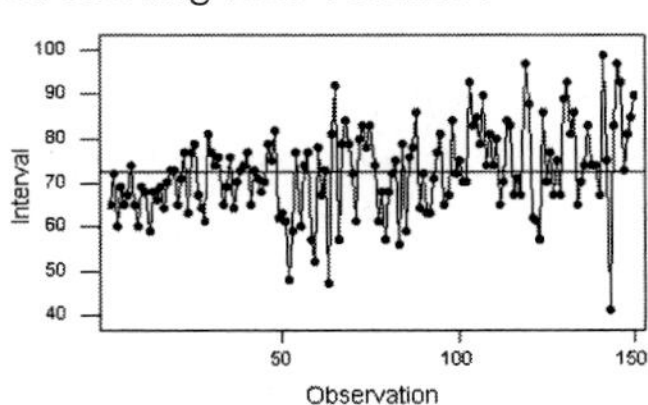

3. There are points lying beyond the upper control limit and there is an upward trend, so the process mean is out of statistical control. An out-of-control process mean indicates that we cannot make accurate predictions.

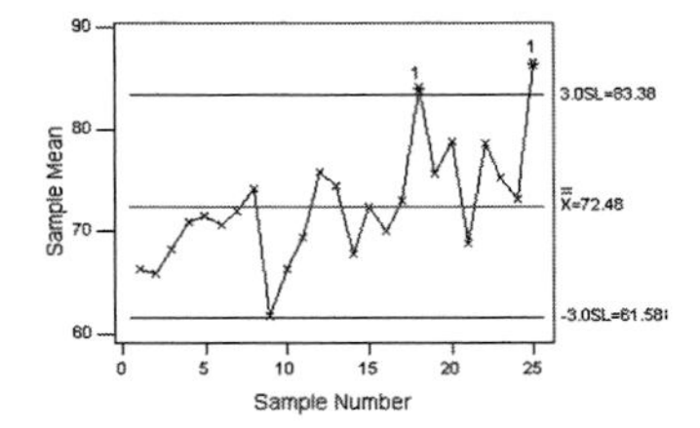

5. There is a pattern of increasing ranges, there are points beyond the upper control limit, and there are eight consecutive points lying below the center line, so the process variation is out of statistical control.

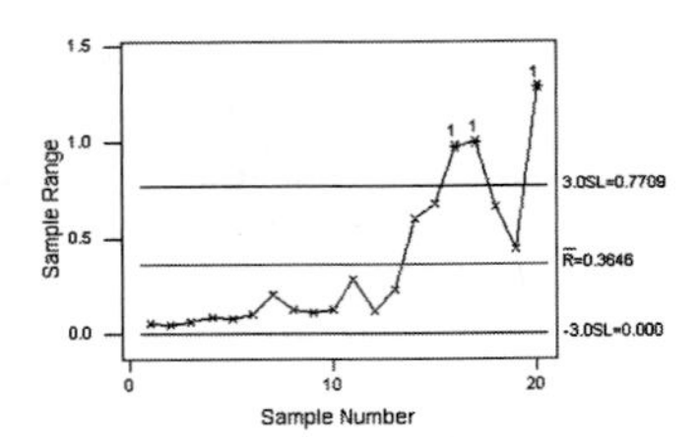

7. There is a cyclic pattern consisting of a rough "W" shape for each year. The pattern can be explained by the temperature changes throughout the year.

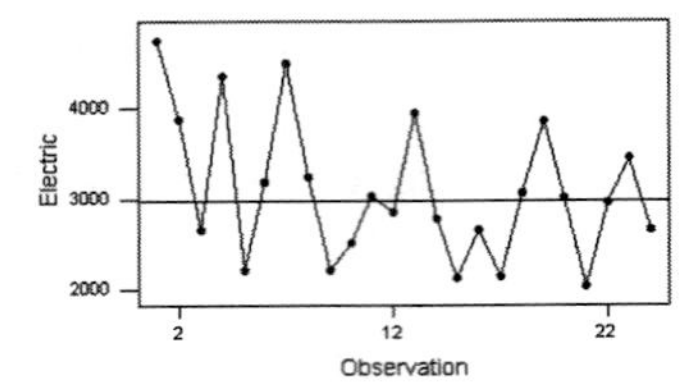

9. The process mean appears to be within statistical control.

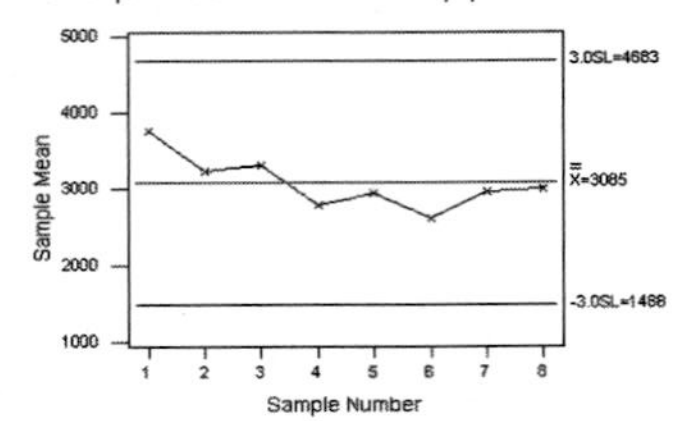

11. The process variation appears to be out of statistical control. There are points that lie beyond the control limits.

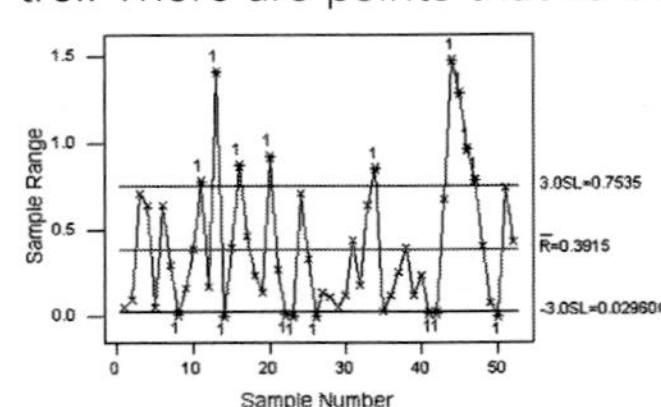

13. Using Table 12-1 values, $\overline{s} = 20.05$, and the control limits are at 2.3659 and 37.7341. The result and conclusions are very similar to those for the R chart.

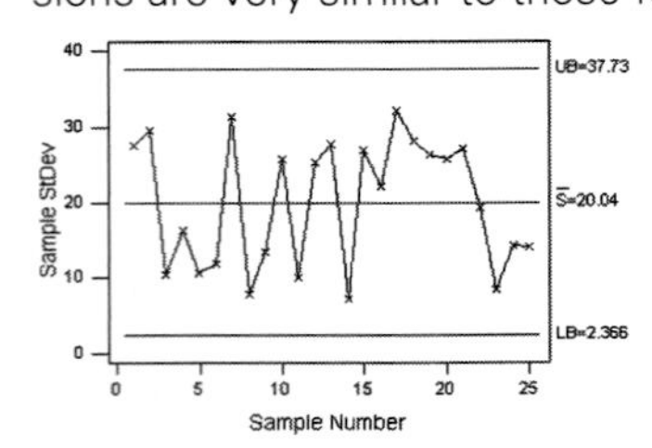

Section 12-3

1. Process appears to be within statistical control.
3. Process appears to be out of statistical control because there is a pattern of an upward trend and there is a point that lies beyond the upper control limit.
5. The process is out of statistical control because there is a downward trend and there are eight consecutive points all lying below the center line. This downward trend is good and its causes should be identified so that it can continue.

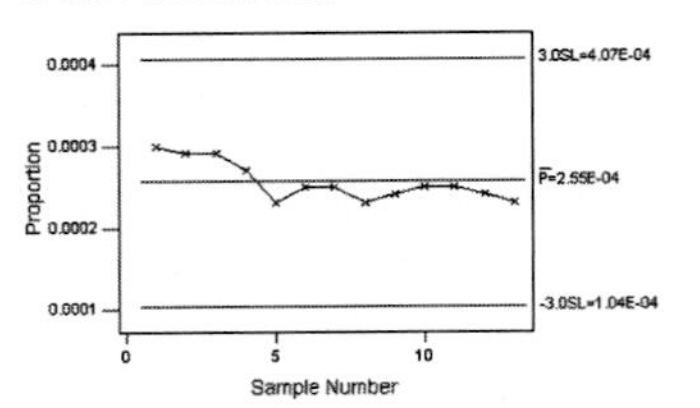

7. The process appears to be statistically stable.

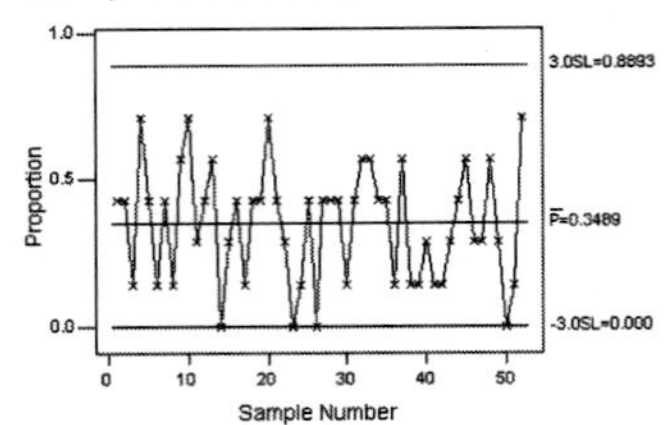

9. Except for the scale used, the charts are identical.

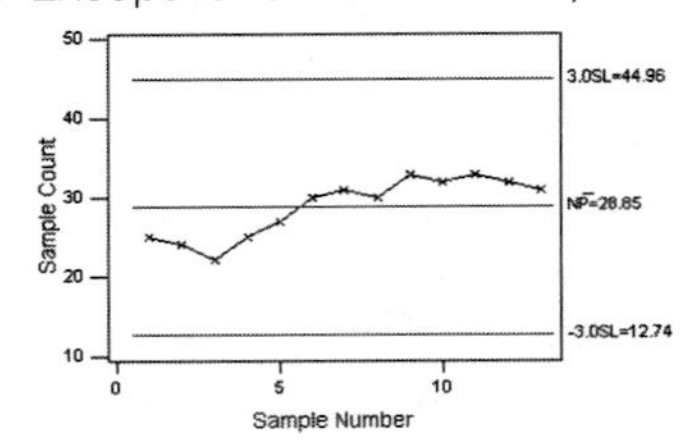

Chapter 12 Review Exercises

1. The process appears to be within statistical control.

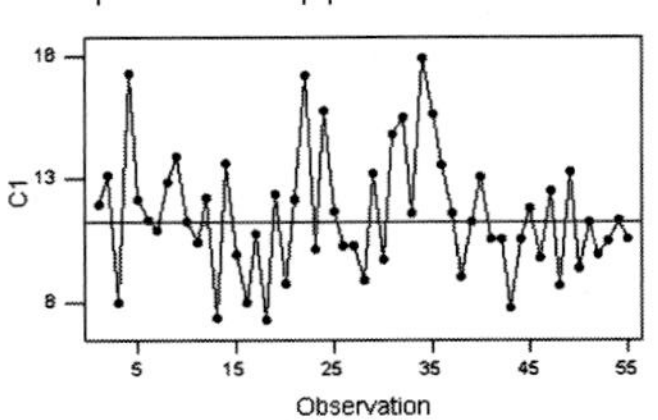

2. The process variation appears to be within statistical control.

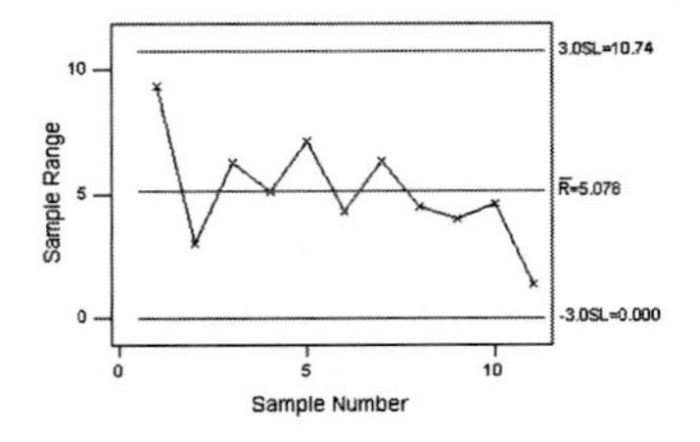

3. Because there is a point that lies beyond the upper control limit, the process mean is not within statistical control.

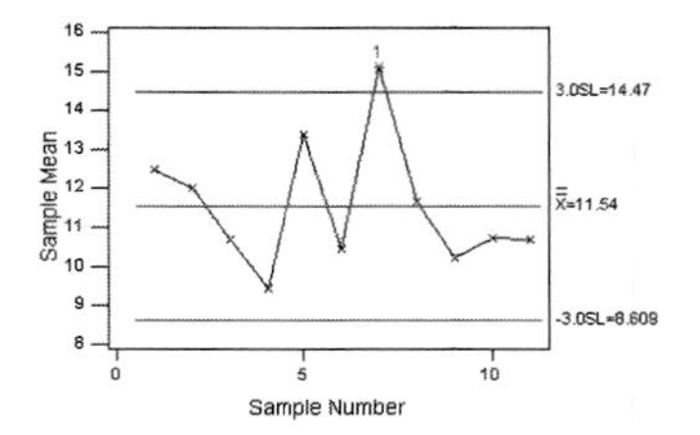

4. The process is out of control because there is a shift up and there are points beyond the control limits.

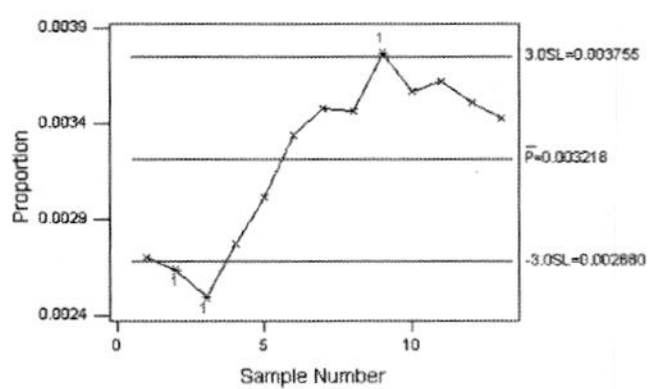

5. The process is out of control because there are points beyond the control limits.

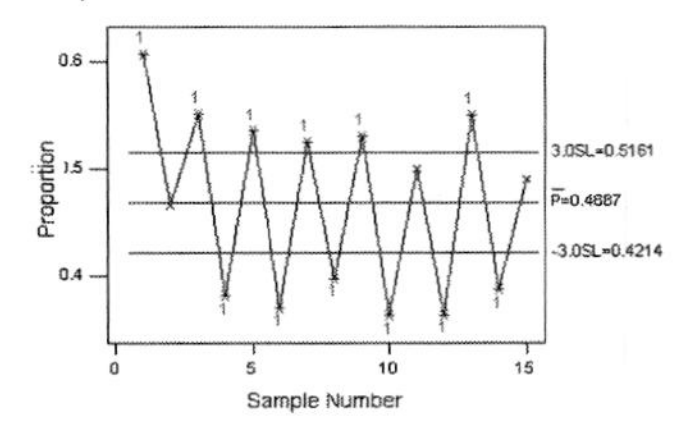

Chapter 12 Cumulative Review Exercises

1. a. The process appears to be within statistical control.

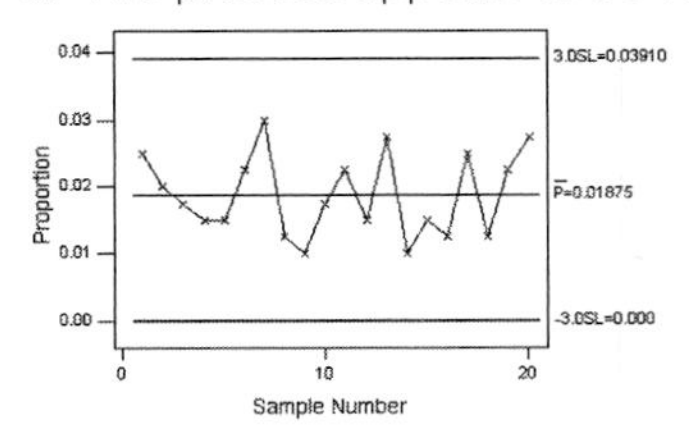

   b. $0.0158 < p < 0.0217$
   c. Test statistic: $z = 5.77$. Critical value: $z = 1.645$. Reject H_0: $p \leq 0.01$. There is sufficient evidence to warrant rejection of the claim that the rate of defects is 1% or less.
2. a. 1/256
   b. 1/256
   c 1/128
3. The run chart reveals very clear cycles, indicating that the process is not statistically stable.

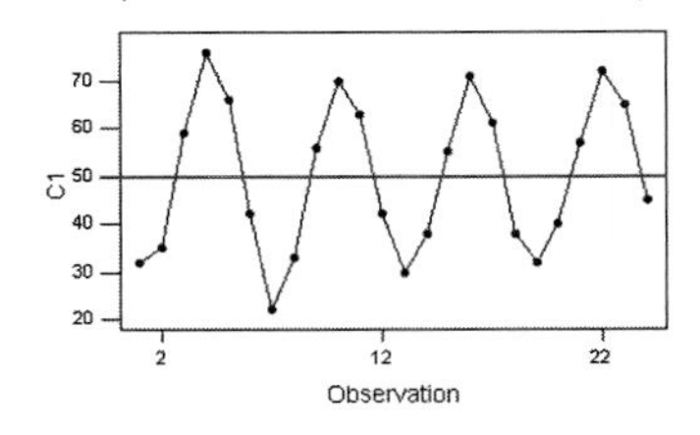

4. a. $r = -0.484$. Critical values: $r = \pm 0.396$ (approximately, assuming a 0.05 significance level). There is sufficient evidence to support the claim of a significant linear correlation between temperature and energy consumption.
   b. $\hat{y} = 4278 - 23.9x$
   c. 2844 kilowatt-hours

Chapter 13 Answers

Section 13-2

1. The test statistic of $x = 3$ is not less than or equal to the critical value of 1. There is not sufficient evidence to support the claim that there is a difference between reported and measured heights of female statistics students.
3. Convert $x = 301$ to the test statistic $z = -12.61$. Critical value: $z = -1.645$ (assuming a 0.05 significance level). There is sufficient evidence to support the claim that the majority of people say that they voted in the election.
5. Convert $x = 1$ to the test statistic $z = -5.32$. Critical value: $z = -1.645$ (assuming a 0.05 significance level). There is sufficient evidence to support the claim that the Coke cans have volumes with a median greater than 12 oz.
7. (Instead of a right-tailed test to determine whether $x = 18$ is large enough to be significant, use a left-tailed test to determine whether $x = 15$ is small enough to be significant.) Convert $x = 15$ to the test statistic $z = -0.35$. Critical value: $z = -1.645$. There is not sufficient evidence to support the claim that the median is greater than 0.9085 g.
9. The test statistic of $x = 1$ is less than or equal to the critical value of 2. There is sufficient evidence to warrant rejection of the claim of no difference between the right- and left-hand reaction times.
11. Convert $x = 6$ to the test statistic $z = -2.83$. Critical values: $z = \pm 2.575$. There is sufficient evidence to warrant rejection of the claim that the drinks had no effect on reaction times.
13. First approach: $z = -1.90$; reject H_0.
    Second approach: $z = -1.73$; reject H_0.
    Third approach: $z = 0$; fail to reject H_0.
15. *, *, *, *, *, 0, 0, 0, 1, 1, 1, 2, 2, 3, 3

Section 13-3

1. Test statistic: $T = 9.5$. Critical value: $T = 3$. Fail to reject the null hypothesis that both samples come from the same population distribution.
3. Test statistic: $T = 4.5$. Critical value: $T = 21$. Reject the null hypothesis that both samples come from the same population distribution.
5. Test statistic: $T = 5.5$. Critical value: $T = 13$. Reject the null hypothesis that both samples come from the same

population distribution. There is sufficient evidence to support the claim that there is a difference between the two times.

7. Test statistic: $T = 0$. Critical value: $T = 14$ (assuming a 0.05 significance level). Reject the null hypothesis that both samples come from the same population distribution. There is sufficient evidence to support the claim that the drug has an effect.
9. a. 0, 27.5
   b. 0, 637.5
   c. 1954

Section 13-4

1. $\mu_R = 577.5$, $\sigma_R = 56.358$, $R = 569$, $z = -0.15$. Test statistic: $z = -0.15$. Critical values: $z = \pm 1.96$. Fail to reject the null hypothesis that red and brown M&Ms have weights with identical populations. They appear to be the same.
3. $\mu_R = 150$, $\sigma_R = 17.321$, $R = 96.5$, $z = -3.09$. Test statistic: $z = -3.09$. Critical values: $z = \pm 2.575$. There is sufficient evidence to warrant rejection of the claim that the two samples come from identical populations.
5. $\mu_R = 1314$, $\sigma_R = 88.792$, $R = 1958$, $z = 7.25$. Test statistic: $z = 7.25$. Critical values: $z = \pm 1.96$ (assuming a 0.05 significance level). There is sufficient evidence to support the claim that the two samples come from different populations. The sugar in regular coke weighs more.
7. $\mu_R = 1330$, $\sigma_R = 94.163$, $R = 1163$, $z = -1.77$. Test statistic: $z = -1.77$. Critical values: $z = \pm 1.96$ (assuming a 0.05 significance level). There is not sufficient evidence to support the claim that new textbook prices are different at the two colleges.
9. The results are the same.
11. $z = 0.74$; the test statistic is the same number with opposite sign.

Section 13-5

1. Test statistic: $H = 14.431$. Critical value: $\chi^2 = 7.815$. There is sufficient evidence to warrant rejection of the claim that the different years have time intervals with identical populations.
3. Test statistic: $H = 4.234$. Critical value: $\chi^2 = 7.815$ (assuming a 0.05 significance level). There is not sufficient evidence to support the claim that left femur loads for the four car weight categories are not all the same. (They are not significantly different.)
5. Test statistic: $H = 6.631$. Critical value: $\chi^2 = 5.991$. There is sufficient evidence to warrant rejection of the claim that the three samples come from identical populations.
7. Test statistic: $H = 2.075$. Critical value: $\chi^2 = 11.071$. There is not sufficient evidence to warrant rejection of the claim that the weights are the same for each of the six different color populations.
9. a. The test statistic H does not change.
   b. The test statistic H does not change.
   c. The value of the test statistic does not change much (because the rank is used instead of the magnitude of the outlier).
11. $\Sigma T = 366$; adjusted $H = 14.431335 \div 0.99668910$, which is 14.479. Change is not substantial.

Section 13-6

1. a. $r_s = 1$, and there appears to be a correlation between x and y.
   b. $r_s = -1$, and there appears to be a correlation between x and y.
   c. $r_s = 0$, and there does not appear to be a correlation between x and y.
3. $r_s = 0.855$. Critical values: $r_s = \pm 0.648$. Significant correlation. There appears to be a correlation between salary and stress.
5. $r_s = -0.067$. Critical values: $r_s = \pm 0.648$. No significant correlation. There does not appear to be a correlation between stress level and physical demand.
7. $r_s = 0.829$. Critical values: $r_s = \pm 0.886$. No significant correlation. There does not appear to be a correlation between the bill and the amount of tip.
9. $r_s = 0.000$. Critical values: $r_s = \pm 0.738$. No significant correlation. There does not appear to be a correlation between age and blood alcohol concentration.
11. a. $r_s = 0.786$. Critical values: $r_s = \pm 0.280$. Significant correlation. There appears to be a correlation between duration and interval.
    b. $r_s = -0.036$. Critical values: $r_s = \pm 0.280$. No significant correlation. There does not appear to be a correlation between intervals and heights.
    c. Eruption duration, because there is a correlation with intervals.
13. a. $r_s = 0.833$. Critical values: $r_s = \pm 0.364$. Significant correlation. There appears to be a correlation between price and weight.
    b. $r_s = -0.329$. Critical values: $r_s = \pm 0.364$. No significant correlation. There does not appear to be a correlation between price and color.
    c. Weight, because there is a correlation with price.
15. a. ± 0.707
    b. ± 0.514
    c. ± 0.361
    d. ± 0.463
    e. ± 0.834

Section 13-7

1. $n_1 = 10$, $n_2 = 5$, $G = 3$, 5% cutoff values: 3, 12; reject randomness.
3. $n_1 = 8$, $n_2 = 12$, $G = 6$, 5% cutoff values: 6, 16; reject randomness.
5. $n_1 = 12$, $n_2 = 8$, $G = 10$, 5% cutoff values: 6, 16; fail to reject randomness.
7. $n_1 = 15$, $n_2 = 13$, $G = 12$, 5% cutoff values: 9, 21; fail to reject randomness.
9. $n_1 = 17$, $n_2 = 14$, $G = 14$, 5% cutoff values: 10, 23; fail to reject randomness.
11. The median corresponds to July 2; $n_1 = 6$, $n_2 = 6$, $G = 2$, 5% cutoff values: 3, 11; reject randomness.
13. $n_1 = 49$, $n_2 = 51$, $G = 43$, $\mu_G = 50.98$, $\sigma_G = 4.9727$. Test statistic: $z = -1.60$. Critical values: $z = \pm 1.96$. Fail to reject randomness.
15. $n_1 = 34$, $n_2 = 31$, $G = 35$, $\mu_G = 33.431$, $\sigma_G = 3.9909$. Test statistic: $z = 0.39$. Critical values: $z = \pm 1.96$. Fail to reject randomness.
17. Minimum is 2, maximum is 4. Critical values of 1 and 6 can never be realized so that the null hypothesis of randomness can never be rejected.

Chapter 13 Review Exercises

1. The test statistic $x = 0$ is less than or equal to the critical value of 2. There is sufficient evidence to warrant rejection of the claim that the drug has no effect.
2. Test statistic: $T = 0$. Critical value: $T = 14$. There is sufficient evidence to warrant rejection of the claim that the drug has no effect.
3. If the drug has absolutely no effect, each "after" score should be equal to the corresponding "before" score, and $r_s = 1$. If the drug is effective and the "after" scores are uniformly lowered, $r_s = 1$. Rank correlation is used to identify the presence of an association between the "before" and "after" scores, and such an association can occur with an effective drug or an ineffective drug.
4. $\mu_R = 188.5$, $\sigma_R = 21.71$, $R = 182$, $z = -0.30$. Test statistic: $z = -0.30$. Critical values: $z = \pm 1.96$. Fail to reject the null hypothesis that the two samples come from identical populations. There is not sufficient evidence to warrant rejection of the claim that the two sample groups come from populations with the same blood pressure levels.
5. Test statistic: $H = 0.144$. Critical value: $\chi^2 = 5.991$. Fail to reject the null hypothesis of identical populations. There is not sufficient evidence to support the claim that the three populations are not all the same.
6. $r_s = 0.713$. Critical values: $r_s = \pm 0.591$. There is sufficient evidence to support the claim of a correlation between the IQs of identical twins.
7. a. The answers of F are followed by four answers of T.
   b. $n_1 = 20$, $n_2 = 5$, $G = 11$, 5% cutoff values: 5, 12; fail to reject randomness.
   c. The pattern shows evidence of a sequence that is not random, but that evidence is not sufficient to warrant the rejection of randomness.
8. Sign test: The test statistic of $x = 35$ is converted to $z = -2.64$. Critical value: $z = -1.645$. There is sufficient evidence to support the claim that the writing course is effective.
9. Test statistic: $T = 5.5$. Critical value: $T = 13$. Reject the null hypothesis that both samples come from the same population distribution. There is sufficient evidence to warrant rejection of the claim that there is no difference between the two times.
10. Test statistic: $H = 2.180$. Critical value: $\chi^2 = 5.991$. Fail to reject the null hypothesis of identical populations. There is not sufficient evidence to warrant rejection of the claim that the three age-group populations of body temperatures are identical.
11. $\mu_R = 201.5$, $\sigma_R = 23.894$, $R = 163$, $z = -1.61$. Test statistic: $z = -1.61$. Critical values: $z = \pm 1.96$. Fail to reject the null hypothesis that the two samples come from identical populations. There is not sufficient evidence to warrant rejection of the claim that the drug has no effect on eye movements.
12. $r_s = 0.190$. Critical values: $r_s = \pm 0.738$. There is not sufficient evidence to support the claim of a correlation between performance and price. Buy the most inexpensive tapes.
13. $\mu_R = 162$, $\sigma_R = 19.442$, $R = 89.5$, $z = -3.73$. Test statistic: $z = -3.73$. Critical values: $z = \pm 1.96$. Reject the null hypothesis that the two samples come from identical populations. There is sufficient evidence to warrant rejection of the claim that beer drinkers and liquor drinkers have the same BAC levels.
14. $r_s = 0.524$. Critical values: $r_s = \pm 0.738$. There is not sufficient evidence to support the claim of a correlation between SAT score and cost per student.
15. Test statistic: $T = 10$. Critical value: $T = 2$. Fail to reject the null hypothesis that both samples come from the same population distribution.

Chapter 13 Cumulative Review Exercises

1. a. $n_1 = 20$, $n_2 = 10$, $G = 15$, 5% cutoff values: 9, 20; fail to reject randomness.
   b. Test statistic: $z = -1.83$. Critical values: $z = \pm 1.96$. Fail to reject the null hypothesis that the proportion of

women equals 0.5. There is not sufficient evidence to support the claim that the proportion of women is different from 0.5.

   c. $0.165 < p < 0.502$
   d. There is not enough evidence to support a claim of gender bias. The sequence appears random.

2. The histogram varies, depending on the choice of class width and starting point, but it is skewed and is very similar to the histogram of the Monday rainfall amounts. The Sunday rainfall amounts do not appear to be normally distributed.
3. a. 0.068 in.
   b. 0.000 in.
   c. 0.00 in.
   d. 1.280 in.
   e. 0.200 in.
   f. 1.28 in.
4. After deleting the last entry for Wednesday, the numbers of days with rain are (from Monday through Sunday) 19, 16, 16, 21, 20, 20, 15. (See Section 10-3.) Test statistic: $\chi^2 = 2.951$. Critical value: $\chi^2 = 12.592$. Fail to reject the null hypothesis of equal proportions. There is not sufficient evidence to support the claim that the proportions of rainy days are not the same for all seven days of the week.
5. a. The test statistic $x = 2$ is not less than or equal to the critical value of 1. Fail to reject the null hypothesis that the course has no effect. The course does not appear to have an effect on SAT scores.
   b. Test statistic: $T = 3$. Critical value: $T = 6$. Reject the null hypothesis that the course has no effect on SAT scores. It does appear to have an effect.
   c. Test statistic: $t = -3.753$.
   Critical values: $t = \pm 2.262$. Reject the null hypothesis that the course has no effect. It does appear to have an effect.
   d. The sign test failed to detect a difference, whereas the t test resulted in the conclusion of a significant difference. Unlike the sign test, the Wilcoxon signed-ranks test did use the magnitudes of the differences, so it is more sensitive than the sign test and it did result in the same conclusion as the t test. (This book does not describe a method for finding P-values with the Wilcoxon signed-ranks test, but the P-value is 0.024, whereas the P-value for the t test is 0.0045. Comparison of these P-values shows that the Wilcoxon signed-ranks test is not as sensitive as the parametric t test.)

Credits

Photographs

Chapter Openers

Chapter 1: Students on Campus © PhotoDisc

Chapter 2: Computer Keyboard © PhotoDisc

Chapter 3: Titanic Lifeboats © Hulton-Deutsch Collection

Chapter 4: Back view of Parents with child walking through woodland © Stone/Joe Cornish

Chapter 5: Navy Fighter Jet © Stone/Chad Slattery

Chapter 6: Thermogram of head and upper torso © Alfred Pasiek /Science Photo Library/ Photo Researchers, Inc.

Chapter 7: Crowd of People © Stone/ Ken Fisher

Chapter 8: Multi-Ethnic College Students © FPG/Ron Chapple

Chapter 9: Red Chair and Red and White Table © Stone/ Robin Lynne Gibson

Chapter 10: Titanic at Bottom of Ocean © Ralph White/CORBIS

Chapter 11: Crash Dummy © Scott Barrow, Inc.

Chapter 12: Aluminum Cans © The Image Bank/Steve Allen

Chapter 13: Crowd of People with Umbrellas © PhotoDisc

The following margin photos are all copyright PhotoDisc, Inc.

Chapter 1
p. 5, The State of Statistics
p. 6, Interesting Statistics
p. 17, Sampling Rejected for the Census
p. 20, Literary Digest Poll

Chapter 2
p. 56, Marion: Average City USA
p. 58, Class Size Paradox
p. 62, Not at Home
p. 70, More Stocks, Less Risk
p. 78, Mail Consistency

Chapter 3
p. 115, Subjective Probabilities at the Racetrack
p. 118, How Probable?
p. 122, You Bet
p. 128, Guess on SATs?
p. 145, Composite Sampling
p. 146, Bayes' Theorem
p. 152, To Win, Bet Boldly
p. 153, The Random Secretary
p. 156, The Number Crunch
p. 157, How Many Shuffles?
p. 162, Boys and Girls Are Not Equally Likely

Chapter 4
p. 184, Is Parachuting Safe?
p. 186, Autism Cluster Exceeds Expected Number
p. 189, Picking Lottery Numbers
p. 195, Prophets for Profits
p. 211, Probability of an Event That Has Never Occurred

Chapter 5
p. 227, Reliability and Validity
p. 281, A Professional Speaks About Sampling Error

Chapter 6
p. 297, Estimating Wildlife Population Sizes
p. 324, Small Sample
p. 330, TV Sample Sizes

Chapter 7
p. 372, Lie Detectors
p. 377, Large Sample Size Isn't Good Enough
p. 410, Polls and Psychologists

Chapter 8
p. 442, Commercials
p. 450, Research in Twins
p. 461, The Lead Margin of Error
p. 464, Does Aspirin Help Prevent Heart Attacks?
p. 470, Do Air Bags Save Lives?
p. 475, Lower Variation; Higher Quality
p. 486, Using Statistics to Identify Thieves

Chapter 9
p. 511, Student Ratings of Teachers
p. 527, Cell Phones and Car Crashes
p. 540, Wage Gender Gap
p. 551, Predictors for Success
p. 552, Model for Alumni Contributions
p. 554, Making Music with Multiple Regression

Chapter 10
p. 595, Home Field Advantage

Chapter 13
p. 701, Gender Gap in Drug Testing
p. 720, Direct Link Between Smoking and Cancer

Misc. photos

Chapter 1
p. 12, Statistics and Land Mines; Copyright Corbis/Nevada Weir

Chapter 2
p. 73, Where are the 0.400 Hitters? Copyright Corbis/AFP

Chapter 7
p. 390, Ethics in Reporting; Copyright Corbis/Robert Maas

Chapter 12
p. 661, Bribery Detected with Control Charts; Copyright Owen Franken/Corbis
p. 659, Don't Tamper! Copyright Davis Lees/Corbis

Chapter 13
p. 730, Sports Hot Streaks; Copyright Bettman/Corbis

Illustrations

The following margin illustrations were rendered by Bob Giuliani.

Chapter 1
p. 11, Detecting Phony Data
p. 18, Hawthorne Effect

Chapter 2
p. 36, Authors Identified
p. 49, Florence Nightingale
p. 61, Six Degrees of Separation
p. 80, Buying Cars
p. 85, Olestra Potato Chips
p. 87, Cost of Laughing Index

Chapter 3
p. 131, Shakespeare's Vocabulary
p. 120, Probabilities That Challenge Intuition
p. 136, Redundancy
p. 151, Monkey Typist
p. 158, Safety in Numbers

Chapter 4
P. 197, Voltaire Beats the Lottery
p. 200, Sensitive Surveys

Chapter 6
p. 298, Captured Tank Serial Numbers Reveal Population Size
p. 303, Estimating Sugar in Oranges
p. 313, Excerpts from a Department of Transportation Circular

Chapter 7
p. 404, Death Penalty as a Deterrent
p. 418, Ethics in Experiments

Chapter 8
p. 451, Crest and Dependent Samples
p. 463, Drug Screening: False Positives

Chapter 9
p. 507, Predicting the Stock Market
p. 528, Pizza Barometer

Chapter 10
p. 592, Delaying Death

Chapter 11
p. 620, Poll Resistance

Chapter 12
p. 656, The Flynn Effect
p. 670, Six Sigma in Industry

Chapter 13
p. 723, Manatees Saved

The following margin illustrations were rendered by James Bryant.

Chapter 1
p. 7, Measuring Disobedience

Chapter 2
p. 75, Data Mining

Chapter 3
p. 137, Independent Jet Engines
p. 139, Convicted by Probability
p. 147, Coincidences?
p. 160, Choosing Personal Security Codes

Chapter 5
p. 250, Survey Medium Affects Results
p. 257, The Fuzzy Central Limit Theorem
p. 270, She Won the Lottery Twice

Chapter 6
p. 332, Push Polls

Chapter 8
p. 460, Polio Experiment

Chapter 9
p. 512, Palm Reading

Chapter 10
p. 581, Safest Airplane Seats

Chapter 12
p. 669, Quality Control at Perstop

Chapter 13
p. 691, Class Attendance and Grades

The following illustrations were rendered by Darwen Hennings.

Chapter 5
p. 242, Queues

Chapter 7
p. 385, Statistics: Jobs and Employers
p. 400, Better Results with Small Class Size

Chapter 8
p. 443, The Placebo Effect

USA Today Snapshots

p. 25, CEO profile. Copyright 1996, USA TODAY. Reprinted with permission.
p. 104, Who can that be? Copyright 1999, USA TODAY. Reprinted with permission.
p. 168, Men from Mars? Copyright 1997, USA TODAY. Reprinted with permission.
p. 215, What's for dinner? Copyright 1997, USA TODAY. Reprinted with permission.
p. 493, Just plane scared. Copyright 1997, USA TODAY. Reprinted with permission.

Tufte Illustration

p. 48, Losses of Soldiers in Napoleon's Army During the Russian Campaign (1812-1813): Edward R. Tufte, *The Visual Display of Quantitative Information* (Cheshire, CT: Graphics Press, 1983). Reprinted with permission.

Index